9th International Scientific Conference "TechSys 2020" - Engineering, Technologies and Systems

IOP Conference Series: Materials Science and Engineering
Volume 878

Plovdiv, Bulgaria
14 - 16 May 2020

Editors:

Valyo Nikolov
Michail Petrov
Nikola Shakev
Sevil Ahmed

ISBN: 978-1-7138-4742-7
ISSN: 1757-8981

Printed from e-media with permission by:

Curran Associates, Inc.
57 Morehouse Lane
Red Hook, NY 12571

Some format issues inherent in the e-media version may also appear in this print version.

Institute of Physics
Dirac House, Temple Back
Bristol BS1 6BE UK

Phone: 44 1 17 929 7481
Fax: 44 1 17 920 0979

techtracking@iop.org

Additional copies of this publication are available from:

Curran Associates, Inc.
57 Morehouse Lane
Red Hook, NY 12571 USA
Phone: 845-758-0400
Fax: 845-758-2633
Email: curran@proceedings.com
Web: www.proceedings.com

TABLE OF CONTENTS

TECHNICAL UNIVERSITY – SOFIA
PLOVDIV BRANCH

9-TH INTERNATIONAL SCIENTIFIC CONFERENCE
"ENGINEERING, TECHNOLOGIES AND SYSTEMS"

TECHSYS 2020

14-16 May, Plovdiv, Bulgaria

IOP Conf. Series: Materials Science and Engineering **878** (2020) 011001 doi:10.1088/1757-899X/878/1/011001

TECHNICAL UNIVERSITY - SOFIA, PLOVDIV BRANCH

Organizing Committee:

Chairman: Prof. Dr. Valyo Nikolov - Director of the Technical University - Sofia, Plovdiv branch

Vice Chairman: Assoc. Prof. Dr. Nikola Shakev - Vice Director of the Technical University - Sofia, Plovdiv branch

Members:

Assoc. Prof. Dr. Georgi Ganev
Assoc. Prof. Dr. Mitko Shopov
Assoc. Prof. Dr. Nikolay Kakanakov
Assoc. Prof. Dr. Krum Kutryansky
Assoc. Prof. Dr. Angel Lengerov
Assoc. Prof. Dr. Valentina Proycheva
Assoc. Prof. Dr. Milcho Tashev
Assoc. Prof. Dr. Valentin Vladimirov
Assoc. Prof. Dr. Hristian Panajotov
Assoc. Prof. Dr. Iliya Chetrokov
Assoc. Prof. Dr. Toni Mihova
Assoc. Prof. Dr. Anton Lechkov
Assoc. Prof. Dr. Krasimir Ambarev
Assoc. Prof. Dr. Nikola Georgiev
Assoc. Prof. Dr. Silvina Ilieva

International Programme Committee:

Honorary Chairman: Prof.DSc Ivan Kralov - Rector of the Technical University - Sofia

Chairman: Prof. Dr. Michail Petrov, TU-Sofia, Branch Plovdiv

Members:

Prof. DSc Georgi Mihov, Bulgaria
Prof. DSc Andre Barraco, France
Prof. DSc. Georgi Todorov, Bulgaria
Prof. DSc. Venelin Jivkov, Bulgaria
Prof. DSc. Vesko Panov, Bulgaria
Prof. DSc Marin Nenchev, Bulgaria
Prof. DSc Emil Nikolov, Bulgaria
Prof. DSc Todor Stoilov, Bulgaria
Prof. DSc Okyay Kaynak, Turkey
Prof. DSc Edmunds Teirumnieks, Latvia
Prof. DSc Mark Himbert, France
Prof. DSc Ivan Jachev, Bulgaria
Prof. DSc Chavdar Damianov, Bulgaria
Prof. DSc Valeri Mladenov, Bulgaria
Prof. DSc Raycho Ilarionov, Bulgaria
Prof. Dr. Nikolay Ganev, Czech Republic
Prof. Dr. Valeri Stoilov, Bulgaria
Prof. Dr. Lubomir Dimitrov, Bulgaria
Prof. Dr. Georgi Venkov, Bulgaraia
Prof. Dr. Grisha Spasov, Bulgaria
Prof. Dr. Ernst Wintner, Austria
Prof. Dr. Veselka Boeva, Bulgaria
Prof. Dr. Mladen Velev, Bulgaria
Prof. Dr. Daniela Goceva, Bulgaria
Prof. Dr. Petar Gecov, Bulgaria
Prof. Dr. Frantisek Zezulka, Czech Republic
Prof. Dr. Ilia Iliev, Bulgaria
Prof. Dr. Nikola Kasabov, New Zealand
Prof. Dr. Petr Louda, Czech Republic
Prof. Dr. Ognyan Nakov, Bulgaria
Prof. Dr. Mihail Todorov, Bulgaria
Prof. Dr. Galidia Petrova, Bulgaria
Prof. Dr. Marsel Popa, Romania
Prof. Dr. Andon Topalov, Bulgaria
Prof. Dr. Sedat Sunter, Turkey
Prof. Dr. Ahmed Hafaifa, Algeria
Prof. Dr. Abdellah Kouzou, Algeria
Prof. Dr. Zdenek Bradach, Czech Republic
Prof. Dr. Iliya Zhelezarov, Bulgaria

Scientific Secretary: Assoc. Prof. Dr. Sevil Ahmed

Technical Secretary: Eng. Tsvetan Petrov, Eng. Christo Christev, Eng. Lalka Boteva

IOP Conf. Series: Materials Science and Engineering **878** (2020) 011001 doi:10.1088/1757-899X/878/1/011001

European Commission

CONGRATULATORY LETTER

On the occasion of the 9th International scientific conference

"TechSys 2020" – ENGINEERING, TECHNOLOGIES AND SYSTEMS

I am pleased to extend my warmest greetings to all gathered today on the occasion of the 9th edition of the International scientific conference "TechSys 2020" - Engineering, Technologies and Systems.

I would like to thank the Technical University of Sofia – branch Plovdiv for hosting this important event, which in the past years, has contributed to the technological advancement of engineering solutions for industry.

In current times, we need engineers more than ever!

The COVID-19 pandemic has hit Europe and the world with serious consequences for many thousands of people and has deeply affected our economy and the life of millions of workers and researchers. We will need the engineering creativity to help with solutions that address both the challenges of the crisis and the recovery process.

If we act together, we can come out of the crisis stronger!

The European Commission is taking robust action to tackle this crisis. We are at the forefront of supporting research and innovation and coordinating global research efforts.

On 4 May, President von der Leyen hosted the Coronavirus Global Response online pledging event to jump-start significant funding to ensure the collaborative development and universal deployment of diagnostics, treatments and vaccines against the coronavirus – the three priorities of the initiative. The Commission registered €7.4 billion in pledges from donors worldwide. My portfolio has made an unprecedented contribution - in total, EUR 1 billion is pledged under Horizon 2020.

The global pledging event was preceded by a number of other decisive actions at EU level. Let me give you some examples.

In April, we launched the ERAvsCorona Action Plan, consisting of ten key measures to coordinate, share and jointly increase support for research and innovation. I warmly welcome the support that all EU Research Ministers gave to the plan.

Manufacturing is part of the Action Plan. We need to work together with engineers on two challenges: access to research data as a matter of solidarity amongst all of us; and, address the most urgent research needs.

IOP Conf. Series: Materials Science and Engineering **878** (2020) 011001 doi:10.1088/1757-899X/878/1/011001

Also in April, I was glad to launch the European COVID-19 Data Platform to enable the rapid collection and sharing of available research data from European research projects. It is a very important concrete measure to support researchers in Europe and around the world in the fight against the coronavirus outbreak.

We need more for the future. We need engineers and scientists in manufacturing to undertake more research.

Manufacturing industry has been hurt greatly during the last weeks, firstly in terms of the disruption of the global supply chains, and then by the indispensable lockdown measures affecting work in factories.

Research and innovation activities dedicated to increasing the flexibility and resilience of European industry are clearly paramount to ensure a prompt response, quick recovery and preparedness of our society to shocks.

Europe has been a major player in supporting the manufacturing world and will continue to be so.

Under the EU R&I Programme Horizon 2020, we invested a total of €1.15 billion in the R&I activities of the Factories of the Future public-private partnership granting funding to more than 200 projects developing innovative production technologies, digitisation, automation, robotics solutions to pave the way to a greener, sustainable and competitive European manufacturing industry.

In 2021, the successor of R&I Programme Horizon Europe will be launched. More than ever, Europe needs the strong commitment from Member States to dedicate an appropriate budget to research and innovation.

I am sure that the Bulgarian scientific community will contribute even more under Horizon Europe, helping us make a competitive and sustainable European industry fit for the challenges of tomorrow.

Be part of this effort, making the most of the huge opportunities available at EU level!

I wish you all a very fruitful conference with many inspiring ideas.

Sincerely,

Mariya Gabriel

Commissioner for innovation, research,
culture, education and youth

Brussels
14.05.2020

IOP Conf. Series: Materials Science and Engineering **878** (2020) 011001 doi:10.1088/1757-899X/878/1/011001

Preface

9th International Scientific Conference: "TechSys 2020" – Engineering, Technologies and Systems is organized by Technical University of Sofia, Plovdiv Branch. It was held on 14-16 May 2020 at Plovdiv, Bulgaria under the frame of "Science Days of Technical University of Sofia" and supported by the University Scientific and Research Sector.

The organization of the conference started very successfully, but the unexpected development of the situation with COVID-19 and the imposed restrictions on holding mass events, travel restrictions and the responsibility for the health of the participants put the organizing committee in a difficult situation. Encouraged by the enthusiasm of the participants and their conviction that the circumstances should not hinder the scientific research, the pursuit of innovation and the sharing of good ideas, the Organizing Committee decided to hold the conference online.

All participants and organizers managed to cope with the limitations in a great way and turn them into an occasion for new activities and creative solutions. This proved very useful experience from which we can draw good ideas and develop them in the future editions of the conference.

The conference was opened with a video address by the Rector of the Technical University - Sofia. All participants in the conference were greeted with a special Congratulatory letter from Ms. Maria Gabriel European Commissioner for Innovation, Research, Culture, Education and Youth.

Thematically, 7 online sessions were held: Techsis 2020 Opening Session, as well as 6 specialized (thematic) sessions. All the sessions were held via video conference and were held with great interest from the scientific community. More than 130 researchers took part. Each of the authors presented their report through video streaming, after which the participants in the forum were given the opportunity for discussions and Q&A sessions. In addition to online presentation, each report was pre-published via posters on the conference website. The sessions were also open to all who wished to take part in this forum.

The main goal of TechSys was successfully achieved - to bring together researchers from academia and industry, scientists and experts in engineering and technology to share ideas and innovations. As a result, selected articles, which represent the significant results of the conference are published in this volume.

IOP Conf. Series: Materials Science and Engineering **878** (2020) 011001 doi:10.1088/1757-899X/878/1/011001

We would like to thank all the authors who have trusted the conference. To thank all those researchers who recognized the TechSys as an appropriate forum to share their recent developments and to get the new trends in the field of technical sciences and engineering.

We must also thank all the reviewers and topic editors for their quiet but very important work, which ensured excellent feedbacks to the authors and a high quality of scientific research.

Last but not least, we sincerely thank the whole team involved in organizing the TechSys 2020 conference.

Editors
Prof. Valyo Nikolov
Prof. Michail Petrov
Assoc. Prof. Nikola Shakev
Assoc. Prof. Sevil Ahmed

IOP Conf. Series: Materials Science and Engineering **878** (2020) 011002 doi:10.1088/1757-899X/878/1/011002

Peer review statement

All papers published in this volume of *IOP Conference Series: Materials Science and Engineering* have been peer reviewed through processes administered by the proceedings Editors. Reviews were conducted by expert referees to the professional and scientific standards expected of a proceedings journal published by IOP Publishing.

IOP Conf. Series: Materials Science and Engineering **878** (2020) 012001 doi:10.1088/1757-899X/878/1/012001

Underwater surveillance of Romanian littoral areas using AUVs - Sea trials results

R G Damian [1,2]

1 Research Center for Navy, Constanta, Romania
2 Military Technical Academy "Ferdinand I", Bucharest, Romania

roxanagabriela.damian@navy.ro

Abstract. One of the many military as well as civilian potential applications regarding autonomous underwater vehicles (AUVs) is detection, localization, tracking and identification (DLTI) of risk factors in littoral areas. Using AUVs in underwater surveillance involves some specific capabilities like underwater data acquisitions, on-board data processing, underwater and surface communication, autonomy aspects and so on. In order for an AUV to be capable of conducting underwater environment surveillance (detection, monitoring) in littoral areas like harbors, coast objectives, ship anchorage areas, mandatory pass points and also to provide warnings about the presence of underwater and surface dangers in the interest areas, it must be first subject to sea trials. This paper presents the theoretical and experimental approaches and results for the surveillance of Romanian Littoral Areas using an AUV.

1. Introduction

The Black Sea is considered to be the eastern maritime border for both the European Union (EU) and the North Atlantic Treaty Organization (NATO). Romania like every member state within EU&NATO holds and will continue to hold responsibilities in the area of maintaining the security and counterattacking eventual threats in the maritime border.

Underwater surveillance, for the purpose of this paper, consists in the task of monitoring the underwater environment for potential risk factors. Risk factors can be defined as internal or external elements, situations or conditions that may, by their nature, affect the security of the country, resulting in adverse effects or attainment of its fundamental interests.

The field of underwater robotics has reached an impressive technological momentum over the past decades by merging areas such as electrical, mechanical and systems engineering. A particular interest is presented by autonomous underwater vehicles (AUV), given the progress in the implementation of revolutionary technologies to overcome the challenges associated with autonomous operation in the underwater environment

Nowadays the concept of using AUVs among complement traditional existing high-value assets as, helicopters, fixed hydrophones systems etc. for ensuring the underwater maritime borders security is being implemented with success. Some of the most common military applications for AUVs are mine countermeasure (MCM), antisubmarine warfare (ASW) and harbor protection.

2. Underwater surveillance scenarios for AUVs

There are several representative features that could be used to draw up scenarios of possible risk factors that will imply the need of surveillance in littoral areas. Based on different combinations of threats, locations, targets and tactics there could be defined plausible scenarios. An example of these scenario features are presented in table 1.

Table 1. Scenarios characteristics.

Threats	Locations	Targets	Tactics
Destruction of harbors, ships, drilling and extraction platforms, etc. Undermining economic activities Destruction of the environment	Harbors Stationary or deployment areas Mandatory crossing points Roads and passes Natural Reserves or Tourist Areas	Military or commercial ships Oil platforms Ferries / cruise ships • Port areas Waterways	Explosives in suicide ships Crash / hitting ships Missiles launched from ships Marine mines and / or improvised explosive devices Diversion divers Underwater vehicles remote / autonomous Intentional sinking of merchant ships

To be able to use AUVs in underwater surveillance first you need to have good knowledge of the environmental conditions (propagation, position of natural obstacles, fund profile, statistical properties of clutter, etc.).

The second aspect to take into consideration in underwater surveillance when using AUVs is the sensors with which it is equipped like sonar (passive or active). Also from the perspective of the sensor used in the detection and identification process there must be considered the characteristics of the target like target strength or target hidroacoustic signature.

Passive sonars have the advantage of covertness but present some challenges when used in littoral areas where the level of ambient noise is high due to ship traffic and biological activity. In passive sonar systems, the searcher listens for the week acoustic signals generated by the target itself.

When using active sonars, the ambient noise issues are overcamed because the sound intensity levels of the echoes tend to be higher than typical ambient noise levels. Active sonar systems create a pulse of sound in the water and then listen for the echo bouncing off the target.

In either case, the structure of the sound speed environment radically affects the propagation paths of the sound energy. Knowledge of the sound propagation model is required to correctly set the sonar system and to choose the appropriate tactics to locate the target.

Table 2 presents such data from the perspective of detection (active or passive).

Table 2. Characteristics of targets from detection perspective using active sonar systems.

Target	Speed [kts]	Action depth [m]	Target strength [dB rel. 1 μPa/1 m]	Generated noise level
Diver	0,5 - 1	1 - 20	25	Low
Torpedo	10 - 50	> 20	-20	High
Mine	-	5 - 10	-15	-
AUV	2 - 5	5 – maximum zone depth	-20	Low

The sound speed depends on temperature, salinity and pressure variation of the marine environment.

The Black Sea Basin presents some specific characteristics like the interaction between the upper and lower layers causing a non-uniform structure of the marine environment. The temperature and salinity of the seawater layers in the area of the continental platform, corresponding to the Romanian shore, does not remain constant during the entire year. It is worth noting the average low salinity in this area with limit values in the vicinity of the mouth of the Danube River (phenomenon of dilution).

All these factors influence the attenuation of the acoustic signal propagation in seawater. The sound propagation in the seawater depends on sound speed profile, sedimentological sea floor structure (seafloor reflections) and marine medium attenuation.

3. Sea trials

The tests were conducted in an area (see figure 1) where:

-noise caused by naval traffic is low;

-no large variations of the thermal and saline environment;

-positioning of the reference buoys and targets can be done easily and with relatively high accuracy;

-depth and bottom type are already known data.

Figure 1. Sea trials place.

The specificity of the testing area is given by:

-naval traffic was only random occasional crossings of leisure craft to/from Limanu Harbour and RHIB boats belonging to RO Navy;

-there were no large variations in the thermal and saline (salinity around 19 PSU, surface temperature around 19 –22 0 C);

-depth of around 11 m, bottom generally composed of peat and covered with mud, only in the southern area having some rocks.

In the sea trials there was deployed a man portable AUV equipped with pressure sensor, magnetic compass, three-ax accelerometer and side scanning sonar. The AUV was integrated by the Research Center for Navy within a national funded research project. The AUV and sonar main characteristics are presented in table 3 and table 4.

Table 3. AUV specification.

Items	Specification
Vehicle weight in air	25kg
Energy and endurance	8h/2Nd
Maximum speed	4Nd
Surface navigation	GPS
Underwater navigation	Magnetic compass
Payload	Side-scan Sonar
Maximum immersion depth	100 m
Connectivity	WiFi (while at surface)

Table 4. Side scan sonar specification.

Items	Specification
Frequency	450 kHz center frequency(430-470 kHz)
Pulse duration	400μs
Signal processing	Compressed pulse(CHIRP pulse)
Hydro-acoustic sensors	2 side sensors attached to the AUV
Dimensions(Length x width x height)	432 x 41 x18 mm
Directivity characteristics of each sensor(VxO)	60∘ x 0,5∘ (-3dB)
Coverage area	1m….100m on channel
Emission power level	<210 dB rel. 1μPa/1m

Theoretical and practical evaluation of the discovery capabilities of underwater targets by the side scanning sonar fitted to the AUV has been conducted for the scenario that involves the detection of two fixed target (mine) with the target strength -20 dB, at a depth of aprox.12 m. The sea trials ware conducted in July and the sea state was SS0 (according to the Sea State Code, sea like a mirror, wind less than one knot, average wave height 0).

The target detection probability for this scenario is presented in figure 2.

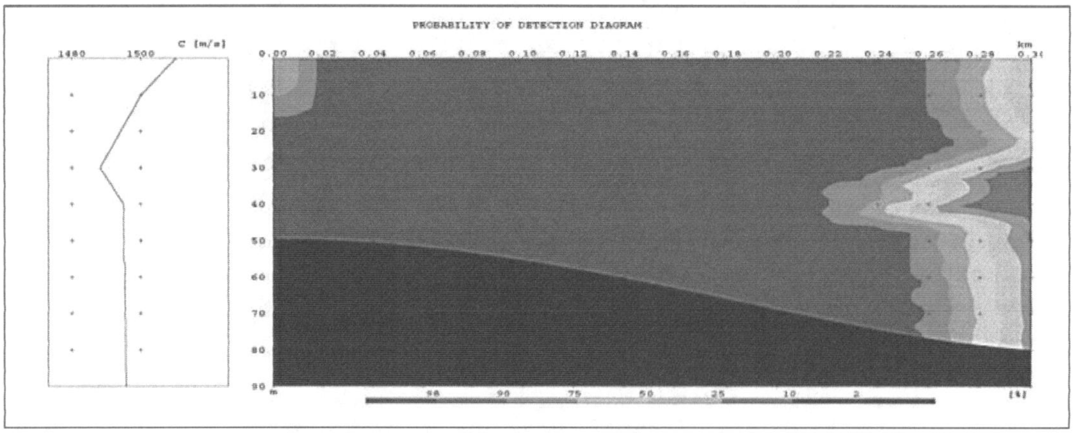

Figure 2. Detection probability distribution.

The equipment used in the sea trials (figure 3) were:

- MP-1 target (portable mine, ballasted version);
- Large target (cylinder Φ = 280 mm L = 700 mm).

Figure 3. Targets used for sea trials.

The AUV performed 7 missions, of which one surface mission and 6 missions on the same route, but with variable dives at different points along the route. A depth variation from these missions is presented in figure 4 and the sonar mosaic from the AUV traveled path is shown in figure 5.

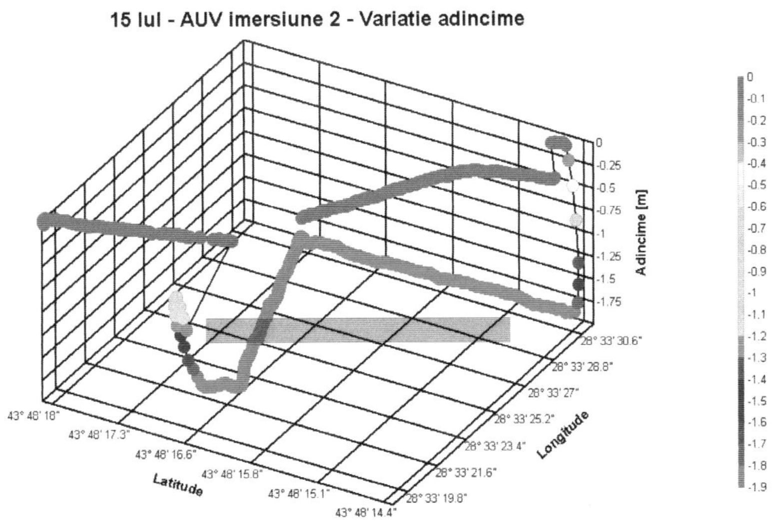

Figure 4. Depth variation in two different missions.

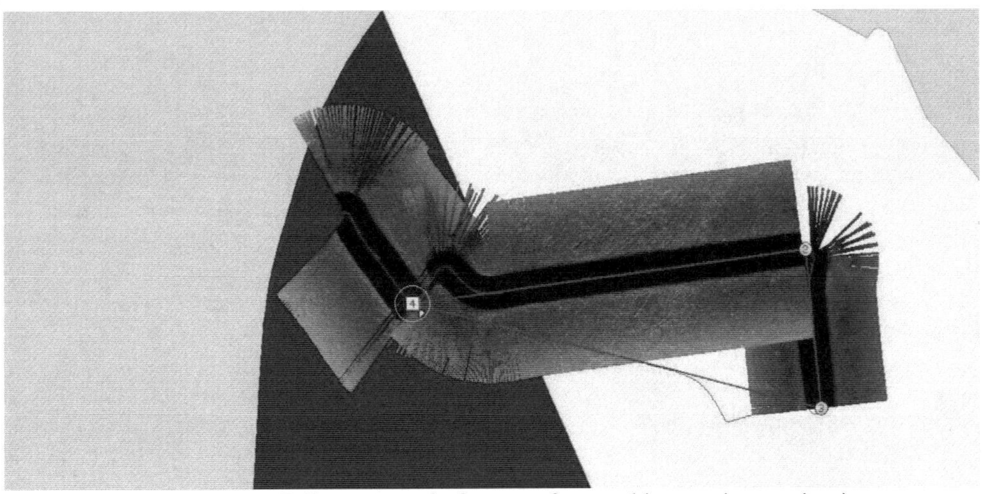

Figure 5. Sonar mosaic from surface and immersion navigation.

IOP Conf. Series: Materials Science and Engineering **878** (2020) 012001 doi:10.1088/1757-899X/878/1/012001

Details about the contact (underwater target) can be obtained through operator interpretation of the data and this is why it is considered that the experience is crucial when trying to obtain data as close to real as possible (figure 6 and figure 7).

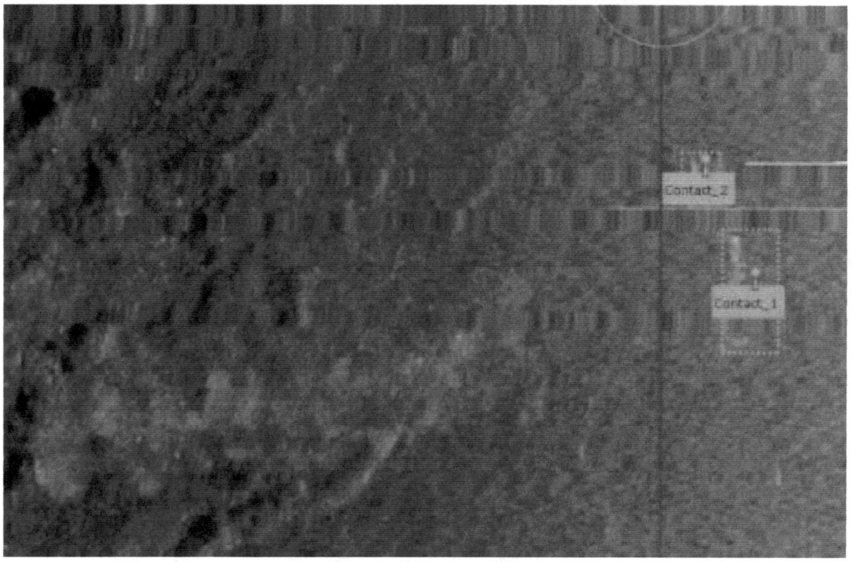

Figure 6. Sonar contacts.

The details of the two contact on the side scan sonar are presented in table 5.

Table 5. Side scan sonar specification.

Contact	Position of Contact	Size of Contact	Size of the selected area	Image
1	: Latitude: 43.8047858200479 Longitude: 28.557099387344	0.3622783 Height(m): 1.233304	Width(m): 2.88 Height(m): 6.17	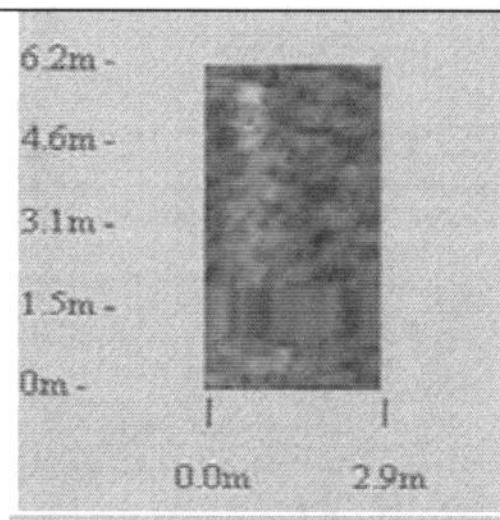
2	Latitude: 43.8048365887958 Longitude: 28.5570673237462	0.2014776 Height(m): 0.9666433	Width(m): 1.97 Height(m): 2.58	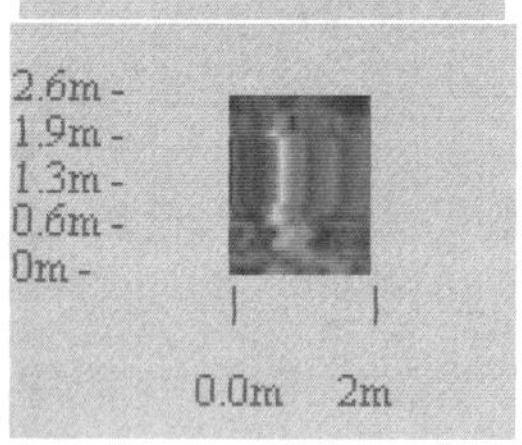

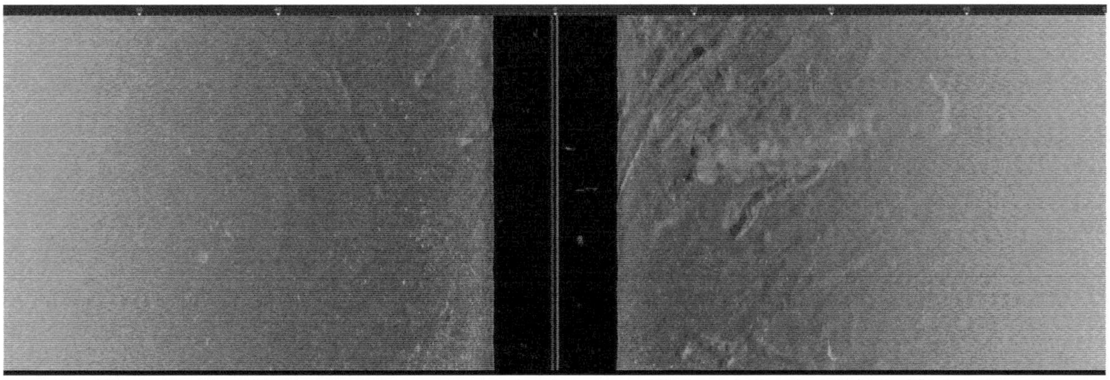

Figure 7. Scanline recording in way point 2 area, with targets in the right half-plane.

Figure 8. RCN team recovering AUV after sea trials.

4. Conclusions

In missions in which the evolution of the AUV is required only at the surface, it follows the programmed route, compensating based on the information from GPS, magnetic compass and accelerometers any deviations due to waves, currents, etc. In immersion developments, in the absence of a Doppler loch (DVL), the compensation of current deviations due to currents could no longer be performed accurately. This fact causes, in perspective, errors in the geo-referencing of the images obtained by means of the sonar with lateral scanning.

Naturally, fluctuations in the speed of travel (imposed by the program or due to the attempt to keep on the road - counteracting the effect of waves, currents, etc. -) occur fluctuations of the instantaneous power consumed.

Analyzing the conducted missions has shown that the sonar has remarkable performance, allowing that at 3-4 meters from bottom to be able to identify reduced dimension targets. We find that the working frequency of the side scanning sonar is low enough so that it reduces propagation losses. Analyzing the characteristics of the side scanning sonar allows us to conclude that it has good performances (very good resolution, relatively low operating frequency).

For TS = having high values (over 10 dB rel. 1 μPa/1m) – the probability of discovery of the target is above 98% within the limit of the sonar maximum range (adjustable between 20 to 100m), no matter what environment conditions, bottom nature, depth and AUV activity conditions (immersions between 10 and 45 m).

It is useful to measure the sound speed velocity in the mission area and to use the value recorded for this parameter in the mission planning software.

It is necessary to carry out a program to assess the capabilities of discovering a wide range of submarine targets (wrecks, predefined targets - exercise mines, torpedoes, etc.) for the autonomous mobile underwater platform. For a better appreciation of these capabilities, it would be necessary to equip the AUV with a video camera or to mount such equipment on the target to be detected / identified.

References

[1] S.Kemna, M. Hamilton, D. Hughes, K. LePage, Adaptive autonoumous underwater vehicles for littoral surveillance, Intel Serv Robotics 2011

[2] S V Paturca, A M Dumitrescu, V B, A M Stefan, A project-based Robotics laboratory, 10th International Symposium on Advanced Topics in Electrical Engineering (ATEE), 2017

[3] M E Mihailov, L Buga, L Dumitrache, L F Constantinoiu, Underwater sound speed features on the western Black Sea shelf, 16th International Multidisciplinary Scientific GeoConference SGEM 2016

[4] A Digulescu, A ŞERBĂNESCU, I Candel, C Ioana, On the extension of acoustic flow metering devices to monitor underwater transient phenomena, OCEANS 2018 MTS/IEEE Charleston

Acknowledgements

This paper presents results from the project "System for detection, Localization, tracking and identification of risk factors for strategic objectives on seaside areas-SIROLC" funded by The Executive Unit for Financing Higher Education, Research, Development and Innovation(UEFISCD) from Romania and coordinated by the Research Center for Navy RCN). The author works as scientific researcher 3rd degree within RCN since 2012 and was part of the project team. The results of the project SIROLC are included in the author's PHD thesis: *Contributions to the development of underwater vehicle, equipped with sensors for monitoring the security risks in the littoral Black Sea area.*

IOP Conf. Series: Materials Science and Engineering **878** (2020) 012002 doi:10.1088/1757-899X/878/1/012002

Nonlinear backstepping tracking control of DC motor driven inverted pendulum

V Petrov

Assist. prof. PhD, Technical University Sofia, Branch Plovdiv, Plovdiv, Bulgaria
E-mail:v.spasovv@tu-plovdiv.bg

Abstract. The paper presents nonlinear backstepping approach for nonlinear control design applied to a DC motor driven inverted pendulum system. The nonlinear closed-loop system designed simultaneously tracks exponentially desired reference trajectory with overdamped response specifications. A simulation example is given for illustration.

1. Introduction

Backstepping approach is a recursive procedure for systematic construction of feedback control laws with respective Lyapunov functions. The designs are related to feedback linearization [10] but backstepping allows flexibilities in the cancelation of the nonlinearities. The designer has the freedom to eliminate only the destabilizing nonlinearities, which leads to robustness of the nonlinear control law. This advantage over the feedback linearization method makes the backstepping approach preferable in industrial applications [8].

The idea of the approach has appeared in the early 1987 [1]. Since then the method has been adopted for development by various authors [5][7][9] in nonlinear [2] and in adaptive control [3][4] techniques. The first book describing the backstepping design [6] discusses SISO systems adaptive and nonlinear control with extensions to MIMO systems.

In this paper the backstepping approach is presented for nonlinear trajectory tracking control design. This allows performance specifications of the nonlinear closed loop system using reference model. The method is applied on a DC motor driven inverted pendulum nonlinear system (figure 1).

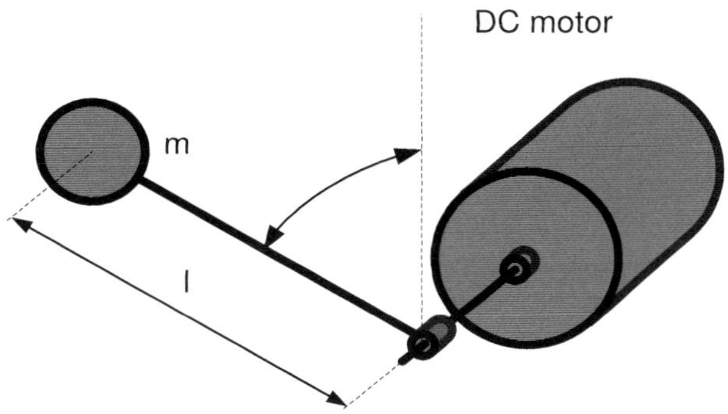

Figure 1. Inverted pendulum system.

The performance of the designed nonlinear control system is illustrated via dynamic simulation and the results are shown graphically.

The paper is organized as follows: the next section gives introductory example to the backstepping design procedure, section 3 is devoted to the inverted pendulum system modelling, section 4 - application of the approach on the objective system, section 5 is for dynamic simulation of the closed loop system and the last section is dedicated to conclusions.

2. Introduction to integrator backstepping

Demonstration of the approach as in [7] is given by considering a second order scalar system of the form

$$\begin{aligned}\dot{x}_1 &= f(x_1) + x_2 \\ \dot{x}_2 &= u\end{aligned} \tag{1}$$

The design objective is stabilization of $x_1 \to 0$ as $t \to \infty$. The only equilibrium point with $x_1 = 0$ is $(x_1, x_2) = (0, -f(0))$ corresponding to $\dot{x}_1 = 0$. The second order system requires two steps recursive design procedure. The backstepping design introduces recursively change of coordinates during the design.

Step 1

The objective is to regulate x_1 to zero, thus the first backstepping variable is chosen as

$$z_1 = x_1 \tag{2}$$

In the first equation of system (1) x_2 will be selected as *virtual control* input. The virtual control is defined as

$$x_2 = \alpha_1 + z_2 \tag{3}$$

where α_1 is called *stabilizing function* and z_2 is a new state variable. Hence the dynamics of (2) reads

$$\dot{z}_1 = f(x_1) + \alpha_1 + z_2 \tag{4}$$

The stabilizing function α_1 must provide the necessary feedback for the z_1-system. Choosing it as

$$\alpha_1 = -f(x_1) - k_1 z_1 \tag{5}$$

yields:

$$\dot{z}_1 = -k_1 z_1 + z_2$$

A candidate Lyapunov function (CLF) for the z_1-system is:

$$V_1 = \frac{1}{2} z_1^2$$

$$\begin{aligned}\dot{V}_1 &= z_1 \dot{z}_1 \\ &= -k_1 z_1^2 + z_1 z_2\end{aligned}$$

with $k_1 > 0$ the feedback gain. Hence system (4) is stabilized.

Step 2

The dynamics of z_2 is computed from (3) by time differentiation:

$$\begin{aligned}\dot{z}_2 &= \dot{x}_2 - \dot{\alpha}_1 \\ &= u - \dot{\alpha}_1\end{aligned}$$

A Lyapunov function candidate for the z_2-system is:

$$V_2 = V_1 + \frac{1}{2} z_2^2$$

$$\begin{aligned}\dot{V}_2 &= \dot{V}_1 + z_2 \dot{z}_2 \\ &= (k_1 z_1^2 + z_1 z_2) + \dot{z}_2 z_2 \\ &= -k_1 z_1^2 + z_2 (u - \dot{\alpha}_1 + z_1)\end{aligned}$$

IOP Conf. Series: Materials Science and Engineering **878** (2020) 012002 doi:10.1088/1757-899X/878/1/012002

Since the control *u* appears in this step no further steps are required. By choosing the control law as:

$$u = \dot{\alpha}_1 - z_1 - k_2 z_2 \tag{6}$$

with $k_2 > 0$, reads

$$\dot{V}_2 = -k_1 z_1^2 - k_2 z_2^2 < 0, \forall (z_1 \neq 0,\ z_2 \neq 0)$$

Instead of using the derivative $\dot{\alpha}_1$ in (6) we can take its analytical solution from (5) as

$$\begin{aligned}\dot{\alpha}_1 &= -\frac{\partial f(x_1)}{\partial x_1}\dot{x}_1 - k_1\dot{x}_1 \\ &= -\left(\frac{\partial f(x_1)}{\partial x_1} + k_1\right)(f(x_1) + x_2)\end{aligned}$$

Then the final expression for the control law is:

$$u = -\left(\frac{\partial f(x_1)}{\partial x_1} + k_1\right)(f(x_1) + x_2) - x_1 - k_2(x_2 + f(x_1) + k_1 x_1)$$

where the z variables are transformed in original coordinates using

$$\begin{bmatrix} z_1 \\ z_2 \end{bmatrix} = \begin{bmatrix} x_1 \\ x_2 - f(x_1) - k_1 x_1 \end{bmatrix}$$

which is called *backstepping coordinate transformation* with inverse

$$\begin{bmatrix} x_1 \\ x_2 \end{bmatrix} = \begin{bmatrix} z_1 \\ z_2 + f(z_1) + k_1 z_1 \end{bmatrix}.$$

3. DC motor driven inverted pendulum modelling

The objective system model is derived using the pendulum kinematics and electromechanical energy conversion of DC motor.

The equation of motion is:

$$J_d \frac{d\omega}{dt} = T_m + T_p, \tag{7}$$

where J_d is motor shaft inertia moment, ω is the angular speed, T_m is the electromagnetic torque of the motor and T_p is the pendulum load torque. The load torque for the objective system has the form

$$T_p = -l^2 m\ddot{\theta} - mgl\sin(\theta) - b\theta$$

where b is viscose friction coefficient, m is the attached ball mass, l is the link length, g is the gravitational constant and θ is the pendulum rotation angle.

The electromagnetic torque of the DC motor is

$$T_m = k_m i$$

where k_m is the motor torque constant.

The pendulum kinematics based on the last three equations reads

$$J\ddot{\theta} = k_m i - mgl\sin\theta - b\dot{\theta} \tag{8}$$

where

$$J = J_d + ml^2$$

is the total moment of inertia.

The electrical part of the system is modelled as current differential equation, which for DC motor is

$$\frac{di}{dt} = -\frac{k_e}{L}\dot{\theta} - \frac{R}{L}i + \frac{1}{L}u \tag{9}$$

where i is the armature current, k_e is the motor back EMF constant, R and L are the motor resistance and inductance and u is the control voltage.

The state space model state variables are selected as

$$x_1 = \theta$$
$$x_2 = \dot{\theta} = \omega$$
$$x_3 = i$$

Replacing the above in equations (8) and (9), yields the state space model of the objective system as

$$\dot{x}_1 = x_2$$
$$\dot{x}_2 = -c_1 \sin x_1 + c_2 x_3 - c_6 x_2 \quad (10a)$$
$$\dot{x}_3 = -c_3 x_2 - c_4 x_3 + c_5 u$$
$$y = x_1 \quad (10b)$$

where y is the system output and the model parameters are

$$c_1 = \frac{mgl}{J}, c_2 = \frac{k_t}{J}, c_6 = \frac{f_m}{J}, c_3 = \frac{k_e}{L}, c_4 = \frac{R}{L}, c_5 = \frac{1}{L}. \quad (11)$$

4. Nonlinear backstepping tracking control design

The control goal is the objective system output (the pendulum angle) (10b) to track exponentially a reference trajectory. This allows via the trajectory settings to control the performance specifications of the objective nonlinear system. The reference trajectory model is defined as

$$\dot{x}_{1r} = x_{2r}$$
$$\dot{x}_{2r} = x_{3r} \quad (12)$$
$$\dot{x}_{3r} = v$$

where

$$v = -k_{3r} x_{3r} - k_{2r} x_{2r} - k_{1r}(x_{1r} - y_d)$$

with y_d the desired pendulum position and $y_r = x_{1r}$. The objective nonlinear system (10 is third order system, hence three recursive steps has to be performed according to the backstepping procedure. Before proceeding with these steps the tracking task is transformed into stabilization task to make the design easier. For this purpose, an error tracking system has to be designed as

$$e_1 = x_{1r} - x_1$$
$$e_2 = x_{2r} - x_2 \quad (13)$$
$$e_3 = x_{3r} - x_3$$

with dynamics

$$\dot{e}_1 = e_2$$
$$\dot{e}_2 = x_{3r} + c_6 x_2 - c_2 x_3 + c_1 \sin(x_1) \quad (14)$$
$$\dot{e}_3 = -c_5 u + v + c_3 x_2 + c_4 x_3$$

Now the task is to stabilize the output tracking error e_1 to zero.

Step 1:

Since the objective is to regulate the error e_1 to zero the first backstepping variable is chosen as

$$z_1 = e_1 \quad (15)$$

In the first equation of system (14) e_2 is the selected *virtual control* input as

$$e_2 = z_2 + \alpha_1 \quad (16)$$

Hence the dynamics of (15) reads

$$\dot{z}_1 = z_2 + \alpha_1$$

The stabilizing function

$$\alpha_1 = -k_1 z_1 \quad (17)$$

IOP Conf. Series: Materials Science and Engineering **878** (2020) 012002 doi:10.1088/1757-899X/878/1/012002

where $k_1 > 0$ is chosen to stabilize the z_1 dynamics as

$$\dot{z}_1 = -k_1 z_1 + z_2$$

A candidate Lyapunov function for the z_1-system is:

$$V_1 = \frac{1}{2} z_1^2$$

$$\dot{V}_1 = z_1 \dot{z}_1$$

$$= -k_1 z_1^2 + z_1 z_2$$

The new state variable z_2 will not be used in the first step, but its presence is important since z_2 has to couple the z_1-system to the system considered in the next step.

Step 2:

The dynamics of z_2 is estimated by time differentiation of (16) as

$$\dot{z}_2 = x_{3r} + c_6 x_2 - c_2 x_3 - \dot{\alpha}_1 + c_1 \sin(x_1)$$

in this case the virtual control is chosen to be x_3 as

$$x_3 = z_3 + \alpha_2 \tag{18}$$

thus the z_2-dynamics yields

$$\dot{z}_2 = c_6 x_2 + x_{3r} - c_2 (z_3 + \alpha_2) - \dot{\alpha}_1 + c_1 \sin(x_1)\,.$$

From this step further the selection of stabilizing functions is done by recursively designed CLF. In this case the CLF is selected as

$$V_2 = V_1 + \frac{1}{2} z_2^2$$

$$\dot{V}_2 = \dot{V}_1 + z_2 \dot{z}_2$$

$$= -k_1 z_1^2 + z_1 z_2 + z_2 \left(c_6 x_2 + x_{3r} - c_2 (z_3 + \alpha_2) - \dot{\alpha}_1 + c_1 \sin(x_1)\right)$$

$$= -k_1 z_1^2 + z_2 \left(z_1 + c_6 x_2 + x_{3r} - c_2 (z_3 + \alpha_2) - \dot{\alpha}_1 + c_1 \sin(x_1)\right)$$

Now the stabilizing function α_2 is selected to make $\dot{V}_2$ negative definite as

$$\alpha_2 = \frac{k_2 z_2}{c_2} + \frac{c_6 x_2 + x_{3r} + z_1 - \dot{\alpha}_1 + c_1 \sin(x_1)}{c_2} \tag{19}$$

with $k_2 > 0$. Then the derivative $\dot{V}_2$ reads

$$\dot{V}_2 = -k_1 z_1^2 - k_2 z_2^2 - c_2 z_2 z_3$$

The $-c_2 z_2 z_3$ term will be eliminated in the next step.

Step 3:

The dynamics of z_3 is estimated from (18) with time differentiation as

$$\dot{z}_3 = c_5 u - c_3 x_2 - c_4 x_3 - \dot{\alpha}_2$$

It is evident that the control u appears in the z_3 dynamics. This shows that the design is in the final step. The last CLF is chosen as

$$V_3 = V_2 + \frac{1}{2} z_3^2$$

$$\dot{V}_3 = \dot{V}_2 + z_3 \dot{z}_3$$

$$= -k_1 z_1^2 - k_2 z_2^2 - c_2 z_2 z_3 + z_3 \left(c_5 u - c_3 x_2 - c_4 x_3 - \dot{\alpha}_2\right)$$

The final goal is to make $\dot{V}_3$ negative definite by selecting the appropriate control u. The choice of

$$u = \frac{-k_3 z_3}{c_5} + \frac{c_3 x_2 + c_4 x_3 + c_2 z_2 + \dot{\alpha}_2}{c_5} \tag{20}$$

turns the CLF to

$$\dot{V}_3 = -k_1 z_1^2 - k_2 z_2^2 - k_3 z_3^2$$

This implies that tracking control system error $\mathbf{z}$ is asymptotically stable. Moreover $\dot{V}_3$ yields

$$\dot{V}_3 = -\mathbf{z}^T \mathbf{K} \mathbf{z} \tag{21}$$

where

$$K = \begin{bmatrix} k_1 & 0 & 0 \\ 0 & k_2 & 0 \\ 0 & 0 & k_3 \end{bmatrix} \text{ and } \mathbf{z} = \begin{bmatrix} z_1 \\ z_2 \\ z_3 \end{bmatrix}$$

Since the matrix $\mathbf{K}$ is positive definite the following relation holds

$$\lambda_{min}[\mathbf{K}]\mathbf{I}_{3x3} \le \mathbf{K} \le \lambda_{max}[\mathbf{K}]\mathbf{I}_{3x3}$$

where $\lambda_{min}[\cdot]$ and $\lambda_{max}[\cdot]$ denote the minimal and the maximal eigenvalues. The slowest component of $\mathbf{z}$ is associated with the smallest eigenvalue, respectively. Considering the last relation, the slowest component of (21) will be

$$\dot{V}_3 = -\lambda_{min}[\mathbf{K}]\mathbf{z}^T\mathbf{z}$$

Thus, the considered slowest dynamics of $\dot{V}_3$ can be represented as a homogeneous equation with respect to V_3 with initial time t_0 as

$$\dot{V}_3 \le -2\lambda_{min}[\mathbf{K}]\, V_3, \ V(t_0) = V_0 .$$

The solution of this homogeneous equation is

$$V(t) \le e^{-2\lambda_{min}[\mathbf{K}](t-t_0)} V_0 .$$

The response $V(t)$ implies that the steady state error will be less than $1\% V_0$ when

$$\lambda_{min}[\mathbf{K}]T_E = \frac{5}{2}$$

The time T_E is the desired response time

$$T_E = \frac{5}{\lambda_{min}[\mathbf{K}]2}$$

Thus, the $\mathbf{z}$ error is exponentially stable with maximum transient response time T_E, which can be controlled by the smallest eigenvalue of the matrix $\mathbf{K}$. The larger $\lambda_{min}[\mathbf{K}]$ leads to faster response.

In order to obtain the final form of the control input the analytical solutions of the stabilizing functions derivatives are required. Differentiating equations (17) and (19) the derivatives read

$$\dot{\alpha}_1 = k_1^2 z_1 - k_1 z_2$$

$$\dot{\alpha}_2 = c_6 x_3 + \frac{(-2k_1 + k_1^3 - k_2)z_1}{c_2} + \frac{(1 - k_1^2 - k_1 k_2 - k_2^2)z_2}{c_2} + \frac{(-c_2 k_1 - c_2 k_2)z_3}{c_2} + \frac{x_2(-c_6^2 + c_1\cos(x_1))}{c_2} + \frac{\nu - c_1 c_6 \sin(x_1)}{c_2}$$

Then by replacing $\dot{\alpha}_2$ in equation (20) the final control input yields

$$u = \frac{(c_4 + c_6)x_3}{c_5} + \frac{(-2k_1 + k_1^3 - k_2)z_1}{c_2 c_5} + \frac{(1 + c_2^2 - k_1^2 - k_1 k_2 - k_2^2)z_2}{c_2 c_5} - \frac{(k_1 + k_2 + k_3)z_3}{c_5} + \frac{x_2(c_2 c_3 - c_6^2 + c_1\cos(x_1))}{c_2 c_5} + \frac{\nu - c_1 c_6 \sin(x_1)}{c_2 c_5}$$

It is evident that the control depends on the backstepping variables z_1, z_2 and z_3. These variables have to be transformed in original coordinates. This is achieved by applying the backstepping diffeomorphism

$$\mathbf{z}=\mathbf{z}(\mathbf{e},\mathbf{x},\mathbf{x}_r)=\begin{bmatrix} e_1 \\ e_2+k_1e_1 \\ -\frac{k_1k_2+1}{c_2}e1-\frac{k_1+k_2}{c_2}e_2-\frac{c_6}{c_2}x_2+x_3-\frac{x_{3r}}{c_2}-\frac{c_1\sin(x_1)}{c_2} \end{bmatrix} \tag{22}$$

By applying the above transformation, the closed loop system control in original coordinates yields

$$\begin{aligned} u=\frac{1}{c_2c_5}\Big[& e_1(c_2^2k_1+k_3+k_1k_2k_3)+e_2(1+c_2^2+k_2k_3+k_1(k_2+k_3))+v+c_2c_3x_2-c_6^2x_2 \\ & +c_6k_1x_2+c_6k_2x_2+c_6k_3x_2+c_2(c_4+c_6-k_1-k_2-k_3)x_3+(k_1+k_2+k_3)x_{3r} \\ & +c_1(x_2\cos(x1)+(-c_6+k_1+k_2+k_3)\sin(x_1))\Big] \end{aligned} \tag{23}$$

With this the task for nonlinear tracking control backstepping design is fulfilled with exponential stability. The transition response time is controllable via the control law coefficients k_1, k_2 and k_3. The overall performance specifications of the closed loop system are set by the reference trajectory.

The analysis of (22) when the error vector $\mathbf{z}$ converges to zero is

$$0=e_1$$

$$0=e_2+k_1e_1 \Rightarrow e_2=0$$

$$0=-\frac{k_1k_2+1}{c_2}e1-\frac{k_1+k_2}{c_2}e_2-\frac{c_6}{c_2}x_2+x_3-\frac{x_{3r}}{c_2}-\frac{c_1\sin(x_1)}{c_2}$$

$$=-\frac{c_6}{c_2}x_2+x_3-\frac{x_{3r}}{c_2}-\frac{c_1\sin(x_1)}{c_2}$$

From the last equation the expected value x_{3d} of the virtual control x_3 can be estimated as

$$x_{3d}=\frac{x_{3r}+c_1\sin(x_1)+c_6}{c_2}. \tag{24}$$

This analysis implies that the tracking errors e_1 and e_2 will converge to zero, while the state variable x_3 selected as virtual control will converge to x_{3d} with the convergence of $\mathbf{z}\to 0$. This will be proven graphically in the next section.

5. Dynamic simulation

Equations (10), (12), (13) and (23) are simulated in the MATLAB environment. The objective system parameters are

Table 1. Inverted pendulum system parameters.

Parameter	Value	Unit
m	0.5	kg
l	0.5	m
g	9.81	$m.s^{-2}$
k_t	0.4775	V.s
k_e	0.02	Nm.s
f_m	0.3	Ω
R	2.7	mH
L	3.7×10^{-3}	$kg.\,m^2$
J_d	0.4775	V.s

IOP Conf. Series: Materials Science and Engineering **878** (2020) 012002 doi:10.1088/1757-899X/878/1/012002

The pendulum system initial conditions are selected as $x_0 = [-\pi/2, 0, 0]^T$. The reference model coefficients are chosen by pole placement to be $k_{1r} = 343$, $k_{2r} = 147$, $k_{3r} = 21$ providing triple pole $p_d = -7$. Its initial conditions are $x_{0r} = [0,0,0]^T$;.The backstepping control parameters are set as $k_1 = k_2 = k_3 = 10$. The closed loop system is simulated for T=4 seconds with desired angle $y_d = \pi/2$ which is changed when $t \geq 2$ to $y_d = -\pi/2$.

The desired output tracking is shown on figure **2** together with its tracking error. It is evident that the

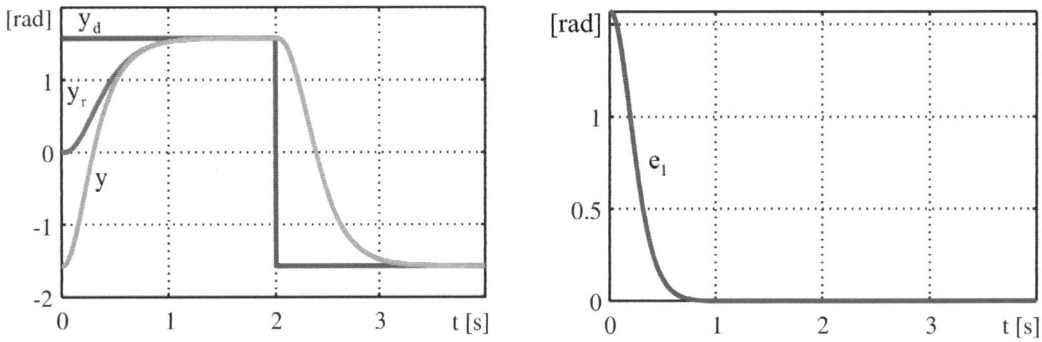

Figure 2. Output angle tracking.

system output y tracks the desired trajectory y_r without error after its convergence even if the desired angle y_d changes. Figure **3** depicts the speed tracking performance which has the same performance.

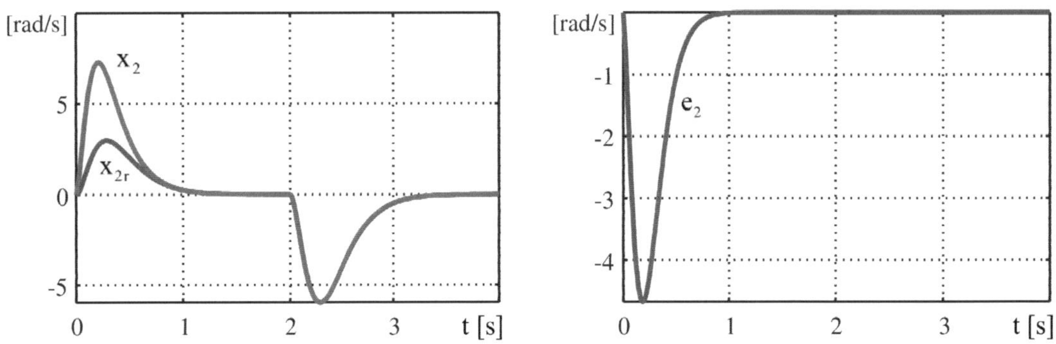

Figure 3. Speed tracking.

On figure **4** we see that the virtual control x_3 tracks it desired value according to (24). Its tracking is the reason of the zero tracking errors of the above variables. On the last graphics **5** the motor control

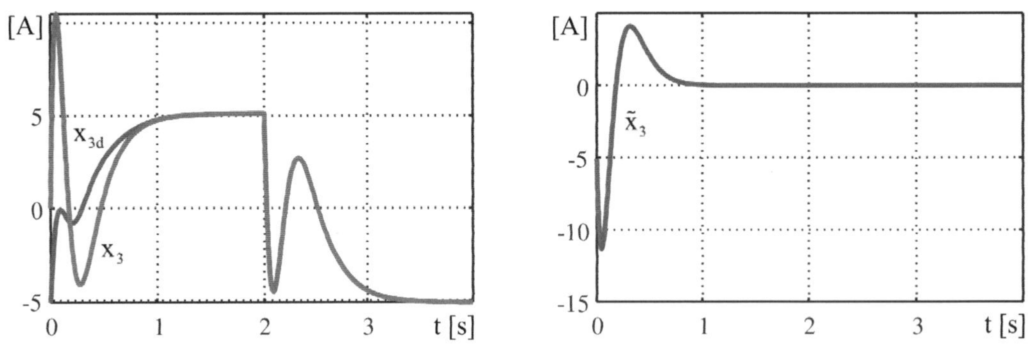

Figure 4. Current tracking.

voltage is shown. The last two figure are useful to check if the current and the voltage to be applied

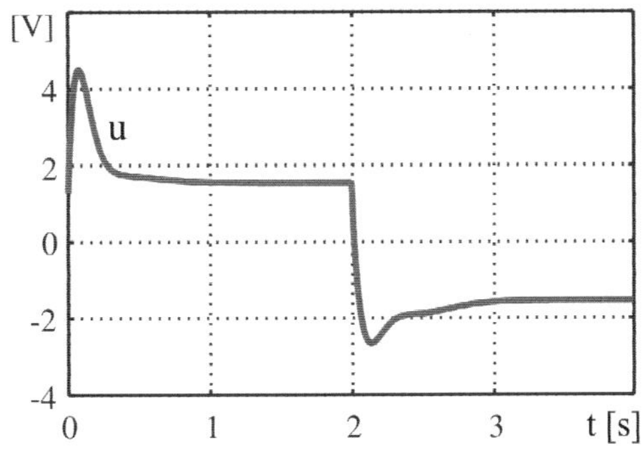

Figure 5. Control input voltage.

are in the motor specification range. For this specific motor the maximum voltage is $V_{max} = 110[V]$ and the maximum current is $I_{max} = 100[A]$ with continuous current $I_n = 12.5[A]$ thus the control is in the motor specification range.

6. Conclusions

The paper has presented trajectory tracking control design procedure for DC motor driven inverted pendulum system using nonlinear backstepping approach. The tracking control allows specification of the controlled system output performance via reference trajectory design.

The designed nonlinear closed loop system is dynamically simulated for two transition responses of the pendulum angle. The simulation results show the overdamped performance of the output specified by the reference trajectory. The control design achieves fast pendulum system output response without exceeding the motor maximum allowed armature current.

Acknowledgement

The author would like to thank the Research and Development Sector at the Technical University of Sofia for the financial support.

References

[1] Koditschek D 1987 Adaptive techniques for mechanical systems *Proc. 5th Yale Workshop on Adaptive Systems* New Haven pp 259–265

[2] Byrnes C and Isidori I 1989 New Results and examples in nonlinear feedback stabilization *Systems & Control Letters* vol 12 pp 437–442

[3] Narendra K and Annaswamy A 1989 *Stable adaptive systems* (New Jersy: Prentice Hall)

[4] Landau Y 1977 *Adaptive control* (New York: Marcel Dekker)

[5] Kanellakopoulos I, Kokotovic P V and Morse A S 1992 A Toolkit for nonlinear feedback design *Systems & Control Letters* vol 18 pp 83–92

[6] Krstic M Kanellakopoulos I and Kokotovic P V 1995 *Nonlinear and adaptive control design* (New York: John Wiley & Sons Ltd)

[7] Fossen T I and Strand J P 1999 Tutorial on nonlinear backstepping: Application to ship control *Modeling, identification and control* vol 20 iss 2 pp 83–134

[8] Merzoug M S and Benalla H 2010 Nonlinear backstepping control of permanent magnet synchronous motor (PMSM), *Int. J. of Systems Control* vol 1 iss 1 pp 30–34

[9] Sepulchre R Jankovic M and Kokotovic P 1997 *Constructive nonlinear control* (Berlin: Springer Verlag)

[10] Astolfi A, Karagianiss D and Ortega R 2008 *Nonlinear and adaptive control with applications* (London: Springer-Verlag)

Design of the H_∞ regulator for the control of glucose concentration in patients with first type diabetes

Sherif Sherif [1], Jordan Kralev[1], Tsonyo Slavov[1] and Veselin Kunchev[1]

[1]Technical University of Sofia, Faculty of Automatics, Department of Systems and Control, Sofia, Bulgaria

E-mail: mr.shark94@abv.bg; jkralev@yahoo.com; ts_slavov@tu-sofia.bg; v.kanchev1993@gmail.com

Abstract. Glucose is a basic source of energy for the living organism and is assimilated by tissues with the help of insulin receptors attached to the cell membrane. A lot of research has focused on improvement the quality of regulation of glucose concentration by applying modern methods of control theory. Mathematical descriptions of glucose and insulin metabolism are nonlinear processes with uncertainty in parameters. Therefore article is purposed at creating a corresponding simulation model, which would allow the research different approaches to controlling blood glucose levels in the presence of disturbances due from the acceptance of nutrients, physical activity and parametric uncertainty.

1. Introduction

Type 1 diabetes is an autoimmune disease in which the immune system attacks and kills beta cells responsible for secreting the hormone insulin, which has an essential role in regulating blood and tissue glucose concentrations. Beta cells are located in an area of the pancreas called Islets of Langerhans. In global, the number of people which are suffering from type 1 diabetes is about 300 million, and in Bulgaria they are about 100 000, however there is also a hidden morbidity and many from cases of diabetes remain undiagnosed.

Until the first administration of insulin treatment in 1922, type 1 diabetes runs as a disease incompatible with life, resulting in a fatal end within a few months. Modern insulin treatment is realized by periodically injecting insulin substitutes several times a day. Patients should also carefully monitor their glucose concentration in blood to prevent other diseases (cardiovascular, blindness, diseases of the nervous system) related to increased concentration (hyperglycemia). It is technically possible to continuously feed insulin through a specialized pump that injects the substance into the organism through a small subcutaneously mounted tubule. Excessively low glucose concentration (hypoglycemia) should not be allowed, in which also develop not so long-lasting, but requiring emergency intervention states (loss of consciousness, speech difficulties, anxiety).

2. Model of Glucose and Insulin metabolism

The processes of regulation of glucose concentration in blood are usually described by pharmacokinetic models for insulin dosage purposes [1-4]. Pharmacokinetic models refer to the rate and extent of distribution of a given the dosage form to different tissues as well as to the rate of elimination (metabolism) of the medicament. Most frequently applies a two-component pharmacokinetic model according to which the body is divided of central and peripheral compartment [4]. The central compartment consists of blood plasma and tissues, in which the distribution of substances is practically instantaneous. The peripheral compartment consists of tissues in which the diffusion of the medicament is slower. The metabolism of glucose and insulin can be decomposed into the following processes - assimilation of subcutaneously injected insulin, assimilation of carbohydrates from the digestive tract and interaction between insulin and glucose. These processes will be examined sequentially and will serve to form a nonlinear model in Simulink®.

a) Model for the absorption of subcutaneously injected insulin

A classic two-component pharmacokinetic model for the species was used:

$$\dot{Q}_{i1}(t) = u(t) - \frac{Q_{i1}(t)}{T_i},$$
$$\dot{Q}_{i2}(t) = \frac{Q_{i1}(t)}{T_i} - \frac{Q_{i2}(t)}{T_i} \tag{1}$$

where $\boldsymbol{Q_{i1}(t)}$ and $\boldsymbol{Q_{i2}(t)}$ are the amounts of insulin in the primary and secondary compartments, $\boldsymbol{u(t)}$ is the flow rate of the introduced insulin substitute from an external source, and $\boldsymbol{T_i}$ is the time constant of the process determined by the time to reach the maximum concentration of insulin in the blood plasma after injection of a single dose of medicine. The concentration of insulin in blood plasma is determined by the expression **(2)**

$$I_p(t) = \frac{Q_{i2}(t)}{T_i w K_{MCR}} + I_b \tag{2}$$

$\boldsymbol{K_{MCR}}$ is the rate of elimination of the insulin substitute, $\boldsymbol{I_b}$ is the average background concentration of insulin, and $\boldsymbol{w}$ is the body weight of the investigated subject. The commented model was implemented in Simulink® as the diagram show in **Figure 1**.

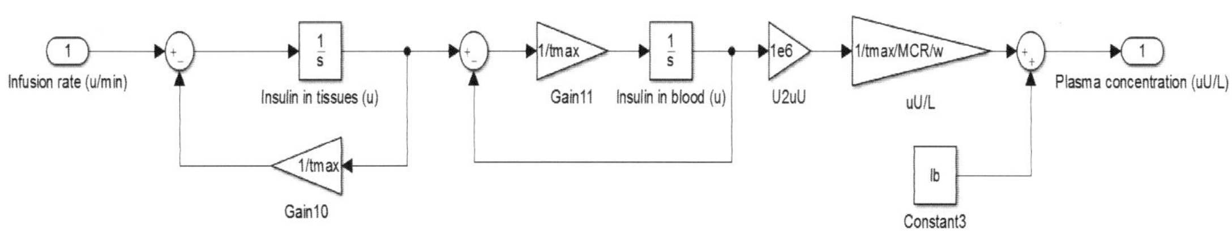

Figure 1. Simulink® model for the assimilation of subcutaneously injected insulin

b) Model of absorption of carbohydrates from the digestive system

The main external disturbance factor on glucose concentration in the body appears the mode on eating, with the predominant influence falls on the consumption of carbohydrates presented in the model as a kinetic value $\boldsymbol{d(t)}$:

$$\dot{Q}_{m1}(t) = K_D d(t) - \frac{Q_{m1}(t)}{T_m},$$
$$\dot{Q}_{m2}(t) = \frac{Q_{m1}(t)}{T_m} - \frac{Q_{m2}(t)}{T_m} \tag{3}$$

IOP Conf. Series: Materials Science and Engineering **878** (2020) 012003 doi:10.1088/1757-899X/878/1/012003

Here, analogous to the previous case $\boldsymbol{Q_{m1}(t)}$ and $\boldsymbol{Q_{m2}(t)}$ are the amounts of carbohydrates in the primary and secondary compartments, and $\boldsymbol{T_m}$ is the time constant of the diffusion between them. $\boldsymbol{K_D}$ determines glucose units obtained from one gram of carbohydrates. **Figure 2** contains model of Simulink® for assimilation on carbohydrate. Accordingly, glucose debit towards organism is described as:

$$U_m(t) = \frac{Q_{m2}(t)}{T_m} \tag{4}$$

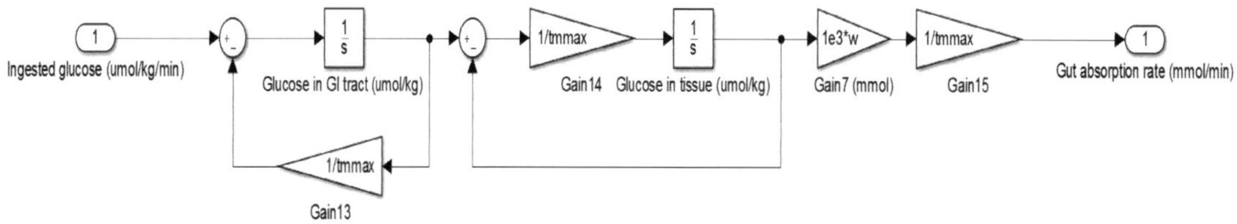

Figure 2. Simulink® model of assimilation of carbohydrates from digestive system

c) Model of the influence between glucose and insulin

The glucose concentration in the organism is again described by a two-component model with states $\boldsymbol{Q_1(t)}$ and $\boldsymbol{Q_2(t)}$, but also taking into account the importance of the following processes - glucose synthesis, assimilation of glucose in tissue and glucose transport. The rate of glucose synthesis $\boldsymbol{x_{EGP}(t)}$ is given by the expression **(5)**

$$x_{EGP}(t) = \begin{cases} x_{EGP,0}(1 - x_3(t)) & ,x_3(t) < 1 \\ 0 & ,x_3(t) \geq 1 \end{cases} \tag{5}$$

where $\boldsymbol{x_{EGP}(t)}$ is the rate of glucose synthesis in the absence of insulin, $\boldsymbol{x_3(t)}$ is a state variable that indicates how much insulin quantity received has affected glucose synthesis. Similarly, the impact of insulin on the processes of glucose transport and absorption are given by state variables $\boldsymbol{x_1}$ and $\boldsymbol{x_2}$, which will be explained below. The glucose concentration model is obtained as:

$$\begin{aligned} \dot{Q}_1(t) &= -F_{01}\frac{Q_1(t)}{1+Q_1(t)} - x_1(t)Q_1(t) + k_{12}Q_2(t) + x_{EGP}(t) + U_m(t) \\ \dot{Q}_2(t) &= x_1(t)Q_1(t) - (k_{12} + x_2(t))Q_2(t) \end{aligned} \tag{6}$$

where $\boldsymbol{F_{01}}$ determines the rate of glucose metabolism to adenosine triphosphate and pyruvate, and $\boldsymbol{k_{12}}$ is the diffusion coefficient between the primary and secondary compartments in the pharmacokinetic model.

d) Activity model of the insulin preparation

The dynamics of the impact of insulin on glucose concentration can be considered in three components [2], which are characterized from specific for each person parameters (**Figure 3**). The impact of insulin on glucose transport is expressed by **(7)**

$$\dot{x}_1(t) = -k_{a1}x_1(t) + k_{a1}S_tI_p(t) \tag{7}$$

and is characterized by time constant $\boldsymbol{1/k_{a1}}$ and sensitivity $\boldsymbol{S_t}$. The impact of insulin on glucose absorption is expressed by the expression **(8)**

$$\dot{x}_2(t) = -k_{a2}x_2(t) + k_{a2}S_dI_p(t) \tag{8}$$

with time constant $\mathbf{1/k_{a2}}$ and sensitivity $\boldsymbol{S_d}$. Analogous, the impact of insulin on glucose synthesis is with parameters $\boldsymbol{k_{a3}}$ and $\boldsymbol{S_e}$.

$$\dot{x}_3(t) = -k_{a3}x_3(t) + k_{a3}S_eI_p(t) \tag{9}$$

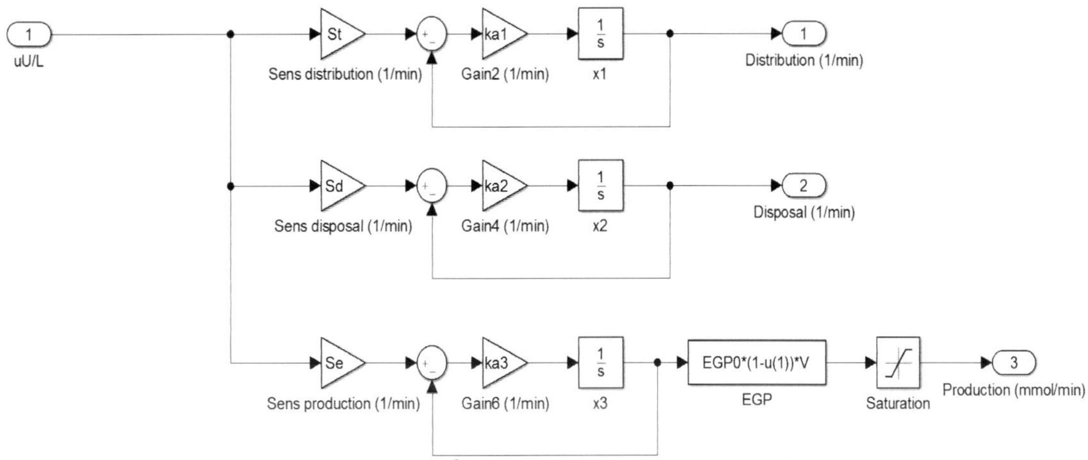

Figure 3. Simulink® model for activity of insulin preparation

Figure 4 shows the complete model of insulin and glucose metabolism, which is created based on the presented equations. For the practical determination of its parameters may be used experimental data obtained from organisms putted on a controlled diet and with suppressed internal secretion of insulin. For the purposes of this research, exemplary values of the commented parameters have been selected, which are given in **Table 1** together with their units of measurement. In the nonlinear model, it is also necessary to include some blocks for conversion between units of measure, as well as for introduction limit values for external signals.

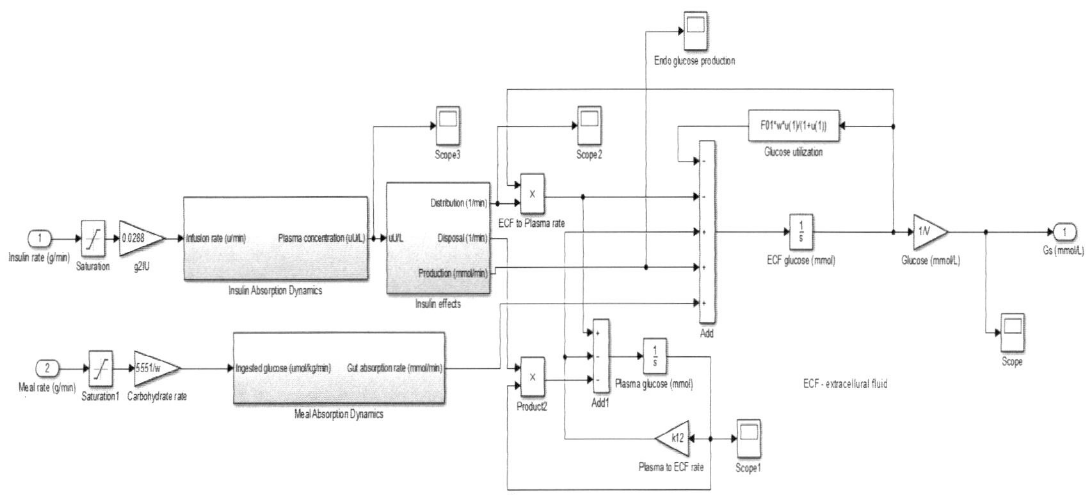

Figure 4. A non-linear Simulink® model of insulin and glucose metabolism

Table 1. Parameters of nonlinear model

Indication	Value	Unit	Indication	Value	Unit
T_i	60	$[min]$	k_{a2}	1	$[min^{-1}]$
T_m	30	$[min]$	k_{a3}	1/40	$[min^{-1}]$
w	80	$[kg]$	S_t	$0.3k_{a1}$	$[L/\mu IU/min]$
I_b	20×10^3	$[\mu IU/L]$	S_d	$0.3k_{a2}$	$[L/\mu IU/min]$
K_{MCR}	0.024	$[L/kg/min]$	S_e	$0.3k_{a3}$	$[L/\mu IU/min]$
F_{01}	1.62/180	$[mmol/kg/min]$	k_{12}	1/30	$[min^{-1}]$
$x_{EGP,0}$	10	$[mmol/L]$	V	7	$[L]$
k_{a1}	0.1	$[min^{-1}]$			

3. Linearized model

The presented model of interaction between insulin and glucose has non-stationary parameters $\boldsymbol{x_1}$, $\boldsymbol{x_2}$ and $\boldsymbol{x_3}$ representing insulin action upon glucose distribution, transportation and generation which multiply the first and second compartment glucose concentration [5]. However do to smooth variation of the insulin action states and considering that model operates around a fixed steady state complying with homeostasis conditions the model can be sufficiently well approximated with its linearization [6,7,8]. Of course the full model can also be employed for direct nonlinear controller design by sliding-mode or feedback linearization techniques. The advantage of using linearized model is that we can employ linear controller, however it has to be tested on the nonlinear model to prove it guarantees closed-loop stability and performance. Additionally the use of linearization is justified if we represent the nonlinear model as a corresponding set of linear models or equivalently as nominal model with uncertainty which describe all input-output behaviors. When the bound of uncertainty is not too large and if the controller is robust with respect to it then we may be sure that the linear controller will perform well in the nonlinear system under all conditions.

The purpose in controlling glucose concentration by injection an insulin substitute is to adjust the glucose level around the operating point of $\boldsymbol{6\,[mmol/L]}$ within $\boldsymbol{\pm 1[mmol/L]}$, so that negative effects of hyperglycemia or hypoglycemia do not occur. Linearization was performed around a selected equilibrium state with parameters given in **Table 2**. When determining the parameters of the equilibrium state, the values of $\boldsymbol{u_{trim}}$ and $\boldsymbol{y_{trim}}$ are fixed, and the remaining values are determined by an optimization procedure so that the derivatives by state are as close as possible to zero, i.e. $\boldsymbol{\dot{x}_{trim}(t) \approx 0}$.

Table 2. Parameters of equilibrium state

Variable	Value	Unit	Variable	Value	Unit
$Q_{1,trim} = x_{1,trim}$	42	$[mmol]$	$x_{3,trim} = x_{7,trim}$	16.1	$[mmol/min]$
$Q_{2,trim} = x_{2,trim}$	4.2	$[mmol]$	$Q_{m1,trim} = x_{8,trim}$	10.2	$[\mu mol/kg]$
$Q_{i2,trim} = x_{3,trim}$	0.173	$[IU]$	$Q_{m2,trim} = x_{9,trim}$	10.2	$[\mu mol/kg]$
$Q_{i1,trim} = x_{4,trim}$	0.173	$[mmol]$	u_{trim}	0.1	$[g/min]$
$x_{1,trim} = x_{5,trim}$	645	$[min^{-1}]$	d_{trim}	0.00488	$[g/min]$
$x_{2,trim} = x_{6,trim}$	6450	$[min^{-1}]$	y_{trim}	6	$[mmol/L]$

After linearization around the equilibrium state, a model in state-space (SS) from a ninth-order:

$$\begin{aligned} \dot{x} &= A(x - x_{trim}) + B(u - u_{trim}) + G(d - d_{trim}) \\ y - y_{trim} &= C(x - x_{trim}) \end{aligned} \tag{10}$$

where the matrices in this description are:

$$A=\begin{bmatrix} -645 & 0 & 0 & -42 & 0 & -70 & 0 & 2667 & 0 \\ 0 & 0 & 0 & 0 & 0 & 0 & 0 & 0 & 0 \\ 0 & 0 & 0 & 0 & 0 & 0 & 0 & 0 & 0 \\ 0 & 26 & 0 & -0.1 & 0 & 0 & 0 & 0 & 0 \\ 0 & 2604 & 0 & 0 & -0.1 & 0 & 0 & 0 & 0 \\ 0 & 0.2 & 0 & 0 & 0 & 0 & 0 & 0 & 0 \\ 0 & 0 & 0 & 0 & 0 & 0 & 0 & 0 & 0 \\ 0 & 0 & 0 & 0 & 0 & 0 & 0 & 0 & 0 \\ 645 & 0 & 0 & 42 & -4.2 & 0 & 0 & 0 & -6450 \end{bmatrix}$$

$$(B\ G)^T=\begin{bmatrix} 0 & 0 & 0.0288 & 0 & 0 & 0 & 0 & 0 & 0 \\ 0 & 0 & 0 & 0 & 0 & 0 & 69.3875 & 0 & 0 \end{bmatrix}$$

$$C=[0.1429\quad 0\quad 0\quad 0\quad 0\quad 0\quad 0\quad 0\quad 0]$$

In **Figure 5** and **Figure 6** shows step responses of the linearized system respectively in the control signal and external disturbance. From these graphics show that the researched system has aperiodic character.

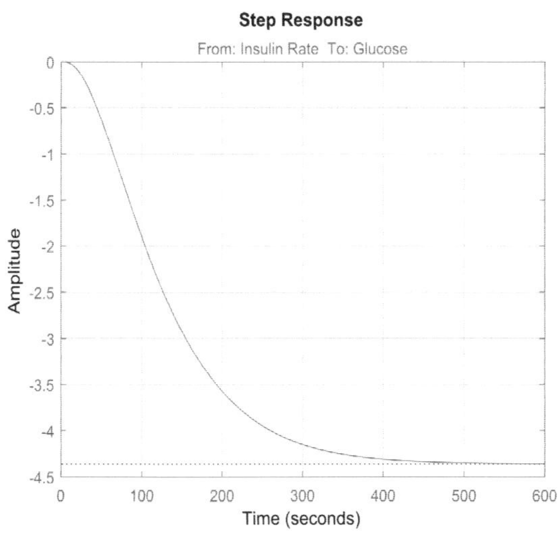

Figure 5. Step response of control input (**debit of insulin**)

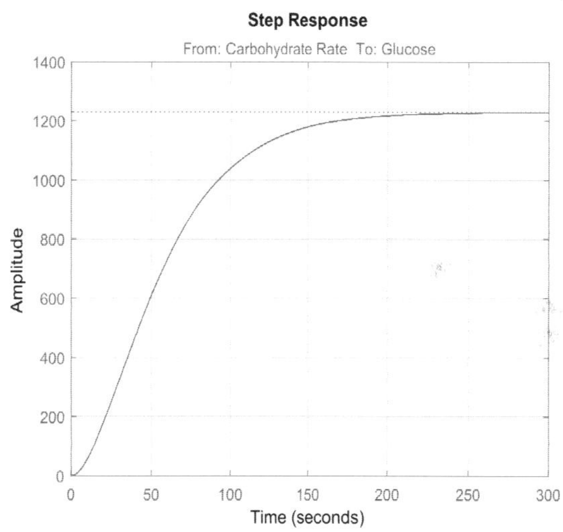

Figure 6. Step response of disturbance (**intake of carbohydrates**)

4. Design of the H$_\infty$ controller

The purpose of artificial pancreas control system is to guarantee that blood glucose is within healthy limits in presence of imprecisely measured disturbances like food ingestion, physical activity, metabolism variation, etc [9,10,11]. The idea behind H-infinity controller is to minimize the magnitude of H-infinity norm of the closed-loop system with respect to external disturbances and performance measures. Let first examine the extended linearized open-loop model on **Figure 7**.

The first input is food ingestion disturbance $\boldsymbol{d}$ with unit norm, which is scaled in frequency domain by the corresponding weighting filter:

$$W_d(j\omega)=0.001\frac{1+j\omega}{1+j0.1\omega} \tag{11}$$

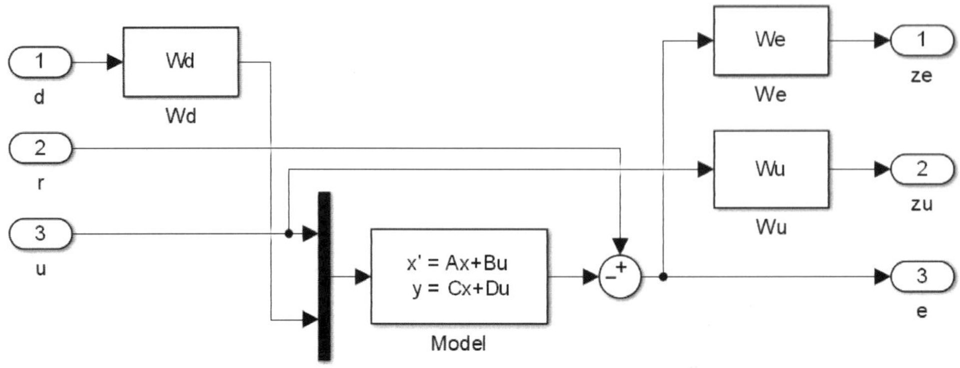

Figure 7. Block diagram of extended plant with weighting filters

Its gain and frequency band are selected with respect to the amplitude of the food intake rate in $\boldsymbol{g/min}$. The food intake is the major driver of blood glucose in the organism, so it should be taken into account during the controller design. The second input is the reference glucose value $\boldsymbol{r}$, which for first-type diabetes patients is typically around $\boldsymbol{6}\ [\boldsymbol{mmol/L}]$. It is very important to regulate glucose to that reference in the whole frequency range as much as possible to limit its negative effects for the health. The manipulated variable is the insulin intake rate in $\boldsymbol{g/min}$ represented by the third input signal $\boldsymbol{u}$.

The first performance index of the system is the weighted error signal between reference and actual glucose levels. The weighting filter is given by the expression **(12)**

$$W_e(j\omega) = 100\frac{1 + j10\omega}{1 + j2000\omega} \qquad (12)$$

is selected to guarantee 100 times attenuation of the external disturbances or $\boldsymbol{0.01}\ [\boldsymbol{mmol/L}]$ glucose error in the low frequency range. With this filter we also specify the desired bandwidth of the closed-loop system which is determinant of the settling time and oscillations around the reference value. The second performance index is the weighted control signal which is filtered.

$$W_u(j\omega) = 0.02\frac{1 + j\omega}{1 + j0.1\omega} \qquad (13)$$

This filter is selected to limit the amplitude and rate of the insulin flow rate to satisfy physical limitations of the insulin pump and to minimize local physiological stress in the tissue. The third output is the measured error signal which is used by the controller to calculated the control signal. The H-infinity design looks to find an optimal controller $\boldsymbol{K}$ such that minimize the H-infinity norm of the closed-loop system with respect to uncontrolled disturbances and performance indices as

$$\min_K \|F_L(T_{zd}, K)\|_\infty = \min_K \max_\omega \bar{\sigma}\big(F_L(T_{zd}, K)\big) < \gamma \qquad (14)$$

where $\boldsymbol{F_L}$ is the lower fractional transform of the controller $\boldsymbol{K}$ with the extended plant $\boldsymbol{T_{zd}}$, $\bar{\boldsymbol{\sigma}}$ is the upper bound of largest singular value of the closed-loop system. A summary of the iterative procedure for synthesis of the H-infinity controller is presented on **Figure 8**. As we can see the achieved value of gamma ($\boldsymbol{\gamma}$) is smaller than unity which means that the closed-loop system satisfies the prescribed requirements.

Figure 9 shows the magnitude response of the controller $\boldsymbol{K}$, where $\boldsymbol{u = K(r - y)}$. The controller dynamics is located in low-frequency range and includes zeros to compensate some of the plant dynamics. **Figure 10** shows the complementary sensitivity of the closed-loop system with the controller. The achieved bandwidth is $\mathbf{0.118}\ [\boldsymbol{rad/s}]$, which is enough for the glucose closed-loop control.

```
gamma     hamx_eig  xinf_eig   hamy_eig   yinf_eig    nrho_xy   p/f
  1.422   2.6e-02   3.5e-20   5.0e-04   -1.5e-11    0.0906    p
  0.961   2.6e-02  -9.8e-20   5.0e-04    0.0e+00    0.7687    p
  0.730   2.6e-02  -7.7e+01#  5.0e-04    0.0e+00    0.4135    f
  0.846   2.6e-02  -4.5e+02#  5.0e-04    0.0e+00    1.6013#   f
  0.903   2.6e-02  -1.0e-14   5.0e-04   -2.2e-11    3.1560#   f
  0.932   2.6e-02  -3.8e-15   5.0e-04    0.0e+00    1.2450#   f
  0.947   2.6e-02  -1.1e-14   5.0e-04    0.0e+00    0.9518    p
  0.939   2.6e-02  -1.8e-15   5.0e-04   -1.1e-11    1.0793#   f

Gamma value achieved:     0.9465
```

Figure 8. Iterative procedure for synthesis of controller using bisection algorithm.

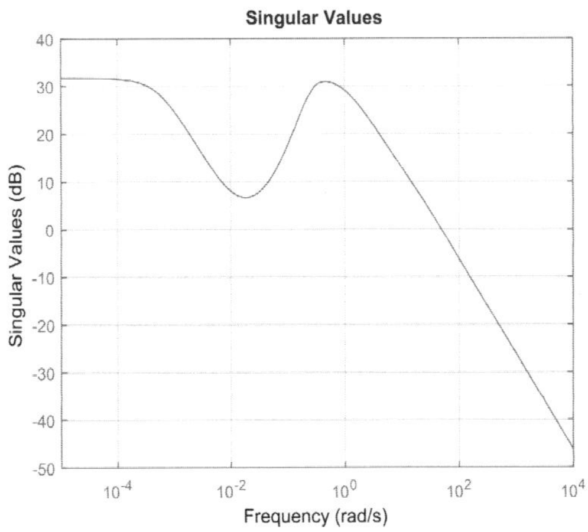

Figure 9. Singular values of H-infinity controller

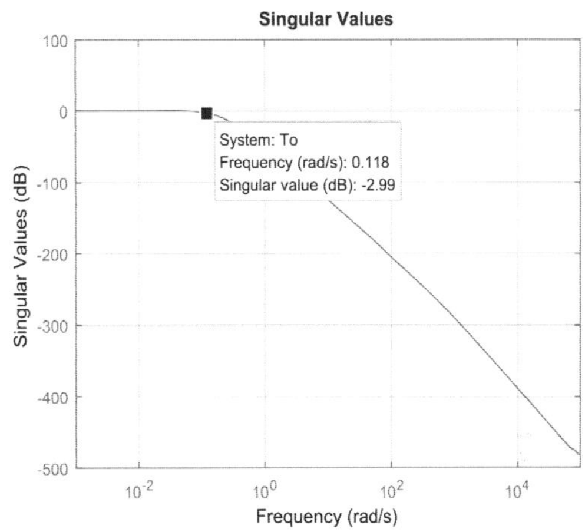

Figure 10. Complementary sensitivity of the closed-loop system with H-infinity controller

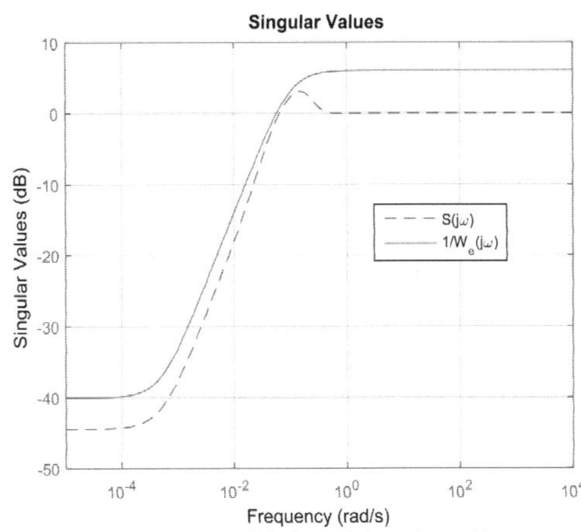

Figure 11. Output sensitivity of the closed loop system compared with its inverted weighting filter

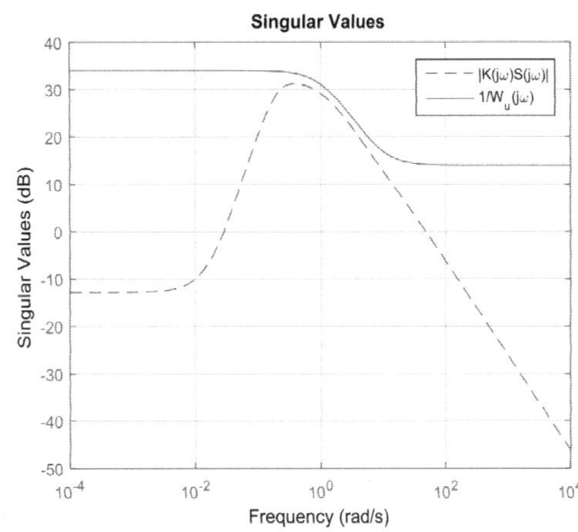

Figure 12. Sensitivity of control signal to output noises compared with its inverted weighting filter

On **Figure 11** is presented the output sensitivity of the closed-loop where we can observe that the system attenuates the output disturbances 100 times. Also the closed-loop system satisfies the requirements specified by the filter $\boldsymbol{W_e}$. **Figure 12** presents the sensitivity of control action to output noise. The system satisfies the prescribed requirements and no amplification of high frequency noise is observed.

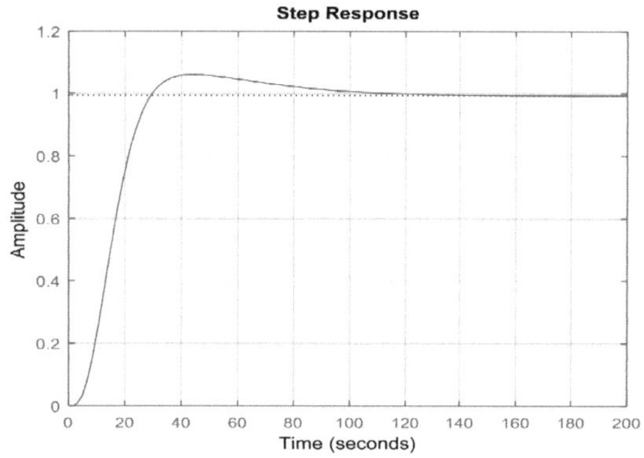

Figure 13. Simulated step response of the closed-loop system with the linearized model

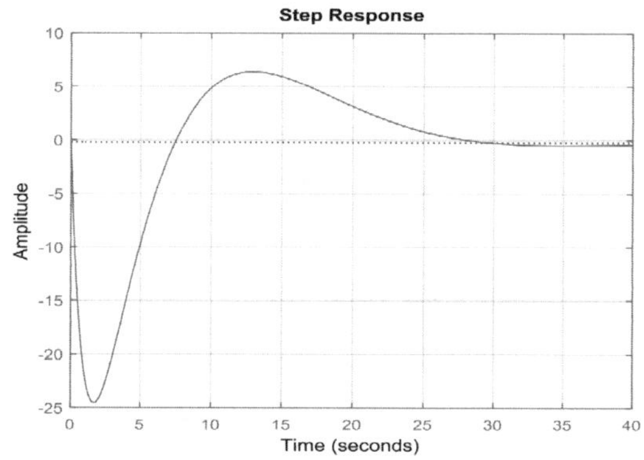

Figure 14. Simulated control signal of the closed-loop system with the linearized model

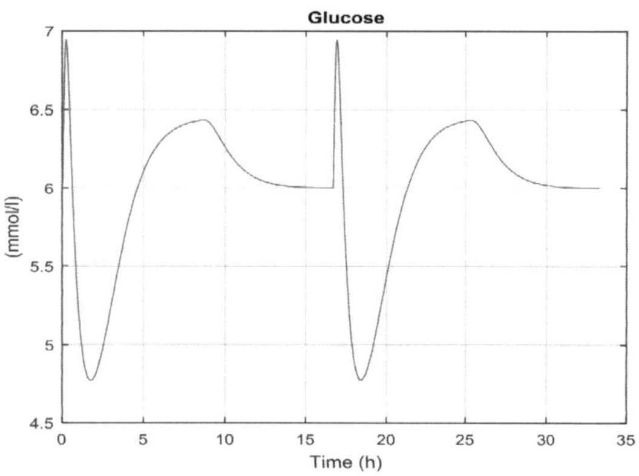

Figure 15. Glucose response of the nonlinear model with the H-infinity controller to periodic ingestion disturbance

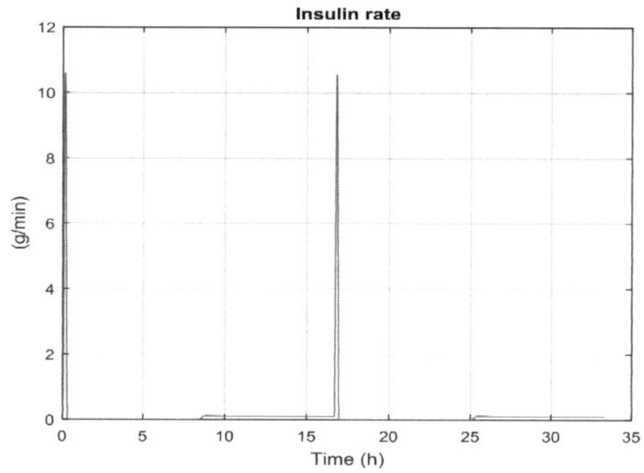

Figure 16. Calculated insulin dosage by the H-infinity controller with the nonlinear model

5. Conclusions

In this article has been reviewed one approach for design H_∞ regulator to stabilization glucose concentration in patients diagnosed with their first type diabetes by subcutaneous injected of an insulin substitute. The main limiting factor for achieving quality control of closed-loop glucose concentration is the slow assimilation of the subcutaneously injected insulin substitute, which creates the risk of an excessive increase in insulin concentration in the tissues immediately after eating, as it is possible the regulator to deliver insulin doses in for 1-2 hours as a reaction to increased glucose concentration. This in turn would lead to an effect of hypoglycemia, which may be for a long period of time.

Acknowledge

The present research was received seed funding by the Bulgarian National Science Fund (BNSF) under contract KP-06-M27/10.

References

[1] A. Haidar 2016 *The Artificial Pancreas* IEEE Control Systems Magazine doi: 10.1109/MCS.2016.2584318.

[2] E. Ferrannini, J.D. Smith, C. Cobelli et al. 1985 *Effect of Insulin on the Distribution and Disposition of Glucose in Men* Journal of Clinical Investication vol. 76 pp. 37-364

[3] B. Kovatchev, W.V. Tamborlane, W.T. Cefalu, et al. 2016 *The artificial pancreas in 2016: A digital treatment ecosystem for diabetes* Diabetes Care vol. 39 pp. 1123-1126

[4] R. Hovorka 2006 *Continuous glucose monitoring and closed-loop systems* Diabetic Med. vol. 23(1) pp. 1-12

[5] P. Biswas, S. Bhaumik, I. Patiyat. 2016 *Estimation of glucose and insulin concentration using nonlinear Gaussian filters* Proc. IEEE 1st In. Conf. Control Measurement Instrumentation pp. 16-20

[6] L. M. Huyett, E. Dassau, H. C. Zisser, et al. 2015 *Design and evaluation of a robust PID controller for a fully implantable artificial pancreas* Ind. Eng. Chem. Res vol. 54(42) pp. 10311-10321

[7] G. M. Steil, K. Rebrin, C. Darwin, et al. 2006 *Feasibility of automating insulin delivery for the treatment of type 1 diabetes* Diabetes vol. 55(12) pp. 3344-3350

[8] C. C. Palerm 2011 *Physiologic insulin delivery with insulin feedback: A control system perspective* Comput. Methods Program Biomed vol. 102(2) pp. 130-137

[9] P. Dua, F. J. Doyle, III, E. N. Pistikopoulos 2006 *Model-based blood glucose control for type 1 diabetes via parametric programming* IEEE Trans. Biomed. Eng. vol. 53 pp. 1478-1491

[10] E. D. Lehmann and T. Deutsch 1992 *A physiological model of glucose-insulin interaction in type 1 diabetes mellitus* J. Biomed. Eng. vol. 14 pp. 235-242

[11] D. R. Worthington 1997 *Minimal model of food absorption in the gut* Med. Inform. (Lond.) vol. 22 pp. 35-45

IOP Conf. Series: Materials Science and Engineering **878** (2020) 012004 doi:10.1088/1757-899X/878/1/012004

Manipulator for upper limbs rehabilitation- ROBCO ®

Nayden Chivarov[1, 2] and Denis Chikurtev[1, 2]

1 Institute of Information and Communication Technologies, Bulgarian Academy of Sciences, Acad. G. Bonchev str., block 2, Sofia, Bulgaria
2 European Polytechnical University, Pernik, Bulgaria

nshivarov@code.bg, dchikurtev@gmail.com

Abstract: This article presents a manipulator for upper limbs rehabilitation Robco ®. The manipulator will be able to allow rehabilitation of the shoulder, elbow and wrist. Manipulator's hardware, software and web-based user interface have been researched and developed to allow the required functions and properties for rehabilitation procedures. Usability experiments have been performed to investigate the functionalities. The results demonstrate precise positioning and repeatability. Finally, we can conclude that the rehabilitation of upper limbs can be robotized, which will reduce time and costs of the rehabilitation procedures.

1. Introduction.

Technological progress poses a number of challenges to medicine. On the way to making progress, a number of treatments have undergone qualitative changes that necessitate a redefinition of the classic methods of treatment and rehabilitation and their adaptation to the new social order [1].

In recent years, robotics has been entering the medical field in various areas [2, 3]. With the implementation of Robots, a variety of manipulations and operations are performed, helping to reduce recovery time. One of the areas where robots are used is rehabilitation [4, 5].

This article addresses the issues of robotic upper limb rehabilitation, using the advances in modern robotic technology.

It is of utmost importance that the principle of controlled autonomy of robots is respected, according to which the initial planning of the treatment and the final decision for its implementation remain always at the discretion of the physician/rehabilitator.

Medical robots will continue to advance precision surgery and repeat procedures and manipulations, and will have the potential to improve rehabilitation outcomes and provide highly efficient logistical support in hospitals [6]. There is great progress and additional potential in the field of exoskeletons that can be worn on the human body and serve both for the rehabilitation and support of the elderly and the disabled [7].

Medical robots have the potential to reduce health care costs by enabling healthcare professionals to focus their attention from treatment to prevention and by providing more budgetary resources to better adapt to the diversity of patients' needs for ongoing care training of medical specialists and research [8].

2. Manipulator for upper limbs Rehabilitation- ROBCO ®.

The Robco upper limb rehabilitation manipulator (Fig. 1) has 2 degrees of freedom, with a stroke volume from -30 to 180 °. It is controlled by PC and microprocessor controller, programming the parameters of the device (extension, flexion, pause, power and speed). It is intended for rehabilitation of the shoulder joint, elbow joints and wrist. The manipulator is versatile and is designed for both left and right hands. Used in a seat position.

The main parameters of the manipulator are the workspace, speed, time, control mode and power.

The angular velocity of joint rotation can be varied on a scale of 1 to 9 as follows:
- at first speed the movement is carried out by 0,5 ° / s, i.e. from -30 to 180 °, run for 7 minutes;
- second: 0,9°/s, for 3 minutes и 8 seconds;
- third: 1,45°/s, for 2 minutes and 41 seconds;
- forth: 1,75°/s, for 2 minutes;
- fifth: 2,15°/s, for 1 minute and 37 seconds
- sixth: 2,5°/s, for 1 minute and 24 seconds;
- seventh: 2,9°/s, for 1 minute and 12 seconds
- eight: 3,3°/s, for 1 minute и 3 seconds
- ninth: 3,75°/s. for 56 seconds

The device has a torque function that indicates the power of operation:

- large - actuators run at 100% power. The system only monitors the trajectory of the movement and seeks to overcome any resistance encountered during movement. In this setting, if a spasm occurs during the therapy process, the system will not respond and will strive to overcome the change in resistance. This can lead to trauma, and therefore is monitor under medical supervision.
- medium - actuators operate at 65% of their capacity and are able to stop and reverse the direction of movement when a resistance greater than engine power occurs.
- small - the smallest power represents 35% of the actuator capacity. When a resistance greater than the power of the device occurs, it will stop movement. This type of adjustment is the most sensitive and guarantees the safety of therapy in patients with contractures, increased muscle tone. For the convenience and safety of the patient, the manipulator has a remote control, which can be stopped and started at any time.

The manipulator has 2 degrees of freedom of rotation type. The manipulator type is "SCARA" [13]. The purpose of the developed model is to allow rehabilitation of the upper limbs and more precisely the shoulder, elbow and wrist.

Figure 1 shows the construction of the robotic manipulator. It consists of 3 units, base, which is static, and two active joints. A specialized pad is mounted on the first joint, on which the user places the forearm. At the end of the second joint, a handle is mounted which user hold with his palm.

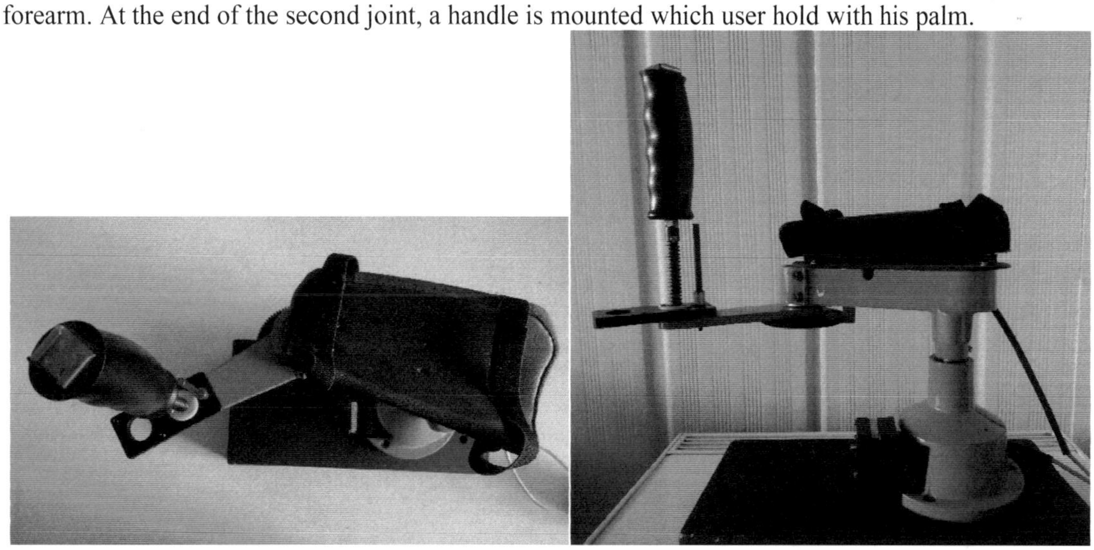

Figure 1. Manipulator for upper limbs Rehabilitation- ROBCO ®.

The maximum rotation angles, which are set as end positions, are: for joints 1 from -120 to 60 and for joints 2 from -60 to 30. The following restrictions are specified to guarantee user safety:

If $\varphi > 45$, then $\theta < 10$;

If $\varphi < -45$, then $\theta > -10$;

Where φ is the angle at the first joint of the manipulator and θ is the angle at the second joint (see figure 2). These restrictions are introduced to limit the movements of the manipulator so that movements that endanger the patient are not performed.

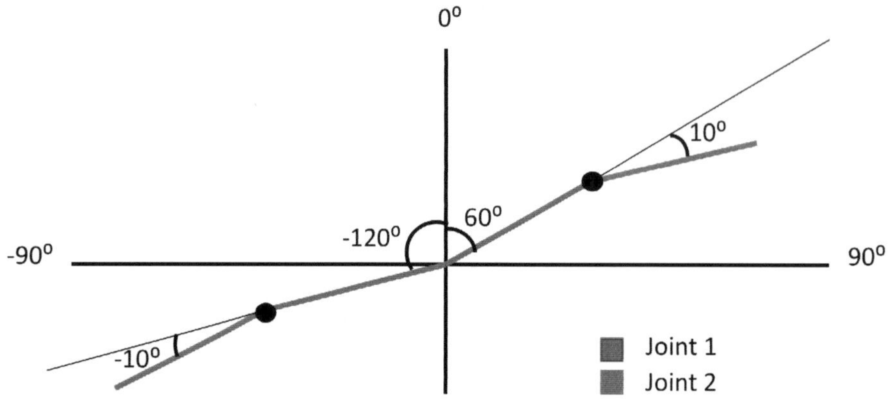

Figure 2. Limitations of rotation angles – maximum end angles.

3. Manipulator Hardware System.

The manipulator is powered by two stepper motors. Depending on the drive gear for first joint, the ratio of the angle of rotation to the number of steps is 1: 3, and for second joint it is 1: 2. In order to operate stepper motors, proper drivers for them have been selected. The motor drivers connect to an Arduino controller that is connected to a computer [9]. Drivers must be provided with additional power according to the controllers used, while the Arduino controller is powered by the computer. The schematic diagram of the connection is shown in Figure 3.

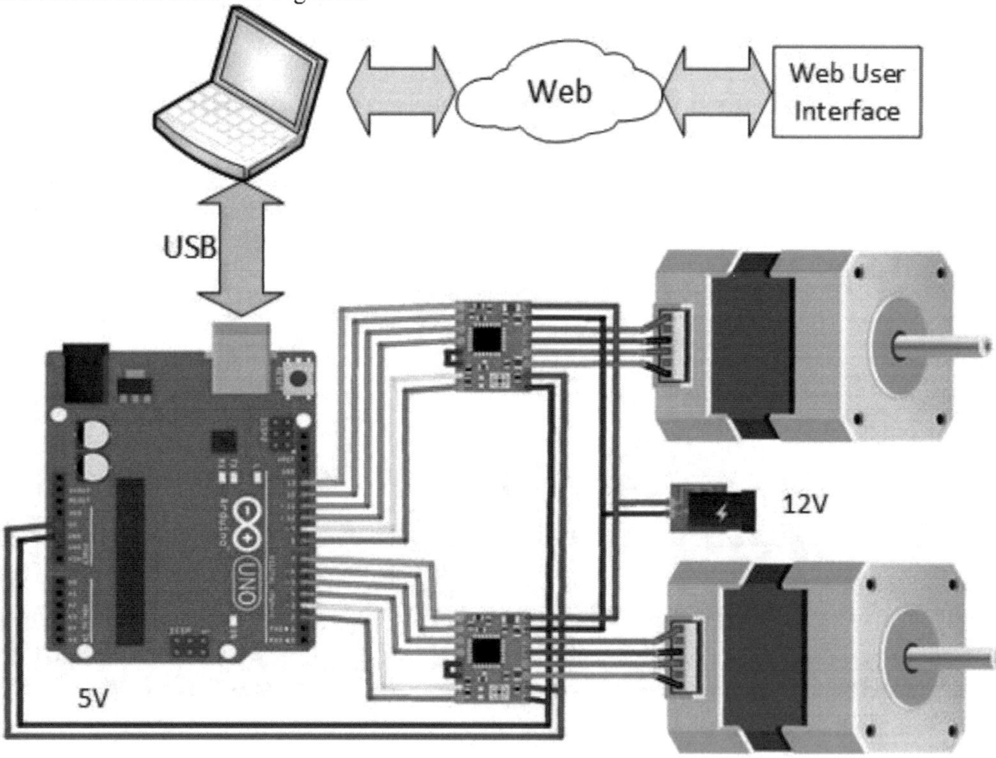

Figure 3. Schematic diagram of Hardware connections.

Since the manipulator is driven by stepper motors, we rely on A4988 drivers to control them [10]. These drivers allow to limit the current to the motor. This is of great importance when it comes to operating a device that will work with humans. In addition, these drivers allow the motor steps to be divided into 1/2, 1/4, 1/8, and 1/16. Thus, accurate positioning can be achieved. The choice of driver is intended to demonstrate the drive mode and capabilities of the robotic manipulator. In fact, most stepper motor drivers are controlled by the same principle as the one chosen. The main parameters are on / off, step resolution, number of steps and direction of rotation.

From a hardware standpoint, these drivers require a minimum of three controller signals: on / off, number of steps, and direction. Additional signals are used for setting the resolution of the steps. An enable input expects a logical 0 or 1, thus activating or deactivating the driver itself. Impulses are used to determine the number of steps as input. Each impulse responds in one step. And the direction input expects a logic 0 or 1 signal. To divide the steps into micro steps, the other 3 inputs (MS1, MS2 and MS3) must be connected to the controller. The steps are set according to the manufacturer's description. These inputs also expect to receive logical 0 or 1. If we have activated the half-step mode, then the number of pulses transmitted corresponds to the number of micro-dull ones.

Arduino controllers are among the most common for controlling robots and other mechatronic systems. They are easy to program and are affordable. In addition, they provide many options for working with different sensors and devices as they have analogue and digital input / output pins [9].

For our purposes, to manage two drivers completely, i.e. to use all their functionality, we need 12 digital input / output pins. This means that we can use an Arduino Uno controller that has 13 such pins. Each pin of the controller must be connected to the corresponding driver pin as shown in Figure 2, and the controller will be connected via USB port to a computer.

The role of the computer is to enable the control parameters of the robotic manipulator and the user interface to be set up and actuated. The scheme thus presented distributes the computing resources between the controller and the computer. For maximum speed, the controller executes trajectories and accelerations calculated from the computer, and in fact takes care of motor control only. And the computer is engaged in complex calculations such as calculating the speed of motion, calculating a trajectory, setting the number of steps to execute, and more. In addition, a user interface will be implemented on the computer, through which the rehabilitator will be able to set parameters for each procedure.

4. Manipulator Software Architecture.

Software control is divided into 3 levels. At the low level is the motor control by the controller. The intermediate level is provided by ROS [11]. And the high level is a web-based user interface.

This type of structure allows remote control of the manipulator via the web. In this way, the parameters of the individual procedures can be set remotely.

4.1. Arduino software.

The low-level software is tasked with executing the received commands from the top level. It receives procedure data such as rotation angle and rotation speed for each joint, then transforms this data into motor-driven data.

The rotation angle for each joint is determined by calculating the required number of steps and / or micro steps.

Steps = deg * motor_steps * microsteps / 360

Where steps are the necessary steps to reach the set angle, deg is the set angle, motor_steps are the number of steps for one complete rotation of the motor and microsteps is the coefficient according to the microstep mode.

The speed of rotation is determined by calculating the pulse rate in milliseconds at a given rpm.

Step_pulse = 60 * 1000000L / motor_steps / microsteps / rpm

Where motor_steps are the number of steps for one complete rotation of the motor, microsteps is the coefficient according to the microstep mode and rpm is the desired rpm.

4.2. ROS.

This level makes the connection between the low level and the user interface. The parameters specified in the interface are processed and sent for execution to the controller [12]. Here the correct execution of the tasks is monitored, the number of repetitions is counted and new parameters are dynamically set. In addition, a stop command can be sent to the controller at any time during operation and to terminate the current procedure.

Figure 4. Software nodes and communications.

In figure 4 is illustrated the general software communication system. The hardware interface between the computer and the controller is USB, using a dedicated library for communication and work with Arduino. Web interface is running on the robot computer and uses a specialized ros-bridge-suite library for communication through web ports.

5. Web-based User Interface.

In order to adequately use the manipulator, a sufficient set of options should be provided for adjusting and monitoring the rehabilitation procedures. We have developed a user interface that will allow physicians and patients to get as much opportunity as possible for easy operation of the manipulator. In figure 5 are shown the features and options provided by the web-based user interface.

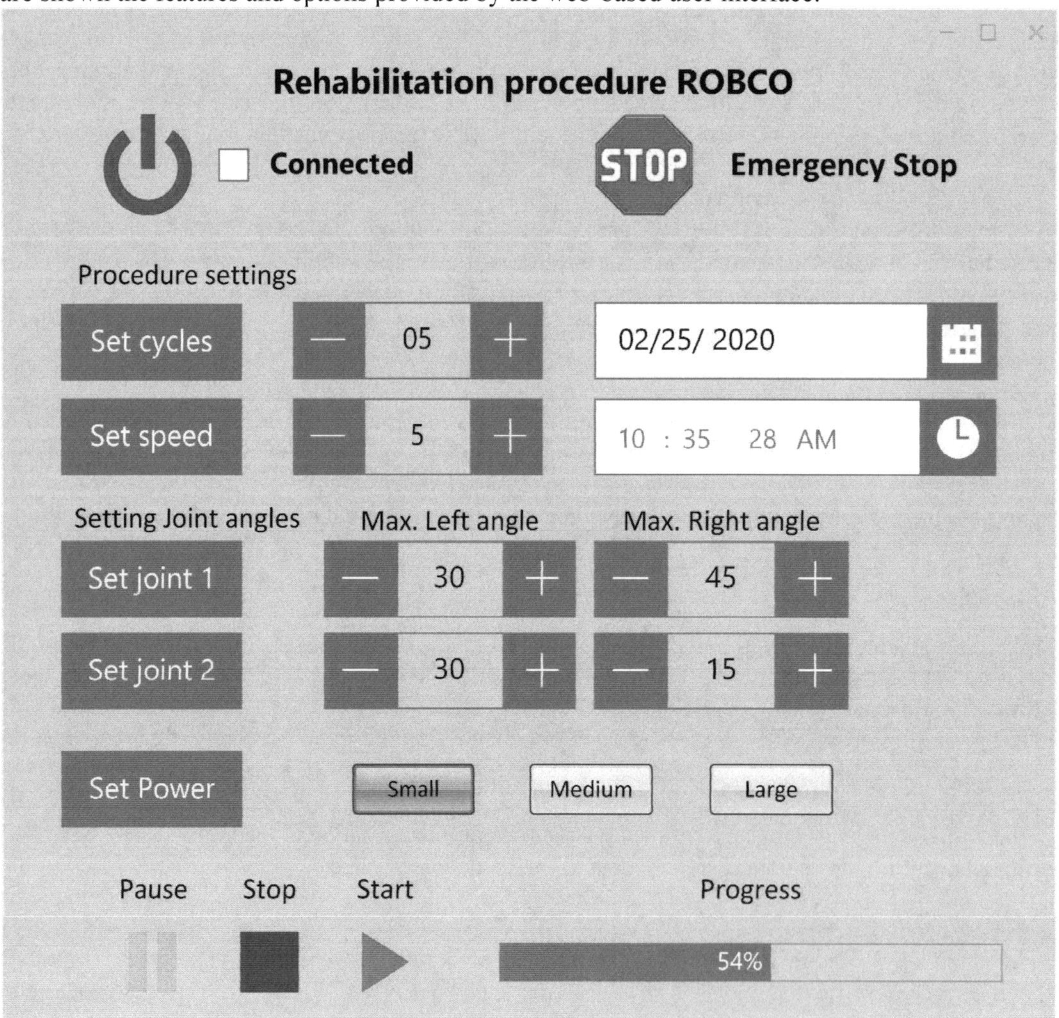

Figure 5. Web-based user interface.

We can divide the main interface menu into five sub-menus. Starting from the top down, these are: connection and emergency stop, settings and timing of the procedure, adjusting the maximum joint angles and managing the procedure.

This arrangement is intended to minimize the likelihood of errors occurring when a new procedure is initiated. The first submenu is represented by two buttons and an empty box where a checkmark appears when the interface has contacted to the manipulator. The connection button is blue and provides a function for connecting and interrupting communication with the robot controller. The emergency stop button instantly interrupts the robot's electric motors. This ensures that the robot does not injure or endanger the health of the patient.

The procedure setting submenu contains buttons for setting the number of cycles and rotation speed. As to set the selected values from the fields, the set buttons must be pressed. In addition, there is a drop-down menu that shows a calendar with a schedule. Here you can trace the history of procedures and upcoming ones. The last menu is for timing and time display.

Entering and storing schedules in the calendar and timer is done using a database where the entered information is stored. These databases are part of the user interface and are not related to the direct operation of the manipulator.

The next submenu is to adjust and set the maximum end angles of rotation. These are the restrictions we set in section 2. Even if the doctor tries to set values that do not meet the set limits, the software will not allow and will set the maximum values according to the restrictions.

There is submenu for power setting. The power can be set to small, medium and large using corresponding button. The active button is colored green after pushing “Set Power” button.

When you press the buttons to set the selected values in the two submenus, these values are sent to the ROS nodes. They, in turn, generate the corresponding parameters for submission to the controller. Changing these parameters is dynamic, but again it is forbidden to change these parameters while the current procedure is running. To change the values, the procedure must be terminated or paused.

Going through all the settings for one procedure follows the submenu for managing the procedure. There are buttons to start, pause and stop the current procedure. In addition, there is a bar that displays the progress of the procedure and the time remaining until its completion.

6. Experiments and Results.

A number of experiments have been conducted to test how the manipulator and control systems perform their work. Initial tests were conducted without a hand on the manipulator. The purpose was to check the correct execution of the tasks without risk to the user. Then, the same experiments were repeated, but this time with the motors loaded, a hand placed by the user. The experiments involved performing three different configurations to reach maximum angles, each test being tested at different speeds from 1 to 9. All experiments were performed for 10 replicates.

During the experiments, it is checked that each joint reaches the specified rotation angle and what is the deviation from the assignment. In addition, the total run time of 10 repetitions is measured and compared to the theoretically calculated times according to the speed selected. The other checks are for the functions of the buttons and whether the system responds adequately when the corresponding button is pressed from the user interface.

The results of the precision positioning experiments show a deviation at idle of 1-2 degrees and with the arm set 4-7 degrees. The results of the speedometer experiments show a minimum delay or acceleration of 1-2 seconds when idling. When working with a raised hand, the difference in speed is within 3-4 seconds. The results of the button functionality tests show that the system reacts instantly when the emergency stop button is activated and the other buttons function normally.

7. Conclusion.

Rehabilitation manipulators can offer significant improvements in the quality and results of rehabilitation procedures. Research in this field can contribute to the creation of modern and smart rehabilitation robots. The researched manipulator and the developed systems for its management mark the successful start of research on manipulators and systems for automated rehabilitation.

Studies show that the model developed can offer good and reliable rehabilitation procedures. In addition, remote control and monitoring capabilities extend its applications and functionality and create great convenience.

The foreseen future work on improving and developing such a rehabilitation manipulator is as follows. Development of a system and algorithm for automatic "STOP" if there is a risk of injury to the patient. Buttons for switching to the previous or next procedure. Preview the set procedure on a special display. Adding a submenu for pre-saving and selecting procedures. Adding a degree of freedom to the wrist rotation. Add translation of elbow pad to adjust to patient need.

Acknowledgment.

This research was carried out as part of the project “Telecontrolled Service Robots for Increasing the Quality of Life of Elderly and Disabled, No. DN 07/23 – 15.12.2016”, financed by the Bulgarian National Science Fund. The work was partially supported by the Bulgarian Ministry of Education and Science under the National Research Programme “Young scientists and postdoctoral students” approved by DCM # 577 / 17.08.2018.

References.

[1] Oujamaa, L., I. Relave, J. Froger, D. Mottet, and J-Y. Pelissier. "Rehabilitation of arm function after stroke. Literature review." Annals of physical and rehabilitation medicine 52, no. 3 (2009): 269-293.

[2] Taylor, Russell H., Peter Kazanzides, Gregory S. Fischer, and Nabil Simaan. "Medical robotics and computer-integrated interventional medicine." In Biomedical Information Technology, pp. 617-672. Academic Press, 2020.

[3] Petrescu, Relly Victoria, Raffaella Aversa, Antonio Apicella, and Florian Ion Petrescu. "Future medicine services robotics." American Journal of Engineering and Applied Sciences 9, no. 4 (2016): 1062-1087.

[4] Krebs, Hermano I., Mark Ferraro, Stephen P. Buerger, Miranda J. Newbery, Antonio Makiyama, Michael Sandmann, Daniel Lynch, Bruce T. Volpe, and Neville Hogan. "Rehabilitation robotics: pilot trial of a spatial extension for MIT-Manus." Journal of neuroengineering and rehabilitation 1, no. 1 (2004): 5.

[5] Van der Loos, HF Machiel, David J. Reinkensmeyer, and Eugenio Guglielmelli. "Rehabilitation and health care robotics." In Springer handbook of robotics, pp. 1685-1728. Springer, Cham, 2016.

[6] Peter S. Lum, Charles G. Burgar, Peggy C. Shor, Matra Majmundar, Machiel Van der Loos, Robot-assisted movement training compared with conventional therapy techniques for the rehabilitation of upper-limb motor function after stroke, Archives of Physical Medicine and Rehabilitation, Volume 83, Issue 7, 2002, Pages 952-959, ISSN 0003-9993, doi.org/10.1053/apmr.2001.33101.

[7] Lu, Renquan, Zhijun Li, Chun-Yi Su, and Anke Xue. "Development and learning control of a human limb with a rehabilitation exoskeleton." IEEE Transactions on Industrial Electronics 61, no. 7 (2013): 3776-3785.

[8] Ryan A. Beasley. “Medical Robots: Current Systems and Research Directions”, Journal of Robotics, Volume 2012, doi:10.1155/2012/401613.

[9] Warren, John-David, Josh Adams, and Harald Molle. "Arduino for robotics." In Arduino robotics, pp. 51-82. Apress, Berkeley, CA, 2011.

[10] Robotic, Pololu. "Electronic. A4988 Stepper Motor Driver Carrier." Pololu Robotics & Electronics.

[11] Quigley, Morgan, Ken Conley, Brian Gerkey, Josh Faust, Tully Foote, Jeremy Leibs, Rob Wheeler, and Andrew Y. Ng. "ROS: an open-source Robot Operating System." In ICRA workshop on open source software, vol. 3, no. 3.2, p. 5. 2009.

[12] Zubrycki, Igor, and Grzegorz Granosik. "Introducing modern robotics with ros and arduino, including case studies." Journal of Automation Mobile Robotics and Intelligent Systems 8, no. 1 (2014): 69-75.

[13] Cox, Damon Keith, Marvin L. Freeman, Jason M. Schaller, Jeffrey C. Hudgens, and Jeffrey A. Brodine. "Methods and apparatus for extending the reach of a dual scara robot linkage." U.S. Patent 8,016,542, issued September 13, 2011.

IOP Conf. Series: Materials Science and Engineering **878** (2020) 012005 doi:10.1088/1757-899X/878/1/012005

Methods for object recognition and classification for tele-controlled service robots

Angel Ivanov[1], Nayden Chivarov[1,2]

1 Institute of Information and Communication Technologies, Bulgarian Academy of Sciences, Acad. G. Bonchev str., block 2, 1113 - Sofia, Bulgaria
2 European Polytechnical University, Pernik, Bulgaria
angel.b.ivanov@gmail.com , nchivarov@gmail.com

Abstract. The need of personal service robots has been increased in the past several years. The tasks assigned to the robots vary in complexity and degree of freedom. In order to assist senior citizens and people with disabilities with their day to day tasks, a greater degree of freedom is required – more autonomy, ability to fulfil more complex tasks and easier for use human-machine interface. In this paper we show some of the methods that can be implemented in personal service robots for object recognition and classification in order to achieve greater autonomy, hence greater degree of freedom.

1. Introduction

The recent growth of aging population and respectively people with disabilities lead to additional care for these people, done by large number of social workers and assistants. To face this issue, the society is forced to use robotics and methods for automation in order to ease the lives of the seniors and people with disabilities. One such solution is the use of household service robot.

The International Federation of Robotics (IFR) has defined in [1] the term "Service Robot" as robot that performs tasks, useful to humans. According to ISO8373 the robots require a degree of freedom, which means that it should have the ability to fulfil the assigned tasks without human intervention and based on current state and sensing. This degree can vary from partial autonomy – the tasks are done with human intervention, to full autonomy – the tasks are done by the service robot alone.

The IFR statistics show [2] that for 2018 the service robots, sold for domestic use and/or household tasks are 16.3 million units. They predict that by 2022, the domestic service robots production will drastically increase to 61.1 million units.

To achieve a higher degree of freedom, the service robot would require having a better "view" of the surrounding environment. At this point, the environment knowledge gathering is divided in 2 major problems:

- Environment mapping;
- Object detection and recognition

The environment mapping is usually done with the help of multiple sensors like ultrasound beacons, radars, GPS, etc. as they gather information about the current physical position of the robot and will help with the route planning and travel.

For object detection, recognition and classification, the only required sensing hardware is a camera. With the help of the camera, the service robot can actually "see" the objects that are in front of it. In order to do the actual recognition and classification, various methods can be used.

In this paper we will present and test 2 different machine learning architectures, with which we will do object recognition in the context of tele-controlled service robot for assisting elderly people and people with disabilities.

2. Object detection and classification

In the recent years a lot of attention is given to the problems of detecting objects in still images and/or video streams. This attention has led to a lot of new machine learning architectures and respectively revisions/improvements to existing ones. Traditional object detection and recognition methods rely on handcrafted features and shallow trainable architectures [3]. Some of the methods use Haar-like features to describe the desired object for recognition, however the task to build the Haar-like feature set is slow and cumbersome.

In [4] was suggested the use of convolutional neural network (CNN) for object detection which yields a very low error rate compared to the current state-of-the-art. In [5] an improvement in terms of speed was proposed with a R-CNN. This type of network scans the image from bottom-up and generates regions of interest (ROI). Each ROI is then parsed through a previously trained large CNN for object detection and classification. This method is said by the authors to improve the mean accuracy (mAP) by 50% compared to traditional CNN.

Although CNNs by themselves are great at object detection and classification in still images, this is not the case for video streams. This issue is easily resolved by adding a backpropagation connection into the neural network architecture. Such architectures are presented in [6] as Recurrent neural networks (RNN) and its special case Long Short-Term Memory (LSTM) networks.

The latest advancements [7,8,9] show the most successful in terms of speed and accuracy architectures. The YOLO and SSD are the architectures of choice in this paper to test the object recognition for service robots.

2.1. YOLO

You Only Look Once (YOLO) [7] is a state-of-the-art architecture that does not require a complex pipeline to process an image. Unlike the sliding window or region-based methods, it sees the whole image during training and test phases, and implicitly encodes contextual information about object's class and appearance.

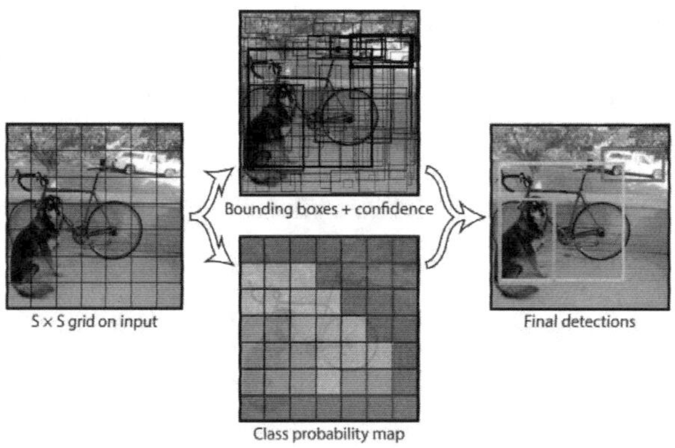

Figure 1. YOLO Model

In figure 1 can be seen the model of the YOLO architecture. It first generates SxS grid on the input image. Then creates prediction bounding boxes B and assigns confidence scores to each box. At the same time creates a class probability map C. These predictions are finally encoded as $S \times S \times (B * 5 + C)$ tensor.

The network architecture is shown in figure 2.

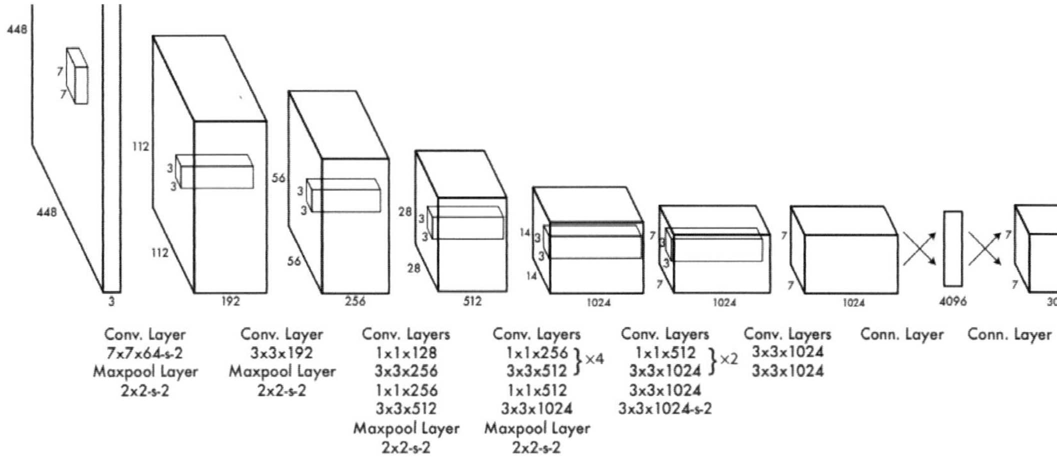

Figure 2. YOLO Network architecture

The network contains 24 convolution layers, followed by 2 fully connected layers. The 1x1 convolution layers are used to reduce the feature set, generated by the previous layers.

The authors claim to achieve speed of 45 frames per second which is more than enough for realtime video processing.

The original YOLO architecture is fast, but it stays behind the competitors in terms of accuracy. Therefor the authors in [8] propose an updated version of the network. The new architecture is shown in figure 3.

	Type	Filters	Size	Output
	Convolutional	32	3 × 3	256 × 256
	Convolutional	64	3 × 3 / 2	128 × 128
	Convolutional	32	1 × 1	
1×	Convolutional	64	3 × 3	
	Residual			128 × 128
	Convolutional	128	3 × 3 / 2	64 × 64
	Convolutional	64	1 × 1	
2×	Convolutional	128	3 × 3	
	Residual			64 × 64
	Convolutional	256	3 × 3 / 2	32 × 32
	Convolutional	128	1 × 1	
8×	Convolutional	256	3 × 3	
	Residual			32 × 32
	Convolutional	512	3 × 3 / 2	16 × 16
	Convolutional	256	1 × 1	
8×	Convolutional	512	3 × 3	
	Residual			16 × 16
	Convolutional	1024	3 × 3 / 2	8 × 8
	Convolutional	512	1 × 1	
4×	Convolutional	1024	3 × 3	
	Residual			8 × 8
	Avgpool		Global	
	Connected		1000	
	Softmax			

Figure 3. YOLOv3 feature extraction architecture

The network now consists of 53 convolutional layers, as opposed to 24 in the original architecture. This boosts the accuracy of the object classification with more than 30% while maintaining the same speed. On higher resolution images, the classification accuracy is improved even more.

For the experiments in this paper, we will use YOLOv3 architecture.

2.2. SSD

The Single Shot Multibox Detector (SSD) [9] is another state-of-the-art architecture that has received a wide publicity because of its speed and accuracy.

The SSD framework is shown in figure 4.

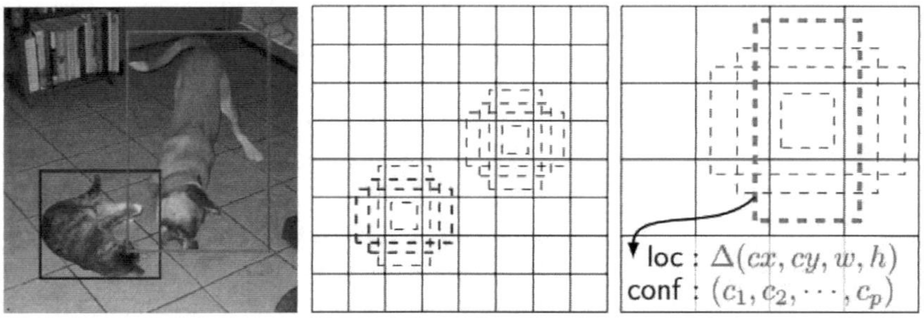

(a) Image with GT boxes (b) 8 × 8 feature map (c) 4 × 4 feature map

Figure 4. Single Shot Multibox Detector (SSD) framework

To train the network, it needs an image with ground truth boxes – figure 4a. Using a convolution layers, small set of default boxes with different aspect ratios are calculated. This is done for each location and in several feature maps with different scales – figure 4 b, c.

The model of SSD is shown in figure 5.

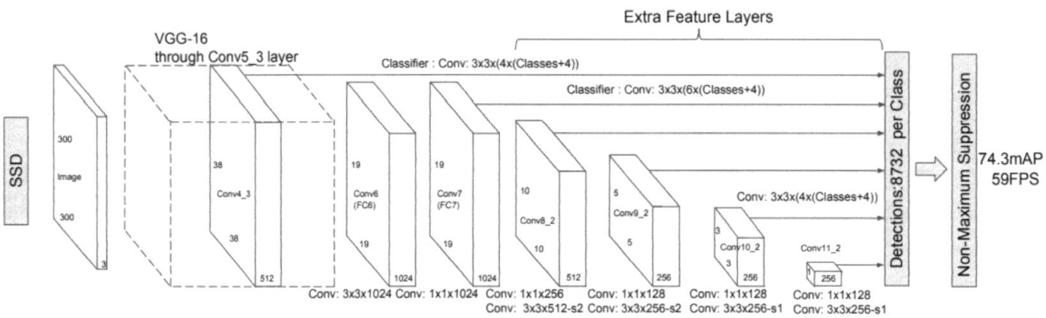

Figure 5. SSD Model

The SSD model is based on a feed-forward convolutional network and produces a fixed size collection of bounding boxes and scores. These scores represent the presence of an object class instance in the respective bounding boxes.

For the experiments in this paper, we will use SSD architecture.

3. Experiments and results

Our goal in this paper is to compare both SSD and YOLOv3 architectures in terms of accuracy and speed. The comparison is within the aspect of usage in autonomous service robots.

For the experiments, we use Intel RealSense D415 [10] stereo camera. For the sake of simplicity, we use only the RGB mode without the additional depth mode.

For video parsing we use the OpenCV library [11][12]. The OpenCV is an open-source computer vision library written in C/C++. It addresses the cases of object segmentation, detection, recognition and tracking. It also supports camera calibration, stereo vision and 2D/3D shape reconstruction.

The OpenCV library is available for use with Python wrappers which accelerates the prototyping speed.

Both SSD and YOLOv3 models are trained using the famous COCO dataset [13],[14]. This dataset contains over 300000 images of 91 object classes. All object instances are manually labelled using extensive crowd worker involvement and that concludes in over 2 mil labelled object instances. Some of the object categories include person, dog, car, bike, chair, sofa, bottle, etc.

Since both SSD and YOLOv3 are single-shot detectors, their performance in terms of speed in frames per second is similar – on CPU processing around 5-7 fps, on GPU processing around 30-40 fps.

Both SSD and YOLOv3 are able to recognize human with high confidence rate of at least 99,95% as shown in figure 6.

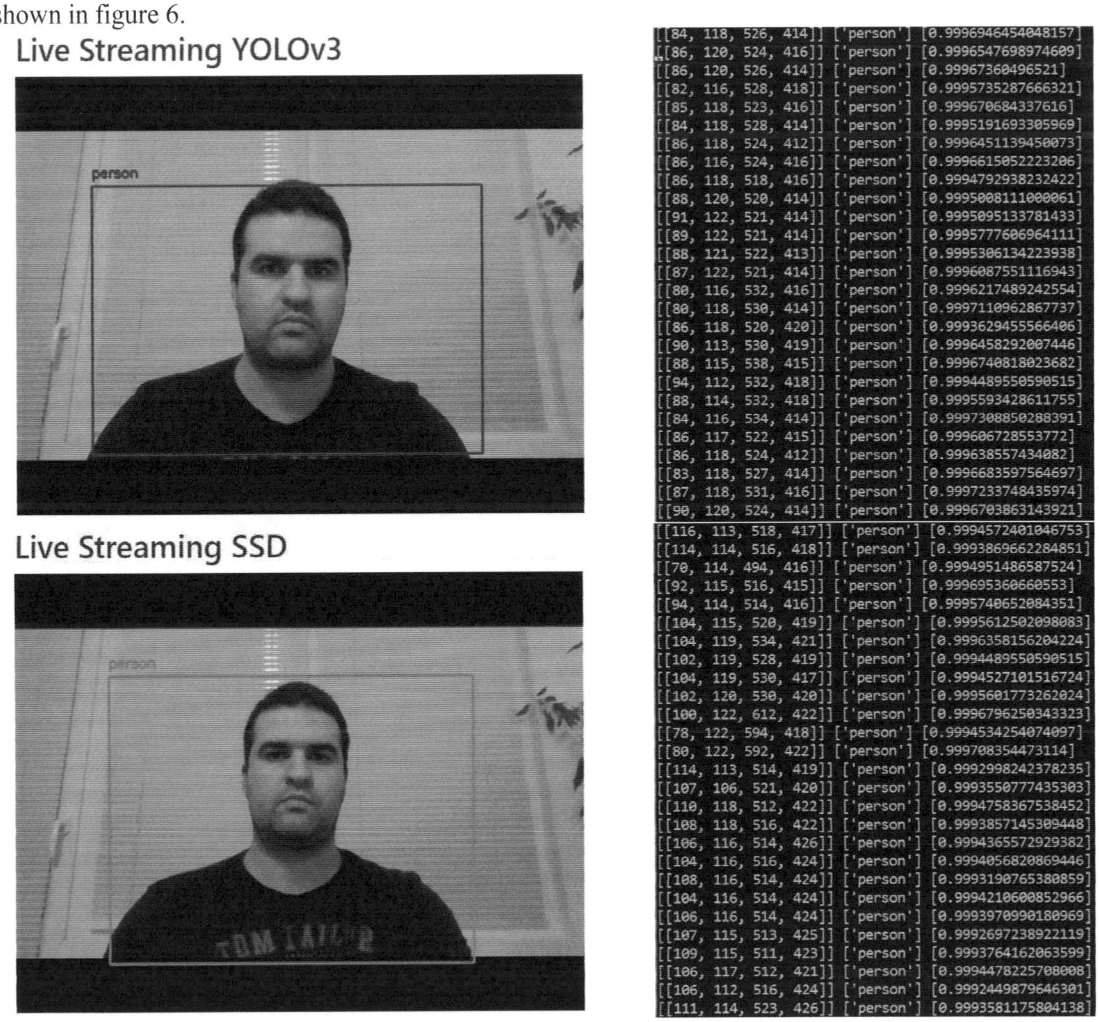

Figure 6. YOLOv3 vs SSD on human detection

Both models can recognize common objects around a typical house with confidence rate of at least 70% as shown on figure 7.

IOP Conf. Series: Materials Science and Engineering **878** (2020) 012005 doi:10.1088/1757-899X/878/1/012005

The difference is observed when multiple objects are presented in front of the camera. In a test case with 10 objects, YOLOv3 is able to detect all of them, while SSD detects and recognizes only 4 that are nearest to the camera.

Depending on the image resolution parsed by OpenCV a wrong classification can occur from both models. Using smaller resolution increases drastically the performance but lowers the accuracy. Higher resolution decreases the performance, but the accuracy is much better. It is a matter of fine tuning for a specific use case to balance the speed/accuracy ratio.

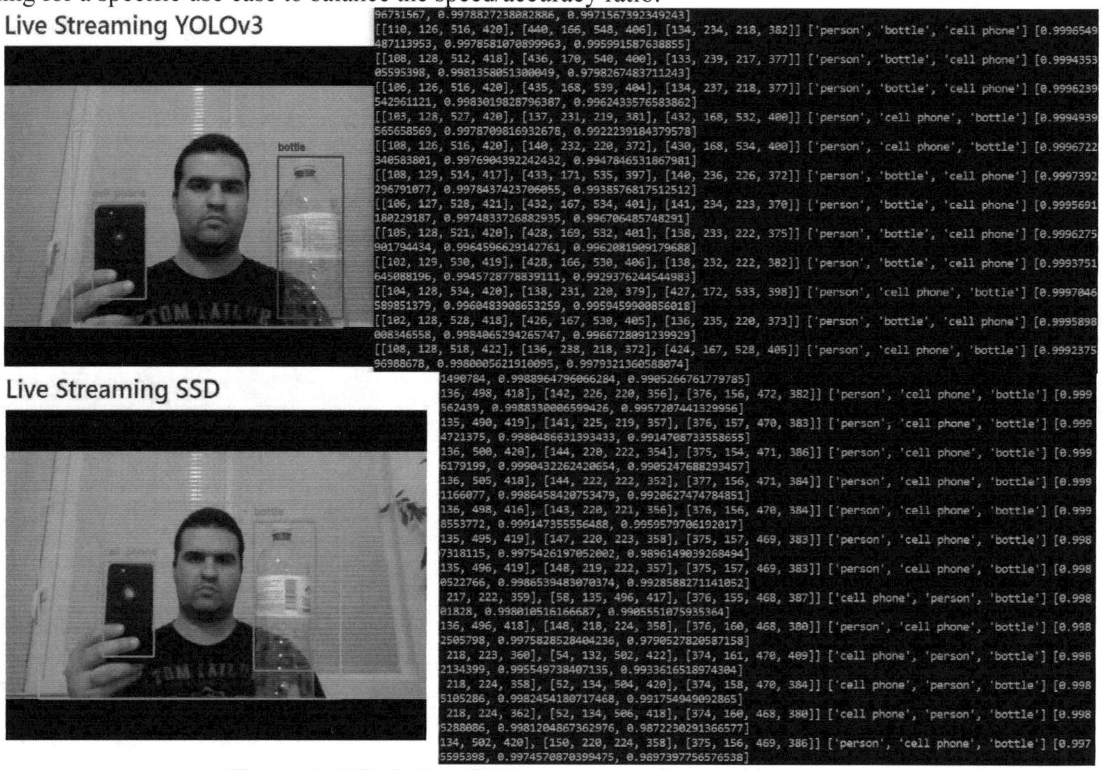

Figure 7. YOLOv3 vs SSD on common house objects

4. Conclusion

The rising need of service robots used as personal assistants for the elderly and people with disabilities require high degree of freedom, hence high level of autonomy. To achieve these requirements a combination of different sensing hardware is needed for navigation through the home of the patients. To achieve the higher autonomy the use of computer vision methods is required to be used.

In this paper a comparison of 2 state-of-the-art machine learning algorithms was made – YOLOv3 and SSD. Both methods are part of the family of single-shot detectors – they go only once through the image/frame.

While both methods provide similar speed/accuracy ratio, the experiments showed that YOLOv3 is able to detect and classify more simultaneous objects in a video stream.

In the use-case of service robots both methods can be used and will provide a similar practical outcome for object detection and classification, so it is up to the researcher to choose which model is more suited for the respective implementation.

IOP Conf. Series: Materials Science and Engineering **878** (2020) 012005 doi:10.1088/1757-899X/878/1/012005

5. Acknowledgement
The research presented in this paper was supported by the Bulgarian National Science Fund under the contract No. DN 07/23 – 15.12.2016 "Tele-controlled Service Robots for Increasing the Quality of Life of Elderly and Disabled".

References
[1] International Federation of Robotics (IFR), **https://www.ifr.org/service-robots/**
[2] International Federation of Robotics (IFR), **Executive Summary World Robotics 2019 Service Robots**
[3] Zhong-Qiu Zhao,Peng Zheng, Shou-tao Xu, and Xindong Wu, **Object Detection with Deep Learning: A Review**
[4] Krizhevsky, A., Sutskever, I., & Hinton, G., **ImageNet classification with deep convolutional neural networks**
[5] Girshick, R., Donahue, J., Darrell, T., & Malik, J., **Region-based convolutional networks for accurate object detection and segmentation**
[6] Alex Sherstinsky, **Fundamentals of Recurrent Neural Network (RNN) and Long Short-Term Memory (LSTM) Network**
[7] Joseph Redmon , Santosh Divvala, Ross Girshick , Ali Farhadi, **You Only Look Once: Unified, Real-Time Object Detection**
[8] Joseph Redmon, Ali Farhadi, **YOLOv3: An Incremental Improvement**
[9] Wei Liu , Dragomir Anguelov, Dumitru Erhan, Christian Szegedy, Scott Reed, Cheng-Yang Fu, Alexander C. Berg, **SSD: Single Shot MultiBox Detector**
[10] Datasheet, **Intel RealSense D400 Series Product Family**
[11] Gary Bradski, **The OpenCV Library**
[12] Kari Pulli, Anatoly Baksheev, Kirill Kornyakov, and Victor Eruhimov, **Real-Time Computer Vision with OpenCV**
[13] Tsung-Yi Lin, Michael Maire, Serge Belongie, Lubomir Bourdev, Ross Girshick, James Hays, Pietro Perona, Deva Ramanan, C. Lawrence Zitnick, Piotr Dollar, **Microsoft COCO: Common Objects in Context**
[14] COCO Dataset Internet Site**, http://cocodataset.org/**

IOP Conf. Series: Materials Science and Engineering **878** (2020) 012006 doi:10.1088/1757-899X/878/1/012006

Description and modelling of robot-manipulator FANUC M-430iA/4FH

E Nikolov, N G Nikolova, M Georgiev

Technical University-Sofia; FA, DIA; 8, Kliment Ohridski str., Sofia-1000, Bulgaria
nicoloff@tu-sofia.bg, ninan@tu-sofia.bg, georgievmg@tu-sofia.bg

Summary: In this work, you can find the results from the description and modeling of the robotic manipulator FANUC M-430iA / 4FH. The basic values that determine the performance of the robot are tested and analyzed. An analytical model was developed on the basis of which it's temporal and frequency characteristics are illustrated in the absence and availability of maximum mechanical load of the working spherical wrist

1. Introduction.

The purpose of this paper is to explore the possibilities of industrial robotic manipulator FANUC M-430iA/4FH. The work tasks are: description, modeling and visualization of the dynamic characteristics of the robotic manipulator. Fig. 1 shows the construction diagram of industrial robotic manipulator FANUC M-430iA/4FH with anthropomorphic arms, spherical working wrist and five independent motion control axes (5-DOF). The electric drive of the robot is carried out with DC motor with permanent magnet (from FANUC-type servo motors β_i or α_i).

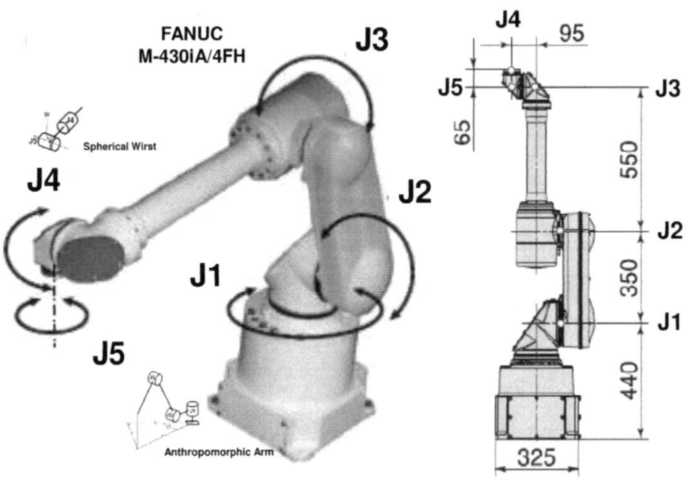

Figure 1. The construction scheme of industrial robotic manipulator FANUC M-430iA/4FH

The mechanical parameters of the robot (rotation ranges and shoulder dimensions, speed, mechanical moments, inertia, maximum lift, energy consumption and power, maximum accuracy, etc.) in the respective dimensions are systematized and presented in Table 1 [12-20].

Table 1.

5-DOF Robot-Manipulator FANUC M-430iA/4FH, Controlled Axis of Motion 5, Payload 4 kg						
Joint	J_i	**J1**	**J2**	**J3**	**J4**	**J5**
Motion Range	$[°]$	$360[°]$	$230[°]$	$383[°]$	$300[°]$	$540[°]$
Robot Maximum Speed	$[°/s]$	$300[°/s]$	$320[°/s]$	$320[°/s]$	$360[°/s]$	$2000[°/s]$
Robot Maximum Speed	$[rad/s]$	$5,24[rad/s]$	$5,59[rad/s]$	$5,59[rad/s]$	$6,28[rad/s]$	$34,91[rad/s]$
Wirst Moment	$[Nm]$	-	-	-	$3,5[Nm]$	$1,5[Nm]$
Wirst Moment	$[kgf.m]$	-	-	-	$0,36[kgf.m]$	$0,15[kgf.m]$
Wirst Inertia	$[kg.m^2]$	-	-	-	$0,064[kg.m^2]$	$0,010[kg.m^2]$
Wirst Inertia	$[kgf.cm.s^2]$				$0,65$	$0,102$
Length Arm	$[mm]$	$350[mm]$	$550[mm]$	$95[mm]$	$65[mm]$	-
Max. Load Capacity an Wirst	$[kg]$	-	-	-	-	$4[kg]$
Repeatability	$[mm]$	$\pm 0,5[mm]$				
Mechanical Weight Robot Mass	$[kg]$	$55[kg]$				
Average Power Consumption	$[kW]$	$1[kW]$				
H-Reach	$[mm]$	$900[mm]$				
Average Power Consumption	$[kW]$	$1[kW]$				

This paper offers and examines a corresponding mathematical model of kinematics of 5-DOF FANUC M-430iA/4FH by Todefine. The kinematic scheme of the axis of the manipulator is presented in Fig. 2 in accordance with the Denavit-Hartenberg method of constructive connection configuration (Fig.1). The following markings are used: z - axis in the direction of the joint; x - axis parallel to the common normal; q_i - the angle from x_{i-1} to x_i along z_{i-1}; d_i - distance from the intersection of z_{i-1} with x_i to the origin of $(i-1)$ - system of axis; a_i - shortcut between z_{i-1} and z_i; α_i angle from z_{i-1} to z_i with length x_i. The kinematic scheme parameters (Fig.2) are shown in Tabl. 2.

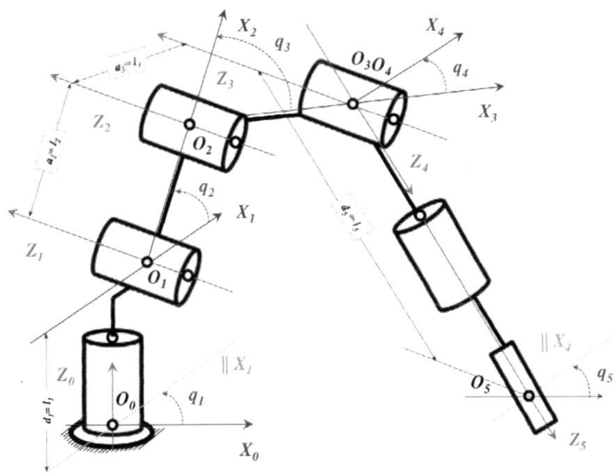

Figure 2. The kinematic scheme of the axis of the manipulator

Table 2.

Denavit-Hartenberg parameters of the 5-DOF robot-manipulator FANUC M-430iA/4FH				
I	q_i	α_i	a_i	d_i
1	q_1	$-\pi/2$	0	d_1
2	q_2	0	a_2	0
3	q_3	0	a_3	0
4	q_4	$\pi/2$	0	0
5	q_5	0	0	d_5

IOP Conf. Series: Materials Science and Engineering **878** (2020) 012006 doi:10.1088/1757-899X/878/1/012006

2. Modeling, parameterization and analysis of the 5-DOF FANUC M-430iA/4FH industrial robotic manipulator.

As an object of control, the industrial robotic manipulator is considered with: input (control) values - control signals μ_{Ji} $(i \in [1,5])$ to the electric motors and output (adjustable) values - positions y_{Ji} $(i \in [1,5])$ of the axes of the 5-DOF manipulator. The structural model (Figure 3) is presented in *"**without loading the wrist**"* mode in Fig. 3.b and in operational ***"with max load an wrist"*** mode in Fig. 3.c, where ς_M indicates the mechanical load of the robotic manipulator by the workpiece being machined, and G_i - the transmission functions of the electromechanical propulsion of the arm of the manipulator.

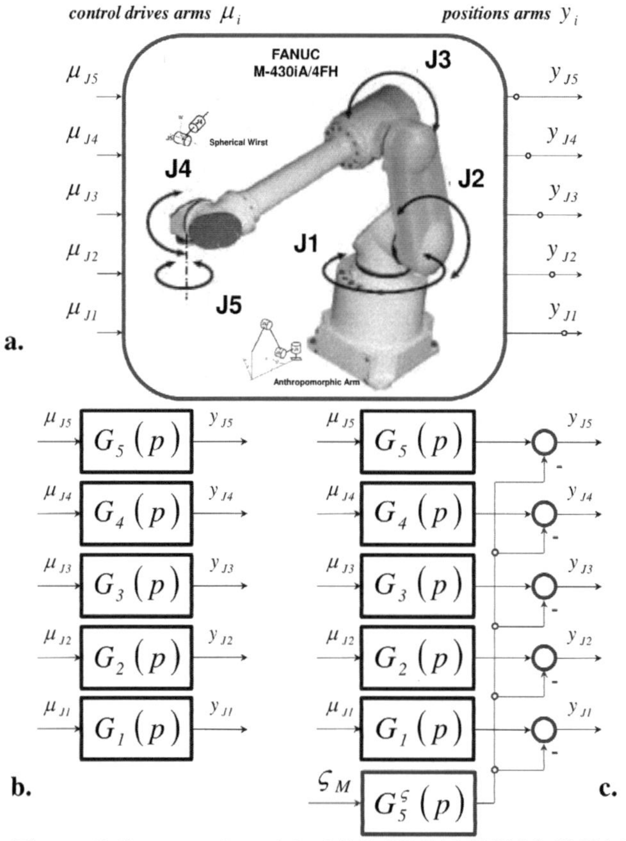

Figure 3. Structured model of 5-DOF FANUC M-430iA/4FH

Literature [1-11] and laboratory studies as well as experimental verification determine the possibilities for analytical modeling of the dynamics of FANUC M-430iA / 4FH in the following sequence. The schematic diagram (Fig. 4) of the electric drive (uniform on all axes of the 5-DOF manipulator) with a DC motor with permanent magnet is considered, where the following markings are used: R_a - anchor resistance; L_a - inductance of scattering; u_a - power supply voltage; i_a - anchor current; T_l - torque of the motor shaft; T_m - electromagnetic moment of the motor shaft; J_m - inertia moment of the motor; K - current-to-moment ratio; B_m - viscous friction coefficient; ω_m - motor shaft speed; θ_m - shaft position.

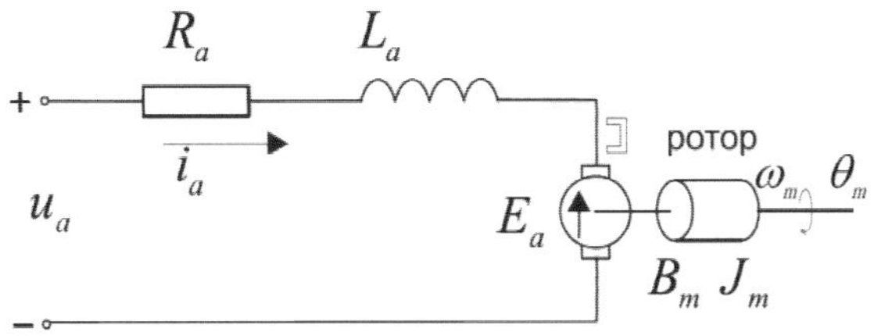

Figure 4. Schematic diagram of the electric drive

The basic equations of analytically describing functional dependencies between the main characterizing variables of the scheme are (1) ÷ (3), and the equations modeling the electrical and mechanical parts are (4) и (5) respectively. After replacing the variables in equations (4) and (5), we reach (6), where T_s is the time constant of the motor and K_{sm} is the gain of the motor. The last two dimensions are determined according to (7):

$$\frac{d}{dt} i_a = \frac{1}{L_a} u_a - \frac{R_a}{L_a} i_a - \frac{K}{L_a} \omega_m \tag{1}$$

$$\frac{d}{dt} \omega_m = \frac{K}{J_m} i_a - \frac{B_m}{J_m} \omega_m - \frac{1}{J_m} T_l \tag{2}$$

$$\frac{d}{dt} \theta_m = \omega_m \tag{3}$$

$$u_a(t) = R_a i_a(t) + K\omega_m(t) \tag{4}$$

$$J\dot{\omega}_m(t) = Ki_a(t) - B_m \omega_m(t) \tag{5}$$

$$T_s \dot{\omega}_m(t) = -\omega_m(t) + K_s u_a(t) \tag{6}$$

$$T_s = \frac{R_a J}{BR_a + K^2} \; ; \; K_{sm} = \frac{K}{BR_a + K^2} \tag{7}$$

With the help of (1) ÷ (7) the transfer functions G_{ω_m} and G_{θ_m} of the robotic manipulator are determined by the speed and by the position of the motor shaft as (8) and (9)

$$G_{\omega_m}(p) = \frac{\omega_m(p)}{u_a(p)} = \frac{K_{sm}}{T_s p + 1} \tag{8}$$

$$G_{\theta_m}(p) = \frac{\theta_m(p)}{u_a(p)} = \frac{K_{sm}}{p(T_s p + 1)} \tag{9}$$

The complete transmission functions (both in speed $G^{\Sigma}_{\omega_m}(p)$ and in position $G^{\Sigma}_{\theta_m}(p)$, $\theta_m \equiv y$) of the electric drive (10), (11) depend both on the supply voltage $u_a \equiv \mu$ and on the moment $M_c \equiv \varsigma_m$ of the mechanical loading of the shaft, which is structurally reflected in Fig. 3.c ., and only by position θ_m - in Fig.5

$$\begin{aligned} G^{\Sigma}_{\omega_m}(p) &= \frac{K_{sm}}{(T_s p+1)} \frac{u_a(p)}{\omega_m(p)} - \frac{k_c(T_a p+1)}{(T_{as} p+1)(T_m p+1)} \frac{M_c(p)}{\omega_m(p)} \\ \omega_m(p) &= \frac{K_{sm}}{T_s p+1} u_a(p) - \frac{k_c(T_a p+1)}{(T_{as} p+1)(T_m p+1)} M_c(p) \end{aligned} \tag{10}$$

$$G_{\theta_m}^{\Sigma}(p)=\frac{K_{sm}}{p(T_s p+1)}\frac{u_a(p)}{\omega_m(p)}-\frac{k_c(T_a p+1)}{p(T_{as} p+1)(T_m p+1)}\frac{M_c(p)}{\omega_m(p)}$$
$$\theta_m(p)=\left(\frac{1}{p}\right)\frac{K_{sm}}{T_s p+1}u_a(p)-\left(\frac{1}{p}\right)\frac{k_c(T_a p+1)}{(T_{as} p+1)(T_m p+1)}M_c(p) \qquad (11)$$

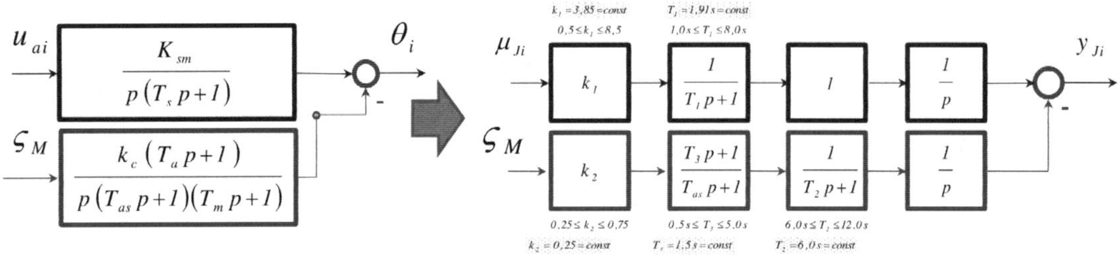

Figure 5. The ranges of parametric fluctuations of the control object

The characteristics (12) ÷ (17) of the control object (Fig.3) are parameterized [1-20], with ranges of parametric fluctuations are indicated and marked in Tabl. 3 and in Fig. 5

Table 3.

	k_{1nom}	T_{1nom}	k_{2nom}	T_{2nom}	T_{3nom}
	$k_1=3,85$	$T_1=1,91s$	$k_2=0,25$	$T_2=6,0s$	$T_2=1,5s$
	k_1 var	T_1 var	k_2 var	T_2 var	T_3 var
1	3,58	0,5 s	0,253	6 s	1,0 s
2	4,58	1,0 s	0,353	7 s	1,5 s
3	5,58	2,0 s	0,453	8 s	2,0 s
4	6,58	3,0 s	0,553	9 s	3,0 s
5	7,58	4,0 s	0,653	10 s	4,0 s
6	8,58	5,0 s	0,753	11 s	4,5 s

$$G_5=G_{\theta_m}(p)=\frac{\theta_m(p)}{u(p)}=\frac{K_s}{p(T_s p+1)}=\frac{3,58}{p(0,91p+1)} \qquad (12)$$

$$G_4=G_{\theta_m}(p)=\frac{\theta_m(p)}{u(p)}=\frac{K_s}{p(T_s p+1)}=\frac{3,58}{p(1,05p+1)} \qquad (13)$$

$$G_3=G_{\theta_m}(p)=\frac{\theta_m(p)}{u(p)}=\frac{K_s}{p(T_s p+1)}=\frac{3,58}{p(1,25p+1)} \qquad (14)$$

$$G_2=G_{\theta_m}(p)=\frac{\theta_m(p)}{u(p)}=\frac{K_s}{p(T_s p+1)}=\frac{3,58}{p(1,55p+1)} \qquad (15)$$

$$G_1=G_{\theta_m}(p)=\frac{\theta_m(p)}{u(p)}=\frac{K_s}{p(T_s p+1)}=\frac{3,58}{p(1,91p+1)} \qquad (16)$$

$$G^{\varsigma_M}(p)=-\frac{\theta_m(p)}{\varsigma_M(p)}=-\frac{k_c(T_a p+1)}{p(T_s p+1)(T_m p+1)}=-\frac{0,253(1,5p+1)}{p(0,01p+1)(6p+1)} \qquad (17)$$

3. Preview of the parameterized performance models of 5-DOF FANUC M-430iA / 4FH industrial robotic manipulator

The analytical models of the electric drive of the robotic manipulator by position $\theta_m \equiv y$ are simulated in the ranges specified in Tabl.3. The transients $y_i(t)$ (at $u_a = 1(t)$) and impulse transients $i_i(t)$ (at $u_a = \delta(t)$) characteristics of the electric drive of the manipulator are presented respectively in Fig.6 and Fig.7 in "***without loading the wrist***" mode and in "***with max load an wrist***" operational mode. The frequency characteristics $G_{\theta_m}(j\omega)$ of the modeled dynamic system by position are illustrated in Fig.8 in Fig.9 in "***without loading the wrist***" mode and in "***with max load an wrist***" operational mode

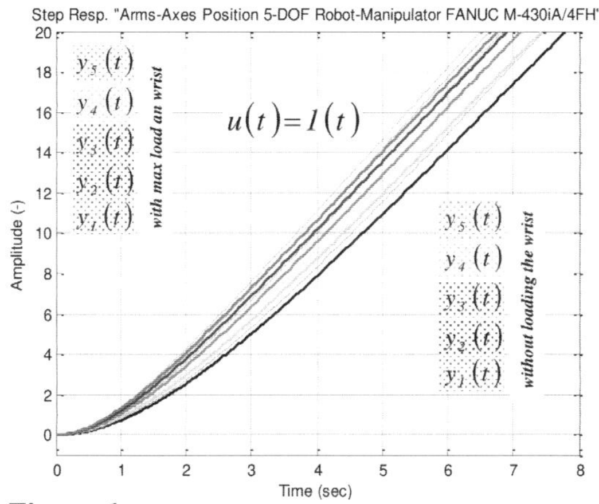

Figure 6.

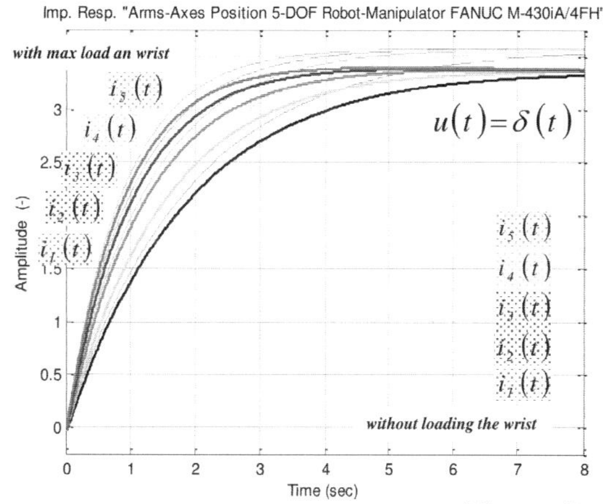

Figure 7.

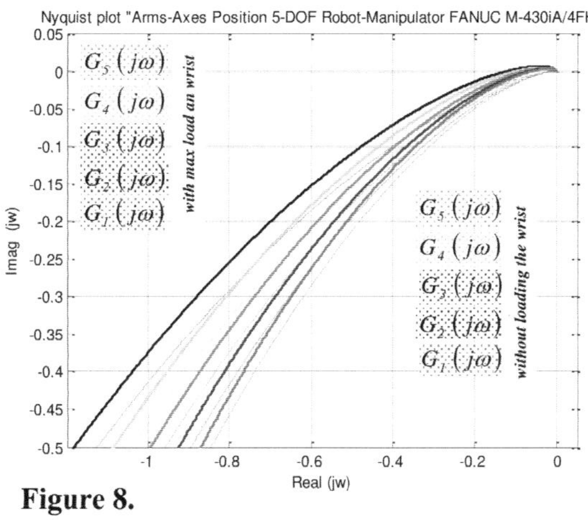

Figure 8.

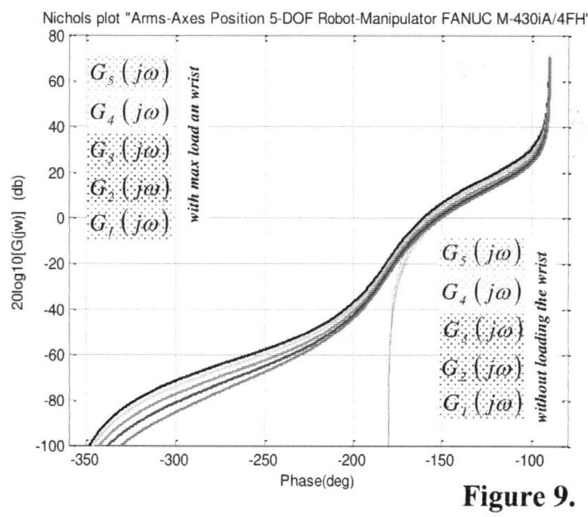

Figure 9.

4. Conclusion

The new and original in this paper is:

■ The proposed description of the structural layout of the 5-DOF industrial robotic manipulator FANUC M-430iA / 4FH, the systematization of the mechanical quantities and parameters of the robot (rotational ranges and dimensions of the arms, speed, mechanical moments, inertia, maximum lifting capacity, energy consumption and power, maximum accuracy, etc.) in the respective dimensions, as well as the developed kinematic scheme of the axes of the manipulator Fig.2 in accordance with the Denavit-Hartenberg method**;**

■ The developed analytical model of the dynamics on the position of the axes of the 5-DOF manipulator in the "without loading the wrist" mode and in the "with max load an wrist" operational mode under the influence of the mechanical load on the robotic manipulator by the workpiece machining during the production, as well as the parameterization of the analytical model developed on the basis of literary and laboratory studies and experimental verification;

■ The transmission functions are determined by speed and by position of FANUC M-430iA / 4FH as an object in a control system, representing the main characteristics of the manipulator in the synthesis of the system;

■ The proposed analytical models of the industrial robotic manipulator are visualized. It presents its transient, impulse transient and frequency characteristics, defining the FANUC M-430iA / 4FH as a static dynamic speed system and an astatic dynamic motion system

Acknowledgment

The authors would like to thank the Bulgarian National Science Fund for the financial support by project КП-06-Н37/17.

References

1. Abdulrahman A.A.Emhemed, Rosbi Bin Mamat (2012), *Modelling and Simulation for Industrial DC Motor Using Intelligent Control*, Procedia Engineering 41 (2012) 420-425
2. Amir Khazaee, Hosein Abootorabi Zarchi, Gholamreza Arab Markadeh (2019), *Loss model based efficiency optimized control of brushless DC motor drive*, ISA Transactions 86 (2019) 238-248
3. F. S. Ahmed, S. Laghrouche, M. El Bagdouri (2010*), Pancake Type DC Limited Angle Torque Motor: Modeling and Identification, Application to Automobile Air Path Actuators*, 6th IFAC Symposium Advances in Automotive Control Munich, Germany, July 12-14, 2010, 978-3-902661-72-2/10/$20.00 © 2010 IFAC, 10.3182/20100712-3-DE-2013.00135, 431-436
4. Fayez S. Ahmed, Salah Laghrouche, Mohammed El Bagdouri (2012), *Analysis, modeling, identification and control of pancake DC torque motors: Application to automobile air path actuators*, Mechatronics 22 (2012) 195-212
5. M. Ruderman, J. Krettek, F. Hoffmann, T. Bertram (2008), *Optimal State Space Control of DC Motor*, Proceedings of the 17th World Congress The International Federation of Automatic Control Seoul, Korea, July 6-11, 2008, IFAC Proceedings, Volume 41, Issue 2, 2008, Pages 5796-5801
6. Masayoshi Tomizuka (1996), *Robust digital motion controllers for mechanical systems*, Robotics and Autonomous Systems, © 1996 Elsevier, 19 (1996) 143-149
7. Mikhov M. (2009), *Mathematical Modeling and Dynamic Simulation of a Class of Drive Systems with Permanent Magnet Synchronous Mo*tors, Applied and Computational Mechanics, Vol. 3, No. 2, pp. 331-338, Czech Republic, 2009, ISSN 1802-680X.
8. Radoslav Vasilev, Dimitar Dimitrov (2012), *Autonomous Mobile Robot for Research and Development of a Perceptual Anchoring System*, Proceedings of the Technical University - Sofia, v. 62, book 2, 2012, ISSN 1311-0829, pp. 313-322
9. Guergov S., Beevski L. (2015), *Study of the Capacity of a Fanuc M430i-A/4FH Robot to Perform Technological Operations*, In Proceedings of Scientific XII International Congress "Machines, Technologies, Materials" 2015, ISSN 1310-3946, Year XXIII, Volume 3 (2015), pp.12-14
10. Vladimir Hristov (2013), *Adaptive Observer of DC Electric Drive*, Proceedings of the Technical University - Sofia, v. 63, book 2, 2013, ISSN 1311-0829, pp. 89-98
11. Xu Chen, Masayoshi Tomizuka (2014), *Transient Enhancement in Add-on Feedforward Algorithms for High-performance Mechatronic Systems*, In Proceedings of the 19th World Congress The International Federation of Automatic Control Cape Town, South Africa. August 24-29, © 2015 Elsevier, 978-3-902823-62-5/2014 © IFAC, 2014, 7214-7220
12. *https://www.robots.com/robots/fanuc-m-430ia-4fh*
13. *http://host118022018037.metio.jp/digianaecatalog/fanuc-430ia/Book/fanuc-430ia-P0000.html*
14. *https://www.flexibleassembly.com/FI-M-430iA-4FH*
15. *https://www.robots.com/images/robots/Fanuc/M-Series/fanuc_M_430IA_4FH_Datasheet.pdf*
16. *https://www.antenen.com/products/industrial-robots/fanuc/501-1400mm/fanuc-m-430ia4fh-r30ia/c-24/c-162/p-16046*
17. *https://www.machinio.com/listings/39560739-fanuc-m430ia-4fh-5-axis-cnc-robot-with-r30ia-controller-in-almont-mi*
18. *https://www.ajsolutions.be/sites/default/files/robot-brochure-en.pdf*
19. *http://robotforum.ru/promyishlennyie-robotyi/fanuc/fanuc-m-430ia-4fh.html*
20. *https://www.fanucamerica.com/docs/default-source/robotics-product-information-sheets/m-430ia-series_17.pdf?sfvrsn=37f0815b_4*

IOP Conf. Series: Materials Science and Engineering **878** (2020) 012007 doi:10.1088/1757-899X/878/1/012007

Design control system of robot-manipulator FANUC M-430iA/4FH

E Nikolov, V Karlova-Sergieva, B Grasiani

Technical University-Sofia; FA, DIA; 8, Kliment Ohridski str., Sofia-1000, Bulgaria
nicoloff@tu-sofia.bg, vaks@tu-sofia.bg, bgrasiani@abv.bg

Summary: This paper offers the results of the control capabilities of the FANUC M-430iA / 4FH robotic manipulator. Classical and DRC robotic position control systems have been synthesized. Both systems are modelled. Their quality has been researched and comparatively analysed in unloaded, maximum loaded and in operation vibration working modes

1. Introduction.

The purpose of the present study is to research the control synthesis capabilities of a FANUC M-430iA / 4FH industrial robotic manipulator. The tasks assigned to this paper is to control the position of the robot in criterion set quality in operational mode

In this paper are presented the main results related to the synthesis of the automatic control system and the analysis of its quality

2. Synthesis of classic linear position control system for 5-DOF FANUC M-430iA / 4FH industrial ic manipulator

This publication offers the design of a classic position control system of the manipulator of 5-DOF industrial robotic manipulator FANUC M-430iA/4FH (Fig.2). It uses linear PID - regulators for controlling the position of each of the five anthropomorphic arms of the manipulator according to the set trajectory (Trajectory Set-Point Generator) of movement and positioning of the spherical wrist. The structure (Fig.2) also takes into account the influence of the disturbing impact ς_m of the workpiece machined in the production on the adjustable variables $y_{Ji} \equiv \theta_{mi}$ of the object, where $\mu_{Ji} \equiv u_{ai}$ are the regulating effects on the electrical movements of the arms (Fig.1).

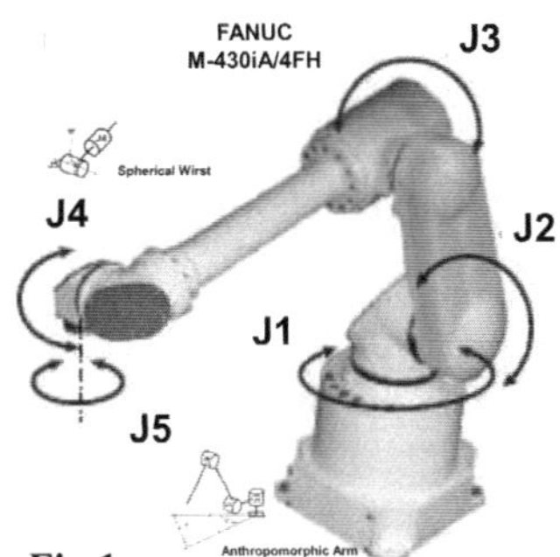

Fig.1

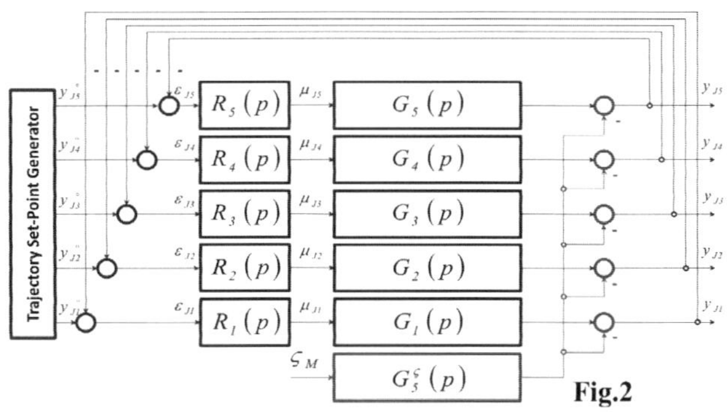

Fig.2

The analytical synthesis of the control system of the manipulator, with a ***local quality criterion*** $\sigma = const$ (boundary aperiodic transient process) and guaranteeing the stability of the nominal system, is based on the use of classical methods for the design of each of the regulators R_i (Fig.2), according to the a priori known dynamics of the respective local object G_i (1) ÷ (6). For example, the result of the synthesis of the regulator R_5 (8) at the position of the spherical wrist is shown.

$$G_5 = G_{\theta_m}(p) = \frac{\theta_m(p)}{u(p)} = \frac{K_s}{p\left(T_s p+1\right)} = \frac{3,58}{p\left(0,91p+1\right)} \tag{1}$$

$$G_4 = G_{\theta_m}(p) = \frac{\theta_m(p)}{u(p)} = \frac{K_s}{p\left(T_s p+1\right)} = \frac{3,58}{p\left(1,05p+1\right)} \tag{2}$$

$$G_3 = G_{\theta_m}(p) = \frac{\theta_m(p)}{u(p)} = \frac{K_s}{p\left(T_s p+1\right)} = \frac{3,58}{p\left(1,25p+1\right)} \tag{3}$$

$$G_2 = G_{\theta_m}(p) = \frac{\theta_m(p)}{u(p)} = \frac{K_s}{p\left(T_s p+1\right)} = \frac{3,58}{p\left(1,55p+1\right)} \tag{4}$$

$$G_1 = G_{\theta_m}(p) = \frac{\theta_m(p)}{u(p)} = \frac{K_s}{p\left(T_s p+1\right)} = \frac{3,58}{p\left(1,91p+1\right)} \tag{5}$$

$$G^{\varsigma_M}(p) = -\frac{\theta_m(p)}{\varsigma_M(p)} = -\frac{k_c\left(T_a p+1\right)}{p\left(T_s p+1\right)\left(T_m p+1\right)} = -\frac{0,253\left(1,5p+1\right)}{p\left(0,01p+1\right)\left(6p+1\right)} \tag{6}$$

$$R_i = k_{pi}\frac{\left(T_{ii}p+1\right)}{T_{ii}p}\frac{\left(T_{di}p+1\right)}{\left(T_{fi}p+1\right)} \underset{\{\sigma = const\}}{\Leftrightarrow} G_i = \frac{K_{si}}{p\left(T_{si}p+1\right)} \tag{7}$$

$$R_5(p) = 0,846\frac{\left(2p+1\right)}{2p}\frac{\left(1,5p+1\right)}{\left(0,025p+1\right)} \underset{\{\sigma = const\}}{\Leftrightarrow} G_5 = \frac{3,58}{p\left(0,91p+1\right)} \tag{8}$$

The designed system (Fig.2) is modeled and the simulation results of model (1) ÷ (6), (7) in "***without loading the wrist***" and "***with max load an wrist***" modes are shown as follows: with transient $y_i(t)$ (at $u_a = 1(t)$) and impulse transient $y_i(t)$ (at $u_a = \delta(t)$) characteristics of the manipulator respectively in Fig.3 and Fig.4: its positional frequency characteristics $W_i(j\omega)$ are illustrated in Fig.5 and Fig.6

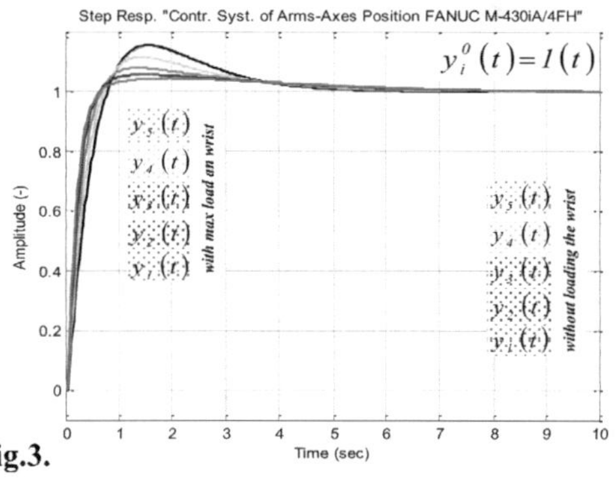

Fig.3.

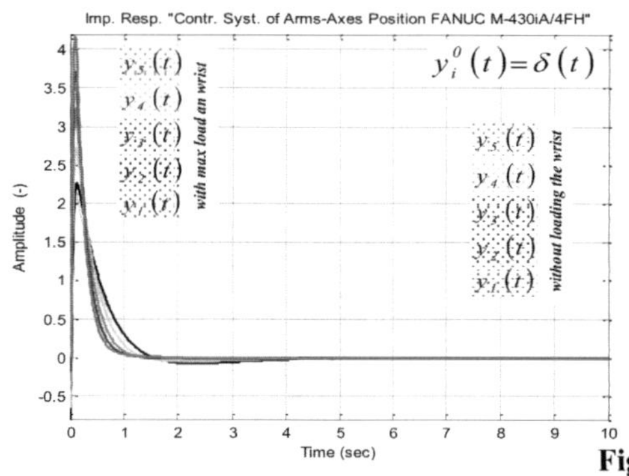

Fig.4.

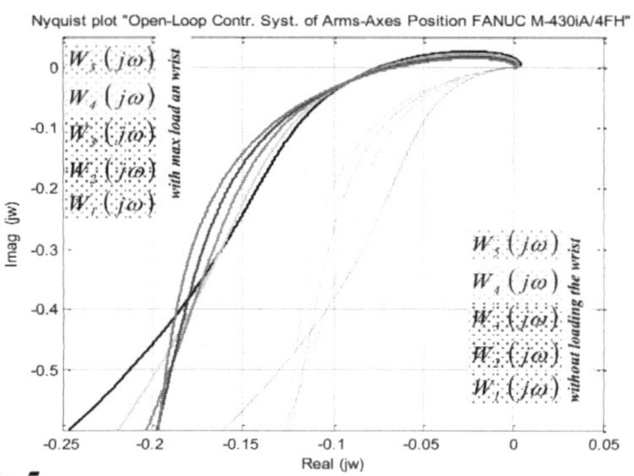

Fig.5.

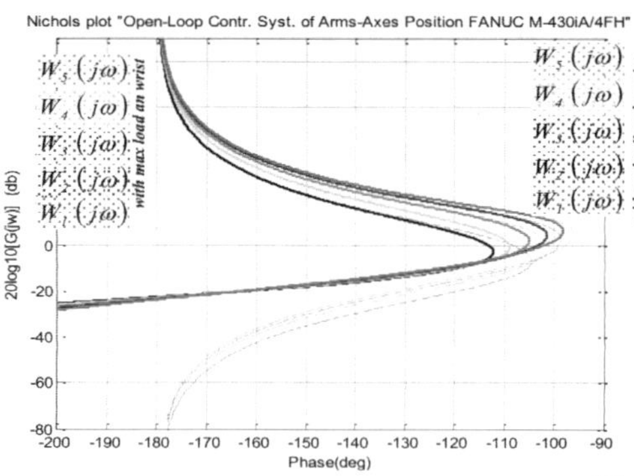

Fig.6.

The simulation results Fig.3 ÷ Fig.6 confirm the reaching of the desired quality of the system according to the synthesis criterion $\sigma = const$ (boundary aperiodic transition process) (7) for optimization, but also show a significant influence in changing the order of the controlled object as a result of the multiplicative disturbance load impact ς_M in "***with max load an wrist***" mode. To overcome this operational dependency, in the paper is proposes to design a robust position control system [15-17]

3. Robust DRC position control system for 5-DOF FANUC M-430iA / 4FH industrial robotic manipulator

In contrast to the discussed (7), Fig.2 classical system with ***PID-regulators*** (symbolically reflected in Fig.7, where $R^* \equiv R_i$) and from the known ones till now [1-14], a robust ***DRC*** position control system is offered for 5- DOF FANUC M-430iA / 4FH. Its structure is illustrated in Fig.8. The ***DRC*** system (***DRC-Disturbance Rejection Control System***) is designed to rectify the influence of parametric and structural disturbances ς_M on both the state and the adjustable variable y_{DRC}, thanks to F_{DRC}. It differs from the classical system by the presence of an internal nominal model G^* on each local contour for controlling the position of the anthropomorphic arms of the manipulator and a corresponding ***DRC***-robust filter F_{DRC} for each of the local contours

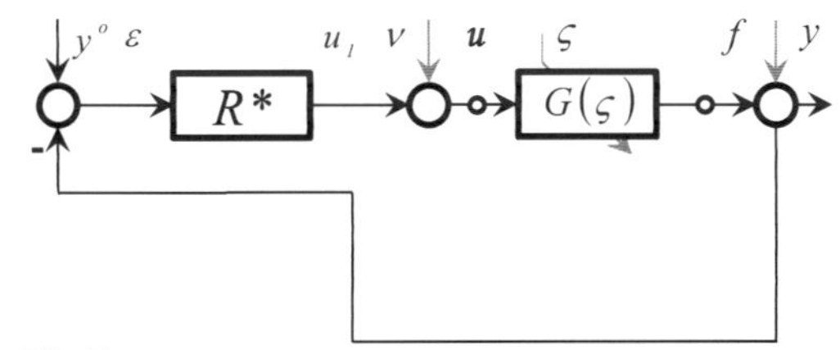

Fig.7.

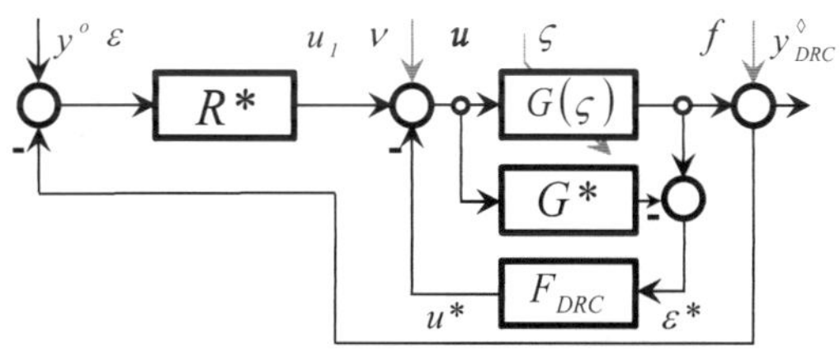

Fig.8.

The synthesis of the main regulators $R^* \equiv R_i$ in the ***DRC*** system (Fig.8) is carried out under a ***local quality criterion*** $\sigma = const$ (boundary aperiodic transition process) and guarantees the stability of the nominal system (7). The analytical design of the ***DRC*** system examines (for each of the control contours) the respective nominal G_i^* and top-of-the-line model $G_i^{■}$ of the local control object. For example, for the spherical wrist position control contour, the corresponding models are shown with (9).

$$G_5^* = G^* = \frac{3,58}{p\left(0,91p+1\right)} \;;\; G_5^{■} = G^{■} = \frac{3,58}{p\left(1,91p+1\right)} \tag{9}$$

The generalized analytical synthesis of the ***DRC*** system consists in designing a filter F_{DRC} (Fig.8) designed to rectify the effect of ς_M on the variable y_{DRC} alone. Synthesis of F_{DRC} is based on the method of the °***balance equation***° (10)

$$y_{DRC}\left(y^0, v, f, \varsigma\right) \hat{\equiv} y_{nom}\left(y^0, v, f\right), \tag{10}$$

at criterion °***zero deviation of the trajectory***° $\left[\, y_{DRC}\left(\varsigma\right)\right]$ (11) ÷ (13) of the parametrically and structurally perturbed ***DRC*** system from the nominal trajectory $\left[\, y_{nom}\,\right]$ (of the corresponding nominal unperturbed system at $\varsigma = 0$), where Φ_i are the transfer functions of the closed systems

$$\left[\boldsymbol{e}_0\right] = \left|\, \left[y_{DRC}\left(\varsigma\right)\right] - \left[y_{nom}\right] \,\right| \hat{\equiv} 0, \; \left(\left[y_{DRC}\left(\varsigma\right)\right] \hat{\equiv} \left[y_{nom}\right] \right), \tag{11}$$

$$\left[\, y_{nom}\left(p\right)\right] = \left[\, \Phi_{y^0 y}\left(p\right) \;\; \Phi_{vy}\left(p\right) \;\; \Phi_{\varsigma y}\left(p\right)\right] \left[\, y^0 \;\; v \;\; 0 \,\right]^T, \tag{12}$$

$$\left[y_{DRC}\left(p\right)\right] = \left[\Phi_{y^0 y}\left(p\right) \;\; \Phi_{vy}\left(p\right) \;\; \Phi_{\varsigma y}\left(p\right)\right] \left[y^0 \;\; v \;\; \varsigma\right]^T . \tag{13}$$

The following relations are valid:

■ If in the structure of the parametric and structurally unperturbed ***DRC*** system ($\varsigma = 0$) the object is determined by (14)

$$G\left(p, \varsigma\right) \hat{\equiv} G^*\left(p\right), \;\; \left(\varsigma = 0\right), \tag{14}$$

the ***DRC*** system will not be differ from the nominal system, it will completely coincide with it, from which follow the equations (15) for the adjustable variables, the transmission functions and the sensitivity and stability functions (16) of the two systems - the nominal and the ***DRC***- system

$$\begin{gathered}[y_{DRC}(\varsigma)] \;\hat{\equiv}\; [y_{nom}] \;\Rightarrow\; [e_0] \;\hat{\equiv}\; 0; \\ \Phi_{DRC}(p,\varsigma) \hat{\equiv} \Phi_{nom}(p)\,;\, W_{DRC}(p,\varsigma) \hat{\equiv} W_{nom}(p)\,,\, (G(p,\varsigma) \hat{\equiv} G^*(p)\,,\, \varsigma = 0)\,,\end{gathered} \tag{15}$$

$$\begin{gathered}(1+R^*(p)G(p,\varsigma))^{-1} \;\hat{\equiv}\; (1+R^*(p)G^*(p))^{-1}\,, \\ GM_{DRC}(\omega,\varsigma) \equiv GM_{nom}(\omega)\,,\; PM_{DRC}(\omega,\varsigma) \equiv PM_{nom}(\omega),\; (G(p,\varsigma) \hat{\equiv} G^*(p)\,,\, \varsigma = 0)\,.\end{gathered} \tag{16}$$

■ If in the structure (Fig. 8) of the ***DRC*** system (parametrically and structurally unperturbed $\varsigma(t) \neq 0$) the object is determined by (17)

$$G(p,\varsigma) \neq G^*(p)\,,\quad (\varsigma = \varsigma(t) \neq 0)\,, \tag{17}$$

the characteristics of the ***DRC*** system will not coincide and will not be equivalent to the corresponding characteristics of the nominal unperturbed system (18), since (19) is valid

$$\begin{gathered}\left(1+R^*(p)\left(\frac{(G(p,\varsigma)-G^*(p))}{1+(G(p,\varsigma)-G^*(p))\,F_{DRC}(p)}\right)\right)^{-1} \neq (\,1+R^*(p)G^*(p)\,)^{-1}\,, \\ GM_{DRC}(\omega,\varsigma) \neq GM_{nom}(\omega),\; PM_{DRC}(\omega,\varsigma) \neq PM_{nom}(\omega),\; (G(p,\varsigma) \neq G^*(p)\,,\, \varsigma = \varsigma(t) \neq 0)\end{gathered} \tag{18}$$

$$\frac{(G(p,\varsigma)-G^*(p))}{1+(G(p,\varsigma)-G^*(p))\,F_{DRC}(p)} \neq G^*(p),\; (G(p,\varsigma) \neq G^*(p)\,,\, \varsigma = \varsigma(t) \neq 0)\,. \tag{19}$$

In this case, the general solution (Fig.10) of the analytical synthesis of F_{DRC} is (20).

$$\begin{gathered}F_{DRC}(p,\varsigma) = \frac{(G(p,\varsigma)-G^*(p))}{G^*(p)\,(G(p,\varsigma)-G^*(p))} - \frac{G^*(p)}{G^*(p)(G(p,\varsigma)-G^*(p))}, \\ (\forall\, G(p,\varsigma) \in \Im,\; G^* \in \Im,\; \forall \varsigma\,)\end{gathered} \tag{20}$$

In this case, the research aims at solving the problem of designing a filter F_{DRC} (Fig.9) with fixed structure and parameters $F^{\diamond}_{DRC}$ (Fig.10) for a real limited range of property fluctuations $G^{\diamond} \in \Im$ (parameters and structure) of the object (wherever $G^* \in \Im$and $G^* \in \Im$ are a priori known (set) nominal and perturbed at the top of the object models) (21)

$$\forall\, G^{\diamond}(p,\varsigma) \in [\, G^*(p) \in \Im,\; G^{\blacksquare}(p) \in \Im \,]\,,\; \forall \varsigma \in [\, 0,\; \varsigma^{\blacksquare} \,]\,. \tag{21}$$

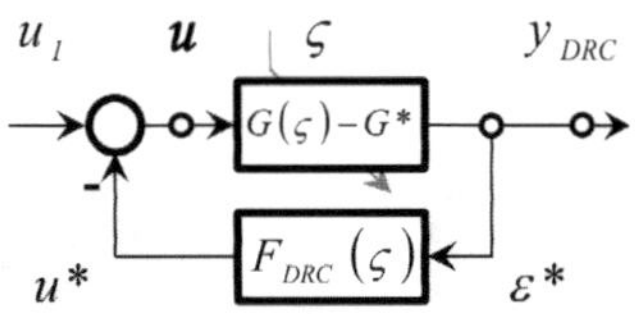

Fig.9.

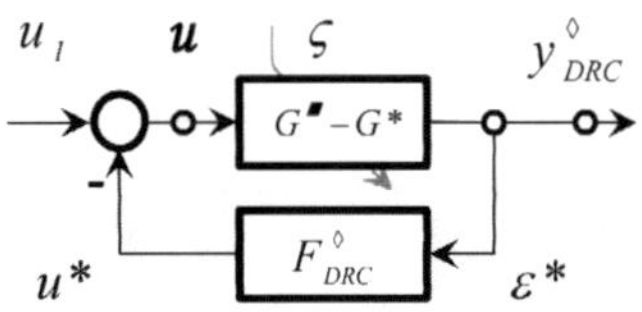

Fig.10.

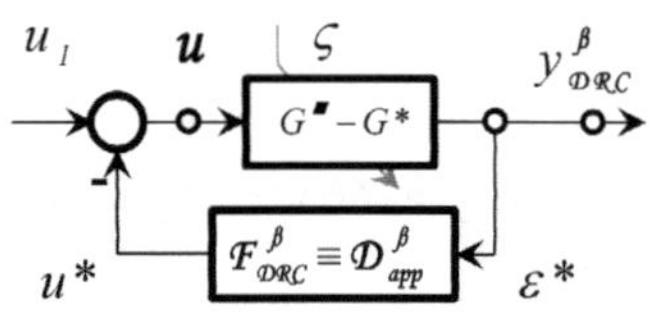

Fig.11.

It is necessary with the help of $F^{\diamond}_{DRC}$ (Fig.10) to achieve the requirements of °***the balance equation***° (22), from which the requirements logically follow (23)

$$\begin{gathered}\frac{(G^{\diamond}(p,\varsigma) - G^*(p))}{1 + (G^{\diamond}(p,\varsigma) - G^*(p))\,F^{\diamond}_{DRC}(p)} \cong G^*(p), \\ (\,\forall\, G^{\diamond}(p,\varsigma) \in [\, G^*(p) \in \Im,\; G^{\blacksquare}(p) \in \Im \,]\,,\; \forall \varsigma \in [\, 0,\; \varsigma^{\blacksquare} \,]\,)\end{gathered}\,, \tag{22}$$

$$\left(1+R^*(p)\left(\frac{\left(G^{\lozenge}(p,\varsigma)-G^*(p)\right)}{1+\left(G^{\lozenge}(p,\varsigma)-G^*(p)\right)F^{\lozenge}_{DRC}(p)}\right)\right)^{-1} \cong \left(1+R^*(p)G^*(p)\right)^{-1},$$
$$GM^{\lozenge}_{DRC}(\omega,\varsigma)\geq GM_{nom}(\omega),\ PM^{\lozenge}_{DRC}(\omega,\varsigma)\geq PM_{nom}(\omega),\qquad,$$
$$\left(\forall G^{\lozenge}(p,\varsigma)\in\left[G^*(p),\ G^{\blacksquare}(p)\right],\ \forall\varsigma\in\left[0,\ \varsigma^{\blacksquare}\right]\right) \tag{23}$$

by the criterion °***Minimum deviation*** e_{min} ***of the trajectory***° $[y_{DRC}(\varsigma)]$ of a perturbed parametric and structural ***DRC*** system from the nominal trajectory and by the criterion $\sigma = const$ defined by (24)

$$[e_{min}] = \left|[y_{DRC}(\varsigma)] - [y_{nom}]\right| \rightarrow 0\ ,\left([y_{DRC}(\varsigma)] \cong [y_{nom}]\right),$$
$$\left(\forall G^{\lozenge}(p,\varsigma)\in\left[G^*(p)\in\Im,\ G^{\blacksquare}(p)\in\Im\right],\ \forall\varsigma\in\left[0,\ \varsigma^{\blacksquare}\right]\right). \tag{24}$$

The analytical design of $F^{\lozenge}_{DRC}$ in the structure of the ***DRC*** system (parametrically and structurally perturbed with ς) for a limited range $G^{\lozenge}\in\Im$ of property fluctuations (parameters and structure) of the object is based on the method of °***the balance equation***° for (15), based on the relations (26), (27), where $\mathcal{G}^*$, $\mathcal{A}^{\blacksquare}$ and $\mathcal{B}^{\blacksquare}$ are the inverse transfer functions of G^*, $\left(G^{\blacksquare}-G^*\right)$ and of $F^{\lozenge}_{DRC}$ respectively (27)

$$\forall G^{\lozenge}(p,\varsigma)\in\left[G^*(p)\in\Im,\ G^{\blacksquare}(p)\in\Im\right],\ \forall\varsigma\in\left[0,\ \varsigma^{\blacksquare}\right], \tag{25}$$

$$F^{\lozenge}_{DRC}(p)=\mathcal{G}^*(p)-\mathcal{A}^{\blacksquare}(p)=\left(\mathcal{B}^{\blacksquare}(p)\right)^{-1},$$
$$\left(\left(G^*\right)^{-1}=\mathcal{G}^*;\ \left(G^{\blacksquare}-G^*\right)^{-1}=\mathcal{A}^{\blacksquare};\left(G^*\left(G^{\blacksquare}-G^*\right)\right)/\left(G^{\blacksquare}-2G^*\right)^{-1}=\mathcal{B}^{\blacksquare}\right),\ , \tag{26}$$
$$\left(\forall G^{\lozenge}(p,\varsigma)\in\left[G^*(p)\in\Im,\ G^{\blacksquare}(p)\in\Im\right],\ \forall\varsigma\in\left[0,\ \varsigma^{\blacksquare}\right]\right)$$

$$F^{\lozenge}_{DRC}(p)=\mathcal{G}^*(p)-\mathcal{A}^{\blacksquare}(p)=\left(\mathcal{B}^{\blacksquare}(p)\right)^{-1},\left([e_0]\underset{(G^*\in\Im;\,G^{\blacksquare}\in\Im)}{\hat{\equiv}}0;\ [e_{min}]\underset{(G^{\lozenge}\in\Im)}{\rightarrow}0\right). \tag{27}$$

The achieved solution (27) for $F^{\lozenge}_{DRC}$ with fixed structure and parameters for a limited range $G^{\lozenge}\in\Im$ of fluctuations by the method of ° ***the balance equation*** ° satisfies the requirements of the criteria for: - °***zero trajectory deviation*** e_0 ° (28) of a perturbed ***DRC*** system of the nominal trajectory for two 'points' of space $\Im$ ($G^*\in\Im$ and $G^{\blacksquare}\in\Im$) and a ***local quality criteria*** $\sigma = const$ for a real limited range of fluctuations $G^{\lozenge}\in\Im$ in the properties of the object

$$[e_0] = \left|[y_{DRC}(\varsigma)] - [y_{nom}]\right|_{(G^*\in\Im;\,G^{\blacksquare}\in\Im)}\hat{\equiv}\ 0\ ; \tag{28}$$

- ° ***minimum trajectory deviation*** e_{min} ° (26) of a perturbed ***DRC*** system of the nominal trajectory and of $\sigma = const$ for a real limited range of fluctuations $G^{\lozenge}\in\Im$ of the properties of the object

$$[e_{min}] = \left|[y_{DRC}(\varsigma)] - [y_{nom}]\right|_{(G^{\lozenge}\in\Im)}\rightarrow 0\ . \tag{29}$$

The analytical design under the constraints of the ***DRC*** system (Fig.8, Fig.10) comes down to the synthesis of $F^{\lozenge}_{DRC}$ and is based on the analytical toolkit of inverse functions [17] and operators of the theory of generalized fractional calculus, such as:

- Takes into account that $\mathcal{B}^{\blacksquare}$ is known (set) "right function" and the desired dynamics of the filter $F^{\lozenge}_{DRC}=\mathcal{B}^{\lozenge}=\left(\mathcal{B}^{\blacksquare}\right)^{inv}_{func}=\left(\mathcal{B}^{\blacksquare}\right)^{-1}$ is corresponding to its "inverse function" $\left(\mathcal{B}^{\blacksquare}\right)^{-1}$;
- On this basis, the analytic design under constraints of $F^{\lozenge}_{DRC}$ is equivalent to the solution of the inverse problem " ***to determine its corresponding inverse function*** $F^{\lozenge}_{DRC}$ ***by a known straight function*** $\mathcal{B}^{\blacksquare}$ ";
- This is possible if the solution for $F^{\lozenge}_{DRC}$ is in the class (Fig.11) of the dynamic systems of fractal differentiators D^{β} by the theory of generalized fractional calculus $\left(F^{\lozenge}_{DRC}\hat{=}D^{\beta}\right)$, where the order β is an integer fractional number

■ The practical applicability of such a solution requires $D^{\beta} = \left(\mathcal{B}^{\blacksquare}\right)^{inv}_{func}$ be a rational, physically and technically feasible function. For these reasons, it is proposed to use a rational approximation of D^{β} with the help of a frequency-limited fractional differentiator $\mathcal{D}^{\beta}_{app}$ of order β (30)

$$\mathcal{F}^{\beta}_{\mathcal{DRC}}(j\omega) \equiv \mathcal{D}^{\beta}_{app}(j\omega) \underset{\{\forall\omega\in[\omega_b,\omega_h]\}}{\cong} F^{\Diamond}_{DRC}(j\omega) \equiv D^{\beta}(j\omega) = \left(\mathcal{B}^{\blacksquare}(j\omega)\right)^{inv}_{func} = \left(\mathcal{B}^{\blacksquare}(j\omega)\right)^{-1}. \quad (30)$$

■ Based on the above, the design of $F^{\Diamond}_{DRC} \cong \mathcal{F}^{\beta}_{\mathcal{DRC}}$ (Fig.11) comes down to the analytical synthesis of a frequency-limited fractal differentiator $\mathcal{D}^{\beta}_{app}$ (31) of order β (32) approximating (Fig. 6) the inverse function $\mathcal{B}^{\Diamond} = \left(\mathcal{B}^{\blacksquare}\right)^{inv}_{func}$ of the straight function $\mathcal{B}^{\blacksquare}$ by the method of ° ***double-transfer frequency-limited approximation*** ° with algorithm (31), (32) as shown in [17].

$$\mathcal{D}^{\beta}_{app}(p) = \left(\frac{1+p\omega_b^{-1}}{1+p\omega_h^{-1}}\right)^{\beta} \prod_{j=1}^{M} \left(\frac{1+p\left(\omega'_j\right)^{-1}}{1+p\left(\omega_j\right)^{-1}}\right) \underset{\{\forall\omega\in[\omega_b,\omega_h]\}}{\cong} \left(\mathcal{B}^{\blacksquare}(p)\right)^{inv}_{func} = D^{\beta}, \quad (31)$$

$$\left(\omega'_j < \omega_j;\ \omega_b < \omega'_1;\ \omega_h > \omega_l;\ n-1<\beta<n;\ \beta = \beta\left(\arg\mathcal{B}^{\blacksquare}, \omega_C^{(\mathcal{B}^{\blacksquare})}\right)\right);$$

$$N \geq 3;\quad \beta = \left|\arg\left(\mathcal{B}^{\blacksquare}\left(\omega_C^{(\mathcal{B}^{\blacksquare})}\right)\right)\right|\left(90^{\circ}\right)^{-1}; \quad (32)$$

In the case of the proposed robust ***DRC*** system (Fig.8), according to the a priori known dynamics of the respective local objects G_i (1) ÷ (6) are synthesized on the basis of (30) according to the algorithm [17] of the corresponding filters $F^{\Diamond}_{i,DRC} \cong \mathcal{F}^{\beta}_{i,\mathcal{DRC}}$. For example, the result (33) ÷ (36) of the filter synthesis $F^{\Diamond}_{5,DRC} \cong \mathcal{F}^{\beta}_{5,\mathcal{DRC}}$ in the local spherical wrist position control system is shown taking into account the relation (9).

$$\begin{gathered} F^{\Diamond}_{5p\,DRC}(p) = \mathcal{G}^{*}(p) - \mathcal{A}^{\blacksquare}(p),\ \left(G^{*}\right)^{-1} = \mathcal{G}^{*};\quad \left(G^{\blacksquare} - G^{*}\right)^{-1} = \mathcal{A}^{\blacksquare}; \\ F^{\Diamond}_{5p\,DRC} = 0{,}2793\left(2{,}648p^{2} + 3{,}82p + 1\right) \qquad ; \\ G^{*}\left(G^{\blacksquare} - G^{*}\right)\left(G^{\blacksquare} - 2G^{*}\right)^{-1} = \mathcal{B}^{\blacksquare} \end{gathered} \quad (33)$$

$$\mathcal{F}^{\beta}_{5,\mathcal{DRC}} = \mathcal{D}^{\beta}_{5,app} = \left(\frac{1+p\omega_b^{-1}}{1+p\omega_h^{-1}}\right)^{\beta} \prod_{j=1}^{M} \left(\frac{1+p\left(\omega'_j\right)^{-1}}{1+p\left(\omega_j\right)^{-1}}\right) \underset{\{\forall\omega\in[\omega_b,\omega_h]\}}{\cong} F^{\Diamond}_{5,DRC} \quad (34)$$

$$\begin{gathered} \beta = \left|\arg\left(\mathcal{B}^{\blacksquare}\left(\omega_C^{(\mathcal{B}^{\blacksquare})}\right)\right)\right|\left(90^{\circ}\right)^{-1} = 3.105;\quad M = 5; \\ \omega_C^{(\mathcal{B}^{\blacksquare})} = 0.008775\ rad/sec;\quad \omega_u = 250\,\omega_C^{(\mathcal{B}^{\blacksquare})} = 2.6325\ rad/sec; \\ \omega_A = 0.1\,\omega_u = 0.26325\ rad/sec;\ \omega_B = 10\,\omega_u = 26.325\ rad/sec; \\ \omega_b = 0.06\,\omega_A = 0.015\ rad/sec;\quad \omega_h = 0.1\,\omega_B = 2.558\ rad/sec; \end{gathered} \quad (35)$$

$$\begin{aligned} \mathcal{F}^{\beta}_{5,\mathcal{DRC}} = \mathcal{F}^{3.105}_{5,\mathcal{DRC}} = 0.33\,&\frac{(0{,}112p+1)}{(7{,}122p+1)}\cdot\frac{(27{,}390p+1)}{(0{,}515p+1)}\cdot\frac{(7{,}620p+1)}{(0{,}143p+1)}\times \\ \times&\frac{(2{,}120p+1)}{(0{,}040p+1)}\cdot\frac{(0{,}589p+1)}{(0{,}011p+1)}\cdot\frac{(0{,}164p+1)}{(0{,}003p+1)} \end{aligned} \quad (36)$$

Together with those of $\mathcal{B}^{\blacksquare}_{5}$ and of $F^{\Diamond}_{5,DRC}$, the transient and impulse transient characteristics $\mathcal{F}^{\beta}_{5,\mathcal{DRC}}(t)$ of the designed filter (36) are shown in Fig.12, Fig.13 and its frequency characteristics $\mathcal{F}^{\beta}_{5,\mathcal{DRC}}(j\omega)$ - in Fig.14, Fig.15. The results confirm compliance with the requirements (30) ÷ (32). They also satisfy the criteria (30), (37) of the dependence between $\mathcal{F}^{\beta}_{5,\mathcal{DRC}}(j\omega)$ and the characteristic of the dynamical system $\mathcal{B}^{\blacksquare}_{5}$ (33) adopted at the synthesis of "straight function" systems.

$$mod\left(\mathcal{F}^{\beta}_{5\,\mathcal{DRC}}(\omega)\,.\,\mathcal{B}^{\blacksquare}_{5}(\omega)\right)_{\{\forall\omega\in[\omega_b,\,\omega_h]\}} = 1\,;$$
$$arg\left(\mathcal{F}^{\beta}_{5\,\mathcal{DRC}}(\omega)\,.\,\mathcal{B}^{\blacksquare}_{5}(\omega)\right)_{\{\forall\omega\in[\omega_b,\,\omega_h]\}} = 0 \quad (37)$$

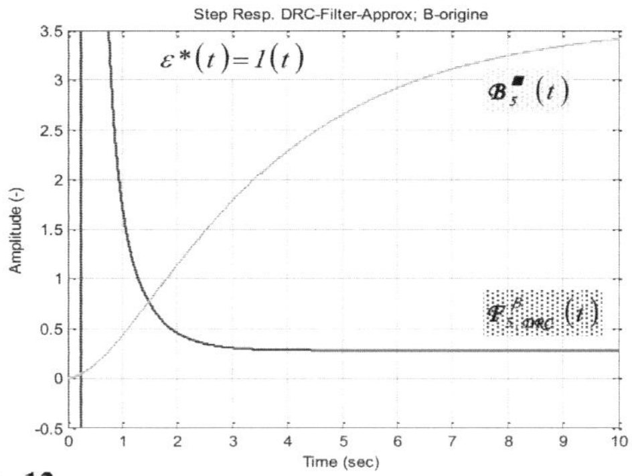

Fig.12.

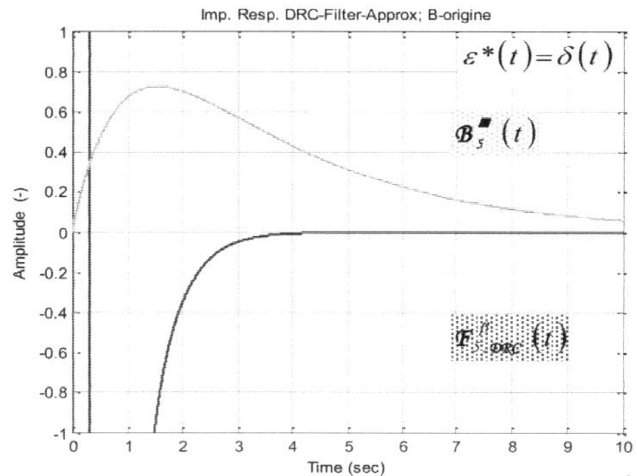

Fig.13.

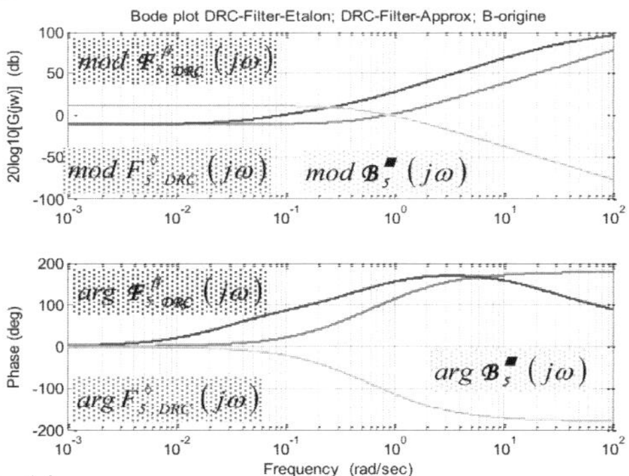

Fig.14.

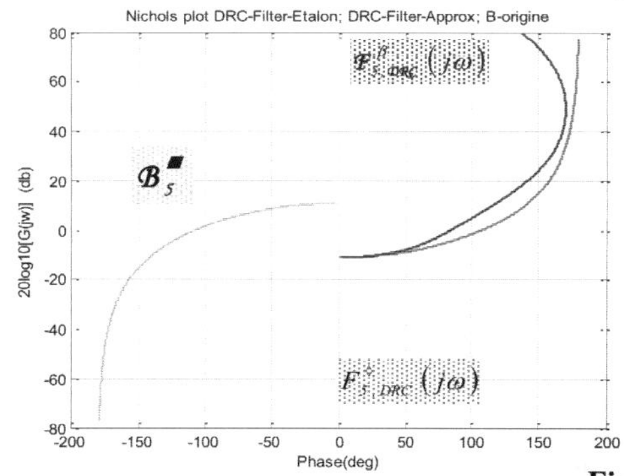

Fig.15.

The designed (10) ÷ (32) robust ***DRC*** system (Fig.8 ÷ Fig.11) is modeled. Parallel to those of the classic (7), (8) system with ***PID***-regulator (Fig.2, Fig.7), its features in "***with max load an wrist***" mode for the model parameter fluctuations indicated and indicated in the illustrations ($k_1 = 3{,}58 \div 8{,}58$; $T_1 = 0{,}5 \div 5{,}0\ s$; $k_2 = 0{,}253 \div 0{,}753$; $T_2 = 6 \div 11\ s$; $T_3 = 1{,}0 \div 4{,}5\ s$) are shown with transients $\mathcal{y}_{5\mathcal{DRC}}(t, k_1, k_2, T_1, T_2, T_2)$ (Fig.16.a. ÷ Fig.19.a.) and with frequency characteristics $\mathcal{W}_{5\,\mathcal{DRC}}(j\omega, k_1, k_2, T_1, T_2, T_2)$ by position (Fig.16.b. ÷ Fig.19.b.). Comparative analysis of the quality indicators of the two (Fig.7, Fig.8, Fig.11) systems ($\mathcal{y}_{5\mathcal{DRC}} \Leftrightarrow y_{5DRC}$ и $\mathcal{W}_{5\mathcal{DRC}} \Leftrightarrow W_{5DRC}$) confirms that for the specified range of parametric ς_M fluctuations the synthesized ***DRC*** system meets the requirements of:

■ the local quality criteria ($\sigma = const$) (7),(38) and maintaining speed (39); - criteria for ° ***minimum deviation from the nominal trajectory*** ° (24) shown for the specific example with $e^{3.105}_{\mathcal{DRC}}$ и $e^{3.105Clegg}_{\mathcal{DRCML}}$ (40).

$$\sigma^{\beta}_{\mathcal{DRC}} \cong \sigma^{\lozenge}_{DRC} \cong \sigma^{*} = const\ ; \quad (38)$$

$$t^{\beta}_{p\ \mathcal{DRC}} \cong t^{\lozenge}_{p\ DRC} \cong t^{*}_{p}\ ; \quad (39)$$

$$\left[e^{3.105}_{\mathcal{DRC}}(\varsigma)\right] = \left|\ \left[y^{3.105}_{\mathcal{DRC}}(\varsigma)\right] - \left[y^{*}\right]\ \right|_{\left(G^{\lozenge}\in\Im\right)} \rightarrow 0\ ;$$
$$\left[e^{3.105}_{\mathcal{DRC}}(\varsigma)\right] \lll \left[e_{DRC}(\varsigma)\right]\ . \quad (40)$$

■ the stability criteria, such as the modular stability stocks *GM* and phase *PM* are illustrated in Fig.16.b., and the ratios in their values being (41)

$$\begin{gathered}\mathcal{GM}^{\beta}_{\mathcal{DRC}}\left(\omega,\varsigma_M\right)\geq GM\left(\omega,\varsigma_M\right);\ \mathcal{PM}^{\beta}_{\mathcal{DRC}}\left(\omega,\varsigma_M\right)\ \geq\ PM\left(\omega,\varsigma_M\right),\\ \left(\forall\, G^{\Diamond}\left(p,\varsigma\right)\in\left[\,G^*(p),\ G^{\blacksquare}(p)\right],\ \forall\varsigma_M\in\left[\,0,\ \varsigma^{\blacksquare}_M\,\right]\right)\end{gathered},\tag{41}$$

■ the requirements for the parametric invariance (42) of the quality indicators of a ***DRC*** system

$$\left(\sigma^{\beta}_{\mathcal{DRC}},\ t^{\beta}_{p\ \mathcal{DRC}},\ [\boldsymbol{e}],\ \mathcal{GM}^{\beta}_{\mathcal{DRC}},\ \mathcal{PM}^{\beta}_{\mathcal{DRC}}\right)\text{ от }\varsigma_M\ .$$

$$\begin{gathered}\sigma^{\beta}_{\mathcal{DRC}}\neq f\left(\varsigma_M\right);t^{\beta}_{p\ \mathcal{DRC}}\neq f\left(\varsigma_M\right);\\ mod\,\boldsymbol{e}^{3.105}_{\mathcal{DR}}\neq f\ \left(\varsigma_M\right),\left(\forall\,\omega\in\left[\,0,\,\omega_h\,\right]\right)\\ \mathcal{GM}^{\beta}_{\mathcal{DRC}}\neq f\left(\varsigma_M\right),\left(\forall\,\omega\in\left[\,0,\,\omega_h\,\right]\right)\\ \mathcal{PM}^{\beta}_{\mathcal{DRC}}\neq f\left(\varsigma_M\right),\left(\forall\,\omega\in\left[\,0,\,\omega_h\,\right]\right)\end{gathered}.\tag{42}$$

It is obvious that the ***DRC*** system is superior to the classic system with ***PID***-regulators, since it is valid (40) ÷ (42), which substantiates the effectiveness of the final solution (34), (35).

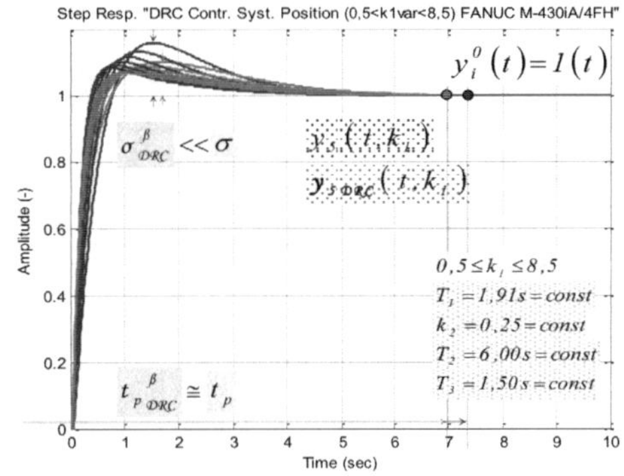

Fig.16.a.

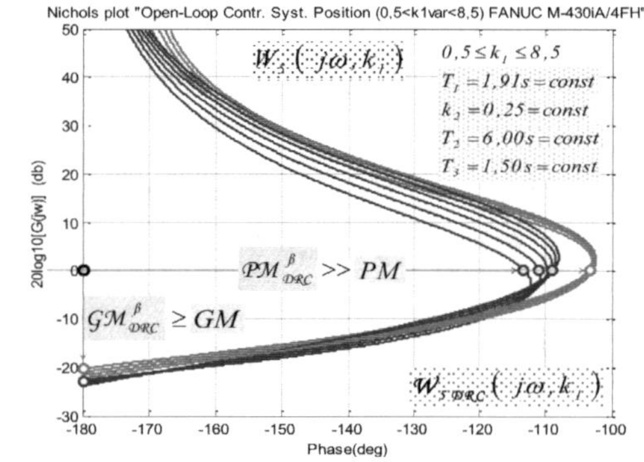

Fig.16.b.

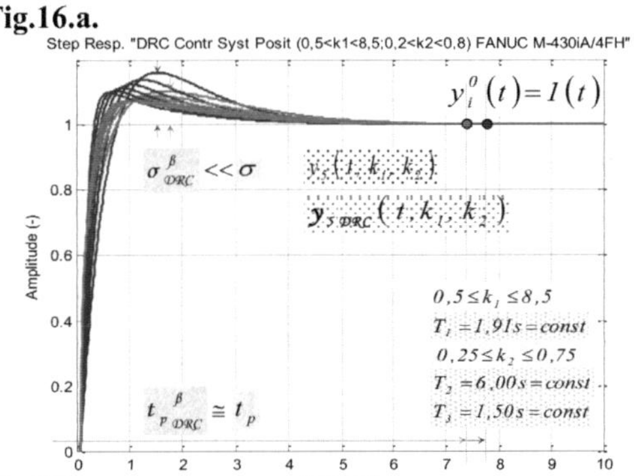

Fig.17.a.

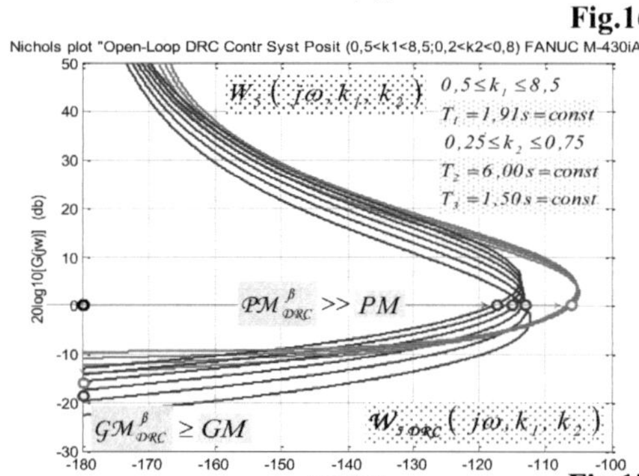

Fig.17.b.

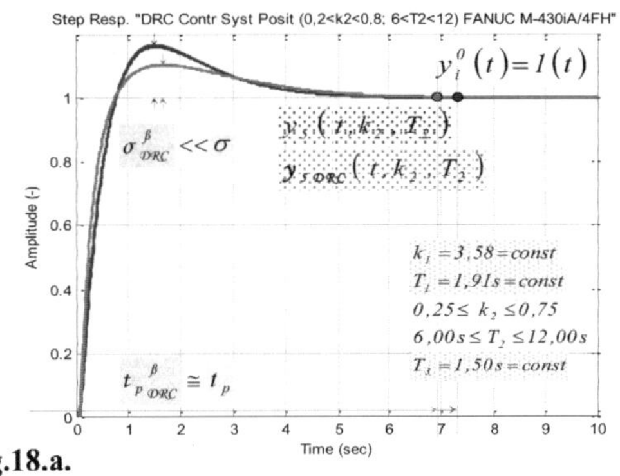

Fig.18.a.

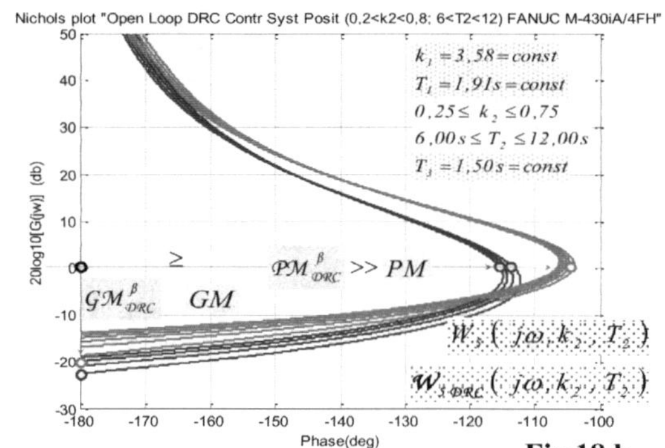

Fig.18.b.

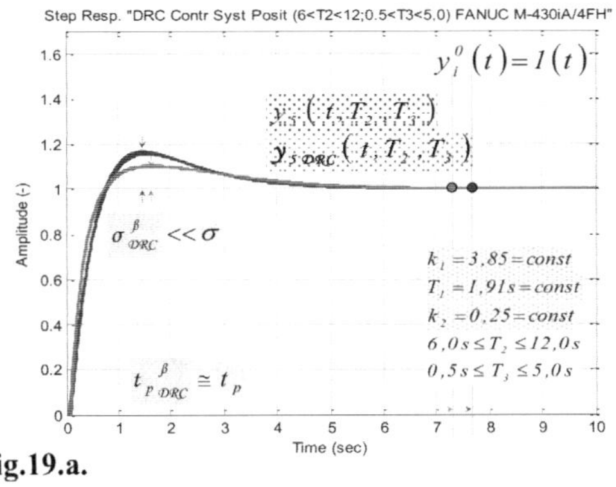

Fig.19.a.

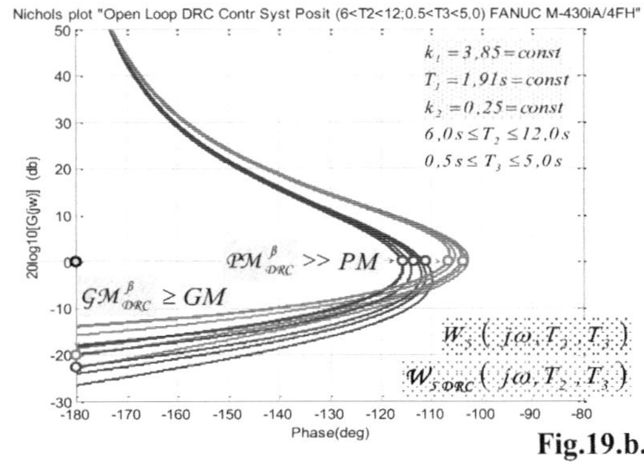

Fig.19.b.

4. Coclusion

New and original in this part of paper are:

■ The strategy of controlling the position of an industrial robotic manipulator with the help of a robust ***DRC*** system, which resects the influence of parametric and structural interferences on the adjustable variable;

■ The proposed methods for ° ***balanced equation*** ° and an appropriate algorithm for the analytical synthesis of a robust ***DRC*** position control system for an industrial robotic manipulator, which is actually used for parameterization and structural interferences regulating the adjustable variable;

■ The results of the structural and parametric synthesis of a classical system with ***PID***-regulators and a robust ***DRC*** position control system for the 5-DOF FANUC M-430iA / 4FH industrial robotic manipulator. Local quality criteria with guaranteed stability and criteria for ° ***minimum deviation from nominal trajectory*** ° were used in the design.

■ The quality analysis of the designed robust ***DRC*** position control system for the robotic manipulator system by modeling and simulating under conditions of multiplicative load perturbation ς_M in "***with max load an wrist***" mode in parallel with the classical system. The comparative analysis of the obtained results confirms that for a wide range of parametric fluctuations, the synthesized ***DRC*** system meets the requirements of: the local quality criteria; the criteria for ° ***minimum deviation from nominal path*** °; the sustainability criteria; the requirements for parametric invariance of quality indicators from parametric fluctuations of a control object

Acknowledgment
The authors would like to thank the Bulgarian National Science Fund for the financial support by project КП-06-Н37/17.

Literature

1. Cong Wang, Minghui Zheng, Zining Wang, Masayoshi Tomizuka (2016), *Robust Two-Degree-of-Freedom Iterative Learning Control for Flexibility Compensation of Industrial Robot Manipulators*, In proc. of 2016 IEEE International Conference on Robotics and Automation (ICRA), © 2016 IEEE, Electronic ISBN: 978-1-4673-8026-3, DOI: 10.1109/ICRA.2016.7487388, 238-2386
2. Cong Wang, Wenjie Chen, Masayoshi Tomizuka (2012), *Robot End-effector Sensing with Position Sensitive Detector and Inertial Sensors*, In Proc. of 2012 IEEE International Conference on Robotics and Automation River Centre, Saint Paul, Minnesota, USA, May 14-18, 2012, 978-1-4673-1405-3/12/$31.00 ©2012 IEEE, 5252-5257
3. Cong Wang, Yu Zhao, Yubei Chen, Masayoshi Tomizuka (2015), *Nonparametric Statistical Learning Control of Robot Manipulators for Trajectory or Contour Tracking*, Robotics an Computer-Integrated Manufacturing, © 2015 Elsevier, 35 (2015) 96-103
4. Guergov S. (2019), *A Method for Determining the Constant DC/AC Motor Power Range and Main Technological Spindle Unit Parameters*, In Proc. IOP Conf. Series: Materials Science and Engineering618 (2019) 012055IOP Publishing, doi:10.1088/1757-899X/618/1/012055, pp.1-6
5. Guergov S., Beevski L. (2015), *A Robot System for Implementation of Technological Operations*, In Proc. of International Scientific Conference "70 years FIT", the 28-th International Scientific Conference of the Faculty of Industrial Technology of TU-Sofia, 11-13 September 2015, Sozopol,© 2019 Publishing House of Technical University of Sofia, ISBN: 978-619-167-178-6, pp.439-444

http://mtf70.tu-sofia.bg/uploads/Downloads/FIT70.pdf
6. Guergov S., N. Nikolova, Gr. Stambolov, (2019), *Robust control application possibilities for machining robots*, 16-th Conference on electrical machines, drives and power systems (ELMA) 6-8 June 2019, Varna, Bulgaria, IEEE Catalog number CFP19L07-USB, ISBN 978-1-7281-1412-5, pp. 562-565
7. Masayoshi Tomizuka (1996), *Robust Digital Motion Controllers for Mechanical Systems*, Robotics and Autonomous Systems, © 1996 Elsevier, 19 (1996) 143-149
8. Masayoshi Tomizuka (2008), *Dealing with Periodic Disturbances in Controls of Mechanical Systems*, Annual Reviews in Control, © 2008 Elsevier, 32(2), doi:10.1016/j.arcontrol.2008.07.002, 193-199
9. Masayoshi Tomizuka (2013), *Control Methodologies for Manufacturing Applications*, Manufacturing Letters, © 2013 Elsevier, 1 (2013), 46-48
10. Shouren Huang, Niklas Bergström, Yuji Yamakawa, Taku Senoo, Masatoshi Ishikawa (2016), *High-Performance Robotic Contour Tracking Based on the Dynamic Compensation Concept*, In Proc. of 2016 IEEE International Conference on Robotics and Automation (ICRA), © 2016 IEEE, Electronic ISBN: 978-1-4673-8026-3, DOI: 10.1109/ICRA.2016.7487577, 3886-3893
11. Xu Chen, Atsushi Oshima, Masayoshi Tomizuka (2013), *Inverse-Based Local Loop Shaping and IIR-Filter Design for Precision Motion Control*, In Proc. of 6th IFAC Symposium on Mechatronic Systems The International Federation of Automatic Control, April 10-12, 2013. Hangzhou, China, © 2013 Elsevier, 978-3-902823-31-1/13/$20.00 © 2013 IFAC, 10.3182/20130410-3-CN-2034.00066, 490-497
12. Xu Chen, Masayoshi Tomizuka (2012), *A Minimum Parameter Adaptive Approach for Rejecting Multiple Narrow-Band Disturbances with Application to Hard Disk Drives*, IEEE Transactions on Control System Technology, © 2012 IEEE, 20(2), doi:10.1109/TCST.2011.2178025, 408-415
13. Xu Chen, Masayoshi Tomizuka (2014), *Transient Enhancement in Add-on Feedforward Algorithms for High-performance Mechatronic Systems*, In Proceedings of the 19th World Congress The International Federation of Automatic Control Cape Town, South Africa. August 24-29, © 2015 Elsevier, 978-3-902823-62-5/2014 © IFAC, 2014, 7214- 7220
14. Youngwoo Lee, Liting Sun, Jun Moon, Chung Choo Chung, Masayoshi Tomizuka (2019), *Reference Modulation for Performance Enhancement of Motion Control Systems with Nonlinear Parameter Variations*, IEEE/ASME Transactions on Mechatronics, July 2019, DOI: 10.1109/TMECH.2019.2930087, pp.1-8
15. N. G. Nikolova, (2019), *Examination of a repetitive process control system, 8th International Scientific Conference* "TechSys 2019" – Engineering, Technologies and Systems, 16–18 May 2019, IOP Conference Series Materials Science and Engineering 618:012013, Plovdiv, Bulgariaq doi:10.1088/1757-899X/618/1/012013
16. N. G. Nikolova, (2019), *Repetitive robust control for systems with uncertain*, (*monography*), Publishing House of Technical University of Sofia, 978-619-167-371-1, pp 177
17. Emil Nikolov (2019), *DRC-Systems - Invers Solutions with Fractional Clegg-Operators - part 1, part 2*, Journal Proceedings of Technical University of Sofia, Volume 69, Issue 3, 2019, © 2019 Publishing House of Technical University of Sofia, ISSN 1311-0829, pp. 47-66

IOP Conf. Series: Materials Science and Engineering **878** (2020) 012008 doi:10.1088/1757-899X/878/1/012008

Expert estimation of leakage losses in the pneumatic systems in the bottling industry

R D Kosturkov, V G Nachev and T P Titova

Department of Automatics, Information and Control Systems, University of Food Technologies, Plovdiv 4002, 26 Maritza Blvd

Abstract. By eliminating leaks in compressed air systems in manufacturing plants, up to 50% of energy can be saved. With a well-designed inspection and maintenance plan, eliminating the majority of leaks can be a routine and effective practice. This paper describes the results of a pneumatic leakage audits in manufacturing facilities from the bottling industry in Bulgaria. The results are analyzed and are defined major groups elements of the pneumatic system, which can be potential sources of leakage leading to significant losses of compressed air and determining energy efficiency in the production.

1. Introduction

Compressed air is a reliable source of energy and is widely used in industry and the service sectors thanks to its easy and safe production, transmission and use. In most industrial facilities, compressed air is required in a significant number of technological operations. The specifics of the production (industry), its scale and the structure of the production equipment, the life cycle stage, make the task of optimizing pneumatic systems multidimensional. With this regards, it is difficult to find a single, generally valid solution for improving energy efficiency when using compressed air [1].

On this basis, it is necessary to link the approach of researching and summarizing the most effective measures to optimize the pneumatic systems to the relevant production or industry. For example, specific industries are subject to analysis in the papers [2,3,4], etc., respectively the woodworking, pharmaceutical and mining industries.

Compress air (CA) leaks are one of the largest sources of energy loss in pneumatic systems and eliminating them in order to improve the energy efficiency of compressed air is of primary importance. In terms of effort and resources, this is also the most effective, conventional measure.

On top of energy loss, CA leaks inflict pressure drops in the system. This effect to the operation of air-powered equipment and tools, and lead to reduction of production efficiency. Leaks cause significant losses, often up to 50% of the CA produced [5,6].

Despite of the recent development and availability of modern equipment, identifying leak points and calculating the flow rate of compressed air losses, caused by them, is a labour-intensive and time-consuming task [7,8].

In the bottling industry like a soft drink, mineral water, beer, etc. production, the main functional modules of bottling lines are similar, regardless of the product manufactured. Most often, they include (Fig. 1) de-palletizer, filling machine, inspection and marking section, automatic labelling machine, automatic multiple shrink wrapper and palletizer. In some industries, depending on the type of individual packaging, a PET blow-molding machine is included in the production line. The main part of the manufactured CA in a bottling plant used in these modules is for linear and rotary actuating, vacuum operations, blowing and drying, transportation, etc. [9]. In these modules are observed greatest losses of CA. Therefore, a systematic approach and prioritized leakage detection and repair is essential to the maintenance.

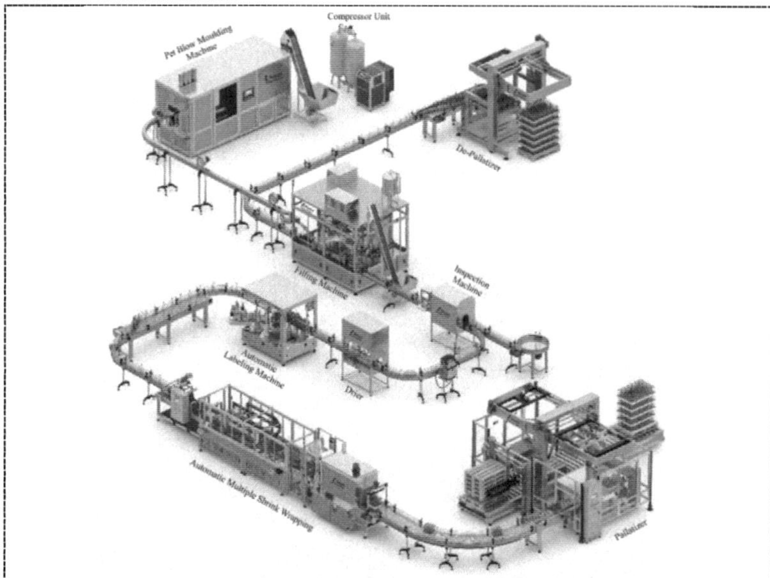

Figure 1. Main functional modules of bottling lines

2. Detection and sizing of leaks - measuring equipment, software

2.1. Soup Liquid Method

In this method, the suspected leakage area is covered with soapy liquid, which, in the presence of leakage, begins to form air bubbles. The size of the bubbles can be used as a basis for determining the size of the leak. This method is cheap and easy to use, but it may be more time-consuming compare to others methods.

2.2. IR Thermography Method

The infrared technology uses IR thermovision cameras to measure thermal energy emitted by an object. These cameras allow very accurate contactless temperature measurement. In almost all compressed air systems, a malfunction or leakage of air lead to change in the temperature. Thermal or infrared energy is a light emission that is invisible to the human eye due to the high wavelength. This is the part of electromagnetic radiation that is popularly known as heat or thermal radiation. Unlike visible light, in the infrared spectrum, any object with a temperature above absolute zero emit heat. Thermography is successfully applied in the audits of pneumatic systems. An infrared camera can record temperature changes in the installation and detect damage or leakage locations [10].

2.3. Ultrasonic Method

Ultrasounds are generated by the turbulence of a flow inherent in the leakage of a CA. In the event of a leakage from a CA system, the friction of the air at the edges of the orifice from which it flows is an ultrasonic source. The effect does not depend on the size of the leak, its flow rate or the size of the opening. Consequently, ultrasonic leak detection is the industry standard and one of the most effective methods for detecting it.

Ultrasonic detectors are relatively easy to use and can detect leaks that cannot be heard with the naked human ear. In addition, during the inspection with such equipment, it is not necessary to interrupt the production cycle, as the background noise does not affect the results. Ultrasonic leak detection is classified as active, passive and vibroacoustic.

This study used passive ultrasound detection to quantify compressed air leakage.

IOP Conf. Series: Materials Science and Engineering **878** (2020) 012008 doi:10.1088/1757-899X/878/1/012008

3. Inspection and analysis of losses of compressed air caused by leakage in bottling industries

The analysis is based on the results of audits of CB leaks at 10 factories in the bottling industry in Bulgaria - bottling of mineral water, soft drinks and production and bottling of beer. The timeframe of the survey is the period 2017-2019.

The leakage audit involves a thorough inspection using an ultrasonic detector.

The elements of the pneumatic systems under examination are classified into 9 groups as follows:

1. Actuators - including all types of pneumatic actuators: cylinders – with rod and rodless, rotary actuators, pneumatic grippers, pneumatic actuators of process valves, etc.
2. FRL (Filter-Regulator-Lubricator) - Including all elements of FRL or individual elements - filters, regulators, lubricators and their manually or automatically shut off valves, pressure gauges, pressure switches and flowmeters (if are part of module of the FRL).
3. Single Valves - this group includes standalone valves, whether used for actuator control or for other purposes (excluding those in the pneumatic preparation group FRLs).
4. Valve islands - includes valves and distributors of block assembly, common base, with or without common power supply and wiring, with or without communication modules and all elements adjacent to the block.
5. Fittings - Includes all Push-in quick connectors regardless where they are mounted.
6. Speed regulators.
7. Tubes and connectors. This group covers leaks in flexible tubes and hoses, as well as the connections between them - fasteners, fittings, clamps, nipples, couplings, etc. without Push-in fittings.
8. Bottle grippers and gripper heads (Fig. 2) - Specific items for bottling industry, mainly used in palletizers and de-palletizer for manipulating bottles. They are separated into a special category because of their specificity and the significant losses of compressed air that generate the leaks detected in them.
9. The last group covers all other leaks detected in the pneumatic system: highways, vacuum elements, etc., which are not included in the first 8 groups.

Figure 2. Bottle grippers and gripper heads

4. Instrumentation

The intensity of the ultrasonic waves generated by the leakage during the examination of the bottling companies was measured with an ultrasonic detector SDT200 [11]. The technical specification is presented in Table.1.

Table 1. Technical specification on SDT200

Central Frequency	38.4kHz
Bandwidth	+-2 kHz
Sensitivity	65 dB/V/μbar in 40 kHz
Measurement range	20dBμV to 120dBμV, ± 0.5 dBμV
Resolution	0.1 dBμV
Response time	>10μs

Two sensors from the SDT200 package were used - the sensor integrated into the unit and a flexible sensor with a flexible body length of 550mm and a focusing tip and a spacer from the sensor - 25mm. The leakage measurement distances are 40cm, 20cm or 2.5cm.

The SDT200 measures the ultrasonic signal, converts it to audio through heterodyne, and amplifies it for audible signalling to headphones. The dimension of the detected signal is dBμV

The conversion of the dBμV signal values recorded by the SDT200 to volumetric flow rate (l/min, m3/h) is analytic using the Field Leak Estimator v.2014, (Fig. 3 and Fig. 4). The reading is calibrated for a flexible 40cm (1.3ft) sensor at 6bar, 20 and 2.5cm (correction with -6 dBμV for each double shortening distance). Which means, for measurements of 20 cm the correction is -6 dBμV, for measurements of 2.5 cm the correction is -24 dBμV.

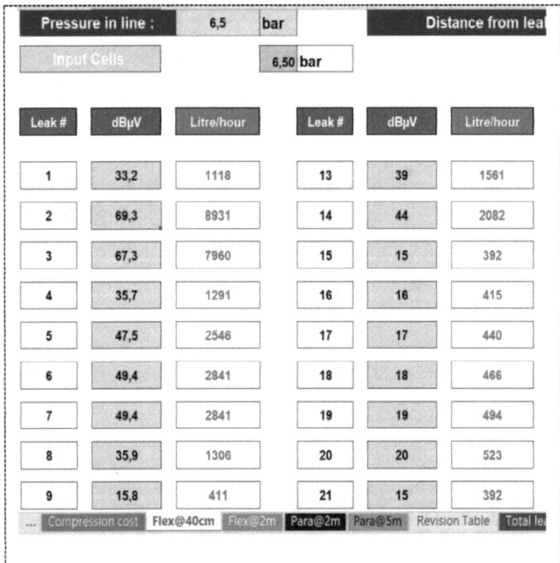

Figure 3. Field Leak Estimator

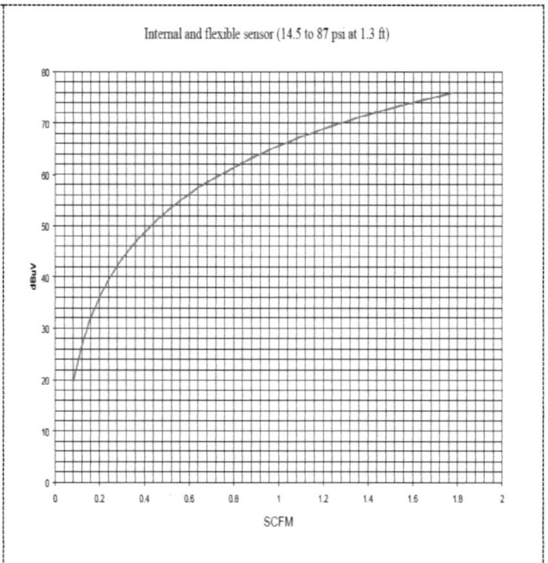

Figure 4. Field Leak Estimator dBmV(Flow)

It should be noted that the conversion is considered approximate. In practice, many factors can influence the result, such as the size and shape of the opening, the profile of the walls, etc. [11].

5. Results and Discussion

During the survey, 1328 significant (over 5 l/min) leakages were detected in the CA systems of the factories examined. The summary results are given in Table 2.

Table 2. Summary results

Summary results	Number	%	Leakage l/min	% of total loss	Average leakage (l/min)
Fittings	457	34,41%	13042,90	29,77%	28,54
FRL	240	18,07%	7547,83	17,23%	31,45
Bottle grippers	125	9,41%	6174,05	14,09%	49,39
Tubes and connectors	106	7,98%	3931,27	8,97%	37,09
Single Valves	99	7,45%	3231,08	7,38%	32,64
Others	94	7,08%	3206,20	7,32%	34,11
Actuators	81	6,10%	3090,93	7,06%	38,16
Speed regulators	75	5,65%	2134,56	4,87%	28,46
Valve islands	51	3,84%	1449,61	3,31%	28,42
Total	**1328**	**100,00%**	**43808,42**	**100,00%**	**32,99**

The distribution of leakages by group of elements is shown graphically in Fig. 5 and 6 shows the ratio of leakages between the groups, respectively, in number and total losses in l/min.

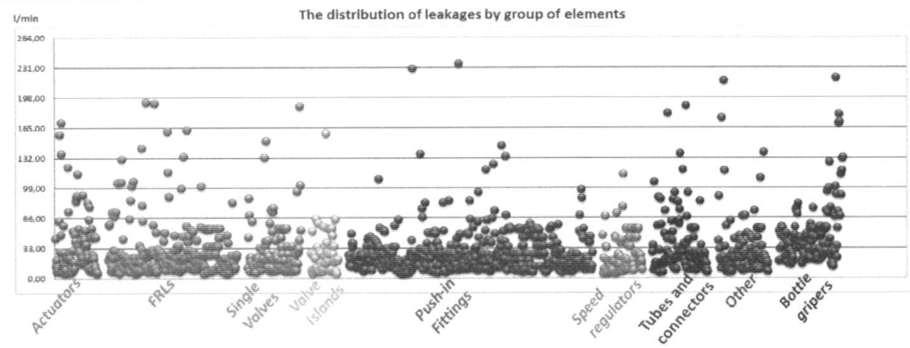

Figure 5. The distribution of leakages by group of elements

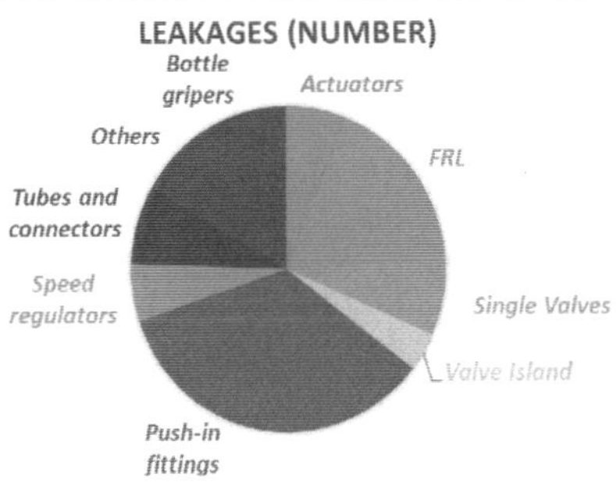

Figure 6. Leakages (number)

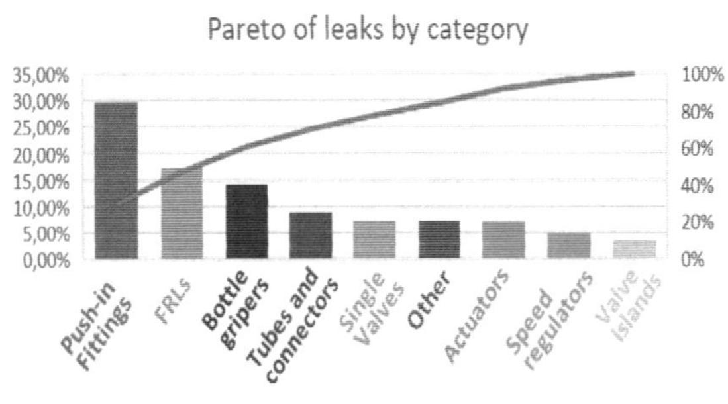

Figure 7. Pareto Leakages by category (l/min)

The following conclusions can be drawn from the results of the survey in bottling companies:

1. Leaks in 4 of the predefined groups - Fittings, FRL, Bottle grippers and Tubes and connectors - represent more than 70% of total losses. The large number of leaks in the Fittings group is not surprising because their number in the pneumatic systems is the largest, and their diversity in quality and prices on the market is also significant.
2. Bottle grippers are a specific element of the pneumatic system in bottling lines and can generate very large losses from compressed air leaks.
3. The commonality between the four groups of items, generating 70% of losses (Pareto chart, Fig.7), is that, unlike actuators, speed regulators, valves, and valve islands, they are easily accessible and quickly replaceable with relatively little down time.

The average leak by group, with the exception of Bottle grippers (Table 2), show that targeting a specific group of components will be not more effective (in terms of air losses) than complex removal. We can see this on Fig.5, where differences are largely determined by the presence of a small percentage of strongly divergent points. On other hand, removing a large number of small leakages can be irrational. However, it can be concluded that about 70% of leaks can be removed by leakage repairmen in the first four groups of components of pneumatic systems.

In this study, we used a passive ultrasonic method for detecting leaks in the compressed air system, which is indirect, non-contact and although accepted as best practice in leak detection, problems with single leak sizing are not neglected. However, many researchers and analysts, including manufacturers of such detectors stands that when detecting and measuring a large number of leaks, the overall results are sufficiently reliable and give a good practical idea of the energy losses generated by leaks in the compressed air system.

The resulting statistics can be used to plan and prioritize the service activities of components in the pneumatic systems.

6. Conclusion

This paper summarizes and analyses the results of 10 compressed air leakage audits at bottling companies. Four main groups of elements of the pneumatic system were defined, contributing to over 70% of losses. The statistics obtained give a realistic idea of the scale of losses caused by leaks. A problem that cannot be permanently resolved over time. The results may find application in training of the maintenance staff, developing a plan for the maintenance of the pneumatic systems, or in the overall task of improving the energy efficiency of production lines.

Acknowledgement

The authors would like to express their gratitude to SMC INDUSTRIAL AUTOMATION BULGARIA for instrumentation support and case studies assembled in this paper.

References

[1] Efficiency, CADDET Energy. 2001. *Compressed air savings by leakage reduction and efficient air nozzles* CADDET Energy Efficiency Result 402

[2] Permaweld PVT LTD. 2017. *Compressed Air System Energy Audit & Leak Survey Report Of Leading Pharmaceutical Industry.* Shivanagar BANGALORE: PERMAWELD PVT LTD

[3] Parkkinen J, Zenger K and Ollila J. 2013. *Statistical Approach To Energy Efficiency Issue In Industrial Pneumatic Systems.* The 13th Scandinavian International Conference on Fluid Power, SICFP2013. Linköping, Sweden: SICFP2013.

[4] Marais J H. 2012. *An integrated approach to optimise energy consumption of mine compressed air systems* Pretoria, South Africa.

[5] Dindorf R and Wos P. 2018. "*Automatic measurement system for determination of leakage*" Metrology and measurement systems, Vol. 25, 159-170

[6] Kosturkov R, Nachev V and Titova T. 2019. "*System Analysis and Opportunities for Optimization of Pneumatic Systems in Manufacturing Plants*" TEM Journal vol. 8 is.3 749-763

[7] Eret P, Harris C, O'Donnell G, and Meskell C. 2011. "*A practical approach to investigating energy consumption of industrial compressed air systems*" Proc. IMechE vol. 226 (Part A: J. Power and Energy) 28-36

[8] Dindorf R, Wos P and Pawelec K. 2017."*Automatic device for indirect measurement of leakage flow rate in compressed air pipeline*" IOP Conference Series: Materials Science and Engineering 233 012030. OP Publishing Ltd.

[9] Harris K, U.S.Department of Energy. 2012. "*Compress air Best Practices Bottling* " The Magazine for Energy efficiency in Compressed Air, Pneumatics, Blower and Vacuum Systems, Jan/Feb .

[10] Kroll A, Baetz W, Peretzki D. 2009."*On Autonomous Detection of Pressured Air and Gas Leaks Using Passive IR-Thermography for Mobile Robot Application*" IEEE International Conference on robotics and automation Kobe Japan. 291-296.

[11] SDT International. 2015. "*Leak Surveyor Handbook*" Brussels (Belgium): SDT International. "SDT200 User Manual." SDT.

IOP Conf. Series: Materials Science and Engineering **878** (2020) 012009 doi:10.1088/1757-899X/878/1/012009

ERP systems and data security

Radoslav Hrischev

Technical University–Sofia, Branch Plovdiv, 25 Tzanko Djustabanov Str., 4000 Plovdiv, Bulgaria

hrischev@tu-plovdiv.bg

Abstract. Enterprise Resource Planning (ERP) integrates all business processes in the organization. Therefore, data security is the main point of security in ERP systems. Most of the biggest ERP developers have already implemented specialized hardware and software security solutions. New ITC technologies – cloud computing, IoT, blockchain are opportunities to make the ERP highly integrated, more intelligent, more flexible, cloud based. The fast growing of the e-business is making ERP systems more open and new decisions are needed to be granted information security. This paper introduces ERP technology regarding the data security point of view. New security solutions for ERP systems are present.

1. Introduction

Enterprise Resource Planning is a business management software by planning the resources of an organization and covering most of business processes. ERP systems are implemented in different types of organizations. The Gartner Group first defined it in 1990 as the next generation of Manufacturing Business System and Manufacturing Resource Planning software [1]. Today, ERP has become from precondition to necessary requirement for successful business development. The mass production with low added value is impossible without effective ERP system.

With ERP, an enterprise can automate its core business applications, reduce the complexity and the cost of the integration. ERP system is a tool for fast and effective business reengineering, optimization of operations and generation of benefits.

The goal of the article is to prepare an overview of the status of the ERP technology and the security risk for information exchange in ERP system. It will be explained the evolution of ERP, its key components – internal and external modules, evolution from a security point of view and the perspective of development. There are many study on this topic, but the fast grown of developers of ERP systems and new information technology requires new approaches to information security. The examples considered are from the biggest developers of ERP systems SAP and IBM, where data security is implemented as standard important part (software and hardware) of the systems.

2. Overview of ERP technology

The ERP system covered all information flow in the organization. It has represented the highest level of information systems. At the base of ERP systems is a client-server technology. Most populate ERP systems at the moment are centralized information systems.

The spread of ERP systems has grown significantly over the years. First systems were developed locally for simple separated business functions such as accounting, controlling, planning, warehouses management and human resources planning. The next step started when companies such as Oracle, IBM, SAP, and Baan developed a set of typical modules, which are developed as applications of ERP systems, intended to manage of main parts of business. The web service is a technology with a serious impact to ERP systems. New systems used secured web-based databases and applications. This technology provided opportunity to use Internet as environment for the data transmission and storage. Modern ERP systems are cloud-based ERP systems such as SAR S/4 HANA. They integrated a result of new technology – Cloud computing, IoT, Big data.

All ERP systems included some core functionalities, realized in similar applications - modules. The modules have a different name, but functionality is the same. These functionalities are presented in following modules: Financial Management /FI and CO modules in SAP/; Human Resource Management /HCM in SAP/; Manufacturing Management /MES or MAP modules/; Sales, Distribution, and Logistics Management /SD module in SAP/; Customer Relationship Management (CRM); Product Lifecycle Management /PP in SAP/; Supplier Relationship Management (MM in SAP); Business Intelligence /BI/; Supply Chain Management (SCM). The SCM includes business processes, part of CRM, SRM, MES, PP modules. Idea is that these business processes have to be connected in enterprise to be granted business success of the supply chain [2].

These functionalities are inter-organizational, communication and data exchange is internally in the company.

The new e-business requires the development and implementation of an e-commerce modules of ERP (e-orders, e-shop, e-store, e-invoice, etc.), focuses on business between companies and customers. The basis of e-business is an electronic way to exchange business information (documents) Electronic Data Interchange /EDI/. Therefore, new external information exchange modules also set new requirements for data security. Depending on the stage of growing, requirements for security also growing.

Historically, in an enterprise, on the first stage some systems have been developed by the enterprise IT team itself. On next step growing most of applications was developed from different external vendors using different databases, languages, and technologies may develop others. All systems have to work together. Systems differ from each other, making it difficult to effectively upgrade the organization's business, strategy, and information technology. Also, applications need ongoing support from different suppliers.

ERP systems are using common standards for communication infrastructure, application, database, data exchange and security. How many and what modules will be implemented in the organization depends mainly on the goal set for the realization, the vision of the leadership, the budget and the available resources of organization. ERP have no alternative in enterprise, but must meet the special requirements [3] – modularity, integrality, flexibility and scalability, profitability. ERP should lead to cost reductions and increased profits, as these are the basic requirements for the organization's existence and the goal of implementing the system.

Data security is a main requirement for ERP systems. Regardless of the quantity and types of modules implemented, the amount and way of information exchange, the system must ensure that information is fully protected. To realize this requirement, the vendors are using special methods of information security, starting from architecture of systems, methods of data transfer, data access, databases. Implements a set of security policy.

3. Information security in ERP

ERP systems are used not only in business - production and commerce, but also in other branches, where information security is critical, such as medicine, finance, defence. First step of ensuring information security begins with the development of a sustainable model beginning with architecture of the systems.

3.1. Three-layer architecture of ERP systems

The ERP system is based on the architecture client-server to distribute by level information processing. Most vendors distributed information processing in three logical layers – Presentation for user connect, Application for data processing and Database for data repository. The three main levels are related each other by communication lines and platform, provided access only to services and information only for authorized people, devices and processes. For example, SAP Net Weaver is information communication server for SAP R/3 [4]. On Figure1 is presented layers' structure of the SAP [4].

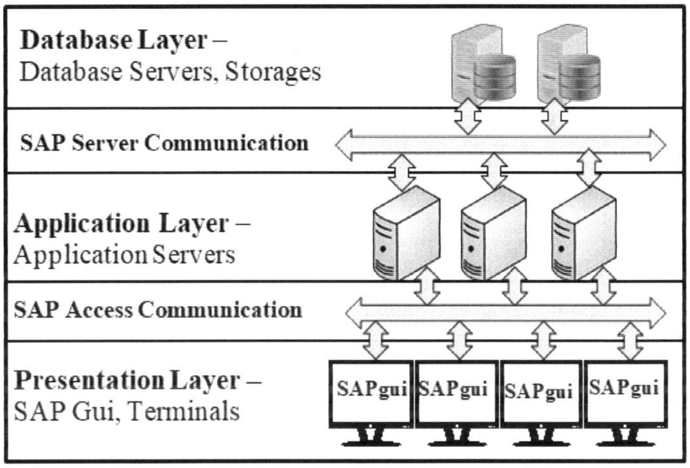

Figure 1. Layers architecture of SAP R/3.

On Presentation Layer (Front End) are remote terminal /hardware or virtual/ with Graphical User Interface (GUI) on desktops or browser that collects input data, transfer information to next layers, generates requests, and returns the results of data processing to the user or machine.

On Application Layer are presented application software which collect the requests from the Presentation layer and process the requests based on the business algorithms and functions.

On Database Layer is Database Management Service that manages the business and operational information. There is stored also information about user access to this information, all operations for every user and service, defined on relevant security policies. This layer included local or cloud based servers and storage of data.

3.2. Security policy for ERP systems

Conception of confidentiality is a main point of security policy, and access control to information and data flow consists following separated elements:

- Every user has access just to this part of information, which he has to know for his daily work.
- Level of sharing information has to be strictly defined depend of functionality, described in communication between modules.
- The information, shared between users, departments, organizations, has to be trusted.
- An information has to be modified just by authorized users, services, applications, and result of this modification has to be visible jus for authorized users.
- During its transmission in communication network information should not be changed.
- If concert implementation included e-modules, information must be encrypted and information should be encrypted and decrypted only by an authorized user.

These securities policies are realized using different methods of user authentication and authorization.

Most of ERP systems are using Role-Based Access Control services [6]. This service granted access to information just to be authorized users with relevant permissions, grouped by his role in organization. Role-Based Access Control (RBAC) includes the following elements – users, roles, permissions, administration.

- Users. The user is a person, service, application, who have access to relevant of him functions in organization information.
- Roles. The role is a described job function within the organization.
- Permissions. Permission is the granted access to part of information, service or application in the system.
- Administrator. The administrator is a user, responsible for granting access of users to this information, described on the table of responsibility. Table of responsibility is a list or matrix distributing the permissions of users by roles in the business process.

The realization of the granted permissions (rights) is possible through a process of authorization. The authorization process is an important part of security and is different for most of vendors. For example, the authorization process in SAP R/3 includes of the following components - authorization administration, authorization object, authorization fields, authorization profile, authorization check, user master record, which enables the users to log on to system and grant limited from permissions access to the information.

For authorization in e-business modules are using also digital signature and security certificates. Digital signature is an e-signature authenticated using encryption and ensures the authenticity of the information. Security certificate is a unique digital ID used for verification of users or services, requesting access to information.

3.3. Methods for secure data transfer in ERP systems

The first ERP systems worked in a closed networks and the most used security technology was isolation and segmentation of the network. For biggest companies were used leased optical lines.
But leased lines are very expensive. Also with the development of web-based ERP systems began using modern methods for information protection such as IPsec (Internet Protocol Security).

IPsec, also known as IP Security protocol, provided a security services for information exchange over IP networks. IPsec is used to provide secured data transfer in ERP systems using technology as IP tunneling. This mean that all data being exchange between two points encrypted by special protocols. Tunneling mean an establish a Virtual Private Network (VPN) connection. The ERP systems are using often public wireless networks for data transfer and VPN technology is a perfect way for secured data transfer over wireless networks.

E-business modules are using also security protocols Secure Socket Layer (SSL) and Secure Hypertext Transfer Protocol (SHTTP) to be granted security of document exchange between customers and web based ERP services over public Internet.

The ERP vendors are developing and implementing more and more strict solutions to improve information security, but also specialized companies are developing software and hardware information security solutions. The most commonly used solutions are the Citrix [5] applications. This company developing a full range of information security solutions. Citrix is software company that developing virtualization technology for desktops, servers and applications and cloud computing.

An ERP system collecting information from terminals, GUI of users, specialized devices and cannot secure the system by limiting access. Therefore, best way to increase security level is using of Citrix infrastructure between the users and the ERP system – Figure 2.

In this case users are using Citrix app for Internet browser to access ERP. They need just additional authentication by user and password. After authentication the users are working on own virtual desktop, no matter the type of access device – PC, terminal, mobile, tablet and Operation System – Windows, iOS, Android, Linux, etc. Citrix infrastructure included secured network, using gateway, and the whole range of IT controllers, services and servers. Citrix also supports server virtualization. At the moment more than 100 million users using Citrix services.

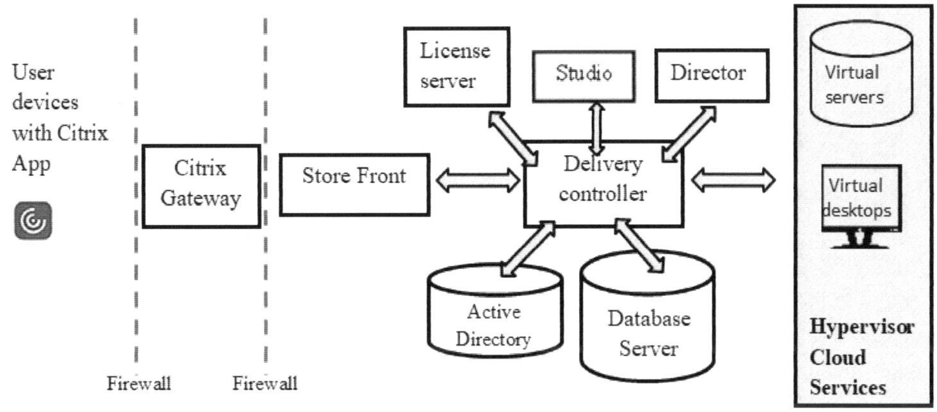

Figure 2. Typical CITRIX infrastructure for ERP.

A lot of companies are developing hardware for secure data transfer directly to ERP systems. For example, company Lantronix [10] offers a wide range of devices and services to provide secure data transfer to control systems.

3.4. Database of ERP systems

All ERP systems currently are using centralized databases, located on the Database layer – figure1. The database is a collection of data. The organization of the data follows a particular model. According to the model, two types of databases can be described - relational and non-relational. Most of the ERP systems are using relational SQL /Structured Query Language/ databases. Mostly are used Microsoft SQL Server and Oracle DB.

NoSQL databases appeared in the early 21st century. They appeared to solve the need to manage large volumes of information in modern web and mobile applications. NoSQL databases are used for big-data and online working web applications. Therefore new versions of ERP systems are using NoSQL databases, SAP HANA, for example.

At the moment biggest vendors of ERP systems SAP [4] and IBM [8] are working on the implementation of new databases based on blockchain technology.

A blockchain is a series of matrices containing encrypted information in the form of records called blocks connected to each other. The blocks form a dynamic and distributed database. The database is decentralized and secure because each block contains an encrypted address of the previous block and transaction data – Figure 3. In its organization, this technology is strikingly similar to the sequential databases used in the first process management systems. By design, a blockchain is resistant to modification of the data and provides an opportunity to build a high-level security database.

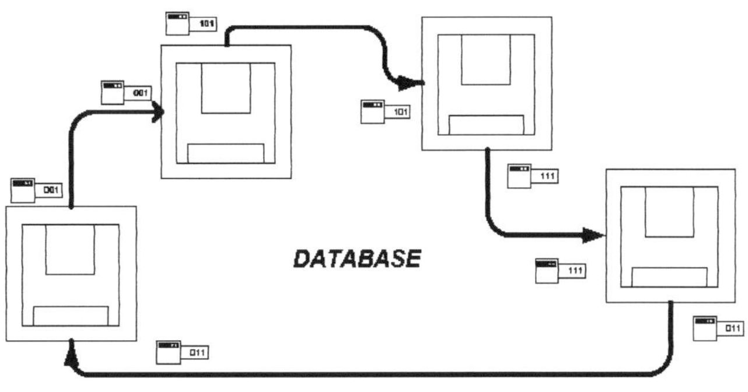

Figure 3. Illustration of new database using blockchain technology.

Blockchain is more than technology. It is a movement to help redefine most important business relationships through trust, transparency and new-found collaboration. It could be the future of the database organization of ERP systems [9]. Blockchain is transforming supply chain management (SCM), deliver confidence and transparency in the process. Blockchain technology is solving trust issues inside and between companies. For the entry of this technology into human resources (HR) management, for example, new legal regulations are needed.

4. A future of ERP systems and information security

The development and implementation of ERP systems that started more than 30 years ago began with mainframe (local PC) structure systems and reached integration with the cloud-based technologies. But the process of development of the ERP systems is not an end in itself, it is conditioned by intensive business growing and the requirement for more manageability and predictability of business performance.

Today ERP is becoming the basis for business management in various sectors and condition for the existence of many organizations. The ERP systems are requirements to benefit generation for the business. Therefore, the security of information going to be of utmost importance.

The future ERP systems depend on three main factors - business growing and benefit generation, implementation of the new modern technologies and provision of information security. With sufficient confidence we can expect evolution in the following directions:

- Intensive developing of the modules, oriented to e-business.
- Collaboration between core ERP modules and e-commerce.
- The ERP systems will be Internet based. This mean that no just databases will be cloud based, but also applications will be placed on the Web.
- Network connectivity will go to wireless data exchange as a result of implementation of new fast speed Wi-Fi networks – 5G, for example.
- The ERP will be more and more intelligent. In the future ERP will include more components for business analysis and decision making. Elements of AI /Artificial Intelligence/ will be integrated into ERP.

5. Conclusion

Enterprise Resource Planning is the technology that reforming business, its condition for economy growing and impact to more and more people and organization. Information security has been a goal for system developers from the beginning, but implementation in newer, sometimes critical areas, also puts new requirements to security. We need new approach, new security technology, new solutions, new tools to grant the security of the systems.

Existing security solutions are based on the features of the implemented currently ERP system, but e-business, the implementation of new e-modules make systems more open and existing solution for information security will become more scarce and systems more vulnerable. The companies, generating more profits will become increasingly subject to cyberattack.

The ERP systems vendor and companies specializing in developing of new security systems have to look for:

- New design of the ERP architecture, granted information security.
- New security policy and authentication and authorization methods in an open environment.
- Systems and devices, protected the data transfer in web based services.

The role of public authorities should also be more proactive for developing adequate legislation to ensure the effective protection of information.

Acknowledgements
The author/s would like to thank the Research and Development Sector at the Technical University of Sofia for the financial support.

References
[1] Bakry A and Bakry S 2005 *Enterprise resource planning – a review and a STOPE view* (Riyadh: International Journal of Network Management 15) p 363-370
[2] Banerjee A 2018 *Blockchain Technology: Supply Chain Insights from ERP* (Amsterdam, Elsevier: Advances in Computers, Volume 111) p 69-98
[3] Hrischev R 2018 *Planning and implementation of the ERP system in packaging production* (Plovdiv: Seventh International Scientific Conference TECHSYS
[4] SAP official site https://help.sap.com/
[5] CITRIX official site https://www.citrix.com/
[6] Wei S and Bhavani T 2007 *Security for Enterprise Resource Planning Systems* (Dallas: Information Systems Security 16) p 152–163
[7] Thuraisingham B 2005 *Database and Applications Security* (Integrating Information Security and Data Management) p 2-19.
[8] IBM official site https://www.ibm.com/blockchain
[9] Forrester 2018 Emerging Technology Projection: *The Total Economic Impact™ Of IBM Blockchain*, Study commissioned by IBM.
[10] Lantronix official site https://www.lantronix.com/

IOP Conf. Series: Materials Science and Engineering **878** (2020) 012010 doi:10.1088/1757-899X/878/1/012010

Research and analysis of modern space heating technologies and management for industrial buildings

V Yosifova[1], D Chikurtev[1] and R Petrov[1]

[1]Institute of information and communication technologies, BAS
Acad. G. Bonchev St., Bl. 25A
1113 - Sofia, BULGARIA

christova@iinf.bas.bg

Abstract. The paper observes different modern space heating technologies suitable for industrial buildings. Their main working principles and applications are researched related to their energy efficiency. The problems about heating of large scale spaces in industry are observed. A compartment analysis between the technologies is carried out with their advantages and disadvantages.

1. Introduction

The industrial sector represents about 26% of total EU-28 energy consumption in 2012. Although industry is only the third largest end-use sector in the EU after buildings with a share of 40% and transport with 32%, in some countries with more energy-intensive industries, such as Austria, Germany, the Czech Republic or Sweden, the share varies between 30 and 35%. In Finland and Slovakia, industry contributes more than 40% of total energy consumption. [1]

In recent years, the industry has been seeking more and more energy efficient solutions for heating and minimizing production costs. One of its main problems is the winter heating of buildings, halls or workshops where the heat goes out because of numerous factors such as: large spaces, high ceilings, flow formation, constantly opening doors, poor thermal insulation, etc. Known heating systems (radiators, convectors, air conditioners, etc.) heat the room on the convection principle, heat the air, it becomes lighter and rises to the ceiling, where the colder surface of the ceiling cools it, where becomes heavier and moves down again to the floor of the room. This cycle is repeated and causes air circulation, causing unpleasant differences between floor and ceiling air temperatures, as well as the spread of dust, smoke, noise and CO_2 emissions.

The temperature of heated industrial workplaces is important issue when it comes to workers health and safety, but also on productivity. It is considered that workers perform better between 16°C and 24°C, depending on working process.

According to the Chartered Institute of Building Services Engineers, (CIBSE) different working environments require different temperatures. For example:

- Heavy work in factories: 13°C
- Light work in factories: 16°C
- Hospital wards and retail: 18°C
- Offices and dining rooms: 20°C

Other problems that may occur due to bad heating systems and their maintenance are employee bad concentration, dizziness, fainting, or even heat stroke. Apart from the social consequences, usually

there are high energy costs and bad environmental footprint because of the use of the wrong heating technologies or lack of proper heater control, poor ventilation and air circulation. [2]
In order to choose the most relevant solution for heating large spaces, the designers should keep up with the modern heating technologies and have the ability to implement them in the best way, based on the building's heating demands. The improvement of existing HVAC technologies, as well as the development of smart building systems and their combination, can contribute in rising the quality of work and working environment, not only in the industry but also where large volumes and open spaces are a complicated test for energy efficient heating like concert halls, sport areas, event centres etc.

2. Heating technologies for industrial spaces

The well known technologies for heating industrial premises, like direct and indirect gas heaters, fan heaters, electrical heaters and boiler systems has few things in common: high energy consumption and convection principle of work, which leads to temperature differences, low heating effect and high consumption costs. In order to make one step closer to energy efficient future, new methodologies with renewable resources and innovative heating technologies should be applied combined with well developed intelligent management systems.

2.1 Heating systems with solar power

Industrial buildings have large and usually flat roof space which is a great opportunity to install solar panels. Industry can significantly reduce their overheads by generating their own electricity from solar energy. Solar heating systems can be applied for heating hot water alone or combined with space heating and can be matched with different type of space heaters. The main components include the collector which collects the solar energy, a thermal storage unit which transports the collected heat to the storage and store it for later use and a heat generator. It is assumed that the solar system is supplementary to the primary heating system, because it is most common to have a solar fraction of less than 50 % of the DHW and space heating demand. The performance is very dependent on especially the size of the solar collectors in relation to the energy consumption. [3]
Solar energy is available at the place, where it can be converted into heat. If it is the same with the auxiliary energy, then it is an autonomous system. Despite that, electricity is always needed for operation of any heating system for the different components such as fans, pumps, compressors of the heat pumps, the automatic control system to operate, thus the solar heating system is not entirely autonomous. In order to be fully autonomous both heat and electricity should be produced on site. For achieving zero energy buildings a well-developed automatic control system should be developed that controls the energy gain and consumption. Their operation depends on the the functions of the system, required temperatures and expected heating load and their adaptation is developed by the consumer heating requirements. [4]
The electricity generated by solar panels first supplies on-site needs, with the grid supplying additional electricity as needed. When the building generates more electricity than it consumes, the electricity is fed back into the grid. It is very effective for administrative buildings, big stores and production factories. There are lots of benefits of solar electricity but the most important one is the carbon free usage — solar electricity is green renewable energy and doesn't release any harmful carbon dioxide or other pollutants.[5]

2.2 Heat pumps

Heat pump technology can deliver great economic, environmental and energy system benefits worldwide. The idea behind of a compression heat pump is the thermodynamic cycle, which is popular also as refrigeration cycle (Fig. 1). It has 5 components: an evaporator (a "liquid-to-gas" heat exchanger), a compressor, a condenser (a "gas-to-liquid" heat exchanger), an expansion valve, a transfer fluid (refrigerant).

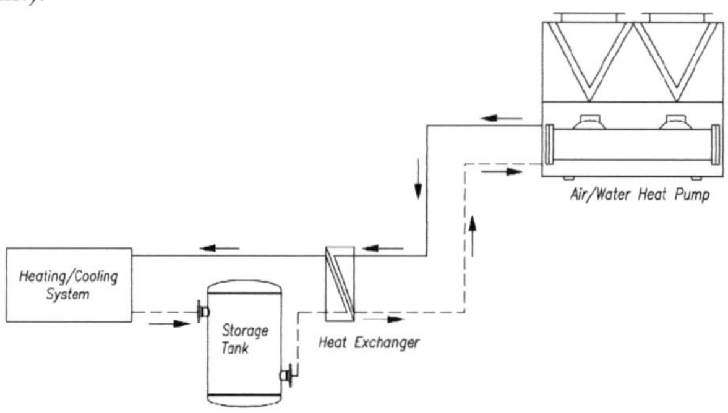

Figure 1. Heat pump working scheme

An electric compression heat pump uses electricity to run the compressor and the pumps; a sorption heat pump uses thermal energy to drive the cycle. But when it counts to the total energy balance, the share of electrical energy needed to drive the cycle is relatively small.

Heat pumps can use renewable energy from air, water or excess energy from industrial processes and infrastructure installations like subway or underground parking. The type of energy used determines the heat pump technology and is also part of the typical naming convention: Air source heat pumps use air as energy source; Geothermal heat pumps use ground energy, which is transferred to a transfer fluid and transported to the heat pump unit; Water source heat pumps use water connected to aquifers, rivers, lakes, sea, and others.

Industrial heat pumps are most often bespoke systems designed to the building's specific needs. The application may be for industrial and production needs, rarely only heating, so it is not preferable choice if it is not included in the production operation. One of the challenges in industrial heat pump design is the selection of components that can operate at higher temperatures on the side of the energy source. [6]

Heat pumps for heating and cooling demand is increasing in most countries. They serve well for cooling in summer peak or heating in winter peak situations, but in very low temperatures(below - 25 C^{o}) they need to be stopped and secondary heating systems should be taken in consideration.

The control of heat pumps is in a similar way as other air-conditioning systems such as electric or gas heating. Usually heat pumps have remote control capabilities for easy operation needs. [7]

2.3 District heating and cooling systems

District Heating and Cooling (DHC) systems increase the overall efficiency of the energy system by recycling heat losses from a different energy conversion processes. It is based on the recovery of heat, that otherwise would be lost, that is delivered for consumption needs of industrial process or building management (Fig. 2). Many renewable sources such as biomass and geothermal energy are difficult to use for other purposes, but in DHC systems they can be efficiently included in the process. The evolution of DHC has three main stages. First, they were using steam as a heat carrier. Until 1970s the systems used pressurised hot water as a heat carrier, with temperatures mostly over 100°C. Modern systems also use water as the heat carrier below 100°C, following the requirements for lower and more flexible distribution temperatures, assembly oriented components, and more flexible materials which leads to s environmentally friendly and customer-oriented solution. [8]

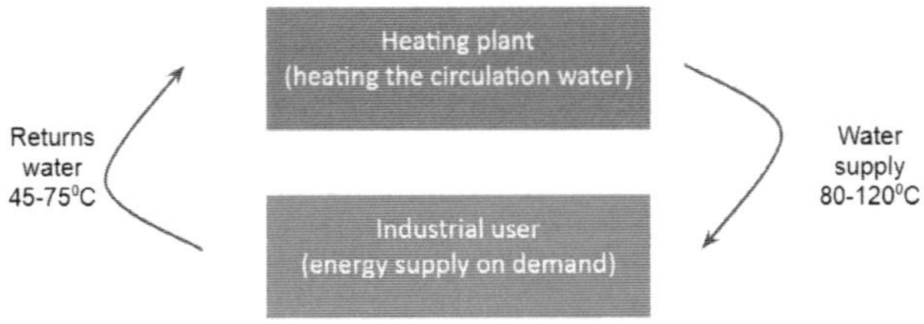

Figure 2. District heating principle

DHC systems are still not very popular in industry heating, but play effective role in emissions reduction and energy savings. They consist of networks of insulated pipes, pumps, energy sources and end users. Heat exchangers transfer heat from a district heating network to a building's own heat and hot water systems, while return water is pumped back to the heating plant. District cooling uses the same principles. Modern DHC can be flexible in terms of source, size and load. Currently they are dominated by fossil fuels such as coal and gas; it can use any fuel including excess heat that is currently being wasted, renewable resources and cogeneration. Each network needs to be developed according to the circumstances and can be adapted to innovation systems like BMS (Building Management Systems) that enable data management relating to temperatures, flows, pressure and leak detection. The repairmen of existing network for construction and expansion of district energy networks and integrating the use of a large share of renewable power, serving as thermal storage can be a major advantage for future smart energy systems. Smart DHC network management would help control the demand and system temperatures, which would reduce heat loss and accordingly energy demand. Smart optimisation and control technologies as well as IoT would also enable better cooperation with service providers and equipment manufacturers. [9]

2.4 Infrared heating

One of the latest technologies entering the market for heating appliances is infrared (radiant) heating. It uses infrared radiation, radiating heat, which concentrates in the areas where it is needed and used, and does not spread throughout the room ort flow through enclosing structures. When the rays reach the earth's surface or other object, they heat up, and this heat warms the air around. Due to the pronounced thermal effect of infrared rays, they are successfully used for heating. This solution is extremely economical compared to mass conventional heating appliances and is considered to be energy efficient thanks to the conversion of up to 98% of electricity into infrared radiation. Due to the fact that air is not used as a heat transporter, heat losses are significantly reduced, since air is a good insulator and not a heat accumulator. It remains at almost the same temperature throughout the entire height of the room. Manufacturers of these types of heating systems claim that energy savings can be as high as 70% when the appliances are properly installed and operated. The best efficiency from heating with infrared panels is registered when they are installed in the ceiling of the respective room, because it has full coverage to the whole room, warms the floor from there it warms the air from the bottom up. After ventilation of a room heated by infrared radiators, the return to the desired room temperature becomes much faster due to the fact that all surfaces in the room are warm. Due to the fact that the surface of objects and bodies in the room is about 1 ° C higher than the air temperature in the room, the possibility of condensation and mould formation in the room is drastically reduced.
With series of experiments in industrial space with thermal camera, was proven the efficiency of the infrared heating comparing to a convection heater. The thermal camera visualized that, although with a lower power, the temperature reached in the radiant heater corpus exceeds three times that in the corpus of the convection heater (Fig. 3). The warm air from the stove was scattered throughout the

room and no significant warming effect was felt during the experiment. The radiant heater heated the area constantly from the start, with warming effect right from the beginning. [10]

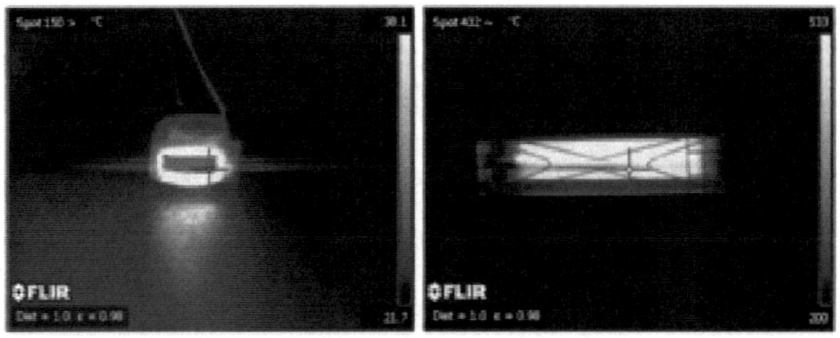

Figure 3. Thermal image of infrared heater and convection heater

Infrared heating systems are very applicable for developing intelligent heating system as part of an overall BMS. Because of coactions between temperature sensors and the control of heating systems, all areas throughout the space or different zones can be maintained with the appropriate temperature even remotely. This helps for several of outside factors such as outside temperature, humidity, natural heat to be taken into consideration in order to reduce energy consumption. Also infrared heating systems can interact flexibly to individual user behaviour to provide safety and comfort.

3. Comparative analysis

In order to achieve the most suitable solution for increasing the energy efficiency, designers and engineers, have to be aware of every method positive and negative sides. The main advantages and disadvantages of the reviewed heating possibilities are listed below in Table 1.

Table 1. Types of heating advantages and disadvantages

	Advantages	Disadvantages
Heating systems with solar power	Uses renewable resources; Suitable for the big scale roofs in industrial buildings; Can use the build in heating/cooling system; No carbon footprint; Return of investment.	High initial cost; Uses the roof space; Depends on the availability of solar radiation at the specific geographic location; Needs energy storage equipment; Needs regular maintenance.

Heat pumps	Uses renewable energy; Effective; Great development opportunities.	Uses electricity Depends on the geographical location; High initial costs; Doesn't work in low temperatures; Complicated design.
District heating and cooling	Easy to install; Reduced maintenance costs; Use small space; Safe; Energy efficiency is improved both on the production and distribution sides. reduced distribution losses; System flexibility; Works with BMS;	Suffers heat losses to the ground; Requires large upfront capital investment; Only available for buildings in areas where a district heat network is available.
Infrared heating	Energy efficient; Heats large and high spaces; Starts to heat up from the moment of switching on; It does not pollute and emit no harmful substances; Noiseless; Easy maintenance; Option for zoned heating; Ability to use remote control, thermostat and smart systems.	Uses electricity; If not installed in suitable height it can cause unpleasant warming effect.

As a future steps, based on the overview of the different heating technologies, a base model for increasing the energy efficiency in industrial buildings will be developed using infrared heating. The model will be based on the heating specifics and building's consumer needs. It will implement modern smart technologies and IoT for maximum results in energy efficient heating.

4. Conclusion

With the new directives for tightening the energy consumption worldwide, the demand of new, energy efficient heating systems increases. This topic is extremely relevant to industrial buildings, characterized by large spaces, wide openings, zoned working processes and special production demands. Most of the systems require large upfront investment to build, or are not suitable for already build premises. Infrared heating has a lot of advantages for these industrial needs and becomes a preferable solution when it comes to easy maintenance, bespoke installation, safe working and lack of heat loss. The ability to integrated it with management systems and IoT helps even more in increasing the energy efficiency, reducing the heating costs and harmful emissions and increasing the comfort of the employees on-site.

Acknowledgments
The paper is partially supported and financed by National Science Fund of Ministry of Education and Science of Republic of Bulgaria, by Project for junior basic researchers and postdocs – 2019, Contract № КП-06-М37/2 from 06.12.2019

References

[1] Energy Efficiency Trends and Policies In Industry , An Analysis Based on the ODYSSEE and MURE Databases, September 2015,Co-funded by the Intelligent energy Europe Programme of the EU
[2] B. Anderson, September, 2019, Heating industrial workspaces, link
[3] N. Pardo Garcia, IET-JRC Kostantinos Vatopoulos, IET-JRC Anna Krook Riekkola, IET-JRC, Luleå University of Technology Alicia Perez Lopez, IET-JRC Lars Olsen, Danish Technology Institute, JRC Scientific and policy reports, Best available technologies for the heat and cooling market in the European Union, EUR 25407 EN --- Joint Research Centre --- Institute for Energy and Transport , ISSN 1831-9424, link
[4] D.A. Chwieduk,2016, Active solar space heating, Advances in Solar Heating and Cooling
[5] V. Christova, 2017, Increasing Energy Efficiency in Existing Buildings., Problems of Engineering Cybernetics and Robotics, vol. 68, ISSN 0204-9848, pp. 58-64
[6] T. Nowak, 2018, Heat Pumps Integrating technologies to decarbonise heating and cooling, European Copper Institute
[7] S. Kärkkäinen, Elektraflex Oy, Heat pumps for cooling and heating, Subtask 5, Report no 3, International Energy Agency Demand-Side Management Programme, Task XVII: Integration of Demand Side Management, Distributed Generation, Renewable Energy Sources and Energy Storages
[8] Colophon DHC+ Technology Platform, District heating & cooling strategic research agenda, March 12, link
[9] L, Lyons., Digitalisation: Opportunities for heating and cooling, JRC Technical Reports, The European Commission's science and knowledge service, Joint Research Centre, ISBN 978-92-76-01438-6
[10] V. Yosifova, M. Haralampieva, 2019"Radiant heating. Types, application area and advantages", XXVIII MNTK „ADP-2019", ISSN: 2682-9584, pg. 327-332;

Proactive compensation of electric power imbalance in day-ahead market

Alexander Angelov[1,2], **Nikola Shakev**[1], **Georgi Ganev**[1]

[1] Technical University of Sofia, Plovdiv Branch, Faculty of Electronic and Automation, Bulgaria

[2] Electricity System Operator EAD, RDS-South, 4000, Plovdiv, Bulgaria

a_angelov.82@abv.bg, shakev@tu-plovdiv.bg, gganev@tu-plovdiv.bg

Abstract. The precise planning of production and demand schedules is essential for the stability of the energy system. For this purpose, various economic incentives have been developed and implemented that encourage accurate adherence to the pre-registered schedules. In order to respond to the strictly following of the schedules a proactive compensation of electric power imbalance in day-ahead market is suggested. An original methodology for parameter estimation of the Imbalance Compensation System is proposed and experimental results are reported.

Key Words. *electric power, imbalance compensation, day-ahead market*

1. Introduction

One of the main features of electricity is the inability to store it, which requires that the balance between the generated and the consumed power has to be maintained at a constant frequency of alternating voltage. When this balance is disturbed, the frequency of the voltage [1] changes immediately. The operational planning, management and control of the electricity system (ES) is carried out by the Transmission System Operator and the operators of each of the electricity distribution grids [2]. The main tasks of the Transmission System Operator related to the real-time control of production capacity is to maintain the balance between consumption and production of electric power [3].

The electricity market in Bulgaria is in a process of gradual liberalization, which started in 2004 and continues to this day. It consists of two segments - a regulated price segment and a freely negotiated price segment, or so-called free market. In the regulated segment, all prices and charges are set by the Energy and Water Regulatory Commission (EWRC), and consumers are served by the Electricity Companies on a territorial basis. Currently this segment includes domestic consumers and business customers connected to low voltage distribution networks. In the free market segment, customers can choose their electricity supplier regardless of their geographical location by paying fixed network transmission charges, as defined by EWRC [4]. According to [2, 5] contracts at freely negotiated prices are concluded between:

- Electricity producers;
- Electricity traders;

• Consumers registered on the free market.

The precise planning of production and demand schedules is essential for the stability of the energy system. For this purpose, various economic incentives have been developed and implemented that encourage accurate adherence to the pre-registered schedules.

In the free electricity market, there are rules for calculating the price of imbalances between the pre-registered schedule and the measured real values of energy provided by producers or consumed by consumers. Settlement periods are defined as distinct one hour time frames for which the imbalance of the responsible parties is calculated.

The pre-registered schedule represents the quantities agreed between trading participants before the actual delivery day. Measured real values are the values reported by the measuring devices of energy producers or consumers. The imbalances are determined on the basis of measured values for production or consumption [L-5]. The value of the amount paid by the free market participant can be determined by the following expression:

$$VPE = VCE + VSuE + VShE = \tag{1}$$

where

VPE is the Value of the Paid Energy

VCE is the Value of the Contracted Energy;

VSuE - the Value of the Surplus Energy;

VShE - the Value of the Shortage Energy;

At the end of each settlement, the imbalance should be kept to a minimum. It does not matter how the consumption has changed over time, the total consumption for settlement is essential. With active management of consumed energy during each settlement period, the imbalance can be minimized.

2. Imbalance Compensation System

The use of an Imbalance Compensation System (ICS) aims to keep the consumer's actual load schedule as close as possible to the contracted one.

The error can be compensated by:

• Switching on and off non-basic electric consumer devices;

• Accumulation and delivery of electricity through an energy storage system (ESS).

When compensating the imbalance error, the consumption is controlled by switching on and off non-essential consumers. If the actions taken are not sufficient, the available ESS capacity can be used.

Various factors can influence the error compensation process. This requires a preliminary analysis of these impacts. The results of the analysis will determine the necessary conditions and parameters under which the error will be compensated by using the ICS.

Some of the major factors that have a significant influence on the work of the ICS have been formulated and substantiated.

2.1. Analysis of the imbalances obtained

The prediction error for each settlement is determined by the expression:

(2)

$$\Delta E(t) = E_{\text{реално}}(t) - E_{\text{прогн}}(t)$$

where $E_{\text{реално}}(t)$ denotes the actual electricity consumption for the settlement, [kWh],

$E_{\text{прогн}}(t)$ indicates the contracted electricity consumption for the settlement, [kWh].

If $\Delta E(t)>0$ there is a shortage and in this case, non-essential consumers should be turned off to compensate the imbalance.

If $\Delta E(t)<0$ a surplus is obtained and in this case, non-essential consumers must be turned on to utilize unused energy.

Consumption, and hence the resulting imbalance of one consumer, can be influenced by different random processes that cannot be predicted at the stage of forecasting.

Experimental data set is required to evaluate forecasting errors. An error histogram is constructed to initially visualize the resulting forecast errors.

In order to determine the appropriate power capacity of the Imbalance Compensation System, a statistical analysis of the errors must be carried out. The result of this will be the basis for determining the parameters of the Imbalance Compensation System.

For statistical analysis, it is recommended to use distribution-free (non-parametric) methods [7-9].

Wilcoxon signed-rank test was used in which the hypothesis was tested by formulating a one-sided criterion.

2.2 Determining the optimal capacity of the energy storage system

An important factor for minimization of the prediction error is the correct power selection of the consumers participating in the ESS. Choosing a smaller capacity may not produce the desired results and, at a higher capacity, there may be significant start-up investments that are ultimately not justified by the results obtained.

Our analyzes indicate that it is recommended that the power capacity of the ESS has to be at least three times the statistical estimate of the imbalance values obtained. This maximum power capacity must be divided into separate steps or levels. We recommend that this number be no less than eight levels.

2.3 Analyse the length of time interval for active error compensation

The main purpose of the ESS is to make the error at the end of each trading period (settlement) as small as possible. Therefore, the following two approaches are proposed:

- Compensation throughout the settlement period;
- Compensation only in the last quarter of settlement (from 45-th to 60-th minutes).

When compensation is made throughout the whole settlement, the ESS monitors continuously (in real time) the error between the real and the agreed value of consumption. If necessary, it switch on or switch off the non-core consumers.

For compensation in the last quarter of each settlement, the approach is as follows. The reference value of consumption in the time t (45<t<60) is:

$$E_{ref}(t) = \frac{t}{60} E_{contracted} \tag{3}$$

And the calculated error $\Delta E(t)$ is

$$\Delta E(t) = E_{ref}(t) - E_{measured}(t) \tag{4}$$

According to $\Delta E(t)$ a control law for feedback control can be applied. A typical solution is the use of linear PID control law.

$$P = k_p \Delta E(t) + k_i \int_0^t \Delta E(t)\mathrm{d}t + k_d \frac{d}{dt}\Delta E(t) \tag{5}$$

where P denotes the additional power which have to be switched on/off to compensate the error $\Delta E(t)$.

For better results a nonlinear control law can be set. It has to take in consideration the limited control time (the time to the end of the settlement) and the limited additional power consumers which can be switched on/off.

2.4 Time delay for switching on and off the non-essential consumers in the Imbalance Compensation System

Consumption may result in short-term abrupt changes in the load schedule caused by the high-power consumers being switched on or off. These changes are random in nature and alternate between positive and negative values. If the frequency of these changes is high, it may cause extraordinary switching on and off a non-essential consumers by the ICS. The consequences of this may be the wear and tear of these facilities as a result of frequent transient start-up and shut-down processes.

This requires exploring the possibility of switching on and off with some timedelay. The timedelay will eliminate short-term peaks. The shorter the delay, the compensation becomes faster and therefore the error is smaller. But this will increase the number of switching cycles. Conversely, if the delay is greater, the compensation will be delayed and therefore the errors will be higher.

These considerations again lead us to an optimization problem for which the following system requirements are formed:

- Minimum number of commutations
- Minimum wage bill for electricity

2.5. Underutilized use of non-core users in the process of error compensation

The application of non-primary consumers for Imbalance Compensation System is related to the following features:

- Since switching them on is not triggered by the need for their use, but depends on the need of the Imbalance Compensation System, their benefits may be reduced. For this purpose, a benefit reduction factor is introduced.

- Since switching them off is not triggered by functional needs but by the need of the Imbalance Compensation System, it is possible that energy savings may be associated with some discomfort. A comfort reduction factor is introduced for this purpose.

Taking into account these peculiarities, correction coefficients are introduced which take into account the degree of real benefit of the non-core consumers' participation in the error compensation process.

The values of the two coefficients depend on the characteristics of the non-basic consumer devices and on the tendency of the user to give priority to efficient use of energy.

IOP Conf. Series: Materials Science and Engineering **878** (2020) 012011 doi:10.1088/1757-899X/878/1/012011

3. Experimental results

For the purposes of this study, the electricity consumption of a consumer - a University building is considered. A mathematical model is used to forecast electricity consumption based on which a forecast is made for the each settlement within the next 24 hours. We will assume that the forecast consumption has been contracted with the electricity provider and there will be financial penalties for deviations between the real (measured) consumption and contracted consumption for each settlement.

To compensate the error during each settlement, the Imbalance Compensation System has been used. It has the following parameters:

- Total power capacity of 18 kW of non-essential consumers that will be switched on and off.
- The consumers are distributed in nine stages of 2 kW each.

Table 1 shows the daily cumulative errors using the Imbalance Compensation System. The daily error has two components negative (surplus) error and positive error. The results shown are averaged on the basis of 3 monthly data.

Table 1. The daily cumulative errors using the Imbalance Compensation System.

Interval for active error compensation	Time delay (min)	Daily cumulative errors			
		-ΔE, (kWh)	-ΔE, (%)	+ΔE, (kWh)	+ΔE, (%)
0 - 60 min	1	-3 461	-1,92%	3 343	1,86%
0 - 60 min	2	-3 651	-2,03%	3 525	1,96%
0 - 60 min	3	-3 811	-2,13%	3 691	2,06%
0 - 60 min	4	-3 951	-2,21%	3 847	2,15%
0 - 60 min	5	-4 079	-2,28%	3 959	2,21%
45 - 60 min	1	-6272	-3,48%	4 847	2,69%
Without Imbalance Compensation System		-17 113	-7,60%	15 995	7,11%

Figure 1 shows series of experimental curves for the negative (surplus) error and positive (shortage) error depending on the power capacity [kW] of the Imbalance Compensation System at different values of the time delay.

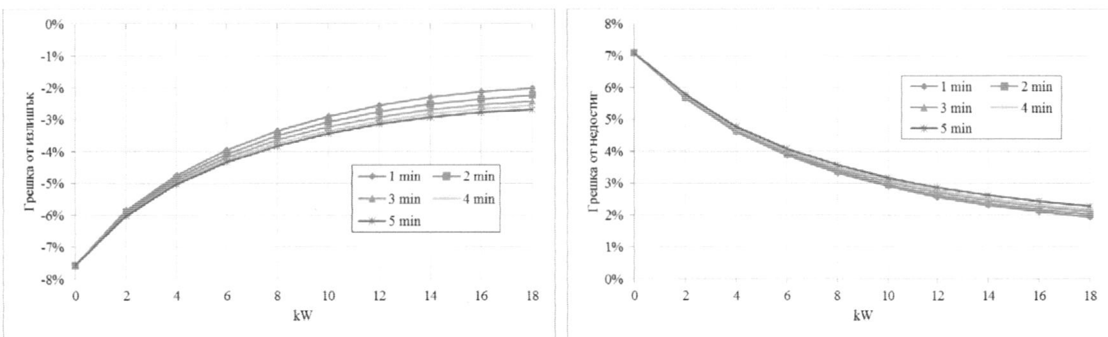

(a) (b)

Figure 1. Compensation for (a) negative (surplus) error and (b) positive (shortage) error depending on the capacity of the Imbalance Compensation System at different time delays

Figure 2 shows the graphs of the change of the total amount paid depending on the capacity of the Imbalance Compensation System at different time delays. Comparison with the data in Figure 1 shows that if the share of the compensated energy error is about 3-5%, then this increased efficiency in terms of money reaches a share of 10%. This is due to the significantly higher energy prices outside the contracted schedule.

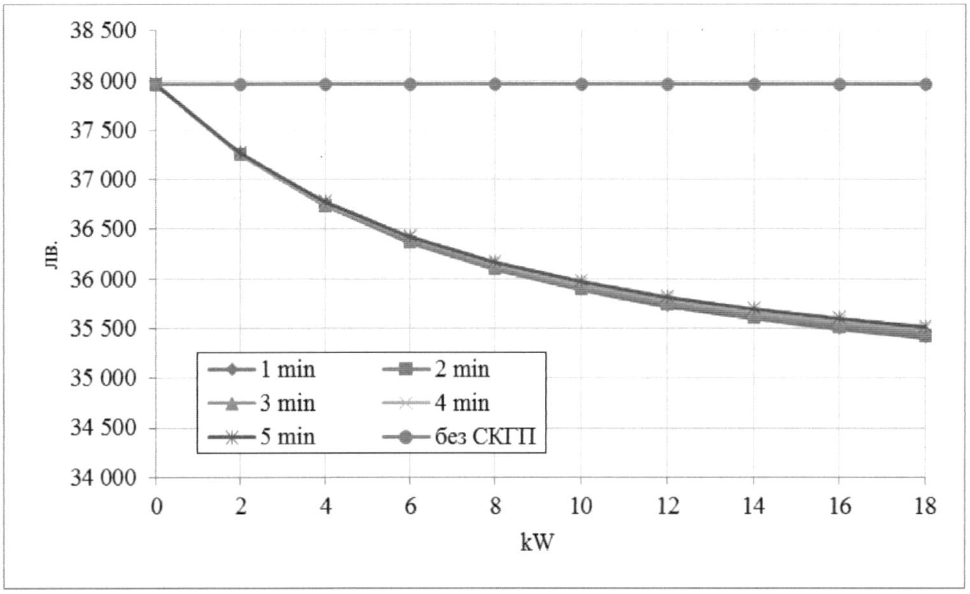

Figure 2. Total amount paid depending on the capacity of the Imbalance Compensation System at different time delays

Figure 3 shows a family of curves of the total amount criterion if we take into account the different levels of benefit reduction factor and comfort reduction factor. The graphs are also calculated in dependence of the power capacity of the Imbalance Compensation System.

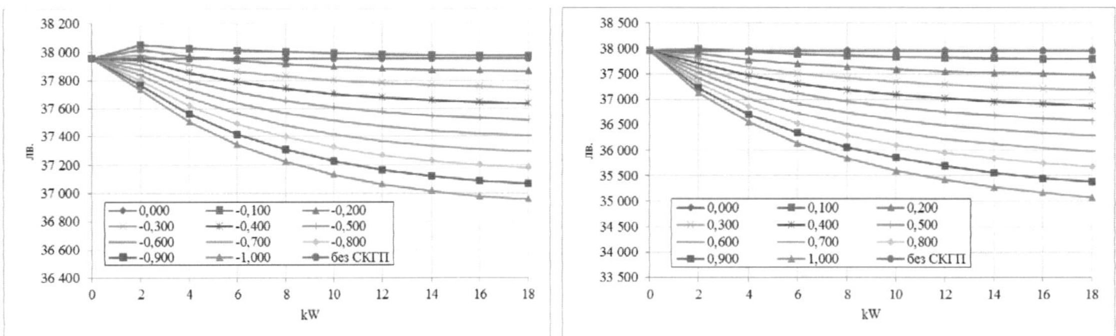

(a) (b)

Figure 3. Total amount criterion if we take into account the different levels of (a) benefit reduction factor and (b) comfort reduction factor.

4. Conclusion

The following conclusions can be drawn from the studies made by the use of non-core consumers for proactive error compensation of electric power imbalance in day-ahead market:
1. The results are better if compensation is made throughout the settlement compared to the case if only part of the settlement (e.g. the last quarter) is used to compensate the error;
2. Increasing the capacity of the Imbalance Compensation System beyond a certain value does not lead to a linear reduction of errors. Saturation trends are observed reaching the error limit;
3. The results at different switching time delays indicate that the obtained values do not differ significantly. For this reason, it can be chosen a time delay of greater value without significantly worsening its economic performance;

From the research we can conclude, the use of non-basic consumers to compensate for the error has a positive economic effect. The identified parameters and factors affecting the algorithm are key to performance and can be explored in more detail.

Acknowledgments

This work was supported by the European Regional Development Fund within the OP "Science and Education for Smart Growth 2014-2020", Project Competence Centre "Smart Mechatronic, Eco-And Energy Saving Systems And Technologies", № BG05M2OP001-1.002-0023.

The authors would like to thank the Research and Development Sector at the Technical University of Sofia for the financial support.

References

[1] Nanchev, St., Automation of electricity systems, 1974
[2] Energy Sector Act, SG. 83/9 Oct 2018
[3] Energy and water regulatory commission, Electricity Power System Management Rules, https://www.dker.bg/
[4] http://ateb.bg/
[5] Energy and water regulatory commission, Electricity Trading Rules, https://www.dker.bg/
[6] Ivanov, Iv., Liberalization of the electricity market in Bulgaria in the context of EU requirements, 2017.
[7] Bojanov, E., Iv., Vuchkov, Statistical decisions in production and research, 1979.
[8] Todorova, S., Statistics in Economics and Business, 2004.
[9] Geshev, A., A. Malinov, I. Geshev, Statistics, 2000.

IOP Conf. Series: Materials Science and Engineering **878** (2020) 012012 doi:10.1088/1757-899X/878/1/012012

Application of the method of dynamic programming for solving the optimal unit commitment problem in distribution networks with local distributed generation

Natasha Dimishkovska[1] and Atanas Iliev[2]

University Ss. Cyril and Methodius in Skopje, Faculty of Electrical Engineering and Information Technologies;
Rugjer Boshkovik 18, 1000 Skopje, Republic of North Macedonia

[1]dimishkovskan@feit.ukim.edu.mk, [2]ailiev@feit.ukim.edu.mk

Abstract. The continuous growth of communities leads to an increase in electrical energy usage. Besides the technological advance of power systems, still, there are outages which lead to an interruption in the power supply. The implementation of distributed generation provides usage of the local renewable energy sources for satisfying local power needs which in good manner complexes the distribution networks. However, because the distributed energy resources depend on the weather conditions, the islanded work is not quite reliable. Grid-connected distributed energy resources have enhanced power supply reliability. In this paper, a dynamic programming model is used for the optimisation of power generation by the local distributed energy resources. Assuming the distributed energy resources have storage systems, the model takes into account the battery charge at the moment and the availability of the distributed energy resources depending on the weather conditions and whether is day or night. It is assumed that the distributed energy resources have equal power capacities. Case study reviews a low voltage distribution network with plenty of distributed energy resources from different renewable energy sources implemented and a cluster of household consumers.

1. Introduction

As renewable energy sources (RES) take a big swing in the power production process, they become easily implemented into the standard power system. In the past decade, they have become quite accessible for commercial use, not just as power plants, but also as a distributed generation. In that way, their implementation is very useful, especially in the isolated distribution networks, where the outages are very common or the electrical grid cannot be upgraded according to the increase of consumption. In such a manner, they provide electrical energy to the local consumers, which manifests with many benefits. One of the beneficial uses of distributed energy resources (DERs) is the opportunity they provide for islanded work. That means that when there is an outage of power lines or in the local substation, the consumers will not be left without power supply.

However, DERs highly depend on the weather conditions, which make them quite unreliable for islanded work only. With storage systems included, they can store the excess power, that can be used whenever needed or it can be sold to the power system when the price for selling it is acceptable. Studies have shown that implementation of storage systems increase the price of DERs, but their purpose justifies the investment made as many communities today have implemented DERs in the local distribution networks.

Distribution networks with implemented DG, work similarly as the standard power system when it comes to optimal DG usage, especially during islanded work. The proper scheduling of the available DERs will provide optimal power usage while decreasing the expenses used for power production and storage.

In this paper, the optimal DERs unit commitment is analysed, considering the availability of the DERs implemented in the local network. For that purpose dynamic programming (DP) is used. The presented approach provides a solution for optimal usage of the local DG, considering their availability in a certain moment, based on the information about weather conditions and battery charge of each of the DERs. Their availability is presented with an availability index, which is a number between 0 and 1. DP method is supposed to give the optimal unit commitment of the DERs available.

2. Related work

DG as one of the solutions to the technical and environmental problems of power supply in modern communities is subject of interest to many research. In this section researches referring to the implementation of DP for resolving the problem with DERs unit commitment are reviewed.

In [1] an optimisation method for optimal energy management in a hybrid system is proposed. The system consists of wind turbines, photovoltaic, diesel generator and a battery. The optimisation is based on minimising the costs by scheduling the DG while taking into account the DERs power generation limitations, emission of greenhouse gasses reduction, balancing the load and production of electrical energy.

In [2] a hybrid DP, genetic algorithm and particle swarm method for optimisation of power systems and unit commitment problem solution is proposed. The analysis made considers system security constraints. The simulation is made on complex power system consisting of hydro and thermal units on IEEE 30 buses system. The proposed method has shown to be quite effective.

In [3] a method for stochastic optimisation is proposed. The method is used for optimisation of power generation in a local grid-connected microgrid with an implemented storage system. The optimisation is made considering the load, which is assumed to be manageable, and the electricity market prices. The simulation on a typical microgrid shows that the proposed method delivers better results when the scheduling uncertainties are taken into consideration.

In [4] an analysis of a microgrid with implemented storage systems is made using DP. The objective function is minimising the emissions from the power production units, especially CO_2 emissions. The constraints are referring to the production and demand power balance, the unit's loading level and the microgrid operation mode. The constraints are different for each mode of operation, and it depends on whether is day or night, the availability of RES and the battery state of charge. The simulation results show that the approach is applicable and effective, providing lower CO_2 emissions, compared to the systems that are not optimised in this manner.

In [4] a DP method is used for computation of the optimal power capacities of four power plants, placed on different locations, if the total power capacity is known. The optimisation is based on decreasing the costs for building each of the power plants, depending on the power capacity installed.

The proposed approach in this paper is based on the DP method used in [4], but it is used for optimal unit commitment in a power network with DERs installed.

3. Dynamic Programming Method

The application of DP is widely known for power system planning and optimal unit commitment in complex power systems. In this paper, DP is shown to be useful for optimal unit commitment in a distribution network with multiple DERs connected.

Usually, the conventional use of DP analyses the costs for running a certain unit, so the optimal unit commitment is based on minimal costs. In this paper, the optimal unit commitment in a complex distribution network with multiple DERs is based on finding the optimal solution depending on the current availability of the DERs.

DP method represents a set of algorithms used for finding an optimal solution to a wide range of input data while maximising or minimising an objective function. The problem is divided into incremental steps, so that, at any step, the expressions are simplified. They are known as sub-problems. The solution to the previous sub-problem is memorised, so eventually, each of the solutions to sub-problems leads to the solution of the main problem.

Presented on a diagram, the solutions are represented with circles, and the paths are the state transitions. An optimal solution is a path which connects the start and the end of the process (in **Figure 1** denoted as 1 and 19, respectively), and at the same time, satisfies some defined conditions.

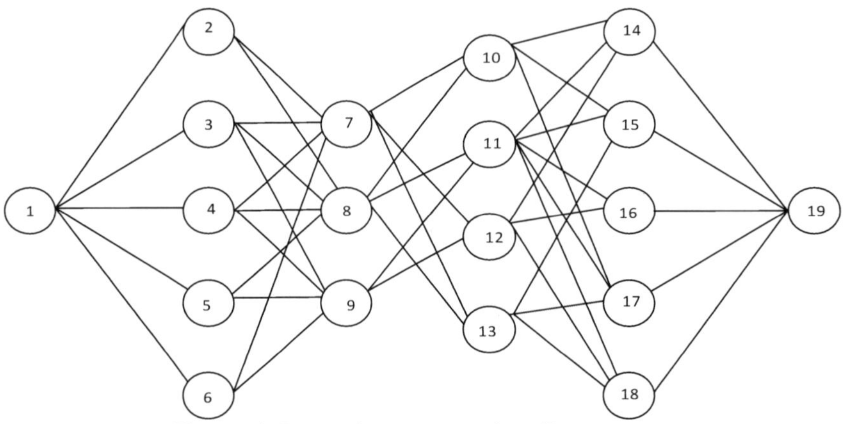

Figure 1. Dynamic programming diagram

4. Distributed Generation problem definition

The implementation of DG into the distribution network has been proven to enhance the power network reliability, especially in cases of an outage of the power lines or even outages in equipment in the substations. Their primary purpose is to serve the local consumers, providing with clean energy in times of power supply interruptions from the main power grid, as shown in Figure 2. However, due to their dependability on weather conditions, they cannot be used as stand-alone systems. Therefore, the preferable solution for the DG is to be grid-connected.

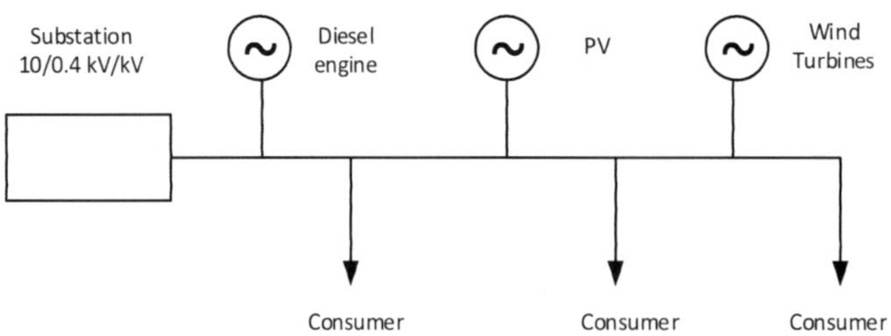

Figure 2. Distributive network with DG implemented

An islanded local distribution network unit commitment can be analysed as a power system with many different units. When analysing a large power system, the optimisation function is based on the costs for unit power generation. However, in the case of DERs, the main problem is their availability, considering that they are mainly RES.

In this paper, the problem with unit commitment is analysed based on units' availability at the time, taking into account the weather conditions.

5. Proposed approach

The problem definition refers to determining which units should be used in a certain moment for power consumption satisfaction, assuming there is an outage, if known the availability of each unit and their power limitations.

The developed algorithm consists of the following stages:

- Current weather conditions data input,
- Current storage data input,
- Calculation of the availability index of each of the implemented DERs,
- Calculation of the DER's power production if there is no energy stored for further usage,
- Computation of the optimal DER's unit commitment.

The power generated from the DERs depends on the availability of the proper weather condition (sun, wind etc.). Applying storage systems provides power storage, in cases when the DERs produce excess power, so it can be used in the future, when there are no proper weather conditions or during night time. Therefore, the power generated from the DERs can be defined by weather conditions and the capacities of the storage systems.

In this paper, the availability ($A_i(P_i)$) is defined as a variable depending on the weather condition index and the capacity and availability of the power stored. The weather condition index denotes as $W_i(P_i) \in [0,10]$, and the capability of operation and stored energy denotes as $C_i \in [1,10]$. The availability function depends on the presence of the renewable energy source at the moment and the stored energy:

$$A_i = \alpha W_i(P_i) + (1 - \alpha)C_i \tag{1}$$

where, α represents the index of importance, which is a subjective decision depending on the one that is applying the proposed approach. The functions of weather conditions depend on the RES used for power generation, and it differs for each DER. For instance, the power generated from the photovoltaic power plant (PV) is not calculated in the same way as the power generated from wind farms.

The zero indexes represent the state of absolute absence of energy. It may occur, for instance, during the night time, when the sunlight is absent, if there is a solar power plant, or if there is a wind farm, the absence of wind. However, it will not be such case if there is a small hydropower plant, or a biogas power plant. Regarding the capability of operation and stored energy, the battery cannot be completely drained, because of the depth of charge limits, and therefore the values for the indexes start from one. The objective function is defined in the following manner:

$$A_j = \sum_i A_i(P_i) \Rightarrow \max \tag{2}$$

where, i is the number of conditions of the DERs, and j is the number of DERs.

The optimisation equation is set up as a multi-step continuous process. First, the power needed is set as a goal. Then, the optimal combination of all the available DERs is done, using the following equation, which takes the form of a recurring relation:

$$A_i^e(P_i^e) = \max \ \{A_i(P_i) + A_{i-1}^e(P_j - P_i)\} \tag{3}$$

The optimal combination is done step-by-step, analysing all of the possible combinations and the power taken from each of the DERs.

6. Case study

The case analysis reviews two different substation configurations with connected distributed generation. The case study analyses a radial distribution network as shown in Figure 3. The system consists of three consumers and five DERs. There are two power lines and the substation voltage rate is 10/0.4 kV/kV. In Table 1 the availability of each the DERs for different power is presented.

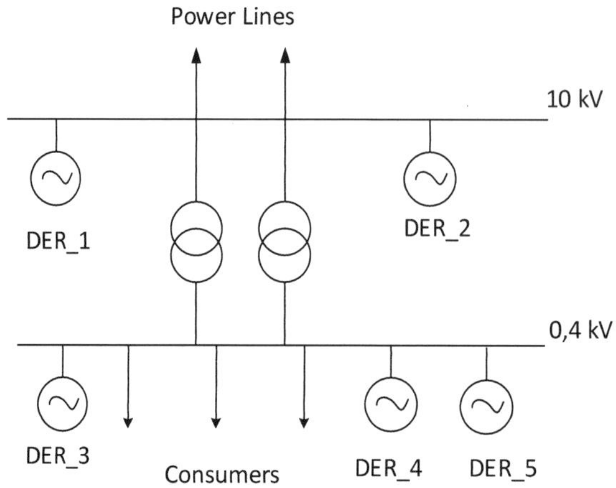

Figure 3. Distribution network with implemented DERs

Table 1 DERs power capacities and their availability

	DGs power capacities [kW]				
	50	100	150	200	250
DER_1	2.3	1.6	5.4	1.8	2.3
DER_2	3.5	6.4	2.9	3.6	3.5
DER_3	8.2	7.3	7.4	6.2	8.2
DER_4	1.5	6.4	1.8	7.8	1.5
DER_5	4.3	4.5	6.8	8.1	4.3

Assuming that 500 kW have to be supplied from the DG at one moment in time, due to an outage in the power lines, the optimal solution calculated using DP is presented in the following. Another assumption made, due to simpler analysis, is that the power generation step is 50 kW.

Following the explanation for applying DP for power system optimisation in [4], in the first step, no optimisation is done, and the data for DER_1 is used. That means that so far, only the first DER will generate power.

In the next steps, the rest of the DERs are loaded. The optimal unit commitment analysis is done as shown with eq. (4). The example refers to a 100 kW power outage, for DER_2.

$$\mathrm{A_2(100)} = \max\begin{cases} \mathrm{A_2(0) + A_1(100) = 1.6} \\ \mathrm{A_2(\mathbf{50}) + A_1(50) = 3.5 + 2.3 = \mathbf{5.8}} \\ \mathrm{A_2(100) + A_1(0) = 6.4} \end{cases} \Rightarrow 5.8 \text{ (at the power of 50 MW)} \quad (4)$$

Then, the function is applied for power of 150 kW, then for 200 kW etc. Then, it continues for the following DERs. The eq. (5) refers to the third DER and the power of 350 kW. This process continues to the desired power of 500 kW. The optimisation is done five times, since there are five DGs. The eq. (6) refers to the fifth DER and the power of 500 kW.

$$\mathrm{A_3(350)} = \max\begin{cases} A_3(0) + A_2(350) = 9 \\ A_3(50) + A_2(300) = 8.2 + 8.3 = 16.5 \\ A_3(\mathbf{100}) + A_2(250) = 7.8 + 11.8 = \mathbf{19.1} \\ A_3(150) + A_2(200) = 7.4 + 8.9 = 16.3 \\ A_3(200) + A_2(150) = 6.2 + 8.7 = 14.9 \\ A_3(250) + A_2(100) = 8.2 + 6.4 = 14.6 \end{cases} \Rightarrow 19.1 \text{ (power of 100 MW) } \quad (5)$$

$$\mathrm{A}_5(500) = \max \begin{cases} A_5(0) + A_4(500) = 27.8 \\ A_5(50) + A_4(450) = 4.3 + 25.5 = 29.8 \\ A_5(100) + A_4(400) = 4.5 + 26.4 = 30.9 \\ A_5(\mathbf{150}) + A_4(350) = 6.8 + 27.7 = \mathbf{34.5} \\ A_5(200) + A_4(300) = 8.1 + 23.3 = 31.4 \\ A_5(150) + A_4(250) = 4.3 + 26.3 = 30.6 \end{cases} \Rightarrow 34.5 \text{ (power of 150 MW)} \quad (6)$$

After the computation is done, the unit commitment process starts. First, the optimal power generation for the fifth DER is determined. In this case, it is the power of 150 kW. Then, the remaining power is transferred to the forth DER, and the process continues until there is no power to be divided among the DERs. The solution gives the information of unit commitment regarding the power injection from all of the DGs.

The information from the results containing the highest availability and the optimal generation for each unit is stored and used in further calculations. The optimal solution to the analysed case is given in Table 2:

Table 2 Unit commitment of the DGs

Distributed Generator	Power generation[kW]
DER_1	0
DER_2	100
DER_3	50
DER_4	200
DER_5	150

The results show that the DER_1 will not be used at that certain moment for a 500 kW power supply.

7. Conclusion

This paper analyses the application of DP for optimal unit commitment in a distributive network with implemented DG. The case study reviewed a distributive network consisting of five DERs connected to different locations in the network, each with the capability of providing up to 250 kW, and a storage system installed. The availability took into consideration the probability of RES power generation and the availability of stored energy. For simpler analysis, a step of 50 kW was considered.

Although the application of DP for solving the unit commitment problem in complex power systems is widely known, in this paper it was shown that it can be applicable for particular part of the power network with local power distribution.

References

[1] An, L. N., & Tuan, T. Q. (2018). Dynamic Programming for Optimal Energy Management of Hybrid Wind–PV–Diesel–Battery. *Energies MDPI.*

[2] Rouhi, F., & Effatnejad, R. (2014). Unit Commitment in Power System by Combination of Dynamic Programming, Genetic Algorithm and Particle Swarm Optimisation. *Indian Journal of Science and Technology*, pp. 2056–2063.

[3] Costa, L. M., & Kariniotakis, G. (2007). A Stochastic Dynamic Programming Model for Optimal Use of Local Energy Resources in a Market Environment. *IEEE Lausanne Power Tech.* Lausanne, Switzerland: IEEE.

[4] Kanchev, H., Francois, B., & Lazarov, V. (2011). Unit commitment by dynamic programming for microgrid operational planning optimization and emission reduction. *International Aegean Conference on Electrical Machines and Power Electronics and Electromotion* (pp. pp. 502-508). Istanbul, Turkey: IEEE.

[5] Iliev, A. (2002). *Selected solved exercises from Planning of Operation and Construction of Power Plants.* Skopje: Faculty of Electrical Engineering and Information Technologies.

IOP Conf. Series: Materials Science and Engineering **878** (2020) 012013 doi:10.1088/1757-899X/878/1/012013

Distributed generation placement in distributive substations analysis using Markov Chain Monte Carlo model considering the reliability of power supply

Natasha Dimishkovska and Atanas Iliev

University Ss. Cyril and Methodius in Skopje, Faculty of Electrical Engineering and Information Technologies;
Rugjer Boshkovik 18, 1000 Skopje, Republic of North Macedonia

dimishkovskan@feit.ukim.edu.mk, ailiev@feit.ukim.edu.mk

Abstract. Proper operation of the power substations is of great importance for power network reliability, stability and uninterrupted power supply. Distributed generation provides higher reliability in power supply, but still, there are contingencies in the electric power production and supply process, which lead to outages in the power supply. In this paper, a method for substations' reliability estimation with distributed generation is presented based on Markov Chain Monte Carlo method. The method considers the possible substation operation states and using random number generator in MATLAB, it simulates faults and calculates the substations' reliability. The method is demonstrated on two cases of 110/35 kV substations, each consisting of two transformers and distributed generator, analysing the best placement for the distributed generation.

1. Introduction

The progress of technology initiates the development of power systems to enhance the quality of life in modern society. Therefore, the need for uninterrupted and quality power supply increases continuously. The power network reliability, i.e. the capability for normal system operation and power supply for the consumers at any time, defines one country's economic and technological development.

The implementation of distributed generation (DG) with storage system, into the distributive network, enhances the reliability of supply, providing the electrical energy needed in conditions of islanded work due to power system outages. However, the substations as part of the power system are infrastructures where the electric power is transformed from one voltage rate to another, and as such, they are objects closest to the consumers and electric power generation facilities. Therefore, the substations, as complex systems that are expected to work with high quality and reliability, have the most important role in the power supply process. The faults that occur in the substations and cause power supply interruption can affect the installed equipment and may cause equipment damages.

The distribution system is part of the power network which is the most prone to failures and disturbances due to contingency, such as equipment failures, human errors or weather-related issues. The implementation of the distributed generation improves the reliability of distribution systems, providing backup power supply during outages. Usually, the distributed generators are renewable energy sources (RES), designed to cater the power consumption to residential consumers or small communities. However, their dependency on the weather conditions aggravates their island work and therefore grid-connected distributive generators are a better solution.

The optimal DG placement requires analysis of the power network where the DG should be connected. It is necessary to know the voltage variations and the technical limits of the installed equipment (rated voltage, current and power). Also, whether the type of DG installed, the uncertainties of load variations have to be taken into considerations. If the DG is a RES generator the uncertainties of the weather conditions should be considered.

Modelling the uncertainties can be done with many deterministic and probabilistic mathematical models. The application of Monte Carlo Simulations for quantitative analysis of uncertain input data has been proven to provide manageable results for further analysis and decision-making processes.

In this paper, an analysis for the best placement of distributed generators is made, considering the reliability of supply. The analysis is based on Markov Chain Monte Carlo (MCMC) method for reliability estimation of distribution substations with implemented distributed generator on the consumers' side. The purpose of the method is to estimate the reliability of a technical solution in the designing process. The model considers normal operation and fault states of the system caused by different states of the installed equipment, evaluating the probability of occurrence of different scenarios which lead to an interruption in the power supply.

The test examples review two different technical solutions, considering equal unavailability rates of the equipment in both cases. The test example is simulated in MATLAB, using a random number generator.

2. Related work

The power network constantly spreads due to the growth of the communities and upgrades with new equipment following the latest technology. Finding a method that estimates the reliability of power networks and takes into account all contingencies, is a challenge. In this chapter, the literature that analyses the network reliability of complex systems with distributed generation based on Markov Chain and Monte Carlo Simulation is reviewed.

In [1], Monte Carlo Simulation technique is used for optimal DG placement. The presented approach analyses implementation of two types of DG, conventional and wind farm. The approach proposed is tested on 15-bus and 33-bus radial distribution systems. The results show the best location (bus) to connect the DG for the two of the examples. Also, the application of Monte Carlo Simulation is justified.

In [2], an integrated approach for evaluation of the impact of distributed energy resources (DERs) have on the reliability performance of power networks. The case study is a power network based on a typical urban medium voltage (MV) and low voltage (LV) networks in the United Kingdom. The potential benefits of the local renewable generation, demand-manageable loads and coordinated energy storage are analysed. The time-variation of electricity demand profiles and failure rates of network components are simulated using conventional MCMC method.

In [3] low and high penetration of distributed energy generators impact on the reliability of a distribution system is investigated. The analysis is made using DGs mathematical modelling and Monte Carlo simulation methods. Also, a Monte Carlo Simulation designed specifically for the investigated cases of low and high penetration of DGs is proposed in detail.

In [4] the reliability enhancement of a distribution system with implemented distributed generation is presented. The paper reviews five different cases, by adding different components in the systems, with and without distributed generation. Also, customers are taken into account. The analysis is done by using analytical methods and the results show reliability enhancement in systems with distributed generation.

In [5] by using the MCMC method the reliability of complex power networks is calculated. The method is based on the random number generator, which generates n numbers from [0, 1] and calculation of the number of variables that satisfies the certain requirement. The requirement is the variables to be smaller or equal to the value of the total unavailability rate of the connected components. By generating a sufficient number of variables, the estimated value i.e. the portability of failure is calculated.

3. Markov Chain Monte Carlo Method

Markov Chain method describes a probability of event occurrence in a stochastic system with a random sequence. It states that the probability of one event occurrence depends only on the previous events. These probabilities are represented in a matrix, known as a transition matrix. The main disadvantage of the Markov Chain model is that the transition matrix significantly enlarges if the number of states increases. This disadvantage can be overcome by dividing the states into smaller independent matrices with smaller dimensions [6].

Monte Carlo simulation is used for different outcomes probability modelling in a process that cannot be easily forecasted due to the random values of the variables. This method is used for better understanding of the influence of risk and uncertainty in the forecast models and it is applicable in the fields of economy, engineering and science [7]. When facing uncertainty in the forecast process or estimation, instead of listing the independent variables' average values, Monte Carlo Simulations offer a better solution.

This method turned out to be inefficient in case of rare events, i.e. events that have probability smaller or equal to 10^{-5}. Therefore, a high number of variables are needed so that the system can be analysed. Monte Carlo method is designed for solving problems with random number generation, but simultaneously for solving deterministic problems tracking the behaviour of random variables an insufficient number of samples [8].

The expected value of the random variable is estimated through a certain number of samples, instead of calculating it mathematically. The determined number of samples is the key factor in Monte Carlo Simulation since it determines precision in the estimation process. The precision can be measured through the coefficient of variation, also known as relative standard deviation. The coefficient of variation is related to the reliability indexes since it represents the degree of accuracy of the indexes estimation.

4. The proposed approach

In [4] a method based on the MCMC method for reliability evaluation of complex systems consisting of serial and parallel-connected components is presented. The method proposed in this paper is based on that method. Calculation of equivalent reliability in a system is based on the theory of series and parallel connected components.

Assuming that the unavailability rates of all components in the system are given, the first step is to calculate the equivalent unavailability of the system. That indicates the probability of system failure.

The analysed system is a complex system which consists of two parallel-connected transformers, two power lines and a distributed generator with storage system, connected to a distribution power network.

Despite the many states of the system, generally, they can be divided into three main states, regarding the installed equipment in the substation and the operation state of the distributed generator:

- State 0: The system is in an outage, there is no power supply;
- State 1: The system runs partially, i.e. only half of the required power is supplied;
- State 2: The system is in a normal operational state.

The application of Markov Chain model on this problem results in a complex solution. However, the solution to the problem can be simplified by performing a Monte Carlo Simulation on the analysed substations. The MCMC method is easily applicable to complex systems since it allows simulation of contingencies that might happen in the installed equipment. For that purpose, random number generators are used, which are defined functions as part of some software. In this paper, a random number generator in MATLAB is used, applying rand(n) function ($\text{rand} \in (0,1)$).

The algorithm of performing the MCMC method is presented in Figure 1.

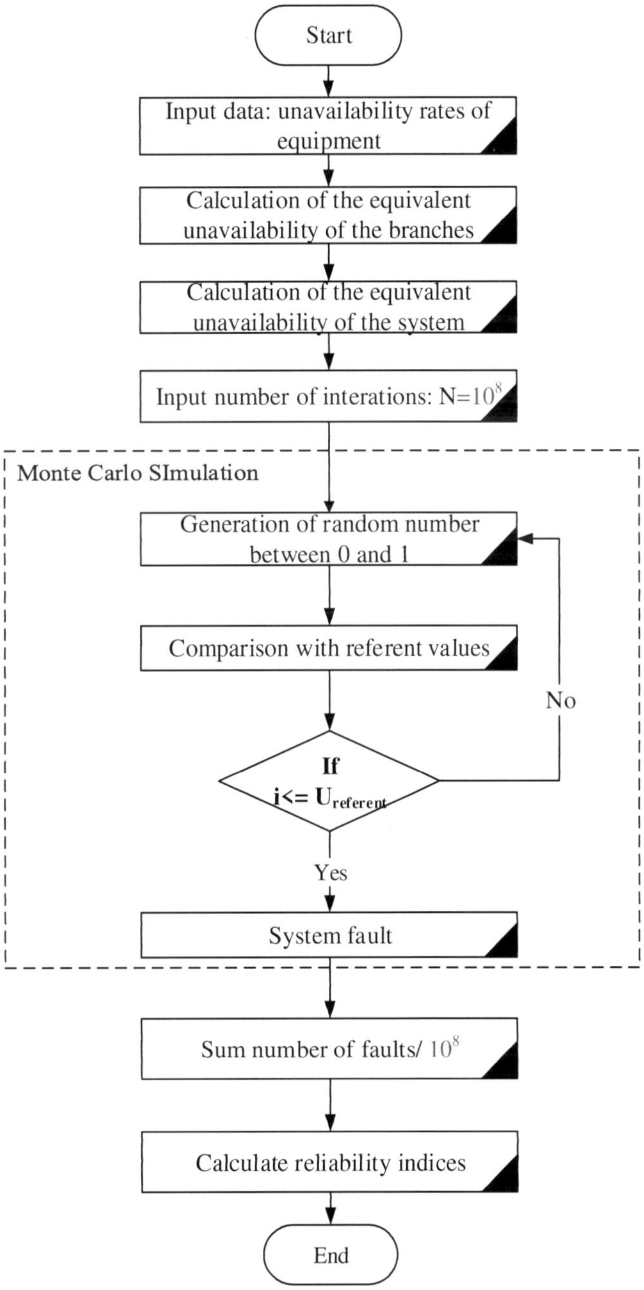

Figure 1. Simple MATLAB MCMC method algorithm for reliability estimation

The MATLAB code consists of three parts:

- Monte Carlo Simulation part: the random number generator,
- Markov Chain part: calculation of partial and total system outage, determining the fixed values that the random numbers are compared to, and,

- Comparison part: determining whether the random number can be analysed as a fault or not.

Firstly, through the input dialogue, the unavailability rates of the components are entered. Then, the equivalent unavailability rates of the power lines and transformer branches are calculated. Finally, using a *for cycle*, the randomly generated numbers are compared with the determined values for faults. This way, the number of partial and total outages is calculated. In the end, the values for the partial and the total number of outages are divided with the total number of randomly generated numbers, which represents the probability of one event occurrence.

5. Case study

The case analysis reviews two different substation configurations with connected distributed generation.

The first substation configurations reviewed consists of a distributed generator connected on the high voltage bus bar. In case of supply interruption due to an outage in the power lines or in one of the transformer branches, part of the load can be supplied by the distributed generator. The single line diagram of the described power system is shown in Figure 2.

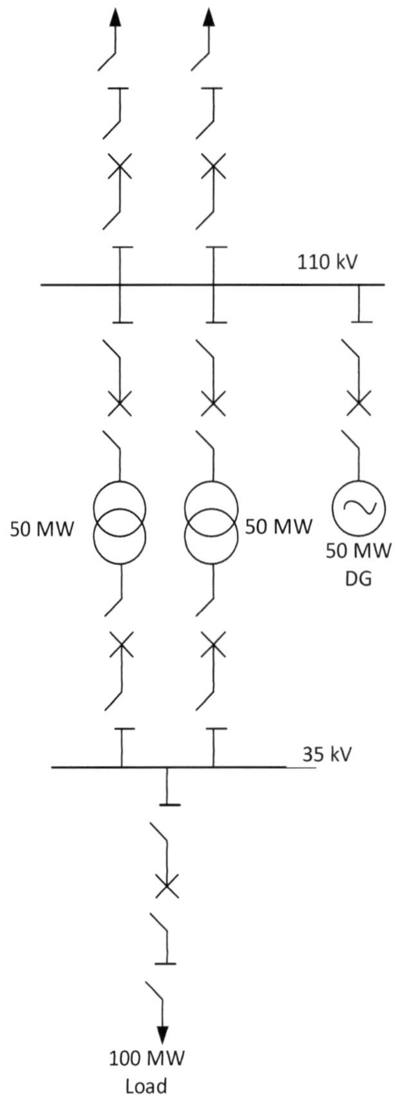

Figure 2 DG connected on the 110 kV bus

Distributed generator represents a large renewable energy source power plant (wind farm, solar power plant, biogas power plant or hydropower plant).

The second substation configuration reviewed consists of multiple smaller distributed generators connected on the low voltage bus. The installed power capacity of each of the generators equals to the installed power capacity of the single distributed generator in the first test example.

In this case, distributed generators provide backup power in case of outage of the both, power lines and transformer branches, at the same time, providing island work and power supply of the consumers. The single line diagram of the described system is shown in Figure 3.

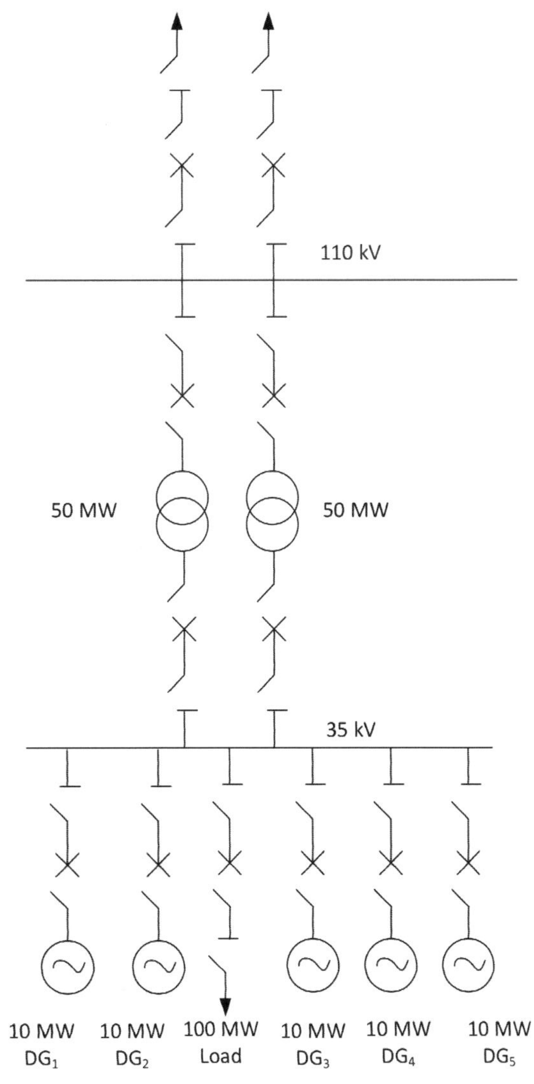

Figure 3 DGs connected on the 35 kV bus

It is assumed that the rating power of each of the transformers in the test examples in both of the substation configurations, is 50 MW, and the power of the distributed generator is also 50 MW. The unavailability rates of the components are given in Table 1. The generator's unavailability rate includes the unavailability of the generators as units and it also takes into account the dependency of the weather conditions.

Table 1. Unavailability rates of the equipment

Components	Voltage rate	Symbol	Unavailability rates
Earth Switch	110 kV	u_{ES}	0.00004
Disconnector	110 kV	$u_{D,110}$	0.00025
Circuit breaker	110 kV	$u_{CB,110}$	0.00038
Bus 110 kV	110 kV	$u_{bus,110}$	0.00007
Power transformer	110/35 kV/kV	u_{TR}	0.0006
Circuit breaker	35 kV	$u_{CB,35}$	0.00024
Disconnector	35 kV	$u_{D,35}$	0.0004
Bus	35 kV	$u_{bus,35}$	0.00001
Generator	35 kV	u_{gen}	0.03

The probabilities of each situation, for different substation configurations, are given in Table 2.

Table 2. Probabilities of normal operation state and outage for different DG configurations

System		Operation state	Partial operation state	Total outage
DG on 110 kV	p	0.99743	0.00175	0.00082
	$\frac{h}{year}$	8737.40	15.35	7.17
DG on 35 kV	p	0.99911	0.00015	0.00074
	$\frac{h}{year}$	8752.16	1.35	6.48
No DG	p	0.99654	0.00264	0.00082
	$\frac{h}{year}$	8729.69	23.14	7.18

The calculated probabilities of the system's different states' occurrence are used to compute the Energy Not Supplied (ENS), which is a power network reliability parameter. The ENS, in the analysed substation configurations, is given in Table 3.

Table 3. Energy Not Supplied for different DG configurations

System	ENS $\left[\frac{\text{MWh}}{\text{year}}\right]$
No DG	3030.96
DG on 110 kV bus	2251.32
DG on 35 kV bus	779.64

6. Conclusion

In this paper, the Markov Chain Monte Carlo method was used for reliability evaluation of distribution 110/35 kV/kV substation with distributed generation. The method is based on the MCMC method, and a simulation of fault occurrence in the presented system was made in MATLAB.

The analysis reviewed two cases of implemented distributed energy generators. The results show that the implementation of distributed generation enhances the reliability in the power networks, especially in the distributive systems with local consumers.

However, there is a slight difference in the calculated reliability of such system. Results show that DG implemented on the low voltage bus provides higher reliability, meaning more operational hours. Besides that, the improvement of power network reliability decreases the penalty costs for ENS.

The MCMC method is applicable for power systems reliability estimation and it provides fast computation and more accurate results compared to the analytical methods. This enables easy calculation of reliability indices, including ENS, and evaluation of the integration of such systems.

References

[1] Rao, B. N., Abhyankar, A. R., & Senroy, N. (2014). Optimal Placement of Distributed Generator using Monte Carlo Simulation. *Eighteenth National Power Systems Conference (NPSC).* Guwahati, India: IEEE.

[2] Ndawula, M. B., Djokic, S. Z., & Hernando-Gil, I. (2019). Reliability Enhancement in Power Networks under Uncertainty from Distributed Energy Resources. MDPI Energies.

[3] Ge, S., Xu, L., Liu, H., & Zhao, M. (2014). Reliability Assessment of Active Distribution System Using Monte Carlo Simulation Method. *Journal of Applied Mathematics*.

[4] Mazidi, P., & Sreenivas, G. N. (2011). Reliability Assessment of A Distributed Generation Connected Distribution System. *International Conference of Power System Operation and Energy Management.* Bangalore, India: Institute for Project Management Pvt. Ltd

[5] Bø, H. S. (2014). *Estimation of Reliability by Monte Carlo Simulations Combined with Optimized Parametric Models.* Trondheim: Norwegian University of Science and Technology.

[6] Rausand, M., & Høyland, A. (2004). *System Reliability Theory: Models, Statistical Methods, and Applications, Second Edition.* Hoboken, New Jersey: John Wiley & Sons, Inc.

[7] Corporate Finance & Accounting: Financial Analysis. (1999). Retrieved July 23, 2019, from Investopedia: https://www.investopedia.com

[8] Amelin, M. (2013). Monte Carlo Methods in Engineering. Stockholm: KTH Royal Institute of Technology

IOP Conf. Series: Materials Science and Engineering **878** (2020) 012014 doi:10.1088/1757-899X/878/1/012014

Study of the technical and economic efficiency and the possibilities for the construction of photovoltaic power plant above the roof structures of transformer power substations 20/0,4 kV and their service areas

S B Stefanov[1], **A P Georgiev**[2]

[1]Departnment of Electrical Engineering, Technical University of Sofia, Branch Plovdiv, 25 Tsanko Dyustabanov Str., Plovdiv, 4000, Bulgaria
[2]Elektrorazpredelenie Yug EAD, 37 Hristo G. Danov Str., 4000 Plovdiv, Bulgaria

[1]glasst@tu-plovdiv.bg, [2]alex_georgiev94@abv.bg

Abstract. This report examines the possibility of constructing photovoltaic systems above concrete complete transformer posts to the medium voltage distribution networks. It has been analysed what capacities could be built depending on the size of the concrete facilities and their service areas, and what would be their economic and electricity efficiency. The main results of the analyses made show that this idea is not devoid of economic and electricity logic.

1. Introduction

In medium voltage electrical distribution networks, transformers housed in concrete cells called concrete complete transformer posts (CCTP) are used to reduce the voltage. These concrete structures have a horizontal roof section without any other structure on it, and in addition they are connected to an adjacent surrounding service area, called service zone, in which no facilities are installed, the entire area being owned by the electricity grid company operating the facility. In view of the size of CCTP and its adjoining areas, horizontal free space above them is available, between 10 – 20 m^2 for CCTP structure and 70 – 100 m^2 for CCTP with the service zone. In this context, it is of interest to study the possibilities of building photovoltaic power systems over these zones and to investigate the effect of the installed capacity.

The study, the results of which are published in this article, was conducted on the basis of the data on the available areas over CCTP and their adjacent areas, owned by an electricity distribution company operating at medium voltage level of 20 kV in the southern part of Republic of Bulgaria. Possibilities for constructing photovoltaic power plants over the CCTP roofs and the area of their service zones are considered, and the connection to the grid should be 0.4 kV in transformer posts.

The power efficiency of the generating capacities has been evaluated in terms of their effect on the loss of voltage in the medium voltage grid and the losses in the transformers when operating in modes close to the real ones. The applied results are for a computer simulated branch of the electricity grid, corresponding in terms of parameters and structure to a real one. The economic estimation has been made on the basis of eligible investment costs for the construction of 1 kWp photovoltaic capacity, adopted by EWRC [1] and on the basis of preferential purchase prices of electricity from producers with total installed capacity of photovoltaic power plants up to 30 kWp.

2. Computing part

The CCTP roof sections are almost horizontal rectangular structures - fig. 1 with the possibility of opening the roof part in order to be able to remove transformers by means of a crane if necessary.

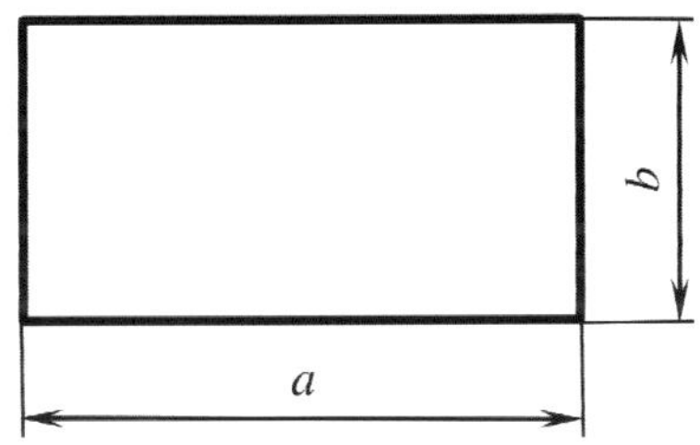

Figure 1. CCTP roof geometry

Two main dimensions of roofs are used a x b - 5.2 x 2.6 m and 5.2 x 5.2 m.

The service zones according to the regulations have the following dimensions, on fig. 2:

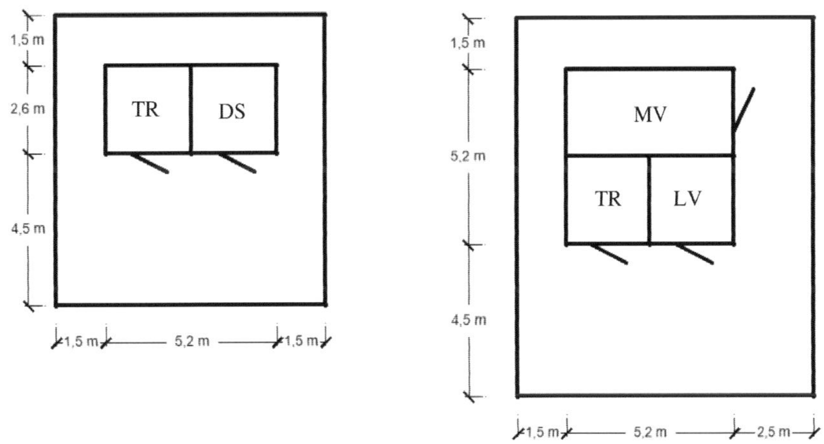

Figure 2. Service zone

The dimensions of one photovoltaic panel with a power of 285Wp and efficiency η = 16.65% are 1.65x0.99m.

According to the mentioned dimensions, it is evident that the square footage of the roof structures is 13.52 m^2 and 27.04 m^2 respectively, and considering the dimensions of the service zones, the available horizontal areas are 70.52 m^2 and 104.04 m^2. Thus, according to the dimensions on fig. 2 of the smaller substation, two rows of photovoltaic panels with 5 panels per row could be assembled - a total of 10 pcs with installed power of 2.85 kWp, and for larger CCTPs the number of panels could reach 20 pcs. with installed power of 5.7 kWp (with additionally added construction, at a slope of 30° in both cases). On the basis of the dimensions of the service zone of the smaller of the two types of Concrete complete transformer posts, above the area by means of a suitable structure there could be placed in 4 rows of 8 a total of 32 photovoltaic panels (of the above-mentioned type) with a total installed capacity of 9.12 kWp without the structure protruding outside the service zone.

The single-line diagram of the electricity grid branch that was used to carry out the electrotechnical and electric power calculations is shown on fig. 3. The diagram shows the numbers of the substations, the rated power of the transformer Sn in each of them, as well as the numbering of the sections with cable power lines at medium voltage level of 20 kV. Table 1 shows data about the cables in the respective sections and their lengths.

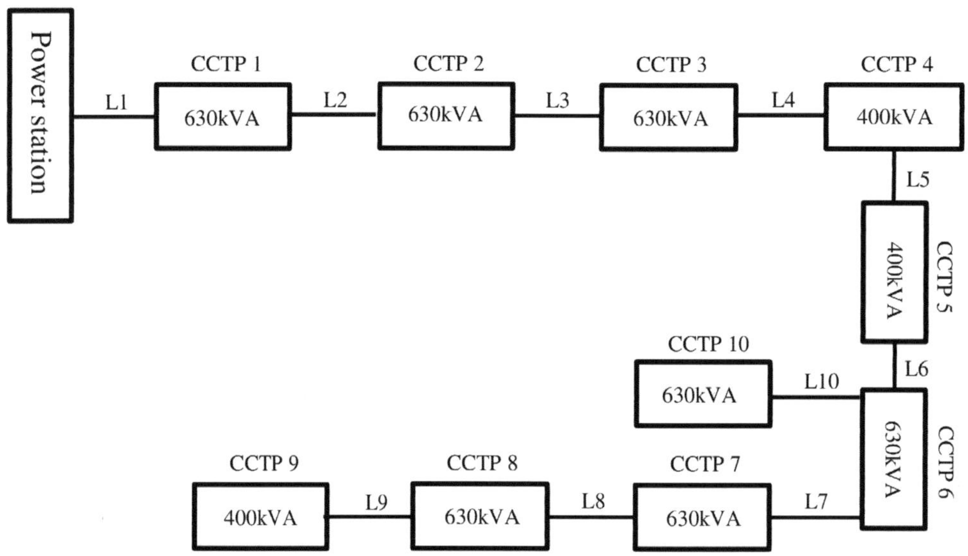

Figure 3. Single-line diagram

Table 1. Power line properties

Power line	L1	L2	L3	L4	L5	L6	L7	L8	L9	L10
distance, m	1826	222	186	64	231	165	72	470	208	159
r_0, Ω/km	0.215	0.215	0.215	0.215	0.215	0.215	0.215	0.215	0.215	0.215
x_0, Ω/km	0.117	0.117	0.117	0.117	0.117	0.117	0.117	0.117	0.117	0.117
R, Ω	0.39	0.05	0.04	0.01	0.05	0.04	0.02	0.10	0.04	0.03
X, Ω	0.21	0.03	0.02	0.01	0.03	0.02	0.01	0.05	0.02	0.02

The loss of active power at no load run in the transformers are ΔP_{o400kW} = 593 W, ΔP_{o630kW} = 855 W respectively, and the loss of active power in the transformer windings at rated load are $\Delta P_{sc400kW}$ = 4567 W, $\Delta P_{sc630kW}$ = 6659 W respectively, according to their rated power S_n.

In three CCTP the dimensions of their roofs are 5.2x5.2 m, and with the service zone the dimensions are 11.2x9.2 m, while the remaining seven CCTP have dimensions of 2.6x5.2 m, and with the service zone they are 8.6x8.2 m. To CCTP 10, on the side 0,4kV an existing 30 kWp photovoltaic power plant has been connected, the energy generated from which influences one of the parameters for which the study was made.

In accordance with the given sizes and number of CCTPs and their service areas, photovoltaic two variants of power plants could be built above them:

1. Variant 1 - When using roof spaces only above CCTP (of course, assuming that photovoltaic plant structures could be installed above them). In this case, with 3 CCTP sized 5.2x5.2 m and 7 CCTPs sized 2.6x5.2 m in size, and using 285 Wp photovoltaic panels, with size 1.65x0.99 m, we

would have the following total installed capacity of photovoltaic power plants on the roofs of transformer posts from the line in question:

$$P_{pv1} = 3x4x5x285 + 7x2x5x285 = 17.1 + 19.95 = 37.05\,kWp \tag{1}$$

2. Variant 2 - using service zones (assuming, of course, that photovoltaic plant structures could be installed above them). In this case, for 3 service zones with dimensions 11.2x9.2 m and 7 zones with dimensions 8.6x8.2 m, and based on the above-mentioned photovoltaic panels with their capacity and size, we would have the following total installed power of photovoltaic power plants:

$$P_{pv2} = 3x6x9x285 + 7x4x8x285 = 46.17 + 63.84 = 110.01\,kWp \tag{2}$$

If these generating capacities are used to produce and sell electricity, they would have the following characteristics for their owner or operator:

1. Annual electricity production from the photovoltaic power generation in Variant 1, based on an annual solar energy of 1580 kWh/m^2, calculated using the software product [2] e 45978.15 kWh/year. At the price of the systems on the individual substations from 2609 BGN/kWp (according to the investment costs recognized for power plants with total installed capacity up to 5 kWp) and preferential purchase price 242.1 BGN/MWh excluding VAT [1], the payback of the investment of photovoltaic systems is 7 years with a total annual revenue from all power plants 14479,62 BGN.

2. The annual electricity production from the photovoltaic power generation in Variant 2, based on an annual solar energy of 1580 kWh/m^2, calculated by software product [2] is 147743.43 kWh/year. At the price of the systems on the individual substations with service zones of 2284 BGN/kWp (according to the investment costs recognized for power plants with a total installed capacity from 5 kWp to 30 kW) and a preferential purchase price of 205.99 BGN/MWh excluding VAT [1], the payback of the investment of photovoltaic systems is 7 years with a total annual income from all power plants 37083.60 BGN.

The following study was carried out in relation to the effect of the photovoltaic systems under consideration on the voltage loss in the ML 20 kV line, including the existing Photovoltaic power plant connected to CCTP 10. The study was carried out using specialized software (used by the electricity grid operator), in which all the data on the cable lines and transformers in the transformer posts are entered for the line/branch under consideration. Two situations were analysed – transformers operating at 60% of their rated power and transformers operating at a percentage capacity corresponding to that at the time of the study. Characteristic for the load on the LV side of the transformers at the moment of the study is that all transformers work unloaded, with currents of about $I_{LV} = (39 \div 43)$A, at $U_n = 400$V and power factor $\cos\varphi = (0.91 \div 0.93)$. Based on the simulations made, the following was established:

1.1. When operating in real mode ($I_{LV} = (39 \div 43)$ A) and without generating photovoltaic power on the substation zones, the loss of voltage at the beginning of the line (20kV buses on CCTP 1) is 0.02%, and at the end it is 0.03% (20 kV buses on CCTP 9).

1.1.1. In Variant 1 of photovoltaic power generators, operating at maximum mode, the loss of voltage at the beginning of the line is 0.01% (CCTP 1) and 0.02% at the end of the line (CCTP 9).

1.1.2. In operation in a real mode, in Variant 2 of photovoltaic power generation, operating at maximum mode, the voltage loss is 0.02% at the beginning of the line (CCTP 1) and 0.02% at the end of the line (CCTP 9), practically not changing anywhere.

1.2. When operating in a mode $S = 60\%S_{nTR}$, at $\cos\varphi = 0.92$ and without the presence of photovoltaic generating power on the substation areas, the voltage loss along the line sections varies from 0.17% at the beginning of the line (20 kV buses on CCTP 1) to 0.26% at the end of the line (20 kV bus systems on CCTP 9).

1.2.1. In Variant 1 of photovoltaic power generation, operating at maximum mode, the voltage loss is changed from 0.16% at the beginning of the line (CCTP 1) up to 0.25% at the end of the line CCTP 9).

1.2.2. In Variant 2 of photovoltaic power generation, 2 operating at maximum mode, the voltage loss is changed from 0.15% at the beginning of the line (CCTP 1) to 0.24% at the end of the line (CCTP 9).

From the data obtained it is shown that the functioning of the photovoltaic power plants connected to the low voltage side of the transformers would reduce the voltage loss, but this would have a negligible effect.

The next study is about the extent to which photovoltaic systems installed above substations and their service areas could offset the loss of active electricity in transformers. It is made for two transformer loads - at $S = 60\%\ S_{nTR}$ and at $S = 90\%\ S_{nTR}$, in $cos\varphi = 0.92$. These two loads have been selected because the operator on which territory the branch in question is located sets such loads on the transformers in their long-running operation schedules. Annual losses of active electricity ΔW in transformers are determined on the basis of the nominal data of transformers for idle losses ΔP_o and the loss at a short circuit ΔP_{sc} by the following formula:

$$\Delta W_{TR} = \Delta P_o . t_{pow.} + \Delta P_{sc} . \left(\frac{S}{S_{nTR}} \right)^2 \tau, \text{kWh} \tag{3}$$

where:

- S is the load on the transformer, kVA;
- S_{nTP} – the rated power of the transformer, kVA;
- $t_{pow.}$ – the time the transformer is connected to the MV network, h. For a period of 1 year it is assumed that, $t_{pow.} = 8760$ h;
- τ – conditional duration of active power loss, h. It is determined based on the hourly usability of the maximum active load on networks MV, at $T = 3000$ h, and at power factor $cos\varphi = 0.92$ [3] – $\tau = 1600$ h.

The following table shows the comparative results from the calculations for the annual losses of electricity in the transformers and for the electricity produced for one year by the photovoltaic systems in the substation areas, determined respectively by formula 3 and [2].

Table 2. Electricity

Loss in transformers, kWh		Photovoltaic power, kWp			
$S_{nTR,}$ kVA 400	$S_{nTR,}$ kVA 630	without service zone		with service zone	
		$CCTP_{2.6x5.2m}$	$CCTP_{5.2x5.2m}$	$CCTP_{2.6x5.2m}$	$CCTP_{5.2x5.2m}$
$S = 60\%\ S_{nTR}$		2.85	5.7	9.12	15.39
7825	11325	Generated electricity ΔW_{pv}, kWh			
$S = 90\%\ S_{nTR}$					
11114	16120	3828	7656	12249	20669

Comparison of results on electricity performance in table 2 shows that the energy produced by photovoltaic systems using only the area above the transformer posts will compensate for about half of the loss in the transformers at maximum load 60% S_{nTP} and about 1/3 of the active energy loss in them at a maximum load of 90% S_{nTP}. But, as it can be seen, if the space above the service zone is also used, these energy losses will be fully offset by the electricity generated by photovoltaic systems. The results in this table raises an interesting case - whether it would be more advantageous for power companies to install such systems above substations and what would be the real economic balance in such a situation with constantly decreasing investment costs of photovoltaic systems, and in the same time rising prices, at which the companies themselves buy electricity. Certainly, one of the effects would be to lower the ambient temperature under the photovoltaic systems due to the shading effect, which would improve the thermal conditions at which transformers operate.

If we compare the preferential prices for the purchase of electricity from photovoltaic systems up to 30 kWp and the prices on the regulated market, it can be seen that there is little difference between them. Also, if we compare the energy production of the photovoltaic systems under consideration with the power consumption of the street lighting systems having medium level of luminous class, having for example LED (30c. x 44 W)/1km = 1.32 kW/km, with an average annual operating time of the lighting system of 2400 h, and an electricity consumption of 3168 kWh/km, we can see that the smallest of the PV systems under consideration can produce energy sufficient to power a one kilometre stretch. Electricity produced and sold at a higher price during the day could be bought at a lower price during the dark part of the day and used for street lighting, in the event of a possible agreement between the municipalities and the electricity grid operators as a form of social service.

3. Conclusion.

The applied calculations and analyses show that the vacant area above the transformer posts, including the service zones, is sufficient to install photovoltaic systems that would produce enough energy to compensate for the energy loss in transformers. It can also be seen that there is some, albeit a slight decrease in the voltage loss values of the studied branch.

It is also evident and should not be neglected the effect of installing photovoltaic systems even only above CCTP, as well as the overall good economic efficiency of such projects, especially in view of the continuously declining investment costs for photovoltaic systems and rising prices of land plots.

Acknowledgements
The authors would like to thank the Research and Development Sector at the Technical University of Sofia for the financial support.

References

[1] Energy and water regulatory commission *Decision № C-9/ 01.07.2018* Republik Bulgaria
[2] http://apexsolar.bg/downloads/Kalkulatory1.htm 06.03.2020 14,30h
[3] Кирчев В, К Янев и М Георгиев 2006 *Електрически мрежи средно и ниско напрежение* Летера България
[4] https://www.evn.bg/Business/Electricity/BusinessElectricityRegulated.aspx 04.03.2020 11,23h
[5] *НАРЕДБА № 3 от 9.06.2004г. за устройството на електрическите уредби и електропроводните линии* МЕЕР ДВ бр. 90/13.10.2004 и бр. 91/14.10.2004 Република България
[6] Routledge Toyler and Francis gorup London and New York 2015 *Solar Photovoltaic Basics: A Study Guide for the NABCEP Entry Level Exam*

IOP Conf. Series: Materials Science and Engineering **878** (2020) 012015 doi:10.1088/1757-899X/878/1/012015

Analysis of earth faults in a wind farm integrated into the electric power system

Ivan Hadzhiev[1], Dian Malamov[2] and Aleksandar Georgiev[3]

[1]Department of Electrical Engineering, Technical University of Sofia, Plovdiv Branch, 25 Tsanko Dyustabanov Str., 4000 Plovdiv, Bulgaria
[2]Department of Communications and Power Engineering, University of Plovdiv Paisii Hilendarski, 24 Tzar Asen Str., 4000 Plovdiv, Bulgaria
[3]Elektrorazpredelenie Yug EAD, 37 Hristo G. Danov Str., 4000 Plovdiv, Bulgaria

E-mail: hadzhiev_tu@abv.bg

Abstract. The aim of this paper is to investigate the processes of earth faults occurrence in a wind farm network, integrated into the electric power system. The wind power farm is realized with squirrel-cage induction generators (SCIG). Each generator has a 0.69/21kV step-up transformer. The connection of the wind power farm to the electric power system is implemented through a step-up substation with a single 21/115kV transformer. The star center of the transformer 21kV winding is grounded by means of a 40 ohm resistor. A synthesized in MatLab/Simulink computer model investigates the processes of earth faults occurrence in the medium-voltage (MV) network of the wind farm. Results have been obtained, concerning the influence of the earth fault location on the transient voltages and currents at both rated torque of the generators and no-operating generators. When there is an earth fault in one of the lines, the voltages and currents of the MV network are asymmetrical and higher harmonics of the currents and voltages are generated. A harmonic analysis of the current and voltage results has been made and the total harmonic distortion (THD) has been determined for several characteristic points of the wind farm network. The parameters of voltage contamination of the power system with higher harmonics in case of an earth fault occurrence in the wind power substation have been determined. It has been found that the earth faults in the MV network of the wind farm generate voltages and currents contamination of the power system with higher harmonics.

1. Introduction

Wind farms are one of the main renewable sources of electricity. The trends in the development of the modern energy systems include an increase in the relative share of electricity produced by wind power farms, as well as increase in their capacity. Wind farms usually work integrated into the region's electricity system. Changes in wind speed and direction over time are largely random, which affects the joint operation of the wind farm and the power system [1]. The wind farms integrated into power systems mainly use squirrel-cage induction generators (SCIG) and doubly-fed induction generators (DFIG). It is known that when taking into account the fluctuation of the parameters of the wind energy these generators have certain advantages if operating in parallel with the electricity system [2]. The reactive power required to excite the magnetic field of the wind farm generators is provided by the synchronous generators in the system or by capacitor banks. The wind farms can consist of a large number of generators (tens, hundreds). There is usually a low-voltage/medium-voltage (/LV/MV) step-up transformer for each generator. Several generators form a group and connect to each other through a main feed cable line. The individual groups are connected by radial lines to a substation that increases the voltage in accordance with the voltage of the electric power system.

The processes in the wind farms are different from those in the traditional power plants with large synchronous generators. The wind farms exert a specific influence on the parameters of the electricity system to which they are connected. This applies both to normal operation and to emergency modes (three-phase and two-phase short circuits, and earth faults). A thorough study of the impact of these processes on the quality of electricity in the electric power systems is of great importance. The wind farms are characterized by a mixed radial-magistral network and presence of elements with nonlinear characteristics (transformers, generators, etc.). Therefore the mathematical models for describing the emergency mode processes in a wind farm integrated into the power system are rather complex [3], [4]. A computer model in MatLab/Simulink, described in [3], examines the normal mode operation processes, as well as the processes in case of a three-phase short-circuit occurrence and in case of a two-phase short-circuit occurrence. In [4], based on a simplified model, presenting the wind farm through an equivalent generator, studies of the effect of both the transformer parameters and the type of generators on the currents at short-circuits and earth faults have been carried out. It is known that the particular method of neutral grounding has a significant influence on the voltages and currents in case of an earth fault in the substation of the MV wind farm. The effect of the different methods of grounding the neutral (on the currents and voltages in case of an earth fault in the MV substation of the wind farm) has been analyzed in [5]. A resistance grounded system is recommended as a compromise option. The studies in [3], [5] show that higher harmonics of voltage and current are generated in the wind farm network in case of earth fault occurrence in it.

The aim of this paper is to study the effect of the earth faults in a wind farm on the voltages and currents of the electric power network. The wind farm under consideration is integrated into the electric power system. The studies have been performed by simulating earth faults by means of an especially designed in MatLab/Simulink computer model.

2. Description of the studied wind farm

The single-line diagram of the wind power farm under study is given in Figure 1. The farm contains eight identical SCIG generators with a step-up LV/MV transformer (0.69/21kV) to each of them. The

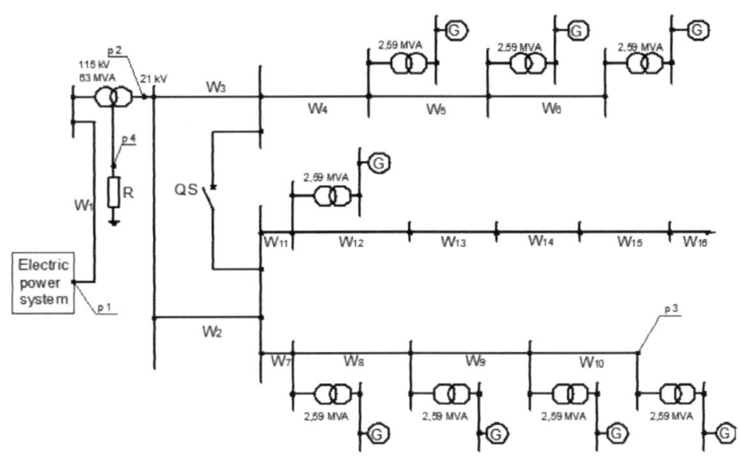

Figure 1. Single-line diagram of the wind power plant.

generators form three groups, which are connected to the input of the medium-voltage/high-voltage (MV/HV) step-up transformer 21/115kV via main feed cable lines. The middle group contains only one generator and the other terminals of the main feed are intended for future expansion of the wind farm. The star center of the 21kV winding is grounded by a 40 Ohm resistor. The output of the MV/HV transformer connects the wind farm to the electric power system via the W_1 cable line in point 1. The MV network is realized by means of NA2XS(F)2Y cables, while the W_1 line uses 2XS(FL)2Y cables. The electrical parameters of the cables, transformers and generators are given in Tables 1÷3.

Table 1. Electrical parameters of the cables.

Number and cross-section of the cable	Resistance (ohm/km)	Reactance (ohm/km)	Capacitance (μF/km)
W_1, 3x1x800 mm^2	0.037	0.302	0.474
W_2, W_3, 2x3x1x400 mm^2	0.078	0.312	0.341
W_4, W_7, W_{11}, 3x1x800 mm^2	0.037	0.302	0.474
W_5, W_6, W_{12}, 3x1x630 mm^2	0.055	0.299	0.402
W_8, W_{13}, W_{14}, 3x1x240 mm^2	0.125	0.334	0.300
W_9, W_{10}, W_{15}, W_{16}, 3x1x95 mm^2	0.32	0.392	0.210

Table 2. Electrical parameters of the transformers.

Parameters	Units	Transformer LV/MV	Transformer MV/HV
Rated power	(MVA)	2.6	63
Primary voltage	(kV)	0.69	21
Secondary voltage	(kV)	21	115
Losses at no-load	(kW)	3.9	24.4
Losses at short circuit	(kW)	14.4	192.9
Relative current at no-load	(%)	0.37	0.74
Relative voltage at short circuit	(%)	7	12
Connection Scheme	-	D/yn-11	Yn/yn-0

Table 3. Electrical parameters of the generators.

Parameters	Units	Values
Type	-	SCIG
Rated voltage	(V)	690
Rated frequency	(Hz)	50
Rated power	(MVA)	2.6
Rated torque	(N.m)	14740
Stator resistance	(mOhm)	1.102
Rotor resistance	(mOhm)	1.497
Leakage inductance of the stator	(mH)	0.065
Leakage inductance of the rotor	(mH)	0.06
Inductance of the magnetizing contour	(mH)	2.14

3. Computer model

The computer model was synthesized in MatLab/Simulink based on the connection scheme in Figure 1 and the equivalent electric circuits of the generators, transformers and cables. The parameters of the equivalent circuits were determined in accordance with the electrical characteristics in Tables 1÷3. The impedance of the electric power system in the point where the wind farm connects to it was taken into account. The model makes it possible to simulate an earth fault and to determine the transient voltages and currents in predetermined points in the wind farm network. In the model the moment of the earth fault occurrence is set by the moment when the current passes through the zero value of the phase in which the earth fault occurs.

The studies were conducted for the following network characteristic points, as in Figure 1:

1) Point 1 – the point of connection of the wind farm to the grid;

2) Point 2 – the line terminals of the MV winding (21kV) of the MV/HV transformer (21/115kV);

3) Point 3 – the line terminals of the MV winding (21kV) of one LV/MV generator transformer (0.69/21kV), which is at the end of one of the main feed cable lines;

4) Point 4 – star center of the MV winding (21kV) of the MV/HV transformer (21/115kV).

The computer model simulated single-phase earth faults in points 2 and 3 of the network, as in Figure 1. The numerical experiments were performed for the cases when all generators operate at the rated shaft torque, on the one hand, and when the generators do not work and are disconnected from the generator transformers, on the other hand. Results about the transient voltages and currents in the described points were obtained at the moment of occurrence of the earth fault in the interval (0÷10) ms from the moment when the current passes through the zero value. Both the harmonic spectrum and the total harmonic distortion (THD%) of the voltages and currents in the indicated points were determined by post-processing of the results. The THD coefficients for voltage were determined by the following formulae:

$$THD_U\,,\% = \frac{\sum_{i=2}^{n} U_i}{U_1}100, \quad (1)$$

$$THD_I\,,\% = \frac{\sum_{i=2}^{n} I_i}{I_1}100, \quad (2)$$

where: n - total number of harmonics; U_1 (I_1) - r.m.s. value of the voltage (current) of the main harmonic; U_i (I_i) - r.m.s. value of the voltage (current) of the respective harmonic.

4. Results

Figure 2 shows the transient voltages for the case of generators, operating at their rated torque and a phase L1 earth fault in point 2. The moment of the earth fault occurrence is when the phase L1 current passes through the zero value.

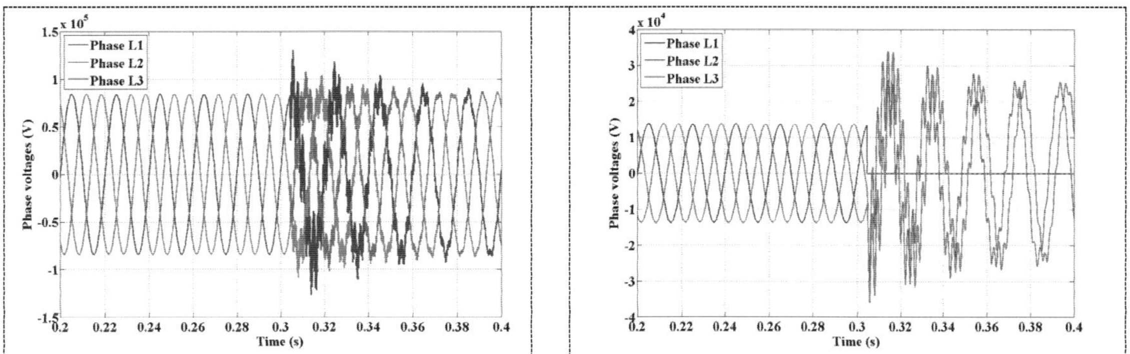

a) b)

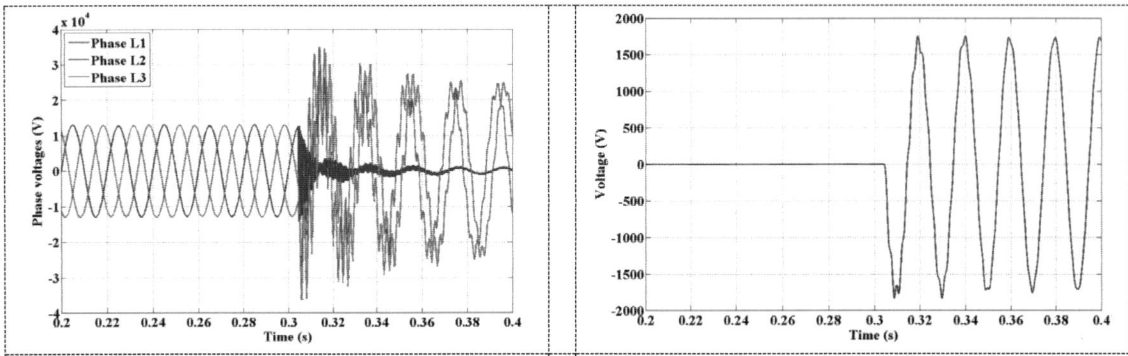

c) d)

Figure 2. Transient voltages in case of an earth fault in point 2 for the following points: a) point 1; b) point 2; c) point 3; d) point 4.

Figures 3÷5 give the results related to the harmonic spectrum and the THD coefficient for the voltages in points 1 and 2 at rated torque of the generators and phase L1 earth fault in point 2.

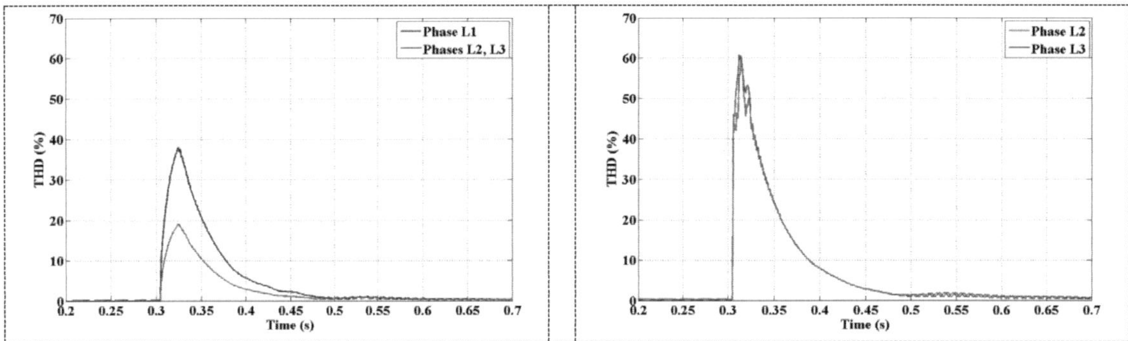

a) b)

Figure 3. Change in THD_U% during the transient process: a) in point 1 for phase L1 (in blue) and for phases L2 and L3 (in green); b) in point 2 for L2 and L3 (in green and red).

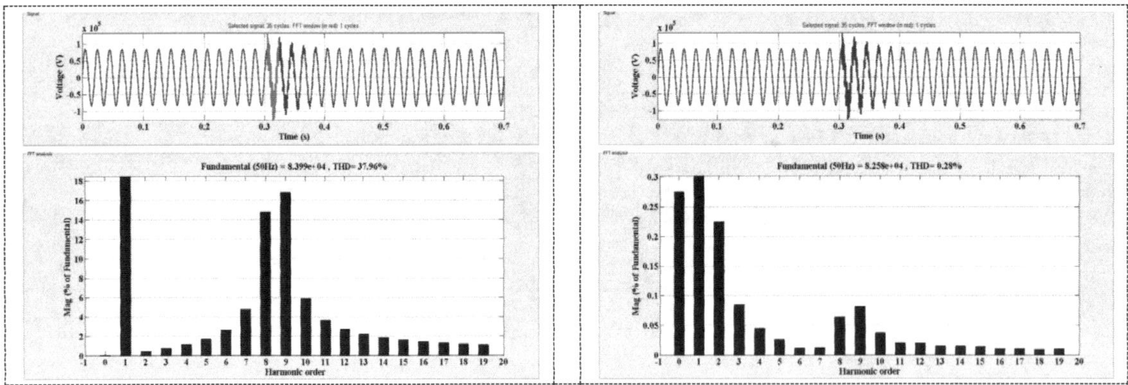

a) b)

Figure 4. Frequency spectrum of the higher harmonics and THD_U for one period of the phase L1 in point 1: a) in the beginning of the transient process; b) in a steady-state mode of an earth fault.

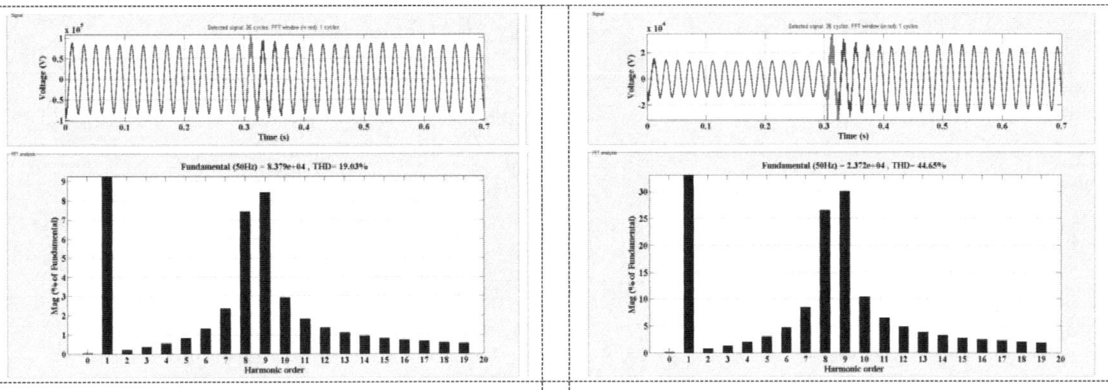

a) b)

Figure 5. Frequency spectrum of the higher harmonics and THD_U for one period of the phase L2 in the beginning of the transient process in: a) point 1; b) point 2.

Figure 6 illustrates the transient processes of the currents in case of a phase L1 earth fault in point 2. The moment of the earth fault occurrence is when the phase L1 current passes through the zero value.

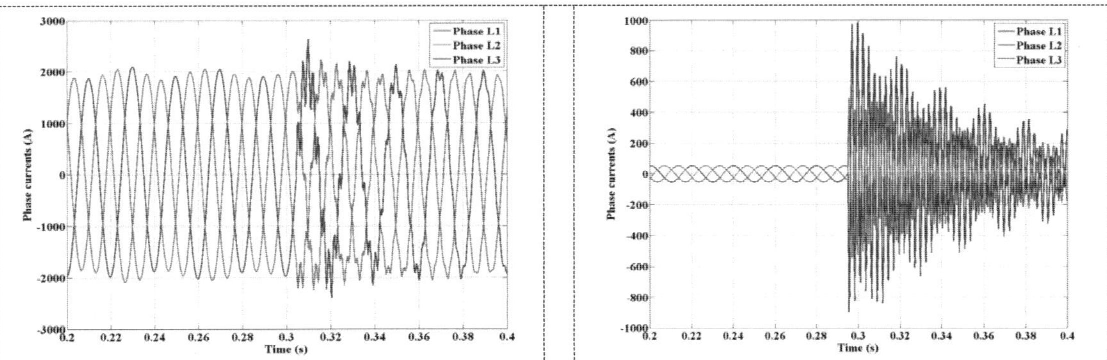

a) b)

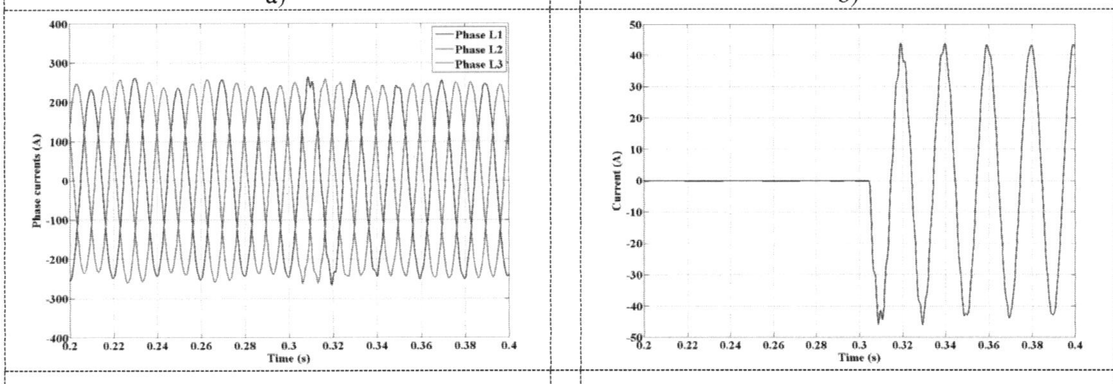

c) d)

Figure 6. Transient currents in case of an earth fault in point 2 for: a) point 2 at rated torque of the generators; b) point 2 for no-operating generators; c) point 3 at rated torque of the generators; d) point 4 at rated torque of the generators.

When the generators operate, the current in point 1 is approximately proportional to the current in point 2. For no-operating generators the current in point 2 is zero. The currents in point 4 are the same for operating and turned off generators.

Figure 7 depicts the transient processes of the currents in case of a phase L1 earth fault in point 3. The moment of occurrence of the earth fault is when the phase L1 current passes through the zero value.

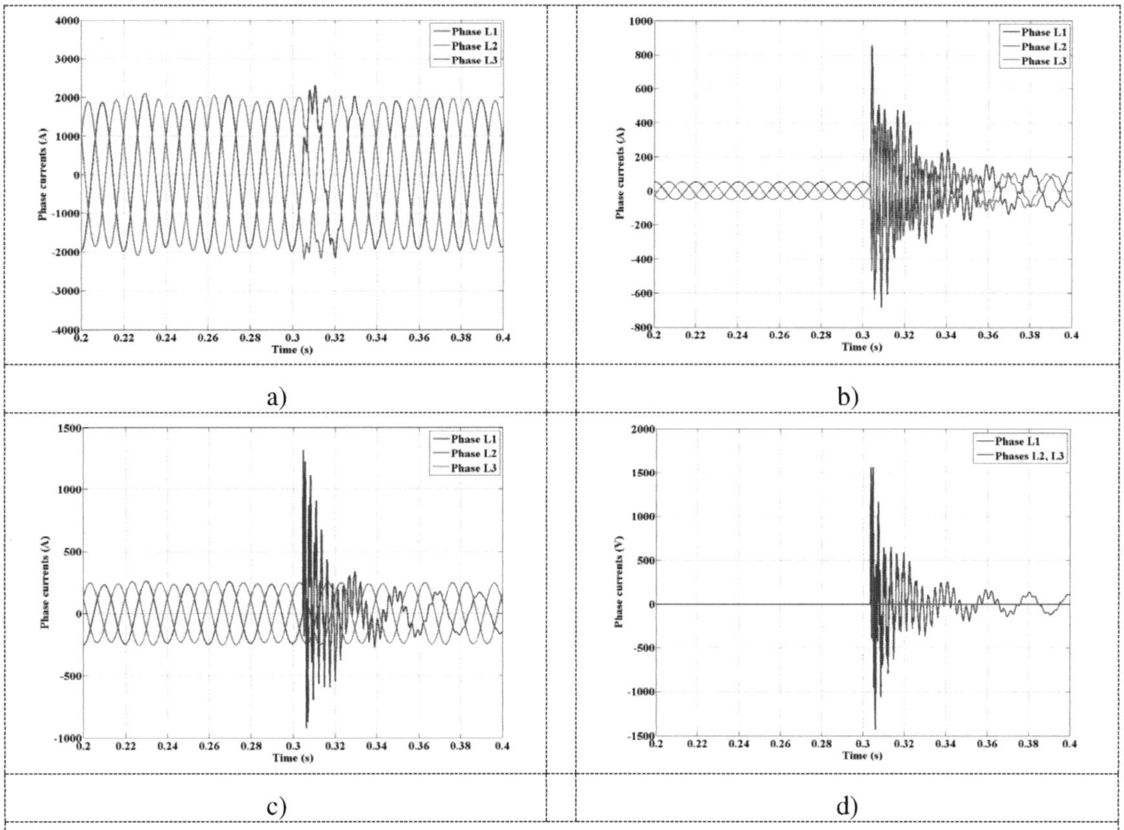

Figure 7. Transient currents in case of an earth fault in point 3 for: a) point 2 at rated torque of the generators; b) point 2 for turned off generators; c) point 3 at rated torque of the generators; d) point 3 for turned off generators.

5. Conclusion

The established r.m.s values of the currents in point 2 of the phases L2 and L3 in case of an earth fault in both point 2 and point 3 at rated torque of the generators are practically unchanged. The observed r.m.s. values of the phase voltages in point 2 of the phases L2 and L3 increase by about 70% for all studied cases of earth faults. This is explained by the method of star center grounding of a MV winding in a 21/115kV transformer through a resistor with an active resistance of 40 Ohm. In the case of no-operating generators, the current in point 2 is 38A (r.m.s.) and it is determined by the idle-run currents of the transformers. In this case, for an earth fault in point 2, the steady-state currents in the phases increase by about 50%.

In case of a single-phase earth fault both in points 2 and 3, network frequency voltage appears over the star center (point 4). Its steady-state value is 1.2kV (r.m.s.) at rated torque of the generators and does not depend on the location and the moment of the earth fault occurrence in the MV network. The current through the grounding resistor in steady state is 30A (r.m.s.). The peak values of voltage and current through the neutral are obtained at the moment of the earth fault of about 6 ms. Higher harmonics in the star center are practically not generated.

In case of a single-phase earth fault, higher harmonics of the currents and voltages with odd and even numbers are generated in both points 2 and 3. The maximum contamination of the network with higher harmonics occurs in the beginning of the transient process when the moment of the earth fault coincides with the moment when the current passes through the zero value.

The earth faults in the MV network contaminate the voltage of the power system with higher harmonics in the point of the wind farm connection to the system.

In case of an earth fault after the transient process has completed, the contamination coefficients (THD_U) for all cases are up to 2%.

Table 4 gives the results, concerning the maximum coefficients of voltage contamination (THD_U) in points 1 and 2.

Table 4. Values of the maximum voltage contamination coefficients (THD_U, %).

Location of the earth fault and generator mode	Phase L1 in point 1	Phases L2 and L3 in point 1	Phases L2 and L3 in point 2
Point 2 – at rated torque of the generators	38	19	45
Point 2 – at no-operating generators	39	20	49
Point 3 – at rated torque of the generators	19	10	38
Point 3 – at no-operating generators	22	11	38

Acknowledgments
The authors would like to thank the Research and Development Sector at the Technical University of Sofia for the financial support. The work of I. Hadzhiev is supported by National program "Young scientists and Postdoctoral candidates" of Ministry of Education and Science.

References

[1] Söder L and Ackermann T 2005 *Wind Power in Power Systems:An Introduction* Ackermann T. (Ed), *Wind Power in Power Systems,* John Wiley & Sons, Ltd, Chichester, England, pp. 25-52.

[2] Noureldeen O, Abdelwahab M and Rashad A 2011 *Operation of Hybrid Wind Farm Based on SCIG and DFIG Under Gust Wind Speed* Journal of Engineering Sciences, Assiut University, Vol. 39, No 3, pp.637-648.

[3] Umashankar S, Kothari D P and Mangayarkarasi P 2010 *Wind Turbine Modelling of a Fully-Fed Induction Machine* S. M. Muyeen (ed.), Wind Power, Intech, Olajnica 19/2, 32000 Vukovar, Croatia, pp. 93-112.

[4] Gevorgian V and Muljadi E 2010 *Wind Power Plant Short Circuit Current Contribution for Different Fault and Wind Turbine Topologies* The 9th Annual International Workshop on Large-Scale Integration of Wind Power into Power Systems as well as on Transmission Networks for Offshore Wind Power Plants, Québec, Canada.

[5] Hansen P, Ostergaard J and Christiansen J S 2007 *System grounding of wind farm medium voltage cable grids* Nicolaos Cutululis and Poul Sørensen (eds.), Proceedings Nordic Wind Power Conference NWPC'2007, Roskilde, Denmark, pp. 164-168.

IOP Conf. Series: Materials Science and Engineering **878** (2020) 012016 doi:10.1088/1757-899X/878/1/012016

Route map for renewable energy sources implementation for household in Bulgaria

Dilyana Gospodinova[1], Kostadin Milanov[1] and Peter Dineff[1]

[1]Technical university – Sofia, Faculty of Electrical Engineering, Department of Electrical Apparatus, 8 Kliment ohridski Blvd, Sofia 1000, Bulgaria
e-mail: dilianang@abv.bg

Abstract. This paper is concerned with the design of a hybrid household system, located in the city of Sofia with the possibility of grid connection. It is comprised of photovoltaic (PV) arrays, wind generator and battery energy storage devices. A multi-objective, non-derivative optimisation is considered in this residential application; the primary objective is the system cost minimisation. The commercial software, HOMER Pro, is utilised to identify the least-cost design among hundreds of options. This paper discusses suitability of renewable energy (RE) applied in one-family household according socio-economic situation in Bulgaria. Finally, a route map of RES implementation options is presented, and economic indicators such as Total net presents cost (NPC), Cost of Energy (COE) and RES fraction are assessed.

1. Introduction

The aim of the European Union's long-term strategy is to keep Europe at the forefront and lead in global climate action, to present a vision that can lead to zero greenhouse gas emissions by 2050 through a socially just transition in a cost-effective way, [1]. Bulgaria adopted the Renewable Energy Act in 2011, and the aim is to achieve national targets for the use of renewable energy sources (RES) in final gross energy consumption. The European Union's Framework Strategy for Energy (2015) aims to achieve an energy transition that is fully European-oriented. The strategy is based on five pillars and aims to achieve the objectives faster and more effectively of the EU in the field of energy climate: reducing territorial greenhouse gas emissions in the EU (by 20 % by 2020 and by 40 % by 2030), increasing the share of energy coming from renewable sources (by 20 % to 2020 and 27 % to 2030) and improving energy efficiency (by 20 % by 2020, by 27 % by 2030), [2].
New renewable energy targets have been set in the new policies of a number of countries where the share of fossil energy, including renewable energy, is expected to reach more 32 % in 2030. Energy consumption from renewable sources in new buildings should represent more than 10% of total energy consumption.
In 2016, household electricity prices in Bulgaria were more than twice below the EU average. When measured in purchasing power standards per unit of energy, Bulgarian retail prices look much closer to the EU average, indicating potential accessibility problems for a particular consumer group. There are no restrictions on the territory of Bulgaria for household consumers to choose their electricity suppliers, but a very large part of them remain with a non-competitive offer of the active suppliers, which are regional. As a result, the switch-over for residential customers and low voltage companies is negligible. In Bulgaria, electricity prices are regulated, and progressive deregulation of prices in the electricity sector is envisaged [3].

1.1. Energy system in Bulgaria

The Bulgarian internal electricity market is highly regulated and there is only one licensed supplier in each geographical region.

The retail electricity market is highly concentrated with a low level of competition and fixed energy prices set by the State Energy and Water Regulatory Commission. Nuclear power and solid fuels are dominated in energy production mix in Bulgaria. In recent years, renewable energy sources have become increasingly important and currently represent about 20% of Bulgaria's electricity production, [4, 5, 6]. Various scenarios related to the electricity mix in Bulgaria from 2020 to 2050 are given in the literature. According to the ambitious decarbonisation objective and the corresponding RES support schemes, Bulgaria will achieve a share of 53÷54 % in electricity production from renewable sources by 2050. This part will be composed mainly of wind, some hydro and solar energy. There are forecasts that, if support for renewable energy is phased out and the carbon dioxide (CO_2) target is not set, the share of RES in electricity consumption will reach around 33 % in 2050 [7].

1.2. Decarbonisation of economy (residential)

Climatic and meteorological conditions affect natural and anthropogenic processes that affect the state of the environment. In recent years, the incidence of extreme meteorological and climatic events in Bulgaria has increased, causing serious material damage to agricultural production, infrastructure, residential and public buildings, including human casualties, [8]. Bulgaria remains the most GHG (*Greenhouse Gas*) - intensive EU Member State. In 2015 in Bulgaria, the largest sectors in terms of GHG emissions were the energy sector, followed by transport (15.2 %), industry (13.9 %) and agriculture (10.5 %). In relative terms, the GHG emissions from residential and commercial sectors of Bulgaria were more than 5 times below the EU average (2.2 % vs 12.1 %) [3].

2. Renewable energy sources implementation and suitability household sector in Bulgarian

It is well known that Bulgaria has long fulfilled its commitments to increase the share of RES in final energy consumption and after the change of the state policy for preferential prices for buying energy from RES, the "boom" in the period 2007÷2012 to build these capacities are sharply reduced. Given the higher targets set in the 2030 climate/energy package for a 27 % share of RES, the Bulgarian policy may need to undergo some changes. The ten-year plan for the transmission system commented on the problems that would arise. Operating capacities with direct dependence on atmospheric intensity, and changes in operating capacities need to be offset by conventional power plants. The connection of new wind and photovoltaic capacities is limited and depends on the available resource for regulation. These considerations are likely to lead to a change in Bulgaria's energy policy and to further decentralization in energy production, especially at a low level [9]. To move to this state of affairs the Bulgarian Government endorses renewable energy sources. These energy sources may deal with restraint about fuel flexibility, efficiency, reliability, economics and emission. The energy mix should revolutionize and start further RES to diminish power generation pollution.

In 2018, renewable energy accounted for 21 % of the total energy used for heating and cooling in the European Union (EU). Increases in industry, services and households have all contributed to the growth in renewable energy used for heating and cooling. According to Eurostat the share of renewable energy in the total energy used for heating and cooling in Bulgaria is 33 % [10].

The development of the RES applied in buildings, no matter the PV, wind or the geothermal heat pump systems (GSHP) application, depend mostly on regional conditions. The dominance of photovoltaic (PV) among renewable energy technologies is owed mostly to its noiselessness, non-toxic emission, and relatively simple operation and maintenance. Taking the application of PV as an example, the commonly used indicators for evaluation of solar radiation condition are the peak sunshine duration hours.

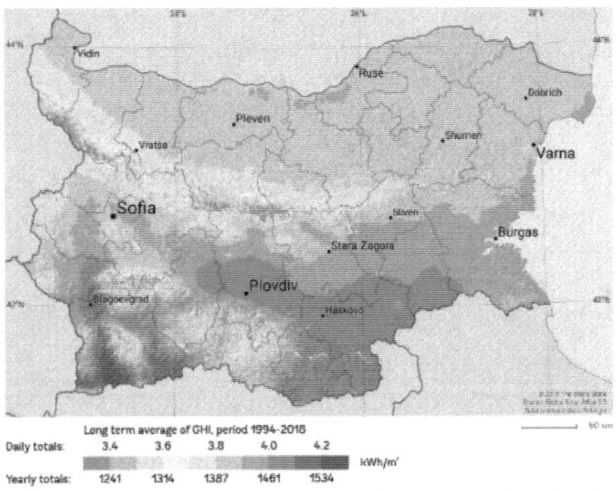

Figure 1. Map of solar radiation for Bulgaria for long term period 1994÷2018 [11].

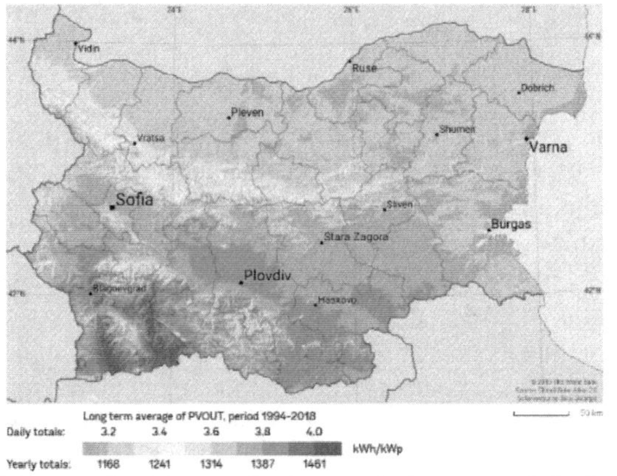

Figure 2. Map of photovoltaic potential for Bulgaria – daily and yearly for long term period 1994-2018 [11].

Having a peak electricity load of 21 kW, being totally dependent on oil and gas provides an excellent opportunity for the generation of solar energy because of available high rate of daily sunshine, Table 1. A Solar Radiation Atlas developed by KACST in collaboration with the US National Renewable Energy Laboratory (NREL) in 2013 to monitor and map Bulgaria solar, wind, geothermal and waste-to-energy resources. Based on data, it has been found that direct normal radiation exceeds 14.4 MJ/m^2 in some regions yearly, Figure 1 and Figure 2.

Table 1. Daylight duration by months and city in Bulgaria, h

Month	Sofia	Plovdiv	Burgas	Varna	Ruse	Pleven	Lom	Sandanski
January	9.26	9.28	9.26	9.22	9.13	9.21	9.18	9.32
February	10.36	10.3	10.29	10.26	10.26	10.25	10.24	10.32
March	11.58	11.52	11.51	11.51	11.51	11.51	11.51	11.52
April	13.27	13.19	13.20	13.22	13.22	13.22	13.23	13.18
May	14.42	14.33	14.35	14.39	14.43	14.40	14.42	14.30
June	15.22	15.11	15.14	15.18	15.20	15.20	15.23	15.07
July	15.03	14.54	14.56	15.01	15.00	15.02	15.05	14.51
August	13.58	13.5	13.51	13.54	13.70	13.55	13.56	13.48
September	12.32	12.26	12.26	12.26	12.24	12.26	12.27	12.25
October	11.07	10.59	10.58	10.56	10.54	10.55	10.54	11.00
November	9.48	9.45	9.43	9.39	9.34	9.38	9.36	9.48
December	9.07	9.09	9.06	9.01	9.00	9.00	8.57	9.13

Bulgaria has good climatic conditions for RES development: great solar radiation > 4 kWh/m^2 per day, Figure 1, and duration of solar lighting > 11 h/day, so the technological possibilities for the utilisation of solar power cannot be neglected, Table 1. According P. Gramatikov, Bulgaria is located at the upper border of the so-called "solar belt" of the world. As a result, the total annual solar intensity in the capital - Sofia, is equivalent to the solar intensity of Madrid in Spain, Athens in Greece, and Sicily in Italy [12]. According EBRD, [13], wind power is one of the RES that has significant growth potential in Bulgaria, where strong winds blow from the coastal area.

3. Simulation and results

Some studies have been conducted to evaluate the RE potential of Bulgaria. Some of these are Government policies for solar investment. It is well known the solar PV and another efficiency improvement for sustainable residential buildings, RE options and economic viability of Solar PV within the residential sector. According some authors the efficient use of renewables through smart grid can reduce oil consumption for electricity and dependency on oil, especially in Southern European countries.

Figure 3. Mean wind speed map of Bulgaria for 10 m high [14].

Renewable energy has become an increasingly competitive way to meet new power generation needs. According to statistics from IRENA (International Renewable energy Agency), the cost of solar modules has been steadily falling, with a decrease of approximately 8 times from 2010 to 2018, [15]. This is a valid reason in this article to propose a technical solution for the introduction of a hybrid system for receiving energy from renewable sources on the territory of Bulgaria. The design of a hybrid power system depends on some important sensitivity variables to optimize system cost and component sizes. The proposed hybrid power system for family house, is flexible and can be easily expanded by adding new photovoltaic panels, wind turbines, fuel cells and rechargeable batteries. The system has the opportunity to add an additional parallel inverter to monitor the added components, [16]. Battery storage can be located at the centralised wind and solar power production site to smooth variable generation output as it is fed into the grid. It can also store excess renewable production for later periods. For many applications, lithium-ion has proved preferable to other chemistries with respect to energy and power density, cycle and calendar life, and cost.

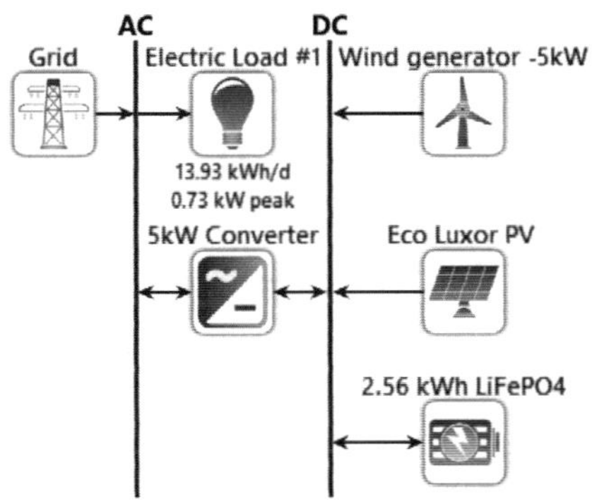

Component	Name	Size	Unit
PV	Generic flat plate PV	5.50	kW
Wind turbine	Aeolos-V5kW	1	ea.
System converter	System Converter EasySolar 48/5000/70-100	5.00	kW
Grid	Grid	999 999	kW
Dispatch strategy	HOMER Load Following		

Figure 4. System architecture and component table of the proposed hybrid power system made by HOMER software.

The capacity of proposed hybrid system estimates include small solar PV (5.5 kW) and wind turbine (1 kW) installations coupled with battery storage at the household level. The data expressed in [17] shows that the li-ion battery with the highest system price actually provides the lowest cost per cycle. Battery storage at a household allows greater self-consumption of electricity produced by solar PV. It can also help relieve local grid capacity constraints. Our results from hybrid system modelling shows that the cost energy produced from battery remains too high: 6.75 BGN/kWh.

The proposed hybrid system may be implemented differently depending on the geographical location and socio-economic situation of the population living there. In some areas of the country, where climatic conditions are not particularly suitable for wind turbines, a hybrid system may be used. There are areas in Bulgaria with wind speed values significantly higher than other areas - the north-eastern

IOP Conf. Series: Materials Science and Engineering **878** (2020) 012016 doi:10.1088/1757-899X/878/1/012016

part of the country near the Black Sea, Figure 3. It is here that the collaboration of solar panels with wind turbines can be successfully used.
The possibility of introducing RES into single-family homes in Bulgaria has been assessed by conducting a feasibility study based on the adopted hybrid system. The analysis was performed using the HOMER Pro software product.
The parameters shown in Table 2 have been loaded into HOMER Pro for optimizing the sizing of the various components. A discount rate of 6 % has been considered for the calculations. The results are valid for a hybrid system located on the territory of the Technical University - Sofia. The total system cost is 22 853 BGN and initial capital of 18 700 BGN. The system lifetime is pre-set at 25 years.

Table 2. Net presents costs of the hybrid system.

Name	Capital, BGN	Operating, BGN	Replacement, BGN	Salvage, BGN	Total, BGN
Aeolos-V5kW	1 600	1 572.00	637.61	- 349.75	3 460.00
Generic flat plate PV	6 100	129.28	0.00	0.00	6 229.00
Grid	0	- 590.83	0.00	0.00	- 590.83
HOMER Load Following	3 000	0.00	0.00	0.00	3 000.00
System Converter EasySolar 48/5000/70-100	8 000	0.00	3 394.00	- 638.82	10 755.00
System	18 700	1 110.00	4 032.00	- 988.58	22 853.00

The hybrid power system has been simulated by HOMER Pro in order to assess its operational and economic characteristics. HOMER Pro is being able to perform thousands of simulations in a few seconds. In this case, 342 feasible solutions were assessed taking into account various system designs to calculate the option with the least net presents cost (NPC). HOMER Pro eliminates all infeasible solutions according to total NPC. Similar models have been made for main cities in Bulgaria, Table 4. In all modelling cases dispatch strategy is Load Following and there is no Unmet Electric Load. Initial capital is 18 700 BGN and the Salvage in the end of system lifetime is constant again – 4 126 BGN.

Table 3. Production summary of the system located on the Territory of Technical University in Sofia.

Component	Production, kWh/year	Percent
Generic flat plate PV LUXOR Eco Line 275 W	6 935	67.9
Aeolos-V5kW	563	5.5
Grid Purchases	2 711	26.6
Total	10 210	100.0

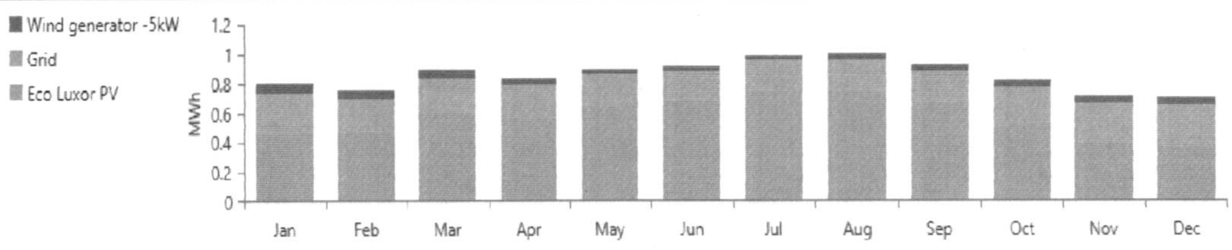

Figure 5. Monthly electric production of the hybrid system located on the Territory of Technical University in Sofia calculated by HOMER Pro.

4. Socio-economic situation in Bulgaria and possibility of RES implementation into the household

Households are the third largest sector in terms of final energy consumption after transport and industry and ahead of services and agriculture. The distribution of electricity consumption of Bulgarian households depends mainly on the heating source they are using – electricity, natural gas, coal, wood or other fuels. In 2015 the electricity consumption of Bulgarian households was distributed as follows: 46 % was used for water heating; 20 % for lightning and small appliances; 16 % for cooling (refrigerator use); 8 % for cooking; and 10 % was used for other purposes. In Bulgaria 20 % of the poorest households spent more than 14 % of their budget on domestic energy [18]. Aaround 67 % of the most socially deprived households were still unable to keep their homes warm. This represents a decrease but it remains significantly above the EU average of 23 % and makes Bulgaria the worst performer in the EU on that metric [3].

Table 4. Compare indicators connected with the financial balance and renewable energy fraction.

City in Bulgaria	Preferred RE source implementation in the hybrid system Solar/Wind	Total net presents cost of the system (NPC), BGN	Cost of Energy (COE), BGN	RES fraction, %
Vidin	Solar	24 051	0.192	72.7
Vratza	Solar	23 923	0.188	73.2
Lovech	Solar	23 758	0.184	73.8
Montana	Solar	23 538	0.181	74.8
Pleven	Solar	23 766	0.185	73.8
Veliko Tarnovo	Solar	23 736	0.184	73.9
Gabrovo	Solar	23 905	0.188	73.3
Razlog	Solar	23 937	0.189	73.2
Ruse	Solar	23 609	0.169	75.6
Silistra	Solar	23 468	0.180	75.0
Varna	Solar and Wind	22 568	0.166	78.9
Dobrich	Solar and Wind	22 711	0.170	78.5
Targovishte	Solar and Wind	22 678	0.169	78.6
Shumen	Solar and Wind	22 574	0.168	79.0
Burgas	Solar and Wind	23 130	0.176	76.7
Sliven	Solar and Wind	22 853	0.172	77.9
Stara Zagora	Solar	23 987	0.191	73.0
Yambol	Solar	23 987	0.191	73.0
Blagoevgrad	Solar and Wind	23 403	0.181	75.6
Kustendil	Solar	24 122	0.191	72.2
Pernik	Solar	24 106	0.190	72.2
Sofia	Solar	24 057	0.189	72.4
Kardjali	Solar	23 073	0.189	72.3
Pazardjik	Solar	23 614	0.183	74.6
Plovdiv	Solar	23 200	0.194	72.0
Smolyan	Solar	24 197	0.194	72.0
Haskovo	Solar	23 582	0.183	74.7

Various sources studying energy efficiency behaviour of Bulgarian households show that energy poverty is a serious issue in the country. According to statistics, 46.6 % of the total population of Bulgaria cannot maintain adequate thermal comfort in their households. The rising electricity and district heating prices in the last years have forced many households towards using coal and wood for heating which further worsens air and living quality. The electricity prices for households in Bulgaria have increased to 9.55 eurocents/kWh in 2017. Although the prices are still the lowest in the whole EU around 444 000 households are claimed to be highly vulnerable to increases in electricity prices [19].

Economic development is a multilayered process endeavors the growth of labor productivity the improvement of the standard of living of the population a highly profitable economy. Overall our country's GDP has been growing steadily since 2013 with wide variations across regions. Bulgaria's GDP per capita is growing slightly faster than the EU average reaching 50 % of the EU average in 2017 but remains rather low [9].

At the district level the highest average annual financial incomes are in the districts of Sofia-city, Varna, Stara Zagora, Gabbrovo, Razgrad, Plovdiv, Burgas, and Ruse (> 1 100 BGN). The incomes in the districts of Vidin, Blagoevgrad, and Smolyan are low (< 950 BGN). The difference between the district with the highest (Sofia-city) and the one with the lowest (Vidin District) average annual household income remains over 2 times financial income Figure 5 [9]. An important economic indicator for assessing the possibility of introducing renewable energy sources in single-family residential buildings is the cost of fixed assets. According to the National Statistical Institute at the national level the Southwestern region of Bulgaria is characterized by the highest costs for the fixed assets.

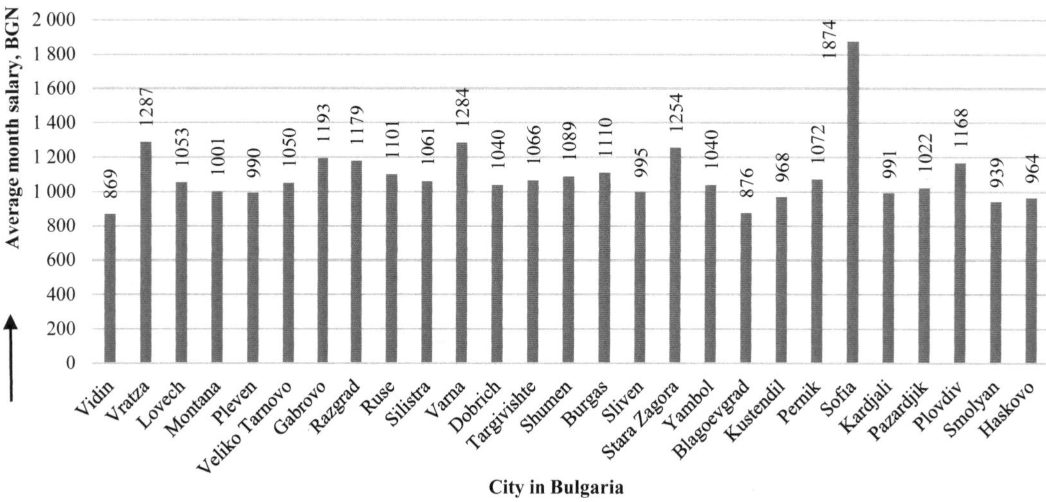

Figure 6. Average month salary in Bulgaria by cities in 2019 [20].

5. Conclusions

Additional efforts could therefore be envisaged to improve energy intensity in various demand sectors. Furthermore the implementation of the national energy efficiency actions and programs needs to be pursued. By 2050 the country's decarbonised electricity mix would include 53÷54 % renewable generation. The rising carbon coal and natural gas prices would lead to an increase of Bulgarian wholesale electricity prices from an average of 34 EUR /MWh in 2016 to over 74 EUR/MWh in the decarbonisation scenario in 2050. The current assessment shows that alternative solutions exist to replace current generation capacity by 2050 with different implications for affordability sustainability and security of supply. The utilizations of the solar PV ST and GSHP are believed to be the most promising technologies in the future. The decentralized production of energy from renewable sources and depending on the local potential and needs is a sector with great prospects in the country. This is a way to avoid the cost of networking and the loss of transmission and distribution of energy. Lower solar PV module prices and ongoing reductions in balance of system costs remain the main driver of reductions in the cost of electricity from solar PV. The proposed solution for supplying electricity to single-family homes is fully in accordance with the national strategy for increasing the share of renewable energy sources in the overall energy mix of our country.

6. References

[1] European commission 2018 A Clean Planet for all A European strategic long-term vision for a prosperous, modern, competitive and climate neutral economy EU Brussels
[2] Vladimirov M Georgiev A Kolarova S 2018 Development of Small-Scale Renewable Energy Sources in Bulgaria Center for the Study of Democracy
[3] European Commission 2017 Energy Union Factsheet Bulgaria Brussels
[4] European Commission 2018 European Network for Research, Good Practice and Innovation for Sustainable Energy
[5] IAEA 2019 Country Nuclear Power Profiles BULGARIA International Atomic Energy Agency
[6] CEICData 2020 Bulgaria Electricity Production CEIC

[7] Szabó L et al 2018 South East Europe Electricity Roadmap Regional Centre for Energy Policy Budapest.
[8] Ministry of Finance 2019 Bulgaria 2030. Part 1 Analysis of the socio-economic development of the country after its accession to the EU Sofia
[9] National center for territorial development 2019 Socio-economic analysis of the regions in the republic of Bulgaria
[10] Eurostat 2020 Eurostat News Eurostat
[11] SolarGis 2020 Global Solar Atlas 2.0 The World Bank
[12] Gramatikov P 2007 *Solar energy utilisation opportunities in Bulgaria* in Solar Desalination for the 21st Century, Rizzuti *et al.* Ed. Springer pp 139-151.
[13] Technical University of Denmark ("DTU") https://globalwindatlas.info/
[14] Technical University of Denmark ("DTU") 2020 Global Wind Atlas 3.0, World Bank Group
[15] International Renewable Energy Agency 2019 Renewable Power Generation Costs in 2018 Abu Dhabi: IRENA
[16] Gospodinova D Milanov K Minchev M and Dineff P 2019 "Techno-economic feasibility analysis of nearly-zero hybrid energy system for the city of Sofia in Bulgaria," *Proc. of 11th Electrical Engineering Faculty Conference (BulEF 2019)*
[17] Kempener R and Borden E 2015Battery storage for renewables: market status and technology outlook, IRENA
[18] Jensen et al 2018 European Network for Research, Good Practice and Innovation for Sustainable Energy
[19] Center for the Study of Democracy (CSD) 2017 A Roadmap for the Development of the Bulgarian Electricity Sector within the EU Until 2050: Focus on Fundamentals Sofia
[20] National statistical institute 2020 Average salary Sofia (in Bulgarian)

Acknowledgments
The authors would like to thank the National Scientific Program NNP "Low Carbon Energy for Transport and Households - Eplus" D01-2014/28.11.2018 for the financial support.

Glow-discharge plasma enhanced coating and functionalization of needle-punched nonwoven PET filter media for textile wastewater decolorization

M P Neznakomova[1], D N Gospodinova[2] and P D Dineff[2]

[1] Department of Material Science and Technology of Materials", Technical University of Sofia, 8 Kliment Ohridski Blvd, 1756 Sofia, Bulgaria
[2] Department of Electrical Engineering, Technical University of Sofia, 8 Kliment Ohridski Blvd, 1756 Sofia, Bulgaria
Corresponding author: diliannag@abv.bg

Abstract. This study reports the use of glow-discharge plasma coated and functionalized PET needle-punched nonwovens as filter media for decolorization of textile dye-containing effluent. The investigation had been aimed at proving the possibility of GDP enhanced coating and functionalization (GDP processing) to produce better results for successfully replace such thermochemical finishing of PEI coat onto the primary PET fibers. The fiber surfaces functionalization with branched polymers such PEI has aroused much interest recently due to their unique properties as dye-immobilization absorber. For this study, the effect of processing time on the dye retention capacity of the functionalized PET nonwovens was investigated. Comparison was made between the most influential parameters of the basic, chemically and GDP processed PET nonwovens, such as: material thickness, air permeability, and puncture strength. The dye-retention capacity of the basic and functionalized nonwovens was determined by two synthetic water effluent – the anionic Orange G and the cationic Methylene Blue test-dyes have been used. The normal glow-discharge plasma processing in air as residual gas at 80 Pa provide the best possible anionic dye retention capacity. Based on this study, it was confirmed that this PEI coated and functionalized PET nonwoven could be used for the immobilization of anionic dyes from colored textile wastewater: the retention capacity from 4.16 to 5.57 mg/g with is comparable to the other available products used as adsorbent in dye-immobilization wastewater treatment.

1. Introduction

A large amount of water, chemicals, and dyes are used at textile wet processes, such as preparation, dyeing, finishing and laundry, with a high load of effluents, causing a high risk of water pollution. Dye removal has recently become of major scientific interest, as indicated by the multitude of related research reports. The increasing occurrence of many synthetic dyes in natural water led to the importance of using adsorption technique as one of the most effective methods of removing dyes from wastewater. Nonwoven filter media designing and developing are capable to improve the adsorption process with very high degree of filtration efficiency [1].

1.1. Nonwoven and functionalized nonwoven filter media

Functionalized nonwoven based filter media were designed to solve this problem. The usage of nonwoven filter fabrics varies according to their end use: they can be manufactured, with different parameters, concerning the nature of material, fiber denier, weight, and needle-punching manufacturing adjustment. The PET fibers have a very low surface free energy and for that reason they have non-polar surface. Their application as dye filter materials was limited due to their lack of a chemical functional group which are averse to applications, such as liquid filtration and dye immobilization. Polyethyleneimine (PEI) is one of the most prominent examples of cationic polymers. Polyethyleneimine coated nonwoven fibers are important for achieving selective adsorption and immobilization of anionic dyes. It's well known that epichlorohydrin (ECH) is a versatile precursor in a thermal finishing synthesis of branched PEI (bPEI) fiber coatings. Radical graft-polymerization of PEI impregnated over porous materials is an effective method for surface modification of inert and hydrophobic nonwovens [2, 3]. The functional PET fiber surfaces and polymers with branched structures such PEI have aroused much interest recently due to their unique properties and prominent applications in various fields of dye filtration and immobilization [4].

1.2. Glow-discharge plasma enhanced coating and functionalization

A. PEI could be effectively bound on the PET-fiber surface after a treatment in cold (non-equilibrium, non-thermal) glow discharge plasma [4]. Plasma-chemical activation introduces new functional groups to the surface. Free radicals are generated in these cold plasma surface treatments inducing a secondary reaction like cross-linking. Depending on the type of chemical groups plasma-chemical activation was used to obtain the desired new properties. A PEI branched monomer which is normally a liquid precursor undergoes a plasma polymerization on the surface. So, PEI could be effectively immobilized on various polymers pretreated by oxygen or air plasma [5]. By generating additional function after both PEI thermochemical and GDP coating and functionalization new material properties can achieve. The combination of both PEI polymer impregnation and GDP graft-polymerization lead to new surface properties of the fibers. In this way GDP processing leads to the plasma-chemical deposition of PEI coating with desired functional group on fiber surfaces. The used monomers (PEI and ECH) include functional groups and radicals such as hydroperoxyl anion radical (-COOH); amino group ($-NH_2$), hydroxyl radical (-OH), epoxide group, etc. proper for dye wastewater immobilization.

B. GDP functionalization (activation) of PET nonwovens can be performed with various gases - argon, nitrogen, oxygen or residual air [2]. Based on the particle density, temperature, pressure, presence or absence of electromagnetic fields, different cold plasma state can be distinguished. The effect of GDP treatment on PET fibres leads to chemical, structural, and physical changes in the surface layers. In this study, we found that bPEI could be effectively immobilized on some PET nonwoven fibers after glow-discharge (air, oxygen) plasma processing of impregnated with PEI/EPI needle-punched nonwovens – GDP enhanced coating and functionalization. At the same time, those of the PET fibers of the non-woven fabric, which are not coated, undergo GDP functionalization (activation). Large complex of ions on the PET non-coated surface appears after an oxidative GDP functionalization (activation, radical and functional group formation) of PET phenyl rings and polyethylene chains of mainly functional group formation: carboxyl (-C(=O)-OH) and carbonyl (C=O) groups, phenolic and alcoholic hydroxyl groups, and hydroperoxyl group (C-OOH) [6]. The GDP processing covers simultaneously as processing: *i)* the GDP enhanced PEI/ECH coating; *ii)* both the new PEI/ECH coating and bare PET surface functionalization. In general, there is a GDP enhanced processing of PET needle-punched nonwoven filter media for anionic and cationic dye immobilization and textile wastewater decolorization.

The effect of GDP processing on PET nonwoven filter media is schematically shown in Figure 1.

The main idea of this paper was based on the concept that the anion/cation active dye will be immobilized on a functionalized filtration medium after the textile effluent containing the residual amount of dye comes into contact with the functionalized part of the PET fiber surface: *a)* the epoxide anion/cation active coating on the PET coating surface; *b)* the GDP chemically activated bare PET surface. Due to the newly formed PEI active centers, amino groups ($-NH_2$), in combination with GDP functionalization of bare PET and the occurrence hydroxyl groups (-OH), the retention of anionic and cationic dye residues can be expected. In addition, the regeneration of a nonwoven filter by changing the pH of the nonwoven filter medium can be expected, due to differences in filtration and dyeing conditions.

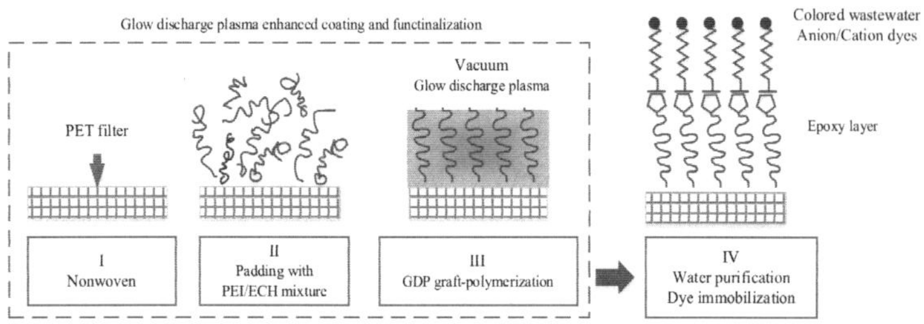

Figure 1. Colorants immobilization onto decolorization nonwoven PET filter media.
I, **II**, and **III** – GDP processing of basic PET nonwoven; **IV** – colorant's immobilization.

2. Materials and methods

2.1. Basic PET needle-punched nonwoven medium

The experimental researches were carried out using PET (polyester) needle-punched nonwovens (*NonWoTex* Ltd., Bulgaria) with *surface density* (mass per unit area) of 500 g/m^2; made by PET filaments with *linear density* (mass per unit length) – 6.7 dtex, and *fiber length* 57 mm. The main parameters of the basic (untreated) PET nonwoven are shown in Table 1.

Table 1. Main parameters of basic (untreated) PET needle-punched nonwoven fabric

Parameter	Value	Unit
Specific surface mass, EN 965	500 ± 8 %	kg/m^2
Minimum average thickness at 2 Pa, EN ISO 5084	4.0 + 0.4	mm
Bulk density, EN ISO 9073-2	125	kg/m^3
Tensile strength, EN ISO 13 934-1	18.1	kN/m
Air permeability, EN ISO 9587	0.21	$m^3/(m^2.s)$
Punch density	260	$needles/cm^2$

2.2. Glow-discharge plasma coating and functionalization

The basic nonwoven sample (BS) was filled and cover (impregnated) on foulard at *load of padder* 43, 68, 91, and 113 kg with a mixture of PEI with *molar mass* 40.04 (repeat unit) and ECH with *molar mass* 92.52 g/mol. The basic samples were cut to a size of 200 x 200 ± 2 mm. After drying at 110°C for 90 min, the impregnated PET samples were treated in the cathode area of GDP at different pressure: 10, 20, 40, 80, and 100 Pa and time: 2.5; 5.0; 7.5 and 10 min. Different glow DC discharge modes – sub-normal (SGD) and normal (NGD) glow, were carried out by varying the residual gas (air) pressure, Figure 2. Comparison was made between the GDP-processed (PFS) and thermochemically processed (CFS) samples. The CFS samples went through the following operations: PEI/ECH mixture padding, drying (105°C), high-temperature condensation (160°C), and ultrasonic washing at about 40°C.

The PET nonwovens changes after GDP processing were determined by measuring the *moisture absorption*, *mass* and thickness *changing*, *puncture resistance* (EN ISO 12236:1999) and *air permeability* (Metrimex). *Retention efficiency* of basic, GDP- and thermochemically processed samples was determined by UV/VIS-spectrophotometry (Shimadzu) with two - cation and anion, synthetic test dyes: *i) Methylene Blue*: $C_{16}H_{18}ClN_3S$, MG 319.85, pH 8 and *ii) Orange G*: $C_{16}H_{10}N_2Na_2O_7S_2$, MG 452.38, pH 3. *Methylene blue* is a synthetic basic **dye.** *Orange G* also called *Acid Orange G* is a synthetic mono-azo dye.

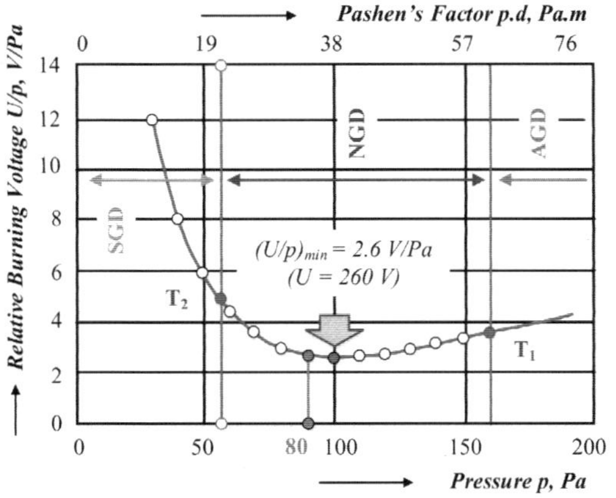

Figure 2. Transitions T_1 and T_2 between subnormal high-voltage (SGD), normal (NGD) and abnormal (AGD) DC glow discharge, are submitted by the change in relative burning voltage *U/p*, V/Pa, and *Pashen's factor p.d*, Pa.m, with the pressure *p*, Pa. The operational gap *g* between the coaxial cylindrical anode and cathode was 0.380 m in size. The inner electrode was a cathode with diameter *D* = 0.825 m. The outer electrode was an anode with diameter *d* = 0.063 m or *D*/*d* = 13.

3. Results and analysis

It was well known that PET fibers have very low surface free energy due to their non-polar surface. Therefore, PET fibers wet slightly, but this is not good for the water filtration. The PET fiber surface free energy was increased by two independent process: the first is the thermochemically bPEI processing and the second - the GDP-enhanced bPEI processing. Functionalized PET nonwovens become the ideal dye filter media. An bPEI nanolayer was formed on the PET surface, which alters the dye retention efficiency. In addition, improvement mechanical bonding and dimensional stability in use of the nonwovens were achieved. A very important feature is that the PET-fiber surface obtains an effective dye adsorption. The application of chemically bounded bPEI layer gives such an opportunity.

Switching PET surface from hydrophobicity to hydrophilicity by GDP processing.

The *moisture absorption capacity B* was determined to select the most effective GDP processing mode Figure 3.

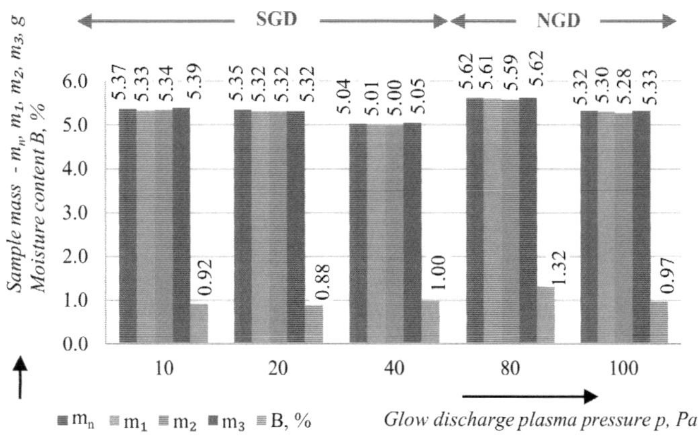

Figure 3. Nonwoven filter media with PEI coated and bare PET fiber surfaces switched by GDP processing from hydrophobicity to hydrophilicity are useful for die immobilization applications - the surface hydrophobicity reduction is illustrated very well by the observed increase of moisture absorption capacity *B*, %, depending on the GDP pressure *p*, Pa, where: m_n - mass of the basic sample; m_1 - mass of the PFS sample two hours after GDP processing; m_2 - mass of a PFS sample after one cycle humidification at 100 % humidity for 24 hours and drying; m_3 - mass of a PFS sample after two cycles of humidification and drying. NGD – normal GDP processing; SGD – subnormal GDP processing.

Samples with a diameter of 30 mm were processed. The GDP time was kept constant and equal to 25 minutes. Sample of basic PET material (m_n) was GDP processed (m_1) and then placed in a desiccator at 100 % humidity for 24 hours and after that was drying at 110°C (m_2) and again the humidification and drying procedure was repeated - sample (m_3). As the increased moisture absorption capacity *B* is a measure of the fiber surface hydrophobicity reduction after GDP processing and the ability to undergo chemical modification – the GDP processing mode at 80 Pa was selected for use. The moisture absorption capacity *B* does not change significantly after moisturizing for 48 hours at 100 % humidity, Figure 3. It can be seen also that after GDP processing, the mass of the treated nonwovens decreases as a result of fibers surface cleaning from different contamination and substances like finishing agents, lubricants, antistatic agents, etc. The moisture absorption capacity maximum coincides with the maximum of mass increase of PET sample at NGD pressure of 80 Pa, Figure 3.

3.1. PEI/EPC liquid mixture padding process

After partly filling the porous medium with PEI/EPC liquid mixture a squeeze processing mode was selected by determining the load on the upper foulard shaft of foulard machine. With less squeeze load, a part of the liquid mixture outflow. The amount of applied film depends on the degree of squeezing. PEI/EPC padding was formed on the fiber surface, Figure 4. The increase in mass *ΔM* of PET basic sample filling with the PEI/ECH liquid mixture before and after thermochemical processing are shown in Figure 4. Under these conditions, the thermochemical processing (condensation polymerization) was carried out at 160°C, after drying in advance at 105°C. This dried PEI/ECH padding was chemically active due to the presence of the amino groups (-NH_2) but after a thermal finishing in the presence of epichlorohydrin a solid insoluble PEI coating was formed. It changes PET surface morphology. The nonwoven becomes more rigid, stabilized in size and increased in puncture strength. These changes were determined by the amount of dry matter present in the porous volume. This amount depends on the degree of mechanically filled and equalized PEI/ECH liquid mixture over the whole height of nonwoven basic sample determined by the load on the upper shaft of squeezing padder (foulard machine). The maximum increase in mass *ΔM* of basic sample after thermochemical processing was implemented at the load of 8.85 kg on the upper shaft of squeezing padder, Figure 4.

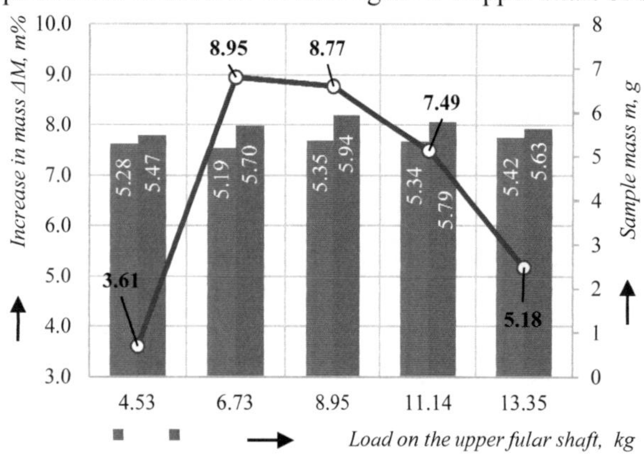

Figure 4. Increase in mass ΔM of nonwoven needle-punched PET basic sample (in blue) after thermochemical processing (in orange) with a liquid PEI/ECH liquid mixture and heat treatment during the following processing: drying at 105°C, condensation polymerisation at 160°C, and ultrasonic washing at about 40°C.

3.2. Glow discharge plasma processing of PET fiber surfaces

Bi-component PET fibers were formed, with bare PET surface and PEI coated PET surfaces. The area ratio of these two PET surfaces depends on the squeeze load magnitude of foulard machine. The area ratio of these two fiber PET surfaces depends on the squeeze machine load. Both PET surfaces acquire different ionic activity due to their functionalization and the appearance of amino (-NH_2) and hydroxyl (-OH) groups. Some of the characteristics that can influence on the removal of textile dye waste are: *adsorption capacity*, *specific surface area*, and pore volume, expressed by *air permeability*. They are so important because if they change the effectiveness of removal also changes [7].

The thickness of GDP processed nonwovens increases and the material becomes larger as a result of the new active centers formation with the same electrical load and their mutual repulsion. After the filling and covering of the nonwoven porous media with PEI/ECH mixture under the squeezing load and the formation of a PEI coating the thickness of the PET nonwoven samples decreases Figure 5a.

The changes of the nonwoven's main characteristics after GDP processing (PFS) compared to the basic PET nonwoven (BS) and thermochemically processed nonwoven (CFS) characteristics are shown in

Figure 5: the increase in the *thickness* (***a***), and *puncture strength* (***b***), and the slight increase in the air permeability compared to CFS (***c***), are well visible.

The *puncture strength* increases compared to the thermochemically processed PET nonwoven in all the investigated cases, but decreses compared to the basic PET nonwoven with one exception at GDP pressure of 10 Pa - the highest puncture strength value was obtained, Figure 5a. This is probably due to the cross-linking process in the deposited layer, which is favoured by the applied GDP processing mode, in the area of the sub-normal glow discharge (SGD). After such SGD plasma functionalization at 10 Pa the basic (bare) PET nonwoven has the highest loss of mass. This means that the largest number of active centers were formed allowing the PEI/ECH layer to be crosslinked and a stable film was formed on the fiber surface. These circumstances leads also to the observed reduction in air permeability, Figure 5b. Probably the reason is the closure of some pores in the nonwoven volume.

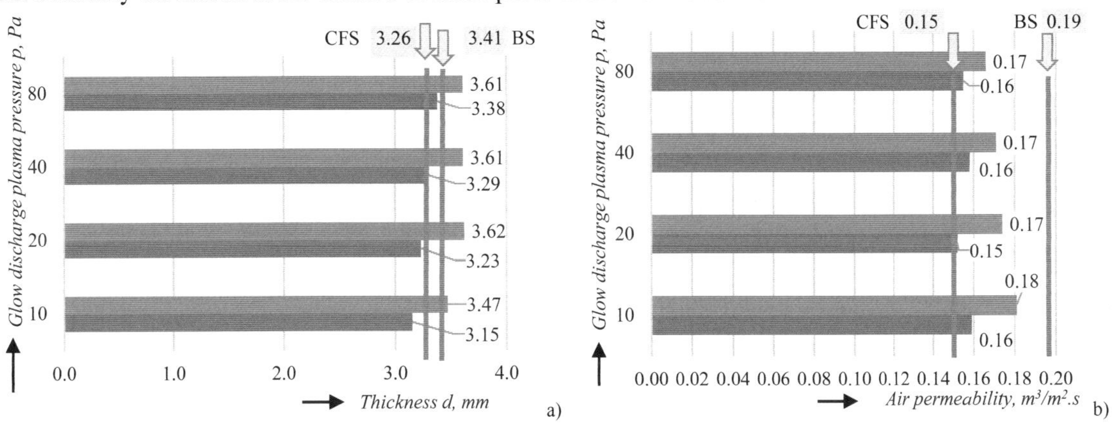

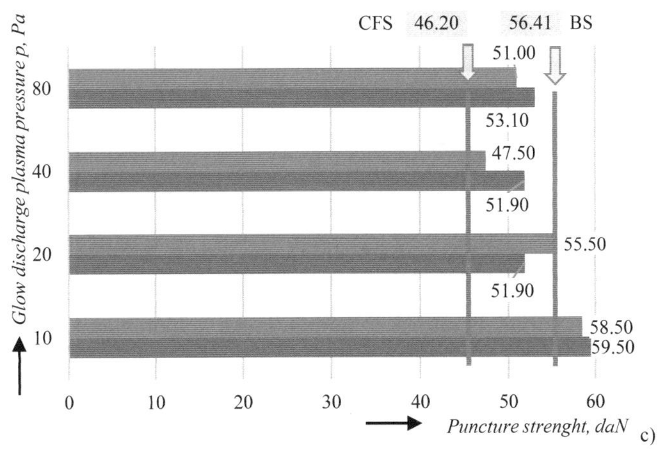

Figure 5. Changes of some basic characteristiques of the *PET* needle-punched nonwoven after *GDP* functionalization (orange columns) compared to *GDP* coating and functionalization (blue columns): ***a)*** thickness, mm; ***b)*** air permeability, $m^3/(m^2.sec)$; and ***c)*** puncture strenght, daN.
BS – basic PET sample; CFS – thermochemically processed sample.

The thickness of the coated and functionalized PET nonwoven decreases (blue columns), Figure 5c. The figure clearly illustrates the difference in PET nonwoven behaviour after GDP bare PET functionalisation and GDP coating and functionalization processing.

The results presented in Figure 6 refers to a single passage of two synthetic dye coloured (*Orange G* and *Methylene Blue*) effluents through a functional PET nonwoven filter. A comparative study was made between the retention anionic and cationic capacity of basic PET, thermochemically and GDP processing PET nonwovens. The basic and processed PET nonwovens are typical cation-functionalized filter media. The maximum anion dye retention capacity (*aRC*) sorts the three-filter media in the following order of increasing values *aRC*: 1.29 (BS); 1.99 (CFS) and 5.86 (PFS) mg/g. The maximum cation dye retention capacity (*cRC*) ranges the three-filter media in the following order of increasing values of *cRC*: 0.040 (BS); 0.044 (PFS) and 0.050 (CFS) mg/g, i.e. the maximum cation retention capacity of PET samples is incomparably smaller than their anion retention capacity: *aRC* >> *cRC* (*aRC*/*cRC* (80 Pa) = 5.76/0.035 = 164.6).

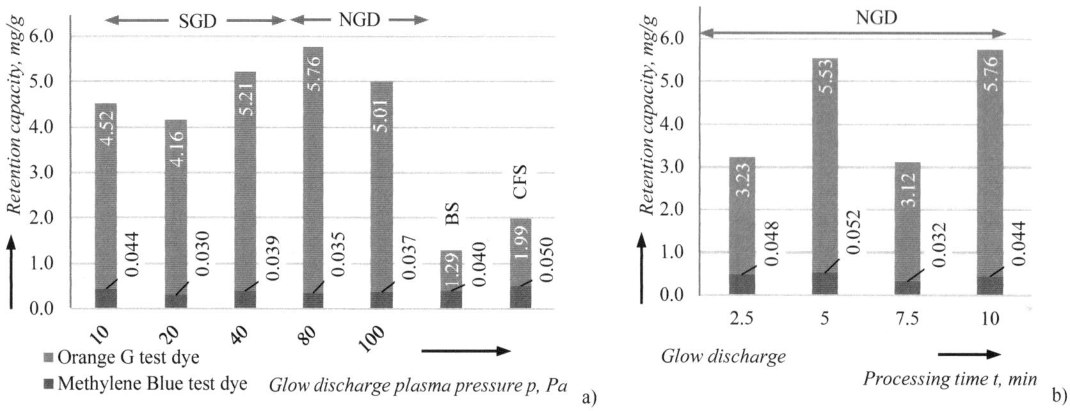

Figure 6. Relationship between the retention capacity RC of the basic PET nonwoven sample (BS), thermochemically (CFS) and GDP- (PFS) processed PET samples and the process factors: GDP-pressure p, Pa; and processing time t = 10 min = Const. The retention capacity was determined after the dye colored test water effluent passing through each sample once of two (anionic/cationic) test dyes – Orange G (orange columns), pH = 3, and Methylene blue (blue columns), pH= 8, onto the functionalized PET nonwoven surface.

The impact of processing time on the NGD plasma processing of PET nonwoven filter media from 2.5 to 10 min at constant in value NGD pressure (80 Pa) changes the results – the *aRC* changes from 3.12 to 5.76 mg/g. From a technological point of view, the smaller the processing time is acceptable regardless of a not so bad result: 5.53 mg/g (t = 5 min) < 5.76 (t = 10 min): 5.53/5.76 x 100 = 96 %. At the same time, the processing time is twice less. These results require further investigation to ensure higher performance of the GDP enhanced coating and functionalization process.
The *aRC* value of SGD plasma treated PET-nonwovens varies from 4.16 to 5.21 mg/g with pressure change from 10 to 40 Pa. The *aRC* value of NGD plasma treated PET-nonwovens varies from 5.01 to 5.76 mg/g with pressure change from 80 to 100 Pa, Figure 6 a. These results for *aRC* are closed to the well-known values of other filter media – 5.5÷5.7 mg/g, [4, 5, 8].

4. Conclusions

Glow discharge plasma enhanced processing (coating and functionalization) was used as a finishing for polyester (PET) nonwoven needle-punched filter media. The aim of this research was to find the possibilities of effective polyester non-woven coating and functionalization in order to manipulate the cation and anion immobilization activities relevant for textile wastewater treatment engineering. It is known that textile dye containing wastewater treatment is a process used to remove contaminants from wastewater and convert it into an effluent that can be returned to the water cycle with minimum impact on the environment:
(1) Glow discharge air (oxidative) plasma enhanced coating and functionalization is an universal tool for bPEI coating and functionalization of polyester filter media. It is a green or environmentally friendly technology for functionalization which refers the improvement in hydrophilicity, wettability, wicking, and ion immobilization activity for textile dye containing wastewater treatment.
(2) This study reports the use of GDP processed needle-punched PET nonwovens as filter media for decolorization of textile dye-containing wastewater. It proves the possibility of GDP enhanced coating to produce better results for successfully replace such thermochemical finishing of bPEI surface functionalized coating onto the primary PET fibers. Replacing the thermochemical process (at 160°C) energy consumption reduces. The low production costs, combined with higher dye retention capacity, make this GDP enhanced processing economical and effective. The PEI coated PET needle-punched nonwoven filter media with increased anionic dye (*Orange G* anionic synthetic water solution) retention capacity of 5.3÷5.76 mg/g is comparable to such filter available products.
(3) The GDP enhanced processing of PET nonwovens produces bifunctional PET fibers since the coating process allows partial surface coverage while leaving bare PET surface: *i)* branched PEI surface

functionalized (-NH_2) coating onto the primary PET fibers, and *ii)* plasma functionalized (-OH; -OOH) bare PET fiber surface. GDP enhanced bare PET functionalization is the first used surface functionalization effect at small processing times. After that the prolonged (5÷10 minutes) GDP activation causes further etching resulting in the saturation of bare PET surface with functional groups. Polyester plasma functionalization does not produce one unique functionality - usually a distribution of different new functional groups and ion exchangers was produced – C-OH, C-O-OH, C-OH, C=O, O=C-O-OH, C-O^-, O=C-O^-, C-H^+, C-H_2O^+, etc. The same glow discharge plasma functionalization procedure produces different concentration of functional groups and ions depending on the used glow regime – sub-normal (SGD) or normal glow (NGD) discharges plasma processing. Plasma activated bare parts of the fiber surface with its immediately increased functional groups and ion exchange capacity become also friendly to the soluble dye containing textile wastewater for physical immobilization on the nonwoven filter media.

References

[1] Abd El-Latif MM Ibrahim AM El-Kady MF 2010 Adsorption Equilibrium, Kinetics and Thermodynamics of Methylene Blue from Aqueous Solutions using Date Palm Leaves *J of American Science* **6** 6 pp. 367-283

[2] Neznakomova M Klotz M-L and Gospodinova D 2019 Non-woven composites intensification properties for air filters by plasma pre-treatment *IRMES 2019* (Kragujevac, Serbia: IOP Science)

[3] Neznakomova M Van Langenhove L and Gospodinova D 2018 Capillar Activity of Non-woven Needle Felted Filters as a Measure of Their Degree of Modification *10th Electrical Engineering Faculty Conference (BulEF)* (Sozopol, Bulgaria: IEEE)

[4] Zhu B Iwata H Hirata I Ikada Y 2000 Hydrophilic modification of a polyimide film surface *J of Adhesion Science and Technology* **14** 3 pp. 351-361

[5] Gospodinova D Dineff P Neznakomova M 2019 Plasma - Aided Modification of Nonwoven Filters for Wastewater Treatment *11th Electrical Engineering Faculty Conference (BulEF)* (Varna, Bulgaria: IEEE)

[6] Al-fawwaz, AT Abdullah M 2016 Decolorization of Methylene Blue and Malachite Green by Immobilized Desmodesmus sp. Isolated from North Jordan *Int. J of Environmental Science and Development* **7** 2 pp. 95-99

[7] Pollard SJT Fowler GD Sollars CJ and Perry R 1992 Low-cost adsorbents for waste and wastewater treatment: a review *Science of The Total Environment* **116** 1-2, pp. 31-52

[8] Hao T 2008 Viscosities of liquids, colloidal suspensions, and polymeric systems under zero or non-zero electric field *Advances in Colloid and Interface Science* **142** 1-2, pp. 1-19

Acknowledgments

The authors would like to thank the Research and Development Sector at the Technical University of Sofia for the financial support.

Study of four-spring electromagnetic harvesters

N P Georgiev[1] **and R P Raychev**[2]

[1]Department of Electrical Engineering, Technical University of Sofia, Branch Plovdiv, 25 Tsanko Dystabanov Street, Plovdiv, Bulgaria
[2]Department of Mechanical Engineering, Technical University of Sofia, Branch Plovdiv, 25 Tsanko Dystabanov Street, Plovdiv, Bulgaria

Abstract. The article considers electromagnetic harvesters, consisting of a metal plate, fixed on four springs at their lower end, and having 2 or 4 magnets. A coil is located above the magnets. Modeling by ANSYS has been done, and the horizontal deviation of the mechanical system has been obtained. With the help of FEMM, the magnetic field of the electromagnetic harvesters has been modeled and their electromotive force has been determined. Experimental studies have been performed to verify the modeling. The influence of the construction parameters on the output electric features of the studied harvesters has been considered.

1. Introduction

Sources, converting mechanical energy into electrical, such as the electromagnetic harvesters, are increasingly replacing the battery-type low power consuming power-supplying electronic devices [1, 2]. There are usually two types of magnet motion in relation to the coil in these harvesters. In the first case, the magnet moves in parallel to the coil [3] and in the second one - vertically with respect to it, often being located in its air gap [4].

Three different structures of four-spring electromagnetic harvesters are considered here, with two different masses, represented by steel plates with 2 or 4 magnets each and two different coils as well. The harvesters have a fixed coil attached to them, in parallel to which the rare earth magnets move. The aim of the present work is to study the effect of the construction parameters (such as the weight of the concentrated mass, the number of turns, the influence of the coil area, the volume of the magnets, and the size of the air gap) on the output electrical parameters of the studied harvesters.

2. Exposition

Figure 1, figure 2 and figure 3 illustrate the construction of the three four-spring electromagnetic harvesters. The following notations are used in the figures: 1 - coil; 2- rare earth magnets, 3- upper base, 4- steel plate, 5- steel springs, 6- lower base and 7 metal fastening. Springs with wire diameter ds = 0.63 mm, spring diameter Ds = 4.63 mm and length ls = 20 mm are used. The steel plates are the same size and with 2 or 4 fixed permanent NdFeB37 magnets, measuring 20x10x2 mm. The first harvester has two magnets and weighs m_1=15 grams - Figure 1; the second is with 2 magnets, spaced from each other by 8 mm - Figure 2; and the third one weighs m2 = 30 grams and has two magnets in two places, spaced from each other by a distance of 8 mm - Figure 3. The coils under study have diameters Dc1 = 20mm and Dc1 = 40mm, thickness lc = 4mm and N1 = 600 and N2 = 1200 turns with conductor diameter dc = 0.1 mm.

IOP Conf. Series: Materials Science and Engineering **878** (2020) 012018 doi:10.1088/1757-899X/878/1/012018

When applying a sinusoidally varying force with a resonant frequency, the mechanical system: "mass (plate with permanent magnets) – springs" begins to oscillate in parallel with the coil. Thus the magnetic flux passes differently through the coil and creates an alternating magnetic field in a different way, thus inducing alternating electromotive force.

The studied harvesters are nonlinear mechanical oscillating systems. Their simulations made by ANSYS R19.1 take into account the fixture, the gravity effect of the plates with permanent magnets and the mechanical characteristics of the used springs. The horizontal deviation x of the mechanical system "mass (plate with permanent magnets) – springs" was obtained while modeling the four-spring electromagnetic harvesters, Figure 4.

The magnetic field distribution of the three electromagnetic harvesters was obtained by means of FEMM 4.2. Figure 5 shows the magnetic field distribution for the third four-spring harvester with two spaced magnets in two places at zero horizontal deflection, and Figure 7 presents the distribution at maximum deviation. Figure 6 and figure 8 illustrate the magnetic flux density changes along the length of the harvester coil at zero and maximum horizontal deviation. From Figure 6 it can be seen that at zero horizontal deviation the normal magnetic flux density is zero, and at maximum deviation the maximum magnetic flux density B_{max} is obtained.

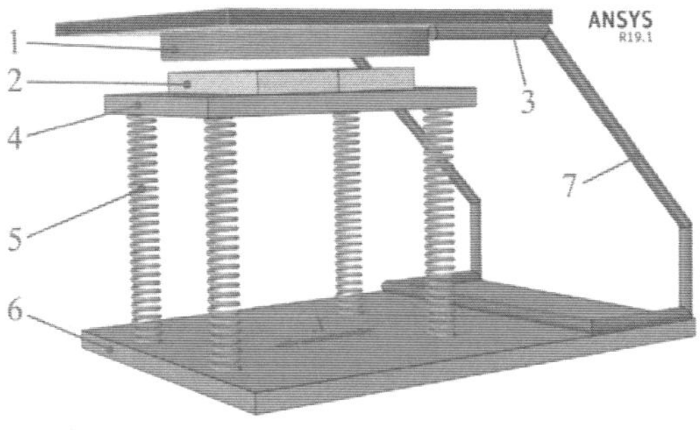

Figure 1. Four-spring harvester with two magnets.

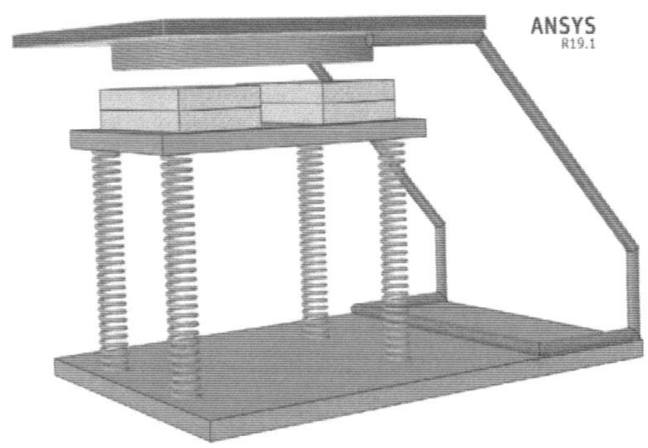

Figure 3. Four-spring harvester with two by two spaced magnets.

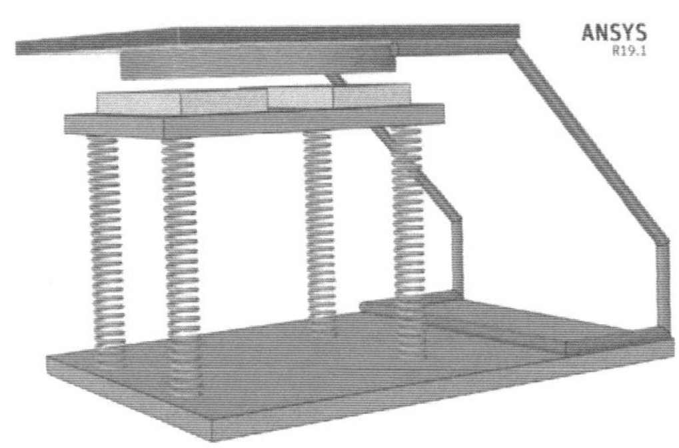

Figure 2. Four-spring harvester with two spaced magnets.

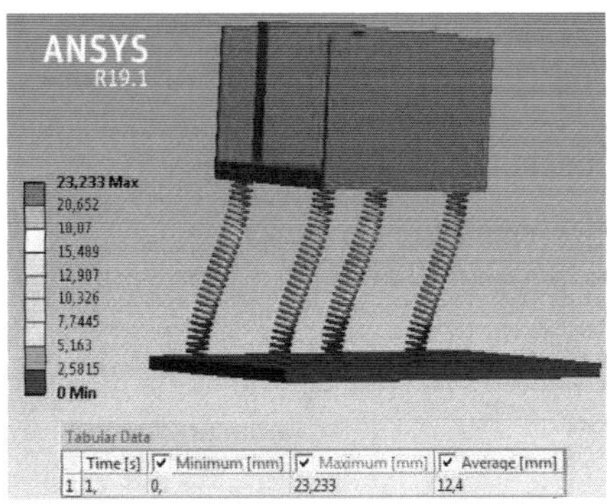

Figure 4. Deviation of the mechanical system.

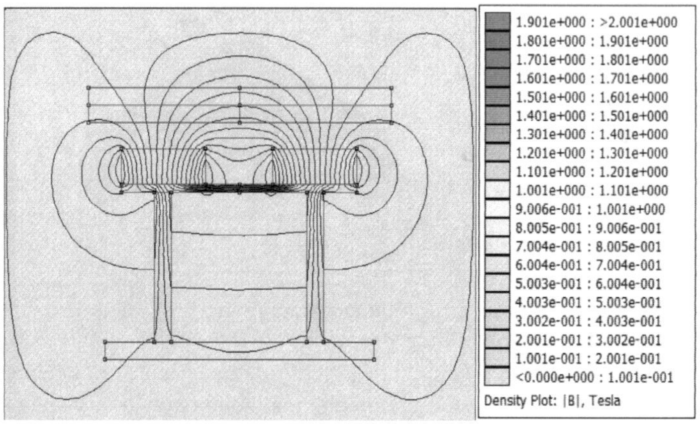

Figure 5. Magnetic field distribution for a four-spring harvester with two by two spaced magnets at zero horizontal deviation.

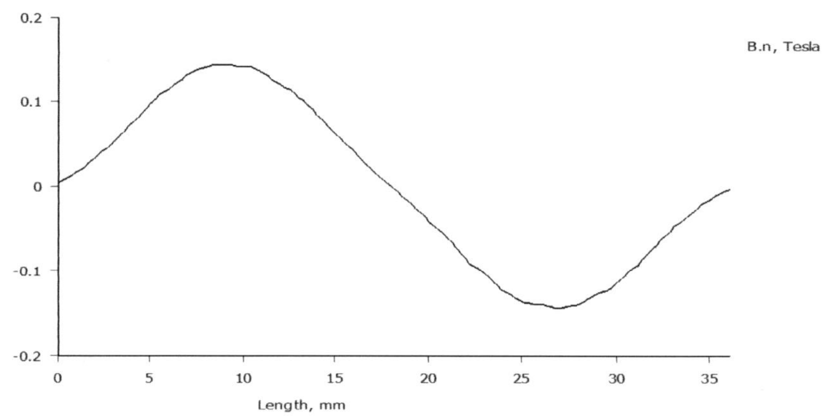

Figure 6. Magnetic flux density along the length of the harvester coil at zero horizontal deviation.

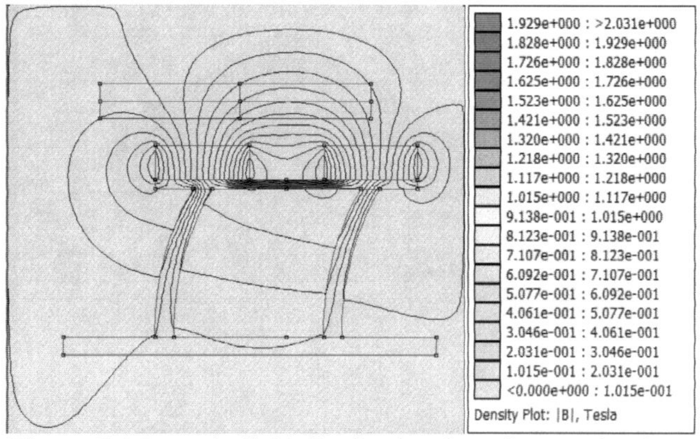

Figure 7. Magnetic field distribution for a four-spring harvester with two by two spaced magnets at a maximum horizontal deviation.

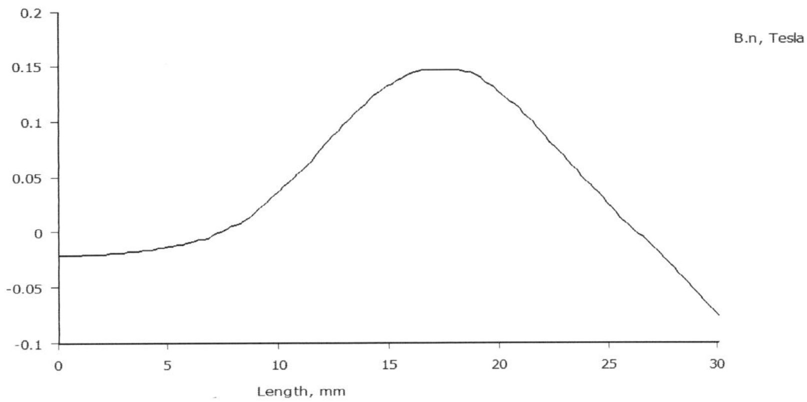

Figure 8. Magnetic flux density along the length of the harvester coil at maximum horizontal deviation.

The horizontal deviation at a sinusoidally changing force with a resonant angular frequency ω is also sinusoidal

$$x(t) = X_m \sin \omega t \tag{1}$$

The horizontal oscillatory speed is equal to

$$v(t) = \frac{d\,x(t)}{d\,t} = \omega X_m \sin\left(\omega t + \frac{\pi}{2}\right) \tag{2}$$

The magnetic flux can be expressed by the change in the cross section of the coil *A(t)*, through which the maximum magnetic flux density B_{max} passes

$$\Phi(t) = B_{max} \frac{dA(t)}{dt} \tag{3}$$

The change in the cross section *A(t)* over time equals the horizontal deviation variation in the time *x(t)* along the diameter D_c of the corresponding coil.

$$\frac{dA(t)}{dt} = D_c\, x(t) \tag{4}$$

From (3) and (4), the magnetic flux through the coil is obtained

$$\Phi(t) = B_{max}\, D_c\, x(t) \tag{5}$$

The induced electromotive force in the coil of the electromagnetic harvester is

$$e(t) = -N \frac{d\Phi(t)}{dt} \tag{6}$$

From (5) and (6) for the induced electromotive force for no-load mode, it is obtained

$$e(t) = -N\, B_{max}\, D_c \frac{d\,x(t)}{dt} \tag{7}$$

The amplitude of the induced electromotive force in the coil is a function of the amplitude of the oscillatory speed ωX_m

$$E_m = N\, B_{max}\, D_c\, \omega X_m \tag{8}$$

The angular frequency of the forced oscillations can be expressed by the frequency *f*

$$\omega = 2\pi f \tag{9}$$

From (8) and (9) for the amplitude of the induced electromotive force in the coil at a resonance it is obtained

$$E_m = N B_{max} D_c 2\pi f X_m \tag{10}$$

Thus, using the magnetic flux density calculated by means of FEMM 4.2 and the given resonant frequency, number of turns, coil diameter and horizontal deviation, the amplitude of the induced electromotive force in the coil for no-load mode can be calculated for the corresponding resonant frequency.

The r.m.s. value of the induced no-load electromotive force is equal to

$$E = \frac{N B_{max} D_c 2\pi f X_m}{\sqrt{2}} \tag{11}$$

Figure 9 shows the equivalent circuit of an electromagnetic harvester with voltage doubling at active load. In it, R_c and L_c denote the active resistance and the inductance of the electromagnetic harvester coil, $e_c(t)$ - the induced electromotive force, and R_L is the active load resistance. U_L indicates the rectified voltage over the load resistance.

The active power, in DC mode, is calculated using the amplitude of the induced no-load electromotive force in the coil at resonance, the voltage on the germanium diode U_D and the parameters of the equivalent circuit (14).

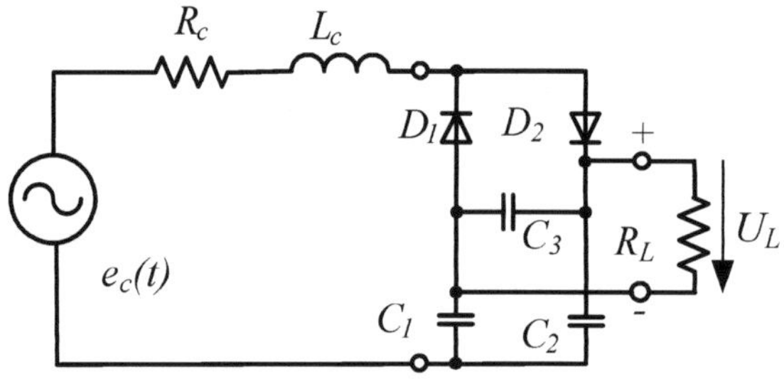

Figure 9. Equivalent circuit.

$$P_L = \left(1,41 \left(\frac{E_m R_L}{\sqrt{\left(\frac{R_C}{b} + R_L \right)^2 + (\omega L_C)^2}} - U_D \right) \right)^2 \cdot \frac{1}{R_L}, \tag{12}$$

In the resulting expression, b denotes the attenuation coefficient, which is determined by the logarithmic attenuation decrement δ [5], where T is the oscillation periodic time and $E_m(t)$ is the amplitude of the measured electromotive force

$$b = 2m\delta \ , \qquad \delta = \frac{1}{T} \ln \frac{E_m(t)}{E_m(t+T)} \tag{13}$$

3. Experimental studies

Figure 10 presents the no-load rectified voltage characteristics with respect to the frequency for the first harvester with two magnets, adjacent to each other and weighing m = 15 grams (denoted as m = 15g 0mm), the second with the two spaced by 8mm magnets (denoted as m = 15g 8mm) and the third one, having two spaced by 8mm magnets in two places, weighing m_2= 30 grams (referred to as m = 30g 8mm).

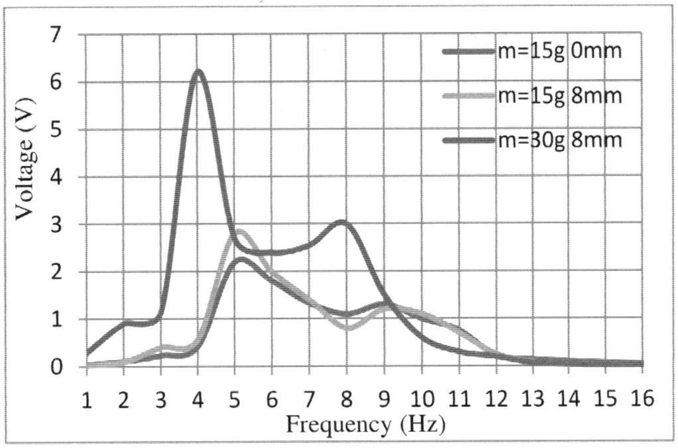

Figure 10. Rectified voltage-frequency characteristics.

From the characteristics it can be seen that as the mass increases, the resonant frequency of the harvesters decreases, while the maximum no-load voltage increases. Both harvesters with equal masses of 15 g each have the same resonance frequencies, and the maximum no-load voltage for the second harvester with spaced magnets is greater by 27% than that of the third harvester with adjacent magnets.

Figure 11 shows the effect of the number of turns on the rectified voltage at load resistances in the range from 0 to 200 kΩ for a two-magnet harvester weighing m = 15 grams. The measured values at 1200 turns are twice as high as those at 600, as it can be seen from the expression for the induced electromotive force (11).

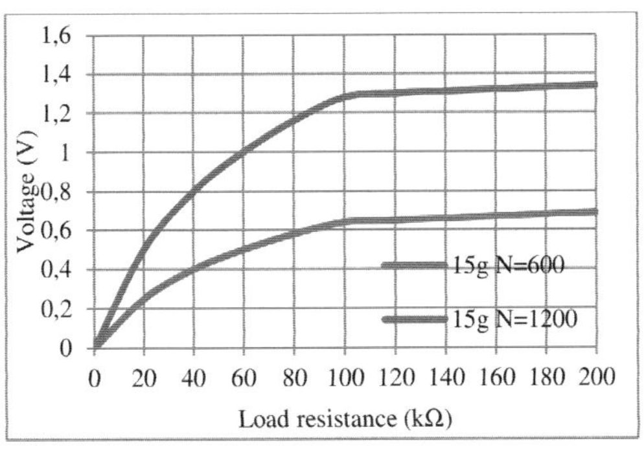

Figure 11. Rectified voltage for a harvester with two magnets, weighing m=15 g.

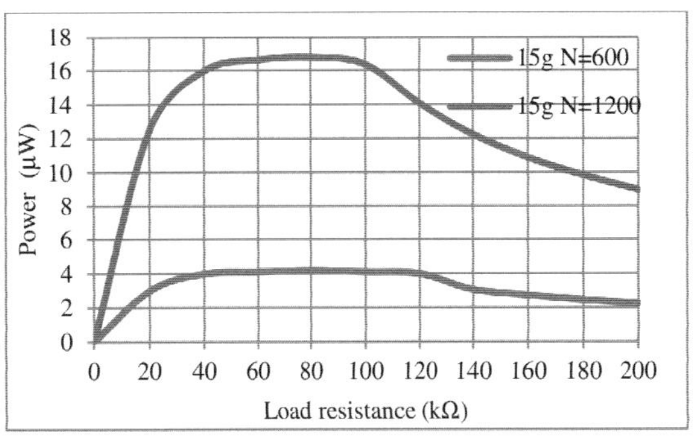

Figure 12. Active power for a harvester with two magnets weighing m=15 g.

Figure 12 shows the effect of the number of turns on the active power. The measured values at 1200 turns are four times as large as those at 600 turns, as it can be seen from the expression (14). The effect of the coil area on the rectified voltage at different load resistances is shown in figure 13. Measured values at 1260 mm^2 are four times larger than those at 315 mm^2, as it can be seen from the expression (10).

IOP Conf. Series: Materials Science and Engineering **878** (2020) 012018 doi:10.1088/1757-899X/878/1/012018

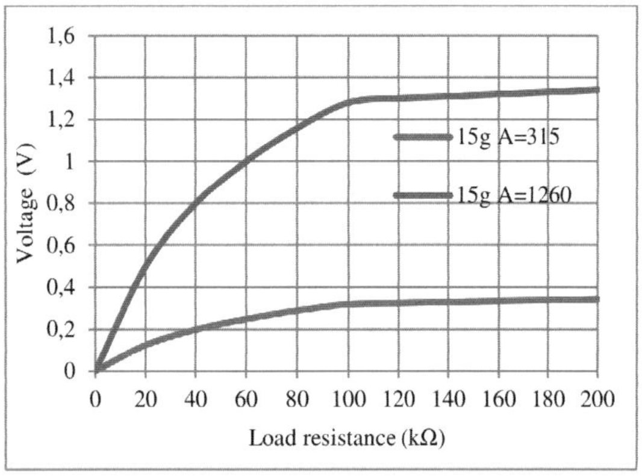

Figure 13. Rectified voltage for a harvester with two magnets.

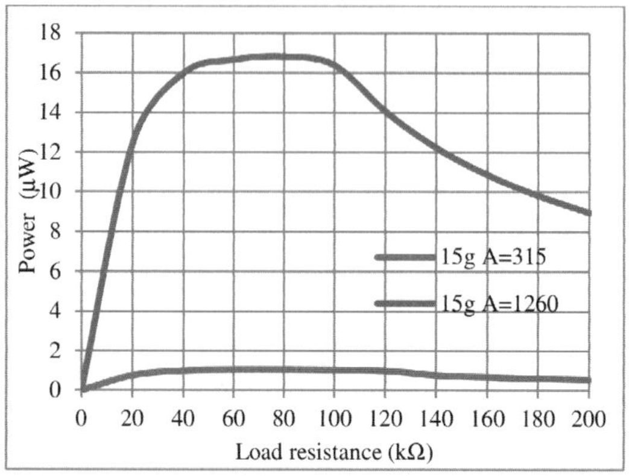

Figure 14. Power for a harvester with two magnets.

Figure 14 illustrates the effect of the coil surface area on the active power. The measured values for an area of 1260 mm^2 are 16 times greater than those for 315 mm^2, as it can be seen from the expressions (10) and (12).

Figure 15 shows the effect of the number of magnets on the rectified voltage at different load resistances. The measured values for the case of two magnets in two places are slightly more than twice as large as for a two-magnet harvester, as it can be seen from the expression (10). This increase of more than 2 times is due to the fact that the air gap between the magnets and the coil has also been reduced.

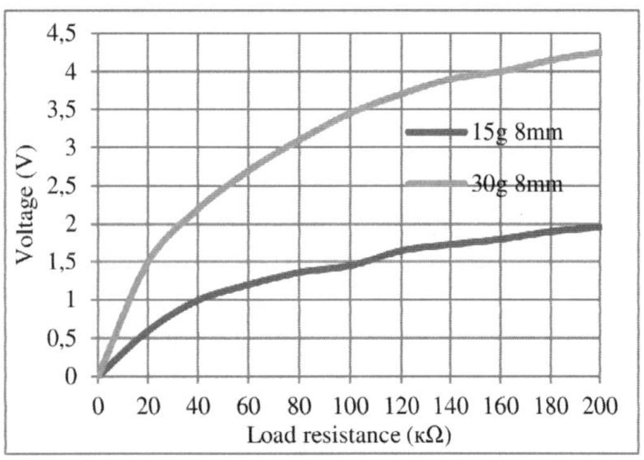

Figure 15. The rectified voltage at different load resistances.

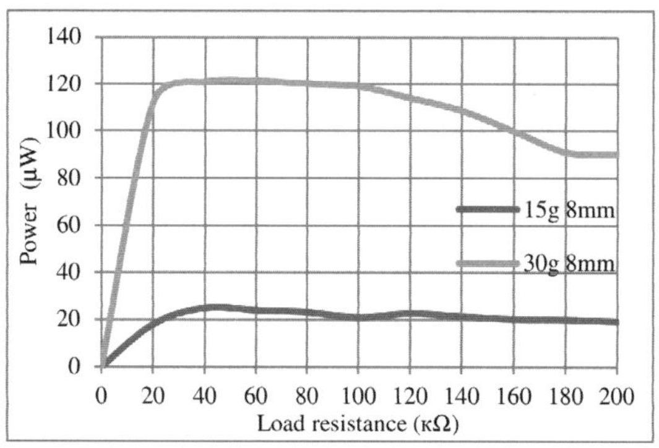

Figure 16. Active power at different load resistances.

The effect of the volume of the magnets on the active power is illustrated by figure 16. The measured values for the case of two magnets in two places are six times larger than for a harvester with two magnets, as it can be seen from the expressions (10) and (12).

Figure 17 shows the three measured powers for the three types of harvester structures and the results obtained from the simulation of the third harvester, having two magnets in two places.

From the theoretical and experimental studies, it can be concluded that by increasing the number of turns (from 600 to 1200), by increasing the area of the coil (from 315 to 1260 mm^2) and the distance between the magnets (from 0 to 8 mm), as well as by increasing the number of magnets from two to two magnets in two places, as well as by reducing the air gap from 3 to 1 mm, significant increase of the rectified voltages and active powers is achieved.

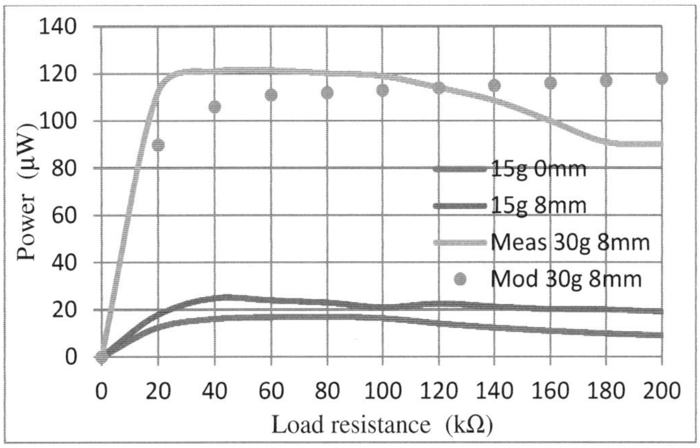

Figure 17. Measured powers for the three types of harvester structures and the results obtained from the simulation.

4.Conclusions

Theoretical deductions, program simulations and experimental studies of three electromagnetic harvester structures have been carried out.

Modeling has been performed (by means of ANSYS R19.1) of the four-spring electromagnetic harvesters and the horizontal maximum deviation X_m of the mechanical system has been found. The magnetic field distribution of the electromagnetic harvesters has been obtained using FEMM 4.2. A simulation has been made to express the induced electromotive force. The equivalent circuitry of the electromagnetic harvesters has been drawn up and an expression has been derived for the DC power at active load.

The frequency characteristics show that as the concentrated mass increases, the resonant frequencies of the harvesters decrease, while the rectified voltages increase.

From the theoretical and experimental studies it can be concluded that by increasing: the number of turns (from 600 to 1200), the area of the coil (from 315 to 1260 mm^2), the distance between the magnets (from 0 to 8 mm), the number of magnets from two to four, and by reduction of the air gap, the rectified voltages and the active powers increase significantly. The third type of construction of the four-spring electromagnetic harvester, having a coil with 1200 turns and 1260 mm^2 area, two spaced magnets in two places (measuring 20x10x2 mm), and an air gap smaller than 1 mm, is the most appropriate type of construction for practical use, due to its best electrical output parameters.

References

[1] Williams C, Shearwood M, Mellor P, Birch S and Yates R 2001 *IEE Proc. Circuits Devices Systems.* **148**

[2] Stephen N 2006 *Journal of Sound and Vibration.* **293**

[3] Saha C 2011 *Score Project, School of Electrical & Electronic Engineering University of Nottingham, NG7 2RD*

[4] Zhu D 2011 *Sustainable Energy Harvesting Technologies - Past, Present and Future. INTECH*

[5] www.fs.vsb.cz/export/sites/fs/330/.content/files/*technical_vibration_1*.pdf.

Acknowledgements

The authors would like to thank the Research and Development Sector at the Technical University of Sofia for the financial support.

IOP Conf. Series: Materials Science and Engineering **878** (2020) 012019 doi:10.1088/1757-899X/878/1/012019

Study of a linear generator with permanent magnets converting sea wave energy into electricity

N P Georgiev[1] and R P Raychev[2]

[1]Department of Electrical Engineering, Technical University of Sofia, Branch Plovdiv, 25 Tsanko Dystabanov Street, Plovdiv, Bulgaria
[2]Department of Mechanical Engineering, Technical University of Sofia, Branch Plovdiv, 25 Tsanko Dystabanov Street, Plovdiv, Bulgaria

Abstract. This paper studies the output electrical parameters of a linear generator with permanent magnets without an iron core for three types of electrical connection of its windings. Software simulations have been performed by means of ANSYS 2019 R1 and FEMM, and experimental studies of the linear generator have been carried out. A model has been obtained, which simulates the operation of the generator under study with good level of precision. The optimal scheme of connecting the generator windings, which is suitable for practical realization, has been determined.

1. Introduction

In the last quarter of a century, there has been a growing interest in devices converting the energy of the sea and ocean waves into electric power. For direct conversion of the sea power into electricity, linear generators with permanent magnets (LGPMs) are often used, which can convert the reciprocation into electrical energy [1]. These generators are divided into two types: without and with an iron core in the windings. The LGPM without an iron core are smaller in size, simpler in construction, and cheaper; their output electrical parameters are lower as well, however [2,3]. The LGPMs with an iron core in the windings, due to the low magnetic resistance, have significantly better electrical parameters [4,5] but higher harmonics are obtained with them because of the availability of an iron core [2,4,5].

The aims of the present work are: to investigate the output electrical parameters of a linear iron generator with permanent magnets without an iron core for three schemes of connecting its windings; and to determine the optimal of the three schemes of connection, most appropriate for practical realization.

The paper studies and models LGPM, which can be used to convert the offshore sea wave energy into near-shore electricity. The proposed generator is a floating element - buoy 1 on wave 2, in which a linear generator 3 is placed. The buoy is secured with a rope 4 to a weight 5, located at the seafloor – Fig. 1. Most LGPM without am iron core in the windings are composed of one or more cylindrical magnets, moving in a non-metallic cylinder, around which several coils are placed. The main disadvantage of this construction is that the magnetic fluxes, created by the permanent magnets, are small, due to the fact that they pass both through the coils and through large sections of air with high magnetic resistance.

In the proposed linear generator, the coils are stationary and placed between movable magnets, which, in turn, are fixed to iron rails as seen in Fig. 2. In this construction, the air gap is significantly reduced and together with the thickness of the coils it forms a layer of only 16 mm. Thus, the magnetic resistances are significantly reduced. At a minimum air gap, the best possible electrical parameters of the LGPM are obtained, as in [6]. Thus the magnetic fluxes across the coils increase, and thence the induced voltages and the resulting power also increase.

Figure 1. Linear generator with permanent magnets.

2. Exposition

The paper studies the output electrical parameters of a linear generator with permanent magnets without an iron core for three different schemes of connecting its four windings.

The proposed LGPM without an iron core (Figure 2) consists of: 1- a movable base on which a movable rail 2 is fixed; two П-shaped aluminum plates 3; two steel plates 4, on each of which five cylindrical permanent magnets (NdFeB32) 5 with dimensions D20xH10 mm are placed. A fixed fiberglass board 7 is attached to the aluminum plate 6, where four coils 8 with a diameter $D`_c$=30 mm, thickness l_c=12 mm and N=2500 turns with a conductor diameter d_c=0,1 mm are located. The movable base 1 is secured to the steel axis 9, which is supported by two steel plates 10, mounted on the lower stationary base 11. To the axis of the motor 12 is mounted a gear-belt washer 13, which transmits the rotational motion of the gear belt 14. By 15 and 16 the magnets of the magnetic brake are denoted. When the DC motor for car wipers is switched on, its rotational motion alternates direction in 1 to 5 seconds and tilts the movable base to the left and right. Thus the sea wave motion is imitated.

The moving part of the LGPM begins to reciprocate and the magnetic fluxes created by the moving magnets intersect the fixed coils and induce electromagnetic voltage in them. When the movable base is tilted to the far left or right position, the acceleration is greatest and a magnetic brake is used to avoid shocks. The magnetic brake consists of two pairs of rare earth magnets 15 and 16, which are turned with the same poles to one another.

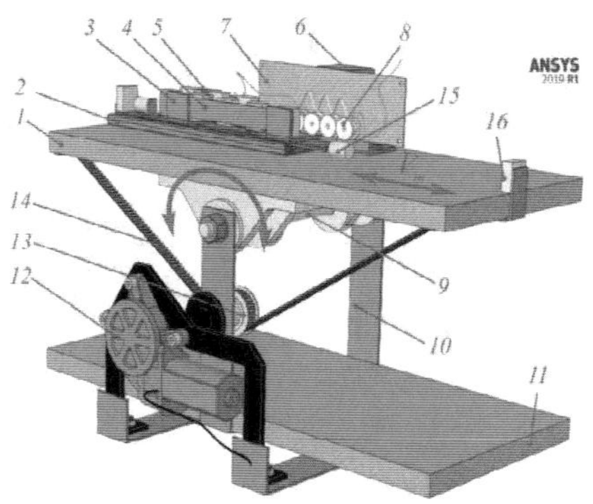

Figure 2. Experimental test rig.

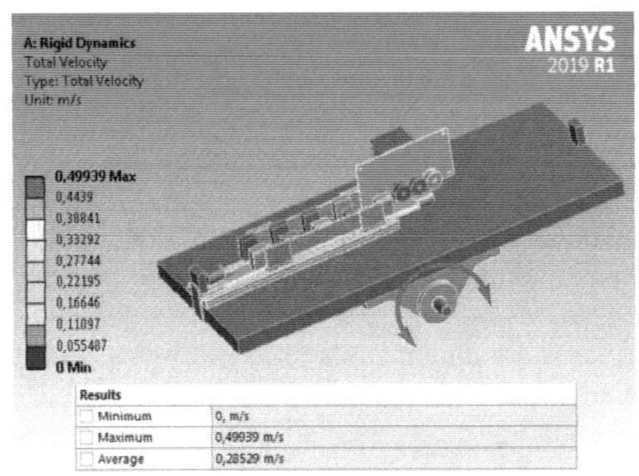

Figure 3. Simulation in ANSYS.

3. Modeling and experimental studies

Modeling was performed by ANSYS 2019 R1 to obtain the maximum linear velocity V_m of the mechanical system - Figure 3. Figure 4 illustrates the distribution of the magnetic field in the considered

LGPM without magnets in its stator windings, obtained by means of the finite elements method Femm 4.2.

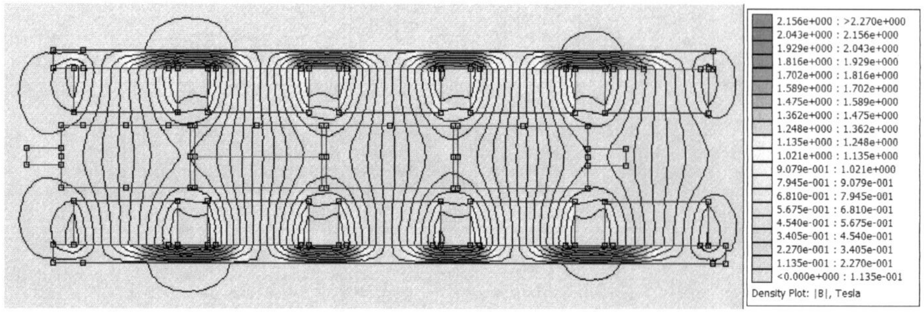

Figure 4. Distribution of the magnetic field.

Figure 5 shows the variation of the magnetic flux density along the length of one winding - B_c. The average magnetic flux density in the perpendicular direction for the stator winding for the considered linear generator is B_c=0,275 T. Figure 6 presents the equivalent circuit of an electromagnetic harvester with a voltage rectifier at active load. In this figure R_c and L_c denote the active resistance and the inductance of the four connected in series windings of the electromagnetic harvester, with induced electromotive voltage c $e_c(t)$ and active load resistance R_L. The rectified voltage over the load resistance is denoted by U_L.

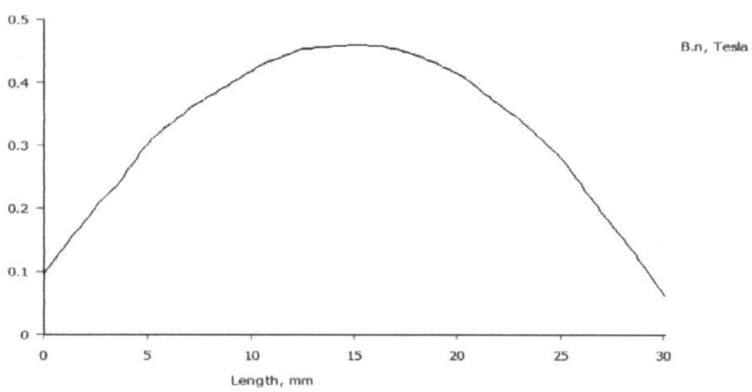

Figure 5. Magnetic flux density along the length of one winding.

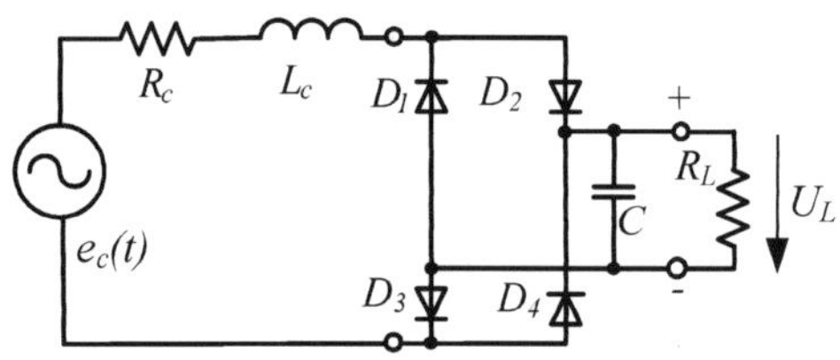

Figure 6. Equivalent circuit of an electromagnetic harvester.

The horizontal linear velocity of the considered LGPM is equal to

$$v(t) = \frac{d\,x(t)}{dt} \tag{1}$$

The magnetic flux change can be expressed by the change in the cross-section of the coil for time $A(t)$, through which the average magnetic flux density B_c passes.

$$\Phi(t) = B_c\, A(t) \tag{2}$$

The change in the cross-section of the coil can be presented by its width b and its moving x in time

$$A(t) = b\; x(t) \tag{3}$$

From (2) and (3) the magnetic flux change in the course of time is obtained

$$\Phi(t) = B_c\, b\, x(t) \tag{4}$$

The induced electromotive force in a single winding with *N* turns of the electromagnetic harvester is equal to

$$e(t) = -N\frac{d\Phi(t)}{dt} \tag{5}$$

From (4) and (5) for the induced electromotive force in one winding it is obtained

$$e(t) = -N\,B_c\,b\frac{dx(t)}{dt} \tag{6}$$

With 4 coils connected in series the total electromotive force $e_c(t)$ will be equal to (7), and its amplitude – to (8), where V_m is the maximum linear velocity

$$e(t) = -4N\,B_c\,b\frac{dx(t)}{dt} = -4N\,B_c\,b\;v(t) \tag{7}$$

$$E_m = 4N\,B_c\,bV_m \tag{8}$$

The active power in DC mode is calculated using the amplitude of the induced electromotive force in the coil at idle-run mode and the parameters of the equivalent circuit

$$P_L = \left(\frac{1,41E_m\,R_L}{\sqrt{\left(R_C+R_L\right)^2+\left(\omega L_C\right)^2}}\right)^2\cdot\frac{1}{R_L}, \tag{9}$$

The paper considers three different connection schemes for the four windings of a linear generator, denoted by the names Circuit 1, Circuit 2 and Circuit 3 in Figure 6. Figures 7, 8 and 9 present the variation of voltage, current and power for the three connection schemes at active loads from 10 Ω to 100 kΩ. Circuit 1 and Circuit 3 have higher than 29 V voltages at a load of 100 kΩ; Circuit 2 has a voltage lower than 24 V. The maximum current for Circuit 2 is the highest of all - 40 mA - at a low-ohm load of 10 Ω; for Circuit 3 it is 30 mA and for Circuit 1 is the least of all - 20 mA. Figure 10 shows that the active power is highest for Circuit 2 - 125 mW – at loads of 700Ω, while for Circuit 1 and Circuit 3 it is the same - 93 mW for each, but for loads of 6 kΩ and 800Ω.

Figure 10 shows that in Circuit 2 the capacities are significantly higher for low-ohm loads of up to 1 kΩ, for Circuit 2 they are lower and for Circuit 1 they are the lowest. From the experimental studies it can be concluded that Circuit 2 has the best output electrical parameters for low-ohm loads of up to 1 kΩ due to the higher current and active power and therefore it is more suitable for the practical realization of a linear generator with permanent magnets without an iron core.

Figure 11 shows that the obtained model simulates with good precision the operation of the studied generator in the range of variation of the load resistance from 10 Ω to 2 kΩ, whereas the maximum relative error is δmax = 16.9%.

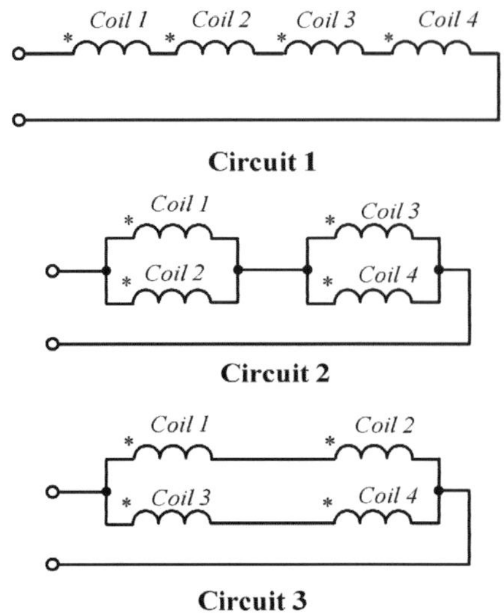

Figure 6.

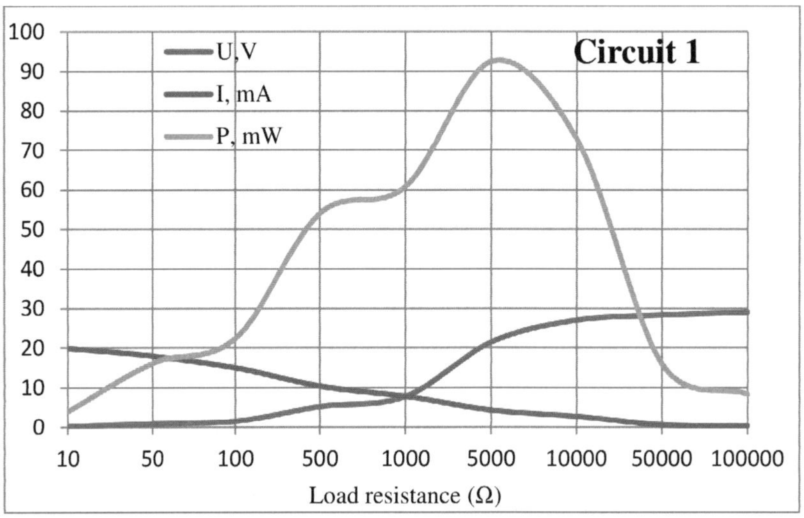

Figure 7. Variation of voltage, current and power for Circuit 1.

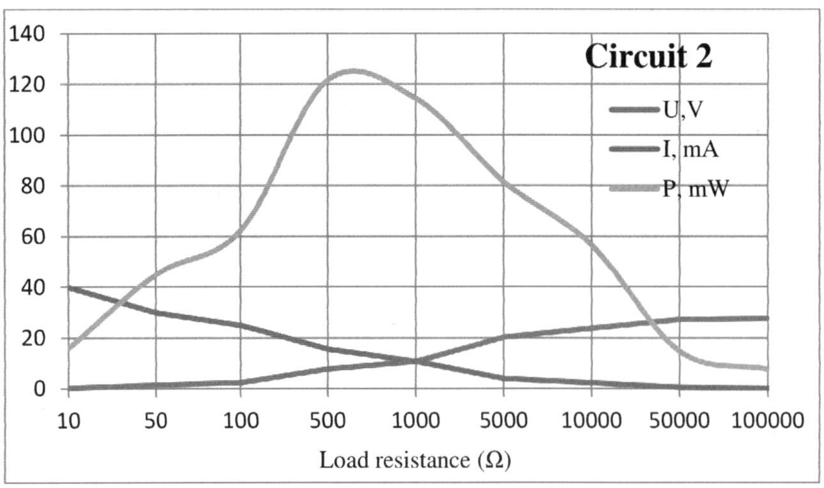

Figure 8. Variation of voltage, current and power for Circuit 2.

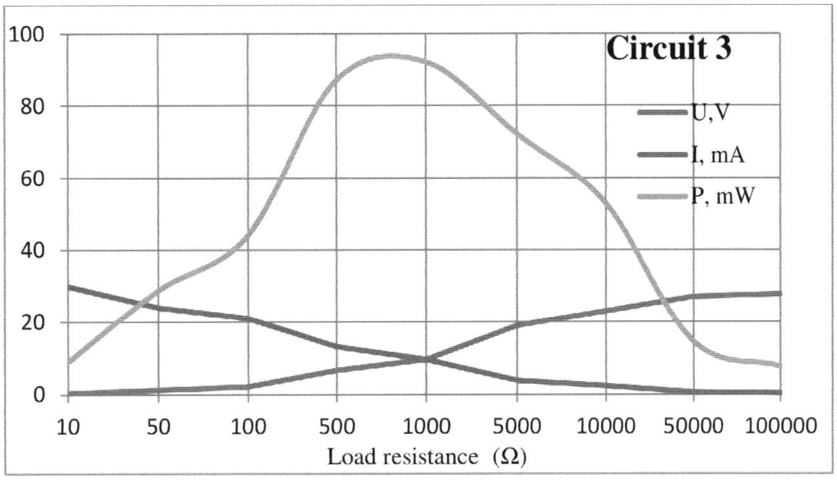

Figure 9. Variation of voltage, current and power for Circuit 3.

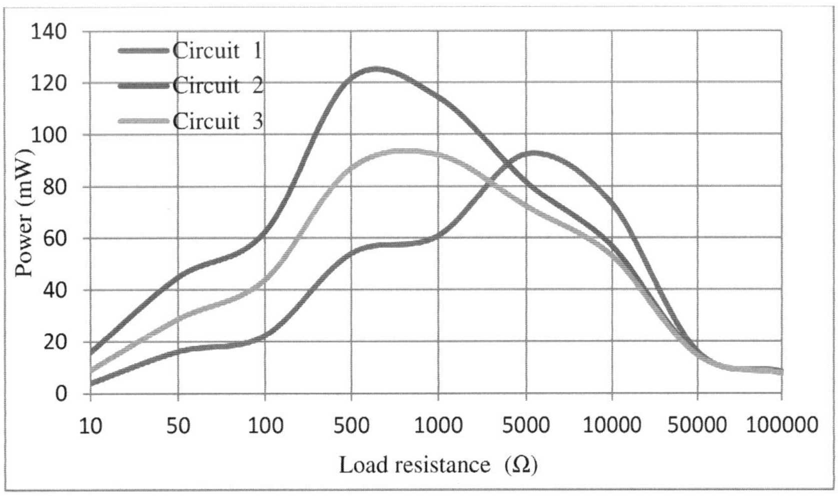

Figure 10. Variation of power for Circuit 1, Circuit 2 and Circuit 3.

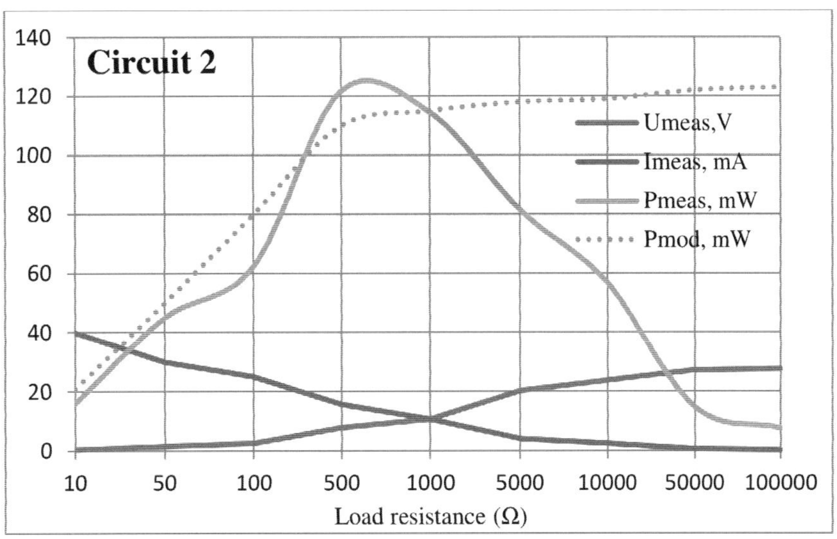

Figure 11. Obtained model simulates and measured powers.

4. Conclusions

Program simulations, theoretical deductions and experimental studies of the proposed linear generator with permanent magnets without an iron core have been performed for three different connection schemes of its windings. Using the ANSYS 2019 R1, the maximum linear velocity of the mechanical system has been found, and by means of Femm 4.2 the distribution of the magnetic field has been obtained. The presented model simulates with good precision the operation of the studied linear generator for the scheme of two by two connected in series windings, in the range from 10Ω to 2 kΩ, where the maximum measured power actually belongs.

The active power is highest for Circuit 2 - 125 mW – at a load of 700Ω; and the current for this circuit is at its maximum of 40 mA at a low-ohm load of 10 Ω.

From the conducted experimental studies it can be concluded that Circuit 2 - the scheme with two by two connected in series windings - has the best output electrical parameters for low-ohm loads of up to 1 kΩ, because of the higher current and active power. It is the most suitable of the three schemes for practical implementation of a linear generator with permanent magnets without an iron core.

References

[1] E. Spooner and M. A. Mueller, "Comparative study of linear generators and hydraulic systems for wave energy conversion", ETSUV/06/00189/REP, 2001.

[2] L. Szabó and C. Oprea, "Wave Energy Plants for the Black Sea – Possible Energy Converter Structures", Proceedings of the International Conference on Clean Electrical Power (ICCEP '2007), Capri (Italy), pp.E3184, 2007.

[3] D. Baa, N. Anha and P. Ngocb, "Numerical Simulation and Experimental Analysis for A Linear Trigonal Double-Face Permanent Magnet Generator Used in Direct Driven Wave Energy Conversion", "2nd Humboldt Kolleg in conjunction with International Conference on Natural Sciences", 2014.

[4] O. Farrok, R. Islam and R. Sheikh, "Analysis of the Oceanic Wave Dynamics for Generation of Electrical Energy Using a LinearGenerator", Hindawi Publishing Corporation Journal of Energy, Article ID 3437027, 2016.

[5] L. Szabó, C. Oprea, I. Viorel and K. Biró, "Novel Permanent Magnet Tubular Linear Generator for Wave Energy Converters", Proceedings of the IEEE International Conference on Electrical Machines and Drives, Antalya (Turkey), vol. 2, pp. 983-987, 2007.

[6] N.Hiron, A. Andang and N. Busaeri, "Investigation of NdFeB N52 magnet field as advanced material at air gap of axial electrical generator", IOP Conference Series: Materials Science and Engineering, 550(1),012034, pp. 1-6, 2019.

Acknowledgements

The authors would like to thank the Research and Development Sector at the Technical University of Sofia for the financial support.

IOP Conf. Series: Materials Science and Engineering **878** (2020) 012020 doi:10.1088/1757-899X/878/1/012020

Investigation of transformers for power electronic converters

K G Milanov

Technical university - Sofia, Department of Electrical Apparatus

e-mail: kmilanow@tu-sofia.bg

Abstract. A transformer is an integral part of power electronic equipment and it is usually the input stage of a high power converter. A good model of the transformer is needed to evaluate the performance parameters of power converters. This work represent 2 new methods for short-circuit test of single-phase transformers. The paper offers measurement methods that do not require powerful adjustable AC sources. Short circuit parameters can be determined by measuring the parameters of the transient when switching on: i) capacitor under zero initial conditions to an idle transformer or ii) by a transient process when connected to a non-powered short-circuit transformer of a capacitor under non-zero initial conditions. In order to accurately determine the period and the transition process in the event of a short circuit, it is suggested to include a parallel connected battery. *PSpice* is being used as a simulation tool for analysis and design of power electric circuits.

1. Introduction

A number of industry applications require Specific Industrial Transformers due to the usage of power (current) as a major resource for production. High-current Rectifier and Converter Transformers need a specific design to supply the necessary power at a low voltage level. These transformer types (EAF Transformers and Ladle Furnace Transformers), as well as converter transformers for large drive applications are called special-purpose or Industry Transformers, whose design is tailor-made for high-current solutions for industry applications [1, 2, 3].

The electromagnetic chokes and transformers are elements in the structure of almost all electronic energy converters.

The increasing utilization of power converters in power electronics requires fundamental analysis and design of power converters. These studies could provide wide opportunities for electromagnetic chokes and transformers at higher frequency exploitation in power electronic circuits [4].

Magnetic components such as chokes and transformers constitute together with the control and the semiconductor components, the main parts in the design of power electronic converters. In today's power electronics, the power electronic converters can withstand high switching frequencies [2, 5].

Comparing the modern power switches used in power supplies with those from older generations, the new switches have significantly reduced switching times, leading to faster and faster rise and fall times for the voltage and current waveforms. These fast edges produce significant energy at surprisingly high frequencies [5, 6].

Transformers supplying converter applications are connected on their secondary side to a converter. The converter input can be a passive rectifier or have an active front end using controllable semiconductor switches. These elements create a lot of harmonic frequencies. Different drive types are quite diverse in the amount of harmonics produced [7].

Therefore, normal transformers, designed and dimensioned for electricity distribution duty do not work satisfactorily in a converter system. Harmonic frequencies increase the thermal, mechanic and dielectric stresses and therefore the transformers must be specially designed for this duty. Over-dimensioning of a normal distribution transformer is not an adequate measure [8].
Aside from power ratings and power losses, transformers often harbor other undesirable limitations which circuit designers must be made aware of. Like their simpler counterparts - inductors - transformers exhibit capacitance due to the insulation dielectric between conductors: from winding to winding, turn to turn (in a single winding), and winding to core. Usually this capacitance is of no concern in a power application, but high-frequency applications may not tolerate this quirk well [9].
The current work proposes not to use the conventional way to determine the short-circuit parameters of transformers and rather use a transitional regime of capacitor's discharging through a transformer. It is also suggested to use a transitional regime in order to determine the frequency of the coils for parameters at elevated frequencies taking into account the capacity of the coils.
In order to determine the parameters of a transformer such as voltage regulation and efficiency, the open circuit and short tests are carried out. These tests are very convenient, as they furnish the required information without actually loading the transformer. Also the power required to carry out these test is very small as compared to full load output of the transformer.
The short circuit test is used to determine the values *Rs* and *Xs* of the series branch of the equivalent circuit and is carry out with current *1.25xIn*. These impedances are usually very low, but appear higher in value when referred to the high voltage side. This test is consequently performed on high voltage side of the transformer in order to keep the current drawn by these impedances at a manageable level [2, 10, 11].
Common short-corcuit test method for power transformers are presented in many in the references. Some testing methods use capacitance and power factor/dissipation factor measurement.
Regularly performing a range of standard electrical tests has proven an effective way to gain a reliable insight into the operating condition of power transformers. Mechanical changes to windings, contact problems in the tap changer or at other connections, shorted windings/coils, as well as interruptions or short-circuit of parallel lines can all be diagnosed early by using conventional testing methods. Severe and costly damage can be thereby prevented. These conventional testing methods include measuring a number of parameters, such as the short circuit impedance, transformers ratio, magnetizing current, etc. Tayeb et al. represents numerical method based on Visual basic language to get the operating characteristics of power transformer, including short circuit parameters [12]. =

2. Experimental investigation

Present work proposed new method for short-circuit parameters determination of transformers for electronic converters to be determined by cause a transient discharge process of preloaded capacitor through the secondary transformer coil.
Work measurement methods that do not require powerful adjustable AC voltage sources are available. The short-circuit parameters are determined by measuring the parameters of the transitional process by switching in:

- a capacitor at zero initial conditions to an idle running transformer;
- or through a transitional process when switching in a non-powered transformer with a capacitor's shading coil at non-zero initial conditions;
- and the third possibility is to determine the short-circuit parameters on basis of measuring processes in commutation phenomenon of converters, more particularly of rectifiers.

The shown equivalent circuit is used to represent the transformer for experimental study. In short circuit mode the equivalent circuit of the transformer is reduced to inductance and equivalent resistance of the windings in a series connected, Figure 1.

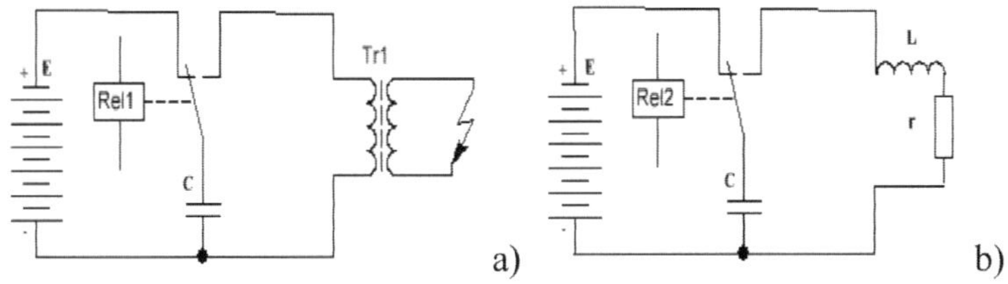

Figure 1. Electrical circuit (a) and equivalent circuit (b) of the transformer used for experimental investigation.

The idea of the proposed method is to cause a transient discharge process of preloaded capacitor *C* through the coil of the transformer when the other coil is short-circuited. For the experimental part electrolytic capacitor was selected: *U = 400 V* and capacitance *C = 1 μF.*

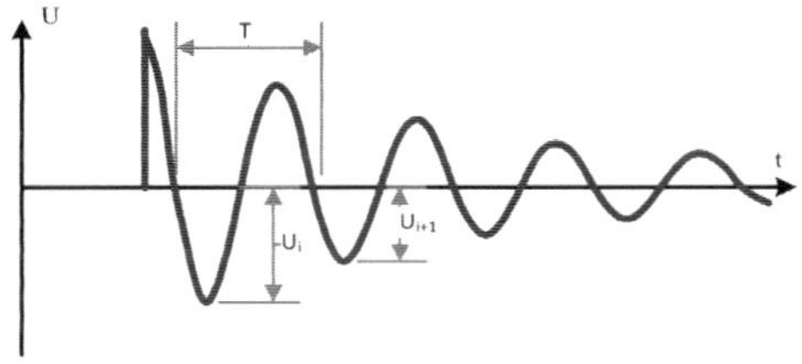

Figure 2. Transient discharge process of preloaded capacitor. *T* is time period and *Ui* is the amplitude.

The parameters of the transitional process, which is a damped sine wave, allow to measure the period *T* of own fluctuations and attenuation - by the ratio of the amplitudes of two consecutive periods *Ui/(Ui+1)*, Figure 2. The measured parameters and the value of the used capacitor *C* determines the inductance of the short circuit. $L_{1,2}$ and the equivalent resistance of the transformer aligned to the used coil.

A similar approach is used in the so called "diagnostics of automation objects", even for more complex objects.

The study was carrying out by using the following formulas and mathematical dependencies:

- Voltage across the capacitor in equivalent circuit:

$$U_C(t) = U_C(0). \cos \beta t, \quad (1)$$

where α- attenuation coefficient of own fluctuations;

- angular velocity $\beta = 2\pi f$. If we use frequency equation: $f = 1/T$ we get next dependence:

$$\beta = \frac{2\pi}{T} = \sqrt{\omega_0^2 - \alpha^2}, \quad (2)$$

where ω_0 is resonance frequency, *T* is time period;

At considered use of transformers for electronic converters, the damping ratio of own fluctuations α is much smaller than the resonant frequency ω_0. Then expression (1) is simplified to:

$$\beta = 2\pi f = \frac{2\pi}{T} = \sqrt{(\omega_0^2)} = \sqrt{\frac{1}{LC}}. \quad (3)$$

The inductance value is calculated by the following formula [1]:

$$L = \frac{T^2}{4\pi^2 C} \quad (4)$$

To account for the impact of the active resistances that cause the attenuation it is suggested to determine the ratio *Ui/(Ui+1)* and after that to determine the attenuation coefficient of the own oscillations α by using the following mathematical expression [2]:

Figure 3. Type of appearance of transformer used in experimental investigations: type ETB 400-10/0.4, voltage 10000/400V, current 23.09/577.4 A, f = 50 Hz, 3-phase step-down transformer.

$P = 400$ kVA; $P_{1F} = 133.3$ kVA;
$P_F = 133.3$ kVA; $U_{2F} = 5780$ V;
$I_{2F} = 23.09$ A $U_{sc} = 3.95$ %;
$Z_{sc} = 9.9\ \Omega$; $L_{sc} = 63.06.10^{-3}$ H

$$\alpha T = ln\frac{U_i}{U_i+1} \quad (5)$$

Experimental tests verification were carried out on a transformer with the following data and type of appearance, shown on Figure 3.
For the studied power transformer, an experiment has been conducted and a waveform of the transitional process has been recorded by oscilloscope, Figure 4. The inductance value have been determined analytically and compared with the experimental data. Also, the error value has been determined: ε = – 2.25%.

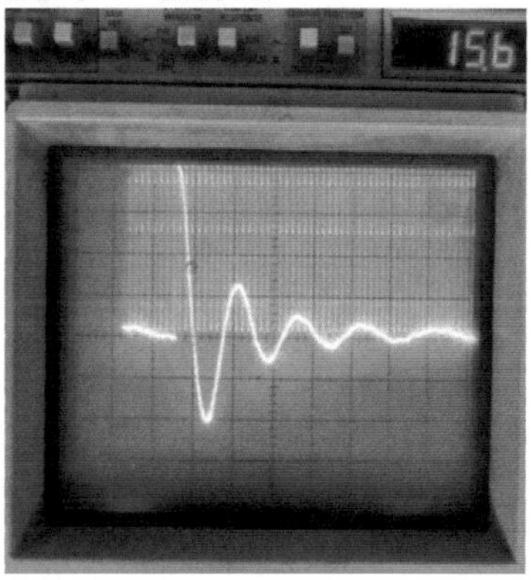

Figure 4. Capacitor discharging waveform connected to power transformer. The waveform was used to determine the period *T*.

$C_1 = 100\ \mu F$ – discharge capacitor connected to the secondary side of a power transformer

$T = 15.6\ ms$ value representing the transient capacitor discharge process through the secondary side of a power transformer

$T = 2\pi\sqrt{LC}$

$L = \frac{T^2}{(2\pi)^2\ C} = 0.06164\ H$ - inductance value

3. Simulation results

This article proposes an analogy between a three-phase medium voltage transformer and a single-phase low voltage transformer for short-circuit testing. For this purpose the article uses a traditional model of a single-phase transformer in a *Pspice* environment, which includes a source of sinusoidal voltage, serial connected resistance and inductance, Figure 5. L1 and L2 are respectively the inductances of the primary and secondary windings of the transformer under test. R1 and R2 are the active resistances of the primary and secondary windings. The inductance L1 models the magnetizing inductance and the resistor R1 models the core losses. The two switches included in the transformer secondary circuit are required to simulate the on and off process. The short circuit mode is achieved by diluting a electrolytic capacitor C1 (U = 400 V, 1μF) connected to the secondary coil of the studied transformer. We use ideal transformer model where the coupling coefficient is set to 1. The *PSpice* schematic is as shown on Figure 5a.
The electrical circuit shown in Figure 5b enables the simulation and the experimental evaluation of the transient and the process of the capacitor to be simulated several times. The selected voltage source has a value of 100 V and a resistance of 10 kΩ.
The article also offers the possibility for precise determination of the period T by parallel connection to the capacitor C1 of an additional voltage source.

IOP Conf. Series: Materials Science and Engineering **878** (2020) 012020 doi:10.1088/1757-899X/878/1/012020

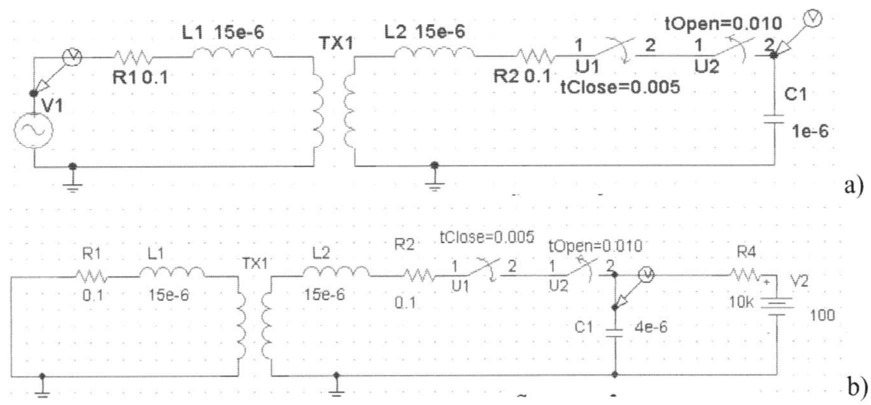

Figure 5. Equivalent *PSpice* circuit model of the transformer (a) and equivalent circuit with additionally included voltage source (b).

The results of the simulation are given in Figure 6. Figure 6 shows voltage waveforms of U1, U2 and voltage across the capacitor C1 obtained when a short circuit occurs.

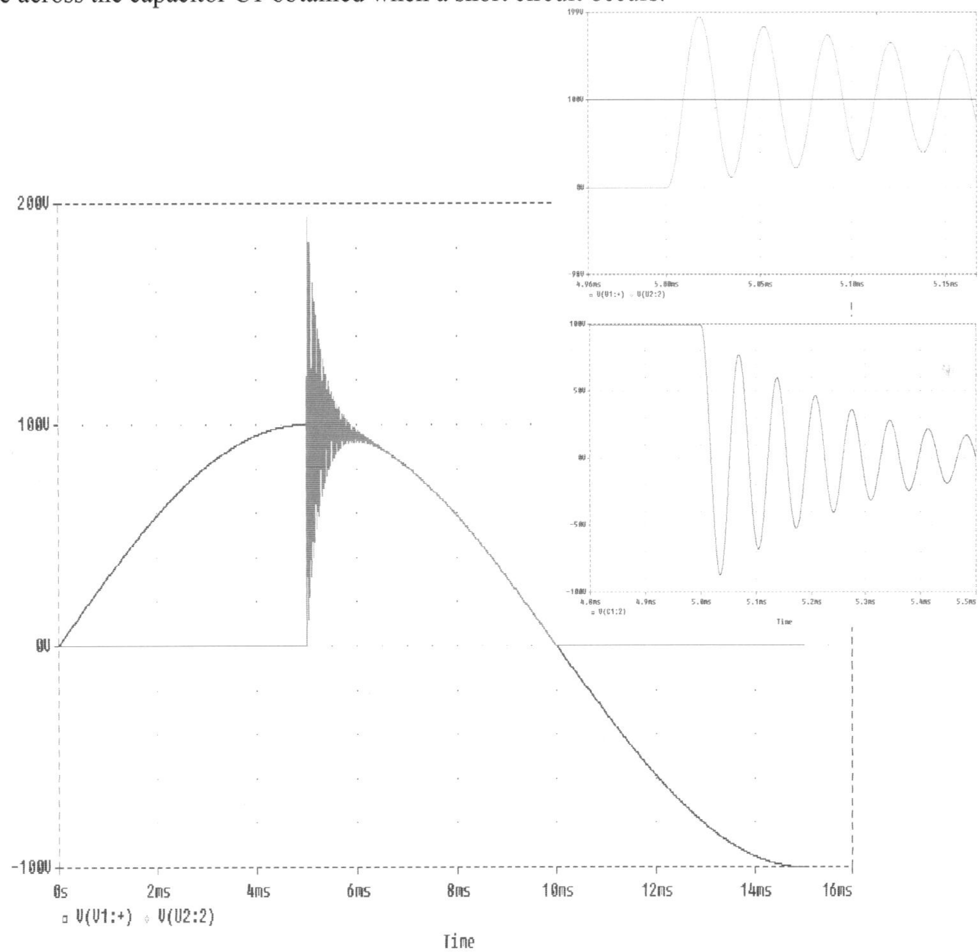

Figure 6. Simulated waveforms of discharge process of preloaded capacitor C1 (U2).

4. Conclusions

Experimental studies of gamma transformers have been carried out and the results of a 400 kVA transformer are presented in the article. The experiment was performed by diluting an electrolytic capacitor connected to the secondary side of the transformer, the primary winding being short-circuited. The paper offers measurement methods that do not require powerful adjustable AC sources. Parameters for short circuit are determined by measuring the parameters of the transient when switching on: *i)* capacitor under zero initial conditions to an idle transformer or *ii)* by a transient process when connected to a non-powered short-circuit transformer of a capacitor under non-zero initial conditions. The proposed method cause a transient discharge process of preloaded capacitor through the coil of the transformer provides sufficient accuracy – 2.25 %. The method is easily applicable to transformers for the needs of electronic converters operating at a wide range of frequencies and powers. Short-circuit testing transformers for high-power should be performed with a high-power source as well. These conditions are difficult to achieve, so the proposed method avoids this disadvantage. In the research laboratories are necessary small investments by applying the proposed method for short-circuit transformer parameters determination. The article proposes two simulation models of a single-phase low-power transformer. The first model, Figure 5a, proposes that a short circuit test be performed with a capacitor simulating the short circuit process. The second model, Figure 5b, adds a voltage source to the capacitor in parallel to charge the capacitor periodically. In conclusion, it can be said that there is a coincidence between the experimentally obtained results and the data provided by the manufacturer for the voltage and inductance of the short circuit. The simulation data obtained confirm the transient process and show the voltage damping by allowing the relationship to be determined $Ui / (Ui + 1)$.

5. References

[1] SIEMENS, "High-current industrial transformers," SIEMENS, [Online]. Available: http://www.energy.siemens.com/hq/en/power-transmission/transformers/special-industry-transformers/converter.htm.
[2] Mintchev M, Shopov I and Ratz E 2008 *Converter technologies* (Sofia: Avangard prima) p. 164
[3] Iman-Eini H and Farhangi S 2006 Analysis and Design of Power Electronic Transformer for Medium Voltage Levels *37th IEEE Power Electronics Specialists Conf.* (IEEE)
[4]Lai J-S, Maitra A, Mansoor A and Goodman F 2005 Multilevel intelligent universal transformer for medium voltage applications *Fourtieth IAS Annual Meeting. Conference Record of the 2005 Industry Applications Conference* (IEEE: Kowloon, Hong Kong)
[5] Bossche A, Valchev V 2005 *Inductors and transformers for power electronics*, (Boca Raton FL 33487-2742: CRC Press Taylor & Francis Group) p. 478
[6] Hesener A 2010-2011 Electromagnetic Interference (EMI) *Power Supplies Fairchild Semiconductor Power Seminar*.
[7] Tahir U, Abbas G, Glavan DO, Balas VE, Farooq U, Balas MM, Raza A, Asad MU and Gu J 2019 Design of Three Phase Solid State Transformer Deployed within Multi-Stage Power Switching Converters, *Applied sciences* **9** 17 pp. 1-24
[8] Reddy BD, Sahoo SK, 2015 "Design of Solid State Transformer *International Journal of Advanced Research in Electrical, Electronics and Instrumentation Energy* 4 (1) pp. 357-364
[9] She X, Huang AQ and Burgos R 2013 Review of Solid-State Transformer Technologies and Their Application in Power Distribution Systems *Journal of Emerging and Selected Topics in Power Electronics* **1**3 pp. 186-198
[10] Popov I 1955 *Electrical mashines*, Second ed., vol. 2 (Sofia: Technika) p. 260 (in bulgarian).
[11] Petrov G N 1974 Electrical mashines (Moskow: Energiya) p. 240 (in russian)
[12] Tayeb Ali A, Ali M, 2013 Extraction of Electrical Power Transformer Parameters," *Int. Journal of Engineering and Computer Science* IS **2** 1 pp. 187-195
[13] Kalantarov P L, Tzeitlin L A 1986 *Inductance calculation* (Leningrad: Energoatomizdat) p. 487

Acknowledgments
The authors would like to thank the Research and Development Sector at the Technical University of Sofia for the financial support.

IOP Conf. Series: Materials Science and Engineering **878** (2020) 012021 doi:10.1088/1757-899X/878/1/012021

The value of the work done by an isotropic vector force field along an isotropic curve

Dimitar Razpopov[1] **and Georgi Dzhelepov**[2]

[1]Department of Mathematics and Informatics, Agricultural University of Plovdiv, 12 Mendeleev Blvd., Bulgaria 4000

E-mail: razpopov@au-plovdiv.bg

[2]Department of Mathematics and Informatics, Agricultural University of Plovdiv, 12 Mendeleev Blvd., Bulgaria 4000

E-mail: dzhelepov@au-plovdiv.bg

Abstract. In the present paper we consider a 3-dimensional differentiable manifold M equipped with a Riemannian metric g and an endomorphism Q, whose third power is the identity and Q acts as an isometry on g. Both structures g and Q determine an associated metric f on (M, g, Q). The metric f is necessary indefinite and it defines isotropic vectors in the tangent space T_pM at an arbitrary point p on M.

The physical forces are represented by vector fields. We investigate physical forces whose vectors are in T_pM on (M, g, Q). Moreover, these vectors are isotropic and they act along isotropic curves. We study the physical work done by such forces.

1. Introduction

The physical work and the physical force on differentiable manifolds have a great application in physics. Vector fields are often used to model a force, such as the magnetic or gravitational force, as it changes from one point to another point. As a particle moves through a force field along a curve c, the work done by the force is the product of force and displacement. There are some papers concerning physical results on light-like (degenerate) objects of differentiable manifolds ([4], [8] and [9]).

The object of the present paper is a 3-dimensional differentiable manifold M equipped with a Riemannian metric g and a tensor Q of type $(1, 1)$, whose third power is the identity and Q acts as an isometry on g. Such a manifold (M, g, Q) is defined in [6] and studied in [1], [2], [3] and [7]. Also, we consider an associated metric f, which is introduced in [7]. The metric f is necessary indefinite and it determines space-like vectors, isotropic vectors and time-like vectors in the tangent space T_pM at an arbitrary point p on M.

We investigate physical forces whose vectors are in T_pM on (M, g, Q). Moreover, these vectors are isotropic with respect to f and they act along isotropic curves. We study the physical work done by such forces.

2. Preliminaries

Let M be a 3-dimensional Riemannian manifold equipped with an endomorphism Q in the tangent space T_pM, $p \in M$. Let the local coordinates of Q with respect to some coordinate system form the circulant matrix:

$$(Q_i^j) = \begin{pmatrix} 0 & 1 & 0 \\ 0 & 0 & 1 \\ 1 & 0 & 0 \end{pmatrix}.$$

Then Q has the property

$$Q^3 = \mathrm{id}. \tag{1}$$

Let g be a positive definite metric on M, which satisfies the equality

$$g(Qr, Qi) = g(r, i). \tag{2}$$

In (2) and further r, i, w will stand for arbitrary vectors in T_pM.

Such a manifold (M, g, Q) is introduced in [6].

It is well-known that the norm of every vector i is given by $\|i\| = \sqrt{g(i,i)}$. Then, having in mind (2), for the angle $\varphi = \angle(i, Qi)$ we have

$$\cos\varphi = \frac{g(i, Qi)}{g(i,i)}.$$

In [6], for (M, g, Q), it is verified that the angle φ is in $[0, \frac{2\pi}{3}]$. If $\varphi \in (0, \frac{2\pi}{3})$, then the vector i form a basis $\{i, Qi, Q^2i\}$, which is called a Q-basis of T_pM.

The associated metric f on (M, g, Q), determined by

$$f(r, i) = g(r, Qi) + g(Qr, i). \tag{3}$$

is necessary indefinite [7].

A vector r in T_pM is isotropic with respect to f if

$$f(r, r) = 0. \tag{4}$$

In every T_pM, for (M, g, Q), there exists an orthonormal Q-basis $\{i, Qi, Q^2i\}$ ([6]). From (1), (3) and (4) we state the following

Lemma 2.1. *Let $\{i, Qi, Q^2i\}$ be an orthonormal Q-basis of T_pM. If $r = ui + vQi + qQ^2i$ is an isotropic vector, then its coordinates satisfy*

$$uv + vq + qu = 0. \tag{5}$$

An isotropic (null) curves $c : r = r(t)$ are those whose tangent vectors are everywhere isotropic, i.e.,

$$f(dr, dr) = 0. \tag{6}$$

The physical forces are represented by vector fields. We investigate physical forces whose vectors are in T_pM on (M, g, Q). Moreover, these vectors are isotropic and they act along isotropic curves. We study the physical work done by such forces.

3. The work in T_pM

We consider an orthonormal Q-basis $\{i, Qi, Q^2i\}$ in T_pM on (M, g, Q).

Let p_{xyz} be a coordinate system such that the vectors i, Qi and Q^2i are on the axes p_x, p_y and p_z, respectively. So p_{xyz} is an orthonormal coordinate system.

The curve c is determined by

$$c : r(t) = x(t)i + y(t)Qi + z(t)Q^2i, \tag{7}$$

where $t \in [\alpha, \beta] \subset \mathbb{R}$.

Let c be an isotropic smooth curve. Thus equalities (1), (3), (6) and (7) imply

$$dxdy + dydz + dxdz = 0. \tag{8}$$

We determine a vector force field

$$F(x, y, z) = P(x, y, z)i + R(x, y, z)Qi + S(x, y, z)Q^2i, \tag{9}$$

where $P = P(x, y, z)$, $R = R(x, y, z)$, $S = S(x, y, z)$ are smooth functions.

Let the vector field F be isotropic. Hence following (5) we get

$$PR + RS + SP = 0. \tag{10}$$

Work A done by a force F, with respect to f, moving along a curve c is given by

$$A = \int_c f(F, dr), \tag{11}$$

where

$$dr = dxi + dyQi + dzQ^2i. \tag{12}$$

Case (A) Let F and c are both isotropic and they are on the same direction. Since c is a trajectory of F we have that the vectors F and dr are collinear. Therefore their coordinates satisfy

$$\frac{dx}{P} = \frac{dy}{R} = \frac{dz}{S} = \frac{1}{k}, \tag{13}$$

where $k \neq 0$ is a function. From (9) and (13) it follows $F = kdr$. Then, having in mind (6) and (11), we get $dA = f(kdr, dr) = kf(dr, dr) = 0$, i.e., $A = 0$.

Case (B) Now, we consider the case when F and c are both isotropic but they are on different directions.

From (3), (11) and (12) it follows

$$A = \int_c [P(dy + dz) + R(dx + dz) + S(dx + dy)], \tag{14}$$

and hence

$$A = \int_\alpha^\beta [P(y' + z') + R(x' + z') + S(x' + y')]dt. \tag{15}$$

- Let $dx + dy = 0$. From (8) we have $dx = dy = 0$ and $dz \neq 0$. Then $dr = dzQ^2i$. Therefore, using (15), we get

  $$A = \int_\alpha^\beta \big(P(k_1, k_2, t) + R(k_1, k_2, t)\big)dt, \tag{16}$$

  where k_1 and k_2 are specific constants.

- Let $P + R = 0$. From (10) we have $P = R = 0$ and hence $S \neq 0$. In this case equalities (11) and (15) imply
$$A = \int_{\alpha}^{\beta} S(x(t), y(t), z(t))[x'(t) + y'(t)]dt. \tag{17}$$
- Let $dx + dy \neq 0$ and $P + R \neq 0$. With the help of (8) and (10) we get
$$dz = -\frac{dxdy}{dx + dy}, \quad S = -\frac{PR}{P + R}. \tag{18}$$
We use (14), (15) and (18) and obtain that the work A is determined by
$$A = \int_{\alpha}^{\beta} \frac{(Py' - Rx')^2}{(P + R)(x' + y')} dt. \tag{19}$$

From Case (A) and Case (B) we state the following

Theorem 3.1. *Let f be the associated metric on (M, g, Q). Let p_{xyz} be a coordinate system such that the vectors i, Qi and Q^2i of the orthonormal Q-basis in T_pM are on the axes p_x, p_y and p_z, respectively. Let F be an isotropic vector force field moving along an isotropic curve c. Let A be the work done by F. Then*

(i) A is zero if F and c are on the same direction;
(ii) A is (16) if F and c are on different directions and $dx + dy = 0$;
(iii) A is (17) if F and c are on different directions and $P + Q = 0$;
(iv) A is (19) if F and c are on different directions and $dx + dy \neq 0$ and $P + R \neq 0$.

4. Work in a 2-plane

Now we consider an arbitrary 2-plane $\alpha = \{i, Qi\}$ in T_pM. We suppose that the angle $\varphi = \angle(i, Qi)$ belongs to the interval $(0, \frac{2\pi}{3}]$. On α we construct a coordinate system p_{xy} such that i is on the axis p_x and j is on the axis p_y, where
$$j = \frac{1}{\sin\varphi}(-\cos\varphi i + Qi). \tag{20}$$

We assume that $\|i\| = 1$ and then p_{xy} is an orthonormal coordinate system.

In [5] it is proved the following

Theorem 4.1. *Let f be the associated metric on (M, g, Q) and let $\alpha = \{i, Qi\}$ be an arbitrary 2-plane in T_pM. Let the vector j be defined by (20) and p_{xy} be a coordinate system such that $i \in p_x$, $j \in p_y$. Then the equation of the circle $c: \ f(w, w) = a^2$ in α is given by*
$$(\cos\varphi)x^2 + \frac{(1 - \cos\varphi)(1 + 2\cos\varphi)}{\sin\varphi}xy - \frac{\cos^2\varphi}{1 + \cos\varphi}y^2 = \frac{a^2}{2}, \tag{21}$$
where $\varphi \in (0, \frac{2\pi}{3}]$.

Let $w = ui + vj$ be an isotropic vector, i.e., $f(w, w) = 0$. Therefore, with the help of (21), we obtain
$$\cos^2\varphi\left(\frac{y}{x}\right)^2 - \sin\varphi(1 + 2\cos\varphi)\frac{y}{x} - (1 + \cos\varphi)\cos\varphi = 0. \tag{22}$$
The discriminant of (22) is
$$D = (1 + \cos\varphi)(1 + 3\cos\varphi).$$

Then we get the following cases:

Case (A) If $\varphi \in (\arccos(-\frac{1}{3}), \frac{2\pi}{3})$, then $D < 0$. There is no isotropic directions in T_pM.

Case (B) If $\varphi = \arccos(-\frac{1}{3})$, then $D = 0$. We have one isotropic straight line $c: \ y = \sqrt{2}x$. Then the force F and the curve c both are on one isotropic direction and the work A of the force F along c is zero.

Case (C) If $\varphi \in \big(0, \frac{\pi}{2}\big) \bigcup \big(\frac{\pi}{2}, \arccos(-\frac{1}{3})\big)$, then $D > 0$. We have two isotropic directions which generate two straight lines:

$$c_1 : y = k_1 x, \quad c_2 : y = k_2 x, \quad x \in [\alpha, \beta],$$

where k_1 and k_2 are solutions of the equation (22) for $\frac{y}{x}$.

- If F is on c_1, then the work of F along c_1 is zero. Similarly, if F is on c_2, then the work of F along c_2 is zero.
- We suppose that F is on c_2 but F acts on c_1. Then

$$F(x, y) = P(x, y)(i + k_2 j), \quad dr = dt(i + k_1 j). \tag{23}$$

Bearing in mind (2) and (20) we calculate

$$\begin{array}{ll} g(i, Qi) = g(Qi, i) = \cos\varphi, & g(i, Qj) = g(Qj, i) = \frac{\cos\varphi - \cos^2\varphi}{\sin\varphi}, \\ g(j, Qi) = g(Qi, j) = \sin\varphi, & g(j, Qj) = g(Qj, j) = -\frac{\cos^2\varphi}{1+\cos\varphi}. \end{array} \tag{24}$$

On the other hand the solutions k_1 and k_2 of (22) satisfy equalities

$$k_1 + k_2 = \frac{\sin\varphi(1 + 2\cos\varphi)}{\cos^2\varphi}, \quad k_1 k_2 = -\frac{1 + \cos\varphi}{\cos\varphi}. \tag{25}$$

Using (3), (11), (23), (24) and (25) we find

$$A = \frac{1 + 3\cos\varphi}{\cos^2\varphi} \int_\alpha^\beta P(t, k_1 t) dt. \tag{26}$$

- Similarly, if F is on c_1 and F acts on c_2 we get

$$A = \frac{1 + 3\cos\varphi}{\cos^2\varphi} \int_\alpha^\beta P(t, k_2 t) dt.$$

Case (D) Finally, the condition $\varphi = \frac{\pi}{2}$ applied to (20) yields $j = Qi$. Then i and j are isotropic vectors. Therefore, from (3) and (11) it follows:

- $F = P(t, 0)Qi$, $dr = (dt)i$. The work is $A = \int_\alpha^\beta P(t, 0)dt$.
- $F = P(0, t)i$, $dr = (dt)Qi$. The work is $A = \int_\alpha^\beta P(0, t)dt$.

The results in Case (A) – Case (D) are summarized in Table 1.

Acknowledgments

This work is supported by project "17 – 12 Supporting Intellectual Property" of the Center for Research, Technology Transfer and Intellectual Property Protection, Agricultural University of Plovdiv, Bulgaria.

Table 1. Work $\boldsymbol{A}$ done by an isotropic vector force field F along an isotropic curve

φ	F acts on	trajectory of F	A
$(\arccos(-\frac{1}{3}), \frac{2\pi}{3})$	-	no is. curves	-
$\arccos(-\frac{1}{3})$	$c:\ y=\sqrt{2}x$	$c:\ y=\sqrt{2}x$	0
$(0,\frac{\pi}{2})\bigcup(\frac{\pi}{2},\arccos(-\frac{1}{3}))$	$c_1: y=k_1x$	$c_1: y=k_1x$	0
$(0,\frac{\pi}{2})\bigcup(\frac{\pi}{2},\arccos(-\frac{1}{3}))$	$c_2: y=k_2x$	$c_2: y=k_2x$	0
$(0,\frac{\pi}{2})\bigcup(\frac{\pi}{2},\arccos(-\frac{1}{3}))$	$c_1: y=k_1x$	$c_2: y=k_2x$	$A=\frac{1+3\cos\varphi}{\cos^2\varphi}\int_\alpha^\beta P(t,k_1t)dt$
$(0,\frac{\pi}{2})\bigcup(\frac{\pi}{2},\arccos(-\frac{1}{3}))$	$c_2: y=k_2x$	$c_1: y=k_1x$	$A=\frac{1+3\cos\varphi}{\cos^2\varphi}\int_\alpha^\beta P(t,k_2t)dt$
$\frac{\pi}{2}$	$c_1: x=0$	$c_2: y=0$	$A=\int_\alpha^\beta P(t,0)dt$
$\frac{\pi}{2}$	$c_1: y=0$	$c_2: x=0$	$A=\int_\alpha^\beta P(0,t)dt.$

References

[1] I. Dokuzova 2017 Almost Einstein manifolds with circulant structures, *J. Korean Math. Soc.* **54** (5), 1441-1456.
[2] I. Dokuzova 2018 On a Riemannian manifolds with a circulant structure whose third power is the identity, *Filomat* **32** (10), 3529-3539.
[3] I. Dokuzova, D. Razpopov and G. Dzhelepov 2018 Three-dimensional Riemannian manifolds with circulant structures, *Adv. Math., Sci. J.* **7** (1), 9–16.
[4] K. L. Duggal and B. Sahin 2010 *Differential Geometry of Lightlike Submanifolds*, Frontiers in Mathematics, (Basel: Birkhäuser), p. 488.
[5] G. Dzhelepov 2018 Spheres and circles in the tangent space at a point on a Riemannian manifold with respect to an indefinite metric,*Novi Sad J. Math.* **48** (1), 143-150.
[6] G. Dzhelepov, I. Dokuzova and D. Razpopov 2011 On a three-dimensional Riemannian manifold with an additional structure, *Plovdiv. Univ. Paisii Khilendarski Nauchn. Trud. Mat.* **38** (3), 17–27.
[7] G. Dzhelepov, D. Razpopov and I. Dokuzova 2010 Almost conformal transformation in a class of Riemannian manifolds, *In: Research and Education in Mathematics, Informatics and their Applications - REMIA 2010, Proc. Anniv. Intern. Conf.* Plovdiv, Bulgaria, 125–128.
[8] T. Korpınar and R. C. Demirkol 2017 Energy on a timelike particle in dynamical and electrodynamical force fields in De-Sitter space, *Revista Mexicana de Fısica* **63**, 560-568.
[9] Robert H. Wasserman 2004 *Tensors and Manifolds: With Applications to Physics*, (New York: Oxford University Press), p. 447.

Study on threshold capabilities of a copper bromide vapor laser

D V Fidanov[1]

[1] Faculty of Mathematics and Informatics, University of Plovdiv Paisii Hilendarski 24 Tzar Assen Street, 4000 Plovdiv, Bulgaria
E-mail: dimitarf@gmail.com

Abstract. This paper presents theoretical and simulation results for the operating parameters of a copper bromide (CBr) laser. The CuBr active medium finds applications as both as a laser source and a vapor amplifier. A methodology is developed to serve as the basis for the solution of a two-dimensional quasi-stationary heat conductivity equation on the cross-section of the gas tube for modeling the temperature of the neutral gas and the electron temperature of the discharge. This methodology makes it possible to determine the maximum electrical power supplied and the maximum electron temperature in terms of thermal ionization and electric ionization resistance of the gas discharge. Analytical results are applied in computer simulations and for the assessment of parameters of an actual CuBr laser.

1. Introduction

Copper and copper compound vapor lasers continue to be the subject of experimental and theoretical research. This type of lasers have numerous unique properties – they are the most powerful sources in the visible spectrum (516.6 nm and 578.2 nm) and have high laser beam coherence and convergence. The development of lasers with high laser output generation (over 100 W) are of particular interest. This extends significantly their range of applications. This type of lasers are also a source of ultraviolet radiation (248.6 nm, 259.2 nm, 260.0 nm, and 270.3 nm) [1].

Copper vapor lasers, and particularly CuBr vapor lasers, have a wide range of practical applications: in medicine and medical research, dermatology and photocoagulation (Tanzi et al., 2003); in industry for microprocessing of various materials - drilling, cutting, marking, etching, etc. [2]; in scientific research for isotope separation of various chemical elements, in studies of magnetic properties of materials, etc.; for pumping titanium-sapphire lasers, dye lasers, etc.; in high-speed photography, show business and advertising, laser microscopy, laser displays and nanotechnology, the defense industry, brightness amplifiers, for air and underwater navigation, and studies of air and ocean pollution, etc. [3, 4, 5].

In recent years, due to the development of high-speed video recorders (CCD and CMOS cameras) for up to 100 fps, there has been renewed interest in metal vapor systems as brightness amplifiers. To this end, CuBr lasers are being used increasingly frequently in some modern laser active optical systems for visual monitoring in real time of various objects and processes protected against intensive wide band background light. Such processes are the interactions of intense energy flows with surfaces, plasma welding, extraction of new materials and nanostructures, processes of thermonuclear synthesis, plasma-induced processes, etc. The experimental results of visualizations of objects and processes using high-speed laser-based monitors with CuBr vapor lasers are presented in [6]. In [7] CuBr brightness amplifiers with 100 kHz pulse repetition frequency are developed and studied. Results obtained by a bistatic laser monitor with two CuBr laser sources at high pulse repetition frequencies are described in [8].

Further development of CuBr vapor lasers poses numerous topical problems both in terms of experiments and research into their physical operating parameters – output power, laser efficiency, laser beam quality, gain and more.

One possible option for the increase of laser output power is to increase the electric power supplied into the gas discharge. This leads to higher electron energy and respectively increased laser generation. The increased electric power supplied to the gas discharge has its natural limits. Thermal overheating occurs in the gas as well as thermo-ionizing instability. This leads to a reduction of laser generation at the center of the tube or its complete termination. In [9] it is determined that the maximum permissible temperature of the gas in copper vapor lasers is $T_g = 3500K$.

Up until now, no methodology has been developed which allows for preliminary assessment the values that can be reached when increasing electron energy from the positions of gas thermal resistance. Such methodology would enable the performance of preliminary planning of the experiment for existing laser sources and its use as the basis for development of new ones.

The objective of this paper is the creation of a new methodology to assess both the increase of gas temperature and electron temperature.

2. Methodology development

2.1. Determination of the gas temperature in the discharge

In order to determine the temperature of the gas in the cross-section of the tube (see Figure 1), it is necessary to solve the two-dimensional quasi-stationary heat conductivity equation:

$$\mathrm{div}\left(\lambda_g \,\mathrm{grad}\, T_g\right) + q_v = 0 \tag{1}$$

where λ_g is the heat conductivity coefficient of the gas, q_v is the volume power density of the internal heat source, and T_g is the temperature in the tube. For gases λ_g usually is approximated by the expression $\lambda_g = \lambda_0 T_g^m$, m and λ_0 are constants.

In order to solve equation (1), we use boundary conditions of the first and second kind

$$T_g(R) = T_w, \quad \left.\frac{\partial T_g}{\partial r}\right|_{r=0} = 0 \tag{2}$$

where $0 < r < R$, R is the tube radius, T_w is the tube wall temperature.

The more general case of a radial distribution of the volume power density is considered. There is insufficient reliable data about the nature of the radial distribution of power along the laser tube cross-section. Research in [10] shows that the most suitable representation of is through a polynomial and in particular the use of a second order polynomial in the form:

$$q_v(r) = Kq_0\left(a + br^2\right) \tag{3}$$

where $K = 1.4383$, $a = 1.0183471$, $b = -0.001077$, q_0 is the average volume power density ($q_0 = \frac{Q}{V}, W/m^3$) , Q is the electric power supplied to the gas discharge, V is the tube volume. When the volume power density is given by (3), the distribution $T_g(r)$ of the gas temperature is presented by the following formula [10]:

$$T_g(r)=\left(T_w^{m+1}-\frac{(m+1)K\ q_0(r^2-R^2)(4a+br^2+bR^2)}{16\lambda_0}\right)^{\frac{1}{m+1}} \tag{4}$$

2.2. Determination of the electron temperature

In order to determine the temperature of the electron gas $T_e(r)$, we need to solve a two-dimensional quasi-stationary equation

$$\nabla\lambda_e\nabla T_e=-\xi.q_v \tag{5}$$

where λ_e is the coefficient of electron heat conductivity of the electron gas, and ξ indicates the portion of the total supplied power that goes to heating the electrons. Under the boundary conditions similar to (2) with $T_e(R)=0$ (the energy of the electrons is zero), the solution to equation (5) takes the form [11]:

$$T_e(r)=\left(T_e(R)^{m+1}-\frac{(m+1)K\dfrac{\xi.q_0}{C_c\bar{\sigma}}(r^2-R^2)(4a+br^2+bR^2)}{16\lambda_0}\right)^{\frac{1}{m+1}} \tag{6}$$

Here quantities C_c, $\bar{\sigma}$ and ξ depend on the specific laser source, its geometric and energy characteristics and these are determined using the method detailed in [11].

3. Subject of investigation

The subject of investigation is a real operational CuBr laser, described in [12]. The principle scheme and the geometric characteristics are given in Fig. 1 and Table 1.

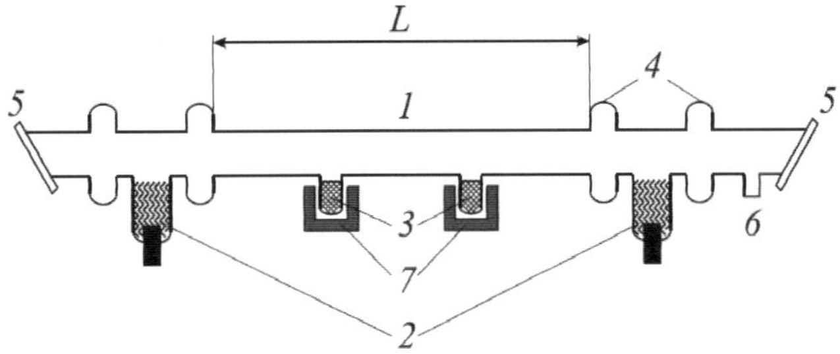

Figure 1. Gas discharge tube (GDT): (1) working channel, (2) electrodes, (3) CuBr containers, (4) traps, (5) exit windows, (6) HBr generator, (7) CuBr container heaters; L is the active zone length.

Table 1. GDT parameters.

Parameter	Value	Measure
GDT diameter ($d=2R$)	0.7	cm
Active zone length (L)	14	cm
Maximum pump PRF	630	kHz
Buffer gas (Ne) pressure	25	Torr

4. Results and discussion

Using the developed methodology, simulations were carried out to estimate the threshold values of CuBr laser parameters. Figure 2 shows the cross-sectional distribution of gas temperature for two values of the supplied electric power in the gas discharge Q = 600 and 800 W, with 50% losses. According to [12], the temperature of the wall is between 500 and $700\ ^{o}C$. We select a value in the specified interval - $T_w = 950K$.

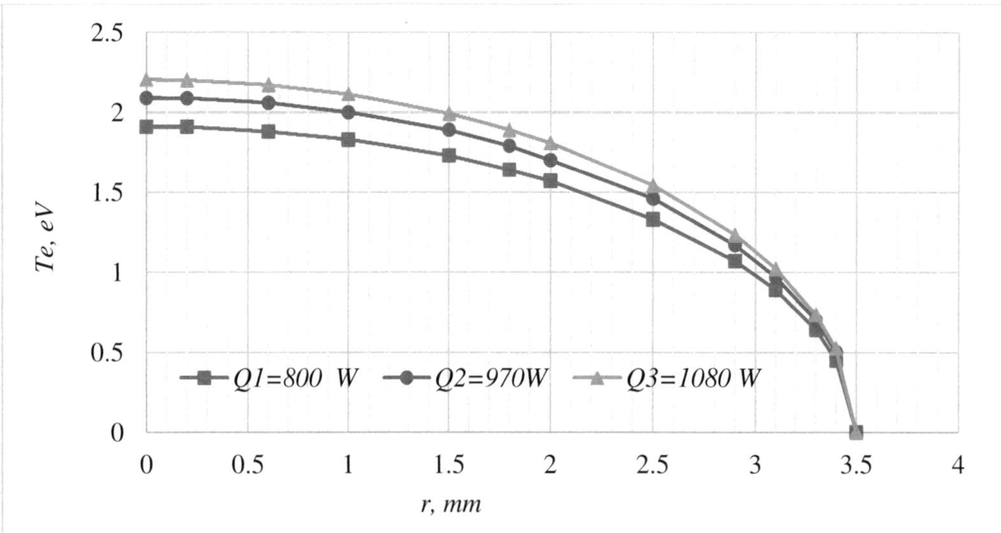

Figure 2. Distribution of gas temperature in the tube's cross-section for two values of the supplied electric power Q = 600 W and 800 W.

Figure 3 shows the value of the gas temperature T_0 at the center of the tube when the electric power supplied to the gas discharge is altered. The temperature of the wall is $T_w = 950K$. From this figure it could be estimated that the maximum permissible electric power supplied to the gas discharge, at which the temperature at the center of the tube reaches its maximum of 3500K is approximately Q =850 W.

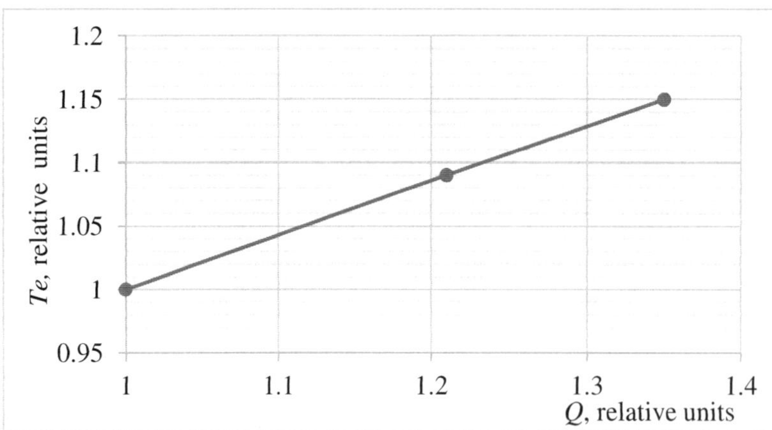

Figure 3. Change in the gas temperature T_0 at the center of the tube when the electric power supplied to the gas discharge is altered. The temperature of the wall is $T_w = 950K$.

In order to determine the distribution of the electron temperature in the cross-section of the tube, in equation (6), it is necessary to pre-determine the constants C_c, $\bar{\sigma}$ and ξ. Following [11], for Q =600W, R=3.5mm, $T_{e,\exp} = 1.5\,eV$, $\bar{\sigma}$ =4.10-5 Ω-1, $C_c = 2.47.10^{-8}$, for ξ we obtain ξ =4.85.10-5.

The *Wolfram Mathematica* software system is used to solve (6) and to run the computer simulations. The distribution of the electron temperature in the tube's cross-section for three values of the supplied electric power (Q 1=600 W, Q 2=800 W and Q 3=1000 W) is shown in Figure 4.

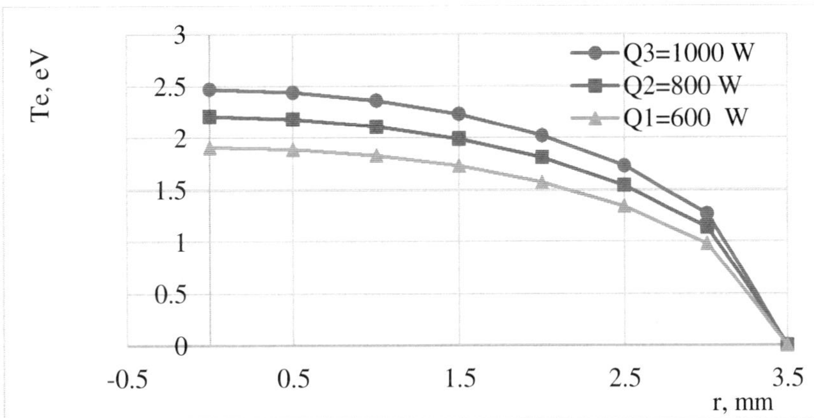

Figure 4. Distribution of the electron temperature in the tube's cross-section for three values of the supplied electric power.

Figure 5 shows in a single graphic the value of the gas temperature, series 1, K and the electron temperature, series 2, eV at the center of the tube as a function of the electric power Q supplied to the gas discharge.

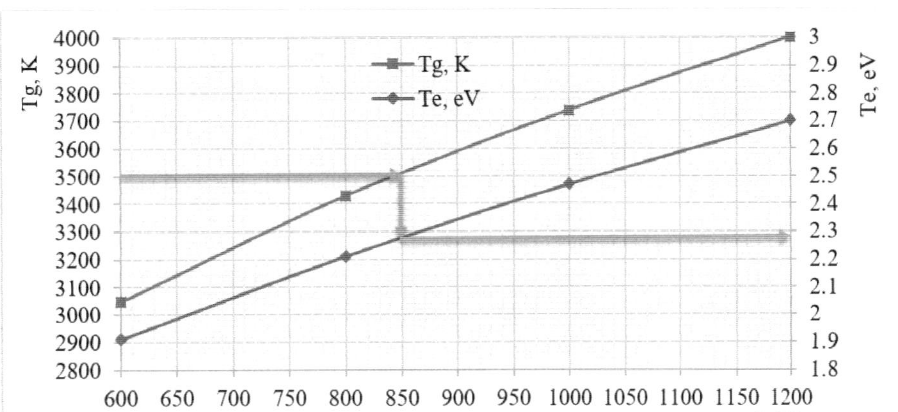

Figure 5. Dependence of gas temperature and the electron temperature at the center of the tube as a function of the electric power supplied to the gas discharge.

The graphic in Figure 5 can be used as a help diagram. For instance, by setting the maximum permissible gas temperature at the center of the tube at 3500 K (left coordinate axis), we calculate that the maximum permissible input power for the gas discharge is Q =850 W, which corresponds to the maximum value of the electron temperature at the center of the tube T_e ≈2.28 eV (right coordinate axis). The obtained results indicate that the increase of the electron temperature resulting from the higher electric power supplied to the gas discharge has its natural limits and it cannot be greater than 2.28 eV. For its further increase, other geometric and thermodynamic methods need to be sought – changes in tube geometry, changes in the shape of the tube's cross-section, changes to the buffer gas type and pressure, various additives, use of forced cooling, etc. It is the first time such a study is published.

5. Conclusion

For the first time a methodology is developed, based on an existing copper bromide vapor laser, which enables the determination of the maximum possible electric power supplied to the gas discharge and the maximum energy of the electrons by the positions of the thermal ionization and electric ionization resistance of the gas discharge. This methodology can be used for a wide range of gas lasers and metal vapor and metal compound vapor lasers to plan experiments for existing lasers and to develop new ones, including for monostatic and bistatic laser optical systems.

Acknowledgement

This work is supported by grant MU19-FMI-010 of NPD at University of Plovdiv Paisii Hilendarski, financed by the Bulgarian Ministry of Education and Science.

References

[1] Webb C and Jones J eds 2004 *Handbook of Laser Technology and Applications* vol 2 (New York: Institute of Physics Publishing)

[2] Steen W M and Mazumder J 2010 *Laser Material Processing* (London: Springer-Verlag)

[3] Zureng X, Guiyan Z and Fucheng L 1992 Applications of the CuBr vapor laser as an image-brightness amplifier in high-speed photography and photomicrography *Applied Optics* **31** 3395–3397

[4] Gocheva-Ilieva S G and Iliev I P 2011 *Statistical Models of Characteristics of Metal Vapor Lasers* (New York: Nova Science Publ Inc)

[5] Kazaryan M A,Batenin V M, Buchanov V V, Boichenko A M, Klimovskii I I and Molodykh E I 2017 *High Brightness Metal Vapor Lasers: Physics and Applications* (Boca Raton: CRC Press)

[6] Evtushenko G S, Trigub M V, Gubarev F A, Evtushenko T G, Torgaev S N and Shiyanov D V 2014 Laser monitor for non-destructive testing of materials and processes shielded by intensive background lighting *Rev Sci Instrum* **85**(3) art no 033111

[7] Trigub M V, Evtushenko G S, Torgaev S N, Shiyanov D V and Evtushenko T G 2016a Copper bromide vapor brightness amplifiers with 100 kHz pulse repetition frequency *Optics Communications* **376** 81–85

[8] Trigub M V, Torgaev S N, Evtushenko G S, Troitskii V O, Shiyanov D V 2016b A bistatic laser monitor *Technical Physics Letters* **42**(6) 632–634

[9] Isaev A A, Kazarian M A and Petrash G G 1974 Possibility of generation of high average laser powers in the visible part of the spectrum *Soviet Journal of Quantum Electronics* **3** 521–523

[10] Iliev I P and Gocheva-Ilieva S G 2010 Model of the radial gas temperature distribution in a copper bromide vapour laser *Quantum Electronics* **40**(6) 479–483

[11] Gocheva-Ilieva S, Iliev I and Fidanov D 2019 Primary analytical model for determining the electron temperature of a CuBr laser *IOP Conference Series: Materials Science and Engineering* **618** 012018

[12] Boichenko A M, Evtushenko G S, Nekhoroshev V O, Shiyanov D V and Torgaev S N 2015 CuBr-Ne-HBr laser with a high repetition frequency of the lasing pulses at a reduced energy deposition in the discharge *Physics of Wave Phenomena* **23**(1) 1–13

IOP Conf. Series: Materials Science and Engineering **878** (2020) 012023 doi:10.1088/1757-899X/878/1/012023

On the use of blockchain technologies in smart home applications

G Pazhev[1], Gr Spasov[1], M Shopov[1] and G Petrova[2]

Technical University of Sofia, Plovdiv branch, [1]Computer Systems and Technologies Department, [2] Department of Electronics, 4000 Plovdiv, Bulgaria,
georgpajev@gmail.com, {gvs, mshopov, gip}@tu-plovdiv.bg

Abstract. The paper presents an implementation of a smart home architecture with two gateways based on collaboration of blockchain technology and Internet of Things (IoT) Message oriented middleware (MOM) architecture for management of internal home appliances. A basic overview of IoT MOM, MQTT (Message Queue Telemetry Transfer) protocol and blockchain architectures is made and the key characteristics and consensus algorithms are presented. Some challenges and the Ethereum smart contracts concept of the blockchain technology are discussed.

1. Introduction

Smart home, in its essence, is a heterogeneous sensor network based on technologies such as Internet of Things (IoT). They are nowadays equipped with many devices such as smart meter, in-home displays, renewable energy sources and storage, and smart appliances such as washing machine, refrigerators, TV, oven, thermostat, HVAC (Heating Ventilation and Air Condition), lights, and plugs for electrical cars. For transparent communication between these devices it is necessary to use gateways for protocol transformations. There is a two-way communication that is utilized in the demand response, advance metering infrastructure, distributed energy generations and storage. The U.S Department of Energy presents two types of home area network architectures – utility-managed and utility and consumer managed. In the utility-managed architecture the utility monitors, controls and manages smart home appliances via his private network using utility gateway connected to this network. The utility and consumer managed architecture extends the utility-managed architecture with a common gateway/hub which acts as an intermediate device between the utility, home owners, third party service providers and the home appliances [1].

On one hand, the presence of many gateways complicates the system in the internal network of sensors and actuators. The presence of additional protocol transformations slows its functionality, and in order to reduce their number, it is necessary to select protocols that are compatible with terminal devices with limited computing capabilities. The presence of MQTT and COAP (Constrained Application Protocol) meet the requirements for devices with limited computational resources and are very convenient to use for building an internal sensor network. On the other hand, the system interface with the outside world must be compatible with different utility providers and their communication protocols. The presence of many heterogeneous devices in IoT leads to compatibility, security and discovery issues, both internally and externally. The blockchain technology, with its unified set of protocols and build-in security, could be used not only as an external link to any home-based network within the house, but also as a distributed database, that stores data contained in each transaction. That data can then be used for further analyses.

In this paper some of the issues and challenges regarding integration of blockchain technology and IoT devices are identified and a use-case implementation of a smart home application is presented.

2. Description of technologies.

2.1. Internet of Things

With the development of sensor networks, wireless mobile communication, embedded system and cloud computing, the technologies of IoT have been widely used in areas such as Smart Cities, public security, Smart homes and so on [2]. There are three essential components of IoT [3]: embedded devices – consisting of both low cost/low power devices and high-end gateways; scalable connectivity – each embedded device should be connected; cloud-based mass device management – centralized management of distributed devices.

The basic idea behind IoT is the pervasive presence around us of a variety of things or objects – such as Radio-Frequency IDentification (RFID) tags, sensors, actuators, mobile phones, etc. – which, through unique addressing schemes, are able to interact with each other and cooperate with their neighbours to reach common goals [4].

2.2. MQTT

Message Queue Telemetry Transfer (MQTT) consists of three components - subscriber, publisher and broker [5], as shown on figure 1. The publisher input messages to a specific topic, the broker dispatches these messages to the subscribed clients. Clients subscribe to particular topics relevant to them and receive each message published to those topics. The broker achieves security by authorizing the publishers and the subscribers.

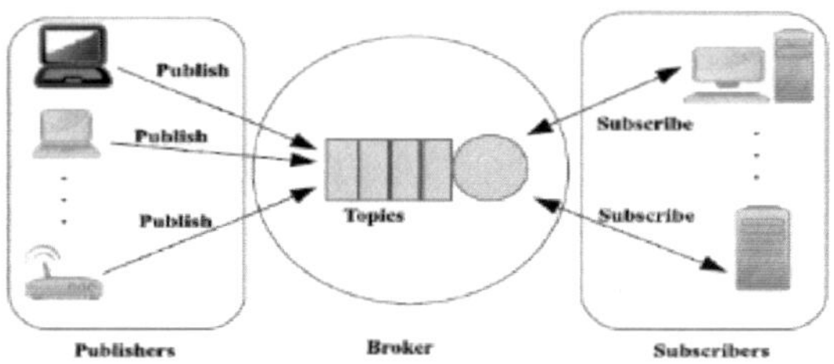

Figure 1. MQTT protocol block diagram [5]

2.3. Blockchain

Blockchain, at its core, is a peer-to-peer distributed ledger that is cryptographically secure, append-only, immutable (extremely hard to change), and can be updated only via consensus or agreement among peers [6]. There are various definitions of blockchain depending on point of view. From a business point of view a blockchain can be defined as a platform whereby peers can exchange values using transactions without the need for a central trusted arbitrator. This is a powerful concept and once readers understand it, they will realize the tsunami potential of blockchain technology. This allows blockchain to be a decentralized consensus mechanism where no single authority is in charge of the database. From a technical point of view blockchain can be thought of as a layer on the distributed peer-to-peer network running on top of the Internet, as shown in figure 2.

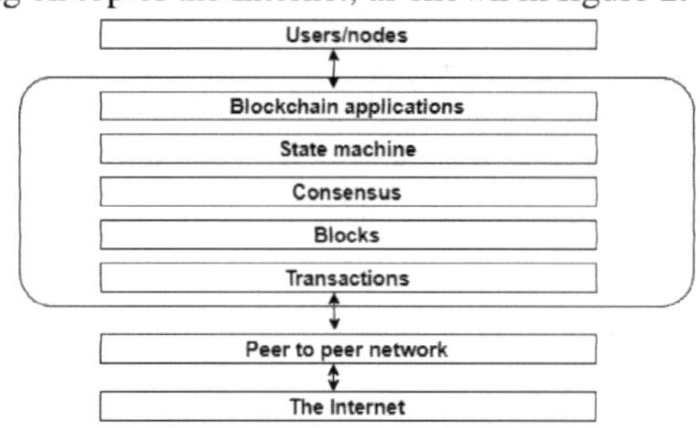

Figure 2. Blockchain as a layer on top of peer-to-peer network [6]

Each user, which participates into the blockchain network, is represented as a blockchain node that has an address, which is represented as a unique identifier. Each node can transfer values to other ones by initiating transactions inside the same network. All initiated transactions are grouped into blocks, which are validated by special nodes in the network called miner nodes. The miner nodes validate the newly created blocks by executing of consensus algorithm. All valid blocks are stored in the distributed ledger as shown in figure 3.

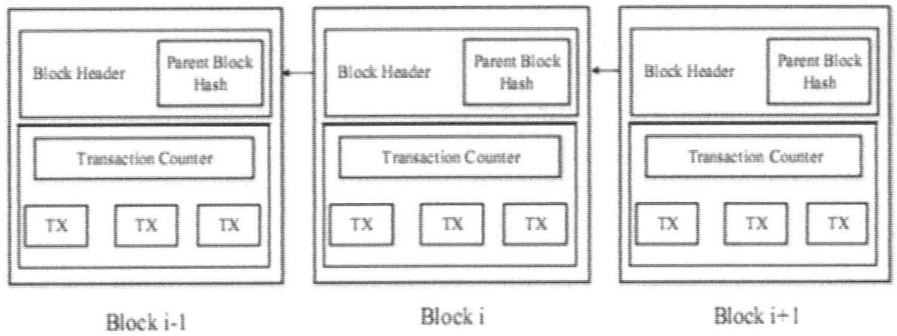

Figure 3. Distributed ledger structure [7]

The ledger is a chain of blocks, which contains all valid transactions (illustrated as TX blocks in figure 3). The first block at the chain is called genesis block. Each block is linked with previous one by Merkle Tree root hash located at his header. Being a decentralized system, blockchain systems do not need a third-party trusted authority. Instead, to guarantee the reliability and consistency of the data and transactions, blockchain adopts the decentralized consensus mechanism [8]. The essential of each consensus mechanism is to make the miner nodes to solve a puzzle to prove the truth of data. In the existing blockchain systems, there are four major consensus mechanisms: PoW (Proof of Work), PoS (Proof of Stake), PBFT (Practical Byzantine Fault Tolerance), and DPoS (Delegated Proof of Stake). The PoW consensus algorithm forces the miners to solve a computationally-intensive ease verifiable task by reaching a target value based on nonce value, hash of previous block and transaction hashes to create a new block. If the task is solved by anyone of nodes inside the network, then this node broadcasts the solution to other miner nodes and all other nodes must mutually confirm the correctness of solution. If the block is validated, other miners would append this new block to their own copy of the ledger. The PoS consensus algorithm is an energy-saving alternative of PoW. This algorithm randomly selects the node responsible to create the new block based on amount of coins that owns. It is based on fact, that the users who contain more coins are more interested the blockchain system to work correctly and these owners are more responsible to provide the ledger to contain valid blocks. The PBFT is proposed by Liskov and Castro as a replication algorithm for toleration of byzantine faults in distributed systems. In this algorithm the replicas move through a succession of configurations called views [9]. In a view one replica is the primary and the others are backups. When the primary receives a client request it starts a three-phase protocol to atomically multi-cast the request to the replicas as shown on figure 4. The three phases are pre-prepare, prepare, and commit. The pre-prepare and prepare phases are used to totally order requests sent in the same view even when the primary, which proposes the ordering of requests, is faulty. The prepare and commit phases are used to ensure that requests that commit are totally ordered across views.

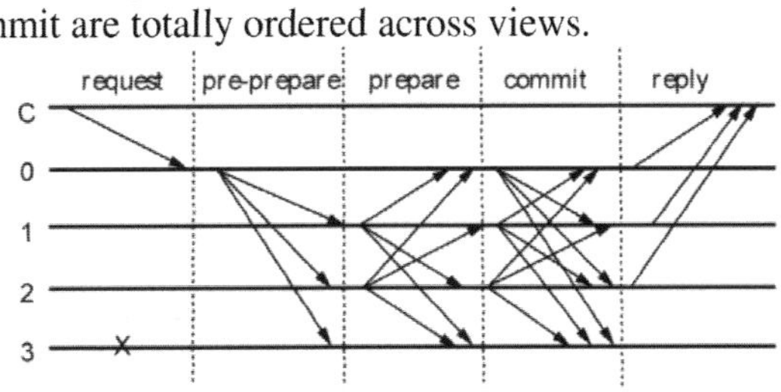

Figure 4. Practical Byzantine Fault Tolerance algorithm [9]

The DPoS consensus algorithm differs from PoS in that PoS is direct democratic while DPoS is representative democratic. In DPoS the stakeholders elect their delegates to generate and validate blocks. The parameters of the network such as block size and block intervals could be tuned by delegates. With significantly fewer nodes to validate the block, the block could be confirmed quickly, leading to the quick confirmation of transactions.

2.4. Ethereum and Smart contracts

Ethereum is an open source software based on blockchain technology, which focuses on providing a platform for building distributed blockchain applications (dApps) [10, 11]. The developer can implement, deploy and execute his application called smart contract in Ethereum network. The smart contract applications are executed by Ethereum Virtual Machine (EVM). Ethereum provides two types of accounts – user and smart contract. The user account does not execute any code. This type of account can only send messages to other accounts by creating and signing transaction using their private keys. The contract accounts always execute the code of any method defined in the deployed corresponding smart contract invoked by transaction. The Ethereum smart contracts are written in domain specific language (DSL) called Solidity. The code of the contract is compiled to byte-code and then is deployed to EVM. Figure 5 illustrates with a real world example how the smart contract acts as trusted intermediary between two users (a seller and a buyer).

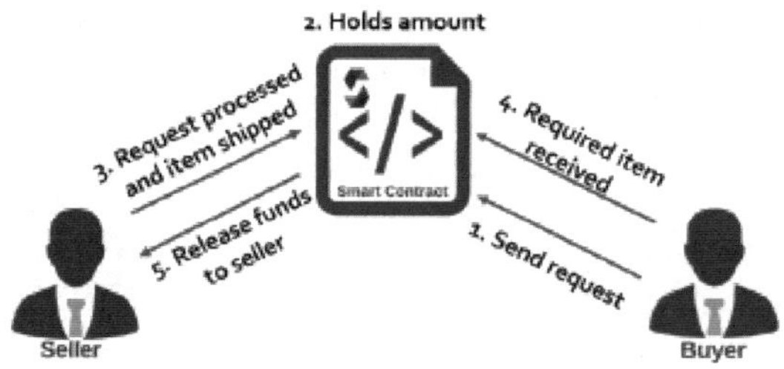

Figure 5. Smart contract execution [11]

The whole transaction which executes smart contract application completes in five steps. In the first step, the buyer sends amount of Ethers to the smart contract's address and this smart contract holds the ethers. In the second step, the smart contract notifies the seller indicating for the sent request by the buyer. In the third step, the seller checks and verifies the buyer's request and if it is valid and the amount of ethers are enough then the seller sends to the buyer the purchased item and inform the smart contract with shipment message. In the fourth step, if the buyer receives the item, then the smart contract is updated with the delivery status. In the last fifth step, after all other steps are successfully executed the hold ethers in the contract are send to the seller's account.

Each smart contract which is deployed on the blockchain network executes variety of elementary computational operations provided by EVM. The complexity of smart contract is evaluated by the number of these operations, which executes. This complexity is measured by the EVM in units named gas. The gas unit is mapped to ether costs as a fee, which is taken over by the user which initiates the request to the contract (the buyer in this case). The smart contract could contain two basic types of methods – methods which set values in their private members (set-methods) and view methods. The invocation of set-methods requires a transaction to be initiated for sending the value to be set in the corresponding private member inside the contract. The execution of this method costs the fee determined by the gas of contract. If this method is set as payable, then the amount of required ether, which will be hold by the contract, should be added to the transaction fee. The invocation of view methods does not require to initiate a transaction to view the value of private member of the contract, however its invocation does not require any fees. If we return to the example with the seller and buyer mentioned above, then the buyer requests items by the seller by invocation of payable set-method provided by the smart contract. In this situation, the buyer sends the amount of ether which will be hold by the contract for the price of items and pays the fee price of transaction determined by the gas of the contract.

3. Challenges

3.1. IoT Challenges

The challenges addressed to IoT and their middlewares are related to dynamic heterogenous resource discovery and composition, scalability, reliability, interoperability, security and privacy [12].

Resource discovery: The nature of the IoT infrastructure nullifies centralized resource registries and discovery approaches. A balance must be achieved between registry distribution and the number of registries. Fewer registries provide consistent and fast discovery of resources under normal circumstances. A problem may arise if there is large number of service discovery queries in IoT applications.

Resource management: Resource conflicts may occur in IoT applications that share resources. A conflict resolution is required to resolve conflicts in resource allocation among multiple concurrent services or applications.

Scalability: The most existing middlewares for IoT are centric around Wireless Sensor Networks (WSNs) and their network level scalability is also limited from WSNs. There is a high possibility that they will perform poorly in IoT's ultra-large-scale network. That's why it is important every component of IoT middleware to be scalable in order to achieve system-wide scalability.

Reliability: To achieve IoT middleware reliability, every component or service of a middleware needs to be easily replaceable. A better understanding of the dependency between reliability and other requirements is needed.

Interoperability: Although network interoperability is supported by most existing middlewares there are some that lack support for semantic and syntactical interoperability. There is a lack of standard in ontologies, which creates a big challenge for IoT. The best support for semantic interoperability is offered by the service-oriented approach, but the support for syntactical interoperability is limited. There is a need for better understanding of IoT services' syntax and semantics.

Security and privacy: Although all the concerns of security, privacy, and trust in all the technologies used in IoT are clearly present in the context of the IoT, they are not fully complete. Most existing middlewares' authentication-based partial security solutions are insufficient for a number of IoT applications. There is a need of research for a holistic security solution that takes care of system as well as middleware level security and privacy aspects.

3.2. Blockchain challenges

The Blockchain technology resolves partly the already mentioned challenges of IoT. In this context, knowledge of the Blockchain challenges is required. Some of them are [13]:

Storage capacity and Scalability: Storage capacity and scalability have been deeply examined in blockchain. In this technology, in Bitcoin for example, the chain is always growing at a rate of 1MB per block every 10 minutes, and there are copies stored among nodes in the network. Although only full nodes (nodes that can fully validate transactions and blocks) store the full chain, storage requirements are significant. As the size grows, nodes require more and more resources, thus reducing the system's capacity scale. In addition, an over-sized chain has negative effects on performance, for instance, it increases synchronization time for new users.

Security: This challenge requires knowledge of some of the most common attacks such as a 51% attack or a majority attack. This attack can occur if a blockchain participant is able to control more than 51% of the mining power. In this situation he can control the consensus in the network. The boom and fast evolution of mining pools, have increased the probability of this attack occurring, which in turn could compromise the integrity of Bitcoin. Another example for attack is the double-spend attack, which consists in spending the same coin twice.

Anonymity and Data Privacy: In public blockchain the ledger is public. However, all transactions are publicly visible for everyone across the blockchain network. Combined with traffic analysis, transactions could be linked back to their source IP (Internet Protocol) addresses, which points to the transactions originator. This is a big concern in privacy point of view. Various proposals have been made to address the privacy issue since the introduction of Bitcoin. These proposals could be divided into two categories: mixing protocols and inherent anonymity.

3.3. Integration of IoT and Blockchain

Another aspect necessary to take into account is related to the IoT interactions, i.e., the communication between the underlying IoT infrastructures. When integrating blockchain, it is necessary to be decided where these interactions will take place: inside the IoTs, through blockchain, or a hybrid design involving IoT and blockchain (figure 6). These three approaches are clearly described by [13]:

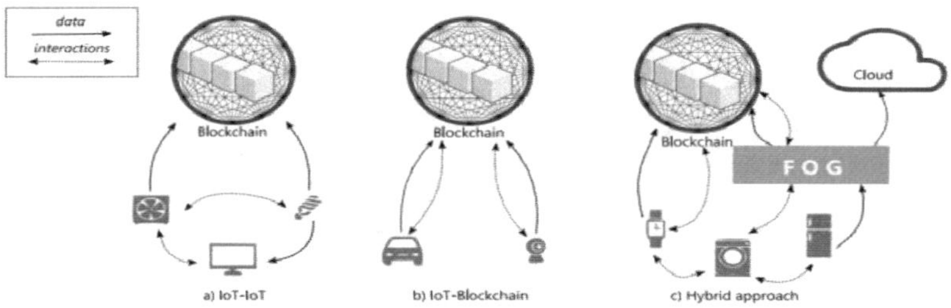

Figure 6. Blockchain IoT interactions [13]

Interactions IoT-IoT: this approach could be the fastest one in terms of latency, and security since it can work offline. The IoT devices have to be able to communicate with each other, which usually involves discovery and routing mechanisms. Only a part of the IoT data is stored in blockchain whereas the IoT interactions take place without using the blockchain (figure 6.a). This approach would be useful in scenarios with reliable IoT data where the IoT interactions are taking place with low latency.

Interactions IoT-Blockchain: in this approach all the interactions go through blockchain, enabling an immutable record of interactions. This approach ensures that all the chosen interactions are traceable as their details can be queried in the blockchain, and moreover it increases the autonomy of the IoT devices. The IoT applications that intend to trade or rent can leverage this approach to provide their services. Nevertheless, recording all the interactions in blockchain would involve an increase in bandwidth and data, which is one of the well-known challenges in the blockchain (figure 6.b). On the other hand, all IoT data associated with these transactions should also be stored in the blockchain.

Hybrid approach: lastly, a hybrid design where only part of the interactions and data take place in the blockchain and the rest are directly shared between the IoT devices. One of the challenges in this approach is choosing which interactions should go through the blockchain and providing the way to decide this in run time. A perfect orchestration of this approach would be the best way to integrate both technologies since it leverages the benefits of blockchain and the benefits of real-time IoT interactions. In this approach fog computing could come into play and even cloud computing (figure 6.c), to complement the limitations of blockchain and the IoT.

4. Use case: A smart home application

A realization of home automation network based on open source hardware is presented in [14]. In this realization CNDEP (Controller Network Data Extracting Protocol) protocol is used for interaction between the gateways for the home appliances and the control gateway. Each device in CNDEP is identified by session id. The control gateway sends command id and sensor function id to extract the sensor data or contact with the actuator. However, with this smart home implementation there are the following challenges that need to be addressed: higher scalability, interoperability and security from the middleware point of view. Although it uses a resurrecting duckling as security policy, this is not quite enough to be highly secure. This home network implementation is not highly flexible concerning the scalability because it requires a new session id to be created to add a new device in the network. This requires the whole network to be re-configured.

Considering these challenges we propose realization of an IoT-based smart home network, using the concept with two gateways and integrating IoT and Blockchain technologies. Using the IoT-IoT

IOP Conf. Series: Materials Science and Engineering **878** (2020) 012023 doi:10.1088/1757-899X/878/1/012023

approach for integrating the two technologies we propose a smart home architecture (figure 7), where each smart home manages its own private sensor network with control gateway, which could share data from this internal network and could provide access to the actuators via its smart contract deployed inside the Blockchain network. The user or utility provider could connect to the corresponding smart home and exchange data with the provided smart contract. Every participant inside the blockchain network needs to create transaction for sending data. Each transaction is digitally signed and encrypted using asymmetric security algorithms. This advantage gives secure connection between the user/utility provider and the home network. From the other hand the transactions are saved in the ledger, however the shared data are also saved at the same place and could be used for further analysis.

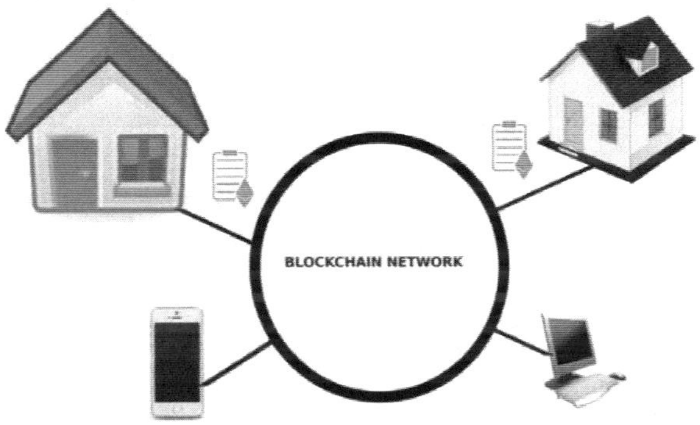

Figure 7. Smart contract execution

The smart contracts provide unified interface, which improves smart homes interoperability with the utility providers and user nomadic agents. Using blockchain network the utility provider does not need to install his own smart meter at each home and to provide own secured infrastructure. The smart home could share the sensor data and the actuators state to its own smart contract. The utility provider could connect to the same smart contract to extract the data. The smart contract also could be used as third-party software for secure connection to the home by the consumer through Internet.

In the blockchain network the participant who initiates the transaction to the smart contract pays the gas for execution of the smart contract. To decrease the number of payments of gas for invocation of the smart contract by the control gateway two approaches could be used. In the first approach, the control gateway could send transaction with all data to the smart contract inside the blockchain network if there are any changes in the sensor data. In the second approach, the consumer or utility provider could initiate a transaction for demanding of the sensor data. The control gateway extracts all the data from the sensors and actuators and creates a transaction to send these data to the blockchain network.

Our test-bed implementation of a smart home application, based on the integration of IoT and blockchain technologies is shown in figure 8.

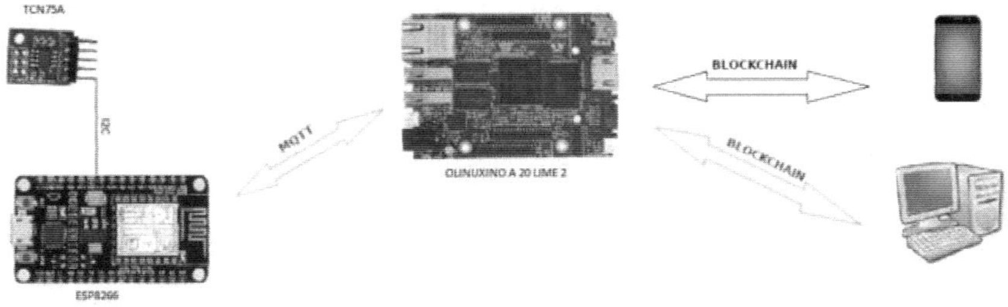

Figure 8. Test-bed realization

The private home network is based on two-gateway architecture – gateway for each appliance inside the home (appliance gateway) and a single control gateway, which provides connection to the outside world by exchanging data with deployed smart contract at the Ethereum blockchain network. Each appliance gateway could contain sensors and actuators. They communicate with the control gateway using MQTT protocol. The control gateway acts as MQTT broker and provides specific topic for each appliance gateway for its sensors and actuators. In our test-bed implementation the appliance gateway is realized by wireless IoT module ESP8266 with connected TCN75A temperature sensor via I2C interface. This platform provides also a WiFi network interface for connection with the control gateway and sends the temperature data in degrees at the specific topic. For implementation of control gateway Olinuxino A20 is used. It has dual core A20 processor with 1GB of RAM and Ethernet interface for connection to the Internet. The control gateway has installed a go-ethereum client (geth) for connection with the blockchain, and mosquito MQTT implementation to work as MQTT broker. Olinixuno A20 platform extracts the data from the specific topic (e.g. temperature topic) and sends these data to the smart contract inside Ethereum Ropsten test blockchain network via JSON-RPC.

5. Conclusion

Every technology has its place within the IoT eco-system. The use of MQTT to build an internal network within a smart home covers a variety of devices with limited computational abilities and it is also characterized by low cost of communication. As a result of the experimental test-bed realization presented in this paper, the following conclusions are reached regarding the use of blockchain in collaboration with MQTT: the use of both technologies reduces the requirements of the control gateway to maintain a large set of protocols for communication with the internal and external networks, as well as the requirements for the use of a database to store the sensor data for further analysis. Used smart contracts provide a unified interface and can also be used as a means of distributed computing. The blockchain integration combined with smart contract feature contributes to the security and interoperability of the house with the external communication. There is a need for further research for exchanging and storage of privacy information, which requires to be visible only for limited range of participants.

6. References

[1] Hafeez A, Kandil N H, Al-Omar B, Landolsi T, and Al-Ali A R, 2014 Smart Home Area Networks Protocols within the Smart Grid Context *Journal of Communications* Vol. **9** No. 9 pp. 665-671
[2] Beng L 2009 Sensor cloud: Towards sensor-enabled cloud services (Intelligent Systems Center Nanyang Technological University)
[3] Geng W, Talwar S, Johnsson K, Himayat N and Johnson K D 2011 M2M: From mobile to embedded internet *Communications Magazine*, IEEE **49**, no. 4 pp 36-43
[4] Giusto, Iera A, Morabito G and Atzori L 2010 The Internet of Things (Springer ISBN: 978-1-4419-1673-0)
[5] Hasan H Evaluation of MQTT Protocol for E-Learning *International Journal of Computer Science & Mobile Computing* vol. **7** issue 11 pp. 57–67
[6] Bashir I 2017 Mastering blockchain (Packt publishing)
[7] Zheng Z, Xie S, Dai H, Chen X and Wang H 2017 An Overview of Blockchain Technology: Architecture, Consensus, and Future Trends *IEEE 6th International Congress on Big Data* pp 557-564
[8] Lie X, Jiang P, Chen T, Xiapu L and Qiaoyan W 2017 A Survey on the Security of Blockchain Systems Future Generation Computer Systems (Elsevier)
[9] Liskov B and Castro M 1999 Practical Byzantine Fault Tolerance *Proceedings of the Third Symposium on Operating Systems Design and Implementation* (New Orleans, USA)
[10] Gencer A , Basu S , Eyal I , Renesse R and Sirer E 2018 Decentralization in Bitcoin and Ethereum Networks", Financial Cryptography and Data Security (http://www.arxiv.org/)
[11] Praitheeshan P, Pan L, Yuy J, Liuy J, and Doss R Security Analysis Methods on Ethereum Smart Contract Vulnerabilities — A Survey (*preprint http:/www./arxiv.org/)*

[12] Razzaque M A, Milojevic-Jevric M, Palade A, and Clarke S 2016 Middleware for Internet of Things: A Survey *IEEE Internet of things journal* vol. **3** No. 1 pp.70-95
[13] Reyna A, Martin C, Chen J, Soler E, Diaz M 2018 On blockchain and its integration with IoT. Challenges and opportunities (Future Generation Computer Systems)
[14] Petrova G, Spasov Gr, Pazhev G 2015 Realization of a Home Automation Network with Two Gateways Based on Open Source Hardware *Annual journal of electronics* pp. 116-119

Acknowledgments
This work was supported by the European Regional Development Fund within the Operational Programme “Science and Education for Smart Growth 2014 - 2020” under the Project Center of Competence "Intelligent mechatronics, eco- and energy-saving systems and technologies“BG05M2OP001-1.002-0023-C0.

IOP Conf. Series: Materials Science and Engineering **878** (2020) 012024 doi:10.1088/1757-899X/878/1/012024

Cycle length optimization through bi-level optimization

Yordanka Boneva[1]

[1] Institute of Information and Communication Technologies – Bulgarian Academy of Sciences, Sofia, Bulgaria

boneva.yordanka@gmail.com

Abstract. This paper describes the formalization of the store-and-forward model in bi-level optimization problem. The study finds the optimal duration of the cycle lengths by given green splits. The queues in front of traffic lights are minimized. Thus, the network allows higher throughput and less congestion with the resulting from this less pollution and better traffic indicators such as delay, density, speed, etc. The results for traffic indicators and pollution are obtained by Aimsun suit – a specialized traffic modelling software product. The optimization is performed by a MATLAB script based on an additional toolbox called YALMIP as well as with a state-of-the-art software product TRANSYT that is compatible with Aismun suit. Results show that it is reasonable to use the script for optimization of traffic as TRANSYT optimizes only green splits and offsets but works with the cycle value form Aimsun.

1. Introduction

Traffic lights are the main mean of control in urban areas. They separate the traffic flow and pedestrians in time. Thus a minimization of conflicts between different traffic flows and traffic flows and pedestrians is achieved.

There are several parameters related to traffic flow. Firstly, the cycle of the traffic light is defined. The cycle is a sum of green, amber and red light. Secondly, an important parameter is the green light also called green split. The amber light may vary between 2-3 seconds depending on the permitted speed in the section of the road. Usually, for urban settings the amber light is set to 3 seconds.

The optimization of these parameters for a single junction as well as for network of junctions is a prerequisite for optimal throughput of vehicles through the junction. Different optimization approaches are explored for the optimization of the cycle length and the green splits. In this paper the bi-level optimization approach is considered. It has the advantage that two objective functions can be optimized at the same time in contrast to classis optimization approach with one objective function. The goal functions are in hierarchical dependence to each other. There are an upper-level objective function and lower-level objective function. The bi-level optimization used in other areas as portfolio optimization, optimisation of railway transport, etc.[1], [2], [3].

A model of queues in front of traffic lights is used called store-and-forward model. It is a simple model based on communication networks and applied to traffic networks. Queues in front of traffic lights are calculated as sum of the cars that were on the queue at a given moment plus entering cars minus cars that are leaving the link. [4], [5], [6], [7].

Based on the bi-level optimization and the store-and-forward model a problem was formulated and solved. The problem is to optimize the duration of cycle length by given green splits. The solution was achieved through the software products MATLAB, Aimsun and TRANSYT. MATLAB uses the tool box YALMIP and the function solvebilevel() to find the cycle length [8]. Further, the results are tested in simulations in Aimsun suit. A comparison is made between the results achieved through MATLAB and the result achieved through state-of-the-art software TRANSYT. Results show that there is a small difference in traffic indicators between the case solved with MATLAB and the one solved in TRANSYT in favour of TRANSYT. Still, because the difference is small and because the cycle length cannot be tackled in TRANSYT as it can be in MATLAB it can be concluded that the bi-level optimization is a promising approach. TRANSYT uses the Hill-Climb optimization method for optimization of green splits and offset and take the cycle length as a given constant. MATLAB uses solvebilevel() to tackle the cycle length and takes the green splits in the network as constants.

The data that are used in this study are acquired at site as the selected network is not equipped with sensors. It is though preferable when data is gathered by sensors and cameras [9].

2. Description of the simulated network

The simulation network consists of four neighboring junctions along Shipchenski prohod Blvd. The junctions are regulated by traffic lights and are situated at relatively small distance of each other which is a good prerequisite for achieving a "green wave". The total distance of the network is 1.5 km.

There are many points of interest in the network like offices, shopping centers, schools, kindergarten, a hospital etc. that generate traffic in the area. Figure 1 is a model of the network in the simulation software Aimsun.

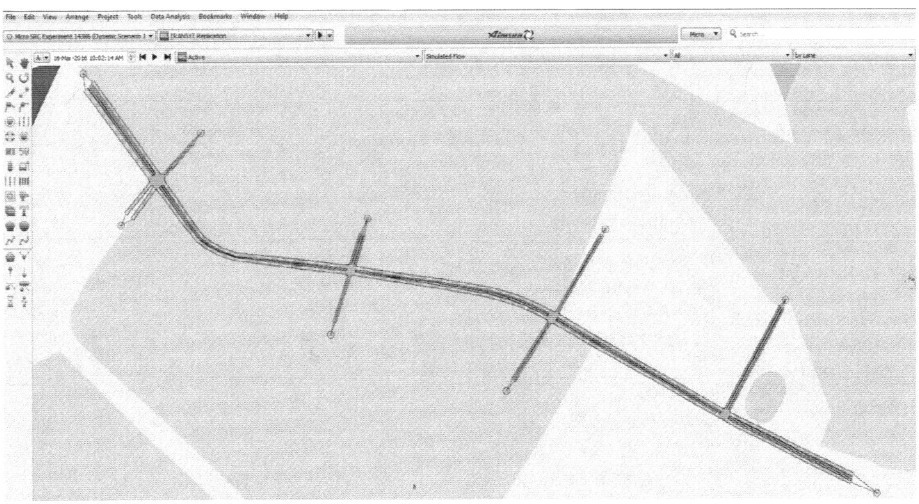

Figure 1. The network of four junctions

3. Description of the experiment

The experiment aims to reveal the potential of bi-level optimization for the optimization of the cycle length in a network of junctions. The function solvebilevel() in MATLAB was used for this purpose as well as the tool box YALMIP.

A script was written in MATLAB that minimizes the cycle lengths of the four connected junctions and at the same time minimizes the queues in front of traffic lights. The duration of green lights are taken as constant control parameters. Green splits are calculated as a part of the cycle length as the cycle comprises of the sum of green, yellow and red light. If the cycle is taken as 1, the yellow is taken as 0.1*cycle length, green is notated with u_i, red is then equal to 0.9cycle $-u_i$. These notifications are important for the understanding of the script in MATLAB and the given model.

The well-known store-and-forward model is applied. The store-and-forward model is a concept taken from computer networks and applied to transport networks. Based on the store-and-forward model the constraints for the lower level problem of the bi-level optimization were defined.

IOP Conf. Series: Materials Science and Engineering **878** (2020) 012024 doi:10.1088/1757-899X/878/1/012024

Bi-level optimization comprises of outer and inner objective. In the case of this study the outer objective is the cycle lengths that are to be minimized. The outer objective is the sum of the queues in the network to minimize. However, a higher flow is achieved than the base scenario.

The notations used by the script are as follows:

OO – outer objective
CO – outer constraints
IO – inner objective
CI – inner constraints

The variables are defined. The variables $x_{1...11}$ are the number of cars in front of traffic lights. The variables $c_{1...4}$ are the cycle lengths.

```
sdpvar x1 x2 x3 x4 x5 x6 x7 x8 x9 x10 x11
sdpvar c1 c2 c3 c4
```

The bounds for the cycle length are defined.

```
cmin = 40/3600;
cmax = 120/3600;
```

Green splt as part of the cycle length is defined.

```
u1  = 0.422535;
u2 =  0.37037;
u3 = 0.446429;
u4 = 0.416667;
```

The flow of cars is given as measured.

```
x1in = 1368;
x2in = 870;
x3in = 246;
x4in = 84;
x5in = 150;
x6in = 90;
x7in = 54;
x8in = 60;
```

The queue length at previous time step is given. In this case the network is accepted as empty.

```
x10 = 0;
x20 = 0;
x30 = 0;
x40 = 0;
x50 = 0;
x60 = 0;
x70 = 0;
x80 = 0;
x90 = 0;
x100 = 0;
x110 = 0;
```

Saturation flow is given as follows:

```
s1 =   624;
s2 =   744;
s3 =   558;
```

IOP Conf. Series: Materials Science and Engineering **878** (2020) 012024 doi:10.1088/1757-899X/878/1/012024

```
s4 =  312;
s5 =  620;
s6 =  30;
s7 =  160;
s8 =  12;
s9 =  72;
s10 =  30;
s11 =  30;
s12 =  360;
s13 =  90;
s14 =  80;
s15 =  40;
s16 =  50;
s17 =  20;
s18 =  40;
s19 =  499;
s20 =  418;
s21 =  32;
s22 =  53;
s23 =  237;
s24 =  28;
s25 =  16;
```

Upper level objective function and upper level constrains are given.

```
OO = c1 + c2 + c3 + c4;
CO = [c1 >= cmin , c1 <= cmax,
   c2 >= cmin , c2 <= cmax,
   c3 >= cmin , c3 <= cmax,
   c4 >= cmin , c4 <= cmax
   ];
```

Lower level objective function and lower level constrains based on the store-and-forward model are given:

```
OI = x1 + x2 + x3 + x4 + x5 + x6 + x7 + x8 + x9 + x10 + x11;
 CI = [ x1 <= x10 + x1in - u1*s2*c1 - u1*s1*c1,
x2 <= x20 + x2in - (0.9 - u1)*s3*c1 - (0.9 - u1)*s4*c1,
x3<= x30 + u1*s2*c1 + (0.9 - u1)*s4*c1 - u2*s5*c2 - u2*s7*c2 - u2*s6*c2,
x4 <= x40 + x3in - (0.9 - u2)*s8*c2 - (0.9 - u2)*s7*c2,
x5 <= x50 + x4in - (0.9 - u2)*s10*c2 - (0.9 - u2)*s11*c2,
x6 <= x60 + (0.9 - u2)*s7*c2 + (0.9 - u2)*s11*c2 + u2*s5*c2 - u3*s12*c3 - u3*s14*c3 -
u3*s13*c3,
x7 <= x70 + x5in - (0.9 - u3)*s15*c3 - (0.9 - u3)*s16*c3,
x8 <= x80 + x6in - (0.9 - u3)*s17*c3 - (0.9 - u3)*s18*c3,
x9 <= x90 + u3*s12*c3 + (0.9 - u3)*s16*c3 + (0.9 - u3)*s18*c3 - u4*s19*c4 - u4*s21*c4 -
u4*s20*c4,
x10 <= x100 + x7in - (0.9 - u4)*s24*c4 - (0.9 - u4)*s25*c4,
x11 <= x110 + x8in - (0.9 - u4)*s22*c4 - (0.9 - u4)*s23*c4

x1 >= 0,
x2 >= 0,
x3 >= 0,
```

```
x4 >= 0,
x5 >= 0,
x6 >= 0,
x7 >= 0,
x8 >= 0,
x9 >= 0,
x10 >= 0,
x11 >= 0 ]

solvebilevel(CO,OO,CI,OI,[c1 c2 c3 c4]);
```

The cycle length is calculated back in seconds.

```
c1 = c1*3600;
c2 = c2*3600;
c3 = c3*3600;
c4 = c4*3600;
```

The solution of the script with approximation for c1 is 40 seconds, c2 is 55 seconds, c3 is 80 seconds c4 is 40 seconds.

4. Results and Discussion

In the previous point the MATLAB solution was given. In this point the TRANSYT solution will be presented as well as a comparison between the base case, the MATLAB case and TRANSYT case. The base case is a simulation in Aimsun suit based on data as it was measured at site.

Results show that the optimization made with TRANSYT give best results but here it should be noticed that TRANSYT optimizes green splits and offsets whereas MATLAB optimizes the cycle lengths. Although optimizing different parameters both MATLAB script and TRANSYT give better results than the simulation of the base case scenario. And comparing traffic indicators after simulation with MATLAB script input for cycle and TRANSYT input for green split and offset it is obvious that these indicators are in favor of TRANSYT but with very small difference. The results are depicted on Table 1.

Table 1. Comparison of traffic indicators for three cases

Traffic Indicators	Base Case	MATLAB script	TRANSYT Optimization	Units
Delay Time	51.16	36.4	35.51	sec/km
Density	7.69	6.91	6.27	veh/km
Flow	4095	4107	4124	veh/h
Fuel Consumption	355.53	338.06	293.98	l
CO2	875112.07	843316.7	768464.08	g
NOx	1311.3	1250.91	1130.48	g
PM (particulate matter)	223.74	210.73	174.77	g
VOC (Volatile organic compounds)	1107.33	997.88	917.66	g
Mean Queue	29.85	20.88	16.28	veh
Number of Stops	0.16	0.14	0.12	#/veh/km
Speed	34.25	37.59	38.8	km/h
Stop Time	38.42	25.48	25.76	sec/km
Total Number of Stops	8322.87	7404.37	6401.28	

Total Travel Time	94.76	85.44	77.96	h
Total Travelled Distance	2925.17	2939.58	2932.89	km
Travel Time	117.76	102.99	102.13	sec/km

5. Conclusion

In this paper a script in MATLAB for the optimization of cycle lengths of traffic lights network along Shipchenski prohod Blvd. in Sofia was presented. The script solves bi-level problem using the store-and-forward method as a lower-level constraint. A comparison is made with the results of state-of-the-art optimization software TRANSYT. Three scenarios were simulated in Aimsun suit – a software product for modeling and simulation of vehicle traffic.

As a conclusion it can be stated base on the result that the bi-level optimization is worth exploring as an optimization approach for the purposes of traffic optimization.

Acknowledgments

This work has been partly supported by two projects: KP-06-H37/6, 6.12.2019 of the Bulgarian National Science fund: "Modelling and optimization of urban traffic in network of crossroads" and project KP-06-M27/9, 17.12.2018 of the Bulgarian National Science fund: "Contemporary digital methods and tools for exploring and modeling transport flows".

References

[1] K. Pavlova,B. Vatchova and E Paunova, 2018. Quantitative estimation of throughput abilities in the transport graph under conditions of limited output data, *Journal of Bulgarian Science*, ISSN: 1314-1031, pp. 45-52.

[2] K.Pavlova, T.Stoilov, 2017. Design of State Rail and Bus Transportation Scheme with Bi-Level Optimization Model. *Journal of Information technologies and control*, N4, 2017, p.2-9, Print ISSN 1312-2622; Online 2367-5357, DOI: 10.1515/itc-2017-0031. http://www.aksyst.com:8081/Sai/Journal/Docum/1-4_17.pdf

[3] T. Stoilov, Krasimira Stoilova, 2012, Portfolio Risk Management Modelling by Bi-level Optimization, *chapter 5 in Handbook in decision making*, vol 2 Risk management in decision making, ed. J.Lu, L.Jain, G.Zhang, Inteligent systems reference library vol.33. Springer – Verlag, Berlin, Heidelberg, ISSN1868-4394, ISBN 978-3-642-25755-1, pp.91-110

[4] D. C. Gazis and R. B. Potts, 1965, *Proc. 2nd Symposium on Theory of Road Traffic Flow*, O.E.C.D.), p. 221

[5] D. C. Gazis, 1964, *Ops. Res*., 12, 815.

[6] D.C. Gazis, 1974.Modeling and optimal control of congested transportation systems, Networks, 4, John Wiley & Sons Inc., pp. 113-124.

[7] T. Stoilov, K. Stoilova, M. Papageorgiou and I. Papamichail, 2015. Bi-Level Optimization in a Transport Network. *CIT*, vol.15, No5, pp 37-49. Print ISSN: 1311-9702; Online ISSN: 1314-4081 DOI: 10.1515/cait-2015-0023.

[8] J. Lofberg, 2004. YALMIP : A Toolbox for Modeling and Optimization in MATLAB. *In Proceedings of the CACSD Conference*. Taipei, Taiwan, IEEE, ISBN: 0-7803-8636-1. pp. 284 – 289, DOI: 10.1109/CACSD.2004.1393890, https://ieeexplore.ieee.org/document/1393890

[9] I. Garvanov and Vladimir Ivanov, 2019. Detection and Parameters Estimation of Moving Objects via Video Surveillance, *Proceedings of the 2019 3rd International Conference on Automation, Control and Robots*, Prague, Czech Republic — October 11 - 13, ICACR 2019, ACM New York, NY, USA, ISBN: 978-1-4503-7288-6, pp. 102-106, DOI: 10.1145/3365265.3366749, https://dl.acm.org/citation.cfm?id=3366749.

Decrease of the portfolio computational workload following the Capital Asset Pricing theory

T Stoilov [1,2,*], K Stoilova[1,2], M Vladimirov [3]

[1]Institute of Information and Communication Technologies – Bulgarian Academy of Sciences, Sofia, Bulgaria
[2] Nikola Vaptsarov Naval Academy, Varna
[3]Varna University of Economics, Varna, Bulgaria

e-mail : todor@hsi.iccs.bas.bg

Abstract. The paper derives an evaluation algorithm for decreasing the computational workload in definition and solution of portfolio optimization problem. The algorithm applies numerical relations, which lead to transformation of the classical portfolio problem to optimization one, which contains parameters from Capital Market Theory. The decrease of the computational workload results from the inclusion of beta coefficients in the portfolio problem instead of evaluation of the covariance portfolio matrix. The algorithm is illustrated by solution of portfolio problem with mutual funds on Bulgarian Stock Exchange.

Keywords: Active investments, mathematical models in portfolio theory, portfolio optimization

1. Introduction

The foundation of Modern Portfolio Theory (MPT) starts with the formal definition of the portfolio problem by Markowitz in 1952. The portfolio problem is defined as optimization one for maximization of the portfolio return, subject to the constraint value of the portfolio variance [1]. The solution of this problem gives the relative values of the investment, which has to be used for buying assets which contribute to the portfolio performance. For the direction of evolution of the MPT one can refer to the overview presented in [2].

The formal presentation of the portfolio return is given by linear relation of expected asses' returns multiplied by the corresponding weights. The portfolio risk is defined as quadratic relation between the assets' weights and the covariance matrix of assets' returns. From practical considerations the evaluation of the covariance matrix requires a big amount of calculations. Nevertheless of the symmetric structure of the covariance matrix, a portfolio with N=100 different assets/stocks would require the evaluation of 5050 covariance components, $\left(\frac{N(N-1)}{2}+N\right)$. Thus, the evaluation of a covariance matrix will insist time consuming resources for finding a set of data which are obligatory for the formal definition of the portfolio problem.

This paper derives an algorithm which constraint the need to find all components of the covariance matrix. The formal ground of the algorithm is the usage of the Capital Asset Pricing Model, launched by Sharpe [4]. Because the MPT has its own formal models and applies portfolio formal relations, below it is given short description of the portfolio problem.

2. Mean variance portfolio problem

The portfolio optimization targets the construction of such allocation of an investment between a set of assets by means to make a trade-off between the portfolio expected return and portfolio risk. The mean variance definition of the portfolio optimization problem targets maximization of the portfolio expected return and minimizing the volatility as measure of portfolio risk [3]. The classical portfolio problem, defined by Markowitz, has two forms:

- maximization of the portfolio return for a given level of risk

$$\underset{\mathbf{w}}{\boldsymbol{max}}\begin{bmatrix}\mathbf{E}^{\mathrm{T}}\mathbf{w}\\ \mathbf{w}^{\mathrm{T}}\boldsymbol{\Sigma}\mathbf{w}\le\sigma_{\max}^2\end{bmatrix} \tag{1}$$

or

- minimization of the portfolio risk for a required level of return

$$\underset{\mathbf{w}}{\boldsymbol{min}}\begin{bmatrix}\mathbf{w}^{\mathrm{T}}\boldsymbol{\Sigma}\mathbf{w}\\ \mathbf{E}^{\mathrm{T}}\mathbf{w}\ge E_{\min}\end{bmatrix}, \tag{2}$$

where

E_i - the mean return of asset $i=1,...,N$, $\mathbf{E}^{\mathrm{T}}=(E_1, ..., E_N)$,

$\sum$ - the covariance matrix between the returns of the assets,

$\mathbf{w}^{\mathrm{T}}=(w_1, ..., w_N)$, w_i –the relative part of the investment, allocated to asset $i=1,...,N$, as a solution (weight) of the portfolio problem,

σ_{max}^2 – given level of portfolio risk,

E_{min}–required level of portfolio return.

The MPT assumes that the portfolio risk can be decreased by holding in appropriate combination the risky assets. The input parameters which have to be evaluated for the classical problem (1) concern the vector of mean returns of the assets, $\mathbf{E}^{\mathrm{T}}=(E_1, ..., E_N)$, where N is the set of returns, participating in the portfolio and the covariation matrix

$$\mathbf{COV}(.)=\Sigma_{N\times N}=\begin{vmatrix}cov_{11}\ \cdots & cov_{1N}\\ \cdots & \cdots\\ covc_{N1}\ \cdots & cov_{NN}\end{vmatrix}.$$

Using the notations from [3,4] these input parameters are evaluated from a historical set of data about the assets' returns $R_i^{(n)}$, $i=1,\ldots,N$ (set of returns), n is the discrete in time in which the assets returns $R_i^{(n)}$ are estimated. Thus, $E_i=\sum_{k=1}^{n}p_i^{(k)}R_i^{(k)}$, where $p_i^{(k)}$ is probability that at time k the asset i has current level of return $R_i^{(k)}$. The classical mean variance (MV) theory assumes normal distribution of the assets returns and each discrete k has equal probability, the mean asset returns is evaluated as an average value

$$E_i=\frac{1}{n}\sum_{k=1}^{n}R_i^{(k)}. \tag{3}$$

Respectively, the covariance components cov_{ij} between couples of assets $i,j\in 1,\ldots,N$ is evaluated as

$$cov_{ij} = \frac{1}{n}\sum_{k=1}^{n}\left(R_i^{(k)} - E_i\right)\left(R_j^{(k)} - E_j\right), \quad \forall i,j \in 1,\dots,N \tag{4}$$

Following [4] it holds

$$cov_{ij} = \sigma_i^2, \quad \forall i,j \in 1,\dots,N \tag{5}$$
$$cov_{ij} = cov_{ji} \ ,$$

which means that the diagonal components of the covariance matrix Σ contain the volatilities (risk) of the assets. The covariance matrix is a symmetrical one, which reduce the computational load for evaluation its components. Unfortunately, with the increase of N, the set of assets, participating in the portfolio it is not practical to implement relation (4) for all $\frac{N(N-1)}{2} + N = \frac{N(N+1)}{2}$ components of the covariation matrix Σ. The decrease of the computational workflow is a prerequisite for implementation of active portfolio management. The research targets the derivation of sequence of computations which result in algorithm for faster solution of the classical portfolio problem (1). It is used additional relations originated from the Capital Market Theory [4]. Then, the solution of the portfolio problem is solved in Excel environment using real data from the Bulgarian stock exchange with actual data of 2019.

3. Relations originated by the Capital Asset Pricing Model (CAPM)

The Capital Market Theory originated a model, titled CAPM [4, 6]. This model introduces two additional formal components in the MPT: the risk free asset and the market point (E_M, σ_M^2). The CAPM introduces relationship between the Market return E_M, the portfolio mean return E_p or asset mean return E_i and the risk free value r_f. This relation is linear, named Security Market Line (SML). This line is derived according to assumption that the assets' returns E_i depend in linear regression to the market returns R_M, respectively E_M . For analytical derivation of SML geometrical considerations are used, given in fig.1.

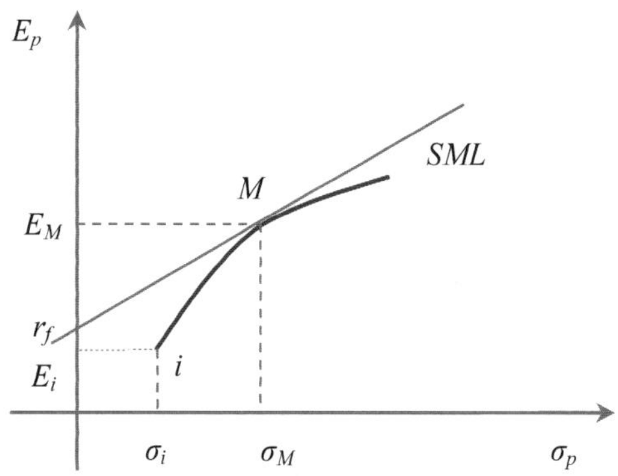

Figure 1. Presentation of asset *i* and Market point

The security *i* and the market has return and risk values respectively (E_i, σ_i) and (E_M, σ_M). The notation σ_i is the standard deviation of the returns R_i around the mean value E_i and it is a measure for the risk of asset *i*. Respectively, σ_M is the risk of the market point M. With the two assets (E_i, σ_i) and (E_M, σ_M) it is possible to make a combination in a portfolio with assets weights, respecting the portfolio constraint

$$w_i + w_M = 1 \ . \tag{6}$$

The resulting portfolio will have mean return and risk is given

$$E_{p_{(w)}} = w_i E_i + w_M E_M,$$
$$\sigma^2_{p_{(w)}} = w_i^2 \sigma_i^2 + w_M^2 \sigma_M^2 + 2 w_i w_M cov_{iM} \quad , \tag{7}$$

where cov_{iM} is the covariance value between the returns of asset i and the market. If the weights w_i and w_M change according to relation (6), the portfolio return and risk (E_p, σ_p) will move to the curve $\widehat{iM}$. Following the CAPM, the point M is a tangent one to the curve $\widehat{iM}$ and passes through the risk free point (r_f, 0). The tangent line in arbitrary point S_i of the curve $\widehat{iM}$ will have slope $s_i = \frac{dE_p}{d\sigma_p}$. To find analytical description of the derivative $\frac{dE_p}{d\sigma_p}$ here it is used the sequential of the differentiation

$$\frac{dE_p(w)}{d\sigma_p(w)} = \frac{\partial E_p/\partial w}{\partial \sigma_p/\partial w} \quad . \tag{8}$$

Using (6), relation (7) can be described only with the weight w_i or

$$w_i = 1 - w_M$$

$$E_{p_{(w_i)}} = w_i E_i + (1 - w_{M)} E_M \tag{9}$$

$$\sigma_p(w_i) = \sqrt{w_i^2 \sigma_i^2 + (1 - w_i)^2 \sigma_M^2 + 2 w_i (1 - w_i) cov_{iM}} \quad .$$

Using (9) the partial derivatives give

$$\frac{\partial E_p}{\partial w_i} = E_i - E_M$$

$$\frac{\partial \sigma_p}{\partial w_i} = \frac{(\sigma_i^2 + \sigma_M^2 - 2cov_{iM}) w_i + cov_{iM} - \sigma_M^2}{\sqrt{\sigma_p}} \quad .$$

Assuming that asset i move to point M, then w_i=0, w_M=1. The ratio of the partial derivatives gives the slope of the tangent line $\overline{pM}$

$$\frac{dE(w_i)}{d\sigma(w_i)}_{w_i=0} = \frac{E_i - E_M}{\left(\frac{cov_{iM} - \sigma_M^2}{\sigma_M}\right)} \; ; \qquad \sqrt{\sigma_p}_{w_i=0} = \sqrt{\sigma_M^2 - \sigma_M} \; .$$

But this tangent line at point M will match with the line $\overline{pM}$ from fig.1, which has a slope equal to $\frac{E_M - p}{\sigma_M}$. Hence, the equality holds

$$E_i = r_f + (E_M - r)\frac{cov_{iM}}{\sigma_M^2} \quad . \tag{10}$$

This relation is the analytical form of the SML. The relations

$$\frac{cov_{iM}}{\sigma_M^2} = \beta_i \tag{11}$$

is named coefficient "beta", which is well known for the portfolio theory. Relation (10) holds not only for asset *i*, but for arbitrary portfolio

$$E_p = r_f + (E_M - r_f)\beta_p . \tag{12}$$

This research uses relations (4), (7), (10), (11) to simplify the evaluations of the components of the covariation matrix $\sum = |cov_{ij}|$, $i,j \in 1, \dots, N$.

4. Evaluation of the portfolio "beta" , β_p

From (11) it follows

$$\begin{aligned} cov_{pM} &= \beta_p \sigma_M^2 \\ cov_{1M} &= \beta_1 \sigma_M^2 \\ cov_{2M} &= \beta_2 \sigma_M^2 \, . \end{aligned} \tag{13}$$

The definition of the coefficient cov_{pM} is

$$cov_{pM} = \sum_{k=1}^{n} p^{(k)} \left(R_p^{(k)} - E_p\right)\left(R_M^{(k)} - E_M\right).$$

By substituting

$$R_p^{(k)} = w_1 R_1^{(k)} + w_2 R_2^{(k)} \quad \text{and} \quad E_p = w_1 E_1 + w_2 E_2$$

it follows

$$cov_{pM} = \sum_{k=1}^{n} p^{(k)} \left(w_1 R_1^{(k)} + w_2 R_2^{(k)} - w_1 E_1 - w_2 E_2\right)\left(R_M^{(k)} - E_M\right)$$

or

$$cov_{pM} = w_1 cov_{1M} + w_2 cov_{2M} \tag{14}$$

Relation (13) shows that the covariance of the portfolio is a weighting sum of the covariance of the assets towards the market portfolio. By substituting (13) in (14) it follows analogical relations for the portfolio "beta"

$$\beta_p = w_1 \beta_1 + w_2 \beta_2 \tag{15}$$

5. Substitution of β in the portfolio goal function

The goal of this research is to decrease the computational workload for the definition and solution of the portfolio problem. This can be achieved by substitution the components of the covariance matrix $\sum = |cov_{ij}|$, $i,j \in 1, \dots, N$ with the beta coefficients. Thus, it is expected to decrease the computations for the definition of the input data for the portfolio problem (1). Problem (1) can be worked out in a new formal view applying relation (11) and (12) of the SML. Thus, in the initial portfolio problem it can be introduced the new coefficients β_i

$$\begin{aligned} &\max_w \left(E_{p_{(w)}}\right) \\ &\sigma_p^2(w) \le \sigma_{max}^2 \\ &\sum_{i=1}^{N} w_i = 1 \, . \end{aligned} \tag{16}$$

To simplify the notations it is assumed *N*=2. Hence, the goal function of (16) is worked out in the form

$$\max_w E_p = \max_w \left[r_f + (E_M - r_f)\beta_p\right] ,$$

IOP Conf. Series: Materials Science and Engineering **878** (2020) 012025 doi:10.1088/1757-899X/878/1/012025

where β_p is evaluated according (15). The goal function does not contain the covariance matrix. For the evaluation of β_p is needed only β_1 and β_2 which are only two components for the portfolio. For the general case of *N* assets in the portfolio the optimization problem (16) needs only N evaluations of coefficients β_i , $i\in 1,\dots,N$ instead of, $\left(\frac{N(N-1)}{2}+N\right)$.

6. Substitution of β in the constraint $\sigma_p(w)$ of the portfolio problem

The portfolio risk analytically is expressed by relatiion (7). This relation does not contain in explicit way the coefficients β_i , $i\in 1,\dots,N$. For that case the next analytical expressions are implemented. The multiplication between the two covariations towards the market gives

$$cov_{1M}cov_{2M}=\sum_{k=1}^{n}P^{(k)}\left(R_1^{(k)}-E_1\right)\left(R_M^{(k)}-E_M\right)\sum_{k=1}^{n}P^{(k)}\left(R_2^{(k)}-E_2\right)\left(R_M^{(k)}-E_M\right)=$$

$$=\sum_{k=1}^{n}P^{(k)}\left(R_1^{(k)}-E_1\right)\left(R_2^{(k)}-E_2\right)\sum_{k=1}^{n}P^{(k)}\left(R_M^{(k)}-E_M\right)^2=cov_{12}\sigma_M^2.$$

Respectively, the square is

$$cov_{iM}cov_{iM}=\sum_{k=1}^{n}P^{(k)}\left(R_i^{(k)}-E_i\right)\left(R_M^{(k)}-E_M\right)\sum_{k=1}^{n}P^{(k)}\left(R_i^{(k)}-E_i\right)\left(R_M^{(k)}-E_M\right)=$$

$$=\sum_{k=1}^{n}P^{(k)}\left(R_i^{(k)}-E_i\right)^2\sum_{k=1}^{n}P^{(k)}\left(R_M^{(k)}-E_M\right)^2=\sigma_1^2\sigma_M^2.$$

From these relations using (13) it holds

$$cov_{1M}cov_{2M}=\beta_1\sigma_M^2\beta_2\sigma_M^2=cov_{12}\sigma_M^2$$

or

$$\beta_1\beta_2\sigma_M^2=cov_{12} \tag{17}$$

Respectively,

$$cov_{iM}cov_{iM}=\beta_i\sigma_M^2\beta_i\sigma_M^2=\tau_i^2\tau_M^2$$

or

$$\beta_i^2\sigma_M^2=\sigma_i^2 \tag{18}$$

After substitution of σ_i and cov_{12} in (7) the expression of the portfolio risk is expressed with the coefficients β_i
or

$$\sigma_p^2=\sigma_M^2(w_1^2\beta_1^2+w_2^2\beta_2^2+2\beta_1\beta_2w_1w_2)=\sigma_M^2(w_1\beta_1+w_2\beta_2)^2 \tag{19}$$

The portfolio optimization problem (1) takes analytical form

$$\max_{w_1,w_2}\left[r_f+\left(E_M-r_f\right)(\beta_1w_1+\beta_2w_2)\right]$$

$$\sigma_p^2=\sigma_M^2(w_1\beta_1+w_2\beta_2)^2\le\sigma_{max}^2\ . \tag{20}$$

This formal description has advantages in comparison with (1) because it is needed to evaluate only the β_i , $i\in 1,\dots,N$ coefficients, which is a set with N number. The classical formulation (1) insists the evaluation of the covariance matrix which has , $\left(\frac{N(N-1)}{2}+N\right)$ components, which are much more in comparison with the case with *N* beta coefficients.

The form of the portfolio problem (20) is implemented in Excel environment with real data from the Bulgarian Stock Exchange.

7. Numerical experiments from Bulgarian Stock Exchange

The numerical experiments were provided with 5 Bulgarian funds [5]: Advance Invest, PFBK Invest, Raiffeisen Bank, Astra Energy, Sky energy. These funds have highest capitalization in comparison with other funds in Bulgaria. Their corresponding parameters $E_i, \sigma_i^2, \delta_i, \beta_i, cov_{iM},\ i = 1, \dots, N$ are given on lines 9 till 14 of the Excel figure 2. The covariances $cov_{iM}\ i = 1,5$ are evaluated towards the Bulgarian Stock market index Sofix (M). All parameters are evaluated on annual base for 2019. The solutions of the portfolio problem are given in cells C17 till C22, fig.2. The weights are the solutions of portfolio problem (1) where r_f=0.02% (G15), σ_M=0.5963 (G12), $\sigma_M^2 = 0.3556\ (G13)$. Only one working column is prepared with the multiplication $\beta_i w_i$ and sell D23 contains the sum $\sum_{i=1}^{5} \beta_i\, w_i$. The portfolio risk is evaluated in cell F23 :

$$\sqrt{\sigma_M^2 (w_1\beta_1 + w_2\beta_2)^2} \le \sigma_{max} = 0.01\ .$$

The goal function is the portfolio return, evaluated in cell F24 : $r_f + (E_M - r_f)(\beta_1 w_1 + \beta_2 w_2)$. The portfolio problem insists nonnegative weights, $w_i \ge 0$ and $\sum_{i=1}^{5} w_i = 1$. The portfolio solution is

$$w^T = (0.325 \quad 0.087 \quad 0.148 \quad 0.186 \quad 0.251)^T,\ \textit{Risk}=0.01,\ \textit{Return}=0.02054\ .$$

The portfolio Return exceeds the risk free value, which is a prerequisite for the implementation of this investment.

	A	B	C	D	E	F	G	H
1								
2								
3								
4								
5								
6								
7								
8	Number	1	2	3	4	5	6	
9	Funds	Advance Invest	PFBK Invest	Raiffeisen BG	Astra energy	Sky energy	market SOFIX	
10	Beta	0.764493954	-11.55652803	-4.943916312	-0.687615472	6.405280265		
11	COV with marketна актив	0.310161365	-0.041321732	-0.02600103	-0.017446157	0.012000765		
12	standard deviation	6.369521703	0.59796473	0.72520321	1.59285801	0.432847987	0.596363311	σM
13	Variation	40.57081	0.35756	0.52592	2.53720	0.18736	0.355649199	σM*σM
14	Mean return	-0.37369	0.08385	0.08693	0.15782	0.07501	-0.012729891	EM
15								
16						rf, risk free	0.02	
17		Beta	Weights	Beta*Weight				
18		0.764493954	0.325961724	0.249195767				
19		-11.55652803	0.087869041	-1.015461034				
20		-4.943916312	0.148048991	-0.731941824				
21		-0.687615472	0.186784666	-0.128436026				
22		6.405280265	0.251335578	1.609874815				
23		SUM(Weights)	1	-0.016768302	Risk	0.01		
24					Return	0.020548825		
25								

Figure 2. Portfolio problem’s solution with Bulgarian mutual funds

This form of the portfolio problem performs less computation in comparison with the classical evaluations of portfolio risk and return according to (7). The benefit comes from fewer evaluations of the components of the covariance matrix as follows

- case with β: evaluation of N values of covariances towards the market points;

- case with relation (4) : $\frac{N(N+1)}{2} + N = \frac{N^2+N}{2} + N$ values are needed for the estimation of the covariance matrix.

The fewer amount of calculations for beta is a prerequisite for the implementation of active portfolio management by means to cope the investment decisions with the fast changing market behaviour.

8. Conclusions

This research makes appropriate formal derivations by means to describe the portfolio optimization problem in new analytical form. The portfolio problem is derived into new form by substituting portfolio parameters from MPT with beta parameters from the CAPM. This form allows to be used less input parameters and respectively less computational workload for the evaluations of the portfolio parameters. The less computational workload allows the portfolio problem to be defined and solved faster in case of dynamical changes of the market. Additionally, the portfolio problem can be solved with less power of the computing system.

Acknowledgments
This work has been supported by project ДН12/10, 20.12.2017 of the Bulgarian National Science fund: Integrated bi-level optimization in information service for portfolio optimization

References

[1] Markowitz H 1952 Portfolio selection *J. of Finance* **7** 77-91.

[2] Kolm PN, Tutuncu R and Fabozzi FJ 2014 60 Tears of Portfolio optimization: Practical challenges and current trends *European Journal of Operational Research* **234** (2) 356-371

[3] Gruber M, Brown S, Goetzmann W and Elton E 2010 *Modern portfolio theory and investment analysis*, 8th ed, Wiley 752 p.

[4] Sharpe W 2000 *Portfolio theory and capital markets* (McGraw Hill 316 p ISBN 0-07-155320-8)

[5] Stoilov T, Stoilova K and Vladimirov M 2019 Financial investments by Portfolio Optimization. *J. Materials Sciences and Engineering* **618** doi:10.1088/1757-899X/618/1/012030

[6] Rossi M 2016 The capital asset pricing model: a critical literature review *J. Global Business and Economics review* **10** (5) 604-617 On-line ISSN 1745-1329

Transportation modelling and solving Travelling Salesman problem

K Stoilova [1,2,*] **and T Stoilov**[1,2]

[1]Institute of Information and Communication Technologies – Bulgarian Academy of Sciences, Sofia, Bulgaria
[2] Nikola Vaptsarov Naval Academy, Varna
e-mail : k.stoilova@hsi.iccs.bas.bg

Abstract. This research addresses the problem of solving the Traveling Salesman Problem (TSP) with wide distributed software suit Excel. The TSP problem has long history and it belongs to a class of optimization problems, which is under continuous attempt for finding appropriate numerical algorithms for its solution. This research does not make attempts for new solution of this problem. The added value of the research is that it gives example how this problem can be solved with wide available software suit. A real problem of transportation in Sofia town is defined. The problem belongs to the class of TSP. It is illustrated how this problem is solved in Excel environment. Appropriate programming in Excel's sheet and in software application Solver is given. This work provides pragmatically example for the definition and solution of TSP transportation problem.

Keywords: optimization, transportation modelling, the shortest path problem, travelling salesman problem

1. Introduction

The transportation problem usually concerns the schedule of transportation of set of products from several source points to number of destinations. The formal definition of such problem in general is given as linear programming optimization problem. Such kind of formalization has wide usage also for problems of assignments and/or scheduling plans. These types of problems in general targets minimization of transportation costs or maximization of some values of profit.

The transportation problems initially have been used practically during the Second World War. Their primary application was the transport of troops from their training places to different battle positions in Europe and Asia. It has been seen the pragmatic and positive outcome of the implementation of the solutions of such problems.

A particular form of the transportation problem is the so-called Travelling Salesman Problem (TSP). This problem is intensively studied due to its internal complexity. The solution of this problem is widely used in many practical applications. This research is oriented to a practical usage of this problem. It illustrates the transportation rules in Sofia by means to minimize transportation costs/distances. The research illustrates the solution of this problem by wide popular software tool like Excel, which is a prerequisite for practical implementation of the results of this work.

2. Classical Travelling Salesman problem

The Traveling Salesman Problem has long history since 18th century. Details about the definition, complexity, classifications, applications one can find in [9]. This paper uses only descriptions about the analytical definition of TSP. Algorithms for solving this problem are not presented here. The goal of the paper is to give pragmatic solution for solving this problem with popular software tool. An illustration how to solve this complex problem is given.

The TSP is a combinatorial optimization problem which makes its solution hard to be obtained. It is mainly used in traffic management for optimization the roots of transport units by means to speed up transportation services and/or increase the efficiency of operation. The content of the problem is that a Salesman has to visit only once a set of destination points by means to minimize the overall length of this trip. At the end of the trip he has to return to his starting point. The distances between the destinations are given. As a mathematical problem it has been defined in1930 and it continues to be one of the intensively studied optimization problems. Following [4, 7] the TSP is used for analyzing the structure of crystals, for handling materials in warehouses, for clustering of data arrays, for scheduling with aggregate deadline, for orienteering problems, for applications in vehicle routing problems.

The wide way to give formal definition of the TSP is to give a description as integer linear programming model [8]. There are cases of symmetrical and asymmetrical definition of the TSP. The analytical definition of the TSP for the symmetrical case can be written as following. A graph $G=(n,A)$ is given where n is a number of nodes, A is a set of arcs between the nodes. Each arc has assigned value of cost (or distance) c_{ij}, $i,j \in \mathbf{1}, \dots, \boldsymbol{n}$. The TSP, written as linear integer problem, has the form

$$\min \sum_{i,j} c_{ij} x_{ij}$$

$$\sum_{\mathrm{i}} x_{ij} = 1, \quad \forall i \in 1, \dots, n$$

$$\sum_{\mathrm{j}} x_{ij} = 1, \quad \forall j \in 1, \dots, n$$

$$x_{ij} - \text{integer}, \quad \forall i, j \in 1, \dots, n$$

$$x_{ij} = \begin{cases} 1 & \text{if arc (i,j) is in the tour} \\ 0 & \text{otherwise} \end{cases}$$

$$x_{ij} \in Z, \text{“sub-tour breaking constraints”}.$$

The notation Z defines that there are not subtours for the travelling salesman except this one, which is the solution of the TSP problem. The analytical description of the set Z has several forms. Following [5, 6]

$$u_i - u_j + n x_{ij} \leq n - 1 .$$

The variables $u_i , i = 1, n$ are dummy ones. Their meaning is the number of visited cities between the city 1 and city i in the optimal tour. These compact relations are named Miller-Tucker-Zemlin formulation of the TSP.

Another well exploited formulation for the TSP is this one, given in [2]. Its form of the breaking sub-tour constraint is given with a set of many additional constraints of the form

$$\sum_{i\in R}\sum_{j\in R} x_{ij} \leq R, \quad R \subseteq [1,\dots,n], \qquad 2 \leq |R| \leq n-1.$$

Thus, the solution contains single big tour and not a set of smaller tours.

The analytical formulation and explicit description in bi-linear integer form is possible to be written only for very small networks, $n<4$. The number of the constraints in the problem is very high, formalizing all possible paths, which can connect the n nodes without making small tours. The peculiarities of the TSP do not allow it to be solved by linear programs. The algorithms, derived for the solution of the TSP have different formal nature: ant colony algorithms [12], evolutionary algorithms [3], heuristics, which decrease the computational workload to polynomial level [1, 10]. For an extensive analysis of the nature of TSP, solving algorithms and application areas one can refer to [11]. According to the accuracy of solving the TSP the algorithms are classified on two sets: exact algorithms and heuristic algorithms [5]. The exact algorithms find optimal solution but in exponential set of calculations. These algorithms are quite complex and need considerable computer power. That is why heuristic algorithms are derived, which give approximate, suboptimal solution but for reasonable computer workload.

This research does not derive new formulation neither algorithms for solving this combinatorial problem. Here it is derived an appropriate application of the TSP and it is illustrated how this problem can be solved with Excel. The peculiarities, applied in this research are the usage of built-in functions in Excel suit, which makes the TSP problem non-analytical with procedural definition. This specific form of the problem later is included in the optimization function Solver. Thus, by making non-analytical constraints and integrating them in optimization function, it is achieved a practical for implementation approach for solving the TSP. This manner of solving the complex TSP has been applied for planning the logistic operation and transportation of goods in the region of Sofia.

3. Case study of the TSP on the example of transportation system in Sofia

The transportation system consists of five stock markets (SM), situated in Sofia, Figure 1. The urban distances between the SMs (in km) are given in Table 1. These stock markets have to be visited by travelling salesman. Particularly, he/it could be boss of a company, supplying vehicle, service company, postman, advertising agent, etc. Each of these nodes has to be visited only once for a round trip. The problem's target is to minimize the travelled path among the five SMs.

The problem will be solved by popular software tool like Excel. At first, the model is designed in Excel and after that the problem is solved by the optimization tool Solver, which is application software of Excel. Table 1 is modeled in the Excel environment in the massive B3:G7, Figure 2.

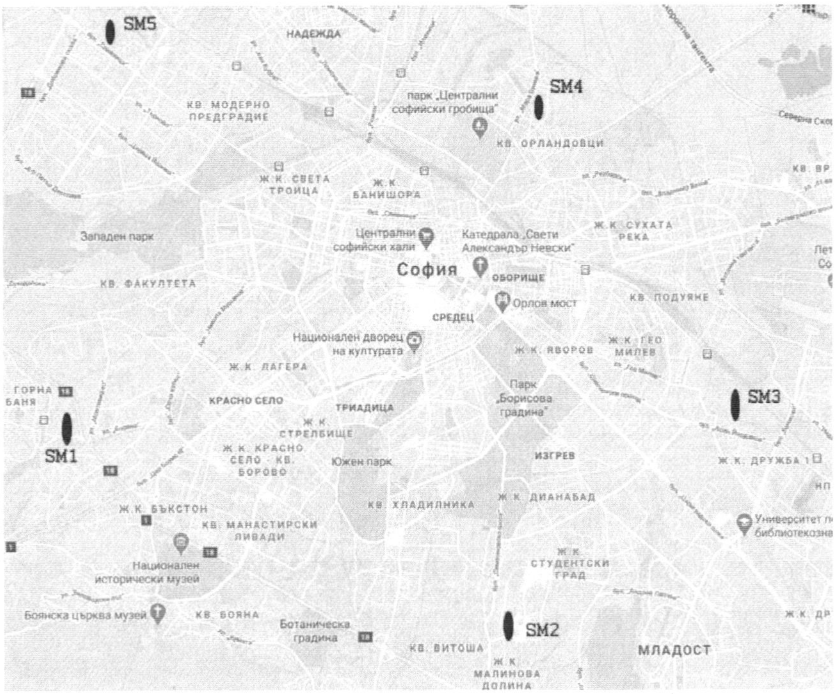

Figure 1. The transportation scheme in Sofia

Table 1. The distance between the SMs

	SM1	SM2	SM3	SM4	SM5
SM1		12.5	18.7	12.3	9.6
SM2	12.5		8.3	12.5	19.5
SM3	18.7	8.3		10.2	16.3
SM4	12.3	12.5	10.2		20.2
SM5	9.6	19.5	16.3	20.2	

The model for solving the problem continues in the area B10:E14. The unknown variables are the sequence of the SMs, each of which has to be visited by the salesman. Column B (B10:B14) represents *the description of the sequence* of the visited SMs. The unknown variables are in cells C10:C14. At the beginning we put arbitrary variables, for instance from the first to the fifth SM, C10:C14. The third column D corresponds to the SMs in the row in correspondence with the variables, D10:D14. These values are obtained by the embedded in Excel function "INDEX". It is a statistical function which finds a cell in massive. This function has three arguments: the area of the values in which we are looking for a cell (in our model this is the massive $B$3:$B$7); the second argument is a cell, which content determines the row of the specified massive; the third argument specifies the column of the massive. For instance, the content of cell D10 is =INDEX($B$3:$B$7,$C10,1). In E10:E14 is the distance from the previous SM. Each of these cells contains again the function "INDEX" as it is shown in Figure 2. The solution of the problem gives the order of the sequentially visiting nodes of the network, C10:C14. These values are the solutions of the optimization problem. The goal function is in cell E16. We have to find the minimum travelling distance. The goal function is formalized in Excel by the function SUM(E10:E15), which minimum has to be estimated.

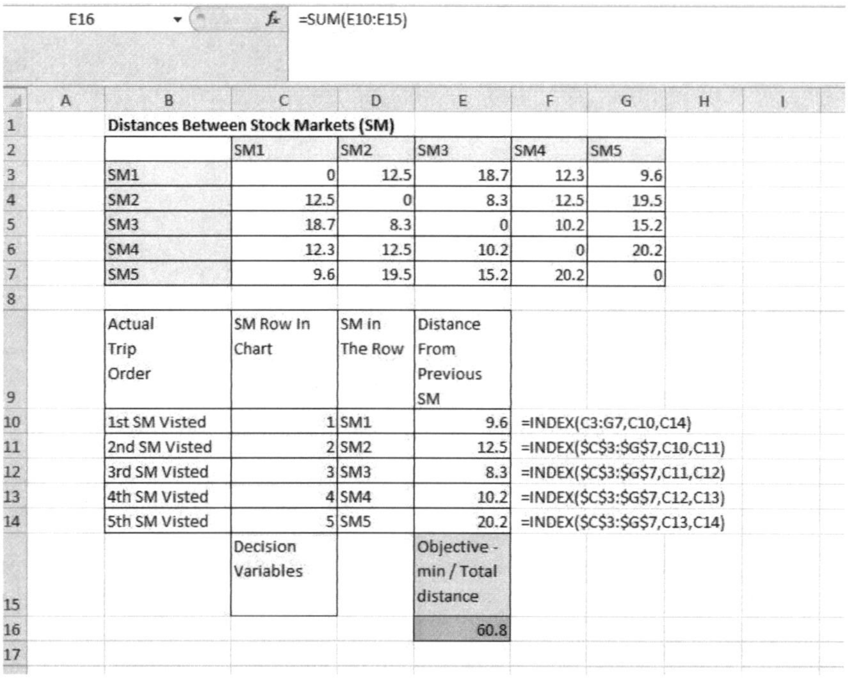

E16 fx =SUM(E10:E15)

	A	B	C	D	E	F	G	H	I
1		Distances Between Stock Markets (SM)							
2			SM1	SM2	SM3	SM4	SM5		
3		SM1	0	12.5	18.7	12.3	9.6		
4		SM2	12.5	0	8.3	12.5	19.5		
5		SM3	18.7	8.3	0	10.2	15.2		
6		SM4	12.3	12.5	10.2	0	20.2		
7		SM5	9.6	19.5	15.2	20.2	0		
8									
9		Actual Trip Order	SM Row In Chart	SM in The Row	Distance From Previous SM				
10		1st SM Visted	1	SM1	9.6	=INDEX(C3:G7,C10,C14)			
11		2nd SM Visted	2	SM2	12.5	=INDEX($C$3:$G$7,C10,C11)			
12		3rd SM Visted	3	SM3	8.3	=INDEX($C$3:$G$7,C11,C12)			
13		4th SM Visted	4	SM4	10.2	=INDEX($C$3:$G$7,C12,C13)			
14		5th SM Visted	5	SM5	20.2	=INDEX($C$3:$G$7,C13,C14)			
15			Decision Variables		Objective - min / Total distance				
16					60.8				
17									

Figure 2. Optimization model in Excel environment

The TSP is formalized in Solver according to Figure 3. As a tool for solving optimization problems, the Solver consists of three main fields, which have to be determined: Target cell or objective function, in our case E16, which minimum has to be obtained. It can be reached by changing variables C10:C14. The third main field is the determination of the problem's constraint – C10:C14 have to be different values because the Travelling Salesman has to visit only once the network's node/SM. The solver has to use nonlinear optimization solving method. That's why in this research the "Evolutionary" solving method is chosen.

The TSP solution is given in Figure 4. The minimum value is 57.9 km and the sequence of travelling is SM5-SM1-SM4-SM2-SM3.

The illustration for the usage and solution of the TSP problem in EXCEL environment can be easily extended for larger set of points for the salesman. One has to take into consideration that the solution of the problem takes time because of its complexity. For this research the TSP solution is found after nearly 3 minutes of computer calculations. If the number of nodes of the problem increases, the solution time is significantly long.

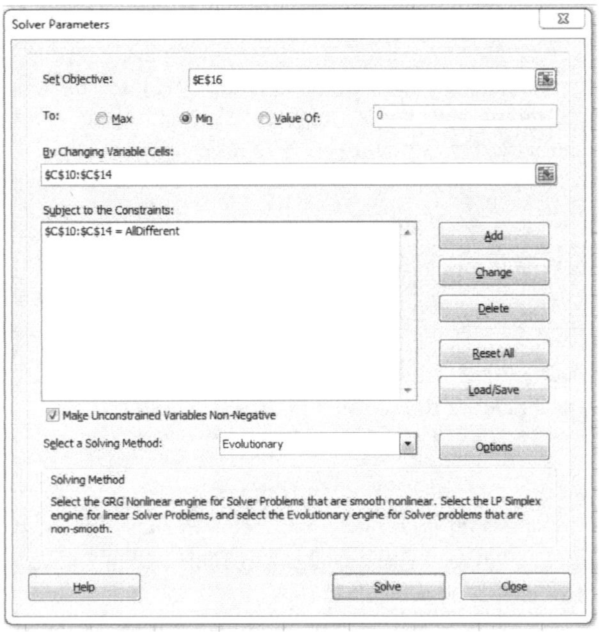

Figure 3. Optimization model in Solver environment

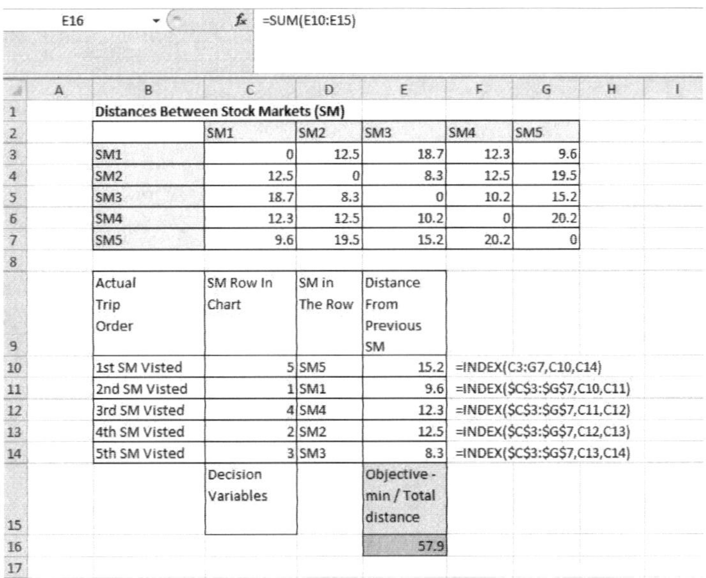
E16 =SUM(E10:E15)

Distances Between Stock Markets (SM)

	SM1	SM2	SM3	SM4	SM5
SM1	0	12.5	18.7	12.3	9.6
SM2	12.5	0	8.3	12.5	19.5
SM3	18.7	8.3	0	10.2	15.2
SM4	12.3	12.5	10.2	0	20.2
SM5	9.6	19.5	15.2	20.2	0

Actual Trip Order	SM Row In Chart	SM in The Row	Distance From Previous SM	
1st SM Visted	5	SM5	15.2	=INDEX(C3:G7,C10,C14)
2nd SM Visted	1	SM1	9.6	=INDEX($C$3:$G$7,C10,C11)
3rd SM Visted	4	SM4	12.3	=INDEX($C$3:$G$7,C11,C12)
4th SM Visted	2	SM2	12.5	=INDEX($C$3:$G$7,C12,C13)
5th SM Visted	3	SM3	8.3	=INDEX($C$3:$G$7,C13,C14)
	Decision Variables		Objective - min / Total distance	
			57.9	

Figure 4. Optimal solution

4. Conclusions

This research presents useful example for definition and solution of practical transportation problem. The transportation problem is defined for the logistics management of a set of nodes/stores in urban environment. A real case for town of Sofia is under consideration. The transportation problem was defined in the class of TSP. Nevertheless the difficulties for solving such kind of problems, the paper illustrates a practical case for its solution. It has been used the software suit Excel, which is widely available. The manner of programming the main sheet, the set of parameters, given by the application Solver, and the usage of specific Excel functions give a useful procedure for the definition and solution of the TSP transportation problem. This research can be used as an approach for solution of hard problems like TSP.

Acknowledgments
The research is funded by Project KP06-H37/6 "Modelling and optimization of urban traffic in network of crossroads" with the Bulgarian Research Fund.

References
[1] Ambati B K, Ambati J and Mokhtar M M 2011 Heuristic Combinatorial Optimization by Simulated Darwinian Evolution: a Polynomial Time Algorithm for the Traveling Salesman Problem *Biological Cybernetics* **65** 31-35
[2] Dantzig G, Fulkerson D and Johnson S 1954 Solution of a Large-Scale Traveling Salesman Problem *J Operations Research* **2** 393–410
[3] Fogel DB 2013 Applying Evolutionary Programming to Selected Traveling Salesman Problems *Cybernetics and Systems* **24** 27-36
[4] Goyal S 2010 A survey on travelling salesman problem. University of North Dakota (pdfs.semanticscolar.org).
[5] Langevin A, Soumis F and Desrosiers J 1990 Classification of Travelling Salesman Problem Formulations. *J. Operational Research Letters* **9** 127–132
[6] Miller C E, Tucker A W and Zemlin R A 1960 Integer programming formulation of travelling salesman problems. *Journal of the ACM* **7** 326–329
[7] Marinakis Y, Migdalas A and Pardalos P 2005 Expanding Neighborhood GRASP for the Travelling Salesman Problem *J. Computational Optimization and Applications* **32** 231-257
[8] Patavi G 2003 Teaching Integer Programming formulation using the Travelling Salesman Problem *J. SIAM Review* **45** 1 116-123
[9] Matai R, Singh S and Mittal M L 2010 Traveling salesman problem: an overview of applications, formulations and solutions, in Traveling Salesman Problem: Theory and Applications, DOI: 10.5772/547, ISBN: 978-953-307-426-9, e-Book(pff) ISBN: 978-953-51-5501-0, [www.intechopen.com].
[10] Okediran O O, Oyeleye C A and Adeyemo I A 2015 A Performance Evaluation of Selected Heuristics for the Travelling Salesman Problem. *International Journal of Advanced Research in Computer Science and Software Engineering* **5** (7) ISSN: 2277 128X 113-117
[11] *Traveling salesman problem, theory and applications* 2010 Ed. by D Davendra INTECH open ISBN 978-953-307-426-9, intechweb.org
[12] Zar Chi Su Su Hlaing and May Aye Khine 2011 An Ant Colony Optimization Algorithm for Solving Traveling Salesman Problem. International Conference on Information Communication and Management IPCSIT **16** IACSIT Press Singapore 54-59

Modelling and estimate of the LTWM noise parameters

Bojan Prlinčević[1], Zoran Milivojević[2]

[1]Higher Technical Professional School Zvecan, St. Nusiceva 6, Zvecan, Serbia

[2]The Academy of Applied Technical Studies, St. Aleksandra Medvedeva 20, 18000 Nis, Serbia

b.prlincevic@vts-zvecan.edu.rs, zoran.milivojevic@vtsnis.edu.rs

Abstract. In the first part of this paper, a statistical analysis of the noise in the image, as a result of the insertion of the watermark by Liu and Tan algorithm, was performed. After that, using the base of test image the noise model was created. Based on the model, an algorithm for estimating the statistical parameters of noise, LTWM algorithm, (μ and σ^2) in the image with a watermark, was created. In the second part of this paper there is a description of the experiment in which a comparative analysis of the accuracy of estimating statistical parameters by the proposed LTWM algorithm and Algorithms from [3], [6] - [8], was performed. Algorithms have been applied on watermarked images from the NRCS base. The results of MAE and MSE between true and estimated variance are shown in tables and diagrams. In addition, the execution time of the tested algorithms was measured. A detailed comparative analysis has shown that the proposed LTWM algorithm was more precise, and has a shorter execution time, compared to the analysed algorithms.

1. Introduction

The efficiency of algorithms for digital image processing (edge detection, resizing, translation, rotation and geometric correction, scaling, improve image quality, image compression, image watermarking, visual pattern recognition, face recognition, object classification,...) [1], [2] can be highly degraded due to noise presence in the image. The degree of degradation of efficiency of the algorithms is directly proportional to the intensity of the noises in the image. In most cases, the noise can be modeled by Gaussian distribution and presented by statistical parameters, the mean μ and the variance σ^2 (e.g. the amplifier noise, the shot noise, the grain noise of photographic film...) [3], [4]. In the image denoising literature, noise is often assumed to be zero-mean additive white Gaussian noise (AWGN) [3], [5]. The first step in denoising is the estimation of noise parameters (e.g. Algorithm from [6], [7], Algorithm from [8], Algorithm from [3], and many others [9] - [11]).

Inserting a digital watermark leads to degradation of the visual characteristics of the image. The degree of degradation is directly dependent on the insertion factor. Decreasing the insertion factor leads to a decreasing of visual image degradation. However, the problem of reliable watermark extraction is related with the watermark low insertion factor. The level of image noise depends of the inserted watermark structure, i.e. of the percentage of white pixels in the watermark. Many transformations are used to insert the watermark into the image: a) *Discrete Cosine Transform* (DCT) [12], b) *Discrete Wavelet Transform* (DWT) [13], c) *Singular Value Decomposition* (SVD) [14], d) *Schur Decomposition*, (SD) [15], and others. A commonly used algorithm for inserting a watermark is the *Liu and Tan* (LT) algorithm, based on the SVD transformation [16].

The authors of this paper tried to answer the questions: a) *whether the noise caused from the insertion of a watermark by LT algorithm could be modeled*? and b) *whether prediction of noise parameters can be made based on the noise model*? The answers to the asked questions were obtained through realization of the experiments. Based on the results of the experiments, it was shown that the images noise caused by inserted watermark, inserted with the LT algorithm (LTWM noise), can be modeled as Gaussian noise. Based on that, in order to predict the statistical parameters (μ, σ^2), an noise (LTWM) model was created. The LTWM model was created based on statistical parameters of standard test images from the base *I*. The accuracy of the estimation of the noise parameters was made by a comparative analysis with the estimation results of: a) the Algorithm from [6], [7], b) the Algorithm from [8] and c) the Algorithm from [3]. The mean absolute error (MAE) and mean square error (MSE) was used as a measure of comparing the estimation accuracy. In addition, a comparative analysis of the execution time (T_e) of algorithms was performed. In the experiment images from the NRCS base (https://photogallery.sc.egov.usda.gov/photogallery/#/) [17], [18] was used.

The paper is organized as follows. In the section 2, analysis of the noise parameters has been performed. Section 3 shows the LTWM model creation proceeding. Section 4 describes the algorithm for estimating the parameters of the LTWM noise. The section 5 describes the experimental results and analysis of the results. Section 6 is a conclusion.

2. The algorithm for estimation LTWM noise statistical parameters

In this chapter an algorithm for estimating the statistical parameters of the noise, which is effect of watermark insertion with LT algorithm, has been described. Confirmation that the LTWM noise can be modeled by Gaussian distribution, was obtained by comparing the true histogram and the histogram obtained based on the estimated parameters μ and σ^2. The algorithm for estimating the statistical parameters of the LTWM noise, is carried out as follows:

Input: Host image *X*, binary watermark *W*, image dimensions $M \times N$, insertion factor α, percentage *p* of white pixels in the watermark.

Output: μ, σ^2.

Step 1: SVD decomposition of the image *X*:

$$X = U \cdot D \cdot V' \tag{1}$$

where *U* is the left orthogonal matrix ($M \times M$), *V* is the right orthogonal matrix ($N \times N$), and *D* is the diagonal singular matrix ($M \times N$).

Step 2: Generating the watermark *W*, with the *p* percentage of the white pixels spatially uniformly distributed.

Step 3: Inserting the watermark *W* into the matrix *D*, with the insertion factor α:

$$D' = D + \alpha \cdot W , \tag{2}$$

Step 4: Creating the watermarked image:

$$X_W = U \cdot D' \cdot V' , \tag{3}$$

Step 5: The LTWM noise matrix is:

$$N_{XW} = X - X_W , \tag{4}$$

Step 6: Statistical parameters of the N_{XW} noise are:

$$\mu = \frac{1}{MN} \sum_{i=1}^{M} \sum_{j=1}^{N} N_{XW}(i, j), \tag{5}$$

and

$$\sigma^2 = \frac{1}{MN-1}\sum_{i=1}^{M}\sum_{j=1}^{N}\left(N_{XW}(i,j)-\mu\right)^2, \quad (6)$$

Step 7: Based on the estimated statistical parameters (μ, σ^2), the Gaussian distribution is modeled:

$$f(x) = \frac{1}{\sqrt{2\pi\sigma^2}} \cdot e^{-\frac{(x-\mu)^2}{2\sigma^2}}. \quad (7)$$

Applying the described algorithm to the test images from base I, the results are obtained and presented by histograms. In Fig. 1 are shown images: a) Lena (Fig. 1.a), b) watermark W (Fig. 1.b), and c) histograms (Fig. 1.c) (the distribution of the noise pixel intensity (4) (marker: '⊖'), and calculated Gaussian distribution (7) (solid line) for the test image Lena). Used watermark (Fig. 1.b) was generated with p = 50% white spatially uniformly distributed pixels. Presented results are for inserted watermark W, with the insertion factor α = 25.

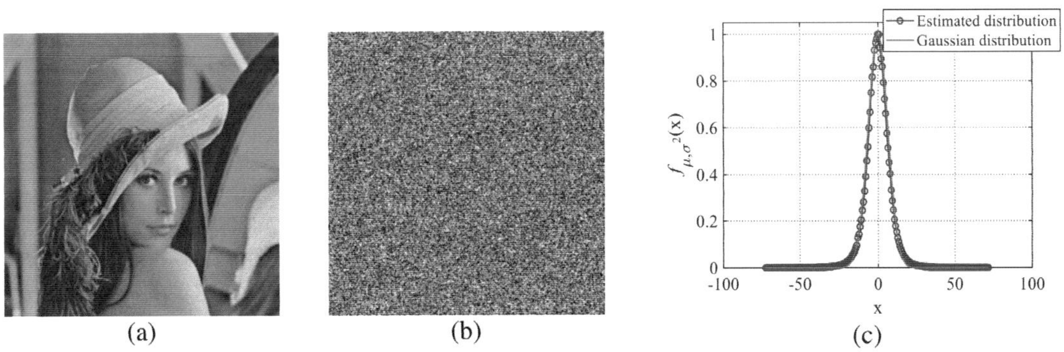

Figure 1. The image Lena (a), Watermark (b), and distribution of the pixel intensity of the LTWM noise and the Gaussian distribution for presented test image (c).

3. Creating of the LTWM model

In this section describes the algorithm for creating the LTWM model. The LTWM model has been created on the basis of the results obtained by Section 2. The algorithm is carried out as follows:

Input: Image base I, with B images, image dimensions $M \times N$, insertion factor $\alpha_{\min}$, $\Delta\alpha$, $\alpha_{\max}$, percentage of white pixels $p_{\min}$, Δp, $p_{\max}$, luminance $Y_{r\_\min}$, ΔY_r, $Y_{r\_\max}$.

Output: LTWM model X_m.

FOR $y_r = Y_{r\_\min} : \Delta Y_r : Y_{r\_\max}$

FOR i = 1 : B

Step 1: Loading the image X_i from the base I.

Step 2: Determining the luminance of the image X_i:

$$y_i = \frac{1}{M \cdot N}\sum_{i=1}^{M}\sum_{j=1}^{N} X_{i,j}, \quad (8)$$

Step 3: Generating the image X with the luminance y_r

$$X = k \cdot X_i, \quad (9)$$

where $k = y_r / y_i$.

FOR $\alpha = \alpha_{\min} : \Delta\alpha : \alpha_{\max}$

FOR $p = p_{\min} : \Delta p : p_{\max}$

Step 4: Generating the binary watermark W with percentage of white pixels p;

Step 5: Creating an watermarked image X_W with the insertion factor α (equations (1) - (3));
Step 6: LTWM noise is $N_{XW} = X - X_W$, equation (4), with statistical parameters μ, equation (5), and σ^2, equation (6), $\Rightarrow Q(\alpha, p, i, y_r) = \mu$, and $S(\alpha, p, i, y_r) = \sigma^2$.

END p

END α

END i

$X_\mu = X_\mu + Q$,

$X_{\sigma^2} = X_{\sigma^2} + S$.

END y_r.

Step 7: Generating LTWM model $X_m = \{ X_{m\mu}, X_{m\sigma^2} \}$, where $X_{m\mu} = X_{m\mu} / B$ and $X_{m\sigma^2} = X_{m\sigma^2} / B$ are components of the model.

Components of the LTWM model X_m, for the image with $Y = 30$, are shown in Fig. 2.a (component $X_{m\mu}$) and Fig. 2.b (component $X_{m\sigma^2}$). The parameters of the algorithm for creating LTWM model (Section 3) are $\alpha_{\min} = 1$, $\Delta\alpha = 1$, $\alpha_{\max} = 50$, $p_{\min} = 1$, $\Delta p = 5$, $p_{\max} = 100$, $Y_{r\_\min} = 10$, $\Delta Y_r = 10$ and $Y_{r\_\max} = 250$.

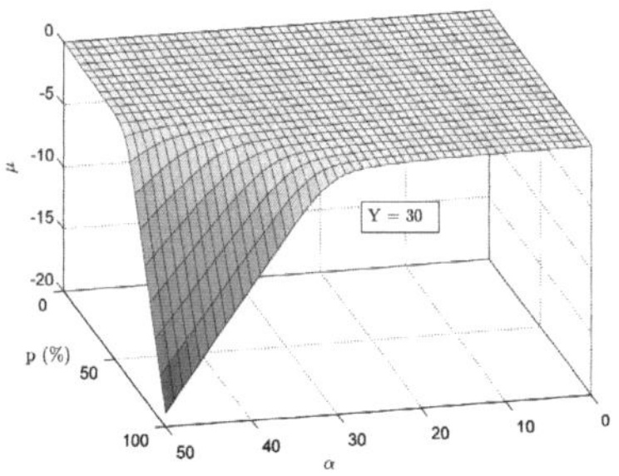

(a)

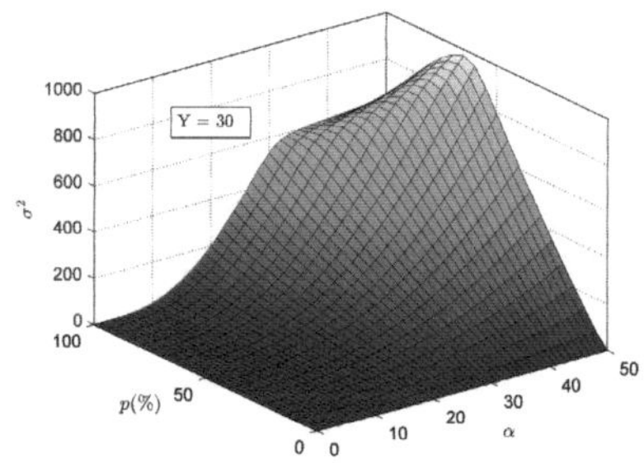

(b)

Figure 2. The components of the LTWM model ($Y = 30$): a) $X_{m\mu}$ and b) $X_{m\sigma^2}$.

4. Estimation of the LTWM noise

The section describes algorithm for estimating of the LTWM noise parameters of the image X, for mean illumination Y, to which the binary watermark W is inserted. The algorithm for estimation LTWM noise (LTWM algorithm) consists of the following steps:

Input: Test image X, binary watermark W, LTWM model $X_m = \{ X_{m\mu}, X_{m\sigma^2} \}$, insertion factor α.

Output: LTWM noise statistic parameters (μ_e, σ_e^2).

Step 1: Luminance of the image X (without a watermark) is $Y = \overline{X}$.
Step 2: Percentage p of the white pixels in the watermark W is:

$$p = (N_W / MN) \cdot 100\ [\%], \quad (10)$$

where N_W – is the number of white pixels in the watermark.
Step 3: Estimation of the LTWM noise parameters by using the components of the LTWM model: $\mu_e \leftarrow X_{m\mu}[p, \alpha, Y]$, $\sigma_e^2 \leftarrow X_{m\sigma^2}[p, \alpha, Y]$.

5. Experimental results and analysis

This accuracy of the statistical parameters estimation of the LTWM noise by using the LTWM model (Section 4) was determined by the following experiment.

5.1. Experiment

The experiment was carried out, within which the estimation of the statistical parameters of the LTWM noise (μ, σ^2) was performed using the algorithm for comparative analysis. The estimation was performed using the LTWM algorithm (Section 4) on the LTWM model (Section 3). The noise matrix N_{XW}, equation (4), was created from the watermarked image X_W, and the statistical parameters were calculated, equation (5) and equation (6), which were considered as the accurate values in the further analysis. After that, the accuracy of the parameter estimation was tested and compared with the estimation results of the noise parameters in the watermarked image X_W, using the algorithms shown in [6], [7], [8] and [3]. As a measure of quality, the the mean absolute error MAE and MSE were used. In addition to the accuracy of the estimation, the analysis of the time T_e of execution of the tested algorithms was performed. The algorithms were performed in Matlab on the personal computer: processor Intel(R) Core(TM) i5 - 3230M CPU @ 2.6 GHz, RAM memory 8 GB, 64-bit OS.

5.2. Comparative analysis

Comparative analysis of the estimation accuracy of the statistical parameters of the LTWM noise consists of the following steps:

Input: Image base $P = \{X_1, X_2,\ldots, X_B\}$, number of tested images B, LTWM model X_m, insertion factor limits $\alpha_{\min}$, $\alpha_{\max}$, iterative step $\Delta\alpha$, percentage of white pixels in the watermark p.

Output: Estimation errors of LTWM MSE, Algorithm from [6], [7] MSE_T, Algorithm from [8] MSE_G, and Algorithm from [3] MSE_W algorithms.

FOR $i = 1 : B$

Step 1: Selecting the image X_i from the image base P: $X_i \leftarrow P = \{X_1, X_2, \ldots, X_i, \ldots, X_B\}$, and determining the luminance Y, equation (8), of selected image.

Step 2: Generating the binary watermark W_p with p [%] of white pixels uniformly spatially distributed,

FOR $\alpha = \alpha_{\min} : \Delta\alpha : \alpha_{\max}$

Step 3: Creating an watermarked image X_W by inserting the binary W_p into the image X_i with the insertion factor α, by using the LT algorithm, equations (1) - (3): $X_W \leftarrow LT\,(X_i, W_p, \alpha)$.

Step 4: Determining variance $\sigma^2_{e\_Tr}$, equation (6), (true value), noise matrix N_{XW}, equation (4).

Step 5: Estimation variance of the LTWM noise using:

a) LTWM algorithm (Section 4) (LTWM model X_m, component $X_{m\sigma^2}$):

$$\sigma^2_{e\_LT}(\alpha) \leftarrow X_{m\sigma^2}(p,\alpha,Y_i),$$

b) Algorithm from [6], [7]:

$$\sigma^2_{e\_T}(\alpha) \leftarrow \text{Algorithm } [6],[7](X_W),$$

c) Algorithm from [8]:

$$\sigma^2_{e\_G}(\alpha) \leftarrow \text{Algorithm } [8](X_W),$$

d) Algorithm from [3]:

$$\sigma^2_{e\_W}(\alpha) \leftarrow \text{Algorithm } [3](X_W).$$

Step 6: Absolute error of estimated variance $\Delta\sigma^2_{\text{alg}}(\alpha) = \left|\sigma^2_{e\_Tr}(\alpha) - \sigma^2_{e\_\text{alg}}(\alpha)\right| \Rightarrow$ a) $\Delta\sigma^2_{LT}$, b) $\Delta\sigma^2_{T}$, c) $\Delta\sigma^2_{G}$, and d) $\Delta\sigma^2_{W}$.

END α

Step 7: Sum of absolute error of estimated variances for all images $\Sigma\Delta\sigma^2_{\text{alg}}(i) = \Sigma\Delta\sigma^2_{\text{alg}} + \Delta\sigma^2_{\text{alg}} \Rightarrow$ a) $\Sigma\Delta\sigma^2_{LT}$, b) $\Sigma\Delta\sigma^2_{T}$, c) $\Sigma\Delta\sigma^2_{G}$, and d) $\Sigma\Delta\sigma^2_{W}$.

Step 8: Sum of estimated variances $\Sigma\sigma^2_{\text{alg}}(i) = \Sigma\sigma^2_{\text{alg}} + \sigma^2_{\text{alg}} \Rightarrow$ a) $\Sigma\sigma^2_{LT}$, b) $\Sigma\sigma^2_{T}$, c) $\Sigma\sigma^2_{G}$, and d) $\Sigma\sigma^2_{W}$.

END i

Step 9: Mean absolute error (MAE) of estimated variance $MAE_{\text{alg}} = \Sigma\Delta\sigma^2_{\text{alg}} / B \Rightarrow$ a) MAE_{LT}, b) MAE_{T}, c) MAE_{G}, and d) MAE_{W}.

Step 10: Mean of estimated variances $\overline{\sigma^2_{\text{alg}}} = \Sigma\sigma^2_{\text{alg}}/B \Rightarrow$ a) $\overline{\sigma^2_{\text{LT}}}$, b) $\overline{\sigma^2_T}$, c) $\overline{\sigma^2_G}$, and d) $\overline{\sigma^2_W}$.

Step 11: MSE of estimated variance $MSE_{\text{alg}} = \frac{1}{\alpha_{\max} - \alpha_{\min}} \sum_{\alpha=\alpha_{\min}}^{\alpha_{\max}} \left(\sigma^2_{Tr}(\alpha) - \sigma^2_{\text{alg}}(\alpha)\right)^2 \Rightarrow$ a) MSE_{LT}, b) MSE_T, c) MSE_G, and d) MSE_W.

5.3. Experimental Base

For purpose of the realization the experiments, image bases were created:

a) the image base of the standard test images I = {*Lena*, *Girl*, *Baboon*, *Peppers*, *Barbara*, *Boat*}, used to create the LTWM model X_m,

b) the image base P, formed from 100 images from NRCS base [17], [18], and

c) binary watermark with p [%] white, uniformly distributed pixels.

5.4. Results

The following results were obtained by applying the algorithm for comparative analysis (Section 5) with the parameters: $\alpha_{\min} = 1$, $\alpha_{\max} = 50$, $\Delta\alpha = 1$, $p = 50\%$ on the image base $P = \{X_1, X_2,..., X_B\}$, $B = 100$ (images from NRCS base P), and by using LTWM model X_m. On the figures are shown results of applied measures, which represents mean value for $B = 100$ images. Fig. 3.a shows the estimated variances $\overline{\sigma^2_e}$ of LTWM noise for: a) True value (marker: '●') (4), b) LTWM (marker: '■'), c) Algorithm from [6], [7] (marker: '▼'), d) Algorithm from [8] (marker: '▲'), and e) Algorithm from [3] (marker: '♦'), for $\alpha = \alpha_{\min} : \Delta\alpha : \alpha_{\max}$. Fig. 3.b shows a part of the diagram from Fig. 3.a, for α = 0 - 10, where LTWM noise are not visible in the watermarked image. Fig. 4.a shows the mean absolute errors MAE of estimated variances: a) LTWM algorithm MAE_{LT} (marker: '●'), b) Algorithm from [6], [7] MAE_T (marker: '▲'), c) Algorithm from [8] MAE_G (marker: '■') and d) Algorithm from [3] MAE_W (marker: '♦')) for $\alpha = \alpha_{\min} : \Delta\alpha : \alpha_{\max}$. Fig. 4.b shows a part of the diagram from Fig. 4.a, for α = 0 - 10, where LTWM noise are not visible in the watermarked image.

Table 1. shows mean square error MSE of estimated variance, for parts of insertion factor: a) α = 0 - 50 ($MSE_{0\text{-}50}$), b) α = 0 - 10 ($MSE_{0\text{-}10}$) where noise is not visible in the watermarked image, and c) α = 11 - 50 ($MSE_{11\text{-}50}$) where noise is visible in the watermarked image. Table 2. shows the execution time for: a) LTWM (T_{LTWM}), b) Algorithm from [8] (T_G), c) Algorithm from [6], [7] (T_T), and d) Algorithm from [3] (T_W) algorithms.

Table 1. MSE of estimated variance.

Algorithm	$MSE_{0\text{-}50}$	$MSE_{0\text{-}10}$	$MSE_{11\text{-}50}$
LTWM	804.94	0.4303	1026.2
Algorithm from [6], [7]	20931	2.4574	26687
Algorithm from [8]	24319	1.4384	31006
Algorithm from [3]	24837	1210.8	31335

Table 2. Execution time T_e.

Algorithm	T_e (s)
LTWM	$T_{e\_LTWM} = 0.0176$
Algorithm from [8]	$T_{e\_G} = 0.0260$
Algorithm from [6], [7]	$T_{e\_T} = 0.3831$
Algorithm from [3]	$T_{e\_W} = 0.61285$

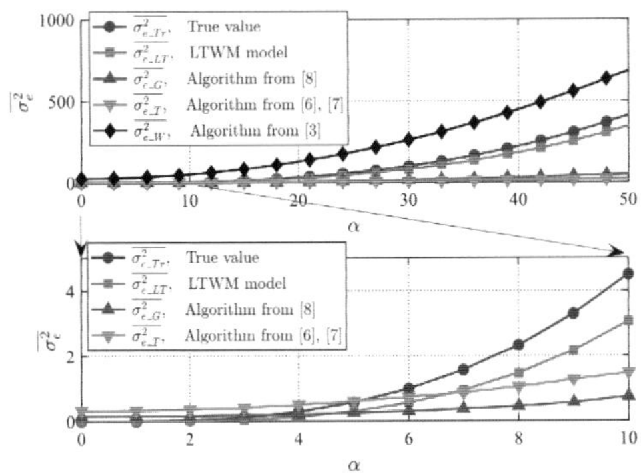

Figure 3. Mean of estimated variances $\overline{\sigma_e^2}$ for insertion factor: a) $\alpha = 0$ - 50 (noise is visible in the watermarked image) and b) $\alpha = 0 - 10$ (noise is not visible in the watermarked image), for True value and LTWM, Algorithm from [8], Algorithm from [6], [7], and Algorithm from [3].

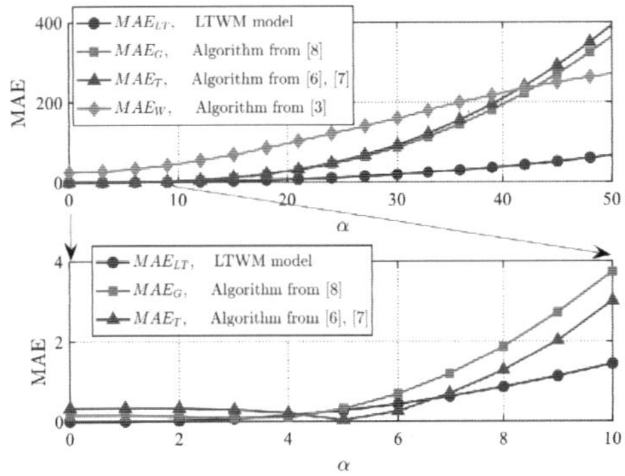

Figure 4. Mean absolute errors MAE of estimated variances for insertion factor: a) $\alpha = 0$ - 50 (noise is visible in the watermarked image) and b) $\alpha = 0$ - 10 (noise is not visible in the watermarked image), for LTWM, Algorithm from [8], Algorithm from [6], [7], and Algorithm from [3].

5.5. Analysis of the results

Based on the results shown in Fig. 3 and Fig. 4 and in Table 1. and Table 2., it is concluded that:

a) the mean estimated variance $\overline{\sigma_e^2}$ curve for the LTWM algorithm is more closed to curve of true value, relative to other three compared algorithms. Mean absolute errors MAE of estimated variances is smaller value for all values α. MSE of estimated variance of LTWM algorithm in relation to other algorithms is 6 (Alg. from [8]), 30.21 (Alg. from [6], [7]) and 3.86 (Alg. from [3] times smaller.

b) in the range where the LTWM noise is not visible ($\alpha = 0 \div 10$), the mean estimated variance $\overline{\sigma_e^2}$ curve for the LTWM algorithm is more closed to curve of true value, relative o other three compared algorithms. Mean absolute errors MAE of estimated variances is smaller value for all values α. MSE of estimated variance of LTWM algorithm in relation to other algorithms is 5.71 (Alg. from [8]), 3.34 (Alg. from [6], [7]) and 2813.3 (Alg. from [3]) times smaller.

c) in the range where the LTWM noise is visible ($\alpha = 11$ - 50), the mean estimated variance $\overline{\sigma_e^2}$ curve for the LTWM algorithm is more closed to curve of true value, relative of other three compared algorithms. Mean absolute errors MAE of estimated variances is smaller value for all values α. MSE of estimated variance of LTWM algorithm in relation to other algorithms is 26 (Alg. from [8]), 30.2 (Alg. from [6], [7]) and 30.5 (Alg.from [3]) times smaller.

d) execution time of the LTWM algorithm in relation to other algorithms is 1.48 (Alg. from [8]), 21.77 (Alg.from [6], [7]), 34.82 (Alg. from [3]), times smaller.

Based on the conducted comparative analysis of the results, it is concluded that the proposed LTWM algorithm has the greatest precision in estimating the variance of the LTWM noise. In addition, the LTWM algorithm has the shortest execution time. The obtained results recommend the implementation of the LTWM algorithm in system for real-time regime.

6. Conclusions

The modelling of the noise, caused by the insertion of the watermark into the image by the LT algorithm, was performed in this paper. It was shown that the noise could be modelled by Gaussian distribution. The LTWM model was described, with which it was possible to estimate the parameters of LTWM noise (μ, σ^2). The detailed analysis showed that *mean absolute errors* (MAE) of estimated variance of LTWM algorithm, proposed in this paper, in relation to other algorithms, is smaller in all range of insertion factor. Further, the detailed analysis showed that *mean square errors* (MSE) in relation to the estimation accuracy in the application: a) Alg. from [8], b) Alg. from [6], [7] and c) Alg. from [3], was smaller than 6, 30.21 and 3.86 times, respectively. In relation to: a) Alg. from [8], b) Alg. from [6], [7] and c) Alg. from [3], it took less *execution time* T_e for the LTWM algorithm, 1.48, 21.77 and 34.82 times, respectively. Based on the above mentioned results, it is concluded that the LTWM algorithm has high accuracy of estimation variance and short execution time. The mentioned performances give a recommendation for the implementation of the LTWM algorithm in systems for real-time regime.

References

[1] Dong L Zhou J and Dai T 2018 Color image noise covariance estimation with cross-channel image noise modeling *IEEE International Conference on Multimedia and Expo (ICME).*
[2] Ding F Shi Y Zhu G and Shi Y 2020 Real-time estimation for the parameters of Gaussian filtering via deep learning *Journal of Real-Time Image Processing,* Vol 17 pp. 17-27.
[3] Liu W and Lin W 2013 Additive white Gaussian noise level estimation in SVD domain for images *IEEE Transactions on Image Processing* Vol 22 Issue 3 pp. 872-883.
[4] Popp A Bock V Sedlmeir F C et all 2019 Investigation of noise sources down to the shot-noise limit in Yb-doped fiber amplifers for TMI investigations *Conference on Lasers and Electro-Optics Europe & European Quantum Electronics Conference (CLEO/Europe-EQEC.*
[5] Russo F 2003 A method for estimation and filtering of Gaussian noise in images *IEEE Trans. Instrum. Meas.* Vol 52 Issue 4 pp. 1148 –1154.
[6] Liu X Tanaka M and Okutomi M 2012 Noise level estimation using weak textured patches of a single noisy image *IEEE International Conference on Image Processing* (ICIP).
[7] Liu X Tanaka M and Okutomi M 2013 Single-image noise level estimation for blind denoising noisy image *IEEE Transactions on Image Processing* Vol 22 Issue 12 pp. 5226-5237
[8] Garcia D 2010 Robust smoothing of gridded data in one and higher dimensions with missing values *Computational Statistics & Data Analysis* Vol 54 Issue 4 pp. 1167-1178.
[9] Yue H Liu J Yang J Nguyen T and Hou C 2017 Image noise estimation and removal considering the bayer pattern of noise variance *IEEE Int Conf on Image Processing (ICIP).*
[10] Ting F F Sim K S and Wong E K 2016 A rapid medical image noise variance estimation method *International Conference on Robotics, Automation and Sciences (ICORAS).*
[11] Kowalski J Mikolajczak G and Peksinski J 2018 Noise variance estimation in digital images using finite differences filter *41st Int Conf on Telecommunications and Signal Processing (TSP).*
[12] Winarno A Setiadi D Arrasyid A Sari Rachmawanto C 2017 Image watermarking using low wavelet subband based on 8×8 sub-block DCT *Int Seminar on Application for Technology of Information and Communication (iSemantic).*
[13] Choudhary R and Parmar G 2016 A robust image watermarking technique using 2-level discrete wavelet transform (DWT) *IEEE 2nd Int Conf on Communication, Control and Intelligent Systems* pp. 120-124.
[14] Su Q Niu Y Liu X and Zhu Y 2012 Embedding color watermarks in color images based on Schur decomposition *Optics Communications* Vol 285 Issue 7 pp. 1792-1802.
[15] B. Prlincevic, Z. Milivojevic, P. Spalevic, and D. Brodic, "Performance of the SD algorithm for inserting watermark in image based on the schur decomposition" *International Scientific Conference* INFOTEH, Jahorina 2015, pp. 640-644, 2015.
[16] Milivojevic Z and Stevanovic Z 2013 Analysis on the robustness SVD-based watermarking algorithms *Int. J. of Computer and Information Technology-IJCIT* Vol 2 Issue 04 pp. 688-693.
[17] Zong H Liu F-L and Luo X-Y 2012 Blind image steganalysis based on wavelet coefficient correlation *Digital investigetion* Vol 9 Issue 1 pp. 58-68.
[18] https://photogallery.sc.egov.usda.gov/photogallery/#.

IOP Conf. Series: Materials Science and Engineering **878** (2020) 012028 doi:10.1088/1757-899X/878/1/012028

Model of process and model of natural language processing system

M Zhekova[1] and G Totkov[2]

[1] PHD student, Department of Computer Science, University of Plovdiv „Paisii Hilendarski", Tzar Asen str.24, Plovdiv, Bulgaria

[2] Prof., Department of Computer Science, University of Plovdiv „Paisii Hilendarski", Tzar Asen str.24, Plovdiv, Bulgaria

m_jekova@uft-plovdiv.bg , totkov@uni-plovdiv.bg

Abstract: Natural language processing is the set of methods for making human language accessibility from computers. An approach that facilitates search in database even for casual non-technical literate users is to use a natural language interface (IEE) for databases. It is important for these IEEs to be intelligent and reliable enough to interpret the most complex sentences most intelligently and effectively enough to deliver results in the shortest possible time. The advantage of natural language interfaces is that they are accessible to a wide range of end users in order to "speak" to the database. Introduction. The solution to the problem of natural-language understanding can be accomplished in two steps: linguistic and database processing. In the first stage of language processing, the natural language query is linked and translated to the corresponding SQL query by appropriate production rules, methods and functions. In the second processing stage, the database is accessed and manipulated and the SQL query is executed by the system. The study presents a research in the field of computational linguistics on a theoretical process model and a natural language processing system model. The process model itself represents knowledge and reflects the conceptual organization of human memory, as well as its flexibility and clarity.

1. Introduction

Natural language processing is a branch of machine learning that focuses on enabling computers to interpret and process human languages in both speech and text forms. The effectiveness of machine learning applications is limited to having the right set of complete, correct, clear and accurate data, as relying on human knowledge base specialists to design, customize and refine models and the inability of these uniquely created models to generalize for each domain area, for all business processes, functions and even geographical regions.

The main purpose of the study is to propose and explore methods and models suitable for creating a system with natural language interface (NLI) for processing user queries in Bulgarian for a given domain area.

Achieving this goal involves developing an NLI for database that upgrades existing systems based on keywords, rules, semantic networks, frames and scripts, making them more automated, improving the quality of the rules. The intent is to create an easy and friendly environment for interacting with computers in natural language and for coping with the task of retrieving a knowledgebase response through a user question in natural Bulgarian.

One of the tasks of natural language processing is to build an abstract conceptual model of the existing database and SQL generating model, which together with the linguistic component and according to the specifications and principles of knowledge modeling, return the answer to the user's query. When creating a computer model of the process of natural language processing of Bulgarian text, it should be taken into account that for the objects and their relations there is already a database (basic data or basic information base) and tools for their processing in order to be able to extract information about them and it does not necessarily depend on the specific methodology proposed. The existing relational-type information data base is the connecting element in identifying and processing text-query in Bulgarian and the tools described below to generate a SQL query.

2. Related work

In most computer information systems, communication with the user is accomplished through a graphical user-friendly interface rather than in a natural way. Any database application interaction requires either a graphical user interface (GUI) or structured query language experience to a database [1]. However, structured query languages are difficult to understand for non-experts. GUI needs to be filled in forms, fields, buttons, menu selections, validation procedures, etc. Instead, NLIs allow the user to communicate directly with the database, in the form of commands that translate a question in their natural language into a database query.

The term 'natural language interface' (NLI) represents a kind of computer-human interface where language forms such as verbs, phrases and clauses act as parameters in the formation of database retrieval requests. Natural language software interfaces allow you to ask questions that are formulated in the user's native language. The NLI must be able to understand human language after typing and translate it into the language that databases understand in order to receive the information sought. Working with NLI is more intuitive than conventional GUI where the user works with menus, tables, and filter fields. Very often menus have inflexible options and sometimes do not meet the specific needs of the user.

In order for a software system to understand a user query, it is necessary to introduce the term "natural language processing". The processing involves converting a sequence of words into a semantic representation that can be used later in the processing. This includes the use of morphology, syntax and semantics. Morphology is the study of the structure and content of word forms. Processing ends with identifying the keywords and generating the meaning of querying the database.

Examples of resource grammars for the Bulgarian language are the syntactic corpus - BulTreeBank; The Chooser system for defining different annotation schemes, used to access detailed information of the proposed meanings through synonymous definitions and semantic relations; The BulNet semantic network of synonymous sets and their semantic relations; knowledge base on semantics of verbs Seminvest et al.

3. Model of a process for natural language processing of Bulgarian text

Figure 1 presents a general model of a dynamic process for natural-language processing of a user-asked-query set in Bulgarian. The development of the process in Figure 1 is presented with possibilities for changing the text of the query and the methodology for synthesizing a SQL query, as a result of the accumulation of new metadata for the received language units and the flow of different subprocesses for natural language analysis and synthesis, following different principles and methods in the domain area in question.

The schematically presented model of interaction between the basic elements of the natural language processing process underlies the design and creation of an automated natural language interface system to the database for any domain area. Designing, creating and using electronic lexical resources (grammars) is a necessary preparatory step in the natural language processing process indicated in Figure 1 with S_Intro. As a result of the digitization of linguistic resources, it can be argued that the linguistic units annotated therein are updated periodically.

The design of a methodology for synthesizing a SQL query to the database depends on the nature of the language units and the results obtained from the next step of the process indicated in Figure 1 with S_Transform. The methodology of the process of forming a pattern-question for generating a response from the system, which is the final stage in natural language processing, denoted by S_Synthese, takes the form of rules, instructions and directions.

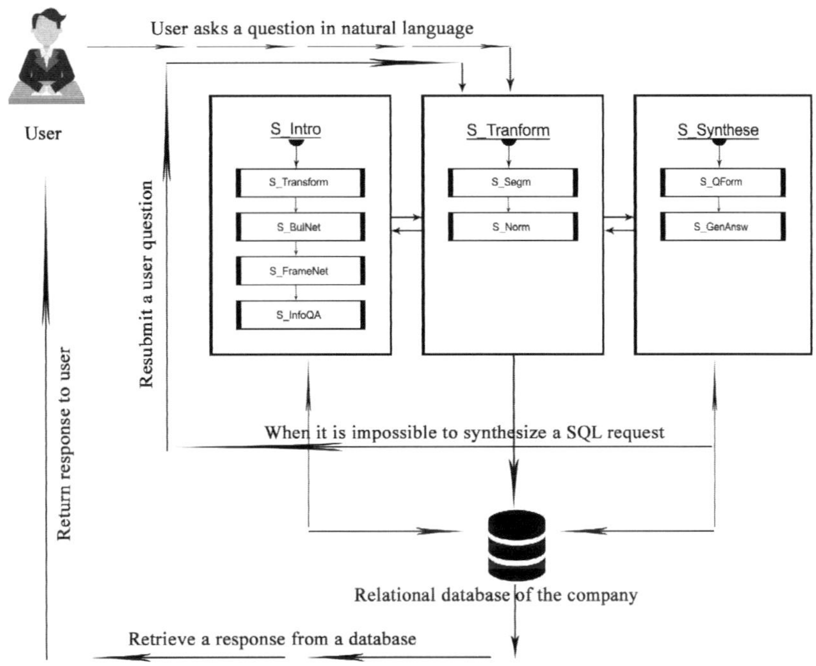

Figure 1. A common model for the natural language request processing process

The correct execution of the process of processing the user request involves performing one of the following actions:

a) After normalization and segmentation of the received request in the first stage S_Transform, and after receiving a list of consecutive values of the specific database and concepts from the specific information base, the process goes to the next stage S_Synthese;
b) If the list of DB values does not correctly generate a question pattern (the corresponding values do not "match"), they do not have matching that will lead to a response from the system, the process returns to the beginning and the user must ask a clarifying question.
c) Upon receipt of the relevant elements of the request, a question-pattern is formed in S_QForm, suitable for the chat-bot question-answer system.
d) Following the question-pattern, a SQL query-pattern related to the database is received.
e) In S_GenAns a reference is generated, extracted from a question in Bulgarian.

This process can be formalized and developed as a theory that combines syntactic, semantic analysis and code generator into one. The article introduces a new approach to semantic analysis based on syntactic-semantic analysis and transformation rules. The algorithm annotates the set of concepts recognized by the input query with their semantic representations. A major innovation of this algorithm is its integration with state-of-the-art natural language comprehension techniques.

4. Model of a natural language processing system

One of the main tasks in the development is to create a model of an automated system for natural language processing in Bulgarian. For this purpose, it is necessary to propose and investigate relevant conceptual and computer models for the elements of Fig. 1. This means that the general model of Fig. 1 will be further developed by incorporating computer models of the linguistic corpus. It is also necessary to propose a software implementation model (computer model) for the S_Analyse modules (with S_Segm and S_Norm sub-modules) and S_Synthese (with the corresponding sub-modules).

The problem of responding to a user's request in natural language is solved using a set of different rules and criteria of the proposed methodology. In the end, the final response from the system is accompanied by a number of tools additional information resources that help to analyze and generate a SQL request to the DB, ie. Response = Request + Resource. It should be noted that preparing a response from the database requires some co-ordination with the other submodules, procedures and processes.

4.1. Creation of a specific (for the given database) information base

The information base (computer model) includes systems and tools that differ in the nature of the processes and processing they perform over their data; technology, maintenance, etc. The existing relational type database is preceded by an analysis of the information structure of the organization within which it operates. The analysis made on the existing information base is made in order to identify the accessible structures of objects, phenomena, processes, relations and tools for their automated processing.

The modeling of each relational database scheme and its relation to the natural language allows:

- interpreting naturally language queries and generating correct SQL requests.
- adding metadata to natural language concepts to enable natural language interpretation of user queries, thus increasing the language coverage of the proposed software.

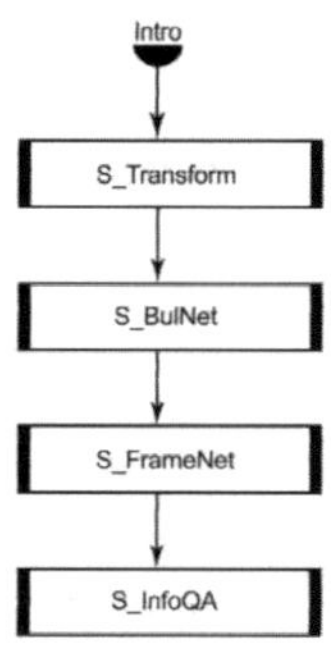

Figure 2. Identification of objects and relationships extracted from the existing database - digitization of language resources

Designing, modeling, and incorporating the elements derived from the analysis is an essential step in the development of the conceptual computer model. At this stage, the process of pairing the linguistic component and creating a computer model is not automated. In the future, however, the need and possibility of automating the matching process between an existing domain-specific database and the proposed compliance database model is discussed.

Terminology, factology and specialized knowledge presentation are important elements in the process of analysis and acquisition of knowledge. Terms such as the linguistic designation of specialized knowledge are inextricably linked to their presentation, storage, transmission, and extraction. Concepts are a core element of both the conceptual model and the knowledge base. Therefore, the concept is crucial to the presentation of knowledge. At any time from the perception of an event, the space around it is also perceived, [2] including objects, participants and objects present in it. Specialist linguists (experts) are involved in the process of collecting, annotating and creating the language component in S_Intro stage.

Let's look at Figure 3. It shows some of the tables in dbUniversity - a relational database in higher education. It consists of objects - 'student', 'teacher', 'discipline', 'specialty', etc. and relationships between them, based on the phenomena and processes taking place in a higher education institution. In the reasoning and examples in the exposition we will be based on this database.

IOP Conf. Series: Materials Science and Engineering **878** (2020) 012028 doi:10.1088/1757-899X/878/1/012028

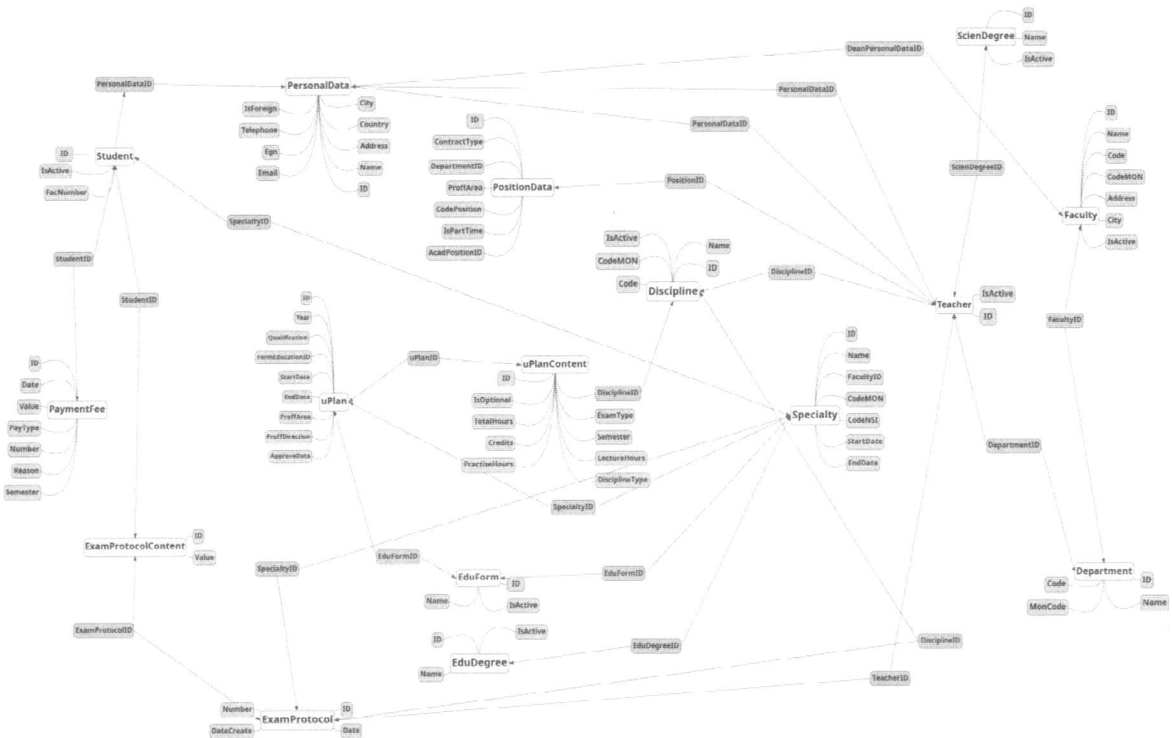

Figure 3. General scheme of relations between objects in the existing database

4.2. Creating a conceptual database model–подмодул_S_Transform

Creating a conceptual model that describes the semantics of the restricted language of a particular database. The existing relational database is transformed into a conceptual network of nodes (concepts) and edges (relationships), which later model query texts from the specific database.

Through the abstract conceptual model model, we schematically represent the structure of the existing relational database. The model is generated automatically by retrieving table names, values, and attributes based on the database schema and is independent of domain area. Initially, the database schema is analyzed to retrieve table names, attributes, data types, and relationships between tables. Subsequently, the structure of all tables and finally the relationships are modeled.

Conceptual modeling is the activity of formalizing aspects of the physical and social world around us for the purpose of understanding and communication. A network of connections is created, mimicking the way the human mind works.

Developing a conceptual data model also includes information about current business processes or workflow analyzes within an organization. After creating a conceptual data model that users are happy with, the next step is translating it into a schema that implements the relevant data structures in the database. This process is often called logical database design, and output is a logical data model, expressed as a schema. Data in a database is usually managed by database management systems (DBMSs). Access to the database is facilitated through a special interaction language called SQL. To overcome the complexity of communication between lay developers and the database management language - SQL, many researchers use natural languages. The idea of using natural language has prompted the development of a new type of processing method called natural language processing to database systems.

In order to model the structure of a table in a database, classes, attributes and properties are defined that describe it. The properties, attributes, and classes used to describe the structure of a relational database are presented in the following table, with the name of each database table representing a TableN class and the columns of each database table representing a ColumnN class. Subsequently, ColumnN class representatives contact the corresponding TableN class representatives using the hasColumn property. The data type name describing each attribute is added to the corresponding attribute via the DataType property [3].

Table 1. Classes and properties for modeling the structure of each database table

Name	Property	Description
TableN	Class	A table name is introduced as a member of TableN
ColumnN	Class	The column name is introduced as a member of ColumnN
hasColumn	Attribute Property	Relationship between a TableN and a ColumnN
isColumnOf	Attribute Property	The inverse property of hasColumn
relatesColToTable	Relation	Relationship between two tables
DataType	Datatype Property	Specifies the data type of each attribute

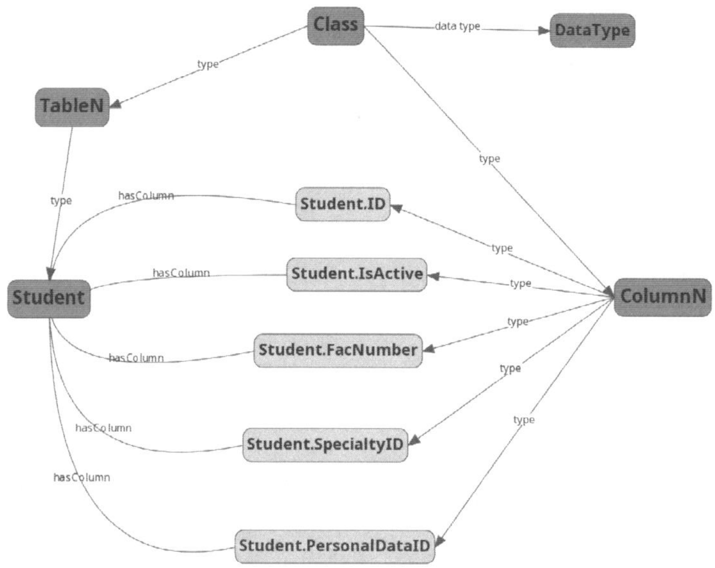

Figure 4. Scheme of the table Student and its attributes

After the structure of all tables is modeled as shown (see Table 1), the relationships between the tables in the database are followed. Figure 5 illustrates the relationship between two tables that is implemented through the relationsColumnToTable property. This property associates an attribute that is a representative of a ColumnN class with a table that is a representative of the TableN class. The TableN class representative must be a primary key in one table, and the ColumnN class representative must be a foreign key in another table (see PersonalData.ID and Student.PersonalDataID in Fig. 5).

The role of the representatives of the individual classes is to classify the objects according to their class. Proper modeling of table-to-table relationships is key to combining multiple table records when generating an SQL query.

Based on the proposed model, a hierarchy of classes and attributes, data types and relationships are dynamically defined for each table and attribute. This necessitates a system with a natural language interface to provide opportunities for:

- creation and management of a basic object (class, attribute property, relation, data type);
- defining roles and identifying objects in the database;
- defining procedures for checking different objects.

Modeling of information extracted from the database schema as a result of the classes and properties described (see Table 1) is performed dynamically and does not depend on the domain area. The user specifies only the database and the configuration is performed automatically. Following the conceptual scheme thus modeled, the user can further develop and customize with a set of natural language tools and metadata to describe each of the objects in the database.

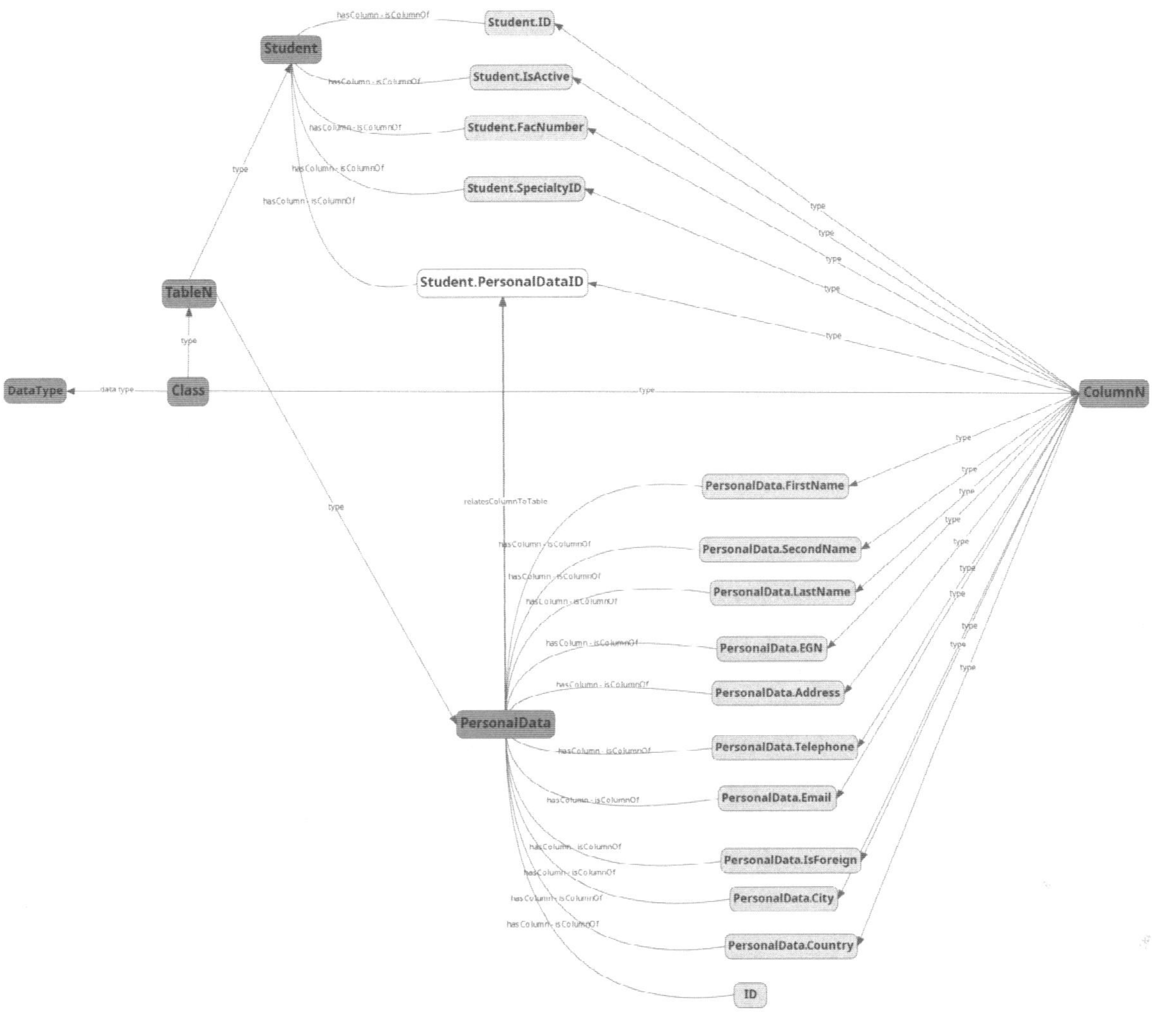

Figure 5. Scheme of relations between two tables - Student and PersonalData, representatives of the class TableN

After modeling the conceptual scheme, the relationship between it and the natural language is determined. For this purpose, a set of natural language words is identified that describes all table and column names in the database that store information that could be requested by users.

The conceptual network is created to answer templates of questions that are centered around a problem that it seeks to solve. This will be accomplished by constructing data structures describing facts, concepts, and valence relationships from the database that match the names of the tables, attributes, and relationships between them, which we will call an abstract syntax tree or concept tree.

The syntax tree corresponding to the statement in Figure 6. will be represented as a tree with vertices, which are arguments describing the concepts Student and PersonalData, and edges, which are relations of syntactic subordination between words. All values in the table PersonalData (Name, EGN, Nationality, etc.) are inherited from the table Student because Student is a type of Person (Student inherits the PersonalData) and all attributes of the class PersonalData class are attributes of the Student class as well.

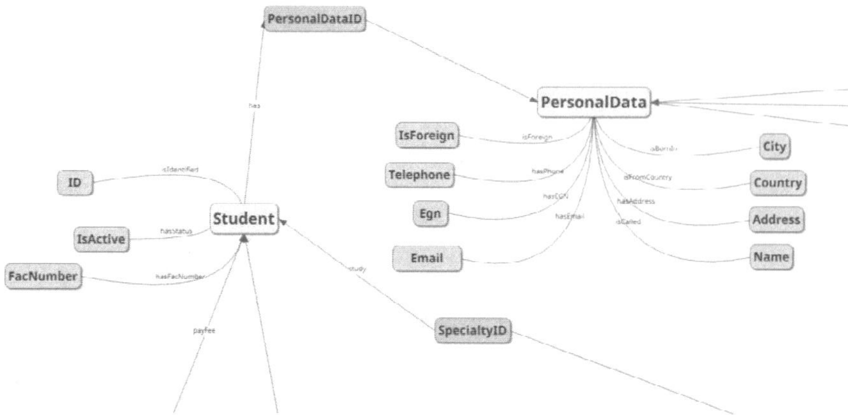

Figure 6. Schematic representation of nodes and edges between two tables in the database

4.3. Creating a Concept Tree for a domain area– подмодул_S_BulNet

From the concepts in the conceptual network to the existing database, builds a tree of concepts described with its synonymous lines (reduced version of BulNet).

At this stage, potential usage information will be collected and concepts will be modeled in the context of the domain area. The resulting knowledge resource will not be static, irrelevant and represented by unrelated data records. The purpose of the_S_BulNet submodule is to present concepts as part of a more abstract concept or situation where the concept is related to other concepts in a dynamic structure that can be optimally and naturally optimized through the interface of the submodule. The resulting abstract syntax tree is a reduced version of the BulNet linguistic corpus.

The submodule_S_BulNet will semantically describe the concepts that make up the conceptual model of each database. Semantic modeling is here to increase the number of correctly answered user queries.

Table 2. A set of attributes describing a Concept of a domain area

Attributes / Metadata	
Concept	Synsets
	Decription / Lexical meaning
	Part of speech
	Semantic class / category
	Core
	Basic form / Lemma
	Hiperonym
	Hiponym
	Принадлежи / Отнася се
	Question words

The feature tables that describe linguistically the concepts at this stage are based on the properties of the language units. The linguistic units are annotated with a syntactic category, part of speech, lexical meaning, synonyms and equivalent values, hyperonyms, hyponyms, etc. (see Table 2). Linguistic annotation data are presented in the form of key-value. Each category receives a single value derived from predefined sets.

A. Morphosyntactic characteristics are part of the morphological annotation. There are two main approaches to automatic morphological annotation.
B. Lexical characteristics describe the information lexical meaning of words, systemic lexical relations - synonyms, hyperonyms, antonyms, meronyms, etc.
C. Semantic characteristics attribute lexical meaning, semantic roles, semantic frames

Attributes of a class are inherited from all its heirs by the type-subspecies relationship, and the successors may have additional characteristics. The_S_BulNet submodule contains basic information about concepts in the domain area, such as semantic and syntactic features, but also contains compatibility (links) with other concepts in the concept tree.

The explicit specification and conceptualization, described by terms organized in taxonomies, their definitions and attributes, as well as related validation methods and value constraints, is intended to answer consumer questions. The conceptual description depends on both the domain and the tasks to be solved. The concepts correspond to the roles played by the objects in the domain area in the process of performing a particular activity.

4.4. Building a conceptual model for correspondences – подмодул_S_FrameNet

From the edges of the conceptual network, a set of control models (reduced version of FrameNet) is obtained.

For a natural language processing system to be effective, it must have the necessary and sufficient knowledge to understand user questions and translate them into question patterns and SQL queries to the knowledge base. In order to interpret natural language user issues, an information system with a natural language interface must know the connection between the natural language (submodule_S_BulNet) and knowledge of the database schema (submodule_S_Transform).

The specialized knowledge resource - submodule_S_BulNet, as part of the specific information base, aims to facilitate the analysis and acquisition of knowledge. Submodule_S_FrameNet, which is the next step in the stage of creating a specific information system, provides descriptions of relationships and content in which the concepts are considered as part of a process or event. Because the acquisition and understanding of knowledge requires a specific situation, therefore horizontal relationships that determine the purpose, task, meaning and outcome of the work and use of an object are as important as vertical and specific relationships [2].

In part 4.2. we have specified that for the creation of a semantic network of concepts structured and grouped by different criteria, nodes are synonymous sets that encode equivalence relations between several linguistic units, and edges express semantic relations between objects located in the nodes. The nodes correspond to the concepts / objects in the knowledge base, and the edges indicate the connections between them. The combination between two nodes and an arc is called a statement or fact. The statement is the relation between different objects. In practice, the semantic tree describes the syntactic behavior of individual words, ie. what words can be attached to them so that the correct language constructions are obtained.

At this point in the S_FrameNet submodule, we build on the concept tree built in part 4.3. and from edges expressing relations we get a set of control models for concepts and verbs related to text queries in Bulgarian. The S_FrameNet submodule will serve to represent the minimum of structured information that defines the types of phenomena, events, situations or processes. We will step on the concept introduced by [4] - frame, as a means of presenting knowledge. Frame models are used to represent relatively independent, logically distinct units of knowledge that can be reused in a variety of situations (including data extraction and aggregation) [5].

Each frame has a unique name and structure consisting of ordered elements called slots [6]. In the frame, each slot has a unique name and is described with a wide range of different features, including dictionary interpretation, classification information about its syntactic and semantic functions, etc. Slots are described with one or more attributes that represent knowledge describing the element in question. Attributes, or so-called facets, represent an expanded knowledge of slot values and allow it to be controlled [7]. The value of a slot has a certain type of data. It can be elementary or composed. The contents of each slot can be text or numeric value, function, formula, relation to another slot, predicate dependency or other frame. The frame structure is dynamic, ie it allows new slots to be added at any time [8]. The purpose of classifying data into frame structures is to maintain and store data, to define relationships between them, and to facilitate access to them.

As elements of the frame system in the S_FrameNet submodule, different concepts are involved with their connections, described by facts and rules in the relevant domain. Structurally, it represents system-bound frame structures with different levels of abstractness.

The purpose of classifying data into frames is to maintain and store the data, to define the links between them, and to facilitate access to them. In this way, overall commitment, hierarchy, and subordination of actions contribute to the complete description of the logical model of the particular domain. Clarification of the basic principles for classifying units and the creation of uniform and consistent categories is a necessary condition for the applicability of classification (Stoyanova, 2012b). This structured knowledge base is reduced to a classification task, where individual frames contain answers to the most frequently asked questions for the domain under consideration.

4.5. Forming a set of question patterns – подмодул_S_InfoQA

The proposed methodology for forming question patterns is illustrated with examples. Each example shows a question pattern and provides the necessary validations and sufficient conditions for the implementation of each specific pattern.

Example 1: Show / Display all semester fees.

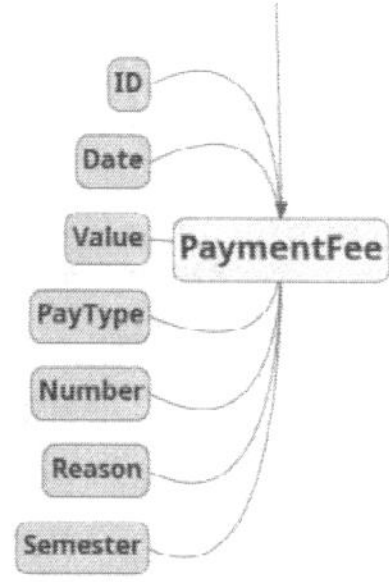

Figure 7. Schema showing the requested TableN instance

Pattern 1. [Which are | Show | Display | Find] [all | every] X, where X is the only TableN class member and no attributes are searched.

The condition for applying this database query pattern to it is that the term is found to be a single TableN class representative. As a result, all records for that particular table are returned.

X – PaymentFee – a specific instance of the TableN class

Therefore, the pattern is undoubtedly applied because no specific attributes were found, and there is no other object in the TableN class. There is only one specific table and the result is all available records in it. The corresponding SQL query for this particular pattern is: SELECT * FROM X

Example 2: Show the type and value of semester fees.

The example shows that in addition to the TableN class representative, there are already ColumnN class representatives, which are specific fields from a specific table.

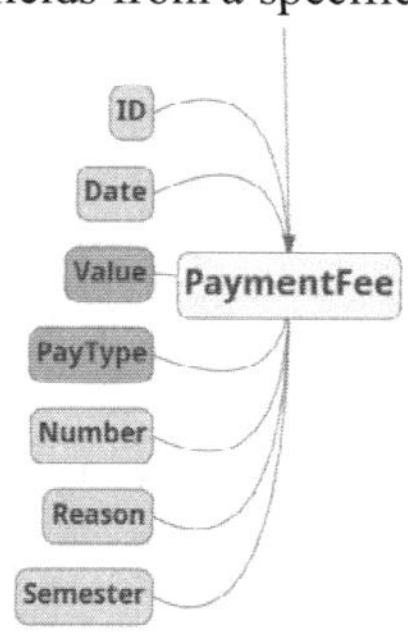

Figure 8. Schema showing the requested attributes of the TableN and ColumnN classes

Pattern 2. [Show | Display | Find] A1,A2,...,An [and | ,] A2 [from | for | in | on] X, where X is the sole representative of the TableN class and A1, A2,.., An are the unknown attributes representing the ColumnN class.

The validations and checks that are required in this case are: Is there another TableN class member in the user string and all the concepts found are eligible to be columns on the only table. As a result, only the columns corresponding to the requested attributes from the table specified in the search string are returned.

X – PaymentFee – a specific instance of the TableN class

A1 – PayType; A2 – Value - are representatives of the ColumnN class and have the property: A = isColumnOf (X)

PayType = isColumnOf (PaymentFee); Value = isColumnOf (PaymentFee)

Therefore, the pattern applies when, after the analysis and refinement of the properties of the attributes (concepts) involved in the query, if all attribute names are from the sole table, ie. the isColumnOf (TableN) property, where TableN is a representative of the same class. Then the corresponding SQL query on the pattern is: SELECT A1, A2,..., An FROM X

Example 3: Show the value of the fee for the second semester.
In the example, the search terms are identified as representing the TableN and ColumnN classes, but there is already an attribute value, i.e. a specific numeric (text) value for a column.

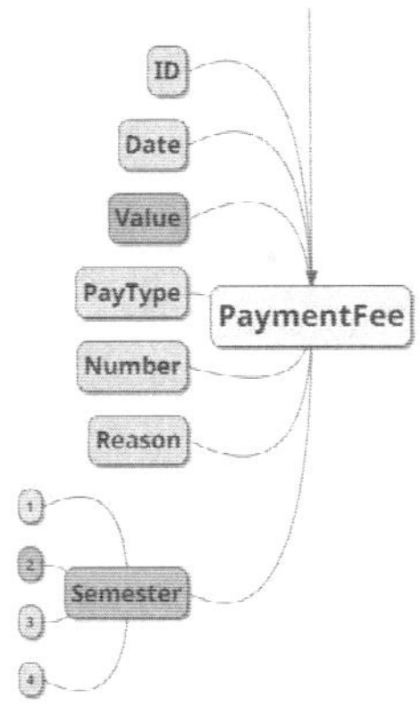

Figure 9. Schema showing the specific attributes of the TableN and ColumnN classes that meet a certain condition

Pattern 3. [Show | Select | Display | Find | Who is | Which are] A1,A2,...An [from | for | in | on] X, [where| when] Ak="value", where X is the sole representative of the TableN class. A1, A2, A3,..., An are the requested attributes, representing the class ColumnN, and 'value' is a specific value for a particular attribute (Ak) in the same table.
For this template, validation methods include: is there another TableN class member in the user string; do all the concepts qualify to be columns of the table you are looking for. As a result, only columns that match the requested attributes from the specified table and that meet the specified condition are returned.

X – PaymentFee - a specific instance of the TableN class
A1 – PayType; A2 – Semester - are representatives of the ColumnN class for which:
A = isColumnOf (X)
A1 – PayType = isColumnOf (PaymentFee); A2 – Semester = isColumnOf (PaymentFee)
Condition – Semester = "2"

Therefore, the pattern applies when, after applying the analysis, it is established that there is only one representative of the TableN class and one or more representatives of the ColumnN class, in addition to the requirement for one of the representatives of the ColumnN class.
Then the corresponding SQL query template is: SELECT A1, A2,...., An FROM X WHERE Ak = "value".

5. Conclusion

This paper presents a new study in the field of computational linguistics - the comprehension and interpretation of texts in natural Bulgarian language by computers. For the purposes of the study, a detailed examination of the results in the field of English language processing was made. Unfortunately, there are no other studies in this field for Bulgarian language so far. Developments for modeling structure, syntax, and semantics in other languages were explored. The subject of the analysis were systems based on pattern matching, keywords, frames, production rules, menus, and question-answer systems. A process model and a model of system with a natural language interface to a database for processing user queries of Bulgarian text is proposed, in the field of higher education.

From the results of the study (a process model and a model of system with a natural language interface to a database) and their general nature, we can conclude that ideas can be transferred to other domain areas. The results of the development may serve as a basis for further research into some promising areas, such as:

- improving the presented models, methods and strategies for accumulating metadata for language units;
- multiplying results for information infrastructures based on different technologies from other domain areas;
- creating tools and applications to describe other types of objects, e.g. work processes and phenomena from activities in other fields of knowledge.

References

[1] Mony M., Rao J., Potey M., An Overview of NLIDB Approaches and Implementation for Airline Reservation System, International Journal of Computer Applications (0975 – 8887) Volume 107 – No 5, December 2014

[2] Faber P., San Martín A., Conceptual modeling in specialized knowledge resources, International Journal "Information Technologies and Knowledge", Vol. 4, Number 2, 2010

[3] González J., Florencia R., Fraire H., Pazos R., Cruz-Reyes L., Gómez C., Semantic representations for knowledge modelling of a Natural Language Interface to Databases using ontologies, International J. of Combinatorial Optim. Problems and Informatics vol.6 no.2 Jiutepec may./ago. 2015

[4] Minsky M, A Framework for Representing Knowledge, MIT, Cambridge, 1974

Ruppenhofer J, Ellsworth M, Petruck M R L, Johnson C R, Scheffszyk J, FrameNet II: extended theory and practice, Technical Report, 2005

[5] Gaftandzhieva S, Doneva R and Totkov G, Frame Models in Programming Training, XI National Conference "Education and Research in the Information Society", 2018

[6] Zhekova M., Totkov G., A Frame Model for Representing Semantic Roles and Processes for Building a Natural Language Interface, ARIO 2019 – Plovdiv, May 30-31, 2019

Fellbaum C, WordNet - An Electronic Lexical Database, MIT Press, 1998

[7] Gaftandzhieva S, Doneva R and Totkov G, Frame representations in e-learning applications and new developments, International Journal on Information Technologies & Security, No 2 (vol. 10), 2018

Zhekova M. Review and Study of Existing Artificial Intelligence Systems, Proceedings of the 21st National Scientific Conference for Students and PhD students "The World is a Open Book", Plovdiv, May 16-17, 2019

[8] Zhekova M., Totkov G., M., Conceptual frame model for the presentation of the concepts and rules in natural language interface for database, 8th International scientific conference "TechSys 2019", Technical university of Sofia, Plovdiv branch 16-18 may 2019

Stoyanova Iv, Automatic recognition and tagging of composite lexical units in the Bulgarian language, 2012 Sofia

IOP Conf. Series: Materials Science and Engineering **878** (2020) 012029 doi:10.1088/1757-899X/878/1/012029

Hybrid deep-learning analysis for cyber anomaly detection

Stanimir Kabaivanov[1], **Veneta Markovska**[2,*]

[1] Plovdiv University "Paisii Hilendarski", Plovdiv, Bulgaria

[2] University of Food Technologies, Plovdiv, Bulgaria

* venetta@abv.bg

Abstract. Cyber threats evolve continuously and so do the detection tools and algorithms. In this paper we analyse the efficiency of hybrid deep-learning analysis as a mean to detect anomalies in computer network traffic. Different deep-learning algorithms are tested against real network intrusion events in an attempt to assess their potential as an early warning system. We suggest a combination of algorithms and rule-based filters as a hybrid system that can improve efficiency and accuracy of cyber anomaly detection.

1. Introduction

Cyber threats are not new and it often seems like we have already done enough to prevent them by installing state-of-the-art protection system and "sanitizing" our online behavior. Recent reports however indicate that although attacks do not grow at a very fast rate they seem to get more sophisticated and indirect ([1]). Taking into account the global move toward even more distributed working environment in the wake of the COVID-19 crisis, the potential for destructive and sophisticated attacks has grown significantly. Some of the possibilities are well known – for example exploitation of cloud vulnerabilities and social engineering attacks. Belani in [2] points out that artificial intelligence, adaptive attacks and machine learning will be major issues in the years to come.

In a distributed environment detecting a malicious software after it has been deployed and activated is a costly and definitely late measure. It makes much more sense to try to monitor the environment and react on anomalies in the normal/expected operation, instead of passively wait for a problem to appear and then try to resolve it. Considering the high amount of data and increasing number of connected software applications such monitoring requires to have dedicated solutions and automated filtering of possible problems. Without proven and solid algorithms that point our attention only to real dangers we will be overwhelmed by false positive alarms that eventually will cause more harm than good.

Bhuyan el. al in [3] classify anomaly detection methods into eight broad groups – statistical, classification, soft-computing, clustering, ensemble, fusion and hybrid. However, when analyzing network intrusion in general, we can suggest a different taxonomy:

- Signature-based detection of network intrusion and misuse.

While being very solid approach, with minimal possibility for a false-positive in case good signature database, signature-based detection requires frequent updates of the threat database and also substantial efforts in maintaining signatures. Exactly because of these significant efforts, there are a lot of databases which are shared and that also increases the chance for a dedicated attacker to test his own malicious activity in advance against these databases and implements exactly the tweaks necessary to go undetected. Signature-based systems can be extended to apply fuzzy-logic ([4], [5]) and machine learning in order to improve their flexibility and responsiveness to new kind of cyber-attacks.

- Anomaly-based detection algorithms.

With sufficiently large amount of network traffic to analyze, it is possible to detect patterns in normal user behavior and detect statistically significant deviations from it. The success of such approach depends on the quality of the patterns being detected/defined and on the predefined deviations boundaries which are considered to be "normal". Taking into account that networks change over time and that user activities are also different, it takes a considerable amount of input data and a careful monitoring of the rules to have a highly accurate anomaly detection. Anomaly-based detection systems can also be combined with flexible AI tools like in [6], [7], [8], [9].

- Hybrid detection algorithms.

The major goal of combining different detection algorithms is to try and detect both well-known as well as new threats. There could be different strategies toward mixing intrusions detection approaches. In this paper we focus mainly on use of machine learning to counter adaptable and AI-base cyber security threats.

2. Hybrid deep-learning model for anomaly detection

Hybrid systems can be built in a different way and for quite different purpose. It is often possible to get similar detection and discovery rates but with dissimilar performance ([9]). When choosing appropriate combination, the goal is to have it working as efficient as possible in both terms of accuracy and resources consumed (e.g. that it can handle as much traffic as possible with the provided hardware resources). Yet it is often necessary to choose between accuracy and versatility, which means that the most-accurate methods may not be suitable in terms of time required to analyze the data or in terms or necessary inputs. In this paper we follow the idea by Wang and Jones in [10] that solutions can be compared in terms of their flexibility, reliability, scalability, robustness and speed. Hybrid solutions typically score higher in the first four categories, compared to statistical and signature-based methods. This comes at the expense of being more complex, thus also harder to maintain and explain, and also requiring more resources and time to analyze the network activities.

We suggest a two-step procedure in building a hybrid anomaly detection. Layering different activities can improve flexibility as each layer can be replaced independently to build pluggable detection systems.

- The first step consists of feature extraction of the underlying traffic, which is adaptable but once executed the feature set remains fixed, until the entire system is reconfigured and re-trained. This makes it possible to remain flexible, while most important features are extracted once and then used. This reduces total calculation time and keeps resource usage low.

Feature extraction step is shown on

Figure 1, where traffic is captured with Wireshark and relevant packets are converted to XML or JSON for further analysis. It should be noted that filtering of packets based on specific protocols and ports is not included in the feature extraction – it can be used to simplify the initial step. The goal of feature extraction is to figure out which packet characteristics are most relevant to be included in the subsequent analysis (for example length, hosts etc.).

- The second step consists of classification algorithm that is responsible for giving hints if analyzed condition should be considered normal or not. This does not mean that output should be only a binary value – different classes of severity can be specified, or the output could be a probability for non-normal situation.

Once feature extraction has been complete, the remaining traffic captured is automatically filtered in accordance with the chosen feature set – e.g. for every packet captured, only features found to be relevant are processed. In case it is necessary to include new features, we can execute again model calibration and run again the initial step. Although this is clearly a limitation for the suggested approach, regularly executing the initial step is also beneficial – it does not have to be done in real time, thus it has no impact on the performance of the whole system. On the other hand – tracking the differences in the important feature subset is a possibility to detect that there has been a change in network use.

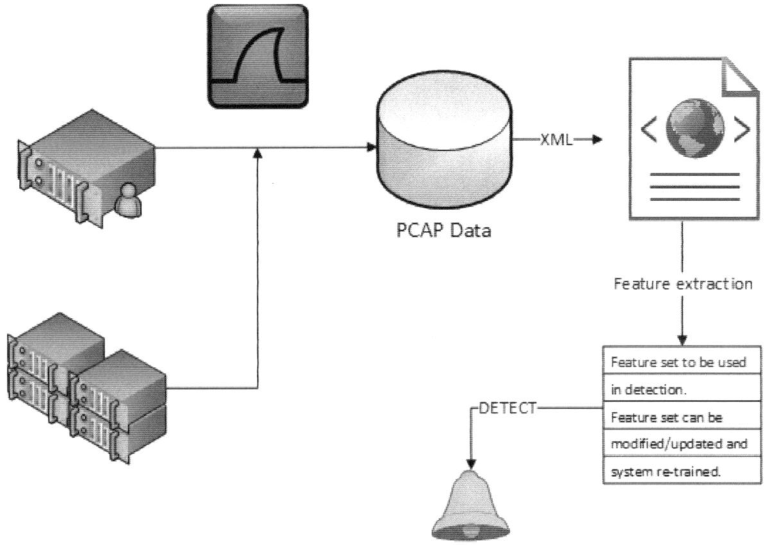

Figure 1. Feature extraction and traffic analysis

For comparing the efficiency of the second step in suggested sequence, we use three different models – generalized linear model, support vector machine and neural network with BFGS optimization.

2.1. Generalized linear model
With a network under attack we assume that there will be no errors of our model that follow normal distribution. Therefore GLM ([11]) can be used to capture these effects, provided that we have a link function such that classification output (probability and/or class) Y, depends on our input characteristics X:

$$E(Y) = f^{-1}(X\beta) \quad (1)$$

In the simplest case where we want to test if the condition being analyzed is normal or not, we assume binomial distribution and use logistic function.

2.2. Support vector machine
Support vector machine ([12]) is used to find the maximum-margin hyperplane between the normal and a-normal cases of traffic. In our case, training vectors contain all features that have been extracted at the first step. Training of the algorithm is done over a training set of inputs, each containing a vector of characteristics and class of the situation (normal or not) - $(x_{1i},..,x_{ki},y_i)$. To allow for non-linear classification a hyperbolic tangent function kernel where $\vec{x_i} = (x_{1i},..,x_{ki})$ and $f(\vec{x_i},\vec{x_i}') = \tanh(a(\vec{x_i}.\vec{x_i}') + c)$. It should be noted that SVM can be used also for novelty and outlier detection, but this is beyond the scope of this paper.

2.3. Artificial neural network

We use a simple neural network with a single layer of hidden neurons. The number of neurons inside is optimized through iterative procedure in a way that maximizes the accuracy and in particular the number of false positives. We focus on false positives as these results would likely lead to higher load for personnel responsible for incident checking and verifying if there has been indeed abnormal activity. Network used for processing packet data has a single hidden layer and 24 neurons inside.

3. Application and results

Data used to test the model consists of 257 673 packets and contains samples of ten different attacks, like denial of service, worms and backdoors. Unlike normal network traffic the amount of abnormal activity is high (39.3% of all cases) and this has been done in purpose in order to keep the test data set small.

Data preprocessing step is kept very simple on purpose, to allow 13 out of 45 network traffic parameters to be included and passed for processing for the second layer. Training set consists of 68% of all cases and the remaining data is used for assessing accuracy of models.

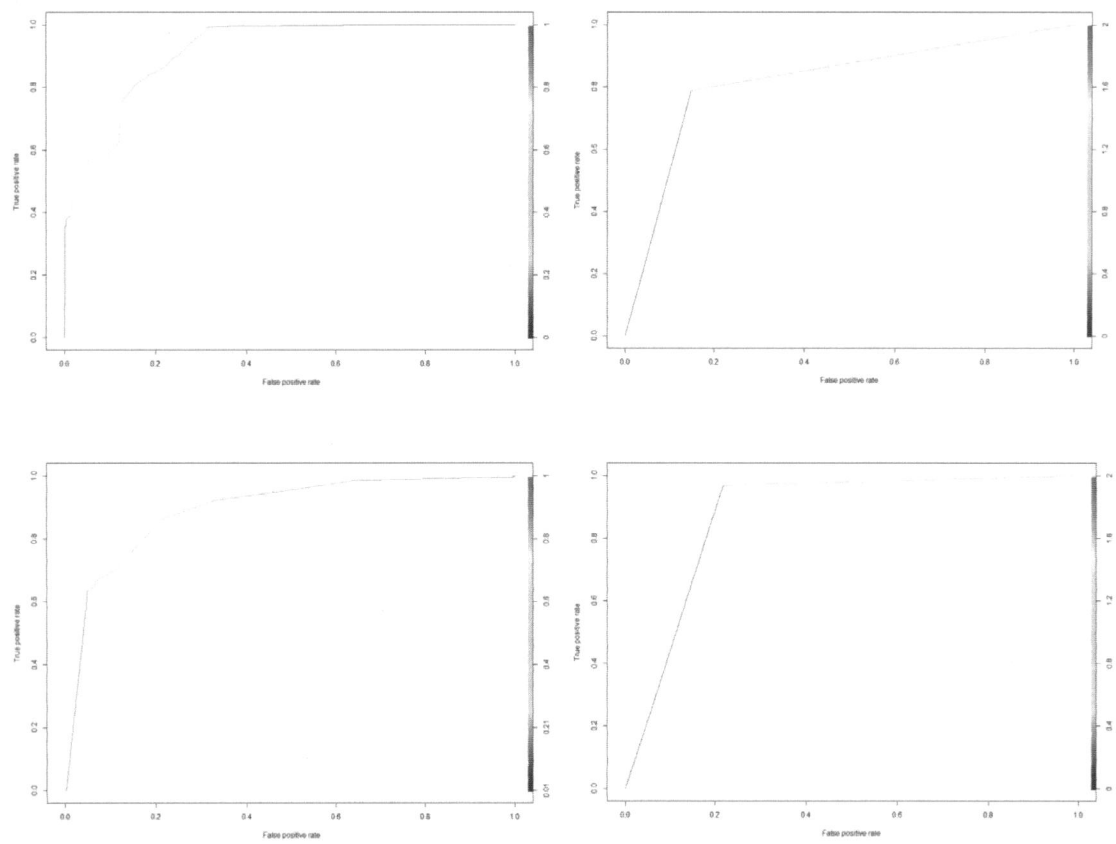

Figure 2. ROC curves for GLM (top left), SVM (top left), ANN (bottom left) and hybrid model (bottom right)

Hybrid model relies on the output of the three models but it is processed again through a simple neural network with single hidden layer containing 3 neurons. This makes it possible to create a dynamic ensemble model, where weights are also modified based on the performance of each participating model. Adding more layers or more neurons is likely to increase the overfitting and actually reduce the accuracy of the combined model use.

Table 1. Accuracy in binary prediction of the test sample

Model	Accuracy in classification (binary)
Generalized linear model	0.881334
Support vector machine	0.808219
Artificial neural network	0.838144
Hybrid model using all three outputs	0.909051

Hybrid model can be extended further by plugging in specific rules that depend on the network topology and services exposed by different nodes. There are no such assumptions made with the current data set, but context-dependent rules can significantly reduce the false-positives and make sure that overall accuracy is improved. Artificial neural networks can be supported by other methods for tracking network activity, for example mean-reverting process modelling in which case normal load is considered to be a starting point and abnormal activities can be checked against their mean-reversion speed [13].

4. Conclusion

Machine learning algorithms can improve accuracy of intrusion detection and checking for anomalies in network activity. We tested several algorithms and found that hybrid methods, supported by additional rules can provide a competitive advantage and increase the overall detection accuracy, while at the same time decrease the amount of false-positives. Taking into account that network diversity and use increases it is especially important to keep track not only at the total correctness, but also at the efforts needed to track cases, which only seem like intrusion but are just normal communication.

Maximum accuracy achieved in the test case is a bit over 90%, which leaves further room for improvement, especially taking into account that means of network intrusion and cyber threats evolve over time. In this context, a major advantage of used detection methods is that they can be adapted and re-trained continuously to respond to changes in the environment. In this paper, out of 45 available features of network traffic only 13 are used, but this set can be extended and more sophisticated methods can be applied in the preliminary feature extraction. Combined with a set of extra rules and context-driven filters, that would significantly improve the detection rate and make it possible to automate further detection of network intrusion and cyber threats.

The research is partially supported by the NPD-Plovdiv University under grant No. FP19-FESS-014 "Contemporary tools and approaches for economic and business analysis".

References

[1] Accenture, "Innivate for Cyber Resilience: Lessons from leaders to master cybersecurity," AccentureSecurity, 2020.

[2] G. Belani, "5 Cybersecurity Threats to Be Aware of in 2020," IEEE Computer Society, 2020.

[3] M. H. Bhuyan, D. K. Bhattacharyya and J. K. Kalita, "Network Anomaly Detection:Methods, Systems and Tools," *IEEE COMMUNICATIONS SURVEYS & TUTORIALS,*, vol. 16, no. 1, pp. 303-336, 2014.

[4] L. Ondrej, M. Manic, T. Vollmer and J. Wright, "Fuzzy logic based anomaly detection for embedded network security cyber sensor.," in *IEEE Symposium on Computational Intelligence in Cyber Security*, 2011.

[5] L. Wei, L. Ci and L. LiPing, "A New Method of Fuzzy Support Vector Machine Algorithm for Intrusion Detection," *Applied Sciences,* vol. 10, no. 3, p. 1065, 2020.

[6] F. Cheng, T. Li and D. Chana, "Multi-level anomaly detection in industrial control systems via package signatures and LSTM networks.," in *47th Annual IEEE/IFIP International Conference on Dependable Systems and Networks (DSN)*, 2017.

[7] A. Kumar, A. Abdelhadi and C. Clancy, "Novel anomaly detection and classification schemes for Machine-to-Machine uplink.," in *IEEE International Conference on Big Data (Big Data)*, IEEE 2018, 2018.

[8] Y. K. Sharma and M. D. Rokade, "Deep and Machine Learning Approaches for Anomaly-Based Intrusion Detection of Imbalanced Network Traffic.," *IOSR Journal of Engineering,* pp. 63-67, 2019.

[9] B. Belchev, A. Veleva and M. Ruseva, "Specialized software product "Statistical Analysis"," in *International conference "Research Methods and Techniques in Economics and Social Sciences"*, Plovdiv, 2012.

[10] S. Otoum, B. Kantarci and H. T. Mouftah, "On the Feasibility of Deep Learning in Sensor Network Intrusion Detection," *IEEE Networking Letters,* vol. 1, no. 2, pp. 68 - 71, 2019.

[11] L. Wang and R. Jones, "Big Data Analytics for Network Intrusion Detection: A Survey," *International Journal of Networks and Communications,* vol. 7, no. 1, pp. 24-31, 2017.

[12] A. J. Nelder and W. R. Wedderburn, "Generalized linear models.," *Journal of the Royal Statistical Society: Series A (General),* vol. 135, no. 3, pp. 370-384, 1972.

[13] C. Cortes and V. Vapnik, "Support-vector networks," *Machine Learning,* vol. 20, p. 273–297, 1995.

[14] S. Kabaivanov and V. Markovska, "Modelling environment changes for pricing weather derivatives," *Scientific Annals of Economics and Business,* vol. 64, no. 4, pp. 423-430, 2017.

[15] A. Adriano, R. Conforti, M. Dumas, M. La Rosa and G. Bruno, "Automated discovery of structured process models: Discover structured vs. discover and structure.," *International Conference on Conceptual Modeling,* pp. 313-329, 2016.

[16] B. F. van Dongen, "BPI Challenge 2018," 4TU.ResearchData, 2018, 2018.

[17] G. Janssenswillen, B. Depaire, M. Swennen, M. Jans and K. Vanhoof, "bupaR: Enabling reproducible business process analysis," *Knowledge-Based Systems,* vol. 163, pp. 927-930, 2019.

Intelligent virtual agent, learning how to reach a goal by making the least number of compromises

Dilyana Budakova, Veselka Petrova-Dimitrova, Lyudmil Dakovski

Technical University of Sofia, Plovdiv Branch, Bulgaria
Technical University of Sofia, Plovdiv Branch, Bulgaria
European Polytechnical University, Pernik, Bulgaria

dilyana_budakova@yahoo.com, vesi_s_petrova@yahoo.com, l.dakovski@gmail.com,

Abstract. The learning process in the Q-learning algorithm is characterized by maximizing a single, numerical reward signal. However, there are tasks for which the requirements toward the way to reach a goal are complex. This paper proposes a modification to the Q-learning algorithm. In order to make the Q-learning agent find the optimal path to the goal by meeting particular complex criteria, the use of measures model (a model of environment criteria), represented as a new memory matrix, is introduced. If the goal cannot be reached by following the pre-set criteria, the learning agent can compromise a given criterion. The agent makes the least possible number of tradeoffs in order to reach the goal. If the criteria are arranged by their level of importance, then the agent can choose more in number and more acceptable compromises. The aim of the modification is to empower the learning agent to control the way of reaching a goal. The modified algorithm has been applied to training smart shopping-cart learning agents. The tests show improvement in their behaviour.

Keywords: Intelligent system, Reinforcement learning, Smart shopping-cart learning agents

1. Introduction

Q-learning is a model-free reinforcement learning algorithm [1][2][3]. The learning process is characterized by maximizing a single, numerical reward signal and by interaction with an unknown environment. The teacher does not point at the actions to be undertaken. Instead, the trainee has to find out those, leading to the greatest reward and then to try to realize them. In the most interesting and challenging cases, not only the immediate reward could be taken into account when choosing an action, but also the further situations and the future rewards.

There are very many ways to improve the reinforcement learning algorithms [4-8]. It is known, for example, that Imitation Learning is a way for their optimization. [4][5][6]. In [7][8] the efforts of the authors are addressed to considering real-life problems that involve multiple objectives, often giving rise to a conflict of interests. The authors propose multi-objective reinforcement learning (MORL) algorithms that provide one or more Pareto optimal balances of the problem's original objectives. In the case the decision maker's preferences are clear and known a priori, single-policy techniques such as secularization functions can be employed to guide the search toward a particular compromise solution [7]. In case the preference of the decision maker is unclear before the optimization process takes place, it might be appropriate to provide a set of Pareto optimal compromise solutions to the decision maker, each comprising a different balance of objectives [7]. A more advanced idea is to simultaneously learn a set of compromise solutions [7]. In other words, criteria are to be set, characterizing a goal, and solutions proposed that lead to another tradeoff goal, optimally balancing between the pre-set requirements.

This paper considers objectives, for which the requirements toward the way of reaching a goal are far too complex to be described by a single, numerical reward criterion. In order to make the Q-learning agent find the optimal path by meeting specific complex criteria, the measures model (a model of environmental criteria), represented as a new memory matrix, is introduced. When it is impossible to reach the goal by following the given criteria, then the agent arrives at compromising a criterion. One option for the agent in the proposed algorithm modification is to make the least possible number of compromise solutions in order to reach the goal. This option has been studied in the present paper. There is another option suitable for the case when there are a lot of requirements toward the way of reaching the goal. The criteria can be arranged by their level of importance then. In such a situation the agent can make more in number and more acceptable tradeoffs. The aim of the proposed modification to the Q-learning algorithm is to empower the agent to control the way of reaching a given goal.

An intelligent system for smart shopping modeling allows for applying and improving the learning algorithms. Therefore the proposed algorithm is implemented to training Smart Shopping Cart Learning Agents (SSCLA). The design of the appropriate intelligent system for smart shopping is described in [9]. The proposed prototype of the system incorporates SSCLA, beacon-based technology, holographic technology, picture exchange communication system, text-to-speech and speech recognizing technology, face recognizing and machine learning techniques. It is envisaged that concrete implementation of the shopping agents will be running on each shopping cart in the shopping centers or on holographic displays [10][11]. The design of the smart shopping cart can be different [10][12]. The k-d decision tree, the best identification tree, and the reinforcement-learning algorithm are used for training the agents. The performance measures, to which both the intelligent system for smart shopping and the shopping agents aspire, include: getting to the correct shop in the shopping mall; getting to the new promotion in the shopping mall; minimizing the path when going through the shops from the shopping list; maximizing passenger comfort; maximizing purchases; and enabling people with different communication possibilities to navigate and shop in the big shopping centers [9-11].

The studies show improvement in SSCLA behavior in result of applying the proposed modified Q-learning algorithm.

The rest of the paper is structured as it follows: in Section II the realization of a Goal-Based Reinforcement Learning Agent is described; in Section III the Q-learning algorithm modification realization is explained; a survey of the performance of the Q-learning algorithm with an introduced measures model (a model of environmental criteria), presented as memory Matrix K is presented in Section IV; the section for discussion and future work is Section V; in the VI-th Section a number of conclusions are drawn.

2. Goal-Based Learning Agent

All reinforcement learning agents have explicit goals, can sense aspects in their environment and choose actions to influence it accordingly. The agent is realized by a program, matching the way the agent perceives reality and the actions it undertakes.

A reinforcement learning algorithms is used for the Goal-based learning agent in [10][11]. The agent receives the shopping list from the customer (this is what the agent perceives) and informs the customer about the sequence of shops he/she can visit in order to buy all the goods needed (these are the actions the agent undertakes). The shortest possible route is suggested, in accordance with the particular shopping list.

Since the goal is to visit all the shops from the shopping list, the particular shopping list can be regarded as a plan or a sequence of goals to achieve in order to fulfill the task completely.

The environment model is a graph (Figure 2) of the different environment conditions. The nodes in the graph are the shops in the exemplary shopping mall. The edges point at the shops, between which there is a transition. Then, this graph is presented by a reward matrix (Figure 1). The number of rows and columns in this matrix is equal to the number of shops in the mall. Zero is put in the matrix in a place where there is a connection between the number of a shop, set by a number of a row, and the number of a shop, given by a number of a column. Values of -1 are placed in the other positions of the reward matrix. The rewards model is needed to set a goal for the agent. Reaching every shop from the customer's shopping list is such a goal. Since the agent is a goal-based one, its behavior can be changed by just setting a new goal, changing the rewards model [1]. A reward is only given when the agent gets to a particular shop.

The agent's memory is modeled by presenting it with the help of an M-matrix (Memory of the agent). The rows in the M-matrix represent the current location of the customer, while the columns are the shops, where he/she can go. It is assumed at the beginning that the agent does not have any knowledge and therefore all elements in the M-matrix are zeros.

The rule for calculating the current location of the customer at the moment of choosing the next shop to visit is as it follows: M (current location of the customer, chosen shop to visit next) = R(current location of the customer, next shop) + ϒ. Max[M(next shop, all possible shops where the customer could go from the next shop)].

The following is taken into account in the above formula: The immediate reward, obtained when the customer decides from the current location to go to a next shop: R(current position, chosen shop to go next); the biggest possible future reward - this is the biggest reward, chosen from among all rewards, which would have been obtained when the customer goes out of the next shop and enters any possible other shop: Max[M(next shop, all the shops where it is possible to go from the next shop)]. The value of the learning parameter ϒ defines the extent, to which the agent will take into account the value of the future reward. The value of the learning parameter ϒ is within 0 to 1 ($0 \leq ϒ < 1$). If ϒ is closer to zero, then the agent will prefer to consider only the immediate reward. Experiments have shown that in this case it is impossible to teach the agent to achieve the goal. If ϒ is closer to one, then the agent will consider the future reward to a greater extent. This is the better option for successful training of the agent. The value of the learning parameter was experimentally chosen to be ϒ=0.8 [11]. One of all shops is chosen, where it is possible to go from the current position. The shop, to which the customer would go next is considered. For this next position now all the shops, to which it is possible to go further are considered. The value of the highest reward is taken. The next position is then set as a current one.

3. Q-learning algorithm modification realization

The task considered here is related to smart shopping realization and it does not allow a teacher to show how to reach the goal in order to achieve better results. Therefore Imitation Learning cannot be applied because of the availability of lots of ways for achieving a particular goal. Besides, the goal is different every time. The shops and stands in the Shopping Center that customers want to reach are different. Some of them look for promotional goods; others need artwork. Some customers use the shopping process as a therapy and want to reach the most frequently visited shops and go through the busiest lobbies and hallways; others want to avoid the crowded zones. So it is important in this task not only to reach the goal. The way of reaching it, together with the criteria, which a certain path meets, are of the same level of importance.

In order to make the Q-learning agent find the optimal sequence of lobbies or hallways by meeting specific criteria, the use of measures model represented as memory K matrix is introduced.

For the purposes of the experiment the Shopping Center is represented by a graph with 17 nodes and 36 edges between them as shown in Figure 2. Every shop in the considered Shopping Center is represented as a node. Every lobby or a hallway, connecting the shops, is represented by an edge. The busiest and most wanted to go through lobbies or hallways are marked in orange (Figure 2) and have a measure of 1 in the K matrix (Figure 1). The secondary, distant, non-desired pathways are marked in blue (Figure 2) and have a measure of 2 in the K matrix (Figure 1). The measures model presented by the K matrix K is similar to the environment reward model, presented by the R matrix (Figure 1). The values of a given criterion, to which each edge in the graph corresponds, are kept in the K matrix. Minus one (-1) in the R matrix and in the K matrix says that there is no edge in this place of the graph. The learning algorithm is changed now. The agent has to go only through those edges in the graph, which have a specific measure value in the K matrix. In other words, an edge to a given node could exist according to the reward model, but this edge should have a precisely defined measure value in the K matrix as well. It means that some nodes remain inaccessible to the agent because the edges that lead to them do not meet the specified criteria in the K matrix.

In some cases it may be found that there is no path to a goal that meets the pre-set criteria. Then the maximum number of compromises the agent can make by selecting edges that have a different measure value from the required one, is set. The rule allows compromises that are less than or equal to (<=) the maximum pre-set number of tradeoffs. As a result, the agent finds the optimal path by making the least number of compromise solutions. When the maximum number of compromises is made but the goal is not reached yet, the agent moves on to searching for another way to reach it.

The software implementation introduces the measures model presented by the K matrix and changes the source code of the function, determining all those possible shops, where the customer could decide to go from the next visited shop. The rule for calculating the current location of the customer at the moment when he/she chooses the next shop to visit does not change.

4. A survey of the Q-learning algorithm performance with an introduced measures model (a model of environment criteria), presented as memory Matrix K

The experiment is conducted in the following way: the goal to reach node 15 in the graph is set in front of the agent (Figure 2); a reward of 100 for going through the edge, connecting nodes 11 and 15 is announced as well in the reward matrix R (Figure 1). The other edges have zero reward points (Figure 1). The black dot line denotes the optimal found path from node 0 to node 15.

First stage. The specific measure values in the K matrix are not used. The desired sequence of edges to reaching the goal is defined based on the rewards model. The optimal found path from node 0 to the goal is given in Figure 3. It can be seen that the path goes through edges with a different value of the criterion, set in the K matrix.

Second stage. The agent is required to reach the goal by going only through edges with a measure value of 1. The optimal found path from node 0 to the goal is given in Figure 4. As it can be seen, the path goes only through edges with a measure value of 1 for the criterion, set in the K matrix.

Third stage. The agent has to reach the goal by going only through edges, having a measure value of 2. The optimal path for this case, starting from node 0 and reaching the goal, is shown in Figure 5, from which it becomes clear that the path goes only through edges with a measure value of 2 for the criterion, set in the K matrix.

In the example under consideration, there are primary and secondary paths that connect all locations in the exemplary Shopping Center. There might be a situation, in which a primary or a secondary path to a given place is missing. Then the algorithm can be modified by allowing the agent to go through a certain number of edges, which do not correspond to the value of the criterion "measure" in the K matrix.

```
//R - reward matrix
public static int [,] R = new int[,] {
{ -1, 0, 0, 0, 0, 0, -1, -1, -1, -1, -1, -1, -1, -1, -1, -1, -1 },
{ 0, -1, -1, 0, 0, -1, -1, -1, -1, -1, -1, -1, -1, -1, -1, -1, -1 },
{ 0, -1, -1, -1, 0, 0, -1, -1, -1, -1, -1, -1, -1, -1, -1, -1, -1 },
{ 0, 0, -1, -1, 0, -1, 0, 0, -1, -1, 0, -1, -1, -1, -1, -1, -1 },
{ 0, 0, 0, 0, -1, 0, -1, 0, 0, -1, -1, -1, -1, -1, -1, -1, -1 },
{ 0, -1, 0, -1, 0, -1, -1, -1, 0, 0, -1, -1, -1, -1, 0, -1, -1 },
{ -1, -1, -1, 0, -1, -1, -1, -1, -1, -1, 0, 0, -1, -1, -1, -1, -1 },
{ -1, -1, -1, 0, 0, -1, -1, -1, 0, -1, -1, 0, -1, -1, -1, -1, -1 },
{ -1, -1, -1, -1, 0, 0, -1, 0, -1, 0, -1, -1, 0, 0, -1, -1, -1 },
{ -1, -1, -1, -1, -1, 0, -1, -1, 0, -1, -1, -1, -1, 0, 0, -1, -1 },
{ -1, -1, -1, 0, -1, -1, 0, -1, -1, -1, -1, 0, -1, -1, -1, 0, -1 },
{ -1, -1, -1, -1, -1, -1, 0, 0, -1, -1, 0, -1, 0, -1, -1, 100, -1 },
{ -1, -1, -1, -1, -1, -1, -1, -1, 0, -1, -1, 0, -1, 0, -1, 0, 0 },
{ -1, -1, -1, -1, -1, -1, -1, -1, 0, 0, -1, -1, 0, -1, 0, -1, 0 },
{ -1, -1, -1, -1, -1, 0, -1, -1, -1, 0, -1, -1, -1, 0, -1, -1, 0 },
{ -1, -1, -1, -1, -1, -1, -1, -1, -1, -1, 0, 0, 0, -1, -1, -1, 0 },
{ -1, -1, -1, -1, -1, -1, -1, -1, -1, -1, -1, -1, 0, 0, 0, 0, -1 }};
```

```
//K- measure matrix
public static int[,] K = new int[,] {
{ -1, 2, 2, 2, 1, 2, -1, -1, -1, -1, -1, -1, -1, -1, -1, -1, -1 },
{ 2, -1, -1, 2, 1, -1, -1, -1, -1, -1, -1, -1, -1, -1, -1, -1, -1 },
{ 2, -1, -1, -1, 1, 2, -1, -1, -1, -1, -1, -1, -1, -1, -1, -1, -1 },
{ 2, 2, -1, -1, 1, -1, 2, 2, -1, -1, 2, -1, -1, -1, -1, -1, -1 },
{ 1, 1, 1, 1, -1, 2, -1, 1, 1, -1, -1, -1, -1, -1, -1, -1, -1 },
{ 2, -1, 2, -1, 2, -1, -1, -1, 1, 2, -1, -1, -1, -1, 2, -1, -1 },
{ -1, -1, -1, 2, -1, -1, -1, -1, -1, -1, 2, 1, -1, -1, -1, -1, -1 },
{ -1, -1, -1, 2, 1, -1, -1, -1, 1, -1, -1, 1, -1, -1, -1, -1, -1 },
{ -1, -1, -1, -1, 1, 1, -1, 1, -1, 2, -1, -1, 1, 1, -1, -1, -1 },
{ -1, -1, -1, -1, -1, 2, -1, -1, 2, -1, -1, -1, -1, 1, 1, -1, -1 },
{ -1, -1, -1, 2, -1, -1, 2, -1, -1, -1, -1, 1, -1, -1, -1, 2, -1 },
{ -1, -1, -1, -1, -1, -1, 1, 1, -1, -1, 1, -1, 1, -1, -1, 2, -1 },
{ -1, -1, -1, -1, -1, -1, -1, -1, 1, -1, -1, 1, -1, 1, -1, 1, 2 },
{ -1, -1, -1, -1, -1, -1, -1, -1, 1, 1, -1, -1, 1, -1, 2, -1, 1 },
{ -1, -1, -1, -1, -1, 2, -1, -1, -1, 1, -1, -1, -1, 2, -1, -1, 2 },
{ -1, -1, -1, -1, -1, -1, -1, -1, -1, -1, 2, 2, 1, -1, -1, -1, 2 },
{ -1, -1, -1, -1, -1, -1, -1, -1, -1, -1, -1, -1, 2, 1, 2, 2, -1 }};
```

Figure 1. Reward matrix R and Measure matrix K.

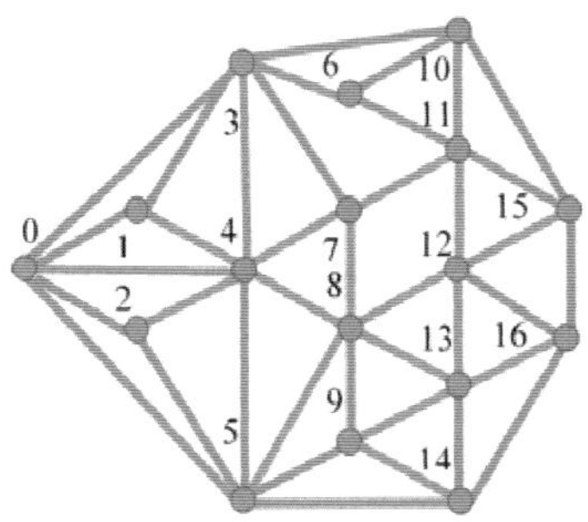

Figure 2. A Shopping Centre with 17 shops and 36 lobbies or hallways between them, presented by an undirected graph.

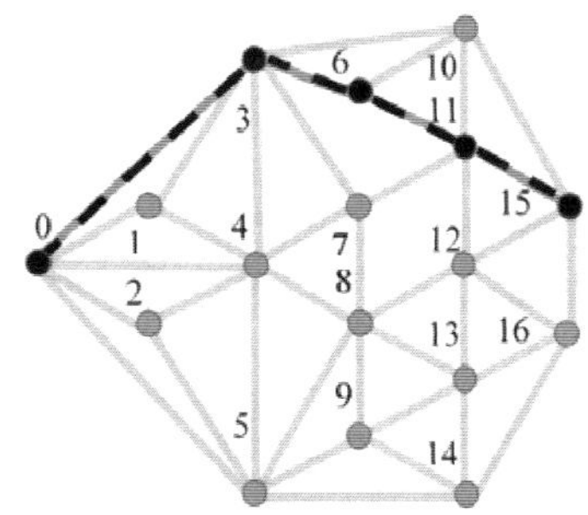

Figure 3. Optimal path from node 0 to node 15. Requirement for measure value not set.

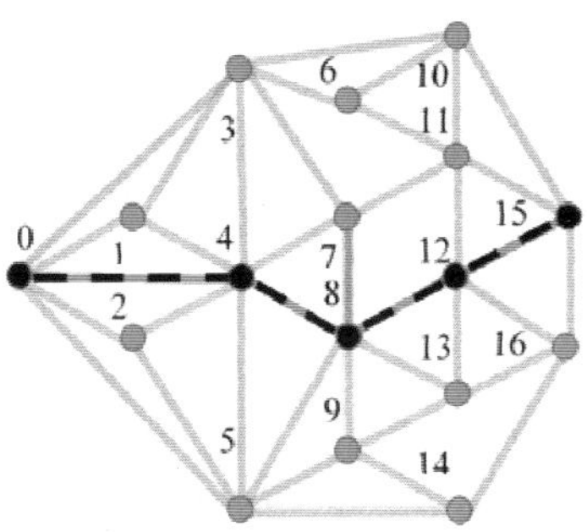

Figure 4. Optimal path from node 0 to node 15. Requirement for measure value: equal to 1.

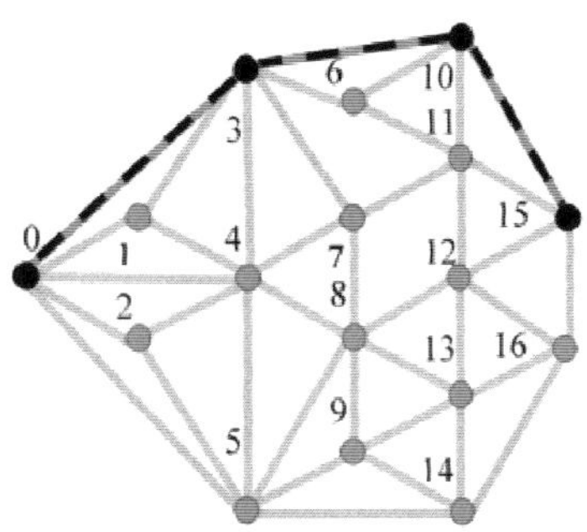

Figure 5. Optimal path from node 0 to node 15. Requirement for measure value: equal to 2.

Forth stage.
For the purposes of the next stages of the experiment the measure values modification of the graph edges is made (Figure 6). There is no path from node 0 to node 15, going only through edges with a measure value of 1 for the criterion, set in the K matrix.

Fifth stage. The agent has to reach the goal by going only through edges, having a measure value of 1. Two tradeoffs at the maximum are allowed, i.e., to achieve the goal the agent can choose two edges at the maximum, having a measure value of 2 as set in the K matrix. Fig.7 shows the found path. It can be seen that the agent has made exactly two tradeoffs on his way to the goal.

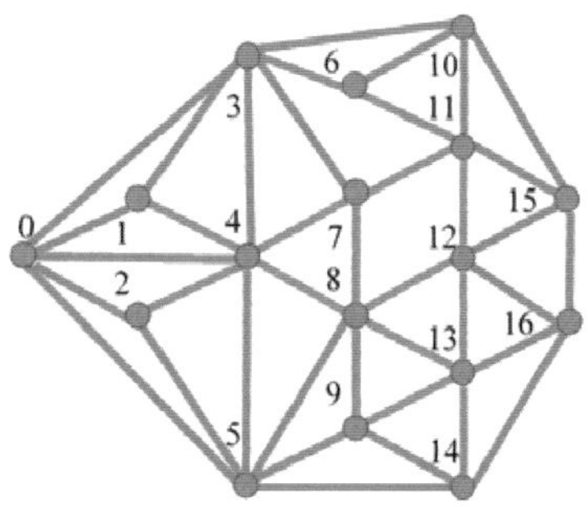

Figure 6. A Shopping Centre with 17 shops and 36 lobbies or hallways between them, presented by an undirected graph. Measure value modification of the graph edges is made.

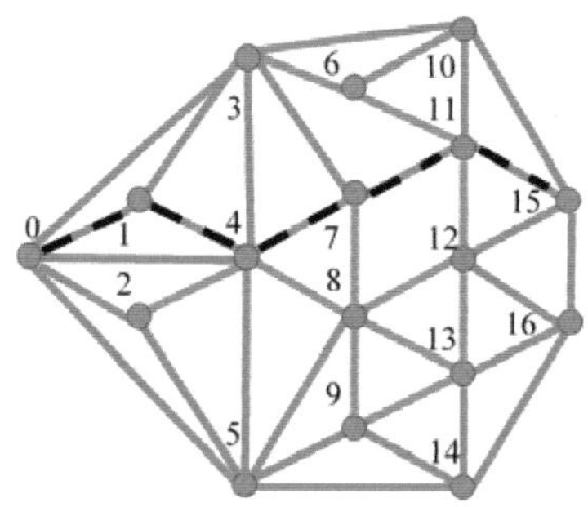

Figure 7. An optimal main path from the start node 0 to the end node 15 is sought, with a maximum of two compromises allowed. It is seen that the agent has chosen an edge with measure value 2 set in the K matrix twice.

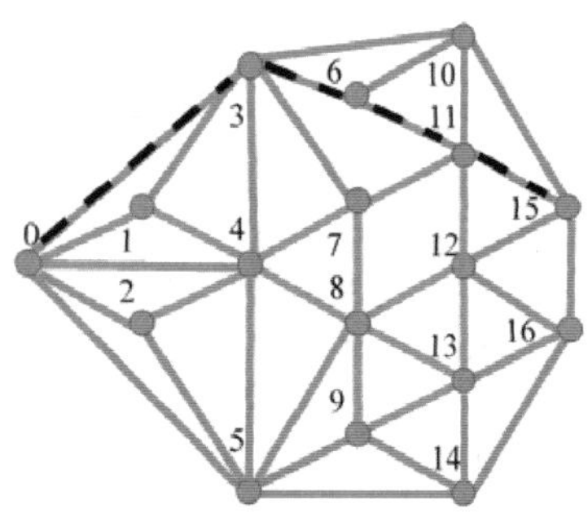

Figure 8. An optimal main path from the start node 0 to the end node 15 is sought, with a maximum of four compromises allowed. It is seen that the agent has chosen an edge with measure value 2 set in the K matrix twice. The agent finds the optimal path to the goal by making only three compromise solutions.

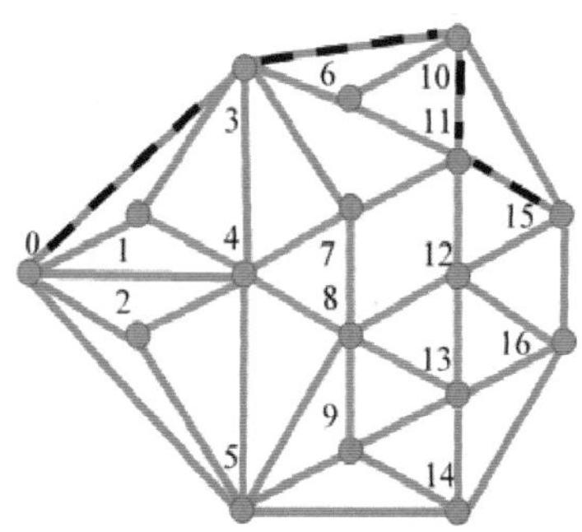

Figure 9. An optimal secondary path from the start node 0 to the end node 15 is sought. No compromises allowed (not allowed to go along main paths to reach the goal). The found path is shown.

Sixth stage. The agent has to reach the goal by going only through edges, having a measure value of 1. Four tradeoffs are allowed at the maximum, i.e., the agent can choose only four edges with a measure value of 2 set in the K matrix to reach the goal. Fig. 8 shows that the agent chooses the same optimal path to the goal as in Fig. 3. Three compromise solutions have only been made to reach the goal, meaning that the agent has made no unnecessary compromise solutions.
Seventh stage. The agent has to reach the goal by going only through edges, having a measure value of 2. No compromises allowed. The optimal secondary path, found by the agent, is given in Figure 9.

5. Discussion and future work

Many problems remain to be solved. In the first place, the work on the development of the Reinforcement learning algorithm will continue.

There are several options when setting requirements for reaching a goal. If possible, the agent will find the optimal path to the goal by meeting all the criteria simultaneously. However, there may not be a path, meeting all the criteria. In addition, some compromises may be more acceptable than others.

One of the solutions is to require from the agent to reach the goal by making as few compromises as possible. Another solution is to allow the agent to choose the compromises to make. If they are graded, the agent can be required to choose to make more in number and more acceptable compromise solutions rather than make fewer but unacceptable ones. These options for modification of the training algorithm under consideration will be useful in more complex social scenarios implementation.

An advantage of the proposed modification of the Q-learning algorithm is that it allows the agent to give explanation of the reasons behind the choice of a given path to a goal. In addition, the proposed modification allows to introduce various criteria for choosing a particular path. If the criteria from Maslow's theory of personality motivation are used, a model of a system of values could be developed using different scenarios.

Opportunities for modeling the training agent's value system will be looked for; efforts will be put to modeling a system for generating explanations by the learning agent. Using a holographic computer, it is possible to model and visualize a virtual advertising agent. It is assumed that the communication with such an agent will be engaging and helpful to customers. As mentioned in [11], there is a lot of interest in modeling a robotic shopping cart to follow the consumer. Efforts will therefore be made to this end. For example, it is important to combine and share intelligent behaviors such as: wander behavior; path following; collision avoidance; obstacle and wall avoidance; patrol between a set of points; flee behavior.

6. Conclusion

The paper describes a modification of the Q-learning algorithm. In order to make the-Q learning agent find the optimal path to the goal by following specific complex criteria, the use of measures model (a model of environment criteria), represented as a new memory matrix, is introduced. If the goal cannot be reached by meeting the set criteria, the agent could just ignore a given criterion and make a compromise solution. In this case, the agent is required to make as few compromises as possible in order to reach the set goal. Experiments have been conducted, illustrating the performance of the modified algorithm. If having the criteria graded by importance, the agent can make more in number and more acceptable compromise choices. This would be useful when developing complex social scenarios. The aim of the modification is to empower the learning agents to: control the way of reaching a goal; better understand the customers; be able to justify their decisions. The modified algorithm has been applied to training smart shopping-cart learning agents. The studies show improvement in their behavior.

Acknowledgments

The authors gratefully acknowledge the financial support provided within the Technical University of Sofia, Research and Development Sector, Project for PhD student helping N202PD0007-19 "Intelligent Cognitive Agent behavior modeling and researching".

References

[1] R. S. Sutton and A. G. Barto, Reinforcement Learning: An Introduction, The MIT Press, Cambridge, London, England, 2014. [Online]. Available from: http://incompleteideas.net/book/ebook/the-book.html, [retrieved: 12, 2019].

[2] A. Gosavi, "Reinforcement Learning: A Tutorial Survey and Recent Advances," INFORMS Journal on Computing, Vol. 21 No.2, pp. 178-192, 2008.

[3] R. R. Torrado, P. Bontrager, J. Togelius, J. Liu, D. Perez-Liebana, "Deep Reinforcement Learning for General Video Game AI," IEEE Conference on Computatonal Intelligence and Games, CIG. 2018-August, 10.1109/CIG.2018.8490422

[4] B. Argall, "Learning Mobile Robot Motion Control from Demonstration and Corrective Feedback", Robotics Institute Carnegie Mellon University Pittsburgh, PA 15213, March 2009.

[5] H. B. Amor, D. Vogt, M. Ewerton, E. Berger, B. Jung, J. Peters, "Learning Responsive Robot Behavior by Imitation," IEEE/RSJ International Conference on Intelligent Robots and Systems (IROS 2013) IEEE, Tokyo, Japan, November 3-7, 2013, pp. 3257-3264.

[6] K. Takahashi, K. Kim, T. Ogata, S. Sugano, "Tool-body assimilation model considering grasping motion through deep learning," Robotics and Autonomous Systems, Elsevier, Volume 91, pp. 115–127, 2017.

[7] K. V. Moffaert, Multi-Criteria Reinforcement Learning for Sequential Decision Making Problems, Dissertation for the degree of Doctor of Science: Computer Science, Brussels University Press, ISBN 978 90 5718 094 1, 2016.

[8] S. Natarajan, P. Tadepalli, Dinamic Preferences in Multi-Criteria Reinforcement Learning, 22nd International Conference on Macine Learning, Bonn, Germany, 2005.

[9] Dilyana Budakova, Lyudmil Dakovski, Smart shopping system, TECHSYS 2019, Plovdiv, 16-18 May 2019, doi:10.1088/issn.1757- 899X; Online ISSN: 1757-899X; Print ISSN: 1757-8981, Scopus.

[10] Dilyana Budakova, L.Dakovski, Veselka Petrova-Dimitrova, Smart Shopping Cart Learning Agents Development, TECIS 2019, 19th IFAC-PapersOnLine, Conference on International Stability, Technology and Culture,Volume 52, Issue 25, 26-28 September, 2019, pp. 64-69, Sozopol, Bulgaria, Web of Science, Elsevier ISSN 2405-8963, https://doi.org/10.1016/j.ifacol.2019.12.447

[11] Dilyana Budakova, Lyudmil Dakovski, Veselka Petrova-Dimitrova, Smart Shopping Cart Learning Agents, International journal on Advances in internet technology, IARIA, pages: 109 – 121, issn: 1942-2652, Vol. 12, nr 3&4. 2019

[12] N. G. Shakev, S. A. Ahmed, A. V. Topalov, V. L. Popov, and K. B. Shiev, "Autonomous Flight Control and Precise Gestural Positioning of a Small Quadrotor," Learning Systems: From Theory to Practice, Springer, pp. 179-197, 2018.

Testing of games through software agents managed by artificial neural networks

Miroslav Trankov[1], Emil Hadzhikolev[2], Stanka Hadzhikoleva[3]

[1, 2, 3] University of Plovdiv "Paisii Hilendarski", Department of Computer Science, 236, Bulgaria Blvd. Plovdiv, Bulgaria

[1]trankov86@gmail.com, [2]emil.hadjikolev@gmail.com, [3]stankah@gmail.com

Abstract. With the development of artificial neural networks, their use in different fields is increasing to optimize different activities. The availability of rich programming languages, various existing machine learning software libraries, along with the virtualization technologies has made intelligent technologies accessible not only to large corporations but also to individual programmers.

Game development is a complex process involving many specialists, experts in various fields. Single-game testing is an iterative process that covers different game development opportunities, different player abilities, as well as player interaction when working in a team. The question is, is it possible to use intelligent software agents in the game testing process in order to optimize the human, time and financial resources needed?

This article introduces our attempt to create an intelligent autonomous agent that plays the Brawl Stars game, managed by an artificial neural network. Initially, the agent is trained by collecting frames from a game that is being played by an experienced player. The frames are then processed and analyzed, and a neural network is trained, which directs the agent to conduct an autonomous game. The developed agent was used in two experiments that confirmed the great potential for the use of artificial neural networks in game testing.

1. Introduction

The fast-growing gaming industry imposes new requirements and rules on game development. The process involves different professionals - designers, programmers, artists, testers, experts in different fields, etc. Artificial intelligence is also used to improve the consumer experience, in its various forms - through autonomous agents [1], neural networks [2], game software templates [3] and so on.

The implementation of AI in games is not a novelty. The first successful attempts to deploy an autonomous agent in a game date back to the early 1970s, and were conducted with the well-known arcade games Computer Space-1971 and Space Invaders-1978. The developers have realized that for better and unique user experience, games must interact with the player independently and adequately, changing the game environment, objects, activities and processes within it. The emergence of AI as an independent field in computer science has an impact on all other industries. In the 1980s, companies actively invested in building expert systems for solving specific problems [4]. A big success in the 1990s was the IBM-created AI Deep Blue chess game application [5]. The second improved version of Deep Blue in 1997 won the match against World Chess Champion Gary Kasparov. Another big challenge for game developers is to develop an application for another strategy game with simple rules - Go. In October 2015, the DeepMind-developed AlphaGo program [6] defeated Fan Hui, the

European Go Champion, five times in a five-game match. In March 2016, an upgraded version of AlphaGo wins four-on-one victories against Go's leading master of the world, Lee Sedol.

Behind the rise of AI applications in recent decades have been major technology and software companies such as Facebook, Google, Amazon, OpenAI, IBM and Intel [7]. They create multiple software libraries for AI and Machine Learning (ML). Most libraries are open source and some have versions in different programming languages, such as Python, C / C ++, C #, Java, and more. Many experts consider Python as the most preferred language for AI and ML due to of its built-in big data capabilities, visualization tools, syntax support for various programming styles (procedural, object-oriented, functional), platform independence etc. That's why Python also has the most AI and ML libraries like TensorFlow [8], Keras [9], Sci-kit Learn [10], Theano [11], PyTorch [12] and so on.

Open source AI and ML libraries provide the opportunity to leverage more powerful and better ways to build their own AI applications, not only to large companies but also to individual developers. Building an artificial neural network for interacting with digital game is nowadays quite accessible and interesting task for developers dealing with AI - GTA 5 [13], FIFA 18 [14], CS: GO [15] and others.

Testing games in the process of their implementing is an important task that can be automated to a great extent by using artificial intelligence techniques. This article describes our experience in creating an automated game playing software agent using neural networks for training and self-study. The object of the experiment is, the well-known among the youngers, multiplayer, strategy and shooting game called Brawl Stars [16]. The agent and the game are running in the Operating System (OS) emulator. The agent's actions are guided by an artificial neural network that monitors the game's development through reading game frames from the screen, and its reactions are transmitted to the game via the OS emulator.

2. Game interaction with autonomous software agents

There are ***two main approaches for an autonomous software agent to communicate with a computer game: direct and indirect interaction***.

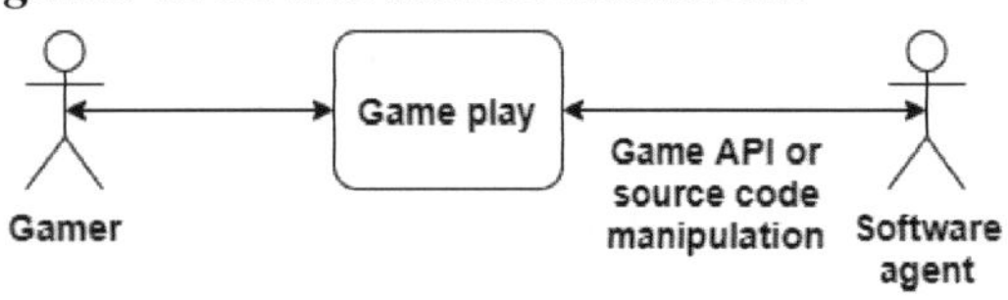

Figure 1. Direct interaction of an autonomous software agent with a game

In ***direct interaction*** (figure 1), the software agent usually communicates with the game through the ***program interfaces*** it supports. For security and copyright reasons, not all applications support this feature. Another option for direct interaction is software ***integration into the code by manipulating the game code***. The difficulties of implementing such an agent are related to learning foreign code without standardized programming interfaces and/or programming techniques.

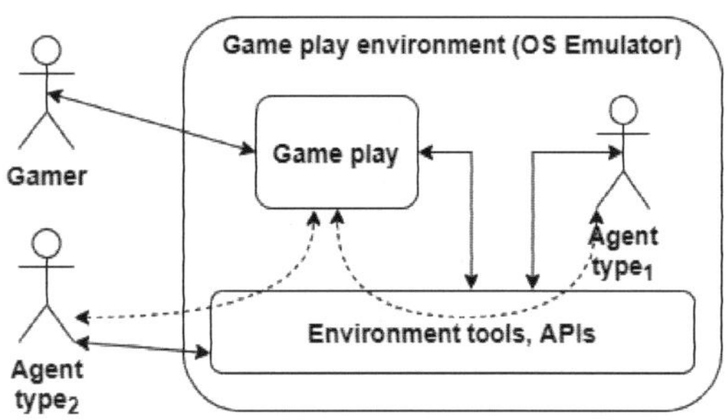

Figure 2. Indirect interaction of an autonomous software agent with a game

Indirect interaction (figure 2) is an approach in which ***the software agent communicates with the environment in which the games are played***, receives information from the environment and through it influences the game. Agents can be located both inside and outside the environment. Programming interfaces and standardized tools come from the environment and are not game-specific. Many existing emulators for different operating systems could be used as an environment. A major drawback of indirect interaction is the inability to use specialized game control capabilities. The advantage is the ability to build universal technology for automated management of games and various other applications, whether or not they provide standard mechanisms.

The inability to directly control the game through indirect interaction requires the search for new approaches. In 2017, Harrison Kinsley [13] used a convolutional neural network to navigate a player in GTA 5. The convolutional neural network uses real-time footage of the game and, after training, begins to play independently. The approach offered by Kinsley is subsequently implemented and adapted across a wide range of games. The main steps in its implementation are (figure 3):

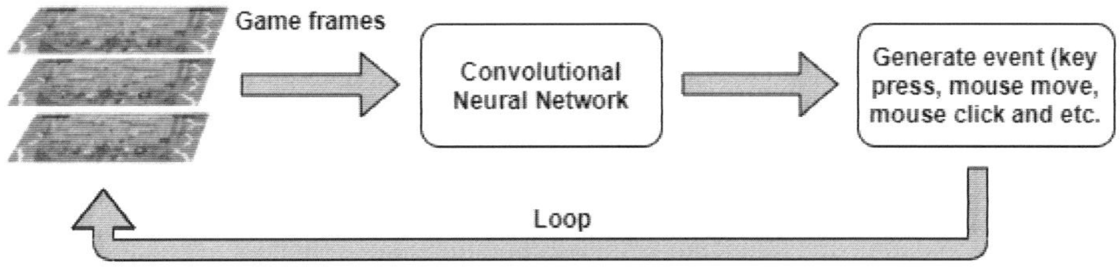

Figure 3. Basic scheme for indirect Kinsley interaction

1. Accumulate enough frames and associated keystrokes from a real player.
2. Training and testing of artificial neural network with the collected frames.
3. Real-time neural network operation using current frames.

3. Model of an autonomous Brawl Stars game agent

Using the approach suggested by Kinsley, we implement an autonomous software agent for the Brawl Stars game [16] (figure 4). The popular among the youngsters multiplayer strategy and shooting game Brawl Stars is an application for iOS and Android-based devices. In the experiments, the game was started in an Android emulator on Windows. The agent is also running on Windows and communicates indirectly with the game using the Python language capabilities to manage Windows input/output operations. ***Important factors for the implementation are the player control capabilities and the ability of the agent to be trained to perform automatically.*** It is controlled by several ***on-screen joysticks:***

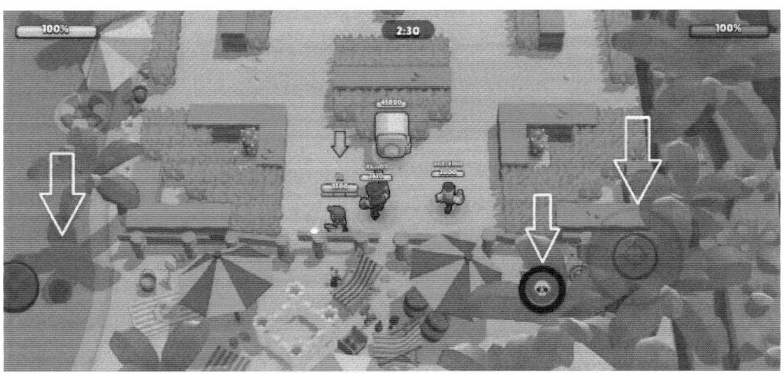

Figure 4. A screenshot of the Brawl Stars game

- ***blue*** (left) - ***to navigate*** the playing field; it is enough to keep it in the right direction; if the opponent's player is within range of the player's weapon, ***the firing is automatically activated;***
- ***red*** (right) - pressed to start ***precise aiming***, and when released, the charge is fired;
- ***yellow*** - appears when the ***Super charge*** accumulates and disappears after it is activated when pressed.

A limitation that exists in firing the weapon is that after three consecutive shots, it is necessary to wait 3 seconds to produce new shots.

Considering the control mechanisms, ***the software agent management model in the first variant developed by us comes down to training and using ANN to navigate the character. In this case, the activity of the neural network is a simulation of pressing the blue screen joystick to move the player in different directions - without movement, left, right, back, forward, as well as intermediate left-up, left-down, right-up, right-down - determined by the location of both the opposing players and teammates***. The restriction that we have imposed on firing is that the shots should be at regular intervals, but not less than 2 seconds. For better results, it is appropriate for the software agent to implement and integrate two neural networks that operate simultaneously but independently of each other: one for navigation and the other for aiming and firing.

The main modules of the software agent that can work independently in different periods (figure 5) are:

- ***The game frames engine,*** with its ***Frames collector*** and ***Frames analyzer*** sub-modules, are responsible for the collection and initial processing of frames during a person-to-person game;
- ***AI training engine,*** through the ***Serialization*** and ***Models training*** sub-modules, processes the collected frames and through NN creates models that can be used in automated control.
- ***Player engine*** - uses the generated models to play in real-time.

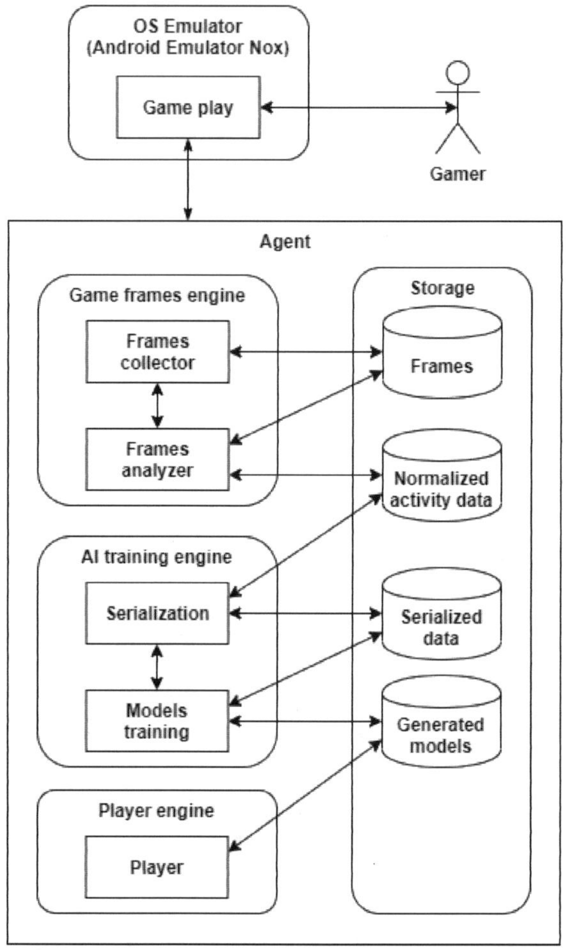

Figure 5. Software Agent Architecture and Core Modules

The first major activity of the software agent is to collect enough screenshots during the course of a person's game. The ***Frames collector*** module is responsible for recording screenshots of the game. Its ***main input parameters are screen size and location, determined by the top-left pixel***. Since the game has many different terrains and characters to play with, ***the terrain and the character are also defined as input parameters***. In our experiments, the game Brawl Stars was launched in OS Emulator, covering a window of 800x600px, the upper left corner of which is (0, 0). The number of screenshots required is determined by the accuracy we want to achieve in neural network training. E.g. our experiments have shown that in the presence of 300 screenshots for each of the possible movements, an accuracy of 99.63% is achieved, and for 200 screenshots - about 95%. As it became clear, the training of the NN is done on the basis of the frames collected during the actual game. For better results, a gamer who plays the game during the training period of the NN must be experienced and skillful.

The ***Frame analyzer*** analyzes and normalizes the collected screenshots, and divides them into nine different folders - for each of the nine movements allowed in the game. Building an effective neural network model requires careful consideration of the network architecture as well as the format of the input data. The most common image data entry parameters are the number of images, the height and width of the image, the number of channels, and the number of pixel levels. Usually, we have 3 data channels corresponding to colors - Red, Green, Blue (RGB). Pixel levels are usually in the range [0,255]. In order to improve the accuracy of the model, the number of photos must be the same in each of the directories. The same number of images for each category is a process of balancing and evenly distributing the input data, which guarantees that the model will not be misled in its training [17]. To test the accuracy of the training model created, a number of random images are extracted.

In the ***Serialization*** module, screenshots, along with their metadata, are saved in stream (HDF5) format for faster access. HDF5 is a cross-platform technology designed to record and quickly access complex big data [18].

For model training, we use a convolutional neural network (CNN) in the ***Models training*** module. CNN [19] is a deep learning algorithm that accepts images as input and then uses different features/aspects in the image to distinguish different objects. The advantage of CNN is the capabilities they offer for image processing and object classification [20]. By examining the color characteristics of two-dimensional images, in particular the frames of the game, CNNs transform them into three-dimensional objects. In addition, CNN can automatically recognize predefined categories such as different types of animals, faces, objects etc. Suitable for research in the development of such applications is the General Game Playing algorithm Monte Carlo Q-learning [21].

The typical CNN has two parts:

- ***A convolutional base that is made up of layers***. The main purpose of the convolutional base is to generate characteristics [22]. Each image can be viewed as a set of specific ***features (structures): points, ovals, lines, edges that form objects. Features can be related to the movement of objects in a sequence of images***, of different sizes and shapes. The concept of features is very general and defining them in a particular computer image classification task can be highly dependent on the particular problem.
- A ***classifier*** that is usually composed of fully coupled neural layers. The main purpose of the classifier is to ***classify the image based on the detected characteristics*** (filters). Any classifier of machine learning algorithms such as Logistic Regression, SVM, Random Forest, Decision Tree, etc. can serve as a classifier with a different performance for different tasks.

The chosen approach for training the CNN in this particular implementation is transfer learning using only gathered game frames [23]. Here, a pre-trained network on a particular problem is used to solve another similar problem. Implementing a ready-made model is a smart solution that saves the need to collect a large array of training data, create your own implementation, and have the serious processing power for NM training. Therefore this gives the advantage to use only frames from the game across the whole process of evaluation (input, training and testing). In this case, in working with images, we can use neural network architectures such as VGG16, VGG19, Xception, ResNet50, InceptionResNetv2 and MobileNet, Inception-v3 [24] in order to speed up the learning process. There are some trained on ImageNet database [25] and available in the Keras [9] library in Python. In our experiments, we use the Inception-v3 model [26]. When configuring it, we set the ***Frames collector*** images that form the user layer, as well as a classification algorithm for determining the nine movements of the player. ***The characteristics of CNN's convolutional base are extracted from the user layer of the model***. We have chosen ***Logistic Regression*** as the most suitable ***classification algorithm***.

Here comes a question regarding how sufficient could be a pre-trained model created only with self-gathered data? There are plenty of academic researches proving how easy and fast are machine learning approaches using transfer learning [27], [28].

In the experiments performed using Logistic Regression, at 459 frames the classification accuracy was 76.67%, while at 2700 frames the model accuracy increased to 99.63%. The time it took to train a pre-trained model using 459 frames was 17 minutes and for 2,700 frames was approximately 53 minutes. This means that time spent on training is approximately 1 sec. for a frame. The hardware specification of the machine used is listed below.

The ***Player*** module is an autonomous player that plays based on the models stored in the ***Storage***. The player navigates through the trained neural network. According to the chosen approach, shooting at opponents takes place at regular intervals by simulating mouse clicks at a point on the screen with certain coordinates.

The main scheme on which the base modules work (figure 6) is the following:

- The user ***Gamer*** starts the game and plays for a certain period of time in order to train the autonomous agent.

- Throughout the game, the ***Software agent*** collects screenshots through the ***Game frames engine***, which analyzes, sorts, and saves the data in ***Storage***, along with automated metadata.
- The NN training process can be started by the ***AI training engine*** when enough screenshots are collected, but it can also be started by the user at any time.

Models created during training are used by the ***Player engine***, which can be started automatically or by a user.

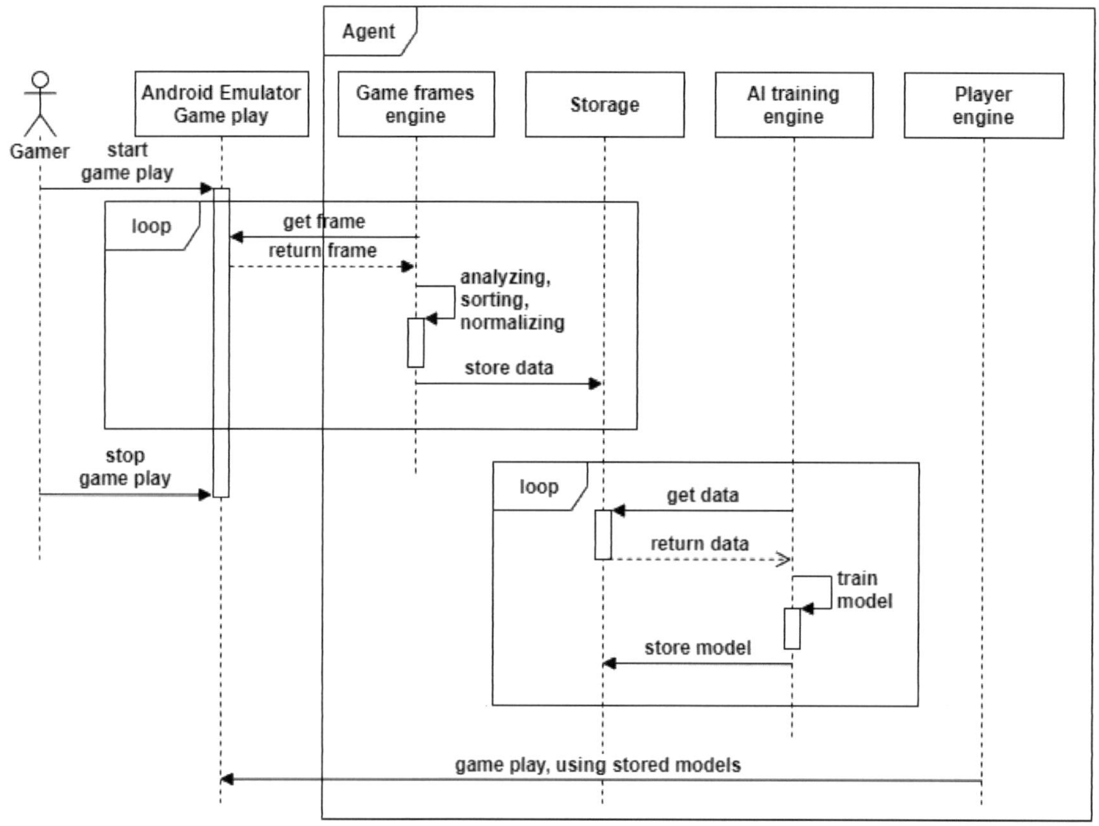

Figure 6. Main activities of the software agent and sequence of their implementation

4. Implementation and experiments

As part of the study, an autonomous Python language agent was developed. The application files are available at https://github.com/snooty7/BrawlStars. The following specialized libraries have been used for the development:

- grab-screen - for taking screenshots;
- numpy - the main library for scientific computing;
- h5py - for working with HDF5 binary data format;
- scikit-learn - a powerful library of different machine learning methods;
- keras - neural-network library;
- pywinauto - for automate the Microsoft Windows GUI, etc.

As Android emulator that plays the game Brawl Stars, we use Nox [29], but other free or paid emulators like Bluestacks, MeMu, and more can be used. The basic parameters of the hardware system on which the experiments were conducted are the following: Intel Core i5-4200M processor (2-core, 2.50 - 3.10 GHz, 3 MB cache), 8 GB RAM (2x4096 MB) - DDR3, 1600 Mhz, video AMD Radeon HD 8750 M Card (2GB) and Kingston 480GB SSD Hard Drive. At peak times, the CPU load reached

75% and running temperature of 62° C. No breaks or other problems with the application modules were observed.

The experiments in the Brawl Stars game were performed with a character called Colt, at an Excellent Escapade terrain in Gem Grab mode, where the team collects gems. The team wins by collecting 10 gems and holding them for 15 seconds. In the current implementation, the ***Agent player*** does not participate in the diamond collection but assists the team in seeking to eliminate the opposing players. The strategy that ***Gamer*** plays in order to train the neural network more effectively is to dynamically navigate the terrain and make it difficult for players on the opposing team to attempt to collect diamonds.

The results of the two experiments are as follows:

- The neural network is trained by a gamer with a personal level of 3,300 cups. The ***Agent player has won 6 of 10 games***.
- The neural network is trained by a gamer with 320 cups. ***Agent player won 8 of 10 games***.

By themselves, these results do not indicate the level of play of the trained agent, since the end result depends on the skills of teammates and opponents, as well as the game strategy through which the training was conducted. The result we found in the game is that ***Agent player*** manages to make relatively correct decisions.

5. Conclusion

The use of different AI technologies in games in order to create more exciting and realistic consumer experiences establishes new, ever-increasing demands on game development. Testing a game in the process of its development requires a lot of time and human resources. Using AI can optimize this activity.

This article presents a study, based on using an autonomous software agent to play as a team in the Brawl Stars game. The agent is trained by collecting, researching and analyzing frames from a real human-player game. He then plays alone, managed by a neural network.

The experiment shows that the use of an autonomous software agent can simulate a player and facilitate the software testing of a multiplayer network game. The implementation performed is suitable for testing any type of navigation games - car racing, arcade games with airplanes, bikes and other vehicles.

Acknowledgements

The work is partly funded by the SP19-FMI-004 project at the Research Fund of the University of Plovdiv "Paisii Hilendarski".

References

[1] Nareyek A 2001 Review: Intelligent Agents for Computer Games *Lecture Notes in Computer Science*, vol. 2063. Springer, Berlin, Heidelberg.
[2] Charles D and McGlinchey S 2004 The past present and future of artificial neural networks in digital games.
[3] Treanor M, Zook A, Eladhari M et al. 2015 AI-based game design patterns *Proc. Foundation of Digital Games 2015 Conf.*
[4] Hart P 1982 Directions for AI in the eighties *ACM SIGART Bulletin* Issue 79 January 1982.
[5] Deep Blue online: https://www.ibm.com/ibm/history/ibm100/us/en/icons/deepblue.
[6] AlphaGo *DeepMind* online: https://www.deepmind.com/research/case-studies/alphago-the-story-so-far.
[7] Patrizio A 2019 Top 45 Artificial Intelligence Companies *Datamation*.
[8] TensorFlow online: https://www.tensorflow.org.
[9] Keras: The Python Deep Learning library online: https://keras.io.
[10] Scikit-learn, Machine Learning in Python online: https://scikit-learn.org/stable.
[11] Theano online: https://github.com/Theano/Theano.

[12] PyTorch online: https://pytorch.org.
[13] Reading game frames in Python with OpenCV - Python Plays GTA V online: https://pythonprogramming.net/game-frames-open-cv-python-plays-gta-v/.
[14] Trivedi Ch 2018 Building a Deep Neural Network to play FIFA 18 *Towards data science* online: https://towardsdatascience.com/building-a-deep-neural-network-to-play-fifa-18-dce54d45e675.
[15] Balsys R 2019 Testing YOLO v3 CS:GO Keras object detection online: https://pylessons.com/YOLOv3-CSGO-detection/.
[16] Brawl Stars online: https://brawlstars.com.
[17] Krizhevsky A Sutskever I and Hinton G 2012 ImageNet Classification with Deep Convolutional Neural Networks *Advances in Neural Information Processing Systems 25 (NIPS 2012).*
[18] The HDF5 Library & File Format *The HDF Group* online: https://www.hdfgroup.org/solutions/hdf5/.
[19] Convolutional Neural Network *UFLDL Tutorial* Stanford University online: http://deeplearning.stanford.edu/tutorial/supervised/ConvolutionalNeuralNetwork/.
[20] Ciresan D Meier U Masci J Gambardella L and Schmidhuber 2011 High-performance neural networks for visual object classification *Technical Report No. IDSIA-01-11* January 2011 Switzerland.
[21] Wang H Emmerich M and Plaat A 2018 Monte Carlo Q-learning for General Game Playing, arXiv:1802.05944.
[22] Chollet F 2017 Xception: Deep Learning with Depthwise Separable Convolutions, *IEEE Conference on Computer Vision and Pattern Recognition (CVPR)*, Honolulu, HI, 2017, pp. 1800-1807.
[23] Transfer learning with a pretrained ConvNet, TensorFlow Tutorial online: https://www.tensorflow.org/tutorials/images/transfer_learning.
[24] Ciaburro G and Joshi P 2019 Python Machine Learning Cookbook: Over 100 recipes to progress from smart data analytics to deep learning using real-world datasets, 2nd Edition, Packt Publishing, ISBN: 1789808456.
[25] ImageNet online: http://www.image-net.org/.
[26] Nguyen L Lin D Lin Zh & Cao J 2018 Deep CNNs for microscopic image classification by exploiting transfer learning and feature concatenation *2018 IEEE International Symposium on Circuits and Systems (ISCAS)* DOI: 10.1109/ISCAS.2018.8351550.
[27] Zabir M Fazira N Ibrahim Z & Sabri N 2018 Evaluation of Pre-Trained Convolutional Neural Network Models for Object Recognition *International Journal of Engineering and Technology*. vol 7, pp 95-98, DOI: 10.14419/ijet.v7i3.15.17509.
[28] Varshney L Keskar N & Socher R Pretrained AI Models: Performativity, Mobility, andChange Preprint arXiv:1909.03290.
[29] NoxPlayer online: https://www.bignox.com/.

IOP Conf. Series: Materials Science and Engineering **878** (2020) 012032 doi:10.1088/1757-899X/878/1/012032

LoRaWAN applications - "Leonardo tasting LoRaWINE"

R Maksimova[1] and K Kolev[1]

[1] University of Food Technologies, Technical Faculty, Plovdiv, Bulgaria

E-mails: rossirm@abv.bg, kr_kolev@abv.bg

Abstract. This paper presents main and specific points of application of LoRaWAN IoT technology as a new idea for the realization of projects in various smart technological systems. In the course of implementing such a task, information technologies like - LoRaWAN, TTN and Python with the kind assistance of a framework for building a real-time wine monitoring web application were used. An example dashboard for monitoring of Food Technology processes was developed and provided using a specially assembled end node with appropriate sensors for the subject's needs. A block diagram of an Internet of Things (IoT) network including The Things Network is presented. A diagram of the end node is synthesized. A flowchart diagram of the dashboard's code is suggested. Different program structures for LoRaWAN applications are built in contrast of ready integrations.

1. Introduction

LoRa (Long Range), as a part of modern low-power wide-area networks (LPWAN), is a proprietary spread spectrum modulation scheme developed for radar applications in the 1940's as well as a radiofrequency (RF) carrier signal technology based in the physical layer (PHY – layer 1 in the OSI model) of a telecommunications network [1, 2]. Therefore, via LoRa modem the data can be converted to this signal. Although there are other, more commonly known radio frequency signal technologies (like Bluetooth and WiFi), LoRa is known for its improved transceiver sensitivity and range - covering up to kilometers under good conditions. This makes it suitable for large-scale networking solutions.

LoRaWAN is a media access control (MAC) protocol developed by the LoRa Alliance for large-scale communication acting as a bridge between signals and an application, allowing transmission of data to devices connected to the application [3]. It is a technology based in the data layer of a telecommunications network. Combined with LoRa RF signal technologies, LoRaWAN has given the possibility for creating low-power, cost-effective, long-term and bidirectional telecommunications solutions used in different situations. These benefits and the ability to reduce the number of Gateway devices needed (compared to other methods of communication such as mobile networks and WLAN) make LoRaWAN suitable for Internet of Things (IoT) networks. Its principles of operation allow battery powered devices to communicate with Internet based applications over long-range wireless connectivity.

Application can be every software package that LoRaWAN devices can communicate with on the Internet. For the purposes of this paper The Things Network (TTN) which is a decentralized infrastructure for the IoT can be used for adding applications. There are different types of applications from custom ones to pre-set integrations provided by TTN that could be used without any need for writing a programming code. Such integrations include functionalities like: syncing and managing LoRaWAN devices in Web Services IoT; designing, prototyping and real-time or historical data visualization IoT solutions; collection of location-ready application programming interfaces (APIs) all allowing free prototyping with instant sign-up; integrations for sending uplink data to an endpoint and receiving downlink data over HTTP; publishing uplink data to brokers for a real-time data exchange for the IoT; storage integrations for persisting data in a database for several days, retrieving the data, using API available over HTTPS and serving it in JSON format (for querying data and listing of devices for which data is stored); uploading coverage information to servers directly from the TTN backend side; aggregating and analyzing live data streams etc. Ready integrations are a good option for beginning from somewhere but for a more flexible implementation we could dive into the world of programming, for example some custom web application. A framework suitable in such a case can be the Python's web application framework - Dash by Plotly.

2. Architecture of IoT Networks with TTN

For the implementation of a basic IoT network general building components are required. Among the main hardware components are LoRaWAN gateways and LoRaWAN end devices. Gateways are the bridge between the end devices and TTN, connecting to the latter via a router. The Things Network backend systems are responsible for routing data from the Internet of Things between devices and applications, i.e. TTN is the bridge between LoRaWAN gateways and applications to which devices are connected.

Non-IP protocols like LoRaWAN require some form of routing and processing before delivery of messages to an application. In figure 1 are presented the basic relations between the different components of an IoT network including The Things Network that performs those routing and processing functions in a decentralized and distributed way. For simplicity, the components are only used once, although there may be many-to-many relations. Messages coming from LoRaWAN end nodes are being received by gateways that are the part of the hardware which transmits the radio signal to the backend systems. They are connected to a router which is responsible for managing the gateway status and scheduling broadcasts. Each router is connected to one or more brokers. Brokers are the core of The Things Network. It is their responsibility to map a device to an application, forward uplink and downlink messages up to the correct application and down to the correct router (which forwards them to a gateway).

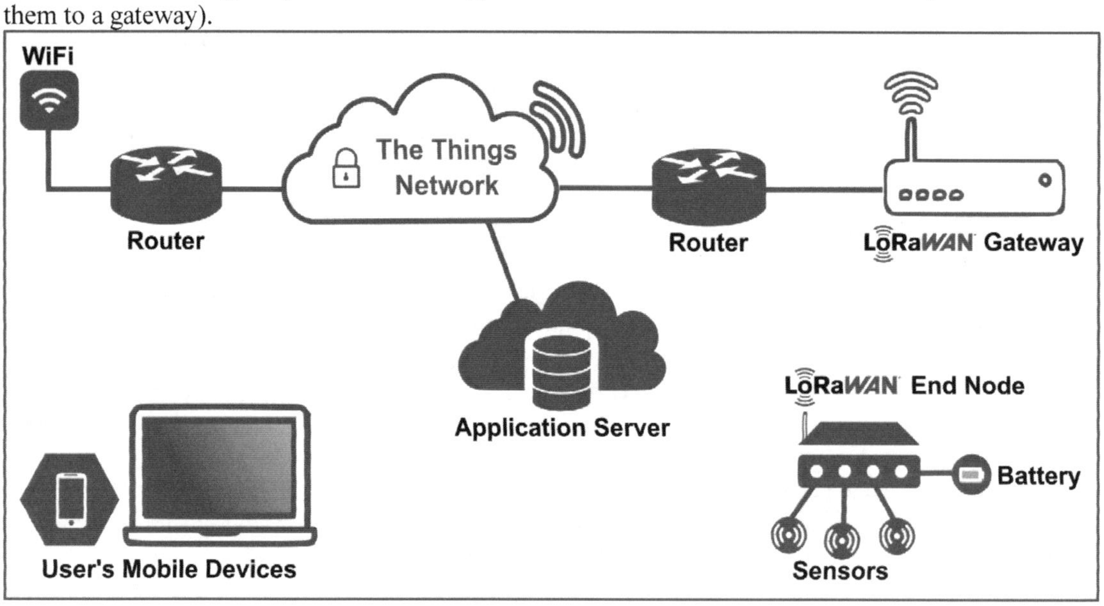

Figure 1. Block diagram of an IoT network architecture

The Things Network aims to be very flexible for deployment options. It prefers the option to contact the community network hosted by The Things Network Foundation or its partners. In this case, the Application connects to a Public Community Network Handler, typically using the Message

Queuing Telemetry Transport (MQTT) API [4]. But it is also possible to deploy private networks performing all of these functionalities in a private environment which keeps the data private, but still being able to use the hosted account on TTN for authentication and authorization.

Network security guarantees the authenticity of end nodes in the network, while the application security ensures that the network operator does not have access to the end-user application data [5]. The Things Network has strong security on the public network that supports end-to-end encryption, mitigations against various attacks and support for different 128-bit encryption keys for every single end device. Users need an account to use The Things Network. On the public community network, user accounts are stored in an account server. These accounts are protected by a password. In private networking setups, users may implement their own means of authentication and authorization. Users can create applications and they can authorize other users to access to applications. Applications are identified by a unique App ID. Each Application has one or more Access Keys to access application data stored on an Application server or to manage devices [6].

Applications and devices can be managed via The Things Network Console, that is a closed-source interface over the open-source TTN command line (CLI) with the same APIs of the open-source server components that applications can use. In addition, TTN provides a User Interface (UI) to its closed-source Network Operations Center to track user's network usage, as well as click-and-run integrations for popular application platforms [7]. There are Brokers and Handlers inside The Things Network which map the traffic of a specific application to be forwarded to an Application server. The Handlers interpret binary data and act like a bridge to higher-layer protocols like MQTT. At the core of things there is a Discovery Server, which is TTN's key to a decentralized architecture. In this server routers, brokers and handlers announce themselves and there they can be looked up [4]. All of these functionalities are performed by The Things Network.

3. Hardware - "Leonardo the taster"

The Things Network supports LoRaWAN for long range (to kilometers, though the strength of the signal is decreased over increasing of the distance and obstacles), low power (months to years on battery) and low bandwidth (51 bytes/message) communication, supports any certified LoRaWAN devices, but also offers its own devices [8, 9]. To connect a device to TTN it needs to have a LoRaWAN module wired or as a shield, which often communicates via serial interface. Currently supported LoRaWAN modules by TTN are the Microchip RN2483 and the RN2903. To save users from the complexity of commands and responses, some modules come with an SDK.

The Things Gateway and The Things Uno board, based on Arduino Leonardo (not Arduino Uno), were purchased and used. The Things Uno (Leonardo in this paper) is fully compatible with the Arduino IDE and existing shields. It is a class A lowest power device (compared to Classes B and C) and supports two-way communication [10]. TTN has built an Arduino SDK which enables transceiving of messages providing an Arduino library which makes it possible devices to communicate with it. Leonardo is assembled with three sensors and acts like a LoraWAN end node. Figure 2 presents a diagram of the main components involved in assembling the end node:

- a DS18S20 Temperature digital sensor - providing 9-bit Celsius temperature measurements, communicating on a 1-Wire bus for data (and ground) with the main microcontroller of the board [11];
- a TCS34725 Color sensor - color light sensor with RGB and clear light sensing elements. An IR blocking filter, integrated on-chip and localized to the color sensing photodiodes, minimizes the IR spectral component of the incoming light and allows color measurements to be made accurately. It communicates via the Inter-Integrated Circuit (I^2C) protocol with the board [12];
- an MQ-3 Gas analog sensor - measuring alcohol. Sensitive material of the sensor is SnO_2 (stannic oxide) which has lower conductivity in clean air. When alcohol gas exists, the conductivity gets higher along with gas concentration rising. Measuring electrode and heater are fixed into a crust made by plastic and stainless steel net. The MQ-3 module has 6 pin, 4 of them are used to fetch signals, and other 2 are used for providing heating current [13].

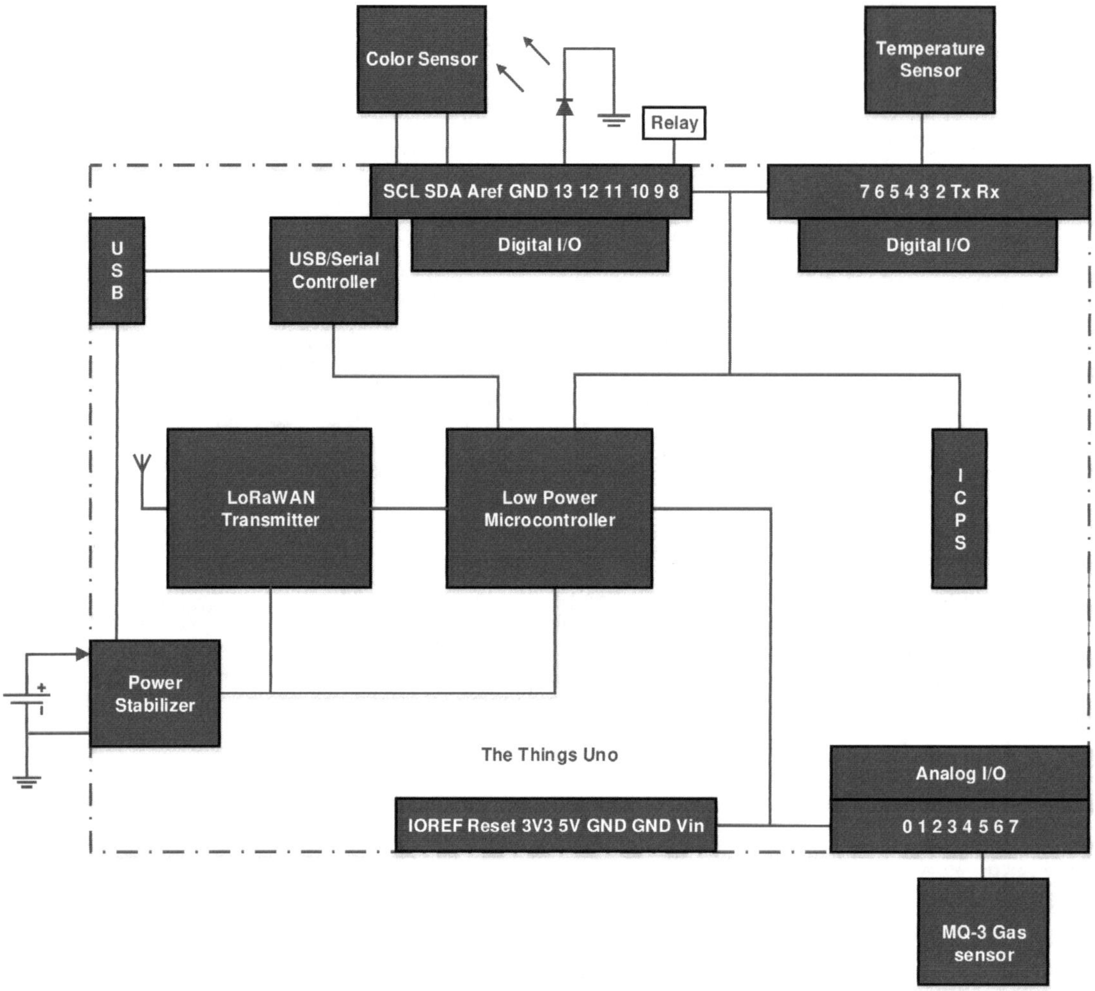

Figure 2. A diagram of our end node

4. Software - "Tasting LoRaWINE"

Before a device can communicate with an application, the latter has to be added to the network and devices to be registered to it. This can be achieved through the TTN Console which is available after creating an account from The Things Network's official website. There are multiple available manners for building applications with TTN:

- working directly with APIs - using standard protocol libraries for direct connection to TTN Handler APIs, sending uplink messages from devices to the application and receiving downlink messages to devices back via the MQTT protocol or managing applications and devices registered to it via the HyperText Transfer Protocol (HTTP);
- via friendly SDKs - providing a layer of abstraction for lighter interaction with the available APIs of TTN;
- via click-and-run integrations - using the same APIs or SDKs an application could use directly. Along with the platform's private or public APIs, they connect the platform-based application to The Things Network. Platform integrations eliminate the need to write code. There are various pre-set integrations available, such as Azure IoT Hub or Amazon Web Services IoT, through which users can manage their applications and devices, making synchronization with The Things Network [14].

Our way for communicating with a LoRaWAN end node via TTN passes through TTN Console:

- Registering The Things Gateway;
- Adding an application;
- Registering devices to the application;
- Building the application.

After registering our end device to our application on TTN we can go on building our web application with the kind assistance of the Python's Dash framework, using TTN provided Python Application SDK which allows transceiving messages to and from IoT devices. According to TTN and not only "Python is powerful, and fast; plays well with others; runs everywhere; is friendly and easy to learn; is Open." Python is a challenge for today's computer engineer [15]. The package ttn for Python provides exchanging traffic with The Things Network via MQTT protocol as a client and a manager of applications (for registering, transceiving uplink and downlink messages, generating callback events etc.). This package can be installed by the command "$ pip install ttn" from a command line.

The MQTT protocol, which was invented in 1999, is a publish and subscribe lightweight messaging protocol for low-bandwidth devices, high-latency or unreliable networks. Such features make the protocol ideal for machine-to-machine (M2M) or IoT world of connected devices, as well as for mobile applications [16].

The web application framework for Python - Dash is based on Flask, React.js and Plotly.js. It is suitable for building data visualization web applications with a highly customized UI (User Interface) in pure Python. Applications of Dash are rendered in the web browser running locally on localhost:8050 port of the user's device. It can be deployed to servers and shared via URLs [17]. Dash is nearly platform independent and mobile ready. It is an open-source library with MIT license and can be installed for Python by the command "$ pip install dash" from a command line.

We have developed an exemplary web application and named it "LoRaWINE Application" that acts as a dashboard. A screenshot of the real-time panel of our application in working mode is shown in figure 3. It is intended for providing the ability to monitor different technological conditions in real time, for example conditions in a wine barrel during wine aging stage from the overall technology of wine making.

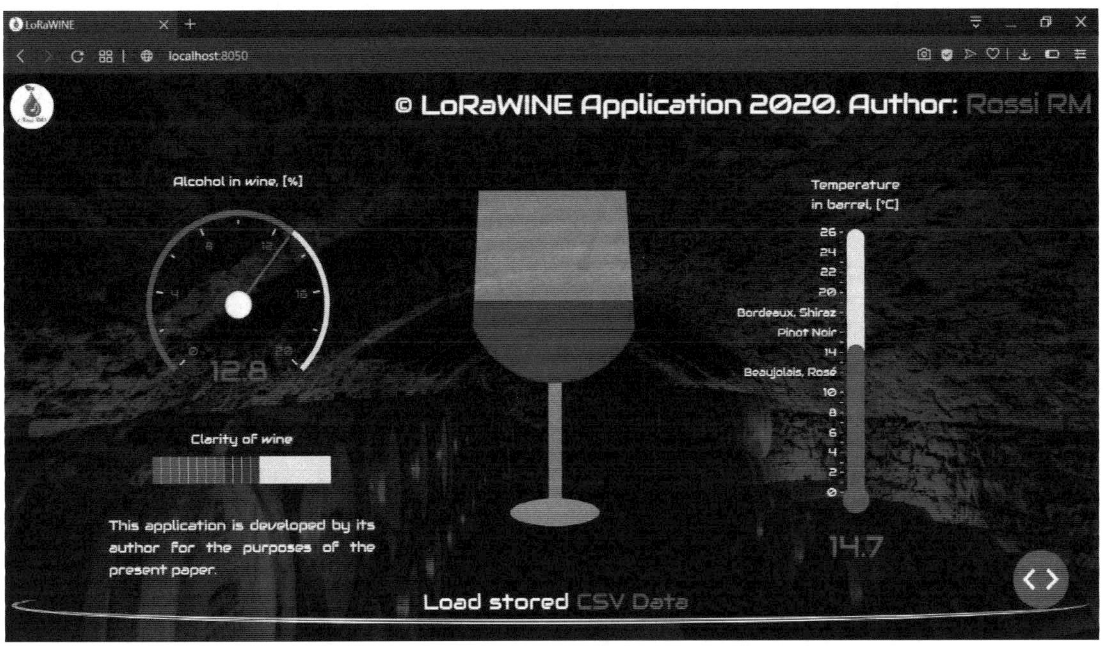

Figure 3. A screenshot of LoRaWINE web application rendered in a web browser

Our dashboard is updating on a predefined interval of time. It listens for callback events on the predefined interval. The data, visualizing on the dashboard, are being fetched by the sensors of our end device put on a suitable place inside the wine barrel. The app pulls live data from TTN via MQTT protocol through the provided Python SDK and parameters like temperature, color, clarity of wine and alcohol content can be updated on panel's display every set updating time. Except the possibility of a real-time monitoring of sensor data, the web application is capable of storing the data into CSV spreadsheets that can be uploaded and previewed any time the user needs some information for past recorded data. This function acts like a database and on the panel is available under the name "Load stored CSV Data" which has the functionality of a button opening a file explorer window for uploading and rendering the created spreadsheet file. Pre-set click-and-run ready integrations of TTN could provide such functionalities but the occasions where components of different nature are combined in a common panel are rare and may cost us a sum when we choose such an option. Figure 4 presents a flowchart diagram tracing the course of logic of our developed web application.

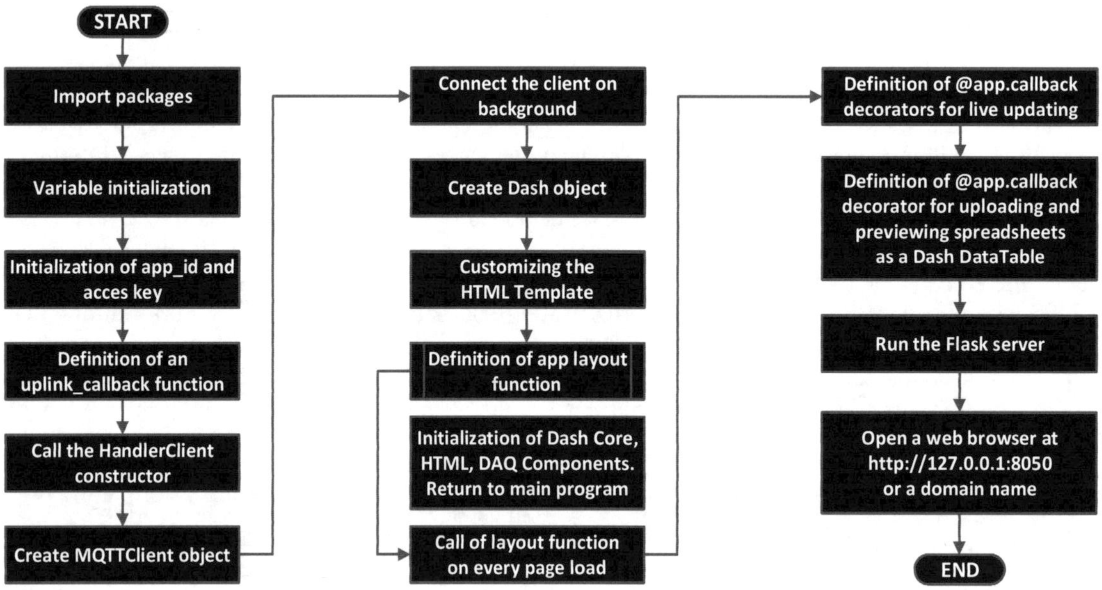

Figure 4. Flowchart diagram of LoRaWINE web application

The structure of a Dash web application's code can be divided to blocks of code, that are clearly distinguishable and act as separate functional units in the web application performing their own callback events. Such a structure looks like this:

```
@app.callback(Output(component_id='my-wine', component_property='style'),
        [Input(component_id='interval-component', component_property='n_intervals')])
def update_color(n):
    global color
    color = rgb_from_color_sensor
    return color

@app.callback(Output(component_id='my-gauge', component_property='value'),
        [Input(component_id='interval-component', component_property='n_intervals')])
def update_gauge(n):
    global alcohol
    alcohol = gas_percentage_from_mq3
    return alcohol
```

5. Conclusion

This subject's idea is a form of integrating LoRaWAN in the world of Food Technologies and not only. It demonstrates the power and the freedom of Dash web applications programmed in Python and is an example for building custom LoRaWAN applications in contrast of pre-set click-and-run integrations. The powerful available components provided by Dash in Python language are very useful in such an implementation with a scientific mark. Similar ideas as the one in this paper are borrowed from the process of developing larger projects by its authors that can be integrated in interdisciplinary fields from the world of science.

The fact that we use a well-prepared basic technology such as TTN would give rise to a lighter and friendlier usage of LoRaWAN networks. TTN is a good platform for developing novel ideas and projects, and in conjunction with Python and Dash, it becomes a powerful and flexible tool for implementing various scientific tasks. The possibility of assembling different customized end devices based on The Things Uno breathes a life to projects like ours giving its users the opportunity to be next-generation "oenologists". Such a dashboard can live not only within the boundaries of Local Area Networks but also on public communities in Internet accessible from every Internet connected node of the world. Such advantages of custom Dash web applications give the sense of the Internet of Things. Among the main advantages of Dash-based web applications, together with LoRaWAN, over ready pre-set dashboards, are the accessibility and flexibility of the source code.

References

[1] Semtech 2005 AN1200.22 LoRa™ Modulation Basics (Available online: http://wiki.lahoud.fr/lib/exe/fetch.php?media=an1200.22.pdf accessed 27 February 2020)

[2] Penkov S and Taneva A 2019 Algorithm development with LoRaWAN for monitoring system *IOP Conf. Ser.: Mater. Sci. Eng.* **618** (2019) 012007

[3] LoRaWAN Architecture The Things Network https://www.thethingsnetwork.org/docs/lorawan/architecture.html (accessed 29 February 2020)

[4] Network Architecture The Things Network https://www.thethingsnetwork.org/docs/network/architecture.html (accessed 27 February 2020)

[5] Kolev K and Maksimova R 2019 An Approach to Infrastructure for Environment Sensor Network *Proc.: IX International Conference Industrial Engineering and Environmental Protection (IIZS 2019 Zrenjanin, Serbia)* pp 428-434

[6] Network Security The Things Network https://www.thethingsnetwork.org/docs/network/security.html (accessed 28 February 2020)

[7] The Things Network Console The Things Network https://www.thethingsnetwork.org/docs/network/console/ (accessed 28 February 2020)

[8] Devices The Things Network, https://www.thethingsnetwork.org/docs/devices/ (accessed 26 February 2020)

[9] Hidayat M, Nugroho A, Sutiarso L and Okayasuet T 2019 Development of environmental monitoring systems based on LoRa with cloud integration for rural area *IOP Conf. Ser.: Earth Environ. Sci.* **355** (2019) 012010

[10] Tokmakov D, Asenov S and Dimitrov S 2019 Research and development of ultra-low power LoraWan sensor node *Proc.: XXVIII International Scientific Conference Electronics - ET2019 (September 12 - 14, 2019, Sozopol, Bulgaria)* pp 1-4

[11] Maxim 2015 DS18S20 High-Precision 1-Wire Digital Thermometer (Available online: https://datasheets.maximintegrated.com/en/ds/DS18S20.pdf accessed 2 March 2020)

[12] Ams AG 2018 TCS3472 Color Light-to-Digital Converter with IR Filter Datasheet (Available online: https://ams.com/documents/20143/36005/TCS3472_DS000390_2-00.pdf accessed 2 March 2020)

[13] Winsen 2014 Alcohol Gas Sensor MQ-3 manual (Available online: https://cdn.sparkfun.com/datasheets/Sensors/Biometric/MQ-3%20ver1.3%20-%20Manual.pdf accessed 2 March 2020)
[14] Build an Application The Things Network https://www.thethingsnetwork.org/docs/applications/options.html (accessed 25 February 2020)
[15] Maksimova R and Kolev K 2019 Modern Approaches with Python in Sensory Fusion *Proc.: IX International Conference Industrial Engineering and Environmental Protection (IIZS 2019 Zrenjanin, Serbia)* pp 58-66
[16] MQTT http://mqtt.org/ (accessed 26 February 2020)
[17] Introduction to Dash https://dash.plot.ly/introduction (accessed 2 March 2020)

Applying technologies in vocational education in Bulgaria

Elena Paunova-Hubenova, Elisaveta Trichkova-Kashamova

Institute of Information and Communication Technologies, Bulgarian Academy of Science, Acad. G. Bontchev str., bl.2, Sofia, Bulgaria

eli_np@abv.bg, etrichkova@abv.bg

Abstract. Information and communication technologies (ICT) give many benefits to every learning stage – from primary school to university and even life-long learning. Vocational education uses the base of the previous stages and aims at preparing students for their future jobs. The present paper focuses on vocational education in Bulgarian high secondary schools and universities (colleges). A short overview of the most used WEB-based educational information systems is presented. The paper also describes an extensive survey among teachers and students on the use of ICT in Bulgarian schools. The profile of responded vocational teachers is briefly presented. Here are presented and analysed their opinions about the available conditions and obstacles for using ICT in the teaching process. The usefulness and the frequency of using the most popular e-resources in the vocational subjects are compared. The paper presents a portfolio optimization approach for the usage of different e-resources. Next, the authors give an overview of the application of new technologies in higher education in Bulgaria. The paper provides information about specific educational systems, created for their own needs of the Universities, according to the specifics of the disciplines.

Keywords: vocational education, e-learning, educational information systems, portfolio optimization

1. Introduction

New technologies change almost all social areas, and education is not an exception. Information and communication technologies (ICT) give many benefits to every learning stage [1] – from primary school to the university and the life-long learning. Vocational education uses the base of the previous stages and aims to prepare students for their future jobs. ICTs can be applied in different activities to achieve different goals, such as interdisciplinarity [2], project-based learning [3], and specific purposes. They are greatly applicable in all subject areas like engineering, medicine, and public administration [4, 2, 5]. Providing the necessary conditions for implementing new technologies in the learning process for the vocational subjects is of vital importance for the efficiency of education.

The present paper focuses on vocational education in Bulgarian high secondary schools and universities (colleges). The next section gives an overview of the role of web-based information education technologies, enabling the e-learning to be implemented and increasing the effectiveness of the educational process. Section 3 presents the methodology of conducting a survey among Bulgarian teachers and the profile of professional ones. Section 4 describes the survey findings for vocational teachers – conditions and obstacles for using ICT in their teaching practice, their opinion about the usefulness, and the frequency of using e-resources. Section 5 describes the need to create their own e-learning systems at universities in Bulgaria. It gives an example of several well-developed and implemented in higher education e-learning platforms - own development. The paper ends with a discussion on the presented issues and a conclusion.

2. WEB-based educational information systems

E-Learning has gained widespread popularity in recent years and is in itself interactive learning that uses various communication technologies as a teaching tool. It provides the learner with easy access to teaching materials at a convenient time and place, thus the trainee receives a quick qualification with minimal costs. E-Learning is both an easy-to-adopt environment that is flexible and high-quality when used in combination with traditional teaching methods to increase the efficiency of the educational process in general.

Along with e-learning, widely used concepts these days are Web-based learning, online training, and distance learning.

Web-based learning can be defined as a collection of specialized resources available via the Internet or Intranet, which allows learners to gain knowledge and skills through self-study. Not only does it provide the necessary teaching material, but in practice, it requires learners to demonstrate the level of mastered knowledge.

The fact is that interest in software products, enabling e-learning to be realized, is growing rapidly. There is a wide range of software products available to provide distance learning [6, 12] which can be classified as:

Authoring Tools (Authoring Packages) – These tools enable educators to develop and publish self-contained learning content. In general, products in this class do not provide opportunities related to the control over the learning process and the success rate, nor do they provide real-time communication.

Learning Management Systems (LMS) – They are designed for the organization of a learning process involving a large number of learners. LMS systems provide opportunities for management, administration, documentation, tracking, and reporting of training programs, classes and online events, e-learning programs, and training content.

Content Management Systems (CMS) – These systems are organized for publishing a large volume of all content-texts, pictures, and multimedia. CMS systems are especially effective in cases where a large number of teachers are working on the creation of the curriculum content, which must use the same fragments of learning materials in different courses.

Learning content Management Systems (LCMS) – A system that creates, stores, assembles, and provides customized learning content in the form of learning objects. LCMS provides the means to create learning materials, to track the training of a large number of people, to store and search for individual elements of the educational content.

Some of the most commonly used systems in education are Content Management Systems. These systems are multifunctional and continually evolve and complicate. The problem of how to structure the Web-based course and present it according to a selected model, how to store it in a properly structured Web-integrated database is particularly important.

Although almost all authors point to the universal nature of the systems they have submitted, there are, in practice, specific differences. The practice shows that individual Universities use information systems entirely tailored to their specifics, justifying, and imposing the need to develop their own systems according to a selected pedagogical model.

3. Methodology

A widespread online survey among Bulgarian teachers form all stages and subjects was conducted in 2017/ 2018. Questionnaire for teachers consisted of 21 questions on several topics: profile, conditions, experience with ICTs and free comments. Survey was held in Bulgarian language, and the hyperlink can be found in the project website [7].

The total number of the responded teachers is 1652 from different settlement sizes and all regions in Bulgaria. The present article focuses on the opinion of vocation teachers, which number is 58, and most of them teach in STEM subjects (Science, Technology, Engineering, and Mathematics). Figure 1 presents their distribution according to their experience in school education. The numbers of responded vocation teachers are almost equal for three of the total four experience groups. The smallest part of them is the teachers with 5 to 15 years of practice in the classroom.

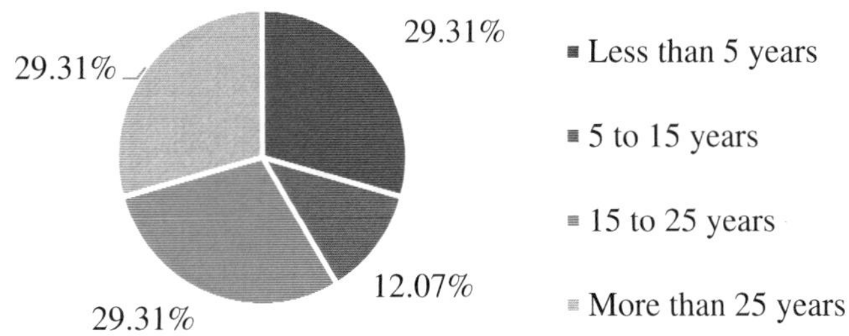

Figure 1. Experience of responded vocational teachers

Figure 2 presents the teachers' distribution according to the size of the settlement they teach in. Most of them work in big cities (38%), followed by teachers from the capital city – about 33%. The rest of them teach in smaller settlements. The most probable reason is that usually high schools, including vocational ones, are situated in the bigger cities, and teachers and students who live in smaller settlements have to travel every day.

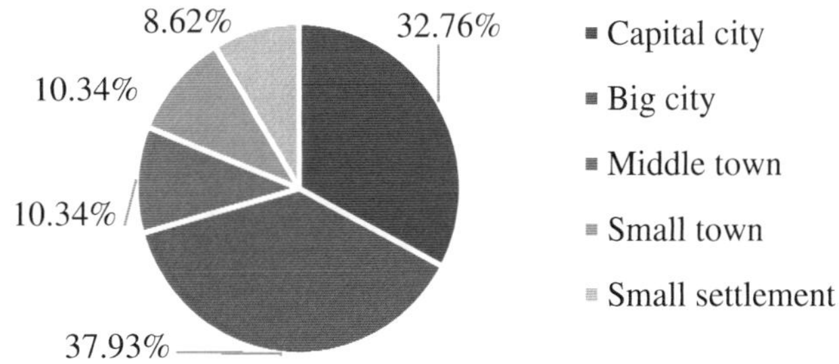

Figure 2. Profile of vocational teachers according to the size of settlement they teach in

4. ICT in professional high secondary school education

Here are presented some outcomes from the above described survey concerning vocational teachers. Authors emphasize on the available conditions in schools and the usability and frequency of using the most popular e-resources.

4.1. Conditions and obstacles for using ICT in the teaching process

This section presents the opinion of Bulgarian vocational teachers about the conditions and obstacles for applying new technologies in the learning process. The most important conditions that the authors took into account are institutional support, and the availability of training courses, technical equipment, appropriate software, and reliable Internet connection. Additionally, teachers commented on the issue in free text.

According to 81% of surveyed vocational teachers, the support from institutions for implementing ICT in the classroom is available. Unfortunately, this is not enough to provide some of the necessary conditions. A relatively small proportion of teachers say that they lack courses, equipment, or software (between 7% and 12%). For a slightly larger proportion, these conditions are only partially available (between 21% and 30%). The survey shows that only 24% of teachers have a good quality Internet connection. 47% of them state that their connection is unreliable, and 24% do not have the Internet in the school at all.

Figure 3 presents the most frequently mentioned obstacles for implementing information technologies in Bulgarian vocational education. Almost 30% of teachers state that there are no obstacles. According to half of the respondents, the e-resources are insufficient. The lack of technical equipment is an obstacle for 47% of them.

About 28% of the vocational teachers point to lack of time in the schedule for including e-resources and the same part of them – the insufficient competence of teachers to use ICT in the learning process. About 20% of the respondents mention a lack of time for preparation and information about appropriate e-resources. Less checked obstacles are not presented on the figure and are as follow: an attitude of teachers (12%), students (9%), and parents (5%); ICTs distract students from the subject (7%) and lack of specialist of educational technologies (7%). Only 2% of the respondents state that ICTs are not appropriate for learning purposes.

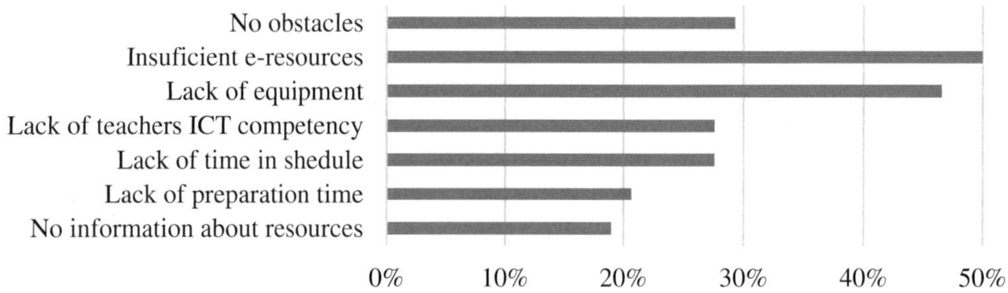

Figure 3. Obstacles for applying ICT in learning process

4.2. Usefulness and the frequency of using e-resources

Teachers had to evaluate the usefulness of the most used e-resources (Figure 4) and how frequently they apply them in the classroom (Figure 5). The most useful e-resources, according to vocational teachers, are learning videos, presentations, and specialized software (respectively, 40%, 38%, and 34% answer with “Extremely useful”). According to respondents, the least useful are educational games and virtual laboratories. Vocational teachers use most frequently presentations (31% for “very often” and 33% - “often”), specialized software (26% for “Very often” and 9% - “often”), and learning videos (19% for “Very often” and 29% - “Often”). The least used resources are virtual labs, educational games, and process simulations. A significant part of teachers did not answer for both issues of e-resources, and the most probable reasons are the lack of such resources for vocational subjects (Figure 4) and the lack of teachers’ experience with them.

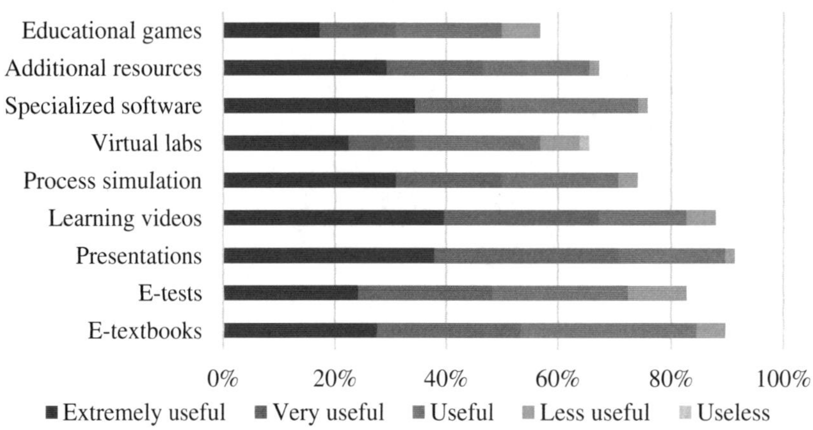

Figure 4. Usefulness of e-resources according to vocational teachers

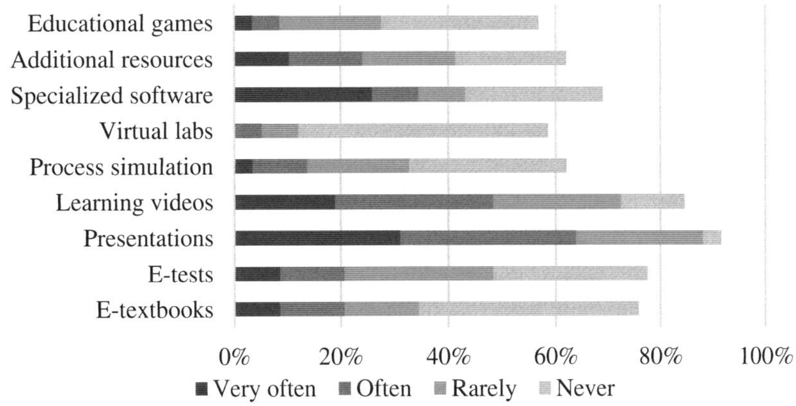

Figure 5. Frequency of using e-resources by vocational teachers

5. Portfolio approach for optimization the usage of e-resources

The idea for an increase of the e-learning performance is the appropriate combination of the e-resources in the educational process. The main e-resources, which are considered in the paper are educational games (x_1), additional resources (x_2), specialized software (x_3), virtual labs (x_4), process simulation (x_5), learning videos (x_6), presentations (x_7), e-tests (x_8), E-textbooks (x_9). These nine tools, x_i, i=1,..., 9 The idea for an increase of the e-learning performance is the appropriate combination of the e-resources in the educational process. The main e-resources, which are considered in the paper are educational games w_1=4, "often" (w_2=3), "rarely" (w_3=2), and "never" (w_4=1).

Following the given data for e-tool "presentation" the mean value of its usage will be

$$E_{presentation}=\sum_i w_i f_i \quad , \quad i=1,\ldots,4 ,$$

where f_i is the frequency parts of the presentations (f_1=31%, f_2=33%, f_3=...).

The portfolio problem for optimal usage of the e-tools will be defined as

$$\max_{x_i} \sum E_i x_i ,$$

$$x_i \geq 0 , \qquad \sum x_i = 1 \quad .$$

The solution of this problem xi will give the relative amount of the corresponding e-tool, which is recommended for usage in the e-learning process.

This portfolio model for optimization of the educational process is in the development stage. The paper presents here only the potential usage of such a portfolio optimization procedure.

6. ICT in professional university education

In today's society, computers have established themselves as an indispensable assistant in all spheres of human activity. Enormously has greatly grown the importance of information which already being treated as a business resource. This was made possible thanks to the computer information systems that provide the main environment for the application of information technologies in practice.

Through good knowledge of the possibilities and trends in the development of computer information systems can increase the efficiency of work at all levels and in all spheres of action.

Development trends in e-learning and greater accessibility of education generally reinforce the need to deploy the ICT at all levels of the educational process and encourage institutions to work in this direction. The development of information systems integrating into itself components for administrative services, communication, e-learning, career development, etc. can significantly improve the quality of educational services offered at all levels [8].

Instead of large application programs with multiple and increasingly difficult opportunities, there are a number of small, efficient, and easily interconnected applications that realize clear unique features [13, 14, 15].

6.1. E-learning platforms - Own development

The rapid development of ICT on the hardware-software-hardware spiral is an advantage for the introduction of new training methods, but also a major challenge for universities, especially when they do not have their own informatics units and teams.

There are a number of reasons justifying the need to create an own e-learning system, such as the high prices of foreign systems for organising and managing training, which is not comparable to our conditions; Language barrier, due to the fact that the systems are oriented mainly to the English-speaking countries; Lack of possibility for the flexible renovation of the system used with the development of new information technologies. The systems purchased or obtained through license may not be altered; Despite the universal nature of most systems, there are, in practice, distinctive differences between engineering, humanitarian, medical and other disciplines, justifying and imposing the development of their own systems, according to the specifics of the disciplines. For example, in medical disciplines, it is important to solve situational tasks (clinical cases).

As an example of some of the best developed and applied in higher education platforms – own development, - can be identified:

eLSe (e-Learning Shell) – The system was developed in the Department of Information Technologies of Rousse University. It is based on open-source products. E-Learning Shell is a platform that allows educators to publish their own WEB-based courses without the need to possess computer knowledge and skills [9].

DeLC (Distributed eLearning Center) – In the Faculty of Mathematics and Informatics of the University of Plovdiv was developed infrastructure called distributed e-Learning Centre (DeLC). This centre is designed to deliver in a context-dependent, adaptive, and personalised way e-learning services and e-learning content located on servers that can be physically separated. DeLC is a dynamic network structure that consists of nodes and relationships. One of the nodes of the DeLC is an educational portal designed to educate both students in higher education and also to be adapted to educate students in secondary education [10].

Virtual educational Environment of Medical University-Sofia – An information system for WEB-based training in medical disciplines has been developed. The system allows the delivery of Web-based courses as a source of basic knowledge. It gives the students a possibility of (self) evaluation of the theoretical knowledge and a possibility of creation and improvement of their professional skills to solve clinical cases. The administrative module allows teachers easily to organize the database necessary to create and edit Web-based courses composed of lectures, interactive tests, and virtual patients [11].

The Courses created currently by using the university Web-based system are a good example of the integration of the latest achievements in the specific discipline (modern content), ICT (Web-environment), Pedagogy (flexible learning), and Medicine (problem-based learning using virtual patients).

6.2. Comparison of eLSe and Web-based system of MU-Sofia with other Bulgarian and foreign systems in the field of e-learning

A comparison of the possibilities between several popular universal e-learning systems and the E-learning platforms - Own development is presented. The purpose of the comparison is not to point out that the systems are better than the above-mentioned universal e-learning systems. The aim is to show that they are commensurate and close to a large extent on their basic qualities of some of the best solutions in the field of e-learning. The comparison was made on criteria based on independent expert studies [9]. The results are visualized through a radar chart in Figure 6. It is evident that the systems possess the necessary means to conduct an effective e-learning.

For years, information technology has had a significant impact on the development of the education system. This is because the curriculum process is a set of participants, means, and methods for conveying the necessary knowledge so that its improvement according to the new technological possibilities is absolutely essential.

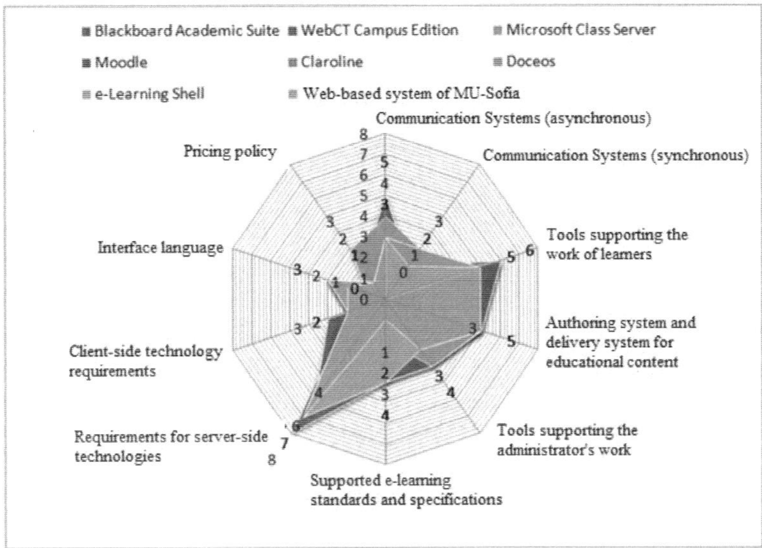

Figure 6. Comparison of platform capabilities with other universal e-learning systems

7. Conclusion

ICT technologies are an inseparable part of the life of our society and thus become indispensable in modern training at all levels – from primary education to university education. The proper use of modern educational technologies enhances the effectiveness of teaching and learning. To this end, it is necessary to use new pedagogical models and approaches that enable collaboration, communication, and mobility. Such learning technologies include virtual educational environments such as content management systems, interactive multimedia technologies, mobile device usage, and wireless technologies. The paper makes an attempt to formalize the usage of e-resources, by the definition of the portfolio optimization problem. The solution to such a problem will give recommendations on how frequently the different types of e-resources have to be used in an optimal way in the educational process.

ICT technologies and resources give a lot of possibilities for diversifying and improving the educational process. Thus, new technologies penetrate in Bulgarian vocational teaching as well. In Bulgarian schools, technical equipment and e-resources are available to various degrees. A positive trend is an increase in the percentage of teachers who claim that they do not encounter obstacles to the use of ICT in education. Unfortunately, in many schools, these conditions are still missing or partially available. Above all, vocational training lacks adequate resources and up-to-date cabinet equipment, as well as a fast and stable Internet connection.

Vocational subjects are considerably diverse, both in the high school and university levels. For this reason, it is difficult to provide the necessary e-resources in all specialized subjects. A good solution to creating a large set of e-resources is to encourage teachers to become actively involved in their development. For this purpose, the educational authorities should ease the curricula and pay the time for generating e-resources.

8. References

[1] Savov T, Terzieva V, Todorova K, and Kademova-Katzarova P 2017 Contemporary Technology Support for Education *CBU International Conference Proceedings* 5 Central Bohemia University ISSN:1805-9961 (online) pp 802-806

[2] Lhotska L and Huptych M 2019 Information Technology and Informatics in Biomedical Engineering Study Programs: Which Topics and What Extend Is Needed? *29th Annual Conference of the European Association for Education in Electrical and Information Engineering*, 4 – 6

September Ruse Bulgaria (in print)

[3] Kakalejčík L and Paľová D 2019 Enhancement of students' skills via project-based learning *International Conference on Computers in Education in MIPRO 42nd International Convention* May 20 – 24 Opatija Croatia pp 769 - 774

[4] Alajbeg T and Sokele M 2019 Implementation of Electronic Design Automation software tool in the learning process,*International Conference on Computers in Education in MIPRO 42nd International Convention* May 20 – 24 Opatija Croatia pp 640 - 644

[5] Klemke-Pitek M 2018 Public Administration Employees Lifelong Learning in Poland co-Financed from the EU Funds in 2014-2020 *10th International Conference on Education and New Learning Technologies* 2-4 July Palma de Mallorca SPAIN pp 10526-10533

[6] Dietinger T 2003 *Aspects of E-Learning Environments* Ph.D. Dissertation Graz University of Technology

[7] *Learning data analytics for ICT resource integration in Bulgarian schools* project Bulgarian National Science fund: http://hsi.iccs.bas.bg/projects/MPIKT

[8] Kasakliev N 2011 Comparative analysis of information systems for higher schools *Proceedings of the National Conference "Education in the Information society"* Plovdiv 26-27 May ISSN 1314-0752 pp139-148

[9] Smrikarov A 2009 National Program for creating a virtual learning environment. Main results and tasks ahead *Youth participation and contribution, Electronic Bulgaria* "Avangard Print" OOD Sofia Bulgaria (in Bulgarian)

[10] Stoyanov S, Doychev E, Valkanova V, and Cholakov G 2012 Education Cluster for Intelligent Provision of eLearning Services, *DBKDA: The Fourth International Conference on Advances in Databases, Knowledge, and Data Applications* 29 February – 5 March Reunion IARIA ISBN 978-1-61208-185-4 pp 45-50

[11] Trichkova E and Kossekova G 2011 "Interactive biochemistry – Sofia" (ibs) – a flexible web-based elearning system, *Proceedings of IADIS Multi Conference on Computer Science and Information Systems, E-learning* 20 - 26 July Rome Italy ISBN 978-989-8533-00-5 pp 3-10.

[12] Stoilova K and Stoilov T 2006 Technologies for integration of e-learning content. *Proceedings of E-learning conference'06: Computer science education* Coimbra, Portugal 7-8 September ISBN 978-989-20-0350-4 pp 2.11-1 - 2.11-6

[13] Stoilov T, Stoilova K and Ivanova Zl 2010 Application of Investment optimization as Web service *International conference Automatics and Informatics'10* Sofia, Bulgaria ISSN 1313-1850 pp II-407-410

[14] Nedyalkova A, Bakardjieva T and Nedyalkov Kr 2019 KM Software Solution at Varna Free University – Comparative Analysis *Proceedings of ICEduTech 2019, International Conference on Educational Technologies* Hong Kong 8-10 February ISBN 978-989-8533-83-8; http://www.iadisportal.org/digital-library/km-software-solution-t-varna-free-university-%C2%96-comparative-analysis

[15] Nedyalkova A, Bakardjieva T and Nedyalkov Kr 2018 KM Software Solution – Case Study at Varna Free University *Proceedings of International University Travnik, XVIII International Conference "Development Trends of Western Balkan Countries Based on Knowledge with a Special Overview on BIH within the Process of Accession to EU* 15-16 December ISSN 2490-2268 pp 25

ACKNOWLEDGMENTS

This work has been supported by project ДН12/10, 20.12.2017, Integrated bi-level optimization in information service for portfolio optimization, funded by the Bulgarian National Science fund.

IOP Conf. Series: Materials Science and Engineering **878** (2020) 012034 doi:10.1088/1757-899X/878/1/012034

System for creation, support and tracking of interactive learning activities

Martin Takev[1], Elena Somova[1], Silvia Gaftandzhieva[1], Miguel Rodriguez Artacho[2]

[1] Dept. Computer Science, University of Plovdiv "Paisii Hilendarski", Bulgaria
[2] Dep. Lenguajes y Sistemas Informaticos, UNED Madrid, Spain

eledel72@gmail.com

Abstract. The paper presents a study in the field of m-learning with interactive learning content. The approach used is to integrate a system (plug-in), which converts learning activities into EPUB learning content, suitable for usage and tracking on any mobile device, with an existing e-learning environment, full with learning courses. Therefore, a plug-in for exporting learning content from the Learning Management System Moodle into an EPUB interactive book is developed. The plug-in can export the following Moodle resources and activities: labels, files, folders, URLs, pages, books, lessons and tests. The data about students' learning activities is collected through xAPI statements, stored in a Learning Record Store. On the basis of the stored data, various types of reports are generated, which help teachers track students' progress. The software is tested with real users and short statistics are derived from their opinions in six areas (practical applicability; design; accessibility; interactivity, support and feedback; communication and collaboration; understanding).

1. Introduction

Without any doubt students prefer using mobile devices. But most of the learning resources at universities are developed and placed mainly in e-learning environments. The Learning Management Systems (LMS) are complex, offer many functionalities and need permanent internet connection. On the other hand, e-books in EPUB format are suitable for all mobile devices and can also be used in offline mode.

One of the main challenges in education is to track the students' learning progress and all of their performed actions, in order to analyze them later and take proper actions to correct learning gaps or failures, especially in performance-centered learning [1]. Since the creation of e-learning standards, this effort has been carried out by several implementations, focused on structuring the learning content and providing mechanisms to track interactions in web-based learning environments [2, 3].

This paper aims to present the results of a study, conducted at the University of Plovdiv "Paisii Hilendarski", in the field of tracking interactive learning activities on mobile devices. Section 2 points some related works that influenced ours – on EPUB standard for e-books and xAPI specification for tracking learning experiences. The methodology of our study is given in Section 3. Section 4 presents an approach how to offer students easier and more comfortable way of learning, using their favorite mobile devices, on the basis of existing e-courses in the university's e-learning environment. Section 5 briefly shows the impact of the developed software to the students' learning process, explained with results from the conducted survey with real students. The conclusion focuses on the obtained results and future objectives.

2. State of the art

E-books in EPUB format are easily usable on all mobile devices like smartphones, tablets and e-book readers, and that is why the standard continues to spread widely.

2.1. EPUB standard

The EPUB standard [4] is an open XML-based format for e-books, which allows copying the static content of printed books into digital ones. The latest version of the standard – EPUB3 introduces mechanisms for interactivity and multimedia integration through JavaScript and HTML5. All these new features of the standard can provide learners' tracking (one of the main objectives in the improvement of learning technologies), like in web e-learning environments.

Some authors identify the main advantages of EPUB as:

- Content is easily shared and reused [5];
- Integration of multimedia and interactive components based on web technologies [5];
- No special metadata related to education [5];
- Mechanism for automatic adjustment and content reformat for devices with small screen sizes [5];
- Simplicity of downloading the learning content [6];
- Access to the educational course content without the necessity of permanent Internet access [5, 7];
- Openness of EPUB [8].

2.2. xAPI standard

The Experience API (xAPI) [9] is a specification for standardizing and collecting formal and informal educational data in the form of statements. XAPI is often described as the next generation of the well-known standard SCORM. The main difference between xAPI and SCORM is the type of learning that they can track. SCORM can store information only for online learning, while xAPI can track wider range of learner activities, both online and offline. Therefore, xAPI gives a more detailed information about the learner's progress in any learning environment. For example, xAPI can track the following learning activities: starting/ reading/ completing a learning resource (webpage), starting/ answering/ completing a test, completing a test question, and etc.

The xAPI collects data about learning experiences, a student has done online or offline, in statements in the form of "Noun, verb, object" (e.g. "I did this"). User actions are saved as statements in a Learning Record Store (LRS) [10] – a repository for storing learning records. The LRS can exist on its own server or inside a Learning Management System, and it can share these statements with other LRSs. Every time the learning content is opened, it goes to the last done activity from the previous session (e. g. last read page) and student can continue from the point where he/she had left the course. This is not device dependent and the student can change devices during the studying without losing the progress made – switching from reading the same course in an e-book on a smartphone to reading it in the e-learning environment on a PC and then again on the e-book, it will always let the student continue from the point, left in the previously used device.

Each statement [11] describes a learning activity, the relevant user action, performed with it, and all of the interconnected objects in the given context. XAPI uses natural language to describe the statements, readable by both humans and machines. They are presented in JavaScript Object Notation JSON as hierarchically structured elements.

For improving the statements, [12] recommends five additional data elements to be incorporated in them: the achievement types, the duration of the learning experience, experience source information, domain learning and forgetting and assessment scores. All of these statements and their corresponding verbs, extensions and contexts are documented into an xAPI profile [13]. The profiles guarantee a semantically interoperable development and successful storage of data, that can be used multiple times for different use cases.

The statements, logged in an LRS, can be previewed and analyzed later. That way students can be monitored, no matter the learning environment or the device they are using – a traditional e-learning system on PC or an e-book on any mobile device.

3. Approach for creating an EPUB e-book

The main goal of our research is the creation of interactive and traceable EPUB content from standard Moodle e-courses. The used approach is to develop a software plug-in, which will be integrated with Moodle, and will automatically export (Fig. 1) all learning resources and activities in EPUB format – in an e-book.

Students can download the e-book and use it online or offline. During the learning process they have to read the learning content and interact with the e-book (e.g. answer to some questions). All student's actions are saved through xAPI statements [14] to an LRS (it does not matter where they occurred – in the LMS or in the e-book), using the same xAPI profile statement templates. This ensures equal tracking of the learner's progress and interactions in both learning places.

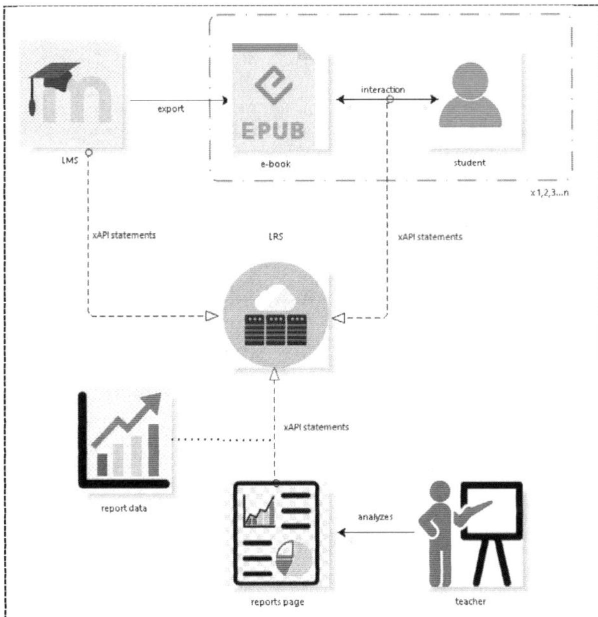

Figure 1. The process of exporting courses to e-books and sending and gathering xAPI statements for reporting.

The data (xAPI statements) in the LRS can be retrieved at any time and processed for different purposes. With it several types of reports can be generated (available in the Moodle reports page) for the teachers to analyze students' behavior and progress.

3.1. A plug-in for creating interactive e-books

The existing adds-in for e-learning platforms, which realize some conversion from/to EPUB format (Lucimoo, custom plugins for Moodle and SHEilds) are presented in [14].

We propose the following functionality in our plug-in: Export, Tracking Learning, Data Collection, Reports and Settings.

Export. An e-book for learning is created through the automated conversion of an e-course from the e-learning environment Moodle. The following Moodle elements can be exported: labels, files, folders, URLs, pages, books, lessons and quizzes. While most of them are static resources and do not require any other specific user action, besides reading them, the last two elements in the list demand students to perform a certain action (like answering a question).

Tracking Learning. Students' actions in the e-book are tracked – completing a test, answering a question, starting to read the book, etc. The xAPI profile used in the e-book contains all learning elements, that are integrated in the e-book, as well as the activities that will be tracked and some additional information (like duration of the activity). The information about all trackable student's activities is passed on by xAPI statements.

Data Collection. Standardized data (from xAPI statements) for student's actions, taken from the e-books, is collected in the LRS.

Reports. Graphical presentations of the collected data from the LRS are prepared for a better analysis of the student's learning progress and behavior by teachers. The system can automatically generate the following reports, divided in categories:

- course reports – show the ratio between the number of people who have started, completed and left a course unfinished or the users' average time spent on a course;
- quiz reports – reports for every question of the quiz, displaying how many users have answered the same, the average time spent on a question and user groups, based on their quiz score;
- assessment reports – show crucial points in the assessment of the students like questions, that got the most wrong answers, or questions, that took the most time to be answered;
- leaderboard reports – give an overall look of users, who performed the best on tests (achieved top scores on a quiz), who have submitted an assessment in record time or who had the most logins or time, spent on a course section;
- learner's path reports – display all of the trackable actions and their sequence for a specific user;
- engagement reports – combine stat measurements like course views, quiz submits and time spent on modules to present how much the users were involved in the learning activities;
- effectiveness reports – compare the results of two user evaluations – one, that's done before the user takes the course and one afterwards, in order to show whether the course was indeed effective;
- time period reports – usually a time period gets selected (month, week, etc.) and a specific metric is evaluated (time spent on a course, users who finished a quiz, the number of activities, performed by the student, grouped by dates, etc.).

Settings. The LRS connection can be set up or changed to correspond to the desired LRS, that's going to be used.

The plug-in is created as a Moodle block. This block can be added on every Moodle page. It shows all courses available for exporting and contains links to download them in EPUB format.

3.2. Exporting Moodle e-learning content

Our plug-in succeeds in the export of the following Moodle elements [14] into EPUB format: file (different formats), folder (with possibility for containing sub-folders), URL (any URL address with learning resources), label (a few words or an image), page (a single, scrollable screen that a teacher creates with the robust HTML editor), book (a multi-page resource with a book-like format), lesson (combination of HTML learning content pages and question pages with adaptive and dynamic navigation) and quiz (a test with different types of questions). The conversion of file, folder, URL, page, label and book builds the static part of the EPUB course.

The Moodle's activities lesson and quiz include interactive parts with questions. Six types of questions are created in the lesson element [15]: Multichoice (or Multiple answers), Essay, Matching, Numerical, Short answer (incl. fill in the blank) and True/false. As for the quiz element, five question types are created. They include the same ones as in the lesson element (excluding the question type Essay), however they are handled differently.

The lesson is probably the most interesting Moodle activity, because it offers the most dynamic content paths (through navigation buttons with predefined activities in the lesson page). The lesson also gives the opportunity for navigating students into different paths (adaptation), according to the correctness of their answers to the questions.

The process of question answering and navigation in the resource is realized with the use of JavaScript. The correct answers are hidden into the e-book files. Student's answers are assessed (by comparison with the correct answers) and sent through respective xAPI statements.

Some of the lesson parts are shown conditionally. The locked parts being open only if a certain prerequisite is met (e.g. answering right to a question bring you to one page, while answering wrong would bring you to another one).

The student's progress is also stored through xAPI – using the endpoint of the State API, available in the same LRS. A statement gets sent every time the student progresses forward in the e-book and it becomes available for retrieval. This ensures that the next time when the student opens the e-book and a connection to the State API is made, the book will open on the last place where the user has left, not from the start (even if the e-book is opened from a different device).

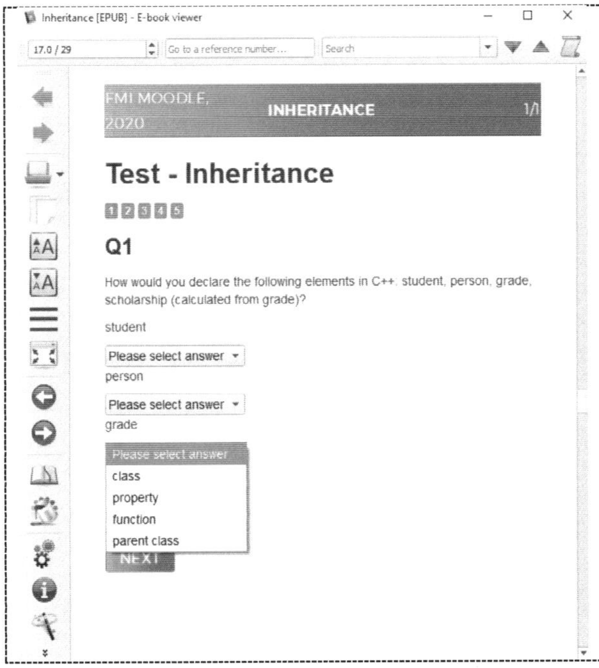

Figure 2. A preview of a test question in the e-book.

4. Experiment

The developed plugin „EPUB+XAPI export" was tested by 42 full-time students at the University of Plovdiv "Paisii Hilendarski" during the "Object-oriented design and programming (C++)" course. Students used the plugin to export the learning materials included in the topic "Inheritance" in an interactive file in EPUB format. After testing the plugin, students had the opportunity to evaluate the developed plugin and its applicability in formal training by taking part in a survey. The questionnaire was prepared in such a way, that without being time-consuming to complete, to give a clear picture of students' attitude towards the use of EPUB books in education. It contains 19 questions divided into six sections (see Table 1):

Section I. Practical applicability (I.1-I.4) – assessing the applicability of EPUB interactive materials in formal training;

Section II. Design (II.1-II.3) – assessing if the learning material in the EPUB files is intuitive and easy to use;

Section III. Accessibility (III.1-III.4) – assessing the ability to access and use materials in EPUB format on various mobile devices;

Section IV. Interactivity, support and feedback (IV.1-IV.3) – assessing how the interactivity in the EPUB file compares to that in an LMS and how fast is the feedback from the interactive actions;

Table 1. Survey questions with summary of students' answers.

Statement	5=SA	4=A	3=N	2=D	1=SD
I.1. The use of EPUB learning resources is easier than the use of learning resources in an LMS (e.g. Moodle).	10 (23.8%)	12 (28.6%)	18 (42.8%)	0 (0%)	2 (4.8%)
I.2. The use of EPUB learning resources is more enjoyable than the use of learning resources in an LMS (e.g. Moodle).	10 (23.8%)	14 (33.3%)	16 (38.1%)	0 (0%)	2 (4.8%)
I.3. EPUB resources are suitable for higher education training.	14 (33.3%)	22 (52.4%)	4 (9.5%)	0 (0%)	2 (4.8%)
I.4. I am inclined to study by using learning resources in EPUB format.	12 (28.6%)	18 (42.8%)	10 (23.8%)	2 (4.8%)	0 (0%)
II.1. The design of learning resources in an EPUB book is good.	8 (19.0%)	22 (52.4%)	8 (19.0%)	2 (4.8%)	2 (4.8%)
II.2. The design of learning resources in EPUB format is intuitive enough.	14 (33.3%)	12 (28.6%)	12 (28.6%)	2 (4.8%)	2 (4.8%)
II.3. I have problems navigating between the pages in some EPUB learning resources.	2 (4.8%)	6 (14.3%)	16 (38.1%)	10 (23.8%)	8 (19.0%)
III.1. Downloading learning resources in EPUB format is easy.	16 (38.1%)	20 (47.6%)	6 (14.3%)	0 (0%)	0 (0%)
III.2. The use of EPUB learning resources is more accessible than the use of learning resources in an LMS (e.g. Moodle).	10 (23.8%)	10 (23.8%)	20 (47.6%)	0 (0%)	2 (4.8%)
III.3. I had trouble opening some EPUB resources.	2 (4.8%)	2 (4.8%)	14 (33.3%)	14 (33.3%)	10 (23.8%)
III.4. I open learning materials in EPUB format faster than it takes me to access the Moodle course.	6 (14.3%)	12 (28.6%)	20 (47.6%)	0 (0%)	4 (9.5%)
IV.1. EPUB learning resources are as interactive as the learning resources in an e-learning course (e.g. in LMS Moodle).	16 (38.1%)	10 (23.8%)	8 (19.1%)	4 (9.5%)	4 (9.5%)
IV.2. I receive fast feedback from the learning resources in EPUB format.	12 (28.6%)	16 (38.1%)	12 (28.6%)	0 (0%)	2 (4.8%)
IV.3. The teacher provides the necessary support for using learning resources in EPUB format.	14 (33.3%)	18 (42.9%)	10 (23.8%)	0 (0%)	0 (0%)
V.1. EPUB encourages communication and collaboration with other students.	12 (28.6%)	10 (23.8%)	18 (42.9%)	0 (0%)	2 (4.8%)
V.2. I find it easy to provide/share EPUB learning resource to other students.	10 (23.8%)	14 (33.3%)	16 (38.1%)	0 (0%)	2 (4.8%)
VI.1. EPUB learning materials are presented in an understandable way and I know what is expected of me.	16 (38.1%)	12 (28.6%)	14 (33.3%)	0 (0%)	0 (0%)
VI.2. Learning with EPUB e-books meets my educational needs.	10 (23.8%)	10 (23.8%)	20 (47.6%)	2 (4.8%)	0 (0%)
VI.3. This learning process develops my skills.	14 (33.3%)	12 (28.6%)	14 (33.3%)	2 (4.8%)	0 (0%)

Section V. Communication and collaboration (V.1-V.2) – assessing whether students are getting the support they need and if they are willing to help others

Section VI. Understanding (VI.1-Vi.5) – assessing how close the training with EPUB files is to the traditional training and whether the tasks in the interactive EPUB files are understandable.

Most of the questions are multiple choice ones. Students should indicate to what extend they agree with the statements in the 5-point Likert-type scale, where 1 means Strongly Disagree (SD), 2 is Disagree (D), 3 is Neutral (N), 4 is Agree (A) and 5 is Strongly Agree (SA). There are two open-ended questions at the end of the questionnaire which allow students to make recommendations for enhancing and extending the functionalities of the plugin.

Table 1 presents summarized results of the survey. For each statement in the table, the number of responses and their percentage are presented (see Column 2 - Column 6).

The results clearly show that students have positive opinion about the developed plugin. Most students were impressed that using EPUB learning resources is easier and more enjoyable than using LMS resources (only 4.8% of students answered SD to statements I.1 and I.2). According to most students, EPUB learning resources are suitable for higher education (85.7% answered SA or A to statement I.3). They are also inclined to study by using resources in EPUB format (71.5% answered SA or A to statement I.4). Most students think that the design of the EPUB book is good (71.4% answered SA or A to statement II.1). The percentage of students who think that the design is intuitive is also high (61.9% answered SA or A to statement II.2). However, many students stated that they have troubles when navigating and switching between modules in learning resources (52.8% answered SA or A and 38.1% answered N to statement II.3). All students stated that downloading learning resources is easy (0% answered SD or D to statement III.1). A small part of the surveyed students does not agree that the use of EPUB learning resources is more accessible than the use of learning resources in an LMS (only 4.8% answered SD to statement III.2). The percentage of students who stated that the opening of study materials in EPUB format is faster than doing so in an LMS is relatively high (only 9.5 responded to SD in statement III.4), although most students had problems opening of some EPUB resources (only 9.6% had opened all resources without problems). Most students shared that learning resources in EPUB format are as interactive as their equivalent in the LMS (61.9% answered A or SA to statement IV.1). Students are satisfied with the speed of the feedback in EPUB learning resources and the teacher's support (only 4.8% answered SD to statement IV.2). Only 2 students do not think that EPUB encourages communication and collaboration with other students and allows them to share learning resources easily. Most students think that EPUB learning materials are presented in an understandable way and they have a clear idea of what is expected from them (66.7% answered SA or A to statement VI.1). When students are asked to agree that training trough EPUB meets their educational needs, almost half of them (47.6%) answered SA or A and N (47.6%). Only 4.8% of students do not agree that the learning process develops their skills and 33.3% of them cannot judge whether this is true (see answers to statement VI.3).

Figure 3 shows the students average score on all sections of the survey. The analysis of the answers and the calculated average score (3.69) shows that students have a positive attitude towards EPUB.

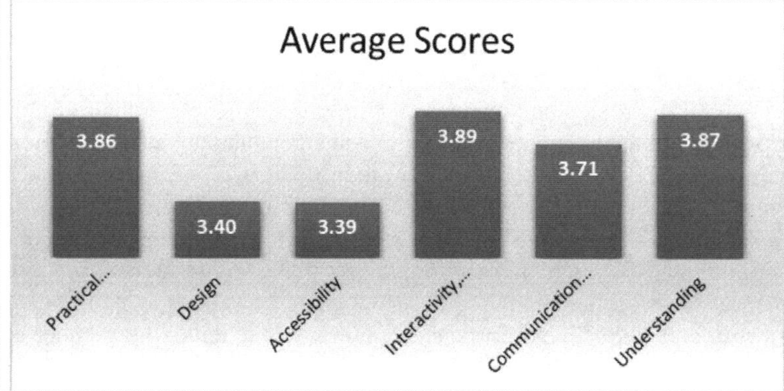

Figure 3. Average scores by sections.

The answers of the open-ended questions indicate that improvements in the learning materials design and navigation are needed. Students also suggest the development of additional functionality, allowing them to write and test programming code.

5. Conclusion

The paper presented a study on the use of interactive e-books in e-learning, where the authors have designed and created a system for creation, support and tracking of interactive learning activities in e-learning. The used approach is to integrate a system (plug-in) in a e-learning environment, that converts learning activities into interactive EPUB learning content, suitable for usage and tracking on any mobile device.

By achieving the objective of the study, the following four results were obtained:

- Researched the EPUB learning content opportunities;
- Creation of an approach to develop, support and track interactive learning activities in an independent mobile environment;
- Software realization of a plug-in for converting learning activities from an e-learning environment into a traceable interactive EPUB learning content;
- Conducted experiment – supported and tracked interactive learning activities in an EPUB course with real users.

References

[1] Stoyanova-Petrova S and Kafadarova N 2011 Performance-centered E-learning and M-learning OFCS Course *ED-MEDIA 2011-World Conference on Educational Multimedia, Hypermedia & Telecommunications* Lisbon, Portugal pp 1978-1983

[2] Artacho M and Verdejo M F 2000 High Level design of Web-Based environments for distance education *Computers and Education in the 21st Century* (Netherlands: Springer) pp. 275-285

[3] Caswell T Henson S Jensen M and Wiley D 2008 Open Educational Resources: Enabling universal education *The International Review of Research in Open and Distributed Learning* vol **9** n 1 pp 1-11

[4] International Digital Publishing Forum (IDPF) EPUB 3.0.1 http://idpf.org/ epub/301

[5] Chang H-P A and Hung J C 2018 Comparison of the features of EPUB e-book and SCORM e-learning content model *International Journal of Distance Education Technologies* vol **16** n 2 pp 1-17

[6] Tse E 2013 Special issue on child computer interaction *Personal and Ubiquitous Computing* vol **17** n 8 (Switzerland: Springer) pp 1573-1575

[7] Murphy A Martin N and Farley H 2012 Using e-readers to increase access to course content for students without internet access *ASCILITE 2012 – Annual conference of the Australian Society for Computers in Tertiary Education* Wellington New Zealand pp. 25-28

[8] Gailer C A Ebner M A Kopp M B Lackner E B Raunig M B and Scerbakov A A 2014 Potential of EPUB3 for digital textbooks in higher education, Lecture Notes in Computer Science

European Conference on Technology Enhanced Learning EC-TEL 2014 Graz Austria pp 564-5659
[9] Kevan J and Ryan P 2016 Experience API: Flexible, Decentralized and Activity-Centric Data Collection *Tech Know Learn* vol **21** (Switzerland: Springer) pp 143–149
[10] Learning Record Store https://en.wikipedia.org/wiki/Learning_Record_Store
[11] Student's Academic Level Evaluation http://www.lib4dev.in/info/nikhildutta333/student-performance-prediction/184567131
[12] Program Interface (xAPI) to Improve Domain Competency Modeling for Adaptive Instruction 2017 pp 265-268
[13] XAPI profiles https://github.com/adlnet/xapi-profiles.
[14] Takev M Rodriguez-Artacho M and Somova E 2019 Creating Interactive and Traceable ePub Learning Content from Moodle Courses *Mathematics and Informatics* vol **4** n 62 pp 459-470
[15] Takev M Somova E and Rodriguez-Artacho M 2019 Converter from Moodle lessons to interactive EPUB ebooks *Mathematics and Informatics* vol **62** n 6 pp 616-625

IOP Conf. Series: Materials Science and Engineering **878** (2020) 012035 doi:10.1088/1757-899X/878/1/012035

Gamification of courses in the e-learning environment

Mariya Gachkova, Elena Somova, Silvia Gaftandzhieva

University of Plovdiv "Paisii Hilendarski", Bulgaria

eledel72@gmail.com

Abstract. The paper presents a study in the field of gamification of learning, in particular – in structural gamification. The four-stages cyclic (Learning – Evaluation – Rewarding – Ranking) gamification learning model is proposed. The used gamified elements (level, story/history, challenge, team, time frame, game progress, status (avatar), bonus, badge, reward, hidden treasures, combo and leaderboard) with their impact to the learner's motivation are explained. Two plug-ins are designed, developed and tested with real users. One of the plug-ins changes the design of the learning course in game-like view, while the other ranks the students in a leaderboard with their avatars, according to their course progress. The short explanation of the experiment with learners is done and part of the statistical results from the conducted survey in five areas (practical applicability; motivation; design and accessibility; interactivity and communication; and intelligibility) are shown.

1. Introduction

The main challenge in the learning process is how to motivate students better. One of the methodological approaches, that gathers speed in recent years – gamification of learning, uses the people passion of playing games. It is an indisputable fact that everyone, regardless of their age, play games. This approach is usually applied on two steps. On the first step, the user is rewarded for task completion and on the second step, the user is encouraged to strive for further improvement by the use of a leaderboard. Gamification relies on user desires for status, achievement, competition and to be part of an inclusive social community.

The paper aims to present the results of the study, conducted at the University of Plovdiv "Paisii Hilendarski", in the field of structural gamification of learning. Section 2 points on some related research that influenced our work. Section 3 presents the study methodology. Section 4 proposes a model of a gamification learning process and two plug-ins for Learning Management System (LMS) Moodle, that add some gamified features. Section 5 briefly shows the impact of the developed software to the students' learning process. The conclusion focuses on the obtained results and future objectives.

2. State of the art

The gamification as an approach has begun to be used in the early 20th century in the stores [1], although the name of the concept was given in 2002 by Nick Pelling. The gamification is part of a wide variety of serious games.

2.1. Definition of gamification

The most popular definition of gamification is "the use of game mechanics and experience design to digitally engage and motivate people to achieve their goals" [2]. [3] point in their definition the resulted perception of users – "gamification is about taking something that is not a game and applying game mechanics to increase user engagement, happiness and loyalty". [4] emphasizes that gamification is a process – "a process of making activities more game-like with involving parts of games".

Some authors, like [2] distinguish gamification from video games and loyalty programs, as gamification uses techniques from behavioral science to "nudge" people into achieving their goals. The others, like [5] describe gamification in broader context as: "incorporation of game-style incentives into every day or non-game activities to enhance customer and employee engagement, boost sales, and cut costs".

Therefore, gamification techniques can be applied not only in learning (where students indeed achieve their goals) but in many different domains like a loyalty card for discounts in a shop, frequent flier programmes, a tourist book for collecting stamps, when visiting historical and cultural places, etc.

According to [6] the gamification of e-learning is two types: **structural** – the application of game elements to propel a learner through content with no changes to the content itself, the content does not become game-like, only the structure around the content; and **content** – the application of game elements and game thinking to alter content to make it more game-like. Our study is devoted to structural gamification.

[7] divided learners in four groups, depending on their specific behaviour and motivation for action: killers, who want to dominate other players by acting on people in the environment; achievers, who like acting and care about the assignments in the environment to win; explorers, who like interacting and exploring the environment and discovering new things there; and socializers, who like interacting with people in the environment and take advantage of the communication to socialize. We propose different elements in our model, motivating all these four groups of learners.

2.2. Gamification in e-learning environments

Some of the existing software systems which implement gamification of learning were investigated.

GENIE [8] is a web-based application, providing game templates or blank ones, where users can add their own game elements. GENIE uses the following game techniques: rewarding with badges and points for achieving learning goals; preparation of leaderboards for stimulating the competitive spirit; putting deadlines to tasks; and learning through gradually passing levels.

Academy LMS [9] is also a web-based system, available as a native application for Android and iOS and implemented with a very colourful design including cartoons and funny pictures. The system supports the following key game elements and techniques - earning badges, earning points after completing certain activities, adding levels, monitoring (by users) the progress of a course and a leaderboard.

TalentLMS [10] is a SAAS (Software as a Service) cloud-based e-learning platform, that implements the concept of serious games with: giving points for performed actions; collecting various badges for passed tests; receiving certificates and awards for course completion; re-certificate on certain time periods; Leaderboards - displayed by charts and diagrams; and passing the course by levels.

Accord LMS [11] have upgraded their e-learning system with specific game elements: Leaderboard ranking, earningbadges upon reaching the required criteria by students, and a points-based reputation system that shows the student's progress together with other student's progress.

Axonify [12] implements gamification as a basic training approach in the system by a very interesting methodology. The software includes game elements such as points, prizes, badges and charts, as well as a variety of up-to-date short games integrated into the application. These games are interrupted by questions triggered by certain game actions.

The most commonly downloaded Moodle gamified plug-ins [13] are LevelUp, Ranking block and Stash. LevelUp enables learners to earn points for various activities they have completed in the course.

The default algorithm is implemented for the progressive number of required points when passing to the upper level. Ranking block captures Moodle events in real-time and rewards points for them. The software offers a ranking of students with their points earned for completed activities. Stash proposes learners to find specific items placed in different Moodle activities or resources. Items can be opened one by one or several, gradually, so that students are encouraged to explore the whole learning material. Stash allows users to set items that can be collected in unlimited number of places.

3. Methodology of the study

The main goal of our study is to propose a model and software tools for carrying out gamified learning in the e-learning environment.

We use the following methodology: a literature review of previous works (learning methodologies and software systems and tools), prepare gamified learning model, design of software tools (plug-ins), development of prototypes, integration of plug-ins in the e-learning environment, and testing the software with real learners.

The paper presents all six steps, however, only one part or a summary of study results conducted are shown, since authors have already published on the topic. The focus is given to the outcomes of software testing, because this part of the research and some updates made, as a result of the testing has not been published yet.

4. Gamification in e-learning environment

We propose a learning model and software plug-ins for gamification in LMS Moodle.

4.1. Gamified learning model

The appropriate game elements from electronic games, which can be used in the gamified methodology of learning are found out in [14] and updated in [15].

We propose a **four-stages cyclic gamified learning model** (see Fig. 1), which uses all found out game elements, to motivate learners better: Learning – Evaluation – Rewarding – Ranking. Students learn using course resources and activities. After an assessment (test assessment, requirements fulfilment, etc.) of some ongoing learning goals, students are rewarded through different means (stimuli): bonus, badge, reward, hidden treasures and combo. As a result, they obtain points and/or new higher level, which directly reflect on the leaderboard ranking. After the processes of assessment, rewarding and ranking students naturally feel motivated to learn more, in order to receive more points, be on a better place in the ranking, obtain badges, receive additional extras, etc.

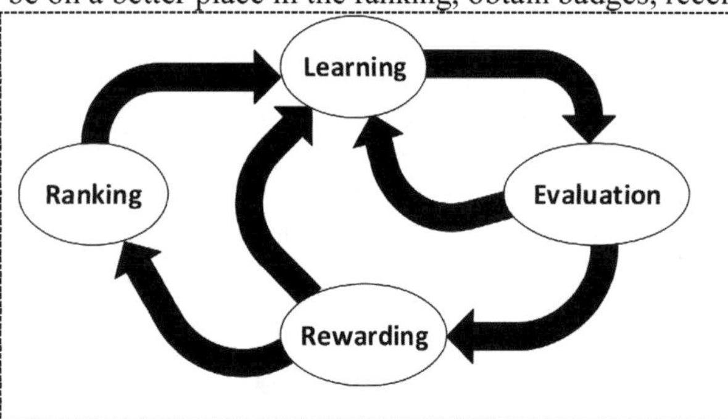

Figure 1. Four-stages gamified learning model.

The learning process is organized in **levels** (locked or unlocked with entry requirements), presenting different learning sections with resources and activities that should be passed. The learning course is introduced with an interesting context (**story/history**) of activities, which have to be done. The learning rules are explained as **game rules**. At each level students have some **challenges** – all learning activities/tasks (reading text file, fulfilling test, writing/doing/solving assignment, etc.), that have to be done for achieving level's learning goals. Some of the quests are individual, but the others are collaborative, where students work in **teams**. Some of the quests have time restrictions for completion of the activity – **time frame**.

For completion of some activities, students receive points/mark – **bonus**. For reaching a set of requirements, students obtain **badge** (Superstar, Adventurer, Explorer, Newbie, etc.). For excellent results or for doing something specific, students receive unexpected **reward** (additional interesting information, points/mark, etc.). For specific activity, students gain advantages as a **combo** (help, recommendations, more detailed examples, etc.). Some of the learning elements are locked (**hidden treasures**) and can be unlocked when students met certain requirements.

Based on collected points and current game level, students are ranking in the **leaderboard**, where leaders are seen. During the whole learning process, students have information about their learning progress as the **game progress**.

Students can choose their **status** in the gamified course as an **avatar** with some name and description.

The detailed diagram of the proposed four-stages learning model is presented in Fig. 2.

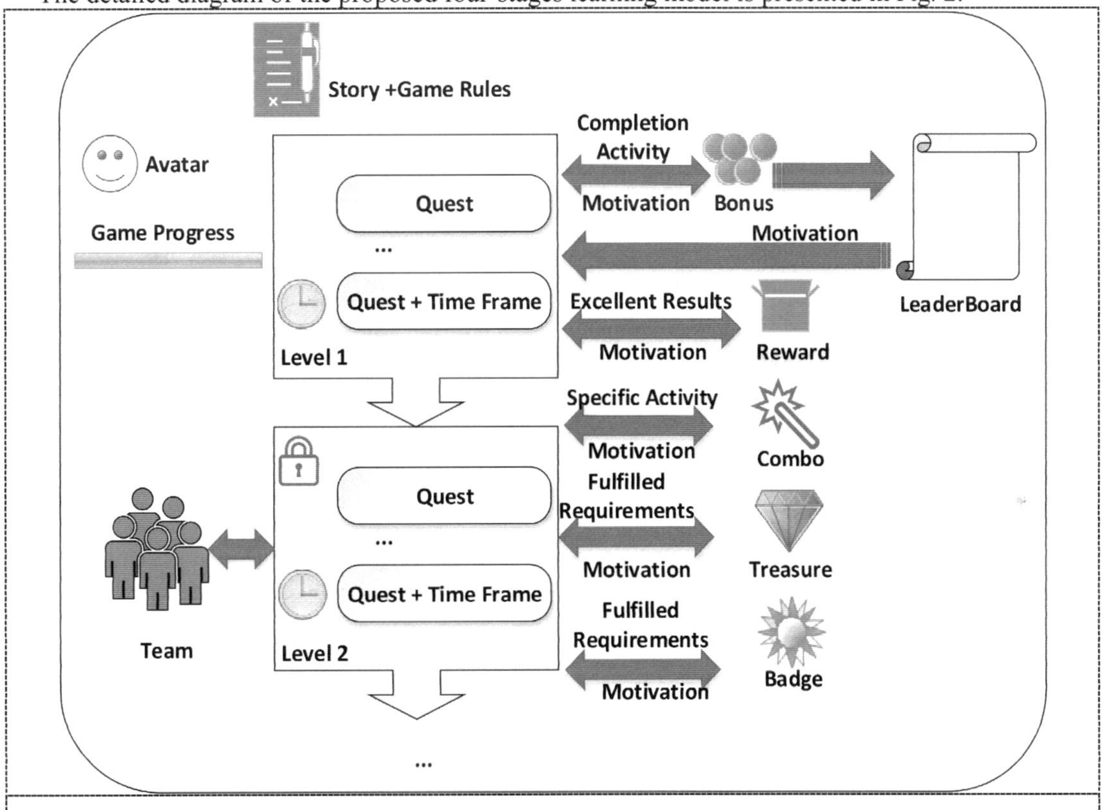

Figure 2. Detailed model of the gamified learning process.

For four user types of [7], the proper motivating game elements are proposed in the learning model [16]: for **killers** – leaderboard, points, course progress, status, combos; for **achievers** – badges, levels, course progress, bonuses; for **explorers** – quests (individual assignments), rewards, history; and for **socializers** – forums, quests (group assignments), history.

4.1.1. Plug-ins for gamification

Moodle platform has a rich variety of learning resources and activities that can be used in the gamification process [17] to realize different gamified elements: **Story/History** – Label, Page, File; **Game Rules** – Label, Page, File; **Challenge** – Page, File, Folder, URL, Book, Lesson, Assignment, Choice, Quiz, Glossary, Workshop, Wiki, Database, Forum, Chat, External tool, Survey; **Hidden Treasure** – Page, File, Folder, URL, Book, Lesson, Glossary, Forum, External tool; **Reward** – Page, File, Folder, URL, Book, Lesson, Glossary, Forum, External tool; **Combo** – Label, Page, File, Folder, Glossary, Database, Book, Lesson, Chat, External tool; **Badge** – Badge; and **Socializing** – Forum,

Chat. Only an avatar and a leader board are missing and can't be realized from the existing Moodle elements. Two plug-ins are designed and developed to cover these missing game elements.

Plug-in for playful design. The first plug-in provides a gamified view of learning courses and changes automatically the design of a standard Moodle course. The course design follows the concept of game levels. User who have a system role Author can use the plug-in to create a new course or to re-design the existing course in a playful design. The software also proposes help for gamified usage of standard Moodle resources and activities (see previous paragraph). If the author follows these recommendations, design a course in multiple section format (like levels) and add some Moodle settings to restrict section/resource/activity access, the developed course will be fully gamified. The plug-in is created in Moodle type course format and implemented as course specific.

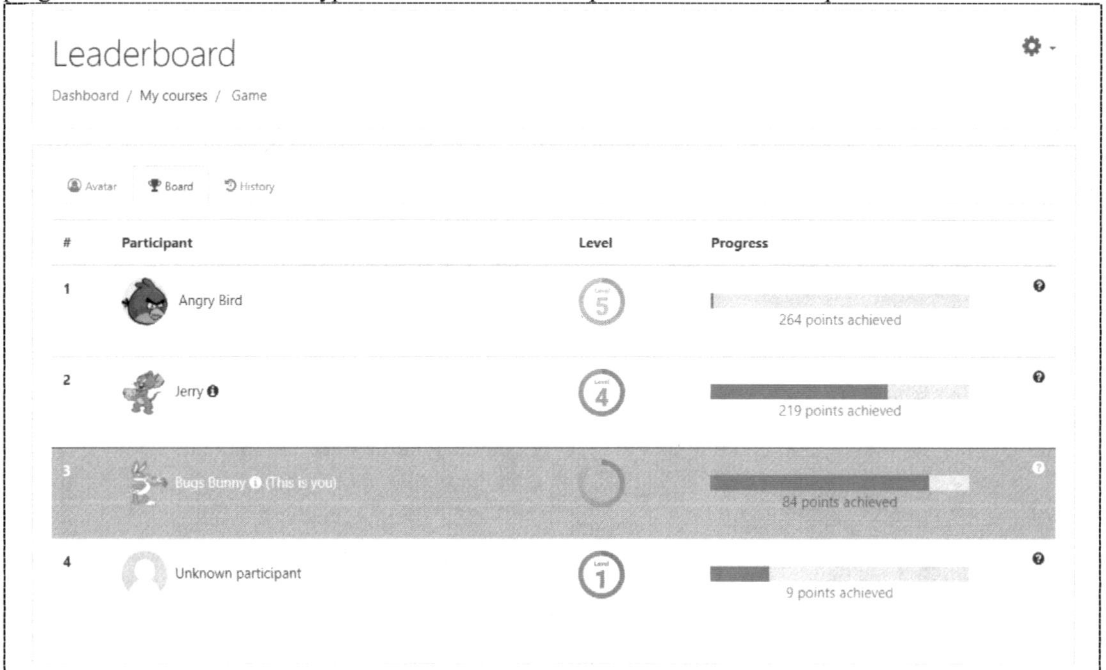

Figure 3. View of the Leader Board ranking with avatars.

Plug-in for specific game elements. The second plug-in proposes for usage game elements that do not exist in Moodle: leader board and avatar. The plug-in can be used in creation of a new course or be added to an existing non-gamified e-course. The software gives the following opportunities for the students:

- To choose an avatar (image), a title (Mr., Mrs. or Prof.), a special name and a short description (the screen in Fig. 3 is taken from account with name Bugs Bunny);
- To see the current LeaderBoard learner ranking (with achieved level and obtained points);
- To see own place in the LeaderBoard learner ranking (Bugs Bunny is on 3rd position);
- To see the reached current level in the course for all learner;
- To see earned points of all learner;
- To see their own current Game Progress;
- To see the logs of her/him actions –actions for which the student has earned the points.

In order to comply with EU laws on concealing sensitive information, each learner participates in the Leaderboard ranking instead of personal name and photo, with a special name and avatar, which may be different for each course.

During the learning process, students collect points (the number of points for each possible learner action is configurated by the teacher) for each activity they have done (entering to read a text learning document, filling in a test, completing an assignment, etc.). They are ranked at the Leaderboard on the basis of current points, despite the fact that everyone moves at their own pace and are currently at a different level.

The plug-in is created in Moodle type block format and implemented as course specific. For the development of both plug-ins are used the following programming languages and tools: PHP, JavaScript, HTML, CSS, Font awesome 4 (for icons) and Bootstrap 4 (for design basis). They are integrated into the Moodle e-learning platform of the University of Plovdiv "Paisii Hilendarski" (http://pdu.uni-plovdiv.bg). Additional tables have been added to the Moodle database via XMLDB. The plug-ins have been implemented with multilingual support in English and Bulgarian.

5. Experiment

The developed plugin has experimented from 38 students studying at the University of Plovdiv "Paisii Hilendarski" during the course "Object-oriented design and programming (C++)". Students had the opportunity to evaluate the developed plugin by taking part in a survey. For the collection of primary data about students' satisfaction a questionnaire was developed. The questionnaire has been prepared in such a way that without being time-consuming to complete to give a clear picture of students' opinion about the plug-ins. The questionnaire contains 19 questions divided into five sections (see Table 1):

Section 1. Practical applicability (Q1-Q4).
Section 2. Motivation (Q5-Q9).
Section 3. Design and accessibility (Q10-Q13).
Section 4. Interactivity and communication (Q14-Q16).
Section 5. Intelligibility (Q16-Q18).

Most of the questions are multiple choice. Students should state how far they agree with the statements on the 5-point Likert-type scale, where 1 means Strongly Disagree (SD), 2 is Disagree (D), 3 is Neutral (N), 4 is Agree (A) and 5 is Strongly Agree (SA). There is an open-ended question at the end of the questionnaire which allow students to make recommendations for enhancing and extending functionalities of the plugin.

Table 1 presents summarized results of the survey. For each statement in the table, the number of responses and their percentage are presented (see Column 2 - Column 6).

Table 1. Survey questions with summary of students' answers.

Statement	5=SA	4=A	3=N	2=D	1=SD
Q1. Training through gamification is more enjoyable than using a standard e-course.	18 (47%)	18 (47%)	2 (6%)	0 (0%)	0 (0%)
Q2. Gamification is suitable for higher education training	18 (47%)	12 (32%)	8 (21%)	0 (0%)	0 (0%)
Q3. Game elements do not overburden or hinder the learning process.	14 (37%)	16 (42%)	8 (21%)	0 (0%)	0 (0%)
Q4. Gamification helps me to achieve educational goals effectively.	14 (37%)	16 (42%)	8 (21%)	0 (0%)	0 (0%)
Q5. Gamification training is more motivating than training trough a standard e-course.	12 (32%)	16 (42%)	10 (26%)	0 (0%)	0 (0%)
Q6. Getting points for each learning activity leads to using more learning resources.	14 (37%)	14 (37%)	10 (26%)	0 (0%)	0 (0%)
Q7. The LeaderBoard element encourages me to get better grades.	16 (42%)	14 (37%)	8 (21%)	0 (0%)	0 (0%)
Q8. The LeaderBoard element encourages me to learn faster.	10 (26%)	20 (53%)	8 (21%)	0 (0%)	0 (0%)
Q9. The Avatar element encourages me to be more active during training.	10 (26%)	12 (32%)	16 (42%)	0 (0%)	0 (0%)
Q10. The proposed gamification design is good.	18 (47%)	12 (32%)	8 (21%)	0 (0%)	0 (0%)

Q11. The use of a gamification design makes the course more accessible than a standard course.	8 (21%)	20 (53%)	10 (26%)	0 (0%)	0 (0%)
Q12. Game elements have enough intuitive design for their easily using.	12 (32%)	24 (63%)	2 (5%)	0 (0%)	0 (0%)
Q13. I had no difficulty with navigation in the gamification course.	14 (37%)	14 (37%)	10 (26%)	0 (0%)	0 (0%)
Q14. I use more interactive learning resources (incl. communication) in the gamification course than in the standard course.	10 (26%)	18 (47%)	8 (21%)	2 (5%)	0 (0%)
Q15. The gamification course encourages communication and collaboration with other students.	16 (42%)	12 (32%)	6 (15%)	4 (11%)	0 (0%)
Q16. Game elements are presented in an understandable way and I have a clear idea of what is expected from me.	18 (47%)	20 (53%)	0 (0%)	0 (0%)	0 (0%)
Q17. The way the training is delivered meets my educational needs.	8 (21%)	18 (47%)	12 (32%)	0 (0%)	0 (0%)
Q18. The gamification course develops my learning skills.	12 (32%)	18 (47%)	8 (21%)	0 (0%)	0 (0%)

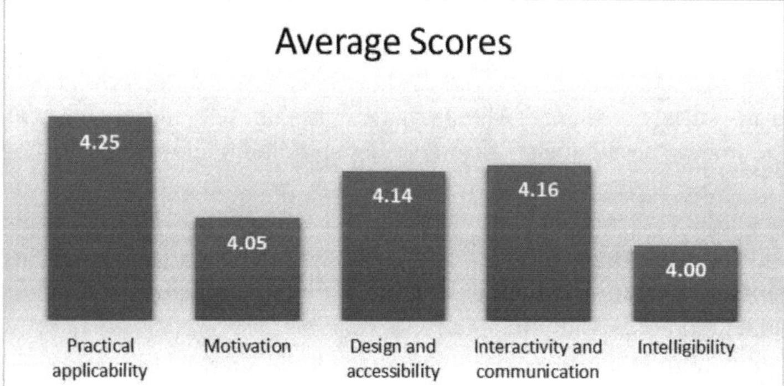

Figure 4. Average scores by sections

The results clearly show that students have positive opinions about the developed plugin. An analysis of the answers show that most students think that training through gamification is more enjoyable than using a standard e-course (84% of the students answered SA or A, and no one student answered D or SD to statement Q1). The percentage of students who think that this teaching approach is suitable for higher education, game elements do not overburden the learning process and that gamification helps them to achieve educational goals effectively is also high (79% answered SA or A to statements Q2, Q£ and Q4 and no one student answered D or SD to these statements). Most students agree that gamification training is motivating and getting points for each activity encourage them to be more active in the learning process (74% answered SA or A to statements Q5 and Q6). The LeaderBoard element encourages students to learn faster and get better grades (79% answered SA or A to statements Q7 and Q8). The lower but relatively high is the percentage of students who agree that Avatar element encourages them to be more active during training (58% answered SA or A to statement Q9). Students approve the proposed design (79% answered SA or A to statement Q10 and 95% answered SA or A to statement Q12) and agreed that the use of a gamification design makes the course more accessible than a standard designed course (74% answered SA or A to statement Q11). 26% of students stated that they sometimes have difficulty when navigating in the gamification course, but the other students stated that they have not such problems (see answers of statement Q13). Most students shared that they use more interactive learning resources in the gamification course than in the standard course (73% answered A or SA to statement Q14) and a small part of them use less interactive learning resources in the gamification course than in the standard course (5% answered D to statement Q14). The lowest degree of agreement students gave to the statements that the gamification course encourages communication and collaboration with other students (11% answered D to statement Q15), but as a whole the percentage of students who agreed with this statement is high (74% answered SA or A to statement Q15). All students think that game elements are presented in an understandable way and have a clear idea of what is expected from them (all answers to statement Q16 are A or SA). When students are asked to agree that this way of training meets their educational needs 33% of them answered N. The percentage of students who cannot judge whether a gamification course develops their learning skills is also high – 21% (see answers to statement Q18).

As a whole, the students are satisfied by the functionalities of the developed plug-ins and think they are useful for them - the average scores on all sections is above 4 (see Fig. 4) and the average score on all statements of the questionnaire is 4.12. According to the students, the functionality of the developed plug-ins can be improved and expanded through adding hints and more badges, removing the fusion of colours in the Level chart (because they associate it with the Progress bar), merging the Progress bar and Level circle functionalities and improving the point accumulation algorithm.

6. Conclusion

The paper presented the study in the field of structural gamification, where the authors aimed to design and develop a system for conducting gamification training in an e-learning environment. The used approach is to integrate a system (plug-ins), that creates and uses a gamified e-course for learning, with an existing suitable learning management system.

In achieving the objective of the study, the following five results were obtained:

- Investigation of game techniques and elements used in gamification and also of existing gamified systems.
- Design of a model of gamified e-learning course.
- Creation of an approach for designing a software system for the implementation of gamified learning in an e-learning environment.
- Software realization of plug-ins for creation and usage of a gamified course.
- Conducted experiment – developed and experimented a gamified course with real users.

References

[1] Christians G 2018 The Origins and Future of Gamification *Senior Theses* 254 University of South Carolina
[2] Gartner https://www.gartner.com/en/marketing/glossary/gamification
[3] What is the definition of gamification? 2018 Growth Engineering https://www.growthengineering.co.uk/definition-of-gamification/
[4] Werbach K 2014 (Re) defining gamification: A process approach. In International conference on persuasive technology *Lecture Notes in Computer Science* vol **8462** (Springer, Cham) p 266
[5] Chappelow J 2019 Gamification https://www.investopedia.com/terms/g/gamification.asp
[6] Kapp K M 2012 The gamification of learning and instruction: Game-based methods for trainig and education (San Francisco: Pfeiffer)
[7] Bartle R 1996 Hearts, clubs, diamonds, spades: Players who suit MUDs *Journal of MUD research* vol **1** n 1 pp 19-58
[8] GENIE http://www.growthengineering.co.uk/genie-content-authoring-tool/
[9] Academy LMS http://www.growthengineering.co.uk
[10] TalentLMS http://www.talentlms.com/
[11] Accord LMS https://evaluate.accordlms.com/learning-management-system/
[12] Axonify https://axonify.com/
[13] Moodle plugins database https://moodle.org/plugins
[14] Somova E and Gachkova M 2016 An Attempt for Gamification of Learning in Moodle, *International Conference on e-Learning (e-Learning'16)* Bratislava, Slovakia pp 201-207
[15] Gachkova M and Somova E 2020 Moodle plug-ins for design and development of gamified

courses *14th annual International Technology, Education and Development Conference – INTED'2020* Valencia, Spain

[16] Gachkova M Takev M and Somova E 2018 Learning and Assessment Based on Gamified e-Course in Moodle *Mathematics and Informatics* vol **61** n 5 pp 444-454

[17] Gachkova M and Somova E 2019 Plug-in for creation of gamified courses in the e-learning environment Moodle *IOP Conference Series: Materials Science and Engineering* vol **618**

IOP Conf. Series: Materials Science and Engineering **878** (2020) 012036 doi:10.1088/1757-899X/878/1/012036

Dynamical stresses in the high class wind turbine blades caused in nonhomogeneous nonstationary wind field. Part 1 Dynamical model

I Kralov, J Genov, I Angelov

Dept. of Mechanics, Technical University of Sofia, 8 Kl. Ohridski Blvd, Sofia 1000, Bulgaria

E-mail: j_genov@mail.bg, kralov@tu-sofia.bg, ivoangelov@gmail.com

Abstract. The wind turbines of a high class are with considerable sizes. For this reason, the consideration of the fluid field as a homogeneous when performing the aerodynamic analysis is not quite correct. In addition, due to the gusts and horizontal turbulence, the wind field is also nonstationary. All this causes dynamic, time-varying loads and fatigue. The determination of these loads is a major aim of this publication. The turbine blades are considered as Euler-Bernoulli beams under the impact of the aerodynamic thrust and torque forces as well as from this of the centrifugal force. The finite element method is applied to solve the partial differential equations describing the model under consideration. Two-node elements with four degrees of freedom for each node and a cubic, polynomial approximation of modal shapes were used. The deformations are relatively small in the separated finite elements and this makes it possible to apply the superposition principle.

1. Introduction

The current study performs dynamic stress analysis directed to the loads caused by the thrust, torque, and centrifugal forces acting on the blades of the wind turbine. In [1-4] the authors proposed modification, that takes into account the vertical wind speed gradient, in the Blade Element Momentum (BEM) theory [5÷9] used for modeling the aerodynamic interaction. Based on this modification it is possible to calculate the distribution of the forces along the blade's length, as function of time. This approach renders an account both the azimuth angle of the blade position (i.e. the vertical wind speed gradient) and the speed turbulence in a horizontal direction. Euler-Bernoulli beam model is used for the determination of the dynamical stresses and deformations in the turbine blades i.e. the section's plane normal to the neutral beam axis remains normal after deformation and the shear deflection is neglectable. The mathematical model takes into account the complex implementation of the turbine blades, consisting sections with different airfoils, variable chords` length, mass, stiffness and damping characteristics distributed along the blade length. Approaches studying different aspects of loads in the turbine blades are discussed in many references [10÷13], but the main aim in this study is the investigation on the influence of the vertical distribution of wind speed combined with longitudinal turbulence, on the dynamic stresses in the blades of the turbines of high class. This is a premise for additional fatigue loads of the turbine's blades and in this manner are important factors for the design and maintenance of the turbines.

2. Dynamical model

The model and the acting forces on the turbine blade and on its element with length *dy* are shown in figure 1. The axes of the used coordinate system, rotating with the turbine blade, are defined as: 'x' is parallel to the hub axis, 'oz' is coincident with the axis of the turbine blade, and 'y' is obtained by cross product of the basis vectors of x and y.

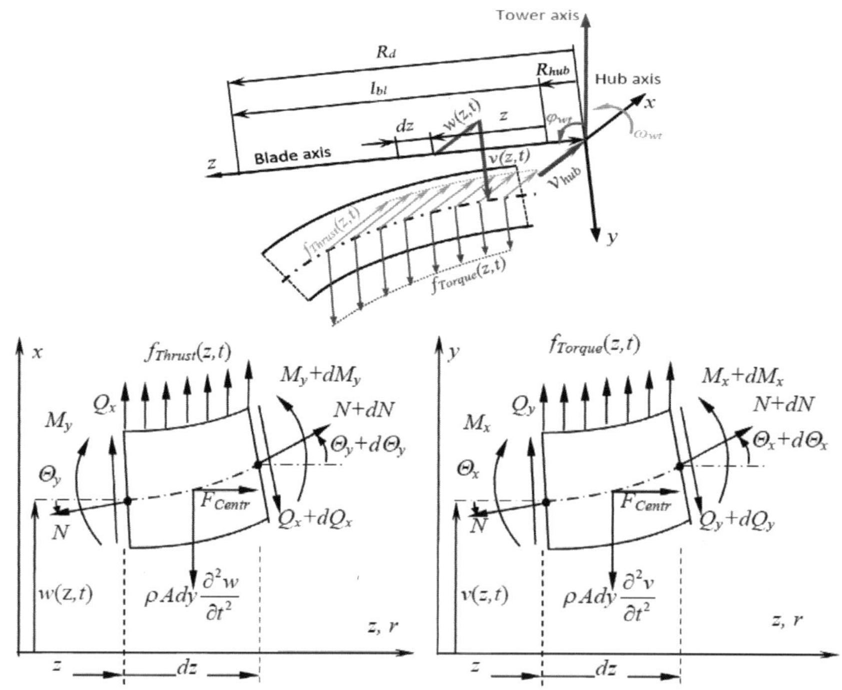

Figure 1. Loads acting on the single blade element considered as Euler-Bernoulli`s beam

The intensity functions of the aerodynamic thrust force - $F_{Thrust}(z,t)$ and torque force - $F_{Torque}(z,t)$ in dependence of the position - z along the turbine blade axis (or radius *r* of the turbine's disk) and of the time - *t*, with considered the vertical and longitudinal turbulence are:

$$f_{Thrust}(z,t) = \frac{\partial F_{Thrust}(z,t)}{\partial z}, \quad f_{Torque}(z,t) = \frac{\partial F_{Torque}(z,t)}{\partial z} \tag{1}$$

The centrifugal force and its intensity function are:

$$F_{Centr}(y,t) = \int_{R_{hub}}^{Rd} \rho A(z)\omega_{wt}^{2}(t) z dz\,,\; f_{Centr}(z,t) = \frac{\partial F_{Centr}(z,t)}{\partial z} = \rho A(z) z \omega_{wt}^{2}(t)\,, \tag{2}$$

where: $\rho(y)$ is the density of the blade`s compose material; $A(y)$ is the cross section`s area and $\omega_{wt}(t)$ the angular velocity of the turbine.

The inertial force is: $\vec{F}_{iner}(z,t) = \vec{F}^{x}_{iner}(z,t) + \vec{F}^{y}_{iner}(z,t) = -\int_{R_{hub}}^{R_d} \rho A(z)\ddot{\vec{w}}(z,t)dz - \int_{R_{hub}}^{R_d} \rho A(z)\ddot{\vec{v}}(z,t)dz$,(3)

where $w(z,t)$ and $v(z,t)$ are the transverse displacements of the elastic line.

The internal loads in the blade's cross section are: $Q_x(z,t)$, $Q_y(z,t)$ - shear forces, $N(y,t)$ - normal force, $M_y(y,t)$, $M_x(y,t)$ - bending moments.

Considering the Euler-Bernoulli`s assumptions, the angular displacements $\Theta_x(z,t)$, $\Theta_y(z,t)$ are relatively small, therefore we can assume:

$$\begin{aligned}\Theta_y(z,t)\approx\tan\left(\Theta_y(z,t)\right)=\frac{\partial w(z,t)}{\partial z}=w^I(z,t),\ \sin\left(\Theta_y\right)\approx\frac{\partial w(z,t)}{\partial z},\ \cos\left(\Theta_y\right)\approx 1\\ \Theta_x(z,t)\approx\tan\left(\Theta_x(z,t)\right)=\frac{\partial v(z,t)}{\partial z}=v^I(z,t),\ \sin\left(\Theta_x\right)\approx\frac{\partial v(z,t)}{\partial z},\ \cos\left(\Theta_x\right)\approx 1\end{aligned} \quad ..(4)$$

Considering the balance of the forces and moments (figure 1), taking into account (4) and eliminating the infinitely small by a second order, it is obtained:

$$\left|\begin{aligned}&\frac{\partial^2}{\partial z^2}\left(EI_y(z)\frac{\partial^2 w(z,t)}{\partial z^2}\right)+\rho A(z)\frac{\partial^2 w(z,t)}{\partial t^2}+2n\rho A(z)\frac{\partial w(z,t)}{\partial t}-\frac{\partial}{\partial z}\left(N(z,t)\frac{\partial w(z,t)}{\partial z}\right)=f_{Thrust}(z,t)\\ &\frac{\partial^2}{\partial z^2}\left(EI_x(z)\frac{\partial^2 v(z,t)}{\partial z^2}\right)+\rho A(z)\frac{\partial^2 v(z,t)}{\partial t^2}+2n\rho A(z)\frac{\partial v(z,t)}{\partial t}-\frac{\partial}{\partial z}\left(N(z,t)\frac{\partial v(z,t)}{\partial z}\right)=f_{Torque}(z,t)\\ &N(z,t)=\omega_{wt}^2(t)\int_{\ell_{bl}}^{z}\rho A(\eta)\eta d\eta\end{aligned}\right. \quad ..(5)$$

where: E – Young`s modulus; $I_y(z)$, $I_x(z)$ – area moments of inertia for the cross section centroidal axes, A(z) – cross sections area, l_{bl} is the length of the blade and n is coefficient of an attenuation.

The boundary and initial conditions are:

$$\left|\begin{aligned}&w(0,t)=0,\ w^I(0,t)=0\\ &w^{II}(l_{bl},t)=0,\ w^{III}(l_{bl},t)=0\end{aligned}\right.,\ \left|\begin{aligned}&w(x,0)=w_0(x)\\ &\dot w(x,0)=\dot w_0(x)\end{aligned}\right.,\ \left|\begin{aligned}&v(0,t)=0,\ v^I(0,t)=0\\ &v^{II}(l_{bl},t)=0,\ v^{III}(l_{bl},t)=0\end{aligned}\right.,\ \left|\begin{aligned}&v(x,0)=w_0(x)\\ &\dot v(x,0)=\dot w_0(x)\end{aligned}\right.. \quad (6)$$

The application the Hamilton's variational approach to (5) derives:

$$\begin{aligned}&\int_0^{\ell_{bl}}\mu_w(x)\left[\frac{\partial^2}{\partial z^2}\left(EI_y(z)\frac{\partial^2 w(z,t)}{\partial z^2}\right)+\rho A(z)\frac{\partial^2 w(z,t)}{\partial t^2}+2n\rho A(z)\frac{\partial w(z,t)}{\partial t}-f_{Thrust}(z,t)-\frac{\partial}{\partial z}\left(N(z,t)\frac{\partial w(z,t)}{\partial z}\right)\right]dz=0\\ &\int_0^{\ell_{bl}}\mu_v(x)\left[\frac{\partial^2}{\partial z^2}\left(EI_x(z)\frac{\partial^2 v(z,t)}{\partial z^2}\right)+\rho A(z)\frac{\partial^2 v(z,t)}{\partial t^2}+2n\rho A(z)\frac{\partial v(z,t)}{\partial t}-f_{Torque}(z,t)-\frac{\partial}{\partial z}\left(N(z,t)\frac{\partial v(z,t)}{\partial z}\right)\right]dz=0\end{aligned},$$

where $\mu_w(z)$, $\mu_v(z)$ are weight functions. (7)

Executing some transformations and integrating by parts leads to:

$$\begin{aligned}&\int_0^{\ell_{bl}}\left\{EI_y(z)\frac{\partial^2\mu_w(x)}{\partial z^2}\frac{\partial^2 w(z,t)}{\partial z^2}+\left[\rho A(z)\frac{\partial^2 w(z,t)}{\partial t^2}+2n\rho A(z)\frac{\partial w(z,t)}{\partial t}-f_{Thrust}(z,t)-\frac{\partial}{\partial z}\left(N(z,t)\frac{\partial w(z,t)}{\partial z}\right)\right]dz-\right.\\ &-\left\{\mu_w(x)\left[\frac{\partial}{\partial z}\left(EI_y(z)\frac{\partial^2 w(z,t)}{\partial z^2}\right)-N(z,t)\frac{\partial w(z,t)}{\partial z}\right]-\frac{\partial\mu_w(x)}{\partial z}EI_y(z)\frac{\partial^2 w(z,t)}{\partial z^2}\right\}\Bigg|_0^{\ell_{bl}}=0\\ &\int_0^{\ell_{bl}}\left\{EI_x(z)\frac{\partial^2\mu_v(x)}{\partial z^2}\frac{\partial^2 v(z,t)}{\partial z^2}+\left[\rho A(z)\frac{\partial^2 v(z,t)}{\partial t^2}+2n\rho A(z)\frac{\partial v(z,t)}{\partial t}-f_{Torque}(z,t)-\frac{\partial}{\partial z}\left(N(z,t)\frac{\partial v(z,t)}{\partial z}\right)\right]dz-\right.\\ &-\left\{\mu_v(x)\left[\frac{\partial}{\partial z}\left(EI_x(z)\frac{\partial^2 v(z,t)}{\partial z^2}\right)-N(z,t)\frac{\partial v(z,t)}{\partial z}\right]--\frac{\partial\mu_v(x)}{\partial z}EI_x(z)\frac{\partial^2 v(z,t)}{\partial z^2}\right\}\Bigg|_0^{\ell_{bl}}=0\end{aligned} \quad (8)$$

In order to solve the system (5)-(6) different methods can be used:

- The Galerkin method is based on the minimization of the error between the deflection function and the boundary conditions. The approach is simple, but it is difficult to form a function that satisfies all of the boundary conditions. The method has slowly convergence for the higher frequencies and modes;

- Rayleigh–Ritz method is based on the energy principles. The deflected shape is express as a linear combination of known functions with unknown coefficients, which are obtained by minimization of the total energy of the system;
- Finite element (FE) method. This method is more convenient in the cases for non-uniform distributed mass-inertial, elasticity and damping characteristics as well in cases of nonstationary acting forces and respectively of a variable angular velocity of the turbine.

Considering all the complexity of the discussed problem the application of Finite Element Method (FEM) is most appropriate approach.

3. Finite element method approach

The blade is divided into n numbers of finite elements. One finite element – figure 2 has two nodes, each node with four degrees of freedom:

- transverse displacements $w(z_i,t)$, $v(z_i,t)$ and
- angular displacements: $\Theta_y(z_i,t) \approx \left.\frac{\partial w(z,t)}{\partial z}\right|_{z_i} = w^{I}(z_i,t)$; $\Theta_x(z_i,t) \approx \left.\frac{\partial v(z,t)}{\partial z}\right|_{z_i} = v^{I}(z_i,t)$, $i=1,\ 2..(n+1)$

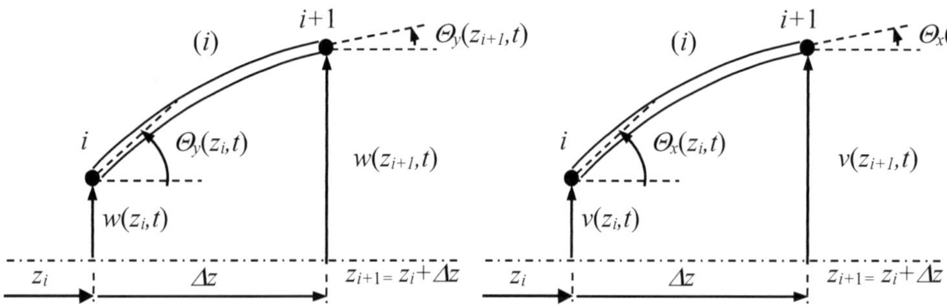

Figure 2. Blade`s finite element

If the generalized coordinates are:

$$\mathrm{q}_i(t)=\begin{bmatrix} q_{i,1}(t) \\ q_{i,2}(t) \\ q_{i,3}(t) \\ q_{i,4}(t) \end{bmatrix}=\begin{bmatrix} w(z_i,t) \\ \Theta_y(z_i,t) \\ w(z_{i+1},t) \\ \Theta_y(z_{i+1},t) \end{bmatrix}\approx\begin{bmatrix} w(z_i,t) \\ w^{I}(z_i,t) \\ w(z_{i+1},t) \\ w^{I}(z_{i+1},t) \end{bmatrix},\ p_i(t)=\begin{bmatrix} q_{i,5}(t) \\ q_{i,6}(t) \\ q_{i,7}(t) \\ q_{i,8}(t) \end{bmatrix}=\begin{bmatrix} v(z_i,t) \\ \Theta_z(z_i,t) \\ v(z_{i+1},t) \\ \Theta_z(z_{i+1},t) \end{bmatrix}\approx\begin{bmatrix} v(z_i,t) \\ v^{I}(z_i,t) \\ v(z_{i+1},t) \\ v^{I}(z_{i+1},t) \end{bmatrix}, \tag{9}$$

$\triangle z = \ell_{bl}/n,\ i=1,2,...n+1.$

the elastic line can be expressed in terms of them as:
$$\left| \begin{array}{l} w(z,t)=\Phi_i(z)q_i(t),\quad z\in[z_i,z_{i+1}], \\ v(z,t)=\Phi_i(z)p_i(t)\quad i\text{=}1,2,..n\text{+}1 \end{array} \right., \tag{10}$$

where the nodal functions vectors are: $\Phi_i(z)=[\phi_{i,1}(z),\phi_{i,2}(z),\phi_{i,3}(z),\phi_{i,4}(z)]$. (11)

At the same time its presentation by separation of variables is:

$$\left| \begin{array}{l} w(z,t)=X(z)Tx_i(t) \\ v(z,t)=Y(z)Ty_i(t) \end{array} \right.,\ z\in[z_i,z_{i+1}],\ i\text{=}1,2,..n\text{+}1, \tag{12}$$

where $Tx_i(t)$, $Ty_i(t)$ are time dependence functions, and $X(z)$, $Y(z)$ are the modal shapes approximated by cubic polynomials:

$$\left| \begin{array}{l} X(z)\approx a_1+a_2z+a_3z^2+a_4z^3=\left[1,z,z^2,z^3\right]\underbrace{\left[a_1,a_2,a_3,a_4\right]^T}_{a} \\ Y(z)\approx b_1+b_2z+b_3z^2+b_4z^3=\left[1,z,z^2,z^3\right]\underbrace{\left[b_1,b_2,b_3,b_4\right]^T}_{b} \end{array} \right.,\ z\in[z_i,z_{i+1}]\Rightarrow \left| \begin{array}{l} w(z,t)\approx\left[1,z,z^2,z^3\right]aTx_i(t) \\ v(z,t)\approx\left[1,z,z^2,z^3\right]bTy_i(t) \end{array} \right. . \tag{13}$$

Considering equations (13), the general coordinates (9) are obtained in the form:

$$q_i(t)=\begin{bmatrix} w(z_i,t) \\ w^I(z_i,t) \\ w(z_i+\vartriangle z,t) \\ w^I(z_i+\vartriangle z,t)\end{bmatrix}=\begin{bmatrix} X(z_i) \\ X^I(z_i) \\ X(z_{i+1}) \\ X^I(z_{i+1})\end{bmatrix}Tx_i(t)=\underbrace{\begin{bmatrix} 1, & z_i, & z_i^2, & z_i^3 \\ 0, & 1, & 2z_i, & 3z_i^2 \\ 1, & z_{i+1}, & z_{i+1}^2, & z_{i+1}^3 \\ 0, & 1, & 2z_{i+1}, & 3z_{i+1}^2\end{bmatrix}}_{B_i}aT_i(t)=B_i aTx_i(t)$$

$$p_i(t)=\begin{bmatrix} v(z_i,t) \\ v^I(z_i,t) \\ v(z_i+\vartriangle z,t) \\ v^I(z_i+\vartriangle z,t)\end{bmatrix}=\begin{bmatrix} Y(z_i) \\ Y^I(z_i) \\ Y(z_{i+1}) \\ Y^I(z_{i+1})\end{bmatrix}Ty_i(t)=\underbrace{\begin{bmatrix} 1, & z_i, & z_i^2, & z_i^3 \\ 0, & 1, & 2z_i, & 3z_i^2 \\ 1, & z_{i+1}, & z_{i+1}^2, & z_{i+1}^3 \\ 0, & 1, & 2z_{i+1}, & 3z_{i+1}^2\end{bmatrix}}_{B_i}bT_i(t)=B_i bTy_i(t) \quad . \quad (14)$$

Substituting (14) in (10) gives:

$$\left.\begin{aligned} w_i(z,t)&=(\Phi_i(z)B_i)aTx_i(t)\underset{(12)}{=}X_i(z)Tx_i(t)=\left(\left[1,z,z^2,z^3\right]\right)aTx_i(t) \\ v_i(z,t)&=(\Phi_i(z)B_i)bTy_i(t)\underset{(12)}{=}Y_i(z)Ty_i(t)=\left(\left[1,z,z^2,z^3\right]\right)bTy_i(t)\end{aligned}\right\}\Rightarrow \begin{aligned}&\Phi_i(z)=\left[1,z,z^2,z^3\right]B_i^{-1} \\ &z\in\left[z_i,z_{i+1}\right], i=1,2,..n+1\end{aligned} \quad (15)$$

Respectively the nodal function vector and its derivations in developed forms are:

$$\left|\begin{aligned} &\Phi_i(z)=[1-3u_i^2+2u_i^3,\ (u_i-2u_i^2+u_i^3)\vartriangle z,\ 3u_i^2-2u_i^3,\ (u_i^3-u_i^2)\vartriangle z] \\ &\Phi_i^{I}(z)=[-6u_i+6u_i^2,(1-4u_i+3u_i^2)\vartriangle z,6u_i-6u_i^2,(3u_i^2-2u_i)\vartriangle z]/\vartriangle z \\ &\Phi_i^{II}(z)=[-6+12u_i,\ (-4+6u_i)\vartriangle z,\ 6-12u_i,\ (6u_i-2)\vartriangle z]/\vartriangle z^2 \\ &\Phi_i^{III}(z)=[12,\ 6\vartriangle z,\ -12,\ 6\vartriangle z]/\vartriangle z^3\end{aligned}\right., \text{ where:}\left|\begin{aligned}&u_i(z)=(z-z_i)/\vartriangle z \\ &u_i(z_i)=0,\ u_i(z_{i+1})=1 \\ &\mathrm{d}z=\vartriangle z\mathrm{d}u_i,\ \mathrm{d}u_i=\vartriangle z^{-1}\mathrm{d}z\end{aligned}\right. . \quad (16)$$

The values at the elements boundaries are:

$$\left|\begin{aligned}&\Phi_i(z_i)=[1,\ 0,\ 0,\ 0] \\ &\Phi_i^I(z_i)=[0,\ 1,\ 0,\ 0] \\ &\Phi_i^{II}(z_i)=[-6,-4\vartriangle z,\ 6,-2\vartriangle z]/\vartriangle z^2 \\ &\Phi_i^{III}(z_i)=[12,6\vartriangle z,-12,6\vartriangle z]/\vartriangle z^3\end{aligned}\right. \quad \left|\begin{aligned}&\Phi_i(z_{i+1})=[0,\ 0,\ 1,\ 0] \\ &\Phi_i^I(z_{i+1})=[0,\ 0,\ 0,\ 1] \\ &\Phi_i^{II}(z_{i+1})=[6,\ 2\vartriangle z,-6,4\vartriangle z]/\vartriangle z^2 \\ &\Phi_i^{III}(z_{i+1})=[12,6\vartriangle z,-12,6\vartriangle z]/\vartriangle z^3\end{aligned}\right. . \quad (17)$$

If the discretization step $\vartriangle z$ is small enough, so the mass, the elastic and the damping characteristics, as well as the external forces for each finite element can be assumed for constant:

$$\left|\begin{aligned}&\rho A(z)\approx const=\rho A(z_i),\ EI_y(z)\approx const=EI_y(z_i),\ EI_x(z)\approx const=EI_x(z_i) \\ &n(z)\approx const=n(z_i),\ f_{Thrust}(z,t)\approx f_{Thrust}(z_i,t),\ f_{Torque}(z,t)\approx f_{Torque}(z_i,t),\ z\in\left[z_i,z_{i+1}\right],\ i=1,.2,..n\end{aligned}\right. . \quad (18)$$

Accounting (5) and (18) the elastic line can be described by next system of differential equations:

$$\left|\begin{aligned}&EI_y(z_i)\frac{\partial^4 w(z,t)}{\partial z^4}+\rho A(z_i)\frac{\partial^2 w(z,t)}{\partial t^2}+2n\rho A(z_i)\frac{\partial w(z,t)}{\partial t}-\frac{\partial}{\partial z}\left(N(z,t)\frac{\partial w(z,t)}{\partial x}\right)=f_{Thrust}(z_i,t) \\ &EI_x(z_i)\frac{\partial^4 v(z,t)}{\partial z^4}+\rho A(z_i)\frac{\partial^2 v(z,t)}{\partial t^2}+2n\rho A(z_i)\frac{\partial v(z,t)}{\partial t}-\frac{\partial}{\partial z}\left(N(z,t)\frac{\partial v(z,t)}{\partial z}\right)=f_{Torque}(z_i,t)\end{aligned}\right., \left|\begin{aligned}&z\in\left[z_i,z_{i+1}\right] \\ &i=1,2,..n\end{aligned}\right. \quad (19)$$

Similarly, to the variance representation (7) with accounting (18) for the single element is obtained:

$$\begin{aligned}&\int_{z_i}^{z_{i+1}}\left\{EI_y(z)\frac{\partial^2\mu_x(z)}{\partial z^2}\frac{\partial^2 w(z,t)}{\partial z^2}+\mu_x(z)\left[\rho A(z_i)\frac{\partial^2 w(z,t)}{\partial t^2}+2n\rho A(z_i)\frac{\partial w(z,t)}{\partial t}-f_{Thrust}(z_i,t)\right]+\right. \\ &\left.+N(z,t)\frac{\partial\mu_x(z)}{\partial z}\frac{\partial w(z,t)}{\partial z}\right]\Bigg\}dz-EI_y(z_i)\left\{\mu_x(z)\left[\frac{\partial^3 w(z,t)}{\partial z^3}-\frac{N(z,t)}{EI_y(z_i)}\frac{\partial w(z,t)}{\partial z}\right]-\frac{\partial\mu_x(z)}{\partial z}\frac{\partial^2 w(z,t)}{\partial z^2}\right\}\Bigg|_{z_i}^{z_{i+1}}=0\end{aligned} \quad (20)$$

$$\int_{z_i}^{z_{i+1}}\left\{EI_x(z)\frac{\partial^2\mu_y(z)}{\partial z^2}\frac{\partial^2 v(z,t)}{\partial z^2}+\mu_y(z)\left[\rho A(z_i)\frac{\partial^2 v(z,t)}{\partial t^2}+2n\rho A(z_i)\frac{\partial v(z,t)}{\partial t}-f_{Thrust}(z_i,t)\right]+\right.$$
$$\left.+N(z,t)\frac{\partial\mu_y(z)}{\partial z}\frac{\partial v(z,t)}{\partial z}\right]\Bigg\}dz-EI_x(z_i)\left\{\mu_y(z)\left[\frac{\partial^3 v(z,t)}{\partial z^3}-\frac{N(z,t)}{EI_y(z_i)}\frac{\partial v(z,t)}{\partial z}\right]-\frac{\partial\mu_y(z)}{\partial z}\frac{\partial^2 v(z,t)}{\partial z^2}\right\}\Bigg|_{z_i}^{z_{i+1}}=0.$$

Establishing the internal loads vector in the nodes

$$\Psi_i(t)=\begin{bmatrix}Q_x(z_i)\\ M_y(z_i)\\ Q_x(z_{i+1})\\ M_y(z_{i+1})\end{bmatrix}=EI_y(z_i)\left[\sum_{j=1}^{4}\left(\frac{\partial^3\phi_{ij}(z)}{\partial z^3}\Bigg|_{z_i}q_{ij}(t)\right),\sum_{j=1}^{4}\left(\frac{\partial^2\phi_{ij}(z)}{\partial z^2}\Bigg|_{z_i}q_{ij}(t)\right),\sum_{j=1}^{4}\left(\frac{\partial^3\phi_{ij}(z)}{\partial z^3}\Bigg|_{z_{i+1}}q_{ij}(t)\right),\sum_{j=1}^{4}\left(\frac{\partial^2\phi_{ij}(z)}{\partial z^2}\Bigg|_{z_{i+1}}q_{ij}(t)\right)\right]^T=$$

$$=\frac{EI_y(z_i)}{\Delta z^3}2\begin{bmatrix}6 & 3\Delta z & -6 & 3\Delta z\\ -3\Delta z & -2\Delta z^2 & 3\Delta z & -\Delta z^2\\ 6 & 3\Delta z & -6 & 3\Delta z\\ 3\Delta z & \Delta z^2 & -3\Delta z & 2\Delta z^2\end{bmatrix}q_i(t) \tag{21}$$

and respectively: $$\varXi_i(t)=\begin{bmatrix}Q_y(z_i)\\ M_x(z_i)\\ Q_y(z_{i+1})\\ M_x(z_{i+1})\end{bmatrix}=\frac{EI_x(z_i)}{\Delta z^3}2\begin{bmatrix}6 & 3\Delta z & -6 & 3\Delta z\\ -3\Delta z & -2\Delta z^2 & 3\Delta z & -\Delta z^2\\ 6 & 3\Delta z & -6 & 3\Delta z\\ 3\Delta z & \Delta z^2 & -3\Delta z & 2\Delta z^2\end{bmatrix}p_i(t), \tag{22}$$

substituting the weight functions: $\mu_{xi}(z)=\mu_{yi}(z)=\phi_{ik}(z),\ k=1,..4$, (23)

and taking into account (18) the equations (20) are transformed into:

$$\begin{cases}\rho A(z_i)\Delta z\sum_{j=1}^{4}\left[\int_0^1\phi_{ik}\phi_{ij}du\right]\ddot q_{ij}(t)+2n\rho A(z_i)\Delta z\sum_{j=1}^{4}\left[\int_0^1\phi_{ik}\phi_{ij}du\right]\dot q_{ij}(t)+\\ +\left\{EI_y(z_i)\Delta z\sum_{j=1}^{4}\left[\int_0^1\phi_{ik}{}^{II}\phi_{ij}{}^{II}du\right]+\Delta z\sum_{j=1}^{4}\left[\int_0^1 N(z)\phi_{ik}{}^{I}\phi_{ij}{}^{I}du\right]\right\}q_{ij}(t)=\Delta z\int_0^1\phi_{ik}du\,f_{Thrust}(z_i,t)+Wx_{ik},\\ \rho A(z_i)\Delta z\sum_{j=1}^{4}\left[\int_0^1\phi_{ik}\phi_{ij}du\right]\ddot p_{ij}(t)+2n\rho A(z_i)\Delta z\sum_{j=1}^{4}\left[\int_0^1\phi_{ik}\phi_{ij}du\right]\dot p_{ij}(t)+\qquad k=1,..4,\ i=1,..n+1,\\ +\left\{EI_x(z_i)\Delta z\sum_{j=1}^{4}\left[\int_0^1\phi_{ik}{}^{II}\phi_{ij}{}^{II}du\right]+\Delta z\sum_{j=1}^{4}\left[\int_0^1 N(z)\phi_{ik}{}^{I}\phi_{ij}{}^{I}du\right]\right\}p_{ij}(t)=\Delta z\int_0^1\phi_{ik}du\,f_{Torque}(z_i,t)+Wy_{ik},\end{cases} \tag{24}$$

where the internal loads are:

$$\begin{aligned}&Wx_{ik}(z_i)=\left[\phi_{ik}(z_i),\ -\phi_{ik}{}^{I}(z_i),\ -\phi_{ik}(z_{i+1}),\ \phi_{ik}{}^{I}(z_{i+1})\right]\Psi_i(t)-\left(\phi_{ik}(z)N(z,t)\frac{\partial w(z,t)}{\partial z}\right)\Bigg|_{z_i}^{z_{i+1}},\\ &Wy_{ik}(z_i)=\left[\phi_{ik}(z_i),\ -\phi_{ik}{}^{I}(z_i),\ -\phi_{ik}(z_{i+1}),\ \phi_{ik}{}^{I}(z_{i+1})\right]\varXi_i(t)-\left(\phi_{ik}(z)N(z,t)\frac{\partial v(z,t)}{\partial z}\right)\Bigg|_{z_i}^{z_{i+1}},\\ &\left[\phi_{ik}(z_i),\ -\phi_{ik}{}^{I}(z_i),\ -\phi_{ik}(z_{i+1}),\ \phi_{ik}{}^{I}(z_{i+1})\right]=diag([1,-1,-1,1]),\end{aligned} \tag{25}$$

and the normal force is:

$$N(z,t)=\omega_{wt}(t)^2\int_z^{\ell_{bl}}A\rho(z)rdr\underset{(20)}{=}\omega_{wt}(t)^2\left[\sum_{j=i}^{n}\left(\rho A(z_j)\int_{z_j}^{z_j+\Delta z}rdr\right)-\rho A(z_i)\int_{z_i}^{z_i+u\Delta z}rdr\right]= \tag{26}$$

$$= \omega_{wt}(t)^2 \left[\underbrace{\Delta z \sum_{j=i}^{n} \rho A(z_j)(z_j + 0,5\Delta z)}_{S(z_i)} - \underbrace{\Delta z \rho A(z_i)}_{m(z_i)} \left(u z_i + 0,5u^2 \Delta z \right) \right], \; z \in [z_i, z_i + \Delta z].$$

Processing (24) gives the next system of differential equations:

$$\begin{vmatrix} \Phi_m(z_i)\ddot{q}_i(t) + \Phi_c^q(z_i)\dot{q}_i(t) + \Phi_k^q(z_i)q_i(t) = \Phi_d f_{Thrust\ i}(t) + Wx_i(t) \\ \Phi_m(z_i)\ddot{p}_i(t) + \Phi_c^p(z_i)\dot{p}_i(t) + \Phi_k^q(z_i)p_i(t) = \Phi_d f_{Torque\ i}(t) + Wy_i(t) \end{vmatrix} \quad i = 1,2,..n+1, \tag{27}$$

where: $$\Phi_m(z_i) = \underbrace{\Delta z \rho A(z_i)}_{m(z_i)} \underbrace{\frac{1}{420} \begin{bmatrix} 156 & 22\Delta z & 54 & -13\Delta z \\ 22\Delta z & 4\Delta z^2 & 13\Delta z & -3\Delta z^2 \\ 54 & 13\Delta z & 156 & -22\Delta z \\ -13\Delta z & -3\Delta z^2 & -22\Delta z & 4\Delta z^2 \end{bmatrix}}_{\bar{\Phi}_m} = m(z_i)\bar{\Phi}_m;$$

$$\Phi_k(z_i) = \underbrace{\frac{EI(z_i)}{\Delta z^3}}_{k_1(z_i)} \underbrace{2 \begin{bmatrix} 6 & 3\Delta z & -6 & 3\Delta z \\ 3\Delta x & 2\Delta z^2 & -3\Delta z & \Delta z^2 \\ -6 & -3\Delta z & 6 & -3\Delta z \\ 3\Delta z & \Delta z^2 & -3\Delta z & 2\Delta z^2 \end{bmatrix}}_{\Phi_{k1}} + \underbrace{\frac{\omega_{wt}(t)^2 S(z_i)}{\Delta z}}_{k_2(z_i)} \underbrace{\begin{bmatrix} 1,2 & 0,1\Delta z & -1,2 & 0,1\Delta z \\ 0,1\Delta z & 2\Delta z^2/15 & -0,1\Delta z & -\Delta z^2/30 \\ -1,2 & -0,1\Delta z & 1,2 & -0,1\Delta z \\ 0,1\Delta z & -\Delta z^2/30 & -0,1\Delta z & 2\Delta z^2/15 \end{bmatrix}}_{\Phi_{k2}} -$$

$$- \underbrace{\omega_{wt}(t)^2 m(z_i)}_{k_3(z_i)} \underbrace{\begin{bmatrix} \frac{6}{35} & \frac{\Delta z}{28} & -\frac{6}{35} & -\frac{\Delta z}{70} \\ \frac{\Delta z}{28} & \frac{\Delta z^2}{105} & -\frac{\Delta z}{28} & -\frac{\Delta z^2}{140} \\ -\frac{6}{35} & -\frac{\Delta z}{28} & \frac{6}{35} & \frac{\Delta z}{70} \\ -\frac{\Delta x}{70} & -\frac{\Delta z^2}{140} & \frac{\Delta z}{70} & \frac{\Delta z^2}{105} \end{bmatrix}}_{\Phi_{k3}} - \underbrace{m(z_i) z_i \omega_{wt}(t)^2}_{k_4(z_i)} \underbrace{\begin{bmatrix} \frac{3}{5\Delta z} & \frac{1}{10} & -\frac{3}{5\Delta z} & 0 \\ \frac{1}{10} & \frac{\Delta z}{30} & -\frac{1}{10} & -\frac{\Delta z}{60} \\ -\frac{3}{5\Delta z} & -\frac{1}{10} & \frac{3}{5\Delta z} & 0 \\ 0 & -\frac{\Delta z}{60} & 0 & \frac{\Delta z}{30} \end{bmatrix}}_{\Phi_{k4}};$$

$$\omega_n(z_i) = \sqrt{\frac{k_1(z_i)}{m(z_i)}} = \frac{1}{\Delta z^2} \sqrt{\frac{EI(z_i)}{\rho A(z_i)}} \ \text{s}^{-1}; \eta(z_i) = \frac{n(z_i)}{\omega_n(z_i)} \Rightarrow n(z_i) = \eta(z_i)\omega_n(z_i) = \frac{\eta(z_i)}{\Delta z^2} \sqrt{\frac{EI(z_i)}{\rho A(z_i)}} \ \text{s}^{-1};$$

$$c(z_i) = 2n(z_i)m(z_i) = 2\eta(z_i)\omega_n(z_i)m(z_i) = \frac{2}{\Delta x}\sqrt{EI(z_i)\rho A(z_i)}\eta(z_i), \ \text{Ns/m};$$

$$\Phi_c(z_i) = c(z_i)\Phi_{k_1}; \ \Phi_d = [2, \Delta z/12, 2, -\Delta z/12]^T \Delta z.$$

The corresponded matrices, for the q and p coordinates, are calculated with the corresponding elasticity coefficients: $EIy(z_i)$ and $EIx(z_i)$.

When we consider the whole blade, we should take into account that the coordinates of the 2nd node of the 1st element coincides with those of the 1st node of the 2nd of the second, the coordinates of the 2nd node of the 2nd element coincides with the those of the 1st node of the 3rd element and so on. The generalized coordinates of the whole blade are 2(n+1):

$$\begin{vmatrix} q_{bl} = \left[w(z_i,t), \Theta_y(z_i,t), \ z_i = (i-1)\Delta z, \ i=1,..n+1 \right]^T \\ p_{bl} = \left[v(z_i,t), \Theta_x(z_i,t), \ z_i = (i-1)\Delta z, \ i=1,..n+1 \right]^T \end{vmatrix}. \tag{28}$$

Considering this relation between the elements coordinates, the system of differential equations can be expressed as:

$$\left|\begin{array}{l}\Phi_m^{gl}\ddot{q}_{gl}(t)+\Phi_c^{gl(q)}\dot{q}_{gl}(t)+\Phi_k^{gl(q)}q_{gl}(t)=\Phi_d^{gl}f_{gl\ Thrust}(t)=U^{gl(q)}(t)\\ \Phi_m^{gl}\ddot{p}_{gl}(t)+\Phi_c^{gl(p)}\dot{p}_{gl}(t)+\Phi_k^{gl(p)}p_{gl}(t)=\Phi_d^{gl}f_{gl\ Torque}(t)=U^{gl(p)}(t)\end{array}\right., \tag{29}$$

where the global matrixes are formed by the locals using the next algorithm:

$$\Phi^{gl}=\begin{bmatrix}\Phi(1{:}2,:)\alpha(1),\ 0_{2x2(n-1)}\\ \left[0_{2x2(i-2)},\Phi(3:4,1:2)\alpha(i-1),\Phi(3:4,3)\alpha(i-1)+\Phi(1:2,1)\alpha(i),\Phi(3:4,4)\alpha(i-1)+\right.\\ \left.+\Phi(1:2,2)\alpha(i),\Phi(1:2,3:4)\alpha(i),0_{2x2(n-i)}\right],\ i=2,..n\\ 0_{2x2(n-1)},\ \Phi(3:4,:)\alpha(n)\end{bmatrix} \tag{30}$$

with substitutions for: $\Phi_m^{gl}\ \therefore \alpha=m,\ \Phi_c^{gl}\ \therefore \alpha=c,\ \Phi_k^{gl}\ \therefore \alpha=k$.

The generalized force matrix is: $U^{gl}=\begin{bmatrix}\Phi_d(1:2,:)f(1)\\ \left[\Phi_d(3{:}4,1)f(i\text{-}1)+\Phi_d(1{:}2,1)f(i)\right],i=2,..n\\ \Phi_d(3:4,1)f(n)\end{bmatrix}$. (31)

The clamped end boundaries are: $\left|\begin{array}{l}w(0,t)=q_{bl}(1)\equiv 0\\ \Theta_y(x_i,t)=q_{bl}(2)\equiv 0\end{array}\right.\ \left|\begin{array}{l}v(0,t)=p_{bl}(1)\equiv 0\\ \Theta_x(x_i,t)=p_{bl}(2)\equiv 0\end{array}\right.$. (32)

This leads to removal of the 1st and 2nd rows and columns of Φ_m^{gl}, Φ_c^{gl}, Φ_k^{gl} and the 1st and 2nd elements of U^{gl}.

Thus the generalized coordinate's matrix size becomes 2n: $\overline{q}_{bl}=q_{bl}\left(3:2(n+1)\right)$ (33)

and respectively the equation's matrix are: $\left|\begin{array}{l}\overline{\Phi}_m^{gl}=\Phi_m^{gl}(3:2(n+1),3:2(n+1))\\ \overline{\Phi}_c^{gl}=\Phi_c^{gl}(3:2(n+1),3:2(n+1))\\ \overline{\Phi}_k^{gl}=\Phi_k^{gl}(3:2(n+1),3:2(n+1))\\ \overline{U}^{gl}=U^{gl}(3:2(n+1),1)\end{array}\right.$. (34)

Defying the state space and the output variables vectors: $X=\left[\overline{q}_{bl}^T,\dot{\overline{q}}_{bl}^T,\overline{p}_{bl}^T,\dot{\overline{p}}_{bl}^T\right]$, $Y=\begin{bmatrix}X\\ \ddot{\overline{q}}_{bl}\\ \ddot{\overline{p}}_{bl}^T\end{bmatrix}$, (35)

the state space presentation of the system of differential equations is: $\left|\begin{array}{l}\dot{X}=AX+BU\\ Y\ =CX+DU\end{array}\right.$, (36)

where the state matrixes are: $A=\begin{bmatrix}0_{nxn}, & E_n, & 0_{nxn}, & 0_{nxn}\\ 0_{nxn}, & 0_{nxn}, & 0_{nxn}, & E_n\\ \multicolumn{4}{c}{-\left(\overline{\Phi}_m^{gl}\right)^{-1}\overline{\Phi}_k^{gl(q)},-\left(\overline{\Phi}_m^{gl}\right)^{-1}\overline{\Phi}_c^{gl(q)},0_{nxn},0_{nxn}}\\ \multicolumn{4}{c}{0_{nxn},0_{nxn},-\left(\overline{\Phi}_m^{gl}\right)^{-1}\overline{\Phi}_k^{gl(p)},-\left(\overline{\Phi}_m^{gl}\right)^{-1}\overline{\Phi}_k^{gl(p)}}\end{bmatrix}$,

$$B=\begin{bmatrix}0_{nxn}\\ \left(\overline{\Phi}_m^{gl}\right)^{-1}\\ 0_{nxn}\\ \left(\overline{\Phi}_m^{gl}\right)^{-1}\end{bmatrix},\ C=\begin{bmatrix}E_n\\ A(n+1:2n,:)\\ E_n\\ A(3n+1:4n,:)\end{bmatrix},\ D=\begin{bmatrix}0_{nxn}\\ B(n+1:n,:)\\ 0_{nxn}\\ B(3n+1:4n,:)\end{bmatrix}.$$

If the nodes displacements are denoted with: q_w=q (from 1, with step 2, to 2n-1), the shear force and torque at the base of the blade are:

$$Q_x(1)=\sum_{i=1}^{N}\left(f_{Thrust}(i)-m(i)\ddot{q}_w(i)\right),\quad M_y(1)=\sum_{i=1}^{N}\left(f_{Thrust}(i)-m(i)\ddot{q}_w(i)\right)z_i\,. \tag{37}$$

The sheared force and torque dependencies for each element can be calculated iteratively:

$$\begin{vmatrix}Q_x(i)=Q_x(i-1)-f_{Thrust}(i)\Delta z+\ddot{q}_w(i)m(i)\\ M_y(i)=M(i-1)-z(i)f_{Thrust}(i)\Delta z+z(i)\ddot{q}_w(i)m(i)\end{matrix},\quad i=2,..n\,. \tag{38}$$

The stresses are: $$\sigma_y(i)=M_y(i)h(i)/I_y(i),\ i=1,..n\,, \tag{39}$$

where $h(i)$ is the thickness of the element.

The calculation of the stress distribution by the other direction is identically.

4. Conclusions

It has been obtained a dynamical model of the loads, stresses, and deformations in wind turbine blades under the action of thrust, torque and centrifugal forces, using the Euler-Bernoulli beam model. Basing on the finite element method the model is discretized and reduced to the system of ordinary differential equations.

Acknowledgements. This work was supported by the Bulgarian Ministry of Education and Science under the National Research Program „Low carbon energy for transport and live", DCM №577/2018.

References

[1] Genov J., 2018, Some aspects of the influence of the vertical wind speed gradient on the blade aerodynamics loads in high class wind generators, *MATEC Web of Conf.* **234** (1), Article 04006, pp.1-5

[2] Genov J., Kralov I., 2018, BEM Theory Adaptation Taking into Account the Wind Speed Vertical Gradient for Wind Turbines of High Class Part 1 A Theoretical Formulation, *AIP Conference Proceedings*, **2048** (1), Article 020015, pp.1-8

[3] Genov J., Kralov I., 2018, BEM Theory Adaptation Taking into Account the Wind Speed Vertical Gradient for Wind Turbines of High Class Part 2 Numerical Analysis of the Aerodynamic Interaction, *AIP Conference Proceedings*, **2048** (1), Article 020016, pp.1-8

[4] Genov J., I. Kralov, I. Angelov, 2019, Dynamic stress analysis of a blade of wind turbine generator taking into account vertical wind speed gradient, *Journal of Environmental Protection and Ecology,* **20** (3), pp.1282–1281

[5] Anderson J., 2011, *Fundamentals of Aerodynamics Fifth Edit.*, The McGraw-Hill Inc., p.1131

[6] Bontempo R., 2014, *The Nonlinear Actuator Disk Method as Applied to Open and Ducted Rotors*, PhD Thesis, Univ. of Naples Federico II, Naples, Itally, p.153

[7] Hansen M., 2015, *Aerodynamics of Wind Turbines - Third edition*, Taylor &Francis, p.186

[8] Sørensen J. N., 2016, *General Momentum Theory for Horizontal Axis Wind Turbines*, Springer Int. Publ. AG Switzerland, pp.183

[9] Branlard E., 2017, *Wind Turbine Aerodynamics and Vorticity-Based Methods Fundamentals and Recent Applications*, Springer Int. Publ. AG Switzerland, p.634

[10] Chortis D., 2013, *Structural Analysis of Composite Wind Turbine Blades - Nonlinear Mechanics and Finite Element Models with Material Damping*, Springer Int. Publ. Switzerland, p.239

[11] Ghorashi M., 2016, *Statics and Rotational Dynamics of Composite Beams*, Springer International Publishing Switzerland, p.240

[12] Ganguli R., 2017, *Finite Element Analysis of Rotating Beams Physics Based Interpolation*, Springer Science+Business Media Singapore, p.283

[13] Sabale A., Gopal N., 2019, Nonlinear Aeroelastic Analysis of Large Wind Turbines Under Turbulent Wind Conditions, *American Inst. of Aeronautics and Astronautics Jour.*, pp.1-19

IOP Conf. Series: Materials Science and Engineering **878** (2020) 012037 doi:10.1088/1757-899X/878/1/012037

Dynamical stresses in the high class wind turbine blades caused in nonhomogeneous nonstationary wind field. Part 2 Numerical simulations

I Kralov, J Genov, I Angelov

Dept. of Mechanics, Technical University of Sofia, 8 Kl. Ohridski Blvd, Sofia 1000, Bulgaria

E-mail: j_genov@mail.bg, kralov@tu-sofia.bg, ivoangelov@gmail.com

Abstract. In this second part of the publication, numerical results for a specific a wind turbine model of a high class – prototype NREL 5MW are presented. A wind speed model is described, taking into account the vertical speed gradient and longitudinal turbulence. Based on the proposed by the authors a modification of Blade Element Momentum theory, the aerodynamic forces, distributed along the length of turbine blades, are calculated as time functions. As a result of this, according to the results of the first part of the publication, the dynamical loads, stresses, and deformations in the blades are calculated.

1. Wind speed model

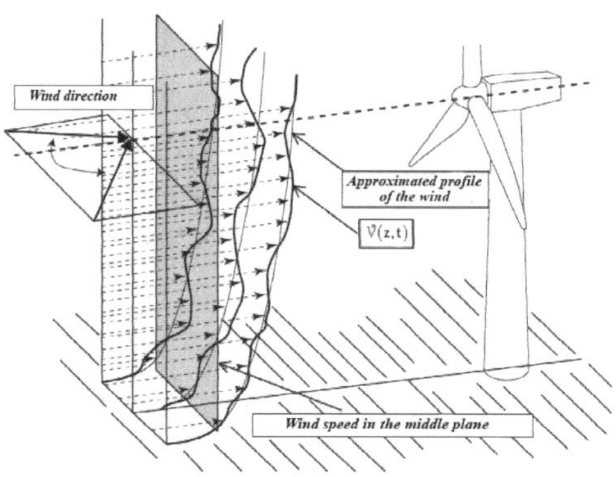

Figure 1. Wind speed profile

1.1 Vertical wind speed distribution

In the present study, the logarithmic wind speed model [1], derived from the Prandtl's mixing length theory is preferred. It takes analytically into an account, many of the physical and meteorological factors causing the vertical movement of the atmospheric masses:

$$V_\infty(z) = k^{-1}V^*\left[\ln\left((z-d)/z_0\right) - \psi\left(z, z_0, L\right)\right], \quad (1)$$

where: $\vartheta^* = \sqrt{\tau_s / \rho}$, m/s – shear (friction) velocity of the air flow;

τ_s, Pa – shear stress of flow; ρ, kg/m^3 – mass density;

k – von Kármán constant for the logarithmic distribution $k \in [0,35 \div 0,42]$ [2];

d – zero-plan displacement, define the height at which the mean velocity is zero due to large obstacles;

z_0 – aerodynamic roughness length is a corrective measure to account the effect of the surface roughness on wind flow;

$\psi(z, z_0, L)$ - integrated form of the Monin-Obukhov's universal function [3].

The values of the parameters of relation (1) are given in Table 1 [4].

Table 1. Values of the parameters for the calculation of logarithmic wind speed distribution

L	Φ_H	ρ	T_0	c_p	T_v	k	h	ϑ^*	τ_s	d	z_0
m	K m/s	kg/m^3	K	J/(kgK)	K		W/(m^{2}K)	m/s	Pa	m	m
59,91	-0,0275	1,225	298	1005	290	0,4	33,09	0,282	0,0975	0,138	0,03

The wind speed profiles, in the diapason from the minimal - 3 m/s to the nominal wind velocity - 11,4 m/s, for the turbine type NREL 5MW [5], are shown in figure 2.

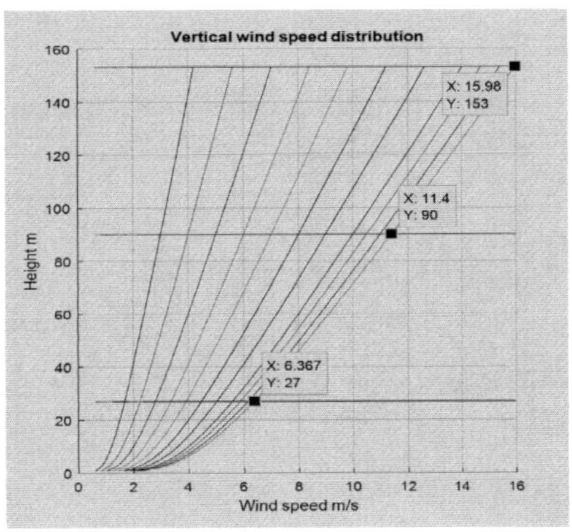

Figure 2. Vertical distribution of the wind speed for NREL 5MW generator in working range: $\vartheta_\infty(z_{hub}) \in [3, 4, 5, 6, 7, 8, 9, 10, 10.5, 11, 11.4]$ m/s

1.2 Longitudinal wind turbulence

The Kaimal's model for the power spectral density (PSD) of the longitudinal wind speed is used. According to this, at the hub height, the PSD has the following form [6]:

$$S_u(f) = \sigma_u^2 \frac{4L_u / \vartheta_{hub}^{av}}{\left(1 + 6fL_u / \vartheta_{hub}^{av}\right)^{5/3}}, \tag{2}$$

where: $\sigma_u = \sigma_{ave} + 1,28\sigma_\sigma$ is the standard deviation of12 the longitudinal fluctuating of the wind speed;

$\sigma_{ave} = I_{ref}\left(a\,\vartheta_{hub}^{av} + b\right)$ - the mean value;

$\sigma_{\sigma} = I_{ref}\left(\alpha\, V_{hub}^{av} + \beta\right)$ - the standard deviation;

I_{ref} - the reference turbulence intensity at 15 m/s;

V_{hub}^{av} - the average wind velocity at the hub height;

a, b, α, β - parameters of the spectral function;

$L_u = 8,1\Lambda_u$ - the longitudinal integral coherence scale parameter,

$\Lambda_u = \begin{cases} 0,7 z_{hub},\ z_{hub} \leq 60\ m \\ 40\ m,\ \ z_{hub} \leq 60\ m \end{cases}$ - longitudinal turbulence scale parameter;

f - frequency in Hz.

The values of the Kaimal's model for the spectral density of the longitudinal wind speed at the hub height are given in table 2.

Table 2. Parameters of the Kaimal's model for spectral density of longitudinal wind speed [7-8]

I_{ref}	a	b	α	β	Λ_u	f_{max}	Z_{hub}	V_{hub}^{av}
-	-	-	-	-	(m)	(Hz)	(m)	(m/s)
0.16	0.75	3.8	0.27	2.75	40	13.65	90	11.4

The obtained wind speed as time function for $V_{hub}^{av} = 11.4$ m/s is shown in Fig. 3.

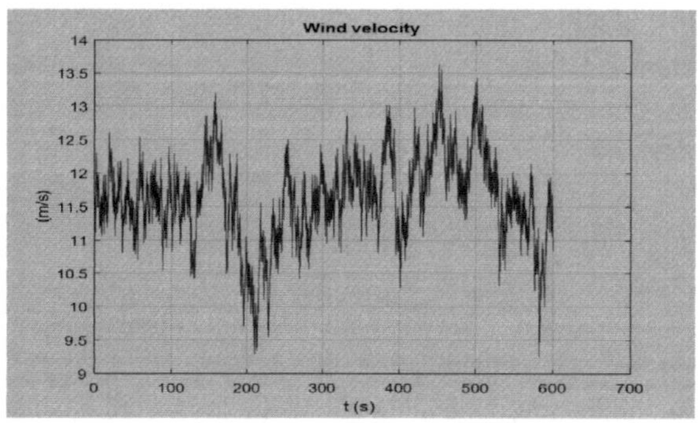

Figure 3. Wind speed as time function for V_{hub}^{av} = 11,4 m/s

2. Aerodynamically loads

2.1 BEM theory adaptation

In [9-12] the authors developed the Blade Element Momentum theory for the cases of turbines of high class, where the aerodynamic interaction is in relation to the azimuth angle. This causes considerable variations of the aerodynamic forces acting on the blades during the turbine rotation and leads to an appearance of cyclic loads, which is a precondition for fatigue arises. In short, the results are:

The thrust and torque forces acting on the blade elements are:

$$\begin{vmatrix} \Delta F_{Thrust}(\varphi_{wt}, r) = 0,5\rho\ell_T(\varphi_{wt}, r)\left(1 - a(\varphi_{wt}, r)\right)^2 K_{Thrust}(\varphi_{wt}, r)\Delta r \\ \Delta F_{Torque}(\varphi_{wt}, r) = 0,5\rho\ell_T(\varphi_{wt}, r)\left(1 - a(\varphi_{wt}, r)\right)^2 K_{Torque}(\varphi_{wt}, r)\Delta r \end{vmatrix} \quad (3)$$

where: $K_{Thrust}(\varphi_{wt}, r) = \sum_{i=1}^{N}\left\{\left[C_L\left(\alpha(\varphi_i, r)\right) cotg\left(\phi(\varphi_i, r)\right) + C_D\left(\alpha(\varphi_i, r)\right)\right]\Psi(\varphi_i, r)\right\}$;

$$K_{Torque}(\varphi_{wt},r)=\sum_{i=1}^{N}\{[C_L(\alpha(\varphi_i,r))-cotg(\phi(\varphi_i,r))C_D(\alpha(\varphi_i,r))]\Psi(\varphi_i,r)\}$$

$\Psi(\varphi_i,r)=V_{\infty}^{2}(\varphi_i,r)F_c(\varphi_i,r)/\sin(\phi(\varphi_i,r)),\ \ \varphi_i=\varphi_{wt}+(i\text{-}1)2\pi/N,\ i=1,..N$

$a(\varphi_{wt},r)=K(\varphi_{wt},r)/(1+K(\varphi_{wt},r))$ - axial induction coefficient for the section;

$a'(\varphi_{wt},r)=\dfrac{1}{\tilde{\lambda}(r)}\dfrac{K_{Torque}(\varphi_{wt},r)}{K_{Thrust}(\varphi_{wt},r)}a(\varphi_{wt},r)$ - tangential induction coefficient for the section;

$$K(\varphi_{wt},r)=0,25r^{-1}\ell_T(\varphi_{wt},r)K_{Thrust}(\varphi_{wt},r)/\int_{\gamma=0}^{2\pi}V_{\infty}^{2}(r,\gamma)d\gamma;$$

$$\tilde{\lambda}(r)=r\omega_{wt}\int_{\gamma=0}^{2\pi}V_{\infty}(r,\gamma)d\gamma/\int_{\gamma=0}^{2\pi}V_{\infty}^{2}(r,\gamma)d\gamma;$$

$V_b(\varphi_{wt},r)$ - a tangential component of the relative flow velocity for the section;

$V_d(\varphi_{wt},r)$ - stream velocity at the position of the turbine disk for the section;

$V_{\infty}(\varphi_i,r)$ - free stream velocity for the section;

$V_{rel}(\varphi_{wt},r)$ - relative (effective) flow velocity for the section;

$C_L(\alpha(\varphi_i,r)),\ C_D(\alpha(\varphi_i,r))$ - lift and drag coefficients;

$\alpha(\varphi_i,r)$ - angle of an attack, $\phi(\varphi_{wt},r)=arctg(V_d(\varphi_{wt},r)/V_b(\varphi_{wt},r))$ - inflow angle;

$\beta(\varphi_{wt},r)=\beta_0(r)+\beta_c(\varphi_{wt})$ – pitch angle for the section; - constructive pitch angle for the section;

$\beta_c(\varphi_{wt})$ - blade pitch angle set by the system for active pitch control (if there is such);

$\varphi_{wt}(t)$ - azimuth angle of the blade, $\omega_{wt}=\dot{\varphi}_{wt}$ - angular velocity of the turbine;

N – number of the turbine blades;

$z(\varphi_{wt},r)=z_{hub}+r\cos(\varphi_{wt})$ - height of the discussed section with respect to the ground surface;

$\ell_T(\varphi_{wt},r)=\ell(r)\cos(\beta(\varphi_{wt},r))$, $\ell(r)$ - section chord; r – distance to the turbine axes.

The parameters of the aerodynamic interaction are shown in figure 4

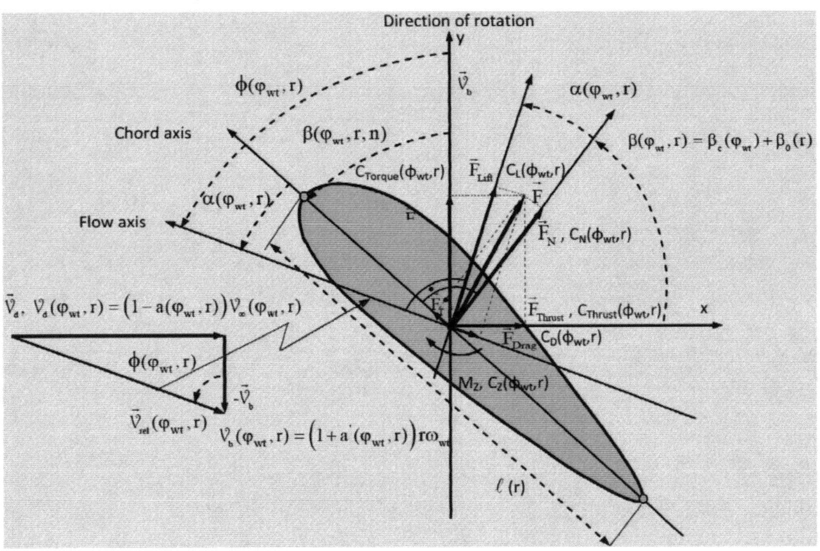

Figure 4. Parameters of the aerodynamic interaction

2.2 Numerical simulations

Table 3. Characteristics of the sections building the turbine blades

Section profile	Section beginning r, (m)	Constructive pitch β_0, (°)	Chord length ℓ, (m)	b/ℓ	Re10^{-6}
Hub	0.000	-	-	-	-
Cylinder	1.500	13.308	3.542	1.000	2.44
Ellipsoid 1	3.400	13.308	3.710	0.930	3.12
Ellipsoid 2	5.544	13.308	3.854	0.780	4.40
Ellipsoid 3	8.631	13.308	4.167	0.610	5.45
DU40-A17	11.750	13.310	4.557	0.477	6.31
DU35-A17	15.850	11.480	4.652	0.383	7.11
	19.950	10.165	4.458	0.331	7.80
DU30-A17	24.066	9.011	4.249	0.290	8.35
DU25-A17	28.150	7.795	4.007	0.260	8.80
	32.250	6.544	3.748	0.238	9.20
DU21-A17	36.351	5.361	3.502	0.219	9.40
	40.450	4.188	3.256	0.202	9.53
NACA 64-618	44.550	3.125	3.010	0.180	9.51
	48.650	2.319	2.764	0.180	9.42
	52.750	1.526	2.518	0.180	9.13
	56.167	0.863	2.313	0.180	8.78
	58.905	0.370	2.086	0.180	8.47
	61.633	0.106	1.419	0.180	8.10
	63.000	0.000	0.779	0.180	7.88

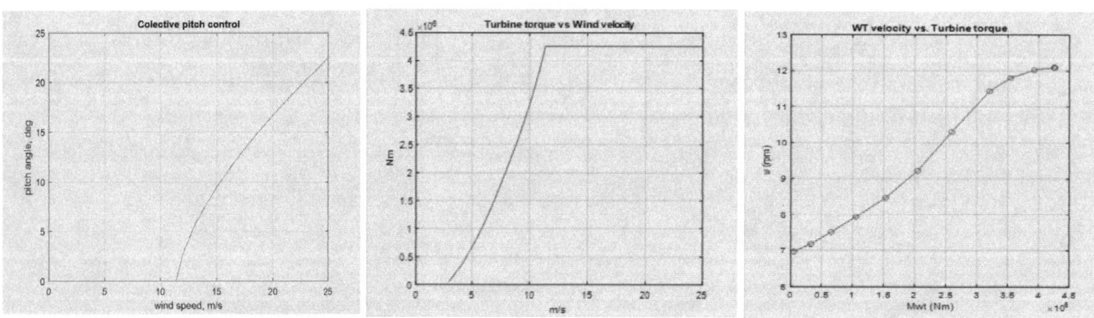

Figure 5. a) Pitch angle vs wind velocity b) Torque vs wind velocity c) Angular velocity vs torque

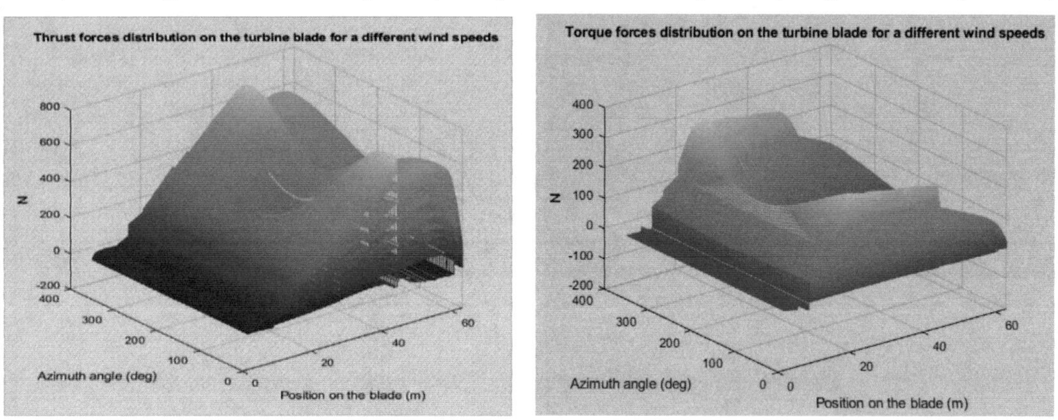

Figure 6. Distribution along the blade: a) thrust force, b) torque force for different wind speeds at the level of the hub axis V_{Hub}=[3, 4,..11, 11.4, 12, 13…25] m/s

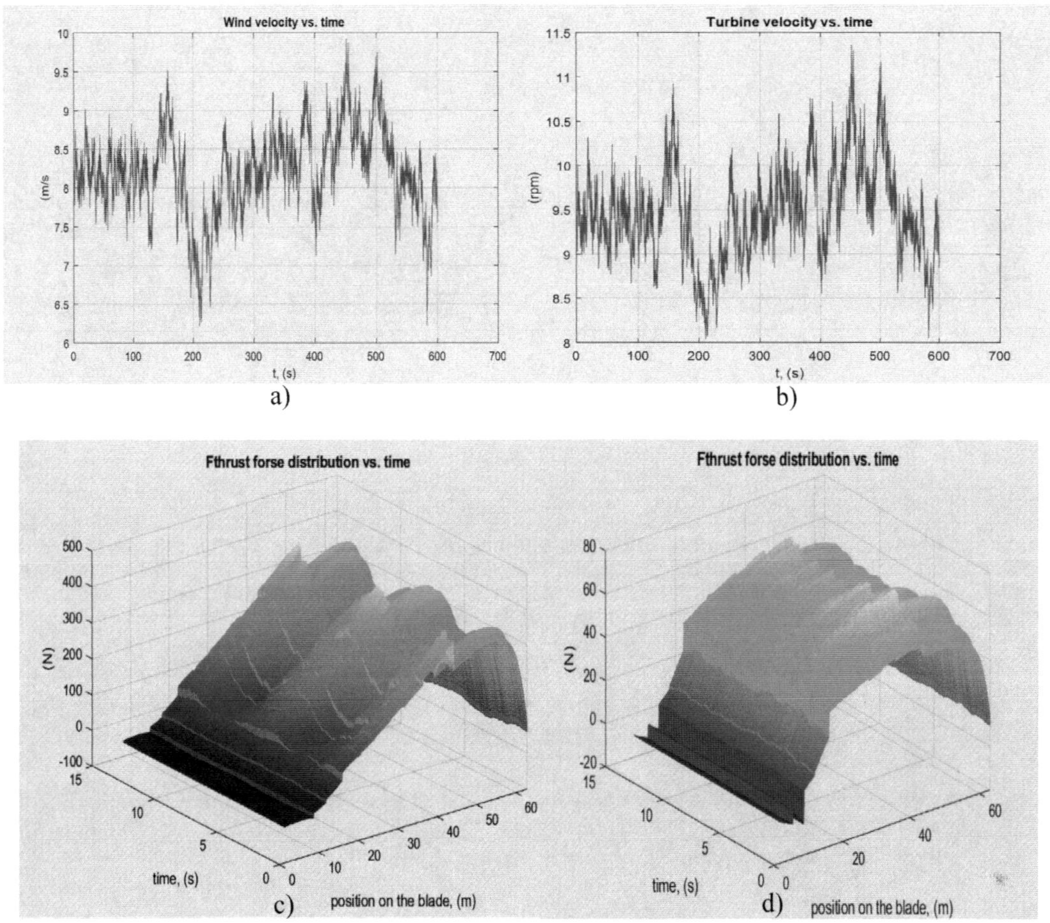

Figure 7. Time responses of: a) wind speed b) turbine velocity c) thrust and d) torque forces distributions for an average wind velocity at the hub axis level $\mathcal{V}_{hub} = 8\ m/s$

Table 4. Mechanical characteristics of the blade of NREL 5MW wind turbine [14]

r (m)	z (m)	ρA (kg/m)	$E_yJ.10^{-10}$ (Nm^2)	$E_xJ.10^{-10}$ (Nm^2)	r (m)	z (m)	ρA (kg/m)	$E_yJ.10^{-10}$ (Nm^2)	$E_xJ.10^{-10}$ (Nm^2)	r (m)	z (m)	ρA (kg/m)	$E_yJ.10^{-10}$ (Nm^2)	$E_xJ.10^{-10}$ (Nm^2)
1.5	0	678.935	1.81100	1.81136	16.7	15.2	349.477	0.238865	0.494849	49.7	48.2	129.555	0.007631	0.070961
1.7	0.2	678.935	1.81100	1.81136	17.7	16.2	346.538	0.227199	0.480802	51.7	50.2	107.264	0.006105	0.051819
2.7	1.2	773.363	1.94249	1.95586	19.7	18.2	339.333	0.205005	0.450140	53.7	52.2	98.776	0.004948	0.045487
3.7	2.2	740.55	1.74559	1.94978	21.7	20.2	330.004	0.182825	0.424407	55.7	54.2	90.248	0.003936	0.039512
4.7	3.2	740.042	1.52874	1.97888	23.7	22.2	321.99	0.158871	0.399528	56.7	55.2	83.001	0.003467	0.035372
5.7	4.2	592.496	1.07824	1.48585	25.7	24.2	313.82	0.136193	0.375076	57.7	56.2	72.906	0.003041	0.030473
6.7	5.2	450.275	0.722972	1.02206	27.7	26.2	294.734	0.110238	0.344714	58.7	57.2	68.772	0.002652	0.0281420
7.7	6.2	424.054	0.630954	0.91447	29.7	28.2	287.12	0.087580	0.313907	59.2	57.7	66.264	0.002384	0.026171
8.7	7.2	400.638	0.552836	0.806317	31.7	30.2	263.343	0.068130	0.273424	59.7	58.2	59.340	0.001963	0.015881
9.7	8.2	382.062	0.498006	0.688444	33.7	32.2	253.207	0.053472	0.255487	60.2	58.7	55.914	0.001600	0.013788
10.7	9.2	399.655	0.493684	0.700918	35.7	34.2	241.666	0.040890	0.233403	60.7	59.2	52.484	0.001283	0.011879
11.7	10.2	426.321	0.469166	0.716768	37.7	36.2	220.638	0.031454	0.182873	61.2	59.7	49.114	0.001008	0.010163
12.7	11.2	416.82	0.394946	0.727166	39.7	38.2	200.293	0.023863	0.158410	61.7	60.2	45.818	0.000755	0.008507
13.7	12.2	406.186	0.338652	0.708170E	41.7	40.2	179.404	0.017588	0.132336	62.2	60.7	41.669	0.000460	0.006426
14.7	13.2	381.42	0.293374	0.7624453	43.7	42.2	165.094	0.012601	0.118368	62.7	61.2	11.453	0.000026	0.000661
15.7	14.2	352.822	0.256896	0.5504896	45.7	44.2	154.411	0.010726	0.102016	63.0	61.5	10.319	0.000017	0.000501
					47.7	46.2	138.935	0.00909	0.079781					

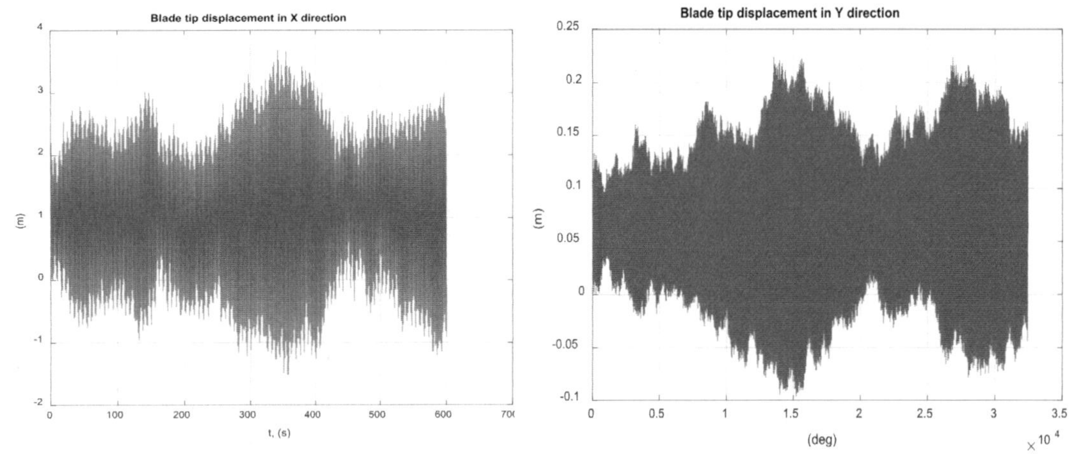

a) b)

Figure 8. Displacement of the blade tip in a) flapwise and b) edgewise directions

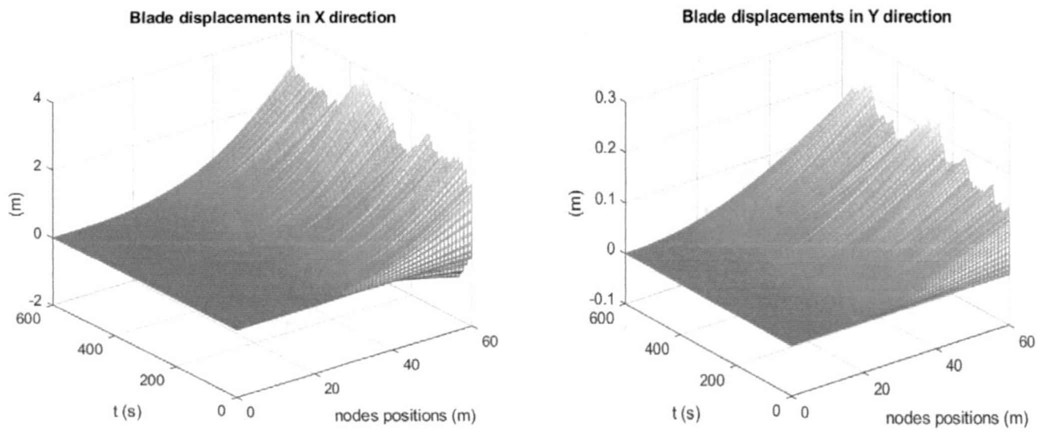

a) b)

Figure 9. Blade displacement distribution in a) flapwise and b) edgewise directions

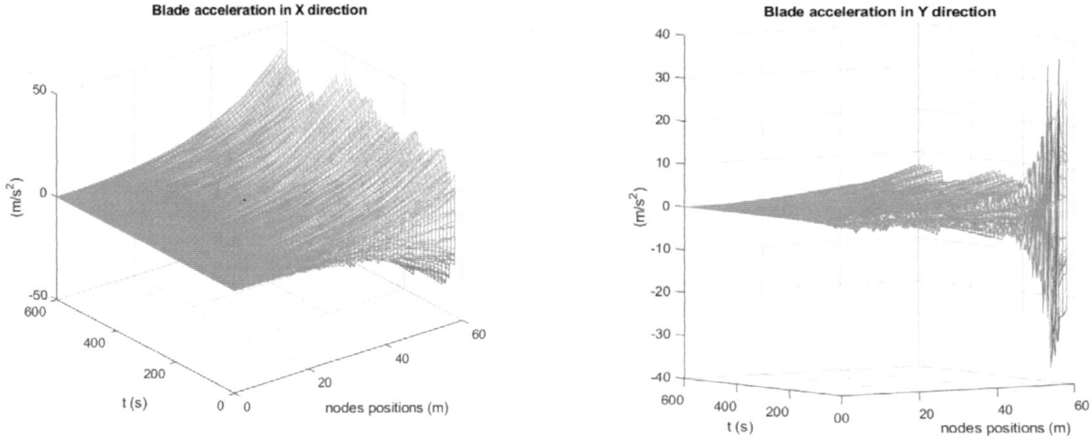

Figure 10. Blade`s acceleration in a) flapwise and b) edgewise directions

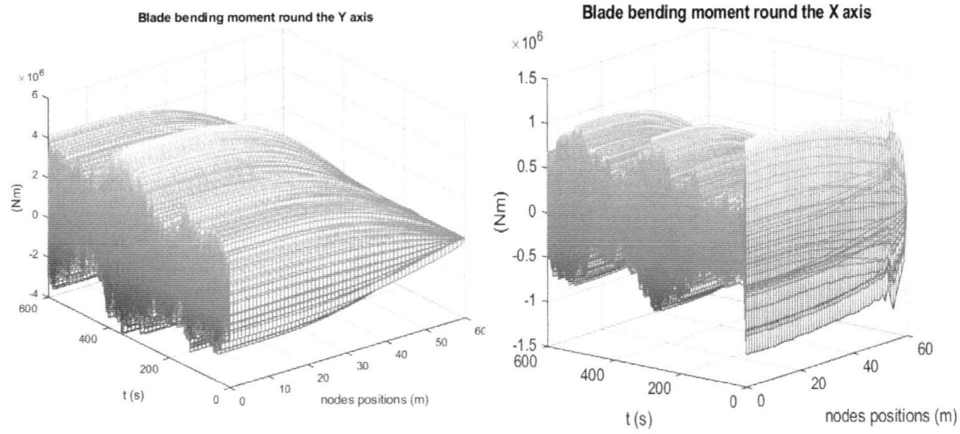

a) b)

Figure 11. Bending moment distribution in a) flapwise and b) edgewise directions

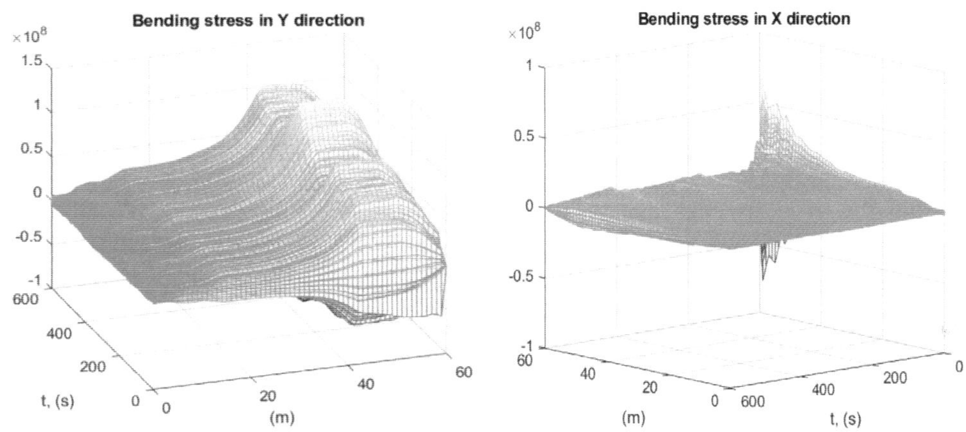

a) b)

Figure 12. Bending stresses in a) flapwise and b) edgewise directions

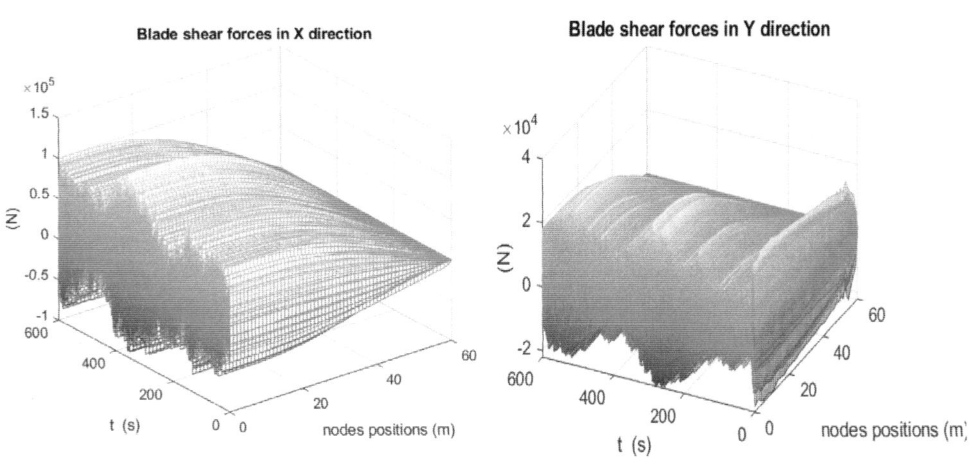

a) b)

Figure 13. Shear force distribution in a) flapwise and b) edgewise directions

Conclusions

Applying the Euler-Bernoulli beam theory, numerical values of dynamic loads, stresses, and deformations in wind turbine blades, under the action of thrust torque and centrifugal forces are obtained. Basing on the finite element method the model is discretized and reduced to the system of ordinary differential equations. Solving the derived numerical model, the influence of the consideration of the vertical wind speed gradient and longitudinal turbulence was evaluated. In the case of high class wind turbine, the impact of the described wind model was proven significant. It leads to appearance of cyclic loads and fatigue.

Acknowledgements

This work was supported by the Bulgarian Ministry of Education and Science under the National Research Program „Low carbon energy for transport and live", approved by DCM №577/17.08.2018.

References

[1] Landberg L., 2016, *Meteorology for Wind Energy - An Introduction*, John Wiley & Sons Ltd., p.215

[2] Zhang Y., Ma J., and Cao Z., 2008, The Von Karman constant retrieved from CASES-97 dataset using a variational method, *Atmospheric Chemistry and Physics* **8**, pp.7045–7053,

[3] A.S. Monin, A.M. Obukhov, 1954, Main characteristics of the turbulent mixing in the atmospheric surface layer, *Trudy Geophys. Inst. AN SSSR*, **24(151)**, pp.153–187

[4] Genov J., Kralov I., 2018, BEM Theory Adaptation Taking into Account the Wind Speed Vertical Gradient for Wind Turbines of High Class Part 1 A Theoretical Formulation, *AIP Conference Proceedings*, **2048** (1), Article 020015, pp.1-8

[5] Jonkman J., Butterfield S., Musial W., Scott G., 2009, *Definition of a 5-MW Reference Wind Turbine for Offshore System Development*, Technical Report NREL/TP-500-38060, p.75

[6] Kaimal J., at al., 1972, Spectral Characteristics of Surface-Layer Turbulence, *Q. J. Royal Meteorol. Soc.*, **98**, pp.563-589

[7] International Electrotechnical Commission: IEC 61400-1, 2010, *Wind Turbines – Part 1: Design Requirements*, p.92

[8] Mücke T., Harkness C., Argyriadis K., 2012, Offshore Wind Turbulence Model versus Measurement, *In: Proceedings of EWEA 2012 Copenhagen 16-19 April*,.pp.1-10

[9] Genov J., 2018, Some aspects of the influence of the vertical wind speed gradient on the blade aerodynamics loads in high class wind generators, MATEC Web of Conferences **234** (1), Article 04006, pp.1-5

[10] Genov J., Kralov I., 2018, BEM Theory Adaptation Taking into Account the Wind Speed Vertical Gradient for Wind Turbines of High Class Part 2 Numerical Analysis of the Aerodynamic Interaction, AIP Conference Proceedings, **2048** (1), Article 020016, pp.1-8

[11] Kralov I., Genov J., I. Angelov, 2019, Modified bem theory application for determining the aerodynamic forces acting on the blade of wind turbine, *Journal of Environmental Protection and Ecology,* **20** (3), pp. 1255–1269

[12] Genov J., I. Kralov, I. Angelov, 2019, Bem theory adaptation taking into account non-uniform, non-stationary wind field, *Journal of Environmental Protection and Ecology,* **20** (3), pp. 1270–1281

[13] Genov J., I. Kralov, I. Angelov, 2019, Dynamic stress analysis of a blade of wind turbine generator taking into account vertical wind speed gradient, *Journal of Environmental Protection and Ecology,* **20** (3), pp. 1282–1281

[14] Jonkman J., 2007, *Dynamics Modeling and Loads Analysis of an Offshore Floating Wind Turbine*, Technical Report NREL/TP-500-41958, NREL U.S. Department of Energy, p.232

[15] Valkov G. and Nikolov V., 2020, Analysis of stresses and deformations in the chassis of terrain forklifts *9th Int. Sc. Conf. TECHSYS 2020 14-16 May 2020 Plovdiv Bulgaria, IOP Conf. Ser.: Mat. Sc. and Eng.* IOP Publ. (in print)

Analysis of stresses and deformations in the chassis of rough terrain forklifts

Georgi Valkov[1] and Valyo Nikolov[2]

[1] Balkancar Record JSC, Department of Design and Technology
17 Kuklensko Shose St, 4004, Plovdiv, Bulgaria
[2] Technical University of Sofia, Plovdiv Branch
Department of Transport and Aviation Equipment and Technologies
25 Tsanko Dyustabanov St, 4000 Plovdiv, Bulgaria

E-mail: : techno@balkancar-record.com[1], vnikolov@tu-plovdiv.bg [2]

Abstract. The outcome of studying the strength and the deformation characteristics of rough terrain forklifts, using a created 3D model applying FEM, is presented herein. The tested chassis design features are described. The external loads of the chassis structure have been estimated throughout two operation modes: handling the load at vertically elevated lifting mast and also featuring load lifted off the ground, using lifting mast in transport mode tilted backwards. Two alternating payloads have been applied for each of the selected modes: featuring reated load and standard size of the centre of gravity, and featuring reduced load at increased height of the centre of gravity. The resulting rates of stresses and deformations of the studied chassis of a rough terrain forklift have been calculated, presented and analysed with regard to the two main operating modes both of which entail two loading alternatives determined by the weight and the location of the payload.

Key Words. rough *terrain forklifts, chassis, stresses, deformations, estimated strength, FEM.*

1. Object and purpose of the study

The manufacture of rough terrain forklifts and the relevant dedicated work equipment has shown a continuous trend of increasing volume in recent years. The reason thereof is the constantly increasing need for mechanising loading and unloading operations and some other types of work on particular varieties of terrain and also an improvement in productivity in the construction business, forestry and agriculture, etc. The fierce competition between the manufacturers of this equipment makes the point of increasing its functionality, strength, and operational reliability increasingly topical [6, 11]. This is very much the case for the chassis design that constitutes a primary assembly of any rough terrain forklift [1, 3, 17, 19].

In view of the fact that the chassis incorporates high degree of metal, weighing from 17% to 20% of the forklift total weight, its optimal design could entail decrease not only in the production costs but also in the operating costs [1, 3, 6, 10, 13].

Given the complex shape of the design components and assemblies, which is characteristic of rough terrain forklift chassis, there is not always an exact analytical solution for estimating stresses and deformations.

Numerical methods, whereof the finite element method (FEM) has become the most widespread, could successfully solve the problem irrespective of the shape and the way of loading and fixing the body [2, 9, 12, 15, 18]. It makes FEM a very appropriate method of studying the phenomenon of stress concentration that is noticeable in abrupt and complex changes to the shape of the component or the assembly, the chassis of rough terrain forklifts being such. The resulting maximum stresses should not exceed the ones that are unsafe for the material.

Regarding resilient and tough materials, the yield limit shall be considered a dangerous stress – R_{eH} ($R_{p0.2}$), and the tensile strength and the compression strength – R_m and $R_{\bar{m}}$ – shall be considered unsafe stresses in respect of fragile materials. The strength characteristics of materials are to be specified experimentally [6, 14]. Regarding the most often used machine-building materials, the reference books provide the values thereof [7].

The object of the study is a chassis of rough terrain forklifts featuring dual wheel drives; it is currently manufactured by the Balkancar Record JSC company [19].

The purpose of the study is to determine the stressed and deformed state of the structure of a rough terrain forklift chassis under two typical load conditions. When developing the model, the chassis has been subjected to the loads resulting from the forces of gravity of the main assemblies and units related thereto, such as engine, tanks, box, lifting mast, counterweight, etc.

The main purpose of the study is to identify the critical points in the chassis design based on the results of the strength and deformation analysis.

An FE model of the chassis and the stresses and the deformations thereto have been made using the SOLIDWORKS Simulation Xpress module within SOLIDWORKS 2019 [9].

2. Load conditions of the chassis of a rough terrain forklifts

The chassis, selected for the study, is used in the R2SR forklifts series, manufactured by Balkancar Record JSC, with wheel drive formula 4X2 and lifting capacities of 30, 40, and 50 kN. [19].

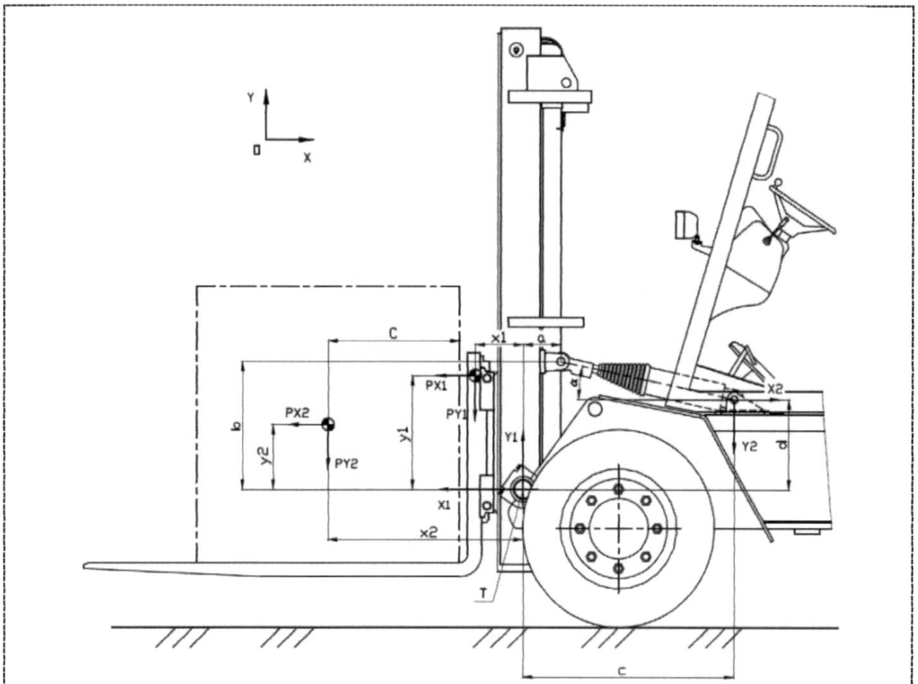

Figure 1. Mode one: Loading by taking the load at vertically elevated lifting mast

The chassis has welded steel structure made of sheet material. The two side plates, left and right, constitute the main carrying elements of this chassis. They are flexed in the form of a Π-shaped section. Since they are linked to the wings and the transverse shield at the front, and to the plates whereto the weight is fixed, at the back, the result is a box-like enclosed form. The side plates thickness is 10 mm, and the mostly used material for these is steel ST355JR. Based on the experience acquired in the perennial company business, the calculations have been made in respect of the two most typical load conditions in the use of these machines [6].

2.1. ***Mode one*** *– taking the load at vertically elevated lifting mast*

2.2. ***Mode two*** *– load lifted 300 mm off the ground, the lifting mast being tilted backwards in transport position*

Calculations involving two positions of the centre of gravity of the load have been made, with regard to each of both load conditions, where C = 600 mm and C = 900 mm concerning the heaviest model representing the series, DV1798, which features lifting capacity of 50 kN [19].

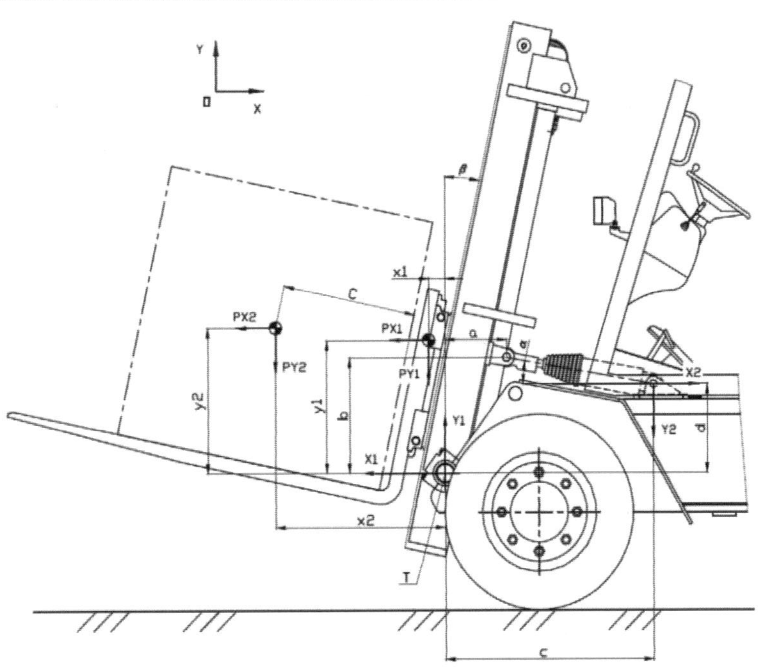

Figure 2. Mode two: Load lifted 300 mm off the ground, the lifting mast being tilted backwards in transport position

It is known from the theory and the design of forklifts [1, 3, 6] that when moving the centre of gravity forward, in respect of the longitudinal axis of the forklift, the reated load should be reduced in order to preserve the machine stability against overturning while handling any load [4, 5, 8].

Therefore, the second purpose of the study has been set – to identify and analyse the stressed state of the chassis structure and its deformed state at various positions of the centre of gravity and using various rates of the payload, which ensure the longitudinal stability of the terrain forklift against overturning [4, 5].

3. Stresses and deformations in the chassis of a rough terrain forklift featuring dual wheel drive

3.1. Stresses and deformations of the chassis under the first load conditions featuring a standard centre of gravity at C = 600 mm and a rated load of Q = 50 kN.

The results of testing the strength under the first load conditions, featuring a standard centre of gravity of the load at C = 600 mm and a rated load of Q = 50 kN, are shown in Figure 3 and Figure 4.

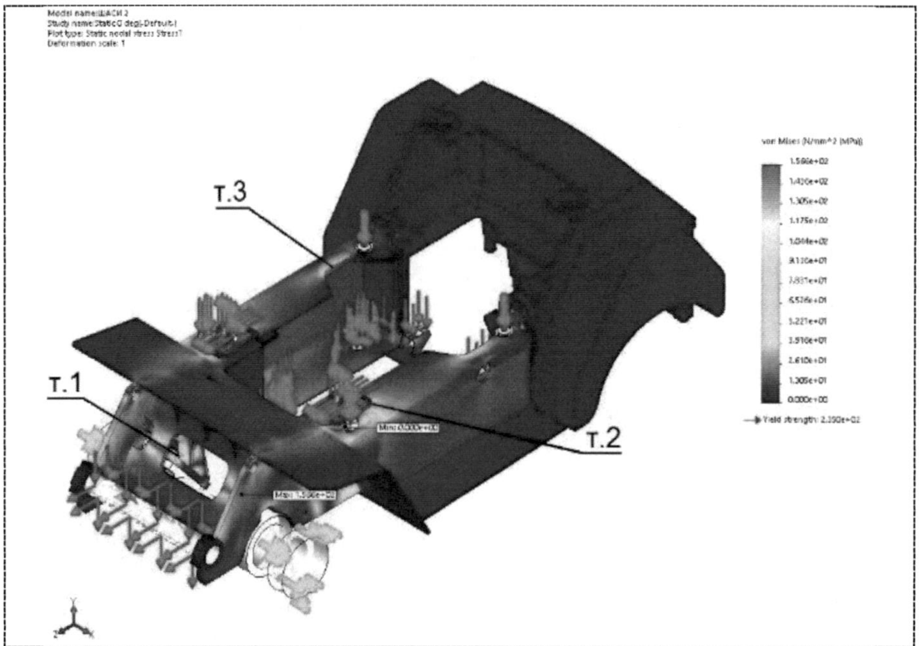

Figure 3. Stresses in the chassis under the first load conditions featuring a standard centre of gravity at C = 600 mm and a rated load of Q = 50 kN

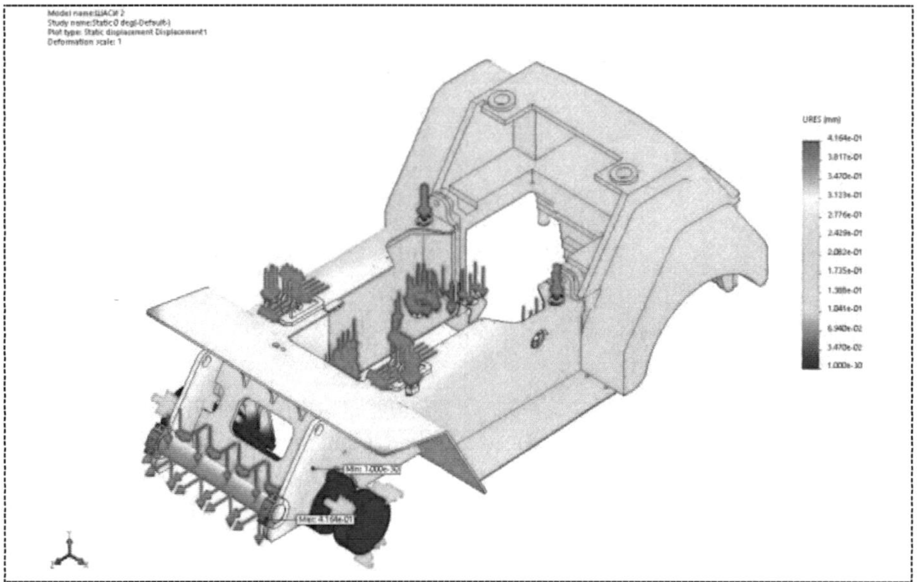

Figure 4. Deformations of the chassis under the first load conditions featuring a standard centre of gravity at C = 600 mm and a rated load of Q = 50 kN

3.2. Stresses and deformations of the chassis under the second load conditions featuring a standard centre of gravity at C = 600 mm and a rated load of Q = 50 kN.

The results of testing the strength under the second load conditions, featuring a standard centre of gravity of the load at C = 600 mm and a rated load of Q = 50 kN, are shown in Figure 5 and Figure 6.

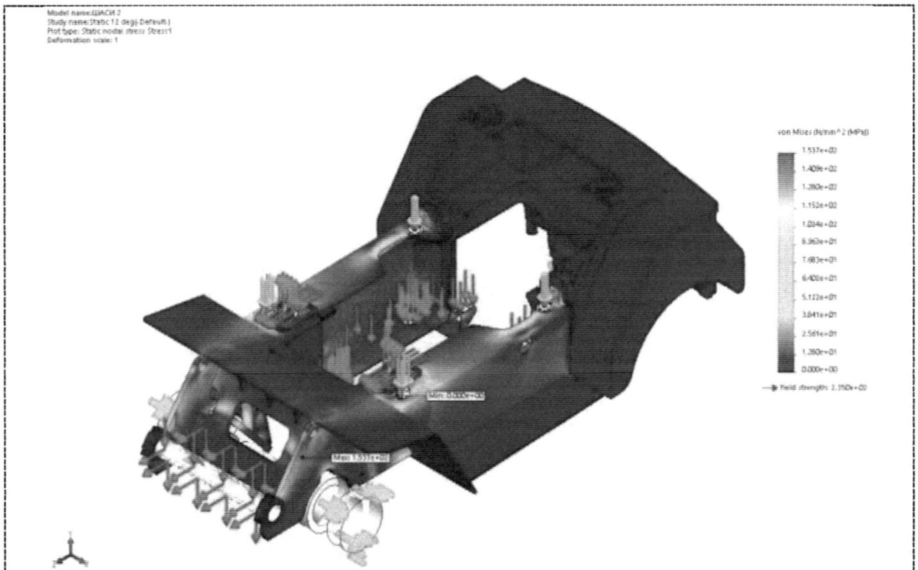

Figure 5. Stresses in the chassis under the second load conditions featuring a standard centre of gravity at C = 600 mm and a rated load of Q = 50 kN

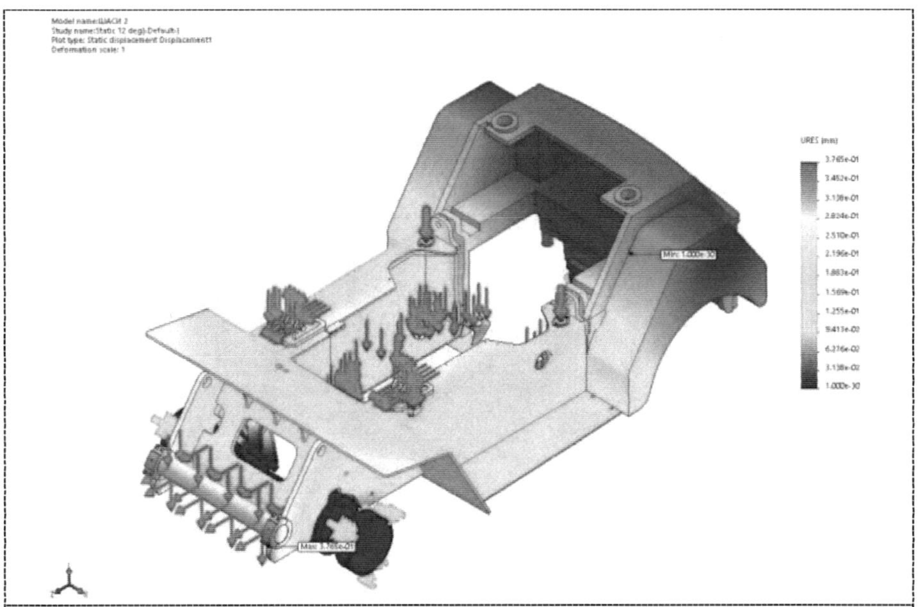

Figure 6. Deformations in the chassis under the second load conditions featuring a standard centre of gravity at C = 600 mm and a rated load of Q = 50 kN

3.3. Stresses and deformations in the chassis under the first load conditions featuring an increased centre of gravity of the load at C = 900 mm and reduced load of Q = 35 kN

The results of testing the strength under the first load conditions and featuring increased centre of gravity at C = 900 mm and reduced load of Q = 35 kN, are shown in Figure 7 and Figure 8.

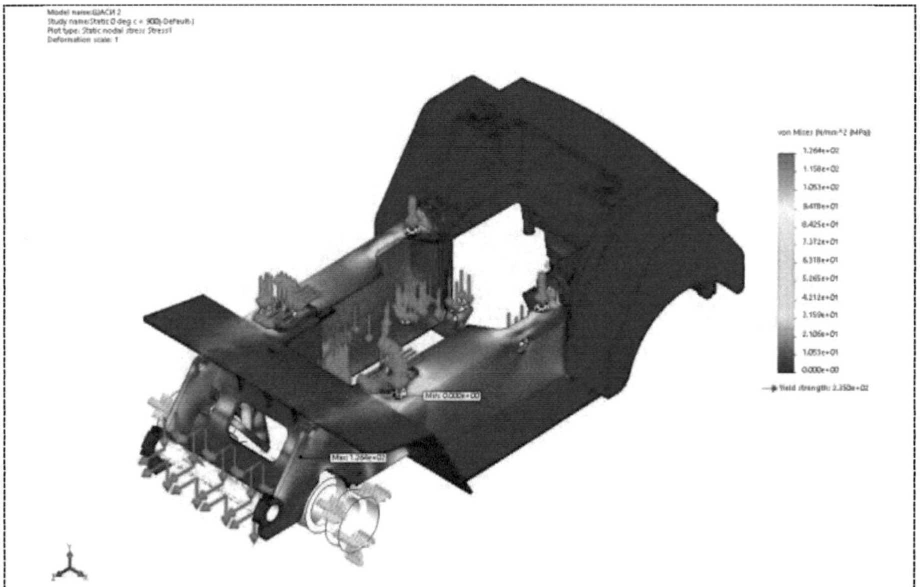

Figure 7. Stresses in the chassis under the first load conditions featuring an increased centre of gravity at C = 900 mm and reduced load of Q = 35 kN

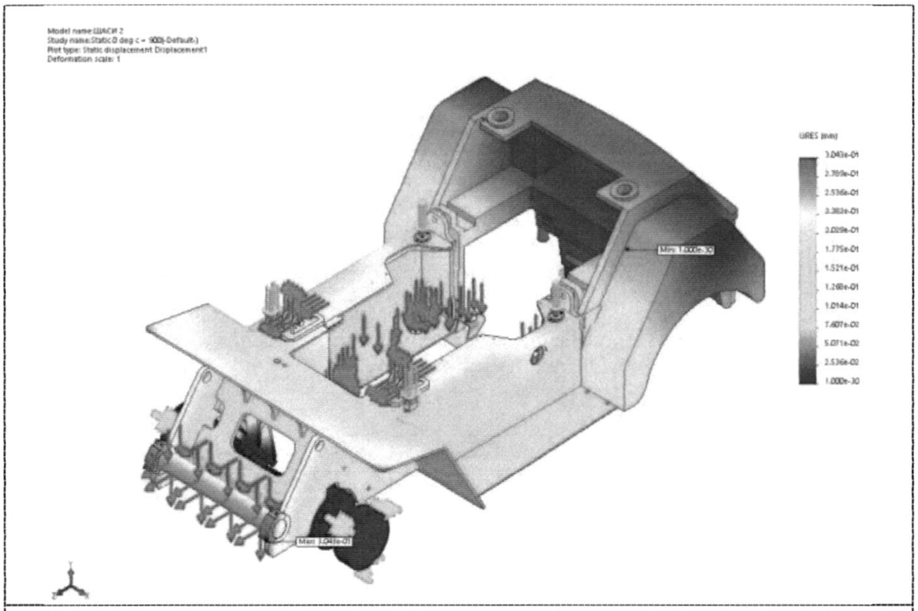

Figure 8. Deformations of the chassis under the first load conditions featuring an increased centre of gravity at C = 900 mm and reduced load of Q = 35 kN

IOP Conf. Series: Materials Science and Engineering **878** (2020) 012038 doi:10.1088/1757-899X/878/1/012038

3.4. Stresses and deformations of the chassis under *the second load conditions featuring an increased centre of gravity at C = 900 mm and reduced load of Q = 35 kN*

The results of testing the strength under the second load conditions and an increased centre of gravity at C = 900 mm and reduced load of Q = 35 kN, are shown in Figure 9 and Figure 10.

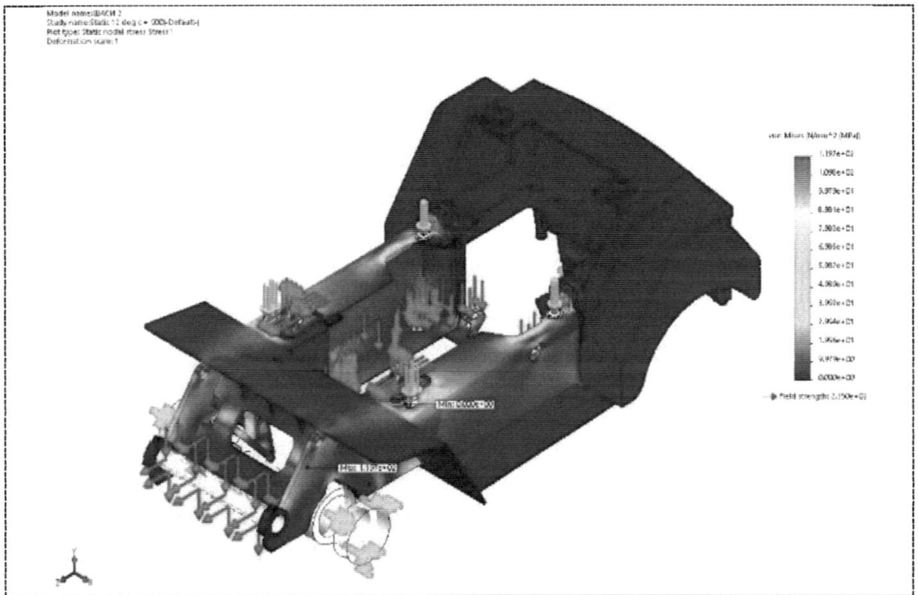

Figure 9. Stresses in the chassis under the second load conditions featuring an increased centre of gravity at C = 900 mm and reduced load of Q = 35 kN

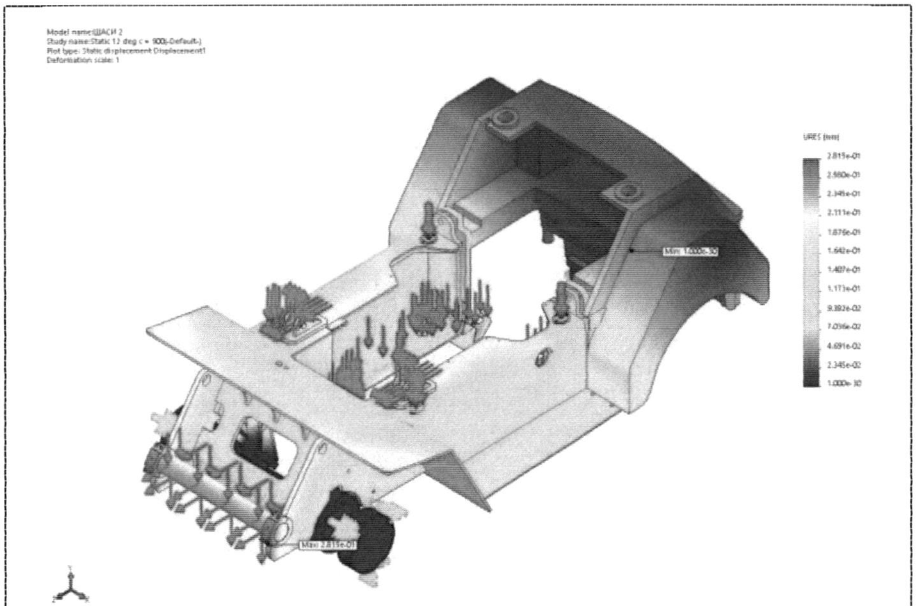

Figure 10. Deformations of the chassis under the second load conditions featuring an increased centre of gravity at C = 900 mm and reduced load of Q = 35 kN

3.5. Results analysis

Based on the obtained results, the following conclusions could be drawn:

1) The critical points in the structure, where the highest stresses under both load conditions have been identified, could be reduced to three, shown on Figure 3: Point 1 – underwing plate used for fixing the drive axle; Point 2 – front upper section of the carrying side plate within the bracket of the tilting cylinder; Point 3 – the rear section of the carrying side plate within the counterweight carrying plate. The estimated maximal static stresses are $\sigma_{Max} = 156$ MPa within the underwing plate (p. 1) in the first instance of loading – taking the load. This value has been confirmed by the conducted strain measuring tests of a chassis of this class of machines at the testing laboratory of Balkancar Record JSC. Pursuant to [3], the dynamism coefficient of chassis and lifting mast of rough terrain forklifts, featuring pneumatic tyres, is $C_d = 1.9$. The maximum stress value multiplied by the dynamism coefficient results in the highest stress value – 296 MPa. Considering the fact that the yield limit regarding the ST 355 JR steel, whereof the chassis components have been made, is 355 MPa, it may be assumed that this section of the construction is optimal.

2) There is a negligible difference in the values of the maximal static stresses regarding both load conditions. When the load is being lifted, the values are 3-6 MPa higher than the ones regarding tilted lifting mast, which shows that operating the forklit in both load conditions does not considerably affect the stresses; it is stability that matters more.

3) Regarding the calculations entailing an increased centre of gravity and reduced load at both load conditions, the registered stresses have been 25% lower compared to the ones entailing reated load and a standard centre of gravity. It practically means that with regard to a forklift optimally designed for rated loads and a standard centre of gravity, compliance with the loads chart specified by the manufacturer shall ensure not only the required stability but also the strength and the reliability of the forklift chassis structure.

4) The achieved results regarding the structure stresses in the other two critical points, 2 and 3, show that there is sufficient reserve for additional lightening the chassis structure by further reduction of the thickness of the carrying side plate, from 10 mm to 8 mm, in respect of forklifts of this series that feature smaller lifting capacity, i.e. 30 kN and 40 kN.

4. Conclusion

The resulting values of the stresses and the deformations in the chassis of rough terrain forklifts have been calculated, presented and analysed using the created 3D model applying the FEM under two main operation modes, each of which has undergone two alternative types of loading determined by the weight and the location of the payload.

A real assessment of the strength and the deformations of the chassis of rough terrain forklifts has been made by analysing the obtained numerical results, and changes to the chassis structure have been proposed, which determines the practicality and the applicability of this study.

Acknowledgments

This work was supported by the European Regional Development Fund within the OP "Science and Education for Smart Growth" 2014-2020, Project Competence Centre "Smart Mechatronic, Eco- and Energy Saving Systems and Technologies", № BG05M2OP001-1.002-0023.

The authors would like to thank the Research and Development Sector at the Technical University of Sofia for the financial support.

References

[1] Georgiev G 1980 Design, construction and calculation of forklifts *Technika Sofia Bulgaria*

[2] Genov J and Kralov I 2019 A linear quadratic regulator synthesis for a semi-active vehicle suspension Part 1-Modeling of the dynamical system *AIP Conf. Proc.* **2172** 110006 1-8

[3] Evtimov I 2014 Electric and motor forklift trucks *University of Ruse, Ruse, Bulgaria*
[4] Iliev I and Nikolov V 2007 System for automatic stability of forklift trucks *Patent BG 62225 B1 November 15 2007*
[5] Iliev I and Nikolov V 2007 System for control of the operations of forklift trucks *Patent BG 62226 B1 November 15 2007*
[6] Kostov K 1982 Electric and motor forklifts. Methodology for strength testing and strength assessment *IEM Sofia Bulgaria*
[7] Nikolov N and Tsonev V 2013 Strength of materials Handbook *Avangard Prima Sofia Bulgaria*
[8] Nikolov V 2013 Mechanical-mathematical modeling of the tilting process of lifting masts of forklift trucks with load *Journal of the Technical University at Plovdiv "Fundamental Sciences and Applications" Plovdiv Bulgaria* **19(2)** 347-350
[9] Rusimov V, Borisov I, Dimitrov N and Angelov P 2019 SOLIDWORKS - Modeling and Drawings *Technologika Sofia Bulgaria*
[10] Sofronov Y, Stoyanova Y, Kopralev N and Todorov G 2019 Kinematic study of the articulated trucks operating layout of turn for articulated vehicles *TechSys 2019 16-18 May Plovdiv Bulgaria IOP Conf. Se.: Mat. Sc. and Eng.* **618** 012044
[11] Stoilov K and Karadimitrov I 1989 Attachment equipment for forklifts *Technika Sofia Bulgaria*
[12] Stoychev G, Chankov E, Dimova B 2010 Stress analysis of a fork-lift truck lifting installation *BulTrans-2010 24-26 September Sozopol Bulgaria Proc.* 205-208
[13] Todorov G and Kamberov K 2017 Virtual prototyping of drop test using explicit analysis *AIP Conf. Proc.*1910 020012
[14] Tsonev V Kuzmanov N Borisov B and Penkov K 2019 System for materials testing at static loading *TechSys 2019 16-18 May Plovdiv Bulgaria IOP Conf. Se.: Mat. Sc. and Eng.* **618** 012048
[15] Tsonev V Stoychev G Chankov E 2017 Stress analysis of a link for slat chain conveyor BulTrans-2017 11-13 September Sozopol Bulgaria Journal MATEC Web of Conferences **133**
[16] Valkov G and Nikolov V 2018 Classification of terrain forklifts 7th *Int. Sc. Conf. TECHSYS 2018 Plovdiv Bulgaria Proc.* 253-257, ISSN online: 2535-0048
[17] Valkov G and Nikolov V 2019 Analysis of the power drives of terrain forklifts. *Proc. of 12th Int. Sc. Conf. "Environment. Technology. Resources" June 20-22 2019 Rezekne Academy of Technologies Rezekne Latvia* **III** 241-248
[18] Yordanov P and Petkov G 2018 Comparative Study of the Bending and Torsional Stresses in Opened Cross Section Thin-Walled Beam. *Sc. Jour. Mechanics of Machines* **121(3)** 49-53 ISSN 0861-9727
[19] https://balkancar-record.com/en-US/Products/Forklifts, [Accessed March 9, 2020]

Application of multi-axis force/torque sensor system for experimental study of small unmanned aerial vehicles propulsion systems – Preliminary results

V Serbezov[1], H Panayotov[2], M Todorov[1], S Penchev[2]

[1] Technical University of Sofia, Department of Aeronautics, blvd. Kliment Ohridski 8, Sofia, 1000, Bulgaria
[2] Technical University – Sofia, Plovdiv Branch, Department of Transport and Aviation Engineering, 4000 Plovdiv, Bulgaria

vserbezov@tu-sofia.bg

Abstract. The experimental evaluation of the performance of small propellers plays important role in the process of multicopter UAV design perfection. This paper presents a small UAV propeller test stand, based on a multi-axis Force/Torque sensor system. The system makes possible to measure the thrust, torque and the pitching moment or lateral force (when in oblique flow) of the tested propeller. The data processing of the experimental results is discussed. Results for static, axial flow and oblique flow conditions are presented. Some conclusions about the measurement methodology are drawn. The results verify that the test bench has good accuracy and is easy for operation.

1. Introduction

The small multirotor unmanned aerial vehicles (UAV) have already taken their place in the society and continue to spread, but they are still far from technological maturity. The propulsion system of such UAV uses small propellers that work at low Reynolds numbers and in very wide range of propeller angles of attack. This features makes the analytical and computational performance evaluation of such propellers quite challenging. For this reason the development of facilities and methodologies for experimental testing plays an important role in the further improvement of the propulsion systems of small UAV. Initial work in this direction was reported in 2011 by Brant and Selig [1], but it covered testing of small propellers only in axial flow. In 2017 efforts to explore UAV propeller characteristics at conditions close to real operation were undertaken in the Plovdiv branch of Technical University – Sofia by Panayotov and Penchev [2]. The test stand was designed with provision to test propellers in wind tunnel at 0 to 90 degrees angle of attack, but it was capable to measure only the thrust of the propeller, while the propeller power had to be estimated from the electric power of the motor. Meanwhile more experimental results by different researcher teams were published with the use of multi-axis force/torque transducers or measuring systems [3,4]. The use of multi-axis measurement systems allowed the measurement not only of the thrust and torque, but also of the pitching moment at nonzero angles of attack of the propeller.

The availability of a six-axis force/torque transducer system at the Department of Aeronautics at Technical University – Sofia made possible the upgrade of the test stand that was previously designed

IOP Conf. Series: Materials Science and Engineering **878** (2020) 012039 doi:10.1088/1757-899X/878/1/012039

in the Plovdiv's branch of the University. The aim of this paper is to describe the new stand and to present the initial test results at static, axial flow and oblique flow conditions.

2. Experimental Setup

The test bench uses ATI-AI Nano-25 six-axis force/torque transducer and WirelessFT data acquisition system (Fig.1). The sensor mount was designed with the following provisions:

- The mount can be fixed to any table in order to perform propeller static testing, as shown on fig.1;
- The mount can be easily attached to the frame of the test bed used in the Technical University of Sofia, Plovdiv Branch (Fig.2);

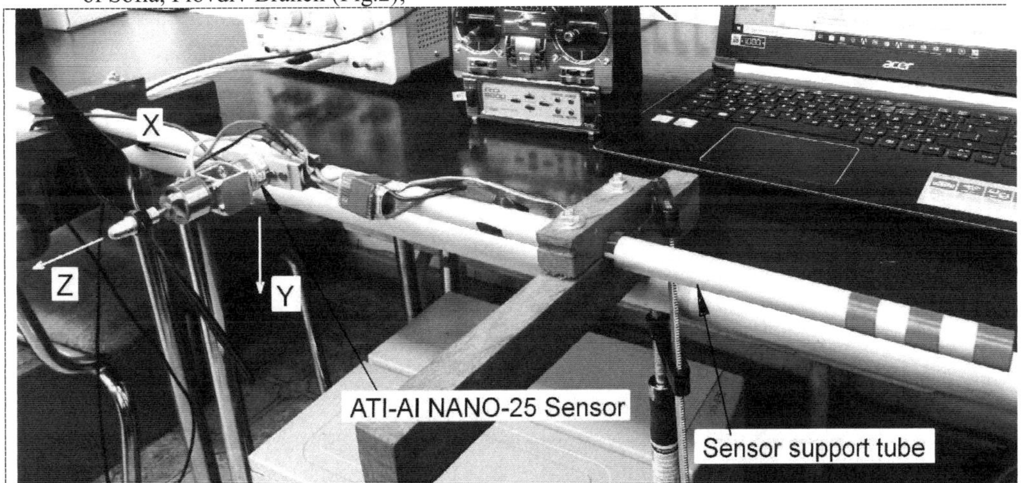

Figure 1. The test bench without sensor thermal isolation. Setup for static testing.

Figure 2. The test bench installed in the ULAK-1 wind tunnel.

- Sensor support tube can be rotated in its holders to allow testing the propeller at different angles of attack, when installed in the wind tunnel;
- The stand can be fitted with any brushless outrunner electric motor with M3 x 16/19 mm mounting holes;
- Tested propellers can be of diameter up to 254 mm (10"). In the particular case APC 10x4.7 propeller was tested (the same as in [2]).

A significant problem that emerged during first system tests was the thermal instability of the transducer, resulting in noticeable zero-load readings drift, caused by the heat coming from the electric motor and conducted through the interconnection structure. The problem was largely eliminated by thermally isolating the transducer from the motor and placing the transducer under an aerodynamic fearing. Even though the thermal drift should be considered by the test methodology.

3. Experimental Results

3.1. *Static Propeller Operation*

For this experiment a stabilized power supply, with precise voltage and current measurement was used, allowing fine calculation of motor power. The propeller rotational speed was measured using an optical tachometer. For every measurement point the following parameters were calculated:

- The propeller power in Watts:

$$P_{prop} = T_z . \frac{2\pi n}{60} \tag{1}$$

where T_z, [N.m] is the torque measured by the transducer along axis OZ, n is the rotational speed in 1/min;

- The electric motor power in Watts:

$$P_e = U.I \tag{2}$$

where U,[V] and I, [A] are the supply voltage and current, measured in the power supply unit;

- The electric motor efficiency (including the efficiency of the electronic speed controller):

$$\eta_e = \frac{P_{prop}}{P_e} \tag{3}$$

The results for propeller thrust are shown on fig.3 and for propeller power on fig.4. The results for the electric efficiency are presented in table1.

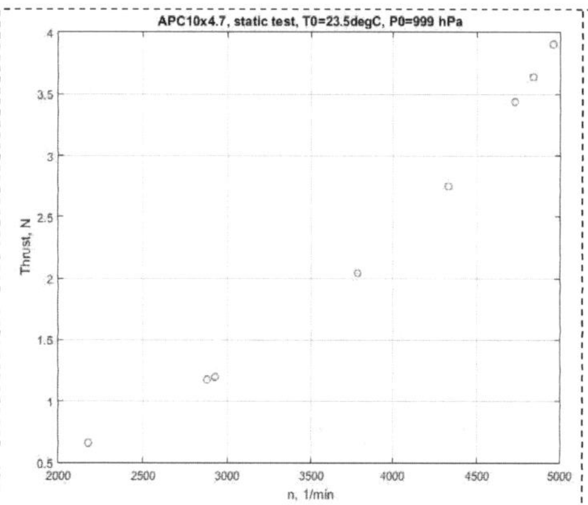

Figure 3. Propeller Static Thrust.

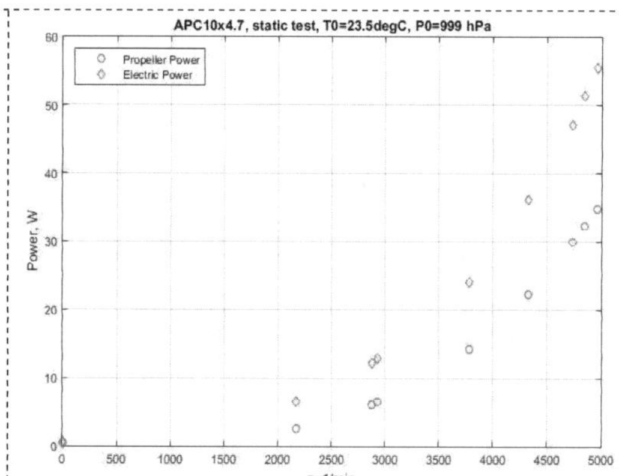

Figure 4. Propeller and electric motor power.

Table 1. Electric efficiency of the APC 10x4.7 / Turnegy D2830/11 propulsion system at static propeller operation.

Propeller rotational speed, [min^{-1}]	2174	2883	2930	3789	4331	4730	4840	4959
Electric efficiency	0.4006	0.5068	0.5058	0.5898	0.6160	0.6374	0.6291	0.6259

3.2. *Axial Flow Conditions*

The experiment was conducted in the ULAK-1 wind tunnel at Technical University – Sofia, Branch Plovdiv. The tests were at airspeeds of 5, 10 and 15 m/s. The propeller rotational speed was varied between 2000 and 8000 min^{-1}. The airspeed and rotational speed were measured with the equipment described in [2]. The obtained data was processed to be expressed in dimensionless form:

$$J = \frac{V}{n.D}, \tag{4}$$

where J is the propeller advance ratio; D, [m] is the diameter of the propeller; V, [m/s] is the speed of the airflow; n, [s^{-1}] is the rotational speed of the propeller;

$$C_P = \frac{\pi.T_Z}{30.\rho.n^2.D^5}, \tag{5}$$

where C_P is the propeller power coefficient; D, [m] is the diameter of the propeller; ρ, [kg/m3] is the measured air density;

$$C_T = \frac{F_Z}{\rho.n^2.D^4}, \tag{6}$$

where C_T is the propeller thrust coefficient;
Finally the propeller efficiency η_p is calculated:

$$\eta_P = J.\frac{C_T}{C_P}, \tag{7}$$

The results are shown on fig.5 for the thrust, fig.6 for the power and fig.7 for the efficiency coefficients. Although some of the data points at low rotational speeds are outliers, the obtained data is sufficient for evaluating the propeller performance parameters and to apply polynomial fit to it. The propeller efficiency is calculated, based on the fitted curves for the C_T and C_P. The results show good resemblance with the theory. Further refinement of the test accuracy can be expected with the improvement of the test methodology.

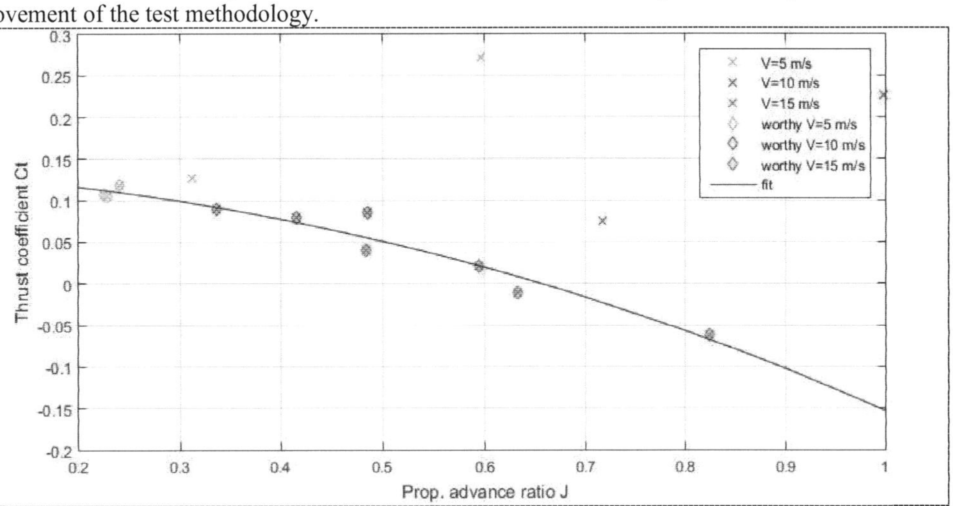

Figure 5. APC 10x4.7 Propeller Thrust Coefficient.

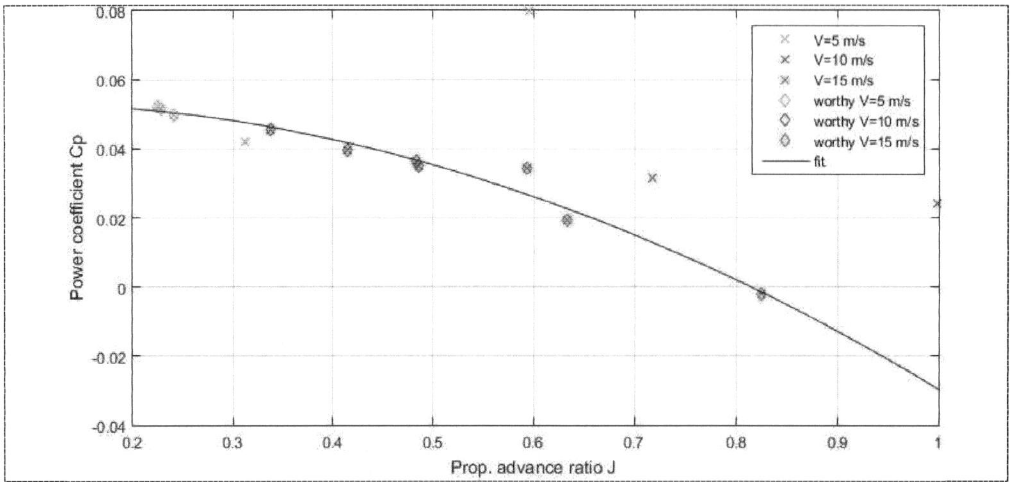

Figure 6. APC 10x4.7 Propeller Power Coefficient.

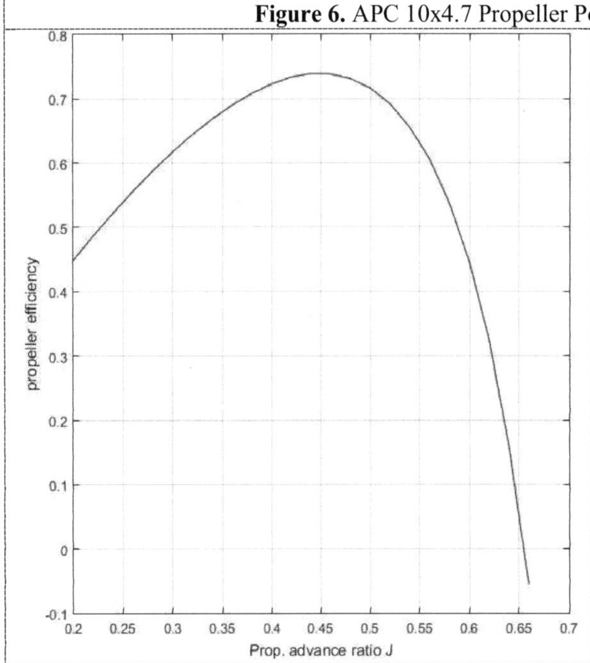

Figure 7. APC 10x4.7 Propeller Efficiency Coefficient.

3.3. *Oblique Flow Conditions*

The test conditions are the same as in axial flow experiment, except the support tube of the sensor is rotated at a given positive angle along axis OX (see fig.1). The main objective of the experiment is to test the ability of the sensor to measure the lateral force, generated by the propeller when operating in oblique flow. The lateral force is measured by the Fy channel of the sensor (giving negative reading for upwards oriented force vector).

For commonality reasons the results are expressed in dimensionless form, by defining propeller lateral force coefficient C_{lat} in analogy with the propeller thrust coefficient:

$$C_{lat} = \frac{F_Y}{\rho.n^2.D^4}, \tag{8}$$

The C_{lat} results for 25 degrees angle of attack of the propeller, airflow speed of 5, 10, 15 m/s and rotational speeds between 2900 and 7800 rpm are shown on fig.8. The results show near-linear relation of the lateral force coefficient and the propeller advance ratio. The results are encouraging for the further exploration of the small propeller performance in oblique flow at different angles of attack.

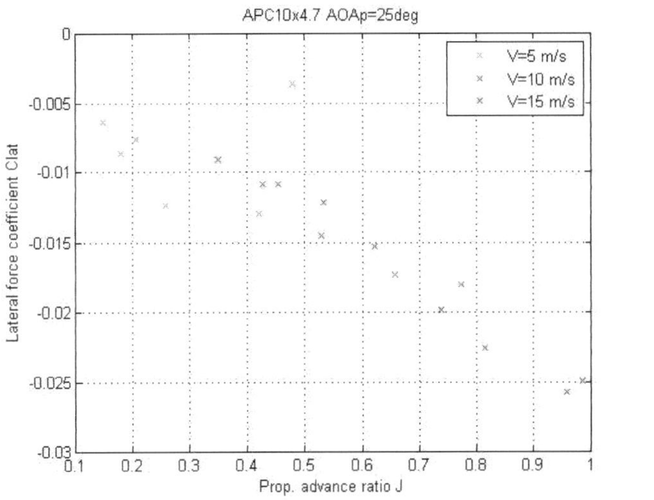

Figure 8. APC 10x4.7 Propeller Lateral Force Coefficient at propeller angle of attack of 25 degrees.

3.4. *Rotational Speed Measurement by Vibrations Frequency Analysis*

The ability of the WirelessFT system to collect data at high sample rates allows the rotational speed calculation to be performed by analyzing the frequency spectrum of the radial vibrations, caused by the propeller unbalance. The vibrations are sensed by the Fy channel of the transducer (see fig.1). The analysis is performed using fast Fourier transformation (FFT) in Matlab. As far as the experiments were performed with sample rate of 100 Hz, the analyzed spectrum was limited to 0 – 50 Hz, corresponding to rotational speeds of 0 – 3000 min^{-1}. Tests were performed with several fragments of the Fy record. The presented example is taken from the oblique flow tests of the propeller at angle of attack of 25 degrees, airflow speed of 5 m/s and recorded rounded rpm of 2900 min^{-1} (fig 9). In this case it is apparent that the propeller vibrations are combined with oscillations aliquot to the sensor support structure natural frequency.

The results of performing FFT and the calculation of the single-sided amplitude spectrum in Matlab are shown on fig.10. The peak at 49.38 Hz is clearly distinguishable, corresponding to the rotational speed of 2963 min^{-1}.

The performed tests show the ability of the multi-axis force/torque transducer to be used for rotational speed measurement. The advantage of this method is the reduction of the number of the sensors in the system. More experiments at higher sensor sampling rates should be conducted to verify the method for the whole rotational speed range of the tested propeller.

4. Conclusions

The preliminary results obtained by the small propeller test bench, based on six-axis force/torque sensor confirm its good accuracy and applicability for applied research purposes. The experiments included cases of propeller static, axial flow and oblique flow operation. In combination with electric measurement equipment the test bench can be used for performance evaluation of the whole UAV propulsion system including the electric motor and power transmission. The possibility of determining the propeller rotational speed by analysing its vibrational frequencies, collected by the force/torque sensor is also examined and confirmed.

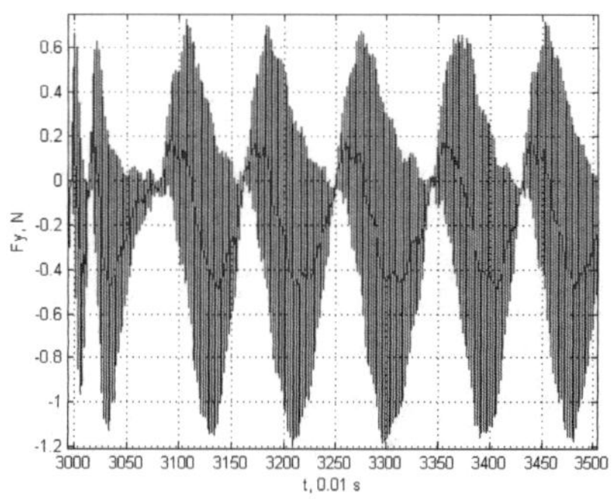

Figure 9. Signal recorded by the Fy cannel of the transducer.

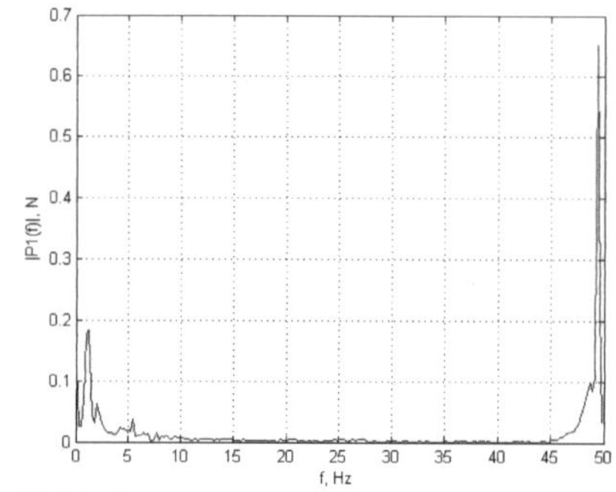

Figure 10. Single-Sided Amplitude Spectrum of Fy(t). Amplitude maximum at 49.38 Hz

Acknowledgements
This work was supported by the European Regional Development Fund within the Operational Program "Science and Education for Smart Growth 2014 - 2020" under the Project CoE "National center of mechatronics and clean technologies" BG05M2OP001-1.001-0008.

References

[1] J.B. Brandt, M.S. Selig, Propeller Performance Data at Low Reynolds Numbers, AIAA Paper 2011-1255, January 2011.

[2] H. Panayotov, S.Penchev, Determination of propeller characteristics for multirotor drone design. *FUNDAmENTAL SCIENCES AND APPLICATIONS*, 2017, 155.

[3] D. Serrano, M. Ren, A. J. Qureshi, S. Ghaemi, Effect of disk angle-of-attack on aerodynamic performance of small propellers, *Aerospace Science and Technology*, *92*, 901-914, 2019.

[4] A. Kolaei, D. Barcelos, G. Bramesfeld, Experimental analysis of a small-scale rotor at various inflow angles, *International Journal of Aerospace Engineering*, 2018.

Applicability of microwave non destructive diagnostics for composite materials used in the aerospace industry

Nikolay L Gueorguiev [1], Atanas Nachev [2], Sergey Ivashov [3]

[1] BAS - Institute of Metal Science, Equipment and Technologies with Hydro- and Aerodynamics Centre "Acad. A. Balevski" – 67, Shipchenskiprohod St., 1574 Sofia;

[2] Technical University of Sofia, Branch Plovdiv, 25, Tsanko Diustabanov St., 4000 Plovdiv, Bulgaria;

[3] Bauman Moscow State Technical University, 2-nd Baumanskaya, 5, 105005, Moscow, Russia.

niki0611@abv.bg

Abstract. The paper offers an analysis of some of the peculiarities associated with the application of the microwaves (MW) for non-destructive testing (NDT) of composite materials used in the aerospace industry. Special attention is paid to the influence of the compound of the materials of which are made composites, their shape and structure on reflection, refraction and attenuation of electromagnetic waves.

1. Introduction

It is well-known that composites (composite materials) have increased resistance to various mechanical, temperature, chemical, electromagnetic and other effects and oftenweight less or have lower manufacturing costs than ordinary materials[1-5]. Therefore, one of the priority areas for composites use is in the aerospace industry. A number of evaluations show that over the last decade, their use in the production of modern aircraft, helicopters, rockets, and spacecraft has grown exponentially. This, together with the high requirements for the reliability of aerospace assets, requires the analysis of various possibilities for composites diagnosis and for the non-destructive control of their quality, taking into account their specific features. The paper offers analysis of some of the peculiarities associated with the application of the microwave diagnostics methods for composite materials used in the aerospace industry. Special attention is paid to the impact of the composition of materials of which the composites are made, their shape and structure, and the system or the device in which it is used to reflect refract and attenuate MW transmissions.

2. General approach

It is known that microwave oscillations are reflected, refracted and damped depending on the electromagnetic characteristics, geometric shape and physical-mechanical structure of the composite materials being diagnosed. The capabilities of the microwave diagnostics, in turn, depend on the particular device in which the composite material is used, its accessibility and the conditions for its operation[6-9]. It is, therefore, appropriate to evaluate the influence of reflection, refraction and attenuation of the ultra-high frequency radiation used for the diagnosis of composite materials, taking into account the composition of materials from which the composites are constructed, as well as their shape and structure.

2.1. Influence of the composition of composite materials used in the aerospace industry on the possibilities for their diagnosis through ultra-high frequency methods.

It is known that the propagation of electromagnetic waves in an medium could be described with the help of the Maxwell equations, which are a function of the dielectric permittivity ε and magnetic permeability μ of the medium in which they propagate. If the medium is homogeneous, these parameters are constant, but in the real medium, they could change.

There are three main types of interconnected energy interactions of microwaves with the composite material – reflection, refraction and attenuation. These interactions depend mainly on the nature of the materials from which the composites are made. As well as the frequency and type of ultra-high frequency radiation[6,8,9]. The reflection and refraction of the microwave oscillations occur from the surfaces between two materials with different dielectric properties and depend on the size and electromagnetic characteristics of these surfaces. Thus, by reflecting on the one hand, it is possible to record the heterogeneities in the composite materials, and on the other hand, the energy of that part of the electromagnetic waves, which continues to propagate in the composite material, is reduced. The refraction does alter the wavefront, which continues to propagate into the composite material. The reflection of the microwave is described upon the basis of unevenness of the surface between the two materials having different dielectric characteristics. In some cases, this surface can be regarded as smooth, in which the reflection is mirror-like; or not smooth, in which case the reflection is diffuse and is determined by the reflections of the individual sections of the surface.

According to the Rayleigh criterion, the reflecting surface can be considered smooth, i.e. reflection is mirror if, in the case of perpendicular irradiation with a plane wave (i.e. at a 90 ° incidence angle), the surface irregularities are less than one sixteenth of the electromagnetic wavelength. This means that in microwave diagnostic methods at frequency of 24 GHz (wavelength is about 1.25 centimeters), the conditions for the reflectance of reflection are border irregularities less than 0.78 millimeters. When the angle of incidence is changed, these irregularities must be smaller (they decrease with the value of the sine of the slip angle of the microwave oscillation).

The refraction of the electromagnetic wave, when passing through medium with variable electromagnetic parameters, is characterized by their refractive indexes (which are the square root of their relative permittivity ε.). According to Huygens' principle, each point on the boundary of the two media is a source of secondary spherical electromagnetic waves. It turns out that the front of the microwave oscillation in them is different, taking into account the different coefficients of refraction of the media. The ratio of the sinuses at the edges of the microwave oscillations in the two media is inversely proportional to the ratio of their refractive indexes, i.e. per square root of the ratio of their relative dielectric constant. If the ε of the second medium is smaller than that of the first medium, then the refraction angle is greater than the incidence angle. Therefore, there is a critical value of the incidence angle at which the broken wave angle becomes 90º. This critical value is equal to the square root of the relation between the relative dielectric permittances of the second and first media. In this case, obviously, if the incidence angle is greater than the critical value, then the electromagnetic wave will be reflected completely from the boundary surface without penetrating the second medium. The later proves that in assessing the possibilities of using microwave radiation, methods for the diagnosis of composite materials must take into account their reflecting and refractive capacities, as well as the arrangement of the installation determining the angles of incidence of the drilling radiation[16,17]. Reflection is the dominant effect of the interaction between ultra-high frequency oscillations and conductive materials – metallic and some non-metallic. For non-conductive materials, the reflection is less pronounced, and the refraction and absorption, as well as the associated damping, are dominant.

The absorption of ultra-high frequency oscillation reduces its energy and is determined by the electromagnetic properties of the materials that make up the composites. In general, for non-conducting materials, the absorption is proportional to the attenuation of the ultra-high frequency oscillations and depends mainly on the dielectric constant of the materials in which it propagates. In practice, the so-called "relative dielectric constant" is usually used for convenience. The relative dielectric constant of the material (ε_r) representing the relationship between its dielectric constant and the dielectric constant of the vacuum.

If the absorbent material is a dielectric, then the damping of the microwave oscillations therein is determined by the dielectric losses, which represent the active electrical power released by the dielectric. Specific dielectric losses have two components. The first one is due to the conductivity of the material and is the same at constant and alternating voltage. The second one is manifested only by the action of variable fields in polar dielectrics and is related to the friction between the particles in their continuous reorientation (it is insignificant in non-polar dielectrics). The second component of the specific dielectric loss is usually much larger than the first and is determined by the tangent of the so-called dielectric loss angle. For some types of dielectrics, the tangent of the dielectric loss angle is very small, in the order of 4 to 10, and they are practically invariant to the frequencies of electromagnetic oscillations.

When analyzing the absorption of microwave oscillations by dielectrics, should be considered that it has a certain maximum value. The maximum density of absorbed power depends on the properties of the material, such as for materials based on rubber or foam polystyrene this power is $0.155 \div 0.465$ W / cm^2, for foam ceramics, it is 7.75 W / cm^2 and for polyurethane, it is 1.3 W / cm^2.

The absorption of microwave radiation from composite materials is uniquely related to their penetration into them. It is obvious, the energy of the microwave oscillation penetrated into the nearest part of the surface of the material is the difference between the oscillation which has fallen on the surface and the reflected oscillation. For its part, the energy of the microwave oscillation at its propagation in a homogeneous isotropic medium at a certain distance from the surface is equal to this distance, multiplied by the absorption (loss) coefficient of the material concerned. Thus, for a material of infinite thickness at some distance from its surface, the energy of the microwave oscillation will become negligible. This distance is called the depth of penetration of the microwave oscillation, and for non-conducting materials it is considerable.

In the case of conductive materials, there is also penetration of ultra-high frequency radiation, but it is limited to a certain critical depth, also called skin depth. This depth is in the order of units to parts of a millimeter and is inversely proportional to the square root of the product of the frequency of oscillation, the electrical conductivity of the material and its magnetic permeability.

The studies conducted to determine the possibility of ultra-high frequency flaw detection of a composite material composed of carbon fibers, show that it is possible that the depth of penetration of some ultra-high frequency oscillations in it may exceed a separate layer of material. As a non-magnet, the magnetic permeability of carbon is equal to that of free space ($4\pi \times 10^{-7}$ H.m^{-1}), whereby the depth of penetration of ultra-high frequency oscillations at it at a frequency of 100 megahertz is about 1 mm and at a frequency of 300 megahertz reaches twice the thickness of one standard layer of 0.25 mm. is on the order of several millimeters.

As already noted, one of the main characteristics of the materials determining the reflection, refraction, and attenuation of the microwave oscillations in them is their relative dielectric constant. It depends on the chemical composition of the materials, their density, shape, temperature, etc. The relative dielectric permittivity values of some materials used in the composites are shown in Table 1.

Table 1. Relative dielectric permittivity values (ε_r) of some materials used in composites at 18 °C (unless otherwise stated)and frequency 50 Hz

material	ε_r	material	ε_r
vacuum	1,0	porcelain	2–6
air	1,0005	epoxy resins	4,3–5,4
styrofoam	1,03	mica	5,4
benzene	2,28	aluminum oxide	7
PTFE and Teflon	2	glass	6–8
polyethylene (PE) (90 °C)	2,4	ceramics	3-11
polypropylene (PP) (90 °C)	2,1	silicon	11,8
rubber	2,5–3	alloyed barium strontium ceramics	100 - 600
quartz	3,8	barium titanate	10^3–10^4
acrylonitrile butadiene styrene (30 °C)	4,3	special ceramics	10^3–10^6

The value of the relative dielectric constant in a vacuum is equal to one, and the increase of the relative dielectric constant of the material through which the electromagnetic oscillation propagates causes its attenuation to increase.

The data shown in Table 1 are for low frequencies of electromagnetic oscillations and can only be used for comparative estimations, and apparently, with increasing frequency of radiation, the relative dielectric capacity will increase significantly. However, it is apparent that a number of materials used for composites for the aerospace industry have relatively good opportunities for transmitting ultra-high frequency oscillations. This makes it possible to use ultra-high frequency diagnostics both to control the uniformity of their structure and to control the quality of the materials beneath them. Of particular interest in this direction is the holographic subsurface radar created by the Moscow Technical University.Thelater is intended for the study of the zones of stable adhesion, the presence of deformations or inhomogeneities of metal housings under thermal insulation coatings up to 50 cm thick.

2.2. Influence of the structure of composite materials used in the aerospace industry on the possibilities for their diagnosis through ultra-high frequency methods.

The structure of the material plays an important role in determining the possibilities for their diagnosis through ultra-high frequency methods. As already indicated, the reflection, refraction, and attenuation of the microwave oscillations depend on the electromagnetic characteristics of the composite materials. If they consist of several elements (eg successive layers, fill matrices with different materials, etc.) with different relative dielectric permissions, then no complex and repeated internal reflections and refractions can be expected. If the construction of the composite material contains elements spaced multiple of one-quarter of the microwave wavelength, the internal reflected wave will be in counter-phase with that propagating. This would lead to an additional "constructive" attenuation of the propagating microwave oscillation, which, depending on the reflection coefficient, could be significant. Overcoming this effect can be achieved by varying the frequency of the microwave oscillation, for example by modulating it in frequency, insisting in the diagnostic process, using multiple frequency microwave means, etc.

If the relative dimensions of the objects are significantly smaller than the wavelength of the microwave oscillation (e.g. in nanocomposites), the concept of geometric optics is not applicable. Then, the principles of wave optics, and in particular those related to diffraction, must be used to evaluate electromagnetic processes. When a flat electromagnetic wave interacts with a finite-size object, two types of non-planar electromagnetic fields emerge – an internal diffraction field within the object boundary and an external diffraction field. The essence of the diffraction wave process is that the object excited by the incident electromagnetic wave is an emitter of an additional wave field.

The consideration of the stated effects of the construction of composite materials on the possibilities of their diagnosis by microwave methods is appropriate to be carried out on the basis of the aims and methods of its use.

In general, the objectives of microwave diagnostics can be classified as superficial diagnostics - for the presence of superficial heterogeneities and/or defects, superficial diagnostics – for heterogeneities and/or defects within the composite material itself, and superficial diagnostics – for non-uniform defects and material behind the composite material or behind a certain layer of composite material.

In addition, depending on the method of diagnosis, it is appropriate to classify it as a reflection-based (reflective) diagnosis or material-based (transient) diagnosis.

Reflective diagnosis is more commonly used and is performed with the ultra-high frequency transmitter and receiver located on the same side of the material. In this case, the heterogeneity and/or the defect is recorded on the basis of the change in the signal received by it. In general, reflective surface diagnostics is applicable to all types of composite materials.

In the transient diagnosis, the transmitter and the receiver of the ultra-high frequency oscillation are located on both sides of the composite material and inhomogeneities and/or defects are recorded on the basis of the change of the signal passed through them.

In the case of composite materials whose surfaces are of an electrically conductive layer, a basic reflective ultrahigh-frequency diagnosis is possible for the presence of surface inhomogeneities and/or defects. In some cases, when the electrically conductive surface layer is twice less than the penetration depth, a reflective diagnosis of the interior of the material is possible, and when this layer is less than the penetration depth, a transient diagnosis is also possible.

For matrix composite materials whose matrices are of electrically conductive material, but their reinforcement is of non-electrically conductive material, the possibilities for subsurface and subsurface ultra-high frequency diagnostics depend on the size and density of the matrix components. If the dimensions of the matrix component are significantly smaller than the depth of penetration of radiation into it, or if the distance between the individual components exceeds $0.25 \div 0.5$ of the length of radiation, it may be possible to perform sub-surface and even sub-surface reflective or transient diagnostics.

If the matrix and its reinforcement are not composed of conductive material, then it is possible to perform sub-surface and sub-surface ultra-high frequency reflection and transient diagnostics.

If the matrix is composed of not electrically conductive material but its reinforcement is of electrically conductive material, then in principle the possibilities of performing sub-surface and sub-surface ultra-high frequency diagnostics are extremely limited.

IOP Conf. Series: Materials Science and Engineering **878** (2020) 012040 doi:10.1088/1757-899X/878/1/012040

If the composite material is multilayered and the individual sheets are composed of conductive material, then there are opportunities to perform sub-surface and sub-surface reflective and transient high-frequency diagnostics. If the composite material is multilayer, but only part of the individual sheets (including the surface) of the composite material is composed of electrically conductive material, then there is a possibility to perform sub-surface ultra-high-frequency reflective diagnostics only up to the first sheet, which is composed of electrically conductive material. If the latter is less than the depth of penetration, it can be assumed that reflective and transient diagnostics could also be made for the layers behind this sheet.

For multilayer composite materials and the use of the reflection diagnostic method, the effect of the summation of the direct ultra-high frequency signal and its reflection from surfaces with different electromagnetic characteristics should be taken into account. If the reflected and direct signals are defaced with a window of 90 degrees (due to the distance between the sheets in proportion to about 0.5 of the wavelength of the signal), then they will offset each other. This effect can be avoided both by selecting the appropriate frequency of radiation for the particular type of composite material and by displacing the transmitter and receiver and providing a definite angle between the broadcasts and the received signal.

2.3. Influence of systems using composite materials on the possibilities for their diagnosis through ultra-high frequency methods.

In general, systems and devices using composite materials predetermine the possibilities of physical access to the technical means for performing ultra-high frequency diagnostics and at the angles at which it is possible to perform microwave radiation. That determines both the limitations associated with the reflection and refraction processes described above, as well as the possibilities of using a particular type of diagnosis.

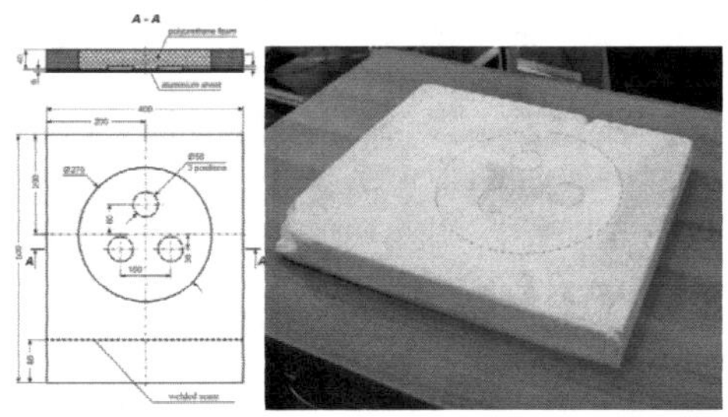

Sketch of the sample N1. View in the top of the sample N 1.

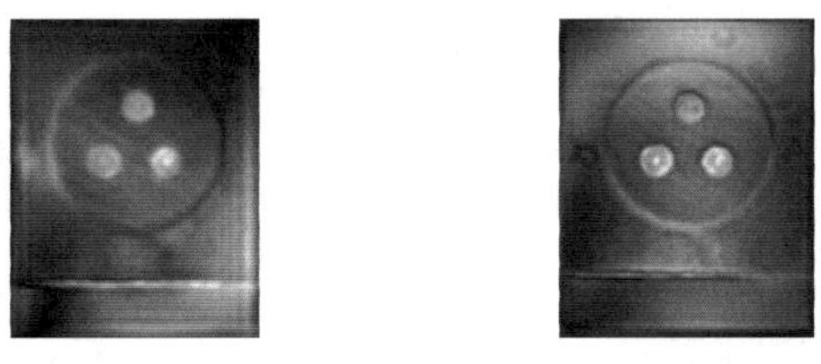

15 GHz (λ≅2 cm) 22.5 GHz (λ≅1.3 cm)

Hologram reconstruction for different frequencies. Sample N 1.

Figure 1. The Sketch and view of the sample N1 and result of the investigation of the polyurethane foam, provided by SPA Tekhnomash, Russia.

In the process of operation, composite materials located on the surface of individual systems and devices are available for various types of reflective ultra-high frequency diagnostics. Some of these could be available for transient microwave diagnostics, such as antenna components, helicopter propellers, etc.

In the course of the various types of repairs, composite materials may be available for one or more types of ultra-high frequency diagnostics, depending on the degree of their dismantling.

Due to the wide variety of possibilities, it is appropriate in the technical, technological and operational documents of the individual aerospace vehicle to determine the composite component and the conditions under which their condition can be investigated and controlled by a method. What is more, including ultra-high frequency diagnostics. Of course, the particularities of the various flaw detectors must be taken into account and the most appropriate cases used. In particular, given the material presented in the presented material, ultra-high-frequency subsurface and subsurface diagnostics is extremely suitable for composites using materials having low acoustic and thermal permeability (e.g., thermal insulation coatings of polyurethane foam), in which acoustic and thermal methods.

One of typical subjects of investigation by the Holographic subsurface radar is the polyurethane foam, used as the thermal insulation. The dielectric permittivity of polyurethane foam that was used as Shuttle tank insulation is essentially close to the same parameters of air, and has very low level of attenuation in microwave range.

The multiple investigations were conducted with the sample N1, made by the polyurethane foam, provided by SPA Tekhnomash, Russia. The Sketch and view of the sample N1 and result of the investigation (figure 1). In the sample are made three circular voids which are subject to registration by the holographic subsurface radar. It is evident that the use of a higher frequency significantly improves the quality of the registration.

It is very interesting to make any comparison of different NDT&E technologies and holographic subsurface radar. Comparative tests were carried out using Thermal (infrared) imager Testo 885 (with high sensitivity of 30 mK at 30°C), X-ray microscope XD75000NT and holographic subsurface radar (HSR). The results are shown in the figure 2.

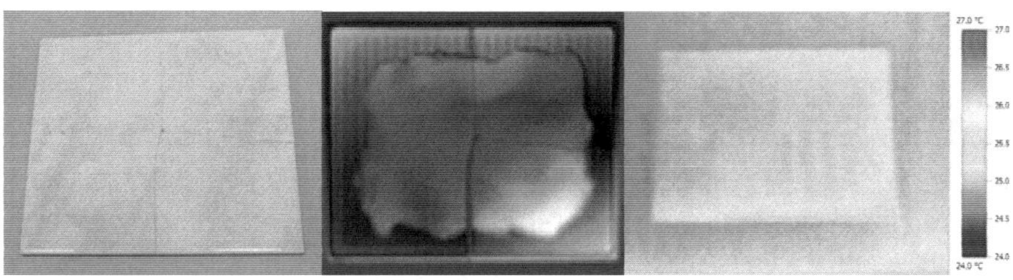

Visual view HSR image Infrared image

(a) Comparison of HSR and Infrared images

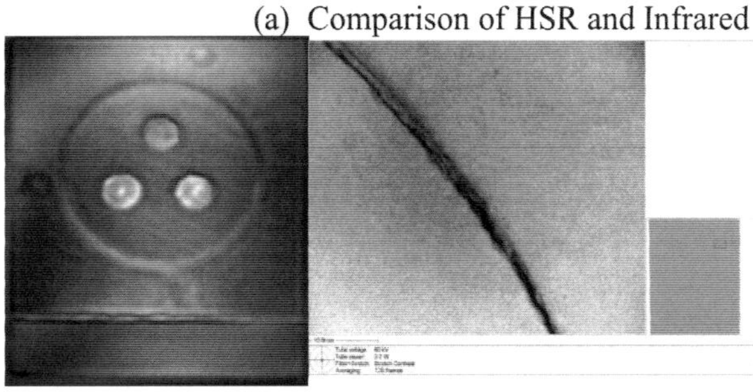

(b) Comparison of HSR **and** X-ray images

Figure 2. Results of the HSR, Infrared and X-ray images.

3. Conclusions

The above paper has offered an approach for analyzing the effect of a composition of the materials of which composite are made, their shape and structure. As well as the system or device in which the composite is used on the basic parameters of the microwave radiations used for their diagnosis. Particular attention has been paid to the analysis of the influence of the indicated characteristics of composite materials on the reflection, refraction, and attenuation of ultra-high-frequency radiation.

The analyzes made can be used in planning and organizing the performance of microwave diagnostics and non-destructive testing, as well as in the preparation of operational and repair documentation of the parts and devices of aerospace vehicles using composite materials.

The results of the study will by used in the Work package 2. Intelligent security systems,Project BG05M2OP001-1.002-0006 - Creation and Development of a Center of Competence "Quantum Communication, Intelligent Security Systems and Risk Management" (Quasar), faunded bythe European Regional Development Fund throughthe Operational programme “Science and education for smart growth”.

Acknowledgments

The results of the study were achieved in the implementation of administrative contract №KP-06-RUSSIA /22 dated September 28, 2019 and RFBR grant 19-57-18001\19, for financing a research project with incl. № NSF 93 with the theme "New Technologies for NDT of composites used in the aerospace industry," designated based financing conducted by Fund "Science" " Competition for projects under programs of bilateral cooperation in 2018 - Bulgaria - Russia 2018-2019". The authors would like to thank the Research and Development Sector at the Technical University of Sofia for the financial support.

References

[1] Borislav G Genov, Nikolaj V Vulkov, Georgi B Genov, Plamen N Chernokojev *Research of the influence of the type jacket of the bullet of the test ammunitions at the testing of the aramid materials*, International conference „HEMUS 2004”, 2004, ISSN: 1312-2916, pp. 269-276

[2] Borislav Genov, Delyan Nedelchev, Mitko MihovskI, Yordan Mirchev *Comprehensive approach for service life assessment of solid-propellant rocket motors*, NDT Days, Volume II / Issue 4 (2019), ISSN: 2603-4018 (print), 2603-4646, pages 467-475

[3] Valeri Panevski, Dimitar Dimitrov, Georgi Nikolov, V Stoichev, St Nanev *Advanced innovation in environmental security*. IMSETHAC-BAS, 2016, ISSN: 1313-8308, 309-313

[4] Nikolova V *Special Means for Individual Protection: Means for Individual Protection in Sites of Critical Infrastructure. Anti-terrorist Activities and in Operations in Crisis Response*. IMSETHAC-BAS, 2014, ISBN:978-619-90310-2-5

[5] Petkov P, Tumbarska A *Optimization of the Design of Fragmentation Warheads Forming an Axial Flow of Preformed Fragments. International Scientific Journal “Security & Future”*, 3, 4, Scientific Technical Union of Mechanical Engineering, 2019, ISSN: 2535-082X, 178-181

[6] Sergey Ivashov, Andrey Zhuravlev, Vladimir Razevig, Binu Thomas, *Frequency Influence in Microwave Subsurface Holography for Composite Materials Testing*, Proceedings of the 17th International Conference on Ground Penetrating Radar, GPR 2018, Rapperswil, Switzerland, June 18–21, 2018, pp. 98-103. DOI: 10.1109/ICGPR.2018.8441592

[7] S Kharkovsky and R Zoughi *Microwave and millimeter wave nondestructive testing and evaluation,* IEEE Instrum. Meas. Mag., April 2007, 10(2), pp. 26–38.

[8] T Lu, C Snapp, T -H Chao, A Thakoor, T. Bechtel, S Ivashov, and I Vasiliev *Evaluation of holographic subsurface radar for NDE of space shuttle thermal protection tiles, Sensors and Systems for Space Applications.* Proceedings of SPIE, vol. 6555, 2007, pp. OS1-8.

[9] Daniels, David, *Ground-penetrating radar*., 2nd ed., The Institution of Electrical Engineers, 2004, pp 761, ISBN O 86341 360 9

Aerodynamic testing of rotors and propellers for small unmanned aerial vehicles at Technical University - Sofia

F Panayotov, V Serbezov, M Todorov

Department of Aeronautics, Technical University of Sofia, blvd. Kliment Ohridski 8, Sofia, 1000, Bulgaria

vserbezov@tu-sofia.bg

Abstract. The recent massive introduction of small unmanned aerial vehicles (UAVs) brought the problem of absence of reliable aerodynamic data and models for propulsion systems that work at low Reynolds numbers. At the same time the new capabilities of modern measuring equipment permit detailed experimental investigations of the aerodynamic characteristics of such systems. This paper presents the results of work in this direction, carried out at, or with serious contribution of the Department of Aeronautics at Technical University of Sofia. Two test benches are presented. The first one is for static testing of small helicopter rotors. It is designed also with provisions for installation in suitable wind tunnels. The second one is a test stand for evaluation of electric power plants for unmanned aircraft vehicles. It can be used statically, or may be installed on the roof of an automobile for mobile testing. As an illustration experimental results are given for a specific model helicopter rotor and a small airplane propeller.

1. Introduction

Unmanned aerial vehicles (UAVs) find ever more applications for both military operations and civilian use in the sphere of observation or in the entertainment and cinema industry. Recent technological advances in the area of propulsion, guidance systems, and microelectronics have made possible commercially applicable small UAVs.

Nowadays, various theories are used to investigate the aerodynamic characteristics of rotors and propellers of UAVs [1]. These are the blade element theory (BET) (classical theory, which can be combined with Glauert theory), vortex methods [2] (the lifting line method and the lifting surface method), and computational fluid dynamics (CFD) methods [3], based on the numerical solution of the Euler/Navier-Stokes equations.

Blade element theory (BET) and vortex methods are fast and adaptive for small angles of attack. The airflow through rotors and propellers is complicated as there are relatively high angles of attack, causing disturbances in the flow-field downstream of the rotor. These methods have low or medium accuracy on the near-flow solution. However, their main advantage is the low-computational power requirements. CFD methods have good accuracy of the solution but they are computationally expensive. Moreover, they are affected by the mesh type and the right choice of turbulent model. Additionally, modeling is complicated by the fact that the small scale rotors and propellers operate at low-Reynolds conditions. Therefore, numerical models need to be validated by experimental data.

In this paper two test benches for the study of the aerodynamic characteristics of rotors and propellers for UAVs are described. The two test benches demonstrate different approaches to the design and construction of such equipment. The first one – for small helicopter rotors, which is designed and produced in the Department of Aeronautics at Technical University of Sofia [4], is made for academic research purposes. The second one – for small propellers, is constructed by researchers from the same department in the interest of Dronamics Ltd. quarter scale prototype program [5]. Specific experimental results from the both stands are also given.

2. Test Benches Description

2.1. *Test Bench for Small Helicopter Rotors*

The test bench is shown in Fig.1. The following measuring devices are used:

- Torque transducer HBM T20WN with nominal (rated) torque of 1Nm, accuracy class 0.2 and output voltage ±10V,
- Force transducer HBM S2M with nominal (rated) force of 10N, accuracy class: 0.02 and output voltage of 2mV/V.

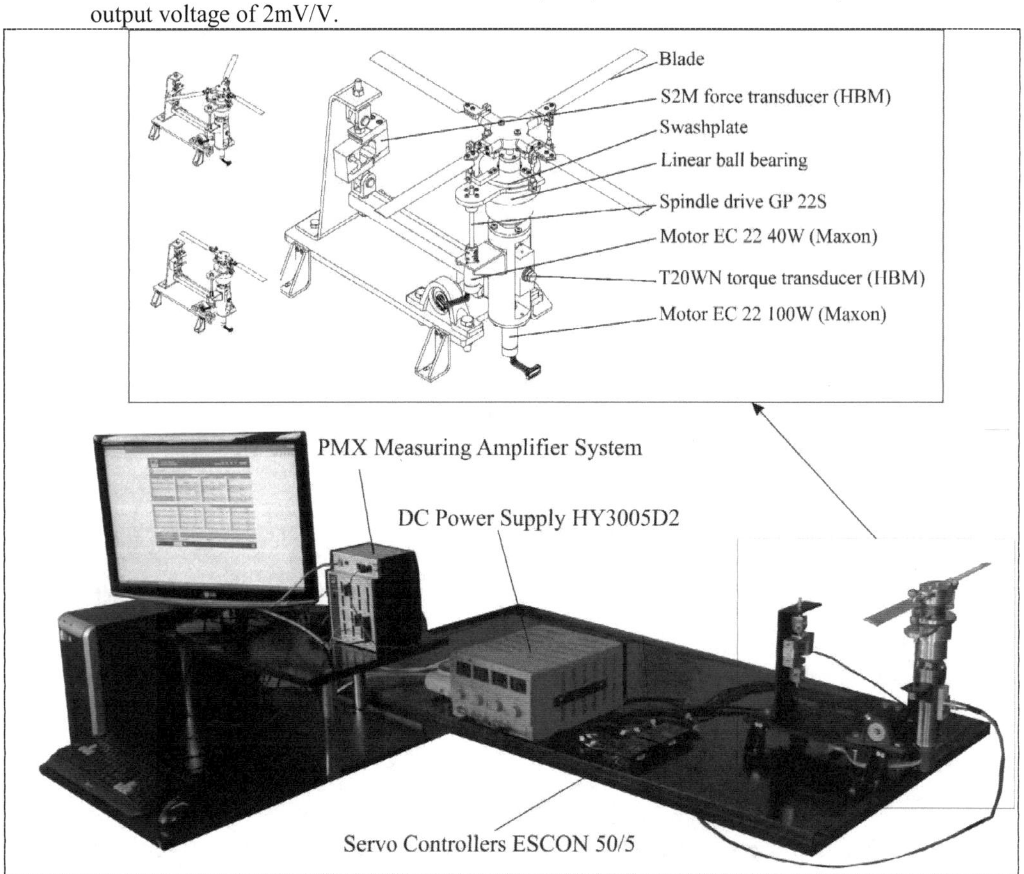

Figure 1. Test Bench for Small Helicopter Rotors.

The model rotors are driven with a brushless electric motor Maxon ECi-40 with nominal (rated) power of 100W. It allows for rotational speeds up to 4190RPM, although the maximum permissible rotational speed is limited to 3000RPM. The electric motor has an in-built Hall sensor, which allow for measurement of the rotational speed. The motor is driven with Maxon ESCON 50/5 servo controller, which allows for precise regulation of the rotational speed.

In order for the pitch angle of the blades to vary, a spindle-driven mechanism is used. It is powered with the combination of Maxon GP22S spindle with gear ratio of 1:370 and the EC-22 electric motor with nominal (rated) power of 40W.

The pitch angle may be varied during rotation of the model rotor. Thus, unsteady transient operation can be studied, in order to evaluate their impact on the aerodynamic characteristics of the rotor. The spindle-driven mechanism is capable to achieve rates of change of the pitch angle of ±2.2deg/s.

Both electric motors can be controlled automatically according to a pre-programmed test schedule. Thus, multiple operational regimes can be evaluated in one test session without the need to constantly stop, re-program and re-start of the test stand. An automatic data acquisition system is organized, which stores the experimental data in a well-structured database.

The data acquisition from the force and torque transducers is ensured by the HBM PMX measuring amplifier system. The test bench is powered by a DC power supply HY3005D2, which provides up to 300W. The test bench provides the possibility to measure the produced thrust T, [N], the torque Q, [Nm] and the angular velocity Ω, [rad/s] of model rotors with diameters of up to 1 meter and pitch angles varying from –12° to +24°.

2.2. *Test Bench for Small Propellers*

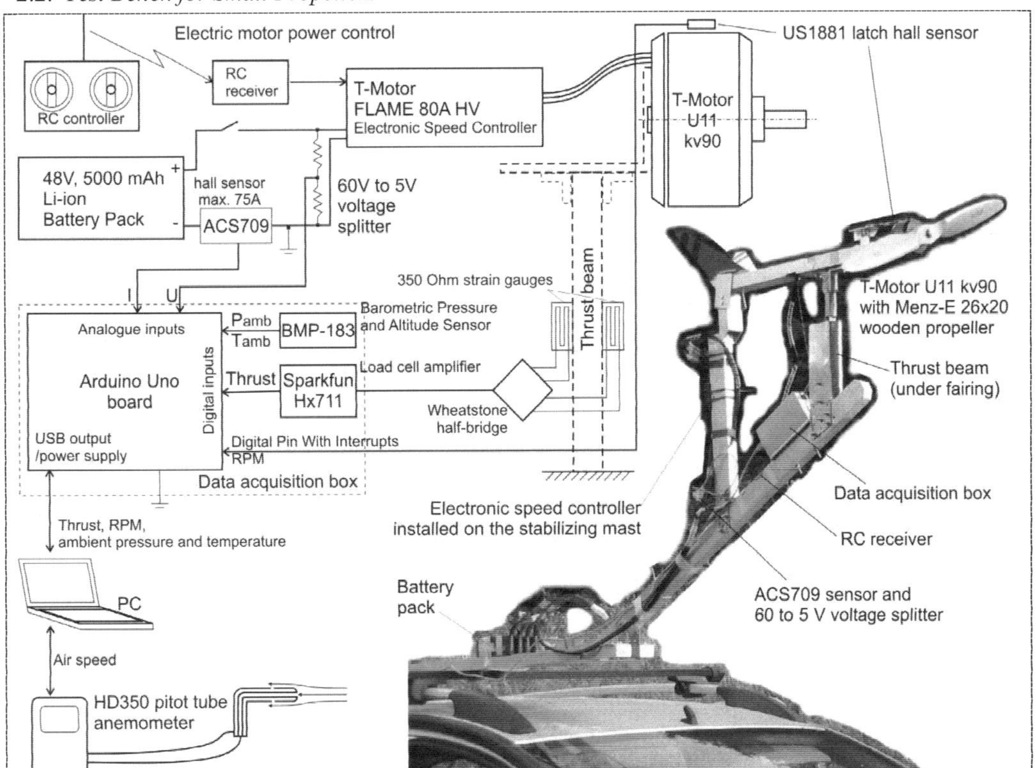

Figure 2. Test Bench for Small Propellers.

The brief description of the test bench and its test results was initially published in [5]. In the present paper some additional details of the measurement and control system of the stand and some conclusions from its operation are given. The system schematic and a picture of the stand, mounted on the roof of a passenger car are shown on Fig.2.

The main objective of the stand was to test the propulsion system of the Dronamics small prototype within a short period and with limited budget. The following requirements were put:

- Measure propellers with thrust up to 100 N and diameter up to 800 mm;
- To be capable of testing propellers in forward motion;
- To use "amateur robotics" hardware and sensors.

It was decided that the objective can be achieved by a mobile test stand, with a provision for mounting on the roof of a passenger car to perform the testing at airfields that are not in use. For this reason the stand was designed with provisions for easy assembly / disassembly and transportation.

Most critical for the small propeller bench are the thrust and the motor rotational speed measurements. To measure the thrust with standard strain gauges, a relatively long and thin thrust beam has to be used. After a number of experiments, a standard I-beam is selected. More detailed information on the thrust beam can be found in [5]. The drawback of this solution is that it is susceptible to vibrations caused by the propeller. For this reason an additional stabilizing mast that limits the lateral vibrations of the motor mount by two side plates is installed.

For the propeller speed measurement it was judged most practicable to use latch hall sensor, placed over the rotating case of the 'outrunner' permanent-magnet brushless motor. The signal of the sensor is led to an interrupt digital input of the Arduino board. As far as the U11 motor has 14 poles on its case, leading to 933 impulses per second at 4000 min-1, the signal rate turns to be challenging for the Arduino Uno board, resulting in inaccurate rpm reading. The problem is overcome by code optimization and restrain from any unnecessary mathematical operations in the Arduino code.

3. Experimental Results

3.1. *Small Helicopter Rotor*

The object of study is a hingeless rotor with inner diameter of 0.162m and an outer diameter of 0.532m. It has two blades which are rectangular and untwisted with a chord length of 0.025m, as it is shown in Fig.3.

The blade profile is Goettingen 417. The measurement methodology is as follows:

- The rotor torque, Q_{hub}, [Nm] is measured without blades to evaluate the internal friction losses in the bearings of the main shaft and the air friction, created by the rotor hub. This measurement is important as this torque later subtracted from the measured torque, Q_m, [Nm], to derive the actual torque of the rotor without overvaluation.
- At the same time the thrust, T, [N] is measured.
- The torque and trust are used to define the Figure of Merit. The Figure of Merit is introduced as a non-dimensional measure of helicopter thrust rotor efficiency, providing a common basis for comparison between rotors with different geometries. The following equations are used:

$$P = Q\Omega \tag{1}$$

$$C_P = \frac{(Q_m - Q_{hub})\Omega}{\rho A \Omega^3 R^3} = \frac{P}{\rho A V_{tip}^3}, \tag{2}$$

$$C_T = \frac{T}{\rho A \Omega^2 R^2} = \frac{T}{\rho A V_{tip}^2}, \tag{3}$$

$$FM = \frac{C_T^{3/2}}{\sqrt{2}C_P}. \tag{4}$$

IOP Conf. Series: Materials Science and Engineering **878** (2020) 012041 doi:10.1088/1757-899X/878/1/012041

where ρ = 1.225, [kg/m^3] is the air density; $A = \pi R^2$, [m^2] is the area of the rotor; Ω, [rad/s] is the angular speed of rotation; R, [m] is the radius of the rotor; $V_{tip} = \Omega R$, [m/s] is the tip speed of the blades; P, [W] is the power consumption due to profile drag.

Furthermore, to ensure comparability between the results for helicopter rotors with different number of blades and thus with different rotor solidity, the following dimensionless parameter is used: C_T/σ – blade loading coefficient. The rotor solidity σ represents the ratio between the combined surface area of all blades and the total surface area omitted by each blade for one complete revolution of the rotor: $\sigma = N_b cR/\pi R^2 = 0.05983$, where N_b, [-] is the number of blades and c, [m] is the chord length of the rotor blades.

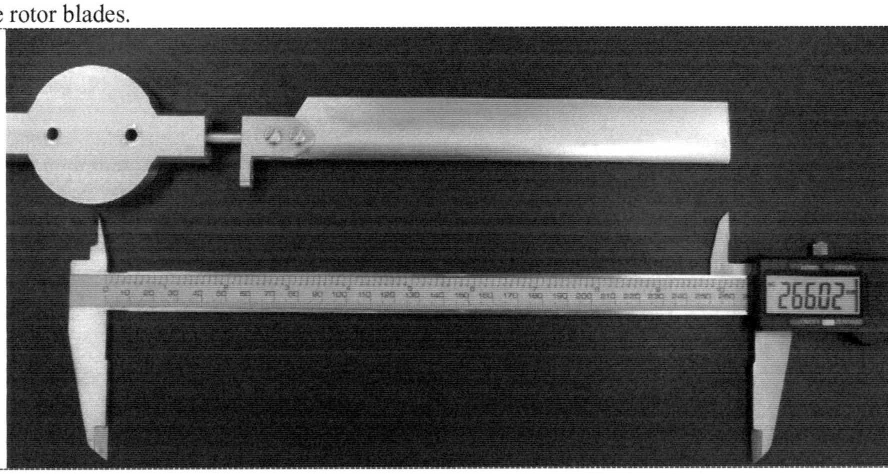

Figure 3. Experimental Two-blade Small Helicopter Rotor.

Fig.4 shows the rotor torque for the two-bladed rotor configuration at various angular speeds of rotation.

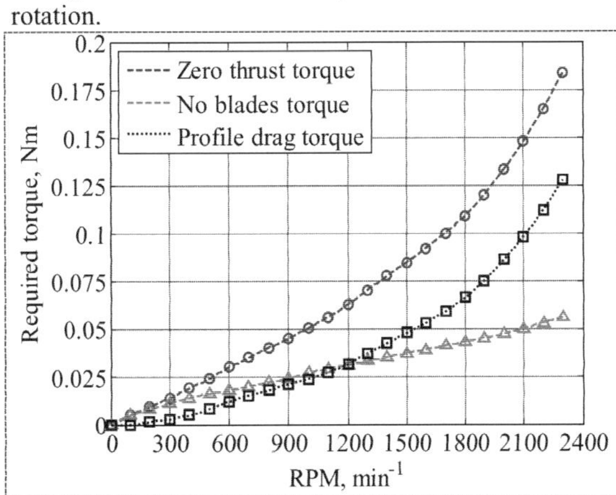

Figure 4. Required torque to drive the test bench with a two-bladed rotor at various rotation speeds.

It can be noticed that initially the "profile drag torque" is negligibly small until 400 RPM, after which it rises substantially. Also, between 800 RPM and 1400 RPM the "profile drag torque" equals the "no blades torque", produced by the internal friction in the ball bearings of the rotor shaft and the aerodynamic friction of the rotating hub without blades. Above 1400 RPM the "profile drag torque" increases as a second-order curve, whether the "no blades torque" increases linearly.

Fig.5. shows the variation of the rotor thrust from the pitch angle of the rotor blades θ. The measurements are done for the following pitch angles θ = 0°, 3°, 6°, 9°, 12° and 15°. Fig.6. shows the hover efficiency against the blade load coefficient.

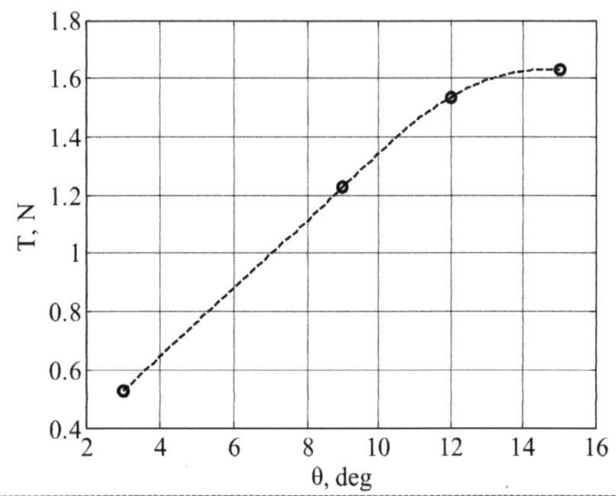

Figure 5. Rotor thrust against the pitch angle of the blades at 900 RPM.

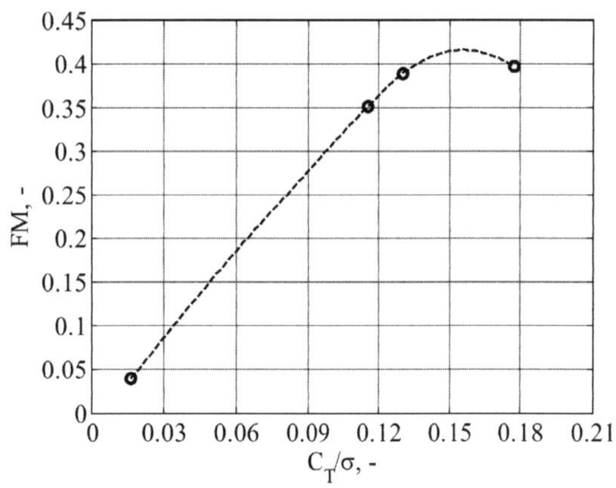

Figure 6. Figure of Merit against the blade loading coefficient for 2-blade rotor configuration at 900 RPM and Re≈29000.

3.2. *Test Bench for Small Propeller*

The object of study is the Menz-E 26x20 propeller, installed on T-Motor U11 KV90 electric motor. The main test results are published in [5]. Due to the schedule restrictions imposed by Dronamics small prototype program, only one round of airfield test runs was conducted. Every test run was carried out at approximately fixed car speed of 10,15,20 m/s and gradual increase of the motor power setting in steps of 20 - 30 s duration.

Although accompanied by technical glitches the testing was successful in general. The rotational speed data is partially corrupted at high speeds due to high vibrations. The dispersion of the thrust and the electric parameters is also significant. For this reason the experimental data had to be heavily processed. In this paper the focus is put on the processing of the test data, complementing the initial result, published in [5]. The data processing generally follows the principles stated in [6].

The applied in Matlab data processing procedure is as follows:

- Merging of the data from Arduino DAQ, the airspeed from Pitot tube anemometer and the GPS speed from Garmin GPSMAP 60CSx handheld system.
- Smoothing of the Thrust, Input current and Voltage data by moving average filter.
- Calculation of the rotational speed rate of change.
- Calculation for every time point of the propeller advance ratio (eq. 5)

$$J = \frac{V}{n.D}, \tag{5}$$

where D, [m] is the diameter of the propeller; V, [m/s] is the forward speed; n, [s-1] is the rotational speed of the propeller.

- Calculation for every time point of the propeller thrust coefficient (eq. 6)
-
$$C_T = \frac{FN}{\rho . n^2 . D^4}, \quad (6)$$

where ρ, [kg/m3] is the measured air density; FN, [N] is the measured thrust of the propeller.

- Creation of set of valid data points (Fig. 7.), by exclusion of data points with:

a) Forward speed bellow 5 m/s;
b) Propeller rotational speed bellow 1500 min-1;
c) Thrust above 100 N - eliminates faulty thrust readings;
d) Excessive rate of change of the rotational speed (dRPM) - eliminates RPM sensor faulty readings;

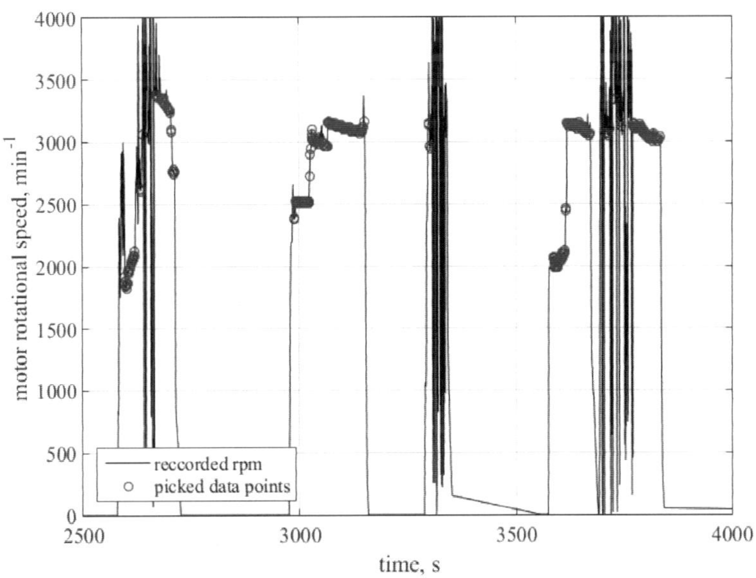

Figure 7. A sample of the propeller rotational speed record with the picked non corrupted data points.

- Curve fitting with second power polynomial of the thrust coefficient in relation with the propeller advance ratio (Fig.8.).

The limit of the rate of change of the propeller speed was varied in order to test the sensitivity of the curve fitting. The final results are presented in Table 1. The results show good stability to input data variations, which verifies the applied data processing procedure and the validity of the experiment, despite the corruption of part of the data.

Table 1. Curve fitting results for data sets with different margin of dRPM.

dRPM, [s-2]	Polynomial coefficients	Rsquare	RMSE
1.6(6)	-0,07172; -0,07930; 0,13849	0.8990	0.0072
0.6(6)	-0,07359; -0,07736; 0,13825	0.9022	0.0070

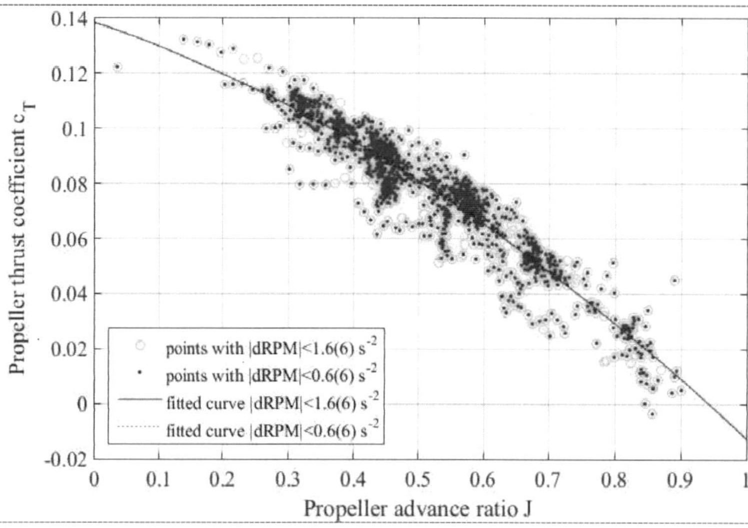

Figure 8. Curve fitting of the propeller thrust coefficient data.

4. Conclusions

The presented test benches show the broad range of solutions that can be used for solving theoretical and practical research problems in the field of UAV propulsion. The first test bench is build at a higher technological level. It uses professional sensors and DAQ equipment, and offers high precision test results, but is somehow restricted in terms of object of study. The second test bench shows the broad capabilities of the low-cost amateur systems and open-air field testing at the price of much more inaccurate experimental data and need for heavy post-processing. If wisely used both approaches can successfully complement each other.

Currently the researchers at the Department of Aeronautics of Technical University of Sofia and Arts et Metiers ParisTech continue their efforts towards improving and enhancing both concepts.

Acknowledgements
This work was supported by the European Regional Development Fund within the Operational Program "Science and Education for Smart Growth 2014 - 2020" under the Project CoE "National center of mechatronics and clean technologies" BG05M2OP001-1.001-0008.

References

[1] J. G. Leishman, Principles of Helicopter Aerodynamics, second ed., Cambridge University Press, New York, 2006.
[2] I. Dobrev, F. Massouh, B. Maalouf, Lifting surface method for prediction of rotor vortex wake, CFM2009, 19 (2009).
[3] C. V. Velkova, Modelling the Dynamic Behavior of Helicopter Rotors, 2013-ENAM-0055, (2013).
[4] F. Panayotov, I. Dobrev, F. Massouh, M. Todorov, Experimental study of a helicopter rotor model in hover, MATEC Web of Conferences, 234 (2018).
[5] V. Serbezov, S. Dimitrov, K. Rangelov, Mobile test stand for evaluation of electric power plants for unmanned aircraft, MATEC Web of Conferences, 133 (2017).
[6] J. Mandel, The statistical analysis of experimental data. Courier Corporation (2012).

Graph-analytic approach to determine the position of the initial impact in the event of pedestrian accidents

Milena Savova-Mratsenkova[1] and Georgi Palagachev[1]

[1]Technical University of Sofia, Faculty of Transport, Department of Combustion Engines, Automobile Engineering and Transport, Bulgaria

savova@tu-sofia.bg

Abstract: In the present research the authors propose a graph-analytic approach to determine the position of the initial impact in the event of pedestrian accidents. When investigating the velocity of movement of the vehicle (participating in the accident), mechanics theorems and parameters are used, which are reported in a scheme and are consistent with the data, gathered from the scene of the accident. The present study uses case law data from a real road traffic accident. The proposed approach can find wide application in the practice of pedestrian accident reconstruction in the absence of specific data that uniquely identify the exact location in the lane in which the initial impact between the vehicle and the pedestrian had happened.

1. Introduction

The vast automobilization in the recent years meets the ever-growing need of the society of vehicles, but it also causes a number of negative consequences, affecting traffic safety. This demands systematic research and analysing the main causes of road traffic accidents. The choice of means to limit the causes of road traffic accidents is necessarily based on extensive research and analysis. An essential step is selecting the right approach to recreate the mechanism of the road traffic accident, which is based on research, analysis and drawing conclusions, we take them into account to determine the causes and events that led to their occurrence [1, 2, 3]. Determining the objective circumstances for the occurrence of road accidents is a key element in establishing the culpability of participants and, accordingly, on a more global scale, to take appropriate preventive measures so as to improve road safety in the future.

In the current research, object of study are accidents involving pedestrians. Most of them occur in settlements and the main pedestrian groups in risk are children under 14 and elderly people [6]. Expert analysis in this group of road traffic accidents is carried out according to certain methodologies, tailored to the specific conditions of the accident [1, 5, 6, 9, 10, 11]. The pedestrian may be injured as a result of the initial impact with the vehicle or during a subsequent collision with the front cover, the front window, the front column, the front part of the ceiling, the fall on the ground, the subsequent overtaking by the vehicle itself or another vehicle. Familiarity with traumatism is required to describe the complete physical process from initial contact to the fall of the victim on the road surface [6,7, 8].

2. Exhibition:

When conducting an expert analysis of a road traffic accident, based on the data provided from the pre-trial proceedings, the mechanism of the investigated incident is restored by assessing the particular road situation and the actions of the individual road users [1, 5, 8]. For this purpose, it is necessary to determine the location of the individual participants at the time of the initial impact, their position relative to the lane, their initial velocity and their velocity at the moment of impact. The key to ensuring accuracy of the analysis is correct determination of the place of initial contact between the participants by fixing it with two coordinates - longitudinally and transversely to the lane. Defining it accurately is essential to carry out the subsequent expert analysis and to adequately address the legal case regarding the technical feasibility of preventing a road accident.

The main output data, while determining the location of the initial impact are the documents, gathered during the investigation: incident inspection control, photographic materials, conclusion of forensic expertise, protocol for conducted investigative experiment, the protocols from the interrogation of the driver, the victim and the witnesses. In determining the location of impact, it is necessary to take into account the trajectory of movement of the vehicle and the position of the detected deformations on it.

Case-law data from a real road traffic accident were used to present the graph-analytic approach developed by the authors in determining the place of initial contact in a pedestrian accident.

The investigated road traffic accident occurred during daylight hours, with good visibility between bus brand "Iveco" model "Mago" and a pedestrian. The roadway in the area of the road traffic accident was open to traffic in both directions with one lane in each direction. The pavement was smooth with clean asphalt. The road section where the traffic accident occurred is straight, no incline. The width of the roadway is 7.50 m, the two lanes are separated by a continuous dividing line – type M1, which near the turnout on the left at RP1, turns into a broken dividing line – type M3. The lane in which the bus was moving is 3.50 m wide, and the lane in the opposite direction is 4.00 m wide. To the right of the right boundary line there is asphalt banquet with unformed edges and a width of 3.75 m. To the left of the left boundary line there is a 2.30 m wide pavement and a curb. There are no walkways in the accident area.

The participants in the investigated road traffic accident are:

- Iveco bus, Mago model, with the following technical characteristics: overall dimensions (l x b) 7,555 x 1,950 m; longitudinal base 4,250 m; width of front track 1,900 m; front overhang 1,20 m; front tires with size 205/75 R16; rear tire size 225/75 R16; own table 4200 kg; maximum mass 6500 kg. During the inspection by the investigating authorities, after the road traffic accident, the following damage was ascertained on the bus: lateral cracks on the lower right side of the front bumper with broken pieces of paint measuring 1 x 2 cm, the first crack being at a height of 0.25 m, the second at 0.37 m and the third at 0.42 m. All are from the front bumper length from 0.30 m to 0.40 m. Exactly, at the right corner of the bumper, at altitudes from 0.48 m to 1.10 m, a layer of old scratches was detected;
- The victim – a 83-year-old pedestrian at the time of the road traffic accident.

During the inspection by the investigating authorities of the accident, the direction of the inspection was chosen to be the direction of the Iveco bus, the landmark was chosen to be a concrete lighting pole, standing on the left walkway in the direction of the inspection and for baseline was used the imaginary line that goes through the landmark perpendicular to the roadway. The following measurements were made relative to this line: in the right lane, in the direction of the bus, two traces of tires are found with trace width - 0.42 m, it starts at a distance of 6.70 m before RP1 and 1 m to the right of the dividing line. The trace ends longitudinally at a distance of 7.20 m after RP1 and 0.62 m to the right of the dividing line. The total length of the trace is 13.90 m. The width of each trace is 0.17 m; rear left outside tire of bus Iveco is found outside the roadway, 0.55 m from the boundary line and 6.41 m from RP1; the front left tire of bus Iveco is found outside of the roadway, 0.55 m to the left of the boundary line and 10.65 from RP1; just before the Iveco bus a blood stain was found and documented in the attached photo material. During the inspection, the bus was oriented with its front in the direction of its original movement. To the inspection protocol photo material is attached.

Based on the data from the inspection protocol a detailed scale scheme of the road traffic accident was drawn up (figure 1).

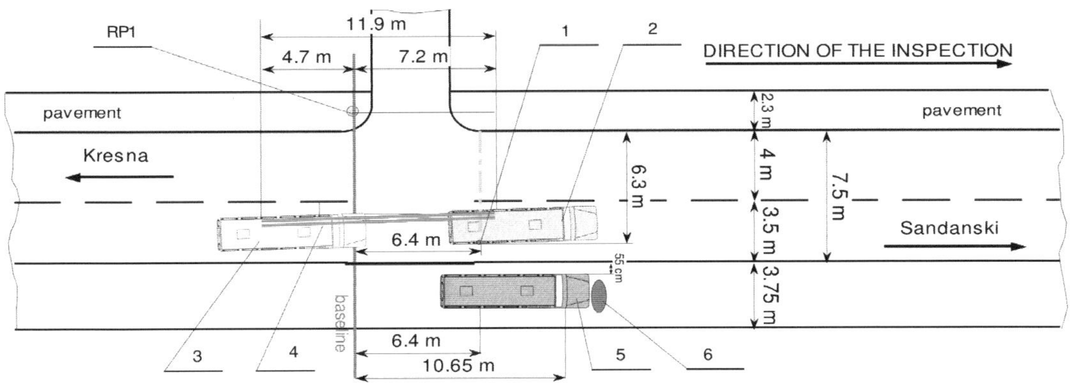

KEY:
1. Position of the bus at the end of the brake traces;
2. Position of the bus at the beginning of the brake traces;
3. Brake traces;
4. Position of the bus, described in the inspection protocol;
5. Blood stain;
6. Reference point – RP1

Figure 1. Scale diagram of the accident.

In order to restore the mechanism of the investigated road traffic accident, it is necessary to determine the location of the impact and the velocity of the bus at the initial moment and at the moment of impact [1, 5]:

- The velocity of the bus at the initial moment in which the driver reacted to the hazardous situation can be calculated technically using the length of the brake trace, described in the inspection of the accident left by the rear left wheels and is 13.90 m long. For this purpose, the following formula is used:

$$V_{Iveko-0.} = 0.5.j.t_{H} + \sqrt{2.S_S.j}, \tag{1}$$

where: $V_{Iveco-0.}$ - the initial velocity of the Iveco bus;

$\mu = 0{,}6$ - coefficient of adhesion;

j = $1{,}2.\mu.g = 7{,}06$ m/s^2 - maximum braking delay according to the specific conditions for a bus equipped with ABS;

tн = 0,8 s – time for increasement of the brake delay of a bus in a loaded condition;

S_S = 13,90 m – length of the brake trace.

$$V_{Iveco-0.} = 0{,}5.0{,}8.7{,}06 + \sqrt{2.7{,}06.13{,}90} = 16{,}83\, m/s = 60{,}60\, km/h \tag{2}$$

The velocity of movement of the Iveco bus in the moment, in which the driver actuated the brake system, was approximately 60,60 km/h.

- The velocity of movement of the bus in the moment of the impact can be defined with the help of different analytical methods. In this particular case two of them have been used.

Method 1 (Appel) – the velocity of movement of the bus in the moment of the impact is defined by the distance in which the pedestrian's body is thrown.

$$V_{Iveko-y} = \sqrt{\frac{S_{\tau}}{0{,}070}}, m/s \tag{3}$$

where $V_{Iveco-y}$ is the velocity of the bus at the moment of the impact, m/s;

S_τ - the distance in which the pedestrian's body is thrown, m.

Method 2 – the velocity of movement of the bus in the moment of the impact is defined by the brake traces after the impact. The following formula is used:

$$V_{Iveco-y} = \sqrt{2.S_y.j}, \text{ m/s} \tag{4}$$

S_y - distance travelled from the point of impact till the bus finally stopped;

The velocity of the bus Iveco in the moment of the impact is defined by using the two methods and a scheme (shown in figure 1). The results from the conducted analysis are presented in table 1.

Table 1. Results from the conducted analysis.

№	S_τ m	S_y m	j m/s^2	$V_{Iveco-y.}$ m/s (Appel)	$V_{Iveco-y.}$ m/s (Brake traces)	$V_{Iveco-y.}$ km/h (Appel)	$V_{Iveco-y.}$ km/h (Brake traces)	Δ km/h
1	5,97	5	7,06	9.235026	8.402381	33.25	30.25	3
2	6,36	5,5	7,06	9.531901	8.812491	34.31	31.72	2.59
3	6,81	6	7,06	9.863352	9.204347	35.51	33.14	2.37
4	7,26	6,5	7,06	10.18402	9.580188	36.66	34.49	2.17
5	7,73	7	7,06	10.5085	9.941831	37.83	35.79	2.04
6	8,17	7,5	7,06	10.80344	10.29077	38.89	37.05	1.85
7	8,59	8	7,06	11.07765	10.62826	39.88	38.26	1.62
8	9.1	8.5	7,06	11.40175	10.95536	41.05	39.44	1.61
9	9.56	9	7,06	11.68638	11.27298	42.07	40.58	1.49
10	10.03	9.5	7,06	11.9702	11.58188	43.09	41.69	1.4
11	10.49	10	7,06	12.24162	11.88276	44.07	42.78	1.29
12	10.97	10.5	7,06	12.51856	12.17621	45.07	43.83	1.23
13	11.45	11	7,06	12.7895	12.46274	46.04	44.87	1.18
14	11.95	11.5	7,06	13.06577	12.74284	47.04	45.87	1.16
15	12.34	11.9	7,06	13.27726	12.96256	47.8	46.67	1.13
16	12.44	12	7,06	13.33095	13.01691	47.99	46.86	1.13
17	12.8	12.5	7,06	13.52247	13.28533	48.68	47.83	0.85
18	13.4	13	7,06	13.83577	13.54843	49.81	48.77	1.03
19	13.98	13.5	7,06	14.13203	13.80652	50.88	49.7	1.17
20	14.56	14	7,06	14.42221	14.05987	51.92	50.62	1.3

The values of the velocity at the moment of the accident, estimated with the help of the two aforementioned methods, are closest to each other when the distance at which the pedestrian's body is thrown is S_τ = 12,80 m and the length of the brake traces left after the accident is S_y = 12,50 m.

Therefore, the velocity of the bus at the moment of the traffic accident is estimated to be approximately $V_{Iveco-y}$ = 48,26 km/h.

The place of the collision is determined when the evidence found and described in the written report from the scene of the accident are taken into account: length and location of the brake trace, deformations in the front part of the bus and location of the blood stain.

Considering the established velocity at the moment of the accident it is determined that the scene of the accident is located along the roadway in the right traffic lane (the lane in which the bus was traveling), at a distance of 0,20 m before the accepted baseline, going perpendicular through RP 1 (figure 2) and along the width of the roadway at around 6,70 m to the right of the left end of the roadway.

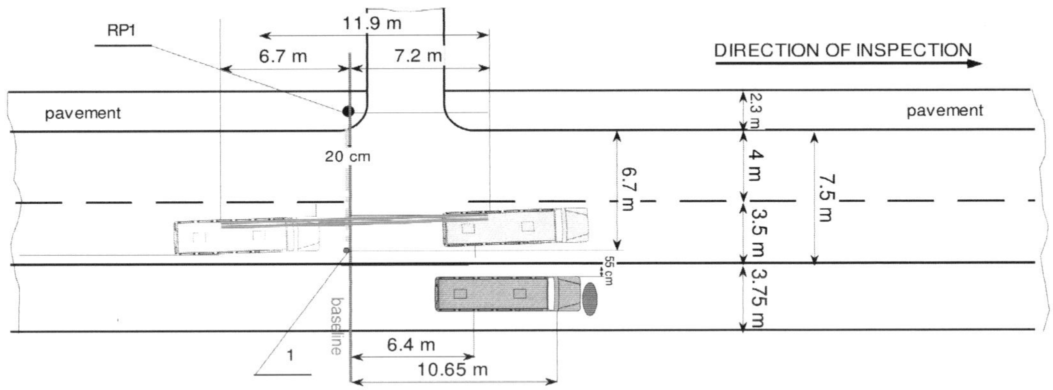

KEY:

1. Place of the initial contact;
2. ------------ Pedestrian movement trajectory.

Figure 2. Scheme of the accident, indicating the location of the impact.

3. Conclusion

Every road traffic accident is characterized with its own specifics, which implies the need of choosing an adequate approach when conducting an expert analysis. The current scientific article proposes a graph-analytic approach for determining the place of the initial impact when analysing road traffic accidents with a pedestrian. In the present research case-law data from a real traffic accident were used. When investigating the velocity of motion of the vehicle that participated in the accident two methods with wide application in the expert practice have been used. The suggested approach can successfully help for the reconstruction of road traffic accidents with pedestrians in situations where there is no specific data in order for the exact place of the initial collision between the vehicle and the pedestrian to be determined.

Acknowledgements

The author/s would like to thank the Research and Development Sector at the Technical University of Sofia for the financial support.

References

[1] St. Karapetkov, Auto-technical expertise. Sofia, Technical University, 2005
[2] M. Savova-Mratsenkova, Djonev G., Influence of the angle of the directory on the accuracy of the measurement when determining vehicles' velocities in case of an impact between them. Conference BulTrans, Sozopol, 11-13 September 2017.
[3] G.Djonev, Savova-Mratsenkova M., Methods for determining the velocity of vehicles in case of an impact between them. Conference BulTrans, Sozopol, 11-13 September 2017.
[4] A. Pisarev, and others. Course in Theoretical Mechanics - Part II. Sofia, Technika, 1988.L.A. Klein, M.K. Mills, D.P. Gibson, in: Traffic Detector Handbook, vol. **I**, third ed., Turner-Fairbank Highway Research Center, McLean, VA, 2006.
[5] V. Ilarionov, Examination of an accident. Moscow, „Transport", (1989).
[6] L.Petrescu, Petrescu Al., Vehicle-pedestrian collisions - Aspects regarding pedestrian kinematics, dynamics and biomechanics, CAR-2017 IOP Publishing: Materials Science and Engineering

252 (2017).
[7] Fang Wang, Chao Yu, Guibing Li, Yong Han, Bingyu Wang, Jikuang Yang, Diandian Lan, A Study on Influence of Minivan Front-End Design and Impact Velocity on Pedestrian Thorax *Kinematics and Injury Risk, Applied Bionics and Biomechanics* Volume **2018**, Article ID 7350159, 8 pages
[8] L. Levulyte et al, Pedestrians' Role in Road Accidents, DOI: http://dx.doi.org/10.7708/ijtte.2017.7(3).04.
[9] World Health Organization. Global status report on road safety 2015/ http://www.who.int
[10] The State-Public Consultative Commission on the Problems of Road Safety. Statistical road safety data / http://dokkpbdp.mvr.bg/default.htm/
[11] National Road Safety Improvement Strategy of the Republic of Bulgaria 2011-2020.

IOP Conf. Series: Materials Science and Engineering **878** (2020) 012043 doi:10.1088/1757-899X/878/1/012043

Use of modern software solutions and systems for analysis and reconstruction of road accidents

I S Damyanov

Technical University of Sofia, Faculty of Transport, Department of Combustion Engines, Automobile Engineering and Transport, Bulgaria

idamyanov@tu-sofia.bg

Abstract. This study examines the possibilities of investigating a road accident and provides initial data for the reconstruction of road accidents using photogrammetric techniques. Compared to the conventional practice of road accident reconstruction, photogrammetry saves time, money and is an effective method for investigating road accidents. The use of unmanned aerial vehicles - UAVs and mapping software enables the collection of collecting the data needed to investigate, analyze and reconstruct road accidents. The measurement results can be considered as initial information for a technique for repairing damage analysis. The study shows that photogrammetry has an increasing potential for application in the investigation and recovery of road accidents.

1. Introduction

"Road safety is a problem that is not paid enough attention to - but it is true that it is one of our great opportunities to save lives around the world," Michael Bloomberg, WHO Global Ambassador for Non-communicable Diseases and Injuries. The 2018 Global Road Safety Report states that road accidents take more than 1.35 million lives annually worldwide [1]. That's 3750 lives every day. In addition, road accidents are already the main cause of death for people between the ages of 5 and 29.

Blocked or closed lanes and congestion are an inconvenience for both drivers and officials responsible for the organization and management of road traffic. By using drones, improving traffic safety and improving the organization is present, but on-site reconstruction and subsequent analysis of road accidents is necessary. Investigating officers should disclose evidence and find out the cause and mechanism of the crash, especially when it comes to physical injuries, death or serious material damage.

Distracted drivers in an emergency situation, delayed or stopped traffic, failing to properly orient themselves, provide a perfect opportunity for secondary road traffic accidents. This issue from the point of view of traffic safety is one of the most important in the investigation and reconstruction of a road accident [2].

The demand for these secondary crashes is that most individuals and employees are quickest to gather evidence and begin to participate in the results of road transport produced [3], but conventional mapping of a fatal accident can be inferred over time. at the time of the severity of the accident.

2. Prerequisites and ways to solve the problem

The actual situation of an accident is only clarified by restoring the actual movement of the objects involved in the incident - vehicles, cyclists and pedestrians [3,4]. The prepared auto-technical expertise is the main source of evidence in the investigation of road accidents [5,6,7]. Conventional data collection techniques required for the analysis and reconstruction of road accidents present risks and problems, the most common of which are:

- the need for a sufficiently large number of highly qualified specialists and employees in the analysis and reconstruction of road accidents to cover the entire road network of the country;

- a long period of time for the operations to collect all the necessary data for the subsequent analysis and reconstruction of the road accident, with the site of the accident completely or partially closed during most of the time;
- high risk of danger for employees involved in the collection of data for the analysis and reconstruction of a road accident;
- the need for inspection of the bodies and staff for the temporary organization of the emergency traffic and the controlled recovery of the overall traffic;
- high risk of danger for people involved in a road traffic accident and all staff involved in its elimination, as well as for all road users in the area of the accident.

Photogrammetry is the science of using images to obtain reliable information about physical objects and the environment through the process of capturing, measuring and interpreting photographic images and models for mapping, measuring distances and other phenomena [8,9]. The minimalist definition of photogrammetry would be: use of photographs to make measurements. The term photogrammetry was first used by the Prussian architect Albrecht Meidenbauer in 1867, however, some of the analysis and study of images is relatively new [10,11]. The term "photogrammetry" itself is Greek by origin and consists of 3 words: photos - light, gramma - recording and metreo - measurements.

The use of unmanned aerial vehicles - UAVs and mapping software for the collecting the data needed to investigate, analyze and reconstruct road accidents will drastically reduce the waste of time and resources [12]. In many cases, even in large road sections, a scene of a car accident takes less than 10 minutes with a drone. The use of UAVs technologies for incident investigations will increase the safety of investigating officers. Once aerial photography is complete, investigating authorities can review office data, reducing the time it takes to stop traffic, which in turn increases the safety of other traffic users and control officers. It also eliminates the need for re-travel in order to collect additional data such as road marking distances, etc. The software programs automatically create a scaled three-dimensional model of the pictures, which allows all measurements to be made by simply rotating the model and by marking two or more points, the measurement can be done with very high accuracy.

The process uses a sufficient number of captured two-dimensional (2D) photographic images, which, after being processed through various softwares, are used to create three-dimensional (3D) digital models, topographic representations or simulations of meteorological models. The hardware to software integration in new technologies is so sophisticated that most photogrammetry operations happen automatically. Photogrammetry is most commonly associated with the production of topographic maps generally than conventional aerial two-dimensional photographs, although digital and satellite images are increasingly used. Photogrammetry using images obtained from aircraft and satellites has been used to create maps at various scales [13]. Recent improvements in the design of cameras and photogrammetric devices have led to an increase in accuracy, allowing mapping at scales up to 1: 200. The photogrammetrically derived 3D information obtained from the images is used to create maps and plans and provide input for terrestrial modeling packages, orthophoto and geographic information systems. Photogrammetry is used to provide national mapping and map review on a small and medium scale. On a larger scale, photogrammetric data form the basis for 3D modeling in a wide variety of applications, including highway design, meteorological anomaly studies, pipeline routing, and more. Photogrammetry is aerial, terrestrial and close-range. Increasingly, UAVs are using sophisticated positioning and control technologies to meet the stability and specific position of the camera during photogrammetry. Aerial photogrammetry requires multiple overlapping images, often up to 80% of the previous image, to produce an accurate, ultra-high resolution large format image. The main difference in terrestrial photogrammetry is that it is usually non-topographical because it is quite difficult to capture and plot large areas of varying height when shooting from the ground. Ground modeling focuses on measuring buildings, engineering structures, film kits, mines, and more.

The mapping process relies on image clarity. When using drones, shooting is best done at around noon when the sun is upright and light shadows are minimal. Depending on the location, noon and vertical sunlight will cause the least visual obstruction [14].

Photogrammetric recordings are made at an angle determined at the angle of the camera of the UAVs, depending on the purpose and object of the study - the camera is perpendicular to the ground or it is necessary to capture the investigated object at an angle. The flight management and position of the drone camera, through modern software products performance are captured from all perspectives as the software uses combined technology for accurate modeling in three dimensions.

3. Results and discussion

An experimental study using an UAVs and three-dimensional photogrammetry software to determine some parameters was carried out in the territory of the Technical University of Sofia, namely the courtyard of block 9. The image processing software used is one of the most widely used and accessible [15], for fast and sufficiently accurate primary processing without the purchase of a license (free access for 14 days). After launching the application, recognizing and establishing communication with the UAVs, all the necessary parameters for the flight area and shooting are determined (Flight Logs Fig. 1). The perimeter of the study area is determined by entering the exact benchmarks (coordinates), the desired altitude for the flight and the shooting, after which the software determines the exact trajectory, the shooting angle and the required minimum number of pictures for the studied area. In the initial experimental study, the choice of altitude for the flight of the UAVs was 50 meters due to the specifics of the surveyed area and the presence of buildings (Fig. 1).

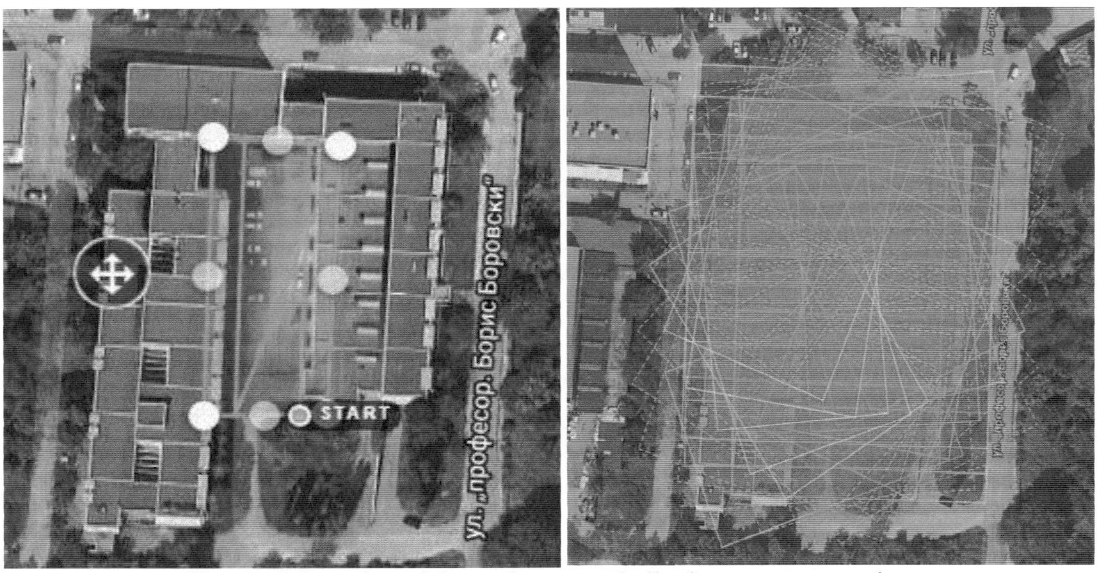

Figure. 1. Location of the study area and pilot image sequence plan.

The figure shows the flight plan, sequence and number of images, the required overlap between consecutive images, the total area of the captured area, the resolution - the resolution of the processed images, represented as the number of centimeters per pixel. After determining the study area and determining the altitude of the operator, the software calculates a minimum number of consecutive shots (in our case 45 frames) and the total time required to complete the flight (2:48 minutes). The status of the battery of the drone battery is checked and, if sufficient, the fields are executed. When exploring larger areas, flight time and trajectory are increased, if the drone's battery capacity is insufficient to perform all of the shooting, the drone lands at a location determined by the software, after replacing the battery by the UAVs recording continues from the moment of interruption. The camera settings during shooting can be in automatic or manual mode, it is provided and it is possible to use the setup through the basic drone software.

After completion of the flight and shooting, images are processed by software. Figure 2 presents image processing information that contains the necessary information for post-processing and downloading. Providing information about the use of the subject to be moved requires a previous site map, which also gives information about the average GPS error of the captured images.

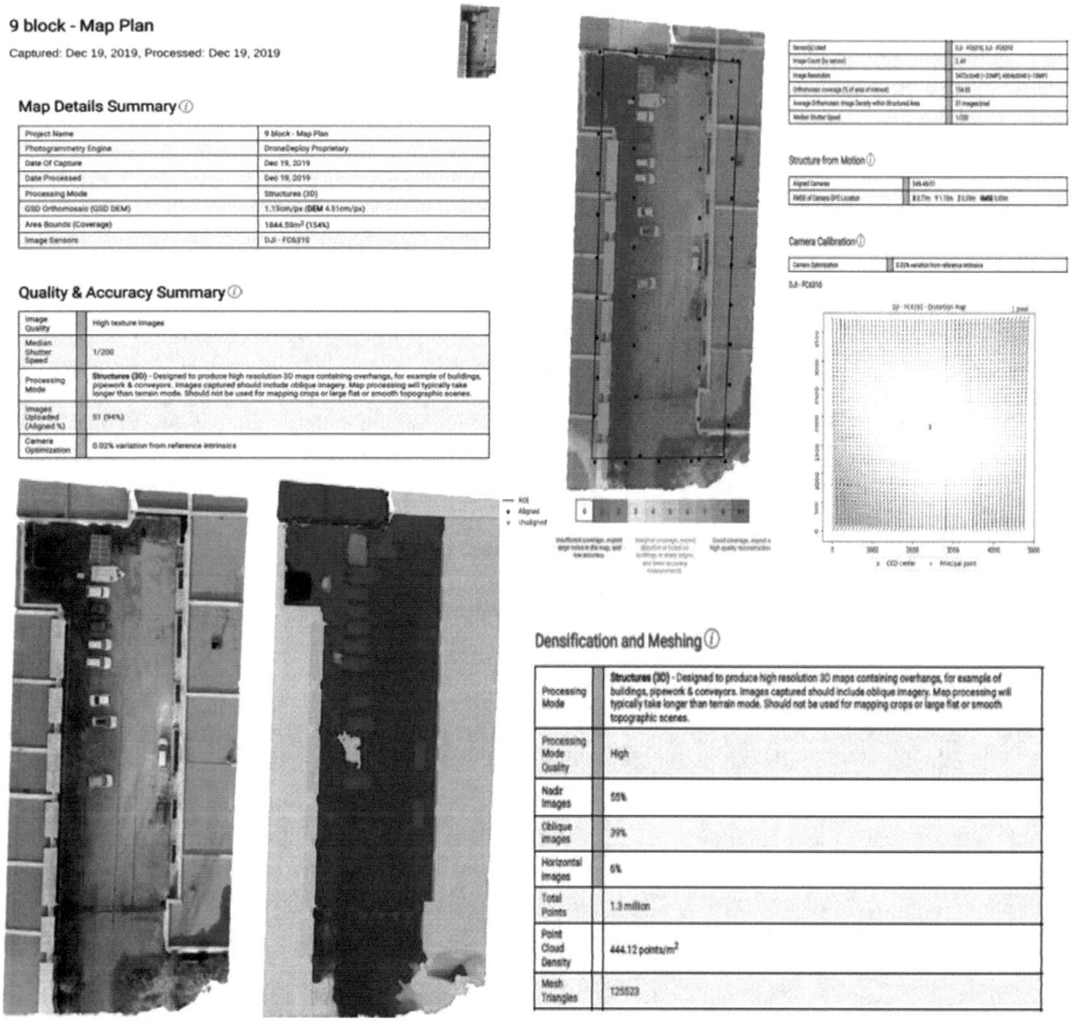

Figure. 2. Area information, sequence of capture, compaction and networking.

The processed images provide an opportunity to determine additional parameters of the study area (terrain, buildings, etc.). The study area includes a building stock, several cars and one trailer, a photovoltaic charging station and one tree. The following figures show some of the data that can be obtained from the processed images. It is possible to determine the coordinates of each selected point in the study area, and can determine slopes and displacements between points of reference. Figure 3 shows the possibility of determining the area of several selected elements (vehicles) from the study area. Several zones have been identified at the same time, with the interior yard of Unit 9 of the Technical University of Sofia being 777 m^2 in the study area, and the area of one of the vehicles that are parked in the study area is defined and shown in the Figure. The figure also shows the possibility of measuring distances between points in a horizontal plane, and measurements in all planes are possible.

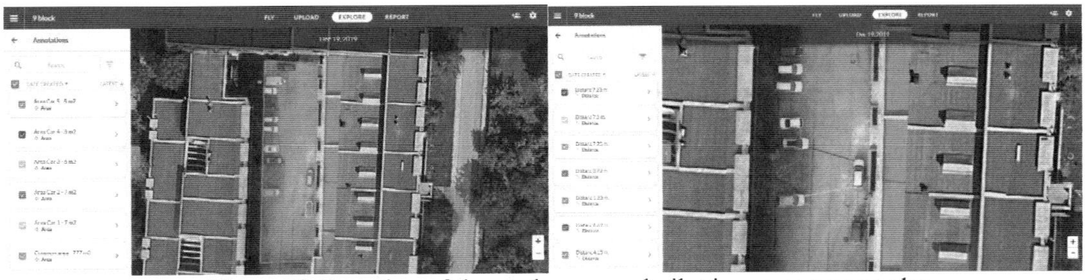

Figure. 3. Location of the study area and pilot image sequence plan.

Processed by software and provided three-dimensional images of the research area, these images are in high definition and allow sufficient accuracy to determine visually and calculate many details.

In order to get a more detailed picture of a particular car using photogrammetric photography, additional shots were taken from a UAV lift height of 10 meters. One on the territory of the 9th building of the Technical University of Sofia (Fig. 4) and the other on the landfill of the Union of the Bulgarian automobilists in Mladost in the city of Sofia (Fig. 5). The software allows accurate and specific shooting of the profile of a car, while calculating its exact area and the volumetric characteristics of the specified car (detail). The quality of the processed images enables accurate road profile capture, unevenness and traces of brake traces, and the listed parameters on the road surface can be determined and measured.

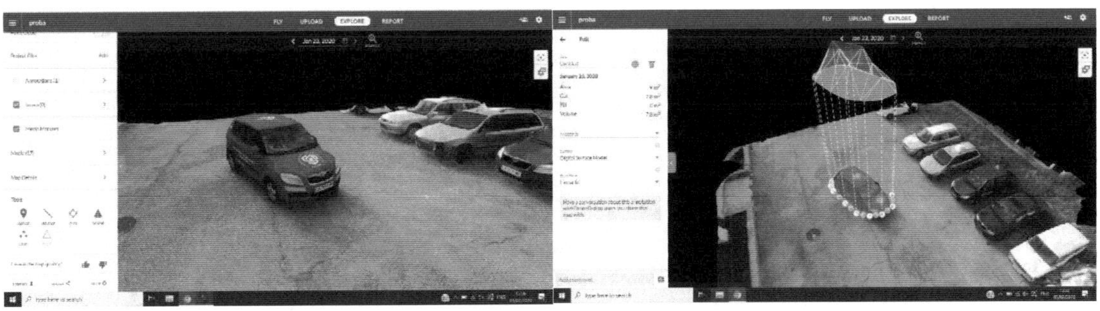

Figure. 4. Three-dimensional detailed image of the examined vehicle.

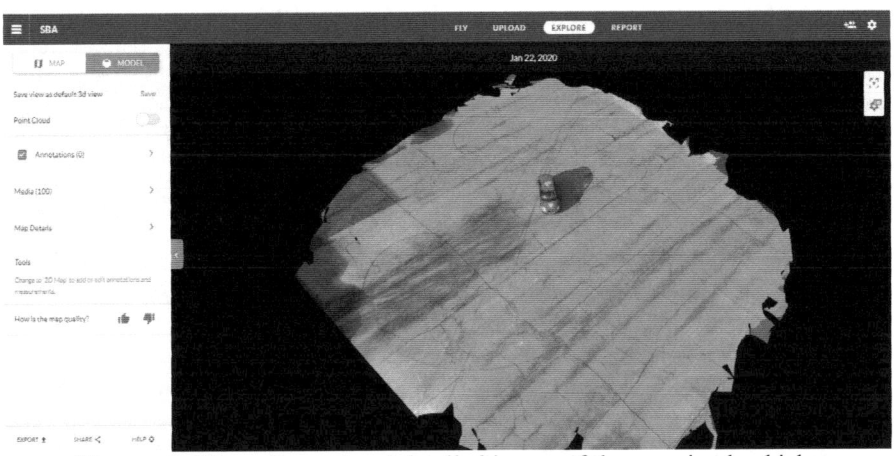

Figure. 5. Three-dimensional detailed image of the examined vehicle.

From the information about the car and the area around it from a lower height (10 meters), it seems that the software processes a two-dimensional image, presented in a three-dimensional picture, presenting a sufficiently accurate, detailed image in scale and a sufficient number of accurate and specific measurements for the vehicle, attack and the space around it.

4. Conclusion

This study presents the potential for road traffic accident investigation and provides initial data for the reconstruction of road accidents using photogrammetric techniques. Compared to the conventional practice of road accident reconstruction, photogrammetry has proven to be a time-saving and effective method for investigating road traffic accidents. The 2D photogrammetry method and the incident trajectory technique can be applied to actual road accidents. Using a portable control system as a reference system, the 3D photogrammetry method can be used in vehicle deformation testing. Measurement results may be considered as initial information on a technique for repairing damage analysis. The study shows that photogrammetry has an increasing potential for application in the investigation and recovery of road accidents.

5. Acknowledgements

The author would like to thank the Research and Development Sector at the Technical University of Sofia for the financial support.

References

[1] World Health Organization, Global status report on road safety 2018, ISBN 978-92-4-156568-4, 2018.
[2] B. Fildes, L. Fechner, A. Linder, Quality criteria for safety assessment of cars based on real-world crashes, Scaling measures and impruvemrnt of data collection. November 2005.
[3] A. Retallack, B. Ostendorf, Current Understanding of the Effects of Congestion on Traffic Accidents, Int J Environ Res Public Health. 2019 Sep; 16(18): 3400, 2019, doi: 10.3390/ijerph16183400
[4] R. Rivers, Evidence in Traffic Crash Investigation and Reconstruction: Identification, Interpretation and Analysis of Evidence, and the Traffic Crash Investigation and Reconstruction Process, Charles C Thomas Publisher, 2006.
[5] G. Djonev, M. Savova-Mratsenkova, Methods for determining the velocity of vehicles in case of an impact between them, BulTrans-2017 Conference Proceedings, Sozopol, 11-13 September 2017, p. 126-128.
[6] D. Fisher, J. Caird, W. Horrey, L. Trick, Handbook of Teen and Novice Drivers: Research, Practice, Policy, and Directions, CRC Press, 2016.
[7] M. Savova-Mratsenkova, G. Djonev, Influence of the angle of the directory on the accuracy of the measurement when determining vehicles' velocities in case of an impact between them, BulTrans-2017 Conference Proceedings, Sozopol, 11-13 September 2017, p. 129-131.
[8] H. Steffan, Accident reconstruction methods, Vehicle System Dynamics 47(8):1049-1073, 2009, DOI: 10.1080/00423110903100440.
[9] L. Cleveland, J. Wartman, Principles and Applications of Digital Photogrammetry for Geotechnical Engineering, GeoShanghai International Conference 2006.
[10] P. Redweik, Photogrammetry, Springer 2012, Retrieved from http://link.springer.com/chapter/10.1007/978-3- 642-28000-9_4/fulltext.html.
[11] D. Stevens, Photogrammetry, Monitoring Building Structures. Springer, Boston, 1992, https://doi.org/10.1007/978-1-4757-5894-8_3.
[12] D. Xinguang, J. Xianlong, Z. Xiaoyun, S. Jie, H. Xinyi, Geometry features measurement of traffic accident for reconstruction based on close-range photogrammetry, https://doi.org/10.1016/j.advengsoft.2008.09.002, 2008.
[13] M. Nolan, C. Larsen, M. Sturm, Mapping snow-depth from manned-aircraft on landscape scales at centimeter resolution using Structure-from-Motion photogrammetry, Published by Copernicus Publications on behalf of the European Geosciences Union, Published: 15 January 2015.
[14] K. Kraus, Photogrammetry, Geometry from Images and Laser Scans, Second edition, Walter de Gruyter, Berlin New York, 2004.
[15] https://www.dronedeploy.com/

IOP Conf. Series: Materials Science and Engineering **878** (2020) 012044 doi:10.1088/1757-899X/878/1/012044

Implementation of advanced software solutions and systems for transport traffic research and control

I S Damyanov

[1]Technical University of Sofia, Faculty of Transport, Department of Combustion Engines, Automobile Engineering and Transport, Bulgaria

idamyanov@tu-sofia.bg

Abstract. This study focuses on the topic of implementation of modern software solutions and systems in transport traffic research and control. The only correct approach to improve traffic organization on the existing road network at minimum cost is to "properly" manage traffic by implementing innovative methods and means to monitor, control, manage and optimize traffic flow indicators. The overall reorganization of road traffic is necessary for a better and safer functioning of the streets and improving not only safety but also the attractiveness and usability of the street network for pedestrians, cyclists and drivers. Ensuring traffic safety and greater security requires more than traditional "engineering, research and implementation".Extensive research is needed to successfully identify and implement the most effective methods and approaches for traffic reorganization and street design.

1. Introduction

The main approach to organizing and management of road traffic is the introduction of innovative methods and tools for monitoring, managing and optimizing traffic flow indicators.

Many studies show that congestion, at least twice a day, kills our productivity, pollutes our environment, damages our health, causes stress, leads to road aggression and turns our dynamic cities into isolated areas. Traffic psychology is a relatively new scientific discipline. Scientists are trying to understand the behavior of people and the mistakes they make that lead to about 90% of road accidents. So far, psychologists have been adamant that you are more likely to become aggressive if you are in your own car than in a public place, since the car's interior creates a false sense of security. On a daily basis we are direct witnesses: when we are behind the wheel and someone crosses our path or takes away advantage; incorrect crossing pedestrian lanes; careless cyclists and drones and more.

The advent of UAVs (unmanned aerial vehicles - drones) into the daily life and rapid development of the technologies in these systems allows them to be successfully used in accounting for the parameters of road traffic flow [1,2,3]. The big advantage of them is that they do not stop the movement in order to install the appropriate receiver (detector). The need to install video cameras to capture traffic at each intersection is eliminated, which is economically feasible and would save a lot of money. The ability to record and directly transmit traffic data in real time to an operator at a traffic management center is essential for good organization and management. Powerful and inexpensive software products are available to handle camcorder data. Mobility of aircraft (drones) is essential [4] because it enables traffic at different intersections to be taken into account, allowing it to be used by different teams and to investigate a sufficiently large number of thoroughfares for a sufficiently short period of time.

2. Prerequisites and ways to solve the problem

In aerial photography, information of a number of road traffic indicators can be obtained: distribution of correspondence between the entrances and exits of the examined thoroughfares; density of traffic flows; traffic flow intervals; time to cross the investigated junction; composition of transport flows; intensity of vehicles; vehicle speed and more. These indicators will cover the study of both road and pedestrian vehicles, cyclists and bicycle lanes. In this way, a large experimental database can be accumulated to enable optimization of the road network.

The advantages of the ability to process video images shot by drones are that they are shot over the road instead of on the road, the placement of vehicle detection areas can be done by the operator, the shape of the detection areas can be programmed for specific applications, and the system can be used for rail vehicles. The disadvantages are the need to overcome obstacles to detection caused by shadows, weather and reflections from the surface of the roadway. Disadvantages can be overcome by designing and installing hardware and designing software algorithms.

The complicated model of a modern road transport network [5,6,7] requires knowing the following parameters:

- Intensity of traffic;
- Vehicle speed;
- Distribution of correspondences between inputs and outputs;
- Density of transport flows;
- Traffic flow intervals;
- Time to pass through the investigated road sections;
- Composition of transport flows;
- Reporting of vehicle registration numbers, by districts in the road network of the given settlement, etc;
- Intensity of pedestrian and bicycle traffic.

All these parameters can be taken into account by video recording [8, 9] of the lane and automated software processing of the motion parameters.

3. Results and discussion

The use of unmanned aerial vehicles for tracking vehicle trajectories and for estimating traffic parameters [10,11,12] is an affordable procedure for determining time-of-flight traffic parameters and providing a useful tool for obtaining data from observations. In fact, the distribution and use of drones, although derived primarily from military applications, has been rapidly expanding for commercial, scientific, leisure, agricultural and other applications [13]. Uses are variable: by the police for surveillance; supply of products; aerial photography; smuggling; racing with drones, etc. [14]. Civilian UAVs now far outnumber military used in urban areas for research purposes to obtain accurate tracking information on video-based vehicles [15]. However, this technology, developed only recently, suffers from problems that have not yet been fully resolved. The level of accuracy of the results of its automotive flow analysis applications has not been adequately tested both under different shooting conditions and modes and under different road conditions.

The elements that form the annotation of the recorded road junctions are the following:

- lanes for traffic, bicycle traffic or pedestrian sidewalks can be defined as entry, exit and other;
- virtual lines at designated locations that register intersecting vehicles, bicycles or pedestrians may be indicated as entry, exit and other;
- geo-referencing of certain points in the picture;
- impact analysis between virtual objects.

Geo-referencing is the internal coordinate system of a map or aerial photo and may be linked to a terrestrial geographical coordinate system [16]. The corresponding coordinate transformations are usually stored in the image file, although there are many possible mechanisms for geo-referencing. The most beneficial effect of geo-referencing is that software can display terrestrial coordinates (such as latitude / longitude or Universal Transverse Mercator (UTM) coordinates) as well as measure terrestrial distances and areas. Geo-referencing means associating a particular aerial imaging area with locations in the physical space.

Experimental studies of traffic flow indicators were performed using an unmanned aerial vehicle. The use of a high resolution camera allows for high quality recording and the ability to track the movement of any pedestrian and vehicle, and hence to obtain the corresponding traffic flows. The software [17] processes the information from the recorded video, links image data (video), geo-referencing, annotations and trajectories of the monitored vehicles, etc. The pursued vehicle can be understood as either an object having an immediate position at any time, or as a complete trajectory in temporal space.

The surveys were conducted on the territory of the city of Sofia at the intersections designated in advance. In the study, the unmanned aerial vehicle was raised to 150 meters in height, with no time to capture traffic at the intersections of more than five minutes, since the software used allows video to be processed without purchasing a license within such a period. The first survey was conducted at the unregulated intersection of Kliment Ohridski Blvd. and Andrey Lyapchev Blvd., and the second survey was conducted at Andrey Lyapchev Blvd. and Andrey Sakharov Blvd. Once the video has been recorded and processed, it is possible to examine and analyze the existing data and manipulate only the view of that data. Figure 1 shows footage of the video footage of the intersections examined.

Figure. 1. Footage of the surveyed intersections.

From the statistical (log. Statistical) of the video footage processed, a total of 432 "elements" involved in the movement were recognized in the five minutes of recording. Type of traffic participant (car or truck, motorcycle, pedestrian, bus, cyclist, etc.), in the analysis is determined and tracked by different parameters: geometric dimensions; shape and speed. The other statistical parameters in the initial analysis are the following: total distance traveled, total follow-up time of all recognized objects, average speed, distance, etc.

Each trajectory of a recognized and identified traffic participant consists of a trajectory line and a marker in the form of a red flag and an identification number. The record can be selected by clicking on its label. The trajectory line is a double-colored green color representing the already traveled part of the trajectory, and the blue color represents the part of the trajectory to be travelled. The Vehicle List shows a list of all that have been detected (recognized) in a given video sequence. Each PPP trajectory can be selected by marking on its recording, providing the following information:

- unique identification ID number of each car;
- vehicle type classification;
- current speed the absolute size of the vehicle's current velocity vector, expressed in kilometers per hour;
- tangential acceleration of the vehicle, expressed in meters per second. A positive value means acceleration and a negative one means slowing down;
- lateral acceleration of the vehicle, expressed in meters per square second. A positive value means acceleration to the right, and a negative value means acceleration to the left;
- total acceleration in absolute magnitude of the direct acceleration vector;
- the average speed of the vehicle at the intersection (between inlet and outlet), expressed in kilometers per hour;
- the duration of the movement of the vehicle spent at the intersection, expressed in seconds;
- the duration of the vehicle spent in the intersection, expressed in seconds;
- the first vehicle ID number of the vehicle ID from the start of its field of capture;
- number of the last image of the vehicle from the beginning of its entry into the field of its capture;
- a unique entry ID through which the vehicle passed when entering the intersection area;
- unique exit ID through which the vehicle passed when exiting the intersection;

In traffic analysis, linear identifiers can be used, divided into three groups: input, output, and neutral. The following information is available for each: ID number; an explanation to it; number and type of vehicles passed through entry / exit (correspondence); time and speed for switching between inputs and outputs (figure 2), etc. It is possible to export data in CSV format with the following content:

- unique record number;
- the numbers of entrances and exits the vehicles pass on when entering / leaving the intersection;
- moment of vehicle passing through the inlet / outlet expressed in milliseconds;
- distance traveled by vehicles, between inlet and outlet, expressed in meters;
- average vehicle speed in the intersection zone (between inlet and outlet), expressed in meters per second;
- list of points in the vehicle path, each point having the following syntax: x [m]; y [m]; speed [m/s]; tangential and lateral acceleration [ms^2];
- position of the vehicle in the x and y coordinate system;
- time marker - the time from the beginning of the video frame when the vehicle was in a certain position.

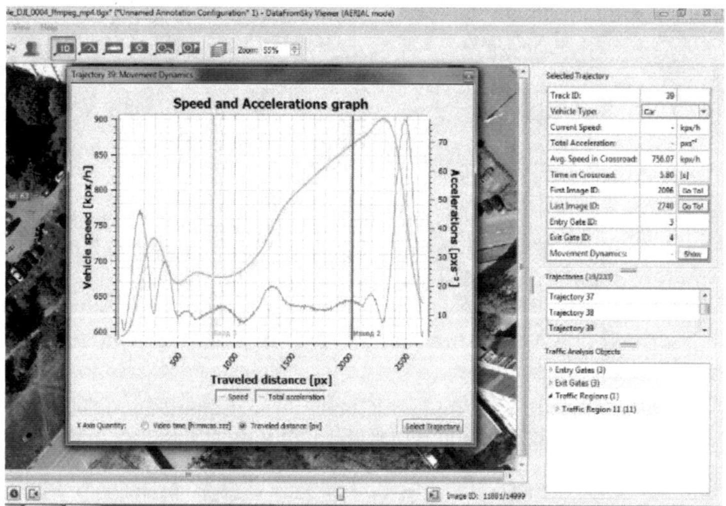

Figure. 2. Motion chart of a specific identified vehicle.

Table 1 presents the statistics of vehicles passing through the entrances and exits of the surveyed intersection for the five minutes interval, with 218 recognized traffic participants passing through the three entrances and 209 passing through the exits respectively.

Table .1. Statistics on vehicles passing through inputs / outputs.

Survey report - Gate Statistics						
	entry 1	entry 2	entry 3	Exit 1	Exit 2	Exit 3
Car count	89	46	57	70	62	56
Medium Vehicle count	9	7	6	6	5	8
Heavy Vehicle count	0	1	0	0	0	0
Bus count	0	2	0	2	0	0
Motorcycle count	0	0	1	0	0	0
Bicycle count	0	0	0	0	0	0
Pedestrian count	0	0	0	0	0	0
Number of all vehicles	98	56	64	78	67	64

In Table 2 are presents the data from vehicles passing through the entrances and exits, as well as from the overpass counting line. It is noteworthy that the difference between the passing vehicles (entry

/ exit) presented in the table is due to the fact that most of the vehicles passing under the bridge are lost by tracing their trajectory. After passing under the bridge, they are re-registered but as new road users, which is why they do not enter the statistics on the correspondence between inputs and outputs.

Table 2. Statistics on vehicles passing through entry / exit / overpass.

Survey report - Gate Statistics									
	entry 1	entry 2	entry 3	entry 4	Exit 1	Exit 2	Exit 3	Exit 4	Overpass
Car count	4	29	10	48	10	40	9	36	22
Medium Vehicle count	3	3	0	4	1	2	1	7	4
Heavy Vehicle count	0	2	1	2	2	0	1	2	0
Bus count	0	0	0	0	0	0	0	0	1
Motorcycle count	0	0	0	0	0	1	0	0	0
Bicycle count	0	0	0	0	0	0	0	0	0
Pedestrian count	0	0	0	0	0	0	0	0	0
Number of all vehicles	7	34	11	54	13	43	11	45	27

It shall be sufficiently detailed to include details of the vehicle identification and type, the time of entry and exit of the designated tail area, the average speed and acceleration in the tail area, and the dwell time (full rest) of vehicles in tail (total and maximum time of each car).

Additional analyzes can be carried out in specific areas: study areas; anonymous perimeter in the study area - unexplored area; traffic area; tail area; traffic area and more. It is possible to export the data in CSV format containing in addition to the above mentioned information and additional such:

- number of traffic participants in the designated area;
- time for movement and stay of each traffic participant in the designated area;
- in the tail area - the length of the tail and all possible times necessary to analyze the time in the queue, first and last in the queue, the acceleration time and the delay of vehicles in the queue (traffic in congestion),
- in the presence of a given anonymous zone in the study area, everything that falls within it is excluded from the analysis.

From the research performed, using an unmanned aerial vehicle on the traffic participants to determine the parameters of the transport process, it seems that the software products to assist the specialists involved in the organization and management of the traffic have evolved radically. It takes a little more time, desire and resource to join all the related units along the chain to achieve optimal operation in the management and optimization of road traffic using modern scientific and technical methods and tools.

4. Conclusion

According to the national strategy for improving the road safety of the Republic. Bulgaria for the period 2011 - 2020 [18], and they are the next ten year period 2021-2030. [19] commitments have been made to the European Economic Community and many tasks are set out to improve both the organization and safety of road traffic, the reduction of road traffic injuries and the optimization of road traffic indicators in order to improve its environmental performance to reduce harmful its impact on the environment and humans. In this regard, the approach to be used in the ever-increasing population and vehicles in major cities is to automatically collect data on traffic parameters and seek optimal solutions to improve them. The approach to solving this multifaceted task, such as optimizing the transport network and avoiding the heavy and expensive projects of deploying roadway detectors, is to use video recording, which will allow for a relatively fast reading of the parameters of transport traffic and its optimization in real time.

5. Acknowledgements

The author/s would like to thank the Research and Development Sector at the Technical University of Sofia for the financial support.

References

[1] M. Khan, W. Ectors, T. Bellemans, D. Janssens,G. Wets, Unmanned Aerial Vehicle-Based Traffic Analysis: A Case Study for Shockwave Identification and Flow Parameters Estimation at Signalized Intersections, Remote Sens. 2018, 10,458, doi:10.3390/rs10030458 www.mdpi.com/journal/remotesensing.

[2] E. Barmpounakis, E.Vlahogianni, J. Golias, Unmanned Aerial Aircraft Systems for transportation engineering: Current practice and future challenges, International Journal of Transportation Science and Technology,Volume 5, Issue 3, October 2016, https://doi.org/10.1016/j.ijtst.2017.02.001.

[3] H. Shakhatreh, A. Sawalmeh, A. Al-Fuqaha, Z. Dou, E. Almaita, I. Khalil, N. Othman, A. Khreishah, M. Guizani, Unmanned Aerial Vehicles: A Survey on Civil Applications and Key Research Challenges, IEEE Access (Volume: 7), DOI: 10.1109/ACCESS.2019.2909530, 2019.

[4] L. Berntzen, A. Florea, C. Molder, N. Bouhmala, A Strategy for Drone Traffic Planning Dynamic Flight-paths for Drones in Smart Cities, The Eighth International Conference on Smart Cities, Systems, Devices and Technologies (SMART 2019), At Nice, France, 2019.

[5] L.A. Klein, M.K. Mills, D.P. Gibson, in: Traffic Detector Handbook, vol. I, third ed., Turner-Fairbank Highway Research Center, McLean, VA, 2006.

[6] Saliev D., “Calculation algorithm for cycle length of signalized intersection”, Machines, Technologies, Materials – International Scientific Journal, ISSN Print: 1313-0226, ISSN Web: 1314-507X, Year XI, Issue 1, 2017, p. 33-34.

[7] Saliev D., “Calculation algorithm for pedestrian green time”, Trans Motauto World – International Scientific Journal, ISSN Print: 2367-8399, ISSN Web: 2534-8493, Issue 1, 2017, p. 17-18.

[8] B. Coifman, D. Beymer, P. McLauchlan, J. Malik, A real-time computer vision system for vehicle tracking and traffic surveillance, Transp. Res. Part C 6 (4) (1998) 777–782.

[9] J.C. Tai, S.T. Tseng, C.P. Lin, K.T. Song, Real-time image tracking for automatic traffic monitoring and enforcement applications, Image Vis. Comput. 22 (2004) 485.

[10] G. Guido, V. Gallelli, D. Rogano, A. Vitale, Evaluating the accuracy of vehicle tracking data obtained from Unmanned Aerial Vehicles, Vol: 5, Issue 3, October 2016, Pages 136-151 https://doi.org/10.1016/j.ijtst.2016.12.001.

[11] B. Coifman, M. McCord, R. Mishalani, K. Redmill, Surface Transportation Surveillance from Unmanned Aerial Vehicles, Conference: Proceedings of the 83rd Annual Meeting of the Transportation Research Board, January 2004

[12] B. Coifman, M. McCord, R. Mishalani, M. Iswalt, Roadway traffic monitoring from an unmanned aerial vehicle, IEE Proceedings - Intelligent Transport Systems 153(1):11- 20 April 2006, DOI: 10.1049/ip-its:20055014.

[13] Tice, Brian P. (Spring 1991). "Unmanned Aerial Vehicles – The Force Multiplier of the 1990s". Airpower Journal. Archived from the original on 24 July 2009. Retrieved 6 June2013. When used, UAVs should generally perform missions characterized by the three Ds: dull, dirty, and dangerous.

[14] Franke, Ulrike Esther (26 January 2015). "Civilian Drones: Fixing an Image Problem?", International Relations and Security Network. Retrieved 5 March 2015.

[15] C. Yahiaa, S. Scott, S. Boylesa, C. Claudel, Unmanned Aerial Vehicle Path Planning for Traffic Estimation and Detection of Non-Recurrent Congestion, https://arxiv.org/pdf/1807.11660.pdf, 2019.

[16] A. Hackeloeer, K. Klasing, J. Krisp, L. Meng, "Georeferencing: a review of methods and applications". Annals of GIS. 20 (1): 61–69. doi:10.1080/19475683.2013.868826, 2014.

[17] https://datafromsky.com/

[18] National Strategy For Improving Road Safety In Bulgaria For The Period 2011–2020.

[19] Staff Working Paper EU Road Safety Policy Framework 2021-2030 – Next steps towards “Vision Zero”

Soil chip thickness of a spading machine with different operating modes

Yordan Stoyanov[1]

[1]Technical University - Sofia, Plovdiv Branch,
Department of Transport and Aircraft Equipment and Technologies,
25 Tsanko Diustabanov str.,4000 Plovdiv, Bulgaria

E-mail: yordan.stoyanov@tu-plovdiv.bg[1]

Abstract. The article uses the SAM software product to synthesize the kinematic model of a spading machine in different operating modes. Dimensions of the kinematic units are set according to the dimensions of the units of real machines. The trajectory of motion of the working body upon which the size and shape of the soil chips depend is analyzed. The soil chip thickness depends on the operating modes of a spading machine. The different operating modes definite different parameters of cross-section area of a soil chip.

1. Introduction

In our country primary soil tillage is done mainly by turning the surface layer at a depth of about 30 cm and is practically used in growing all filed crops requiring such tillage. For this purpose, mainly ploughshares are used and the operation is carried out after harvesting the previous crop, that is, during the summer-autumn season. Surface treatments preceding sowing are carried out during the spring season [2].

In growing vegetable crops outdoors or in greenhouses intensive methods are used and the time between primary tillage and sowing or transplanting is minimal. In these cases it is preferable to carry out the primary tillage with machines with active working organs to provide a grain-size soil composition suitable for the subsequent operations. The primary tillage with them can be done at elevated soil moisture without distorting the microrelief, which is important if watering is by gravitation. According to Coleman [1], primary tillage machines with active working organs are suitable for organic farming and according to their working principle, they are: with rotary working organs (rotary ploughs) and with reciprocating movement (spading machines). Spading machines virtually imitate manual turning of the soil by using a soil with a spade.

According to some authors [3], the advantages of the spading, i.e. the better residues incorporation and the absence of the soil compacted layer, cannot be easily evaluated, because in this case the crop yield and quality should be checked for a significant time period.

The time needed for cutting the soil chip is determined depending on the constructive parameters and the operating mode of the performing unit [4].

According to some authors, the spading machine [5] with active drive ensure maximum improvement of soil structure by intensive crushing, aeration of the cultivated soil layer and its separation in fractions. In coarse soil aggregates and residues from grown vegetables remain on the surface, preventing soil from claying, drying, wind and water erosion. Provides excellent levelling of the soil surface, a prerequisite for precise execution of subsequent technological operations.

IOP Conf. Series: Materials Science and Engineering **878** (2020) 012045 doi:10.1088/1757-899X/878/1/012045

The objective of the present paper is to study the soil chip thickness of a spading machine by kinematic model created by a specialized engineering software.

2. Material and methods

The synthesis of kinematic models of the spading machine and determination of the soil chip thickness has been done by the software product SAM [7]. The software product allows drawing of a mechanism, setting operating parameters of that mechanism, and outputting velocity and acceleration graphs relative to the path of selected points of the mechanism.

The following unit dimensions were used to draw the mechanism of the spading machine (Figure 1):

- knee length 1 (crankshaft rotation radius) – 0,15 m;
- arm length for mounting the working body 19+20+23 – 0,37 m;
- length of working body 22 (spade) – 0,3 m;
- distances between the crankshaft axis and the hinge for mounting the timing bar on the frame 3+10 – 0,4 m;
- length of the timing bar 7 – 0,26 m.

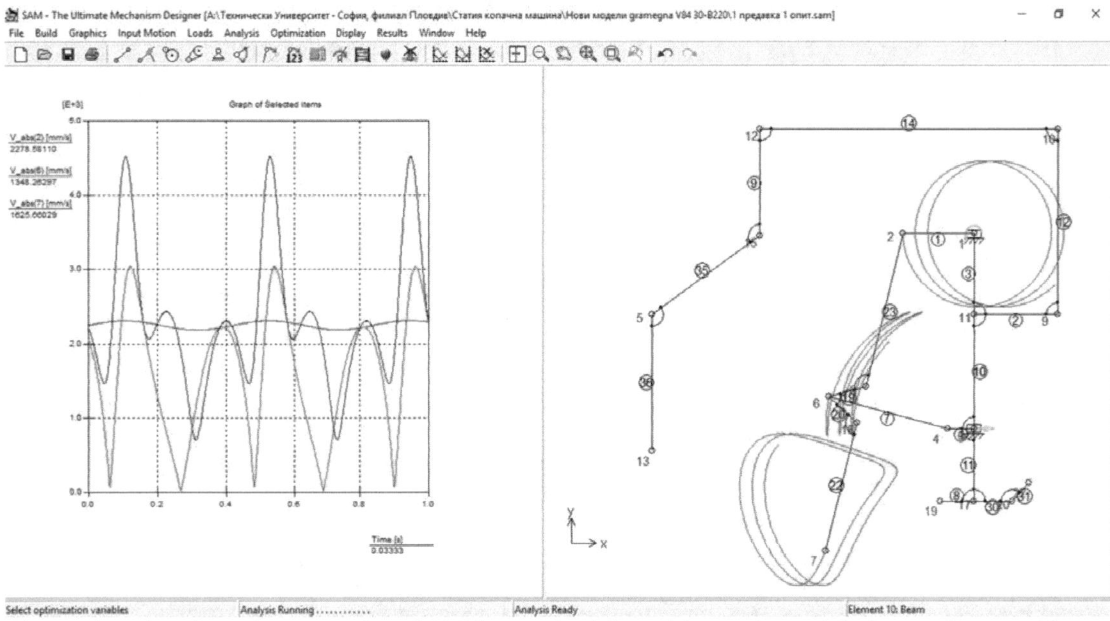

Figure 1. Using the software product SAM for synthesizing a kinematic model of a spading machine

It is assumed that the tractor with which the machine is aggregated moves in one gear and the changing of engine rotations variations the velocity of movement (working velocity *Vw*) which takes the following values 0,063 m/s, 0,147 m/s, 0,231 m/s. Since the velocity of rotation of the independent power take-off shaft (PTOS) of the tractor is proportional to its velocity (within one gear), at the relevant velocities of movement the velocity of rotation of PTOS is 460 min^{-1}, 500 min^{-1} and 540 min^{-1}. The gear ratio of the spading machine reductor is 3,21, therefore the crankshaft velocity of rotation of the machine ranges from 143 min^{-1} to 168 min^{-1}.

The trajectories of movement of selected points in the machine mechanism have been built, the absolute velocity of the working body and the parameters of the soil chips at operation of the spading machine have been determined.

3. Results and discussion

Fig. 2 presents a diagram of a spading machine created by using the software product SAM. The trajectories of point B at the crank of crankshaft 1 of the machine and point M of the working body (spade) 3 of the machine are shown. The trajectory of point B is divided into sections at 45° by the crankshaft rotation angle. The trajectory of point M is also divided into sections that correspond to the given rotation angle of the machine crankshaft. It can be seen that the individual sections of the trajectory of point M, corresponding to the same angle (45°) of rotation of the machine crankshaft, are of different length. Accordingly, the absolute velocity at point M of the working body has different meanings at different rotation angle of the machine crankshaft.

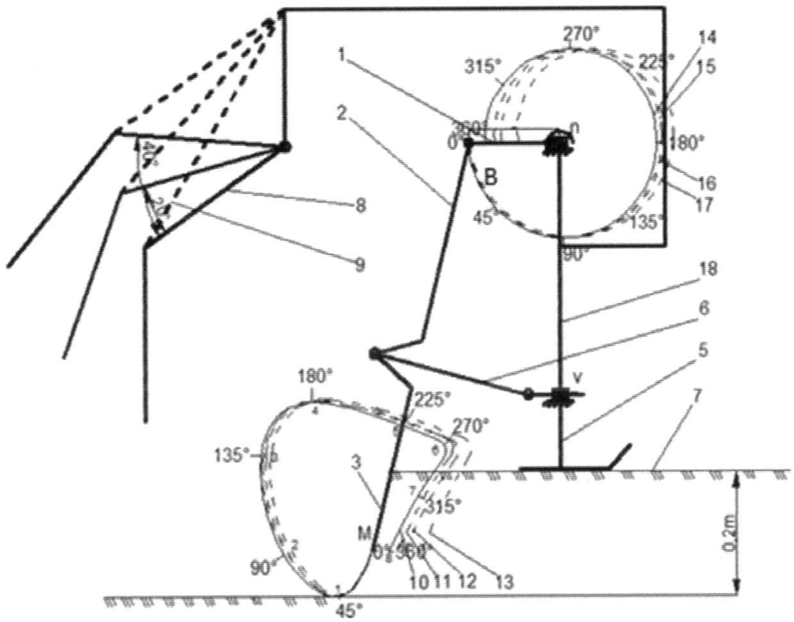

Figure 2. Diagram of a spading machine:
1 – crankshaft; 2 – arm; 3 – working body (shovel); 4 – arm; 5 – sliders; 6 – timing bar; 7 – soil layer; 8 – deflector cap; 9 – chain for cap adjustment chain;10, 11, 12, 13 – trajectory of movement of point M of the working body (the shovel); 14, 15, 16, 17 – trajectory of movement of point B of the crankshaft.

Fig. 3 is a graphical representation of the variation of the absolute velocity of the M (lower end of the working body) relative to the crankshaft rotation angle. It can be seen that within a small interval of rotation of the machine crankshaft the velocity of point M increases significantly. In this interval the soil is overthrown by the working body, and therefore higher velocity, respectively greater kinetic energy of the overthrown layer is needed in order to obtain better fragmentation of the soil.

The values of the trajectory and the absolute velocity of point M of the working body of the machine are shown in table 1. It shows that the total length of the road traversed by point M of the working body of the machine does not change and is about 1 m per revolution of the crankshaft. The absolute velocity of the working body increases with an increase in PTOS rotation frequency from 5,203 m/s at 460 min^{-1} to 6,165 m/s at 540 min^{-1}. The absolute velocity of the working body reaches its maximum in the range between 90° and 135° of the crankshaft rotation.

The trajectory covered by point M on the spade on the vertical plain OXY (Figure 4) can be divided depending on the process carried out by the spade into the following sections:

- A – B – inserting the spade into the soil;
- B – C – overthrowing the soil;
- C – D – moving the spade for the next insertion.

The soil chip is the area surrounded by points AD0B. It can be seen that it has almost Rectangular cross section (Figure 5). The thickness of the chip is equal to the distance between points A and D, which is the step of the working body. In individual sections, the absolute velocity value changes as follows: when the spade penetrates into the soil it is the lowest; when the soil aggregates are dropped off the spade, the speed reaches its maximum; when shifting the spade for subsequent penetration into the soil, the speed is slightly higher than that at insertion.

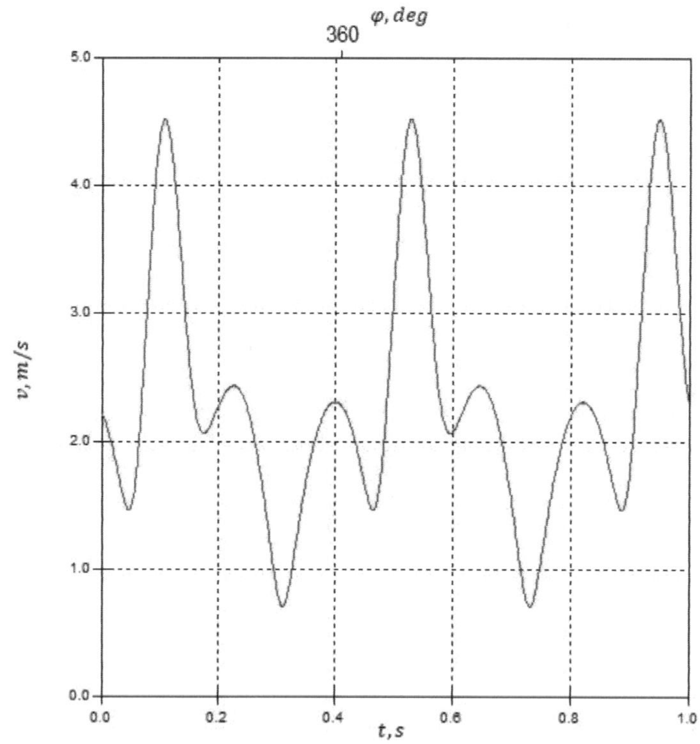

Figure 3. Diagrams for absolute velocity V at point M in the mechanism of the spading machine according to the angle φ of rotation of the machine crankshaft

Figure 4 shows the step of the working body of the spading machine - the distance between two identical points of two consecutive trajectories. At the accepted operating parameters it is 0.088 m. The step of the machine is determined analytically by the expression

$$X = \frac{60.V_w.i}{n}, m, \quad (1)$$

where *Vw* is the progressive speed of the machine, *m/s*;
i – the transmission ratio of the reductor of the spading machine;
n - rotation speed of the tractor power take-off shaft, min^{-1}
The rotation speed *n* of the tractor PTOS at the same tractor gear changes in pro rata to the crankshaft rotation frequency of the tractor engine, and the speed of movement *Vw* also changes accordingly. For this reason, the machine step remains constant when changing the rotation frequency of the PTOS within the gear used.
From Fig. 4, it can be seen that the bottom of the furrow in operation of the spading machine is uneven, the height of ridge depending on the size of the step of the working body. At a smaller step Fig. 4 the height of ridges along the bottom of the furrow decreases, and a bigger step Fig. 5 the height of ridges along the bottom of the furrow increases. The soil chips are with different thickness because the step X (1) of the working body of a spading machine is different. The soil chip is the area by points ADOB in the figures.

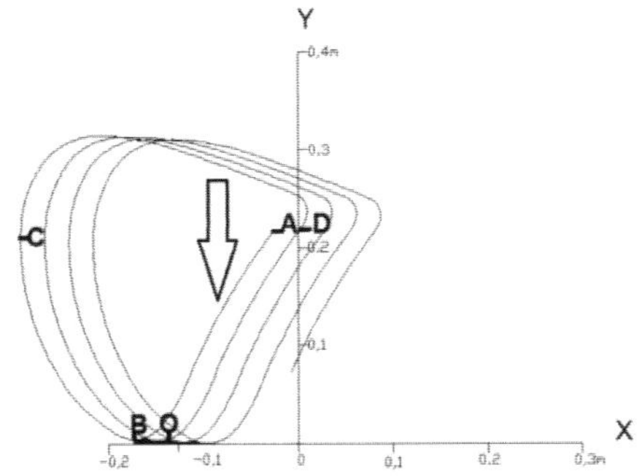

Figure 4. Trajectory of point M on the working body of the spading machine for smaller step.

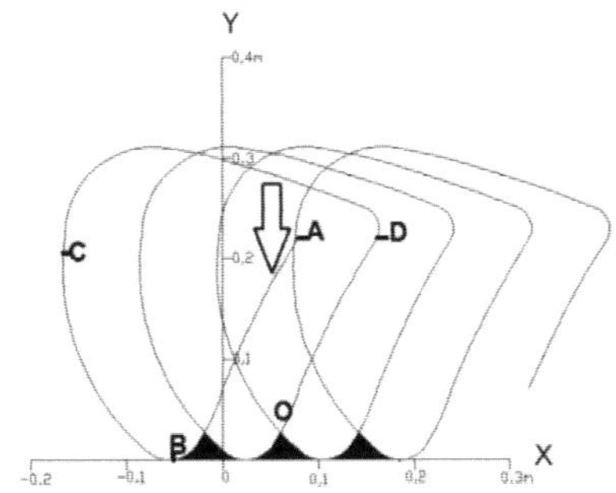

Figure 5. Trajectory of point M on the working body of the spading machine for bigger step.

Table 1. Values of absolute velocities and the distance covered by point M of the spade depending on the crankshaft rotation angle of the machine

Crankshaft rotation angle, °	Distance covered by point M, m	Absolute velocity V, m/s
At rotation frequency of PTOS n = 460 min^{-1}		
45	*0,116*	*2,006*
90	*0,233*	*4,410*
135	*0,350*	*3,968*
180	*0,467*	*2,082*
225	*0,584*	*2,361*
270	*0,700*	*1,608*
315	*0,817*	*2,086*
360	*0,935*	*2,242*
Max		***4,514***
Min		***0,695***
At rotation frequency of PTOS n = 500 min^{-1}		
45	*0,116*	*2,224*
90	*0,233*	*4,762*
135	*0,350*	*4,271*
180	*0,467*	*2,242*
225	*0,584*	*2,584*
270	*0,700*	*1,797*
315	*0,817*	*2,266*
360	*0,935*	*2,433*
Max		***4,961***
Min		***0,753***
At rotation frequency of PTOS n = 540 min^{-1}		
45	*0,116*	*2,324*
90	*0,233*	*5,168*
135	*0,350*	*7,509*
180	*0,467*	*2,435*
225	*0,584*	*2,767*
270	*0,700*	*1,895*
315	*0,817*	*2,470*
360	*0,935*	*2,601*
Max		***5,314***
Min		***0,825***

The study has been conducted at the farm of agricultural producer. The soil aggregates dimensions are determined according the Kachinsky classification [6]. During each experiment from the experiment plan a soil sample is taken to working depth, which is divided by sieves into fractions of different aggregate sizes. Sieves with openings 0,001, 0,01, 0,025, 0,05 and 0,1 m are used, these are weighed to the nearest 1g and the values are averaged. The relative share of fractions in the sample is determined by the dependence [8]. The fraction size is with no difference when the change the operating speed rotation frequency of the PTO. The step X (1) determinate by tree operating speeds is 0,0225m (Fig. 4) 0,0524m 0,0824 m (Fig. 5). Step size 0,0225m including fraction size from 0,001 to 0,01 mm. Aggregates with this sizes up to 0,01 m are the most desired in preparation soil tillage. When using the machine for pre-sowing soil tillage it is preferable the share of that fraction to be until 50 percent of total aggregates. Step size 0,0524 m including fraction sizes from 0,001 to 0,01 m up to 40 percent and fractions sizes from 0,01 to 0,25 and form 0,25 to 0,50 m up to 40 and 25 percent. Step size 0,0824 m including fractions sizes above 0,1 m up to 95 percent of total aggregates. The position of the deflector cover helps to better crushing of the aggregates. When the deflector cover is on 0 degrees crushing increase and decrease with the increase degrees.

4. Conclusion

An analysis of the kinematics and the soil chip thickness of a spading machine has been made by using a specialized engineering software. It has been found out that the speed of the working body varies considerably depending on the rotation angle of the machine crankshaft. The trajectory of the working body and the shape of the soil chip shredded by it is determined. The soil chip thickness increase with the working speed of a spading machine. The soil fractions of certain size have been drawn in relation to operating speed, frequency of rotation of the PTO of the tractor and the position of the deflector cover. The granulometric composition of the soil when cultivated with a spading machine is influenced by the machine operating speed and the position of the cover.

Acknowledgments
The authors would like to thank the Research and Development Sector at the Technical University of Sofia for the financial support.

5. References

[1] Coleman, E. (1995). The New Organic Grower. Chelsea Green, White River Jct., VT.
[2] Delchev N., Trendafilov K., Stoyanov Y. (2017). Study of the kinematics of a spading machine by using engineering software. ARTTE Vol. 5, No. 4, ISSN 1314-8788 (print), ISSN 1314-8796 (online), doi: 10.15547/artte.2017.04.001.
[3] Giordano D. M., Facchinetti D., Pessina D. (2015). The spading machine as an alternative to the plough for the primary tillage, Journal of Agricultural Engineering, volume XLVI: 445.
[4] Guglev D. (2009). THEORETICAL RESEARCH ON A VERTICAL ROTARY SOIL TILLAGE UNIT. Agricultural sciences, Plovdiv, volume I, issue 2, ISSN 1313 – 6577.
[5] Guglev D. (2011). Defining the critical kinematic parameters of rotary harrow with vertical axis of rotation. AGRICULTURAL SCIENCE AND TECHNOLOGY, VOL. 3, No 3, pp 237 – 239.
[6] Kachinsky N. (1965). Physics of the Soil. Part 1, 1st ed., High School, Moscow, p. 323 (Ru).
[7] SAM Mechanism design by Artas Engineering Software [Online]. Available: http://www.artas.n0l [2017-12-18].
[8] Tenu I., Carlescu P., Cojocariu P. and Rosca R. (2012). Impact of agricultural traffic and tillage technologies on the properties of soil. Resource Management for Sustainable Agriculture Vikas Abrol and Peeyush Sharma, http://dx.doi.org/10.5772/47746.

IOP Conf. Series: Materials Science and Engineering **878** (2020) 012046 doi:10.1088/1757-899X/878/1/012046

Design of an automated railway crossing system with Verilog language in CPLD

M Zhilevski[1] **and V Hristov**[1]
[1] Faculty of Automatics, Technical University of Sofia, Bulgaria

E-mail: mzhilevski@tu-sofia.bg

Abstract. In this paper the basic requirements for the automatic railways crossing devices are formulated. A block diagram, which consists of elements that make the automatic railways crossing systems are presented and their working principles are discussed. Based on the hardware description language - Verilog, a digital railway crossing control system is designed and is implemented with complex programmable logic device, taking into account the requirements. As an opportunity for practical application the controller CoolRunner II could be used. Simulation results with set input parameters are shown to represent the functionality of the designed system. The results obtained can be used for design, implementation, simulation and development of such types of systems in railway crossing automatics.

1. Introduction

The rail transport is one of the most energy efficient means of mechanized land transport, characterized by relatively high safety, comfort and cheapest modes of transportation. As a result of these benefits, rail transport is a major form of public transport in many countries [1, 2, 3].

But accidents related with railways are more dangerous than other transportation accidents. Therefore more efforts are necessary for improving safety. There are many railway crossings without automatic systems. Hence many accidents occur at such crossings since there is no one to take care of the functioning of the railway gate when a train approaches the crossing. That's why different automated railway crossing systems are used [4, 5].

The railway crossing represents the passing through a railway line of a road for other road transport, tram and trolley lines, as well as with other railway lines. The safety of traffic through crossings for both rail and road transport is achieved by means of automatic railway crossing devices (ARCD). They provide maximum possibilities for the two modes of transport with a minimum time to close the crossing before the railroad vehicle arrives and open it immediately after the clearance zone of the last axle is released [1, 2].

The complex programmable logic devices (CPLD) like the controller Xilinx's CoolRunner II are appropriate option in designing and implementing a control of many different types of electromechanical objects [6, 7, 8, 9]. These devices comprises of a group of programmable functional blocks. The inputs and the outputs of these functional blocks are connected together by a global interconnection matrix. This interconnection matrix is reconfigurable, so that we can modify the contacts between the functional blocks. There will be some input and output blocks that let us unite CPLD to the external world. Thanks to the hardware description language like Verilog, it is relatively easy to create digital schemes for control. The main advantage of Verilog is that it allows the system to be described (modeled) and verified (simulated) prior to implementation [10, 11, 12, 13].

In this paper the main requirements for the automatic railway crossing devices are formulated. The schematic of the elements that make the automatic railways crossing devices are presented. Based on the language Verilog, an automated railway crossing control is designed and implemented with complex programmable logic device. The CPLD controller is used like practical application for control of barriers, signaling to road vehicles, and so on. Some simulation results of railway crossing systems are shown and discussed.

2. Automated railway crossing devices

The automated railway crossing devices are the main elements used in designing and implementing the railway crossing system.

2.1. Basic requirements

The control system of the ARCD must be located in a cabin or other suitable room and protected from unauthorized access and inappropriate weather. ARCD apparatus executes the algorithm of action and performs control functions of all components. It can be implemented with electronic and / or relay elements [14].

The supply voltage required for the normal functioning of the ARCD is provided by two independent sources (industrial network and contact network) that switch automatically to ensure the safe functioning of the ARCD.

The basic requirements for the technical parameters of the ARCD can be formulated as follows [15]: to be connected with a special technical specification; power supply must be with the following parameters: 220V, 50Hz; to meet the safety level requirements - not lower than SIL3; the time of turning on the red flashing light of the cross railway traffic lights (CRTL) after an overlay closing command should be 1 second; the time of the closing command should be 6 ÷ 9 seconds after switching on CRTL; horizontal position of the beams should be 16 seconds; the ARCD contains a diagnostic system – an information board, which provides details for the position of the railways at one of the close stations.

The basic elements and the principle of operation of a railway crossing system must be taken into account when designing, implementing and simulating.

2.2. Main blocks

Main blocks of the automatic railway crossing devices are presented in Fig.1. Basic notations are as follows: CE – cabin for equipment; CRS – cross railway signals; 1, 2, 3, 4 – traffic sensors (axle counters); BM – barrier mechanisms; BS1 and BS2 – block sections.

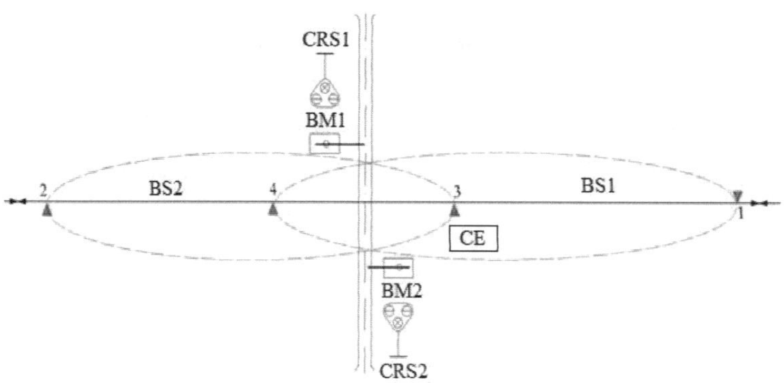

Figure 1. Main blocks of the automatic railway crossing devices.

The cabin for equipment is used for accommodation and installation of relays, fuses, rectifier, converters, axle counters, etc. related to ARCD.

The Cross Railway Signals are two pieces situated on the left and right side of the train. When a railway vehicle is crossing the traffic lights, they should be lightening consecutively with a red light of 60 ± 5 flashes per minute and the audible alarm should be activated.

Four traffic sensors are used and the railway vehicle should activate the ARCD for at least 30 seconds before entering the area of the railways. The axle counters are used to control the movement of railways before and after the passing of the vehicle.

One configuration of a railway crossing is shown in Fig. 2.

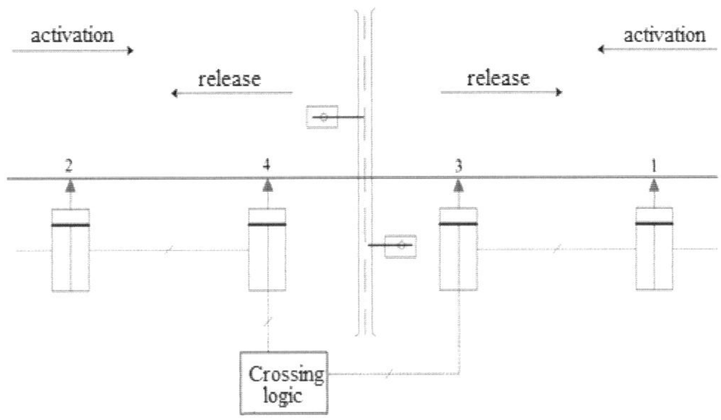

Figure 2. Configuration of the railway crossing.

Automatic barriers are falling down only on the one side of a road with two-way movement (the barrier end may not reach the center of the roadway by at most 1.00 m). When the traffic lights turn red, the beams begin to fall and a maximum of 16 seconds occupy a horizontal position.

2.3. Operation principle

Traffic information on the train is transmitted by four road sensors. The used Xilinx CoolRunner-II controller is used to address the operational-technical requirements for the automatic overlay devices.

2.3.1. Output state of the apparatus

In the non-active state of the ARCD when a railway vehicle is not in the area of the railways, the traffic of vehicles is allowed - road traffic lights blink in white - 30 pulses per minute at a pulse / pause ratio of 1:1 ± 10% [16], the barriers are in the vertical position (85 ÷ 90° to the horizontal). This informs the driver that there is no train in the area of the overturning zone.

The controller receives information about: free block sections BS1 and BS2; control of road traffic lights; superior control of the barrier mechanisms and integrity of the beams. It provides information that the traffic is permitted when there is a white flashing light on CRS.

2.3.2. Automatic closing of the level crossing

Upon the arrival of the railway vehicle over the first axle of the railway, a command from CoolRunner-II is generated to close the level crossing. The white traffic lights switch to red lights; the buzzers in CRTL turn on; after 6 seconds a command is generated to close the overlay and the barriers are lowered to horizontal position (maximum 16 seconds); after obtaining a lower control of the barrier mechanisms, the buzzer is turned off.

2.3.3. Open the railway crossing

When the railway vehicle leaves the crossing (passes a counting point) the overhead device opens for the road vehicle - the barrier beams are raised to the vertical position (max. 16 seconds) after releasing the area from the railroad vehicle. Upon receiving upper control of the barrier mechanisms, the red flashing lights of the CRTL switch to white ones.

IOP Conf. Series: Materials Science and Engineering **878** (2020) 012046 doi:10.1088/1757-899X/878/1/012046

3. Synthesis of the control program

The synthesized program in Verilog is divided into four main parts: defining inputs and outputs; determination of registers, parameters and clock generator; initialization, clock generators and axle counters; logic control.

Some of the used inputs and outputs are: "clk" – an input for generating a clock signal; "in1", "in2", "in3", "in4" – 1-bit inputs for axle counters; "barriers_up" and "barriers_down" – 1-bit inputs for extreme top and bottom positions respectively; "lights_error" and "barriers_error" – 1-bit inputs showing that there is a fault with the lights and barriers respectively and others.

Some of the used registers, parameters and clock generator are: **reg** [7:0] count1, **reg** [7:0] count2 – 8-bit registers for internal counters 1, 2 respectively; **reg** r_barrier1, **reg** r_barrier2 – 1-bit registers related to the outputs „barrier1" and „barrier2"; **reg** r_white1, **reg** r_white2 – 1-bit registers related to the outputs „white1" and „white2".

The first part of the Verilog code development initializes the signals in such way that when the program starts it will be in an open state and without any errors:

```
initial
begin
  count1<=0;
  count2<=0;
  clk_lights<=0;
  clk_lights_delay<=0;
  clk_sounds<=0;
  clk_sounds_delay<=0;
  r_barriers_alert<=0;
  r_lights_alert<=0;
  r_barrier1<=0;
  r_barrier2<=0;
end
```

The next part of the code generates a clock generator that is used to control the flashing of the lights from CRS:

```
always @ (posedge clk)
begin
  if(clk_lights_delay==149)
  begin
    clk_lights<=~clk_lights;
    clk_lights_delay=0;
  end
  else
  begin
    clk_lights_delay=clk_lights_delay+1;
  end
end
```

With each increase of the signal "clk", the variable "clk_lights_delay" will increase by 1 until it reaches the set value in the "if" block, which in this case is 149. Then "clk_lights_delay" will reset and the clock generator "clk_lights" will change its status from logical "0" to logical "1" or inverse.

In a similar way the creation of a clock generator is realized, which is used to control the signaling sound and axle counters in both directions.

The logic of operation of the railway crossing system is described by the following Verilog code:

```
always @ (posedge clk)
begin
```

```
if (lights_error==0 && r_barriers_alert==0)
begin
```

For the implementation of the opening of the railway crossing the following code is used:

```
if (count1==0 && count2==0)
begin
  r_barrier1<=0;
  r_barrier2<=0;
  r_red11<=(barriers_up)?0:clk_lights;
  r_red12<=(barriers_up)?0:~clk_lights;
  r_red21<=(barriers_up)?0:clk_lights;
  r_red22<=(barriers_up)?0:~clk_lights;
  r_white1<=(barriers_up)?clk_lights:0;
  r_white2<=(barriers_up)?clk_lights:0;
  r_sound1<=0;
  r_sound2<=0;
end
```

When the signals "count1" and "count2" are in logical "0", this means that the railway crossing is free and the passage of vehicles is allowed. The signals "barrier1" and "barrier2" become in logical "0", which means that the barrier mechanisms have been signaled to be opened. Until logic 1 receives the signal "barriers_up", the lights "r_red11", "r_red12", "r_red21", "r_red22" flash with the clock frequency "clk_lights", and the lights "r_white1" and "r_white" are off. When the signal "barriers_up" becomes in a logical "1", it means that the barriers are in the highest position and the vehicles can pass. Then the lights "r_red11", "r_red12", "r_red21", "r_red22" go out and the lights "r_white1" and "r_white" flash with the frequency of the clock generator "clk_lights", which indicates to the vehicles that they can pass freely. The buzzers "sound1" and "sound2" are off in this process.

For the implementation of the closing of the railway crossing the developed code is similar like in the opening of the railway crossing.

Protection against damage to light signals and barrier mechanisms has been synthesized in Verilog code.

4. Simulation results

In the Verilog Test Bench Code a 4-axle passing locomotive is simulated in one direction. Some results from the computer simulations are presented in the next figures.

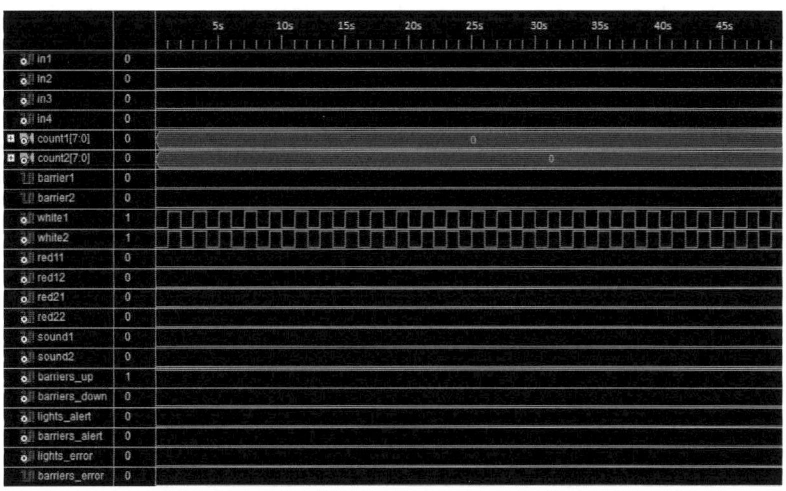

Figure 3. Initial initialization.

Fig. 3 presents the results obtained in the initiative initialization of the program. At this initialization of the program the railway crossing is open. Axle counters are reset. Barriers are lifted. Feedback is obtained that the barriers are in the extreme top position. The white lights blink with 30 pulses per minute at a pulse/pause ratio of 1:1 ± 10%. Red lights and buzzers are off. Lights and barriers as well as their alarms are turned off.

Fig. 4 shows the results obtained when the railway crossing passes from open to closed state. When the locomotive enters the section in one direction of the movement, the corresponding counter increases its value by as many as the number of wheels passed. The counter is different from 0. The white lights stop flashing. The red lights start flashing with 30 pulses per minute in a pulse/pause ratio of 1:1 ± 10%. After 6 seconds the barrier signal is switched. While the barriers are closed, the buzzer starts sounding. When feedback is received that the barriers are in the extreme lower position, the buzzer stops sounding.

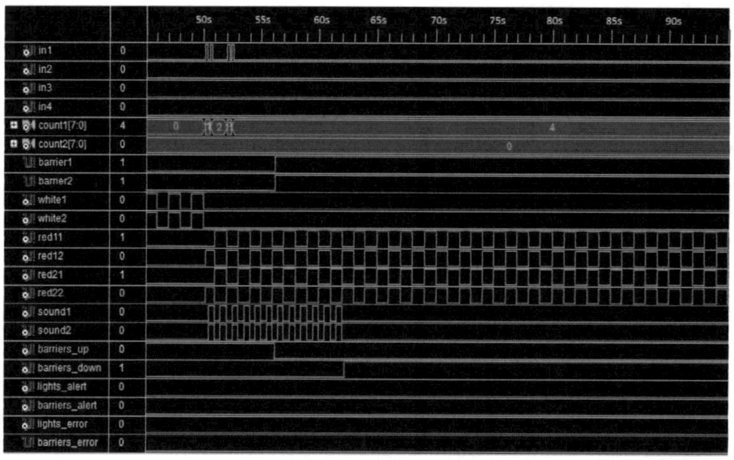

Figure 4. Passage of railway crossing from open to closed state.

Fig. 5 presents the results obtained during the passage of the crossing from the closed to the open position. Upon exiting the locomotive from the section on one direction of movement, the corresponding counter decreases its value by as many as the number of the wheels passed. When the counter is set to 0, a barrier signal is displayed. When the feedback is received that the barriers are in the extreme position, the red lights stop blinking. The white lights start flashing with 30 pulses per minute at a pulse/pause ratio of 1: 1 ± 10%. The buzzers are turned off.

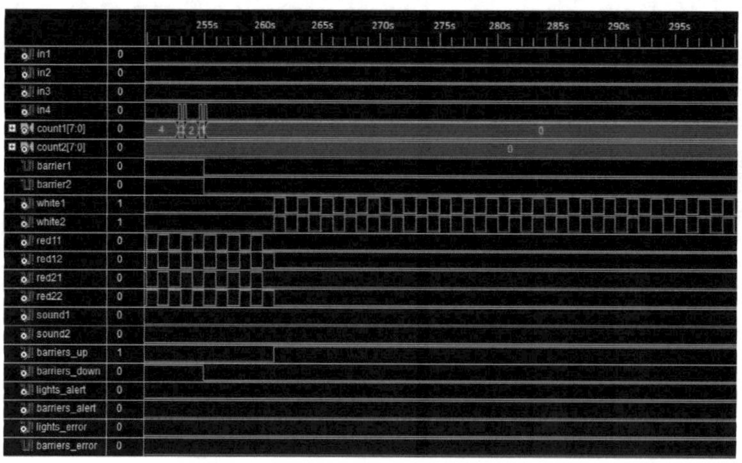

Figure 5. Passage of railway crossing from closed to open state.

IOP Conf. Series: Materials Science and Engineering **878** (2020) 012046 doi:10.1088/1757-899X/878/1/012046

Fig. 6 shows the results obtained in case of failure in CRS.

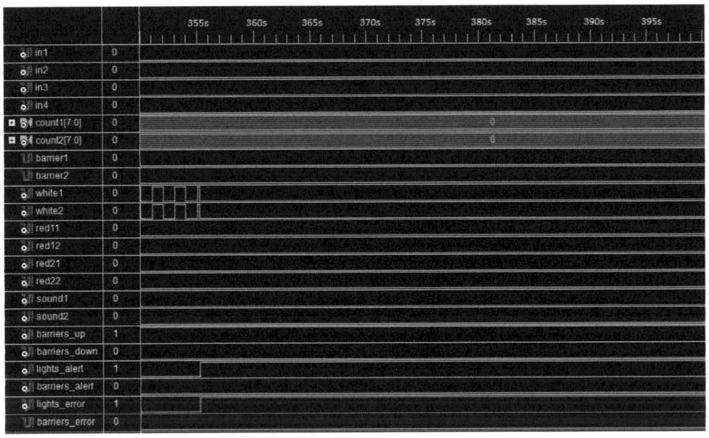

Figure 6. CRS failure.

The RTL Schematic has been generated using Xilinx ISE Version 14.6. Fig. 7 shows RTL schematic of the design railway system. This schematic shows the hardware generated inside the complex programmable logic device. The design is successfully implemented on CoolRunner II board, after testing the design by simulation.

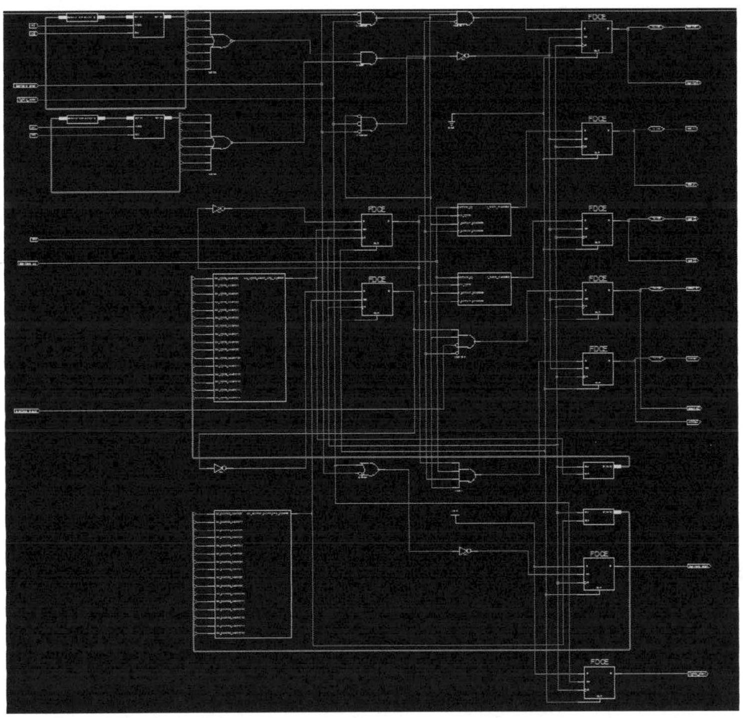

Figure 7. RTL schematic of the design railway crossing system.

5. Conclusion

Complex programmable logic devices are used for high performance and different control applications. They can be applied in digital designs to perform the functions of boot loader. CPLD are used: for loading the configuration data of a field programmable gate array from non-volatile memory;

many applications like in cost sensitive, battery operated portable devices due to its low size and usage of low power.

In this paper an automated railway crossing system is designed and implemented through a complex programmable logic device-CPLD based on a hardware programming language - Verilog HDL. The controller of Xilinx company is used to industrial control barrier mechanisms, vehicle alarms, etc. Some results from the simulation are shown and discussed. The results obtained can be used for the design, implementation, simulation and development of such types of systems in railway crossing automatics.

Acknowledgments

The authors would like to thank the Research and Development Sector at the Technical University-Sofia for the financial support in this conference.

References

[1]. Enokela, J, K. L. Oladejo, An Automated Railway Traffic Station Control System, Nigerian Journal of Technology (NIJOTECH), Volume 36, No. 1, pp. 138 – 147, 2017.

[2]. Leena G., C. Vidawat, N. Jha, Automatic Railway System, International Journal of Computer Applications, Volume 159, No 8, pp. 30-33, 2017.

[3]. Yordanov, R., Miletiev, R., Damyanov, Il., Iontchev, E., Measurement and analysis of the road traffic, Proc. XXVII International Scientific Conference Electronics - ET2018, 2018.

[4]. Al-Zuhairi, A. S. M., Automatic Railway Gate and Crossing Control based Sensors & Microcontroller, International Journal of Computer Trends and Technology (IJCTT), Volume 4, Issue 7, pp. 2135-2140, 2013.

[5]. Krishna, S. Yadav, Nidhi, Automatic Railway Gate Control Using Microcontroller, Oriental Journal Of Computer Science & Technology, Volume 6, Issue 4, pp. 435-440, 2013.

[6]. Marella, V., CPLD Based Design and Implementation of Pipelined 32-bit RISC Processor with Floating Point Unit, International Journal of Emerging Technology and Advanced Engineering, Volume 5, Issue 12, 2015.

[7]. Hristov, V., M. Zhilevski, Extension of Programming Possibilities for Programmable Logical Devices through Matlab, Proceedings of Technical University of Sofia, Volume 68, Issue 2, pp. 79-88, 2018.

[8]. https://www.xilinx.com/publications/products/cpld/coolrunnerii-product-brief.pdf.

[9]. https://www.elprocus.com/complex-programmable-logic-device-cpld-architecture-applications/.

[10]. Samir Palnikar, "Verilog HDL – A Guide to Digital Design and Synthesis", 2nd Edition, IEEE 1364-2001 Compliant, published by Pearson Edition.

[11]. Pavlitov, C., Blocks and Elements of Electrical Drives Implemented by Programmable Logic Devices, Monograph, Sofia, 2013.

[12]. Hinova, A., Synthesis of logic circuits with programmable logic controller, KNOWLEDGE – International Journal, Natural and Biotechnical Sciences, Volume 31.3, pp. 747-752, 2019.

[13]. Zhang Yi, Guo Xiaojuan, Development of intelligent traffic control system based on FPGA and single chip microcomputer technology, Journal of Chemical and Pharmaceutical Research, 2014, 6(6):1156-1159p.

[14]. Damyanov, Il., Investigation and analysis of transit road traffic through specific residential area with the use of an integrated system of mobile lpr cameras, Academic Journal of Mechanics, Transport and Communications, Volume 17, Issue 1, 2019.

[15]. Regulation No. 4 for the rail crossings, Railway Administration Executive Agency, Mart 1997.

[16]. Technical specification – "002-2005-Automatic level crossing", National Railway Infrastructure Company.

Logistics 4.0 and supply chain 4.0 in the automotive industry

Krasimir Markov[1] and Pavel Vitliemov[2]

[1] University of Ruse, Department of Engines and Vehicles
[2] University of Ruse, Department of Management and Business Development

krmarkov@uni-ruse.bg

Abstract. The Industry 4.0 technologies give real benefits to the plants where they are implemented such as performance improvement, costs and delays reductions. The automotive plants are among the factories that could implement most of these technologies. Industry 4.0 can affect all of the processes which happen in the automotive industry manufacturing plants – engineering, production, logistics, management, etc. Analysis of all of the processes in one article is too complex task and due to this the paper is focused only on logistics and supply chain. The aim of this article is to investigate the potential possibilities for implementation of Logistics 4.0 and Supply Chain 4.0 in the automotive industry plants, in particular in Bulgaria. Specially designed survey was used as investigation method. It consists of two groups with proposals and assumptions. The survey was conducted by 12 automotive industry logistics experts who are on 6 different logistics positions. The results show that the production visibility (assembly lines connected to internet) and blockchain technologies have the biggest values to the logistics and supply chain. But the cybersecurity and faulty data are the main challenges in front of them.

1. Introduction

The automotive industry is one of the most engaging and dynamic industries. It is developing extremely strongly and it is open to new technologies which would rise the production rates and which would provide better control. Industry 4.0 is one of these new technologies. Industry 4.0 includes additive manufacturing, advance robotics and cobots, artificial intelligence, augmented reality, human-machine interfaces, machine-to-machine communication, blockchain, internet of things, cloud stored data, internet of services, digital transformation, autonomous vehicles, drones, etc.

Key factor in it is the usage of connectivity and communication among of devices. The industry 4.0 technologies together with real-time data would transform the whole manufacturing process in every industry including also the automotive industry.

The goal of this article is to investigate the potential possibilities for implementation of Logistics 4.0 and Supply Chain 4.0 in the automotive industry plants, in particular in Bulgaria based on unique survey that has been conducted by 12 automotive industry logistics experts.

The paper is structured as follows: In the section in terms of related work are listed the main lessons following the smart manufacturing and smart factory and the challenges in the implementation of Logistics 4.0. Section 3 shows our approach to conduct a unique survey by automotive industry experts in terms of possibility for implementation of Logistics 4.0. Section 4 describes an analysis of the results of the survey. The conclusion of the investigation is placed in the end of the paper.

2. Related work

Characteristic feature of industry 4.0 are the intelligent networks based on cyber-physical systems. Cyber-physical systems are physical and engineering systems which operations can be monitored, coordinated, controlled and integrated by computing and communication systems [1]. It relies on interaction between physical and network devices such as sensors, actuators, control units and communication devices. The general structure of cyber-physical system is shown in figure 1. The implementation of Internet of Things can affect the interaction between the cyber-physical systems and the way of their monitoring, controlling and managing. By this way it could contribute to more effective and cost efficient production [1][10]. There are three features of the Internet of Things – context, ubiquity and optimization [11].

There are two more terms in the literature connected with Industry 4.0 – smart manufacturing and smart factory. The smart manufacturing is the use of cyber-physical systems together with Internet of Things. The smart factory refers to the close connection between products, machines, transport systems and people [10]. The smart factories could give numerous opportunities and benefits including highly flexible mass production, real-time coordination and optimisation of the supply chain, reduction of delays and costs, emergence of brand new business models and services [4].

Industry 4.0 can affect all of the processes which happen in the automotive industry manufacturing plants – engineering, production, logistics, management, etc. Analysis of all of the processes in one article is too complex task and due to this the paper is focused only on logistics and supply chain. The terms Logistics 4.0, Smart Logistics and Supply Chain 4.0 have been introduced according to Industry 4.0. These terms refer to the usage of logistics together with cyber-physical systems and Internet of Things. From technological point of view the implication of Logistics 4.0 is not aiming to replace the humans but to avoid inaccuracies and to have faster processes with easy shared information in real time [1].

The logistics could be define as to deliver the right product/s to the right customer/s in the right location/s and time. The introduction of Logistics 4.0 can change the process in different ways – faster speed with using of autonomous vehicles and drones, higher reliability, lower operating costs by using inventory monitoring and refilling smart sensors (smart shelves with weight sensors), improve efficiency because of delays reducing, smart transportation with delivery time reduction and vehicles emissions reductions [8]. The main applications of Logistics 4.0 can be in the resource planning, warehouse management systems, transportation management systems, intelligent transportation systems and information security [1]. Logistics 4.0 application model is shown in figure 2. Some of the sensors that could be used in Logistics 4.0 applications include RFID, LIDAR, accelerometers, GPS, cameras, humidity sensors, etc. [10]. The existing logistics concepts such as JIT/JIS and Kanban could be improved with the implementation of Logistic 4.0 [4].

The supply chain 4.0 has the following features – extensive use of Internet of Things, constant visibility of the complete supply chain from all participants, a possibility of accurate verification of the supply chain coherence by blockchain technology, a possibility of dynamic optimisation of supply chains and suppliers [10]. The blockchain technology has significant potential to improve supply chain processes [5]. The main reason for this is the possibility for simultaneously sharing of the information and documents between all companies in the supply chain. Additionally it could reduce the delays and the incorrect data in the documents.

The introduction of Supply Chain 4.0 reduces the lead-time of the whole chain [3]. It can also affect the Engineer-to-Order industries which also include the project lead-time [7].

Sustainable Supply Chain 4.0 can be formed only if the entire business is connected digitally [6]. Dossou (2018) suggests a reference model for transition to supply chain 4.0 of small and medium enterprices where the sustainability is placed as kernel for improvement.

According to Torbacki et al. (2019) there are ten key performance indicators divided into three perspectives – Indusry 4.0 with the three indicators (supply chain, production planning and reconfigurable manufacturing), Logistics 4.0 with four indicators (Internet of Things in logistics, warehouse, intelligent transportation systems and data safety) and Sustainability with three indicators (economy, environment and society).

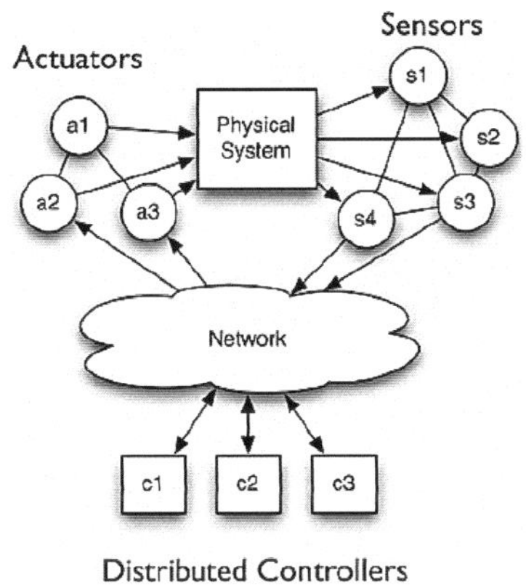

Figure 1. General structure of cyber-physical system [1].

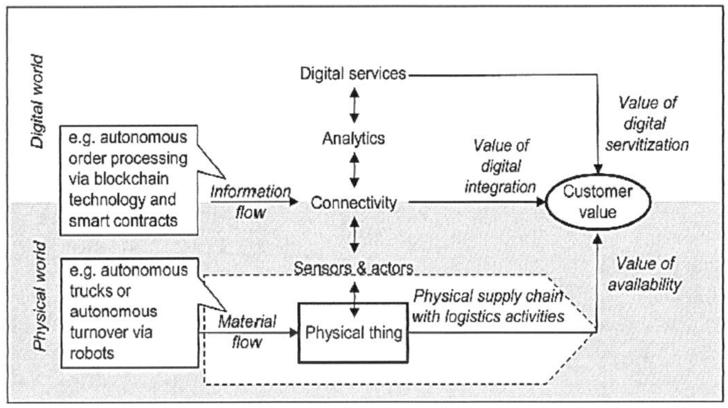

Figure 2. Logistics 4.0 application model [4].

Some of the challenges in front of the introduction of Industry 4.0 are the cyberattacks, faulty data, safety regulations and privacy issues [8]. It has to be also noted that some of the Industry 4.0 technologies would have negative impact on the performance [9]. This depends from the context of the company.

3. Resources and methods

The first thing which had to be done in the investigation was the selection of Industry 4.0 technologies which are more likely to be implemented in the logistics of the automotive plants. The selected technologies are advance robotics, artificial intelligence, machine-to-machine communications, Internet of Things, blockchain, cloud stored data and autonomous vehicles.

Based on the literature analyze in the previous section, the survey is common used method for evaluation of the possibility for implementation of Logistics 4.0. Taking this into account, the investigation uses a specially created survey in order to achieve the purpose. The survey consists of two groups. One of the group considers only the possible potentials and has seven proposals. The other group considers the possible challenges and has seven assumptions. Each proposal has answers with points from 0 to 2 (0 point to "disagree", 1 point to "agree", 2 points to "already implemented"). Each assumption has answers with points from 0 to 2 (0 point to "disagree", 1 point to "agree", 2 points to "already is problem"). The sum of the points gives the result of the respective proposal or assumption. The survey proposals with the linked to them technologies are given in table 1. The survey assumptions with the linked to them technologies are given in table 2.

Some of the proposals are opposite to the assumptions – proposal 1 is opposite to assumption 1, proposals 4 and 5 are opposite to assumption 2, proposal 6 is opposite to assumptions 3, 4, 5 and 6, proposal 7 is opposite to assumption 7.

The survey is conducted by 12 automotive industry logistics and supply chain experts from 2 automotive plants in Bulgaria. They have six different work positions and all of them were with different work experience in the logistics and supply chain field. The experts were asked for their Industry 4.0 level of knowledge before the creating of the survey. This was done in order to create more clear proposals in the survey and to get accurate answer from the experts. In the end of the investigation the logistic managers were asked additionally which technology would give the best delay reduction and the percentage of the reduction. The experts work position, work experience and knowledge are shown in table 3.

Table 1. Logistics 4.0 potential proposals.

№	Industry 4.0 technology	Proposal
1	IoT/Blockchain/Cloud stored data	The internet connection of the assembly line to the customer would increase his satisfaction due to production visibility.
2	Machine-to-machine communications	The internet connection between production linked assembly lines would increase the costs savings due to production traceability.
3	Machine-to-machine communications	The smart shelves with sensors would increase the production rates due to assembly line refilling shorter time.
4	Autonomous vehicles	The autonomous forklift would increase the production rates due to the shorter time for reaching and delivering the required pallets with parts.
5	Autonomous vehicles/Robots/AI	The autonomous milkrun would increase the production rates due to shorter time and flexibility for delivering the required parts to the assembly line.
6	Blockchain/Cloud stored data	The blockchain and use of internet portals for all companies in the supply chain would increase cost saving due to real-time information and documentation shared between all participants.
7	IoT/Blockchain/Cloud stored data	Real-time GPS route optimization would increase cost savings due to transportation shorter time.

Table 2. Logistics 4.0 challenges assumptions.

№	Industry 4.0 technology	Assumptions
1	IoT/Blockchain/Cloud stored data	Cyberattacks over internet connected assembly line could break the production process.
2	Autonomous vehicles/Robots/AI	The use of autonomous vehicles and robots could result in employee's replacement.
3	Blockchain	The different standards among the companies in the blockchain could result in increase of the costs.
4	IoT/Blockchain/Cloud stored data	The use of blockchain and cloud computing could result in confidential data security leakage.
5	Blockchain/Cloud stored data	Faulty data inputed in the blockchain portals could increase the costs.
6	IoT/Blockchain/Cloud stored data	Old version software could result in blockchain costs increase due to missing of information or security problems.
7	IoT/Blockchain/Cloud stored data	GPS connection missing could result in cost increase due to transportation delays.

Table 3. Automotive Industry logistics and supply chain experts.

№	Position	Experience	Industry 4.0 knowledge
1	Logistic Manager	21 years	moderate
2	Logistic Manager	18 years	moderate
3	Disponent	10 years	little
4	Disponent	7 years	basic
5	Material Planner	8 years	little
6	Material Planner	14 years	moderate
7	Forwarder	5 years	basic
8	Forwarder	8 years	basic
9	Purchaser	9 years	little
10	Purchaser	12 years	little
11	Seller	11 years	moderate
12	Seller	10 years	little

4. Results and discussion

The results of the survey are shown in table 4.

Table 4. Results of the survey.

№	Proposal result	Assumption result
1	14	11
2	9	4
3	8	4
4	7	8
5	4	12
6	12	5
7	8	4

The results show that the proposal 1 has the biggest value. This is because the fulfilment of the customer orders has big impact on the customer satisfaction. Combining of order fulfilment with real-time production visibility has the potential to raise his satisfaction even more. That is why some of the automotive plants have already implemented the connection of some assembly lines by internet directly with their customers. The blockchain technology (proposal 6) gives additional advantage to this and it is also implemented in some of the automotive plants. In the same time the assumption 1 has one of the highest values. This indicates that if there are not any cybersecurity activities, the benefits could be totally lost. The other big problem which could happen is the faulty data which could be inputted in the blockchain portals (assumption 5). This would have impact over the all participants in the supply chain resulting with big delays and costs.

Most of the proposals have moderate values. There are two reasons for these values. The first reason is the high costs for the implementation (proposal 4). The second reason is the low/moderate benefits of the proposed technologies (proposals 2, 3, 7). It has to be noted also that the safety regulations for the autonomous vehicles have impact over the results for the respective proposals.

There is just one assumption with moderate value (assumption 4). Although all of the companies from the automotive industry use some internet portals where the documents are shared, there is high attention to the cybersecurity. It has to be carried out cybersecurity activities in the internet systems in order to avoid confidential data security leakage.

The lowest value in the proposal has the autonomous milkrun (proposal 5). This is mainly because even if the milkrun is autonomous, a worker would still be needed to take the parts from the milkrun to the assembly lines. This could change only if the milkrun is combined with robots. Another reasons are the higher costs for autonomous milkrun and the safety regulations.

There are several assumptions with low values. The reasons are as following: the employees would monitor the autonomous vehicle and robot's work (assumption 2), the standards are not problem for the experts mainly because the automotive plants are certified to the automotive standards (assumption 3), old software version is not normally used in high quality production factories such as the automotive plants (assumption 6), the short-time lack of GPS connection would not have significant impact over the transportation time (assumption 7).

According to the opposite proposals and assumptions it can be seen strong correlation between proposal 1 and assumption 1, proposal 5 and assumption 2, proposal 6 and assumption 5.

The autonomous vehicle was pointed from both logistics managers as the technology which could give the best delay reduction in the logistic process in the automotive plants. According to them there would be around 20% time reduction of the process. It should be noted that this does not mean 20% increase of the production rates.

5. Conclusion

The Industry 4.0 technologies give real benefits to the plants where they are implemented such as performance improvement, costs and delays reductions. Industry 4.0 can affect all of the processes which happen in the automotive industry manufacturing plants – engineering, production, logistics, management, etc. Analysis of all of the processes in one article is too complex task and due to this the paper was focused only on logistics and supply chain.

The aim of the article is to investigate the potential possibilities for implementation of Logistics 4.0 and Supply Chain 4.0 in the automotive industry plants, in particular in Bulgaria. The investigation approach consisted with specially designed survey which was conducted by 12 logistics experts from the automotive industry in Bulgaria who are on 6 different logistics positions. After the survey, some of the expert were asked additional questions.

The results showed that the production visibility (assembly lines connected to internet) and blockchain technologies have the biggest values to the logistics but the cybersecurity and faulty data are the main challenges in front of them. The results also indicate strong correlation between them. Although the autonomous vehicles have the lowest result in the proposals, they were pointed from both logistics managers as the technology which could give the best delay reduction in the logistic process in the automotive plants (around 20%). It has to be pointed that this does not mean the same increase of the production rates. The challenges in front of the autonomous vehicles are the higher costs for implementation and safety regulations.

Future work which will be conducted is the implementation in the automotive plants of some of the technology which has high or moderate result in automotive plant. The next work will be focused on assessment of the performance improvement and delays reduction in real production environmental.

Acknowledgement

The investigation is conducted with the support from project **ФНИ РУ 2020-ФБМ-01**.

References

[1] Barreto L, Amaral A and Pereira T. 2017. Industry 4.0 implication in logistics: an overview. *Procedia Manufacturing* **13**, 1245-52.

[2] Dossou PE. 2018. Impact of Sustainability on the supply chain 4.0 performance. *Procedia Manufacturing* **17**, 452-59.

[3] Dossou PE and Nachidi M. 2017. Modeling supply chain performance. *Procedia Manufacturing* **11**, 838-45.

[4] Hofmann E and Rusch M. 2017. Industry 4.0 and the current status as well as future prospects on logistics. *Computers in Industry* **89**, 23-34.

[5] Issaoui Y, Khiat A, Bahnasse A and Ouajji H. 2019. Smart logistics: Study of the application of blockchain technology. *Procedia Computer Science* **160**, 266-71.

[6] Manavalan E and Jayakrishna K. 2019. A review of Internet of Things (IoT) embedded sustainable supply chain for industry 4.0 requirements. *Computers and Industrial Engineering* **127**, 925-53.

[7] Muller JM and Voigt KI. 2018. The Impact of Industry 4.0 on Supply-Chain in Engineer-to-Order Industries – An Exploratory Case Study. *IFAC PapersOnLine* **51-11**, 122-27.

[8] Tang CS and Veelenturf LP. 2019. The strategic role of logistics in the Industry 4.0 era. *Transportation Research Part E* **129**, 1-11.

[9] Tjahjono B, Esplugues C, Ares E and Pelaez G. 2017. What does Industry 4.0 mean to Supply Chain? *Procedia Manufacturing* **13**, 1175-82.

[10] Torbacki W and Kijewska K. 2019. Identifying key performance indicators to be used in Logistics 4.0 and Industry 4.0 for the needs of sustainable municipal logistics by means of DEMATEL method. *Transportation Research Procedia* **39**, 534-43.

[11] Witkowski K. 2017. Internet of Things, Big Data, Industry 4.0 – Innovative Solutions in Logistics and Supply Chains Management. *Procedia Engineering* **182**, 763-69.

Clearance speed study for intergreen time determination

D N Saliev

Department of combustion engines, automobile engineering and transport, Technical university of Sofia, City of Sofia, 8 St Kliment Ohridski blvd, Bulgaria

durhan_saliev@tu-sofia.bg

Abstract. Intergreen is the period between ends of a green period of one phase and the beginning of a green period of the next phase at signalized intersections. Its determination requires data which are in relation of geometrical parameters of junctions and that of dynamic parameters of the entering in and passing through the crossroads vehicles. One of them is clearance speed of a last passing vehicle of a phase which is necessary to calculate its clearance time. The present study aims to determine values of clearance speed for straight ahead and turning vehicles and find a relations between them and curves radii, slope of the road etc. The study use field data bring at three traffic lights junctions in Sofia city. The obtained results and the conclusions reached would be used in intergreen time determination for phasing design optimization at these junction and the other one in the same condition includes clearance distance and curves radii of turning flows.

1. Introduction

Intergreen periods are very important for traffic safety in cases of crossing a traffic light junction. In the international scientific community, the *inter-green* time is defined as the time between the end of green of one stream and the onset of green of a conflicting stream [1]. They are secure the necessary time last passing car of a flow of a stage to leave conflict zone whit first entered cars of a flow in next stage, which will cross or merge the way of first one.

These times are parts of time-spacing diagram of signalized intersections. There are various methods to calculate their duration around the world. Main of them is German method [2] that is applied in many countries all over the world directly or in some modifications or supplements. Intergreen calculation principles in Germany, United States of America, Switzerland, Austria, France and some other country are reviewed by Wolfermann in [1].

In Bulgaria intergreen times are calculated according German model with Austrian acceptances in accordance in (1). Differences are in that the value cannot be less than three seconds (Austrian models says less than four seconds recommended). Intergreen time in Bulgaria may be adjusted for special local situation, if justifiable as permitted by the Swiss model [1].

$$t_{ig} = t_y + t_c - t_e + t_{ad}, s \qquad (1)$$

Where t_{ig} is intergreen time, s

t_y – Amber after green signal (its value is equal to 3 s, normatively defined);

t_c – Clearance time, s

t_e – Entrance time, s

t_{ad} – Additional time by the specialists view, s

$$t_c = \frac{L_c + L_v}{V_c}, s \quad (2)$$

Where L_c is clearance distance of last departure car of a phase, m
L_v – Vehicle length, m
V_c – Clearance speed, ms^{-1}.

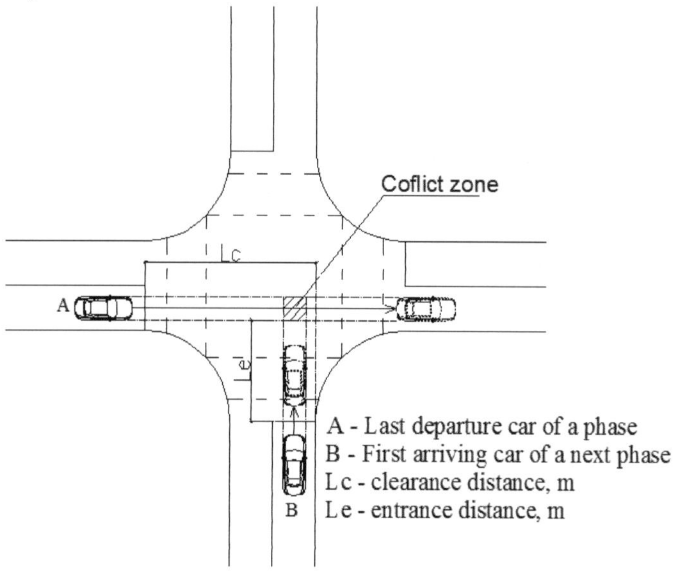

Figure 1. Clearance distance and entrance distance for the timing of intergreen periods.

A part of necessary data for intergreen time determination is a clearance speed in accordance of equation (2). This is the speed of last passing car of a departure flow (Figure 1). This paper gives the value of that speed which are result of field study in Sofia city.

2. Prerequisites and means for solving the problem

The applied values of clearance speed of a car in case of intergreen time determination varies in all over the world despite the similar approaches and adoptions in question in different countries. Austria applies 10 ms^{-1} for clearance speed of a car turning in curve radius smaller and 20 meters and 7 ms^{-1} of a car turning in curve radius smaller and 10 meters. Germany used values are 10 ms^{-1} for straight ahead moving and 7 ms^{-1} for turning cars. Also applied a value of 5 ms^{-1} of a car turning in curve radius smaller and 10 meters [3]. Retzko and Boltze in [2] were applied maximum value for clearance speed equal to the speed limit at an intersection. The minimum value they were considered 23 fts^{-1} (approximately equal to 7 ms^{-1}). Also they were assumed for vehicles that pass small curves clearance speed 16.4 fts^{-1} (approximately equal to 5 ms^{-1}) for R<49,2 ft (about 15 meters). Keshuang and etc. in [4] were determined free flow speed of right-turn traffic (6.60 to 8.00 ms^{-1}) and the shock-wave speed for the discharge flow of right-turn traffic (about 5 ms^{-1}). Guberinic in [5] assumed the speed of the vehicles leaving the conflict areas in range of 25-30 kmh^{-1} (6.9-8.3 ms^{-1}) independent of their directions. They were not explained about curves radii of the observation. Other scientists in this relation were not used clearance speed to determinate intergreen time. They were used specific values for yellow and all-red periods of different phases of time-spacing diagram of signalized intersections [6]. These values depend of designed speed and the distance in meters measured from the departure stop line to the furthest point of conflict with vehicles or pedestrians in the next or subsequent phase, taking into account the longest distance of any straight or turning movement within the phase.

In aims of study presented in this paper research was conducted at three traffic light junctions in the Bulgaria capital – Sofia city. The junction are formed by crossing the one of main boulevard “St Kliment Ohridski” whit “Osmi dekemvri” street (junction “T”), “Traycho Stanoev” street (junction “U”) and

“”Dr G. M. Dimitrov” boulevard (junction “D”), located in Darvenica district (Figure 2). Schemes of the crossroads are shown in Figure 3, Figure 4 and Figure 5.

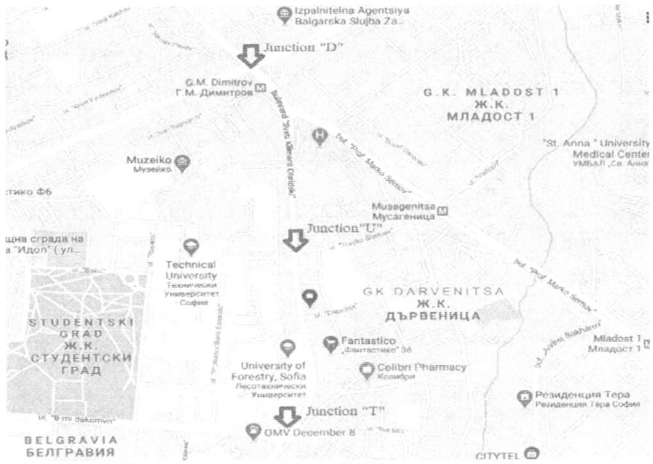

Figure 2. Location of observed junction.

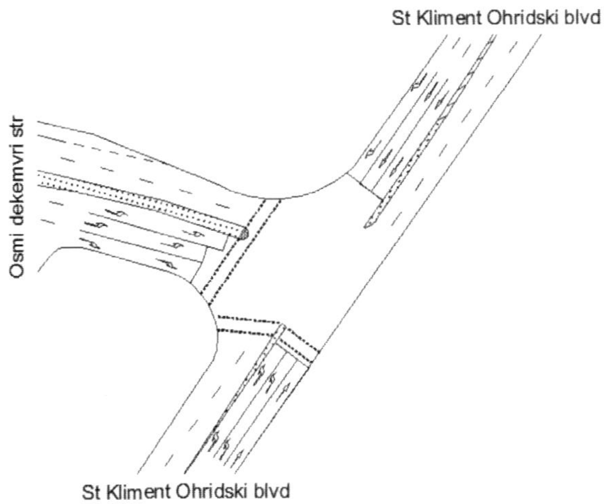

Figure 3. Scheme of junction “T”.

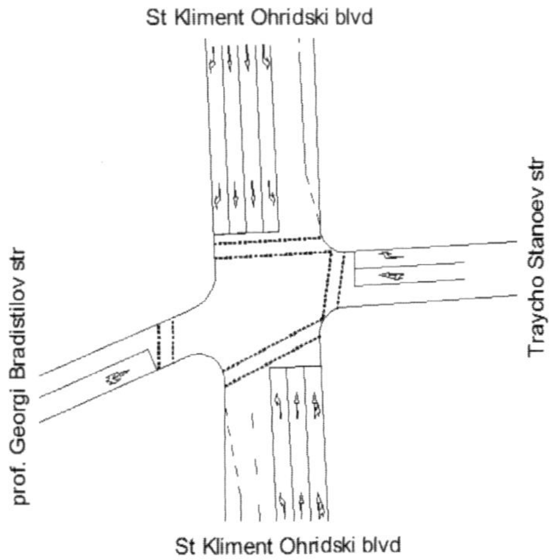

Figure 4. Scheme of junction “U”.

Figure 5. Scheme of junction “D”.

The research include determine clearance distance for every traffic flows cross the junctions and time to pass the distance from last clearing car of a departure flows of a stage. The purpose set require defining curves radii of turning flows wherefore also were measured. The data were obtained by hand measurement from video captured whit unmanned aerial vehicle known as a drone. Its type, principles of work and its opportunities were presented by Hristov, Damyanov, etc. in [7] and [8]. Usage of this type of aircrafts were described in detail in [9] and [10]. The authors were shown different applications of that technique which had been used in this study.

3. Results and discussion

For the purpose of the study traffic flows of each junction were indicated as shown in Figure 6 for junction "T", Figure 7 for junction "U" and Figure 8 for junction "D".

Necessary data about clearance distance, average time to pass the distance by flows, curves radii and result about average clearance speed by flows of each junction were presented in Table 1 for junction "T, Table 2 for junction "U" and Table 3 for junction "D".

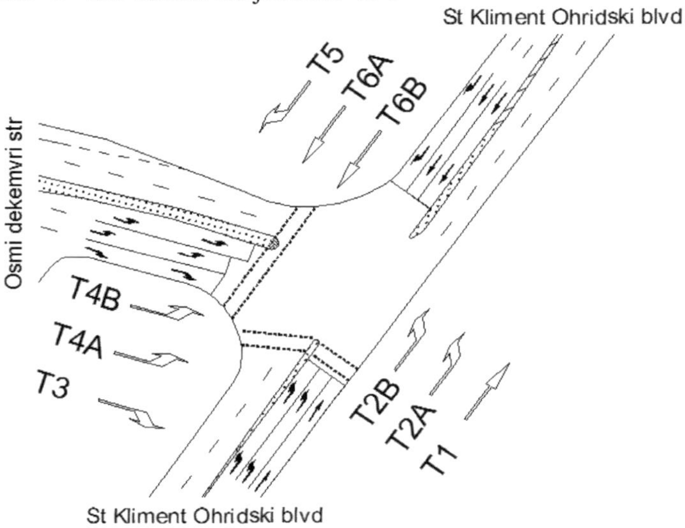

Figure 6. Traffic flows indication at junction "T".

Table 1. Necessary data and obtained results for junction "T".

Traffic flows	Passed distances, m	Turning radii, m	Average of last passed car		
			t, s	V, kmh^{-1}	V, ms^{-1}
T1	30	-	2.54	44.06	12.24
T2A	40	18	5.53	26.59	7.39
T2B	30	16	4.56	24.38	6.77
T3	17	12	3.18	19.78	5.49
T4A	40	37	5.45	28.14	7.82
T4B	31	33	4.68	24.92	6.92
T5	20	14	3.11	24.68	6.86
T6A	45	-	3.84	44.32	12.31
T6B	45	-	3.33	50.40	14.00

Results reading in Table 1 about junction "T" can see that the straight ahead flows run from 12.31 ms^{-1} to 14.00 ms^{-1} for last passing car of a departure flows. Last passing car of a turning flows had clearance speed from 5.49 ms^{-1} to 7.82 ms^{-1}. It is clear that the turning flows on minimum curve radius (12 m) had a minimum clearance speed at the junction. It gives an impression that the other turning flows had approximately equal values although a big differences of their curves radii. That can explain because the flows T4A and T4B despite of curves radii values have to decrease the speed at the end of passing which is determined of geometry of the crossroad.

The analysis shows increase of clearance speed of flow T6B compared with other straight flows. Its value is above 2.00 ms^{-1} from others. This cannot be explained by the junction geometry or traffic conditions furthermore there is a slope and flow was moving uphill. Obtained values show exceeding of speed limit about in 15% of cases.

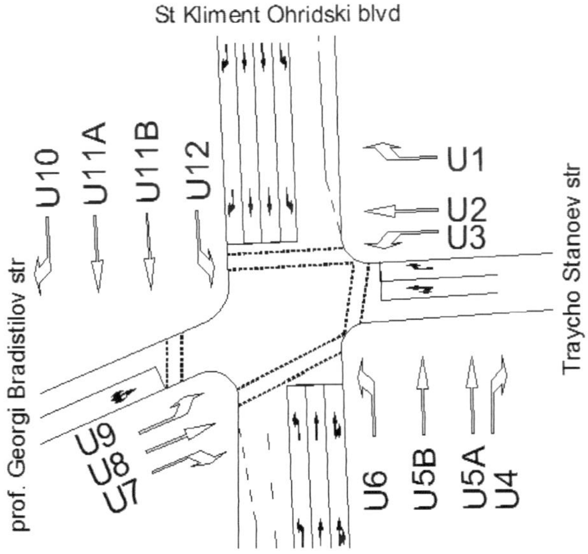

Figure 7. Traffic flows indication at junction "U".

Results about junction "U" show that the straight ahead flows had a speed from 6.43 ms^{-1} to 14.56 ms^{-1}. Flows U8 and U2 had a low speed in comparison with other straight flows. That explain with the conflict between flows U1, U2, U3, U7, U8 and U9 which runs in one phase. Approaches to the crossroad of these flows do not offer good visibility, straight ahead flows must to change their trajectory (it can be seen at Figure 4 and Figure 7) also contribute to decrease clearance speed of these flows. The same reasons are valid in the analysis of the speed when comparing flow U3 and U9 which had 10 meters higher curve radius but above 1.3 ms^{-1} lower clearance speed than flow U3.

Turning flows at this junction had a very various values for clearance speed. This is due to differences of curves radii of them. The analysis shows differences of clearance speed not only in different radii but in the equal values of this metric. For example flows U3 had about 1.2 ms^{-1} higher clearance speed than flow U12 which is had the same curve radius. It is the same at comparison about flow U1 and U4 which had above 1.00 ms^{-1} lower clearance speed than U1.

It gives an impression that the flow with minimum clearance speed are not that with minimum curve radius. This is flow U7 with curve radius 7 m and clearance speed of 4.91 ms^{-1}. Reason in this case is specific right turn conditioned of its geometry (approach, turn and exit condition).

Average clearance speed of last passing car of flow U11B was obtained like highest of observed ones. It value is above 3.00 ms^{-1} compared to flow U11A which is the same flow as regards to approach, traffic condition, direction and other conditions in relation for passing the junction of a car with this movement which is the same like situation about flow T6B at the junction "T". The values of the observed speed of flow U11B show exceeding the speed limit in about 30% of cases. This remains unexplained to the moment since it cannot be bound by the technical parameters of the intersection or traffic conditions.

Table 2. Necessary data and obtained results for junction "U".

Traffic flows	Passed distances, m	Turning radii, m	Average of last passed car		
			t, s	V, kmh^{-1}	V, ms^{-1}
U1	9	6	1.37	23.94	6.65
U2	38	-	4.58	33.07	9.19
U3	28	15	3.91	27.72	7.70
U4	13	6	2.42	19.61	5.45
U5A	22	-	2.06	42.71	11.87
U5B	22	-	1.94	42.11	11.70
U6	25	15	3.61	25.43	7.06
U7	18	7	3.80	17.69	4.91
U8	37	-	5.86	23.15	6.43
U9	36	25	6.03	22.89	6.36
U10	22	10	3.80	21.35	5.93
U11A	25	-	2.42	40.54	11.26
U11B	24	-	1.79	52.43	14.56
U12	16	15	2.93	19.96	5.55

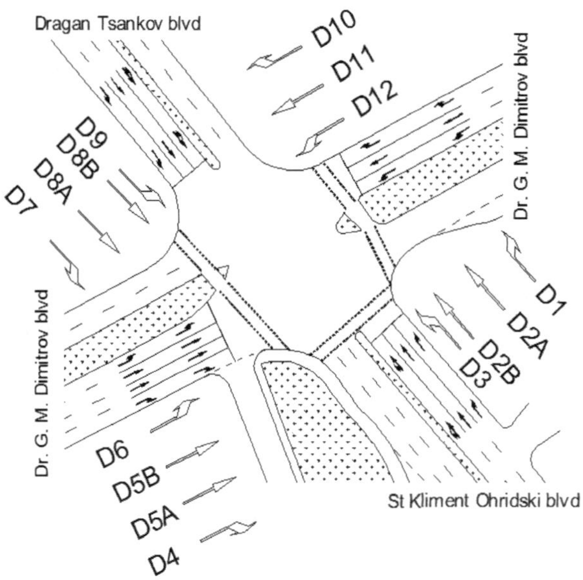

Figure 8. Traffic flows indication at junction "D".

Table 3. Necessary data and obtained results for junction "D".

Traffic flows	Passed distances, m	Turning radii, m	Average of last passed car		
			t, s	V, kmh^{-1}	V, ms^{-1}
D1	12	10	2.17	21.60	6.00
D2A	48	-	4.43	39.75	11.04
D2B	48	-	4.68	37.24	10.35
D3	50	37	5.20	35.26	9.79
D4	21	16	2.92	26.63	7.40
D5A	41	-	4.97	30.20	8.39
D5B	41	-	4.16	35.87	9.96
D6	38	30	5.14	27.50	7.64
D7	14	14	2.04	25.35	7.04
D8A	50	-	4.77	38.41	10.67
D8B	50	-	4.86	37.47	10.41
D9	48	38	4.95	35.14	9.76
D10	23	16	3.37	24.95	6.93
D11	40	-	3.66	39.91	11.09
D12	37	26	5.37	25.49	7.08

Shown results in Table 3 make clearly view of that the clearance speed of straight ahead flows are not much different from that of left turning flows (about 10 ms^{-1}). That can explain with their curves radii values which are than 30 meters expect the flow D12. Independently of his radius value (26 m) this flow have a behavior like flows with smaller curves radii. This is proven by his clearance speed (7.08 ms^{-1}) which is common for flows D7 of this intersection, U3, U6 of intersection "U" and others with curve radius about 15 meters that were been subject of the study.

It can be seen that the clearance speed of flows D5A and D5B is deferent form that of flow D12 although the flows D5A and D5B have a downhill slope. Similar is situation about approximately equal values for clearance speed of flows D2A, D2B, D8A and D8B, where the last two flows have large uphill slope. That show steadily driver behavior to keeping approximately constant speed in the same condition independently of road slope.

4. Conclusions

Presented study has given results about clearance speed used field data at three traffic light junction in Sofia city. Their analysis had shown that the clearance speed value depends on many factors which cannot be considered individually but always in the aggregate. That are the direction of the flow, curve radius in case of turning flow, traffic condition, organization, etc. The results showed the relation not so much between curves radii of the turning flows but between directions of turning. Was observed relation between left turn flows, right turn flows and straight flows. Study has shown speeds which were been above of speed limit at the area of junction. This result must not to be taken into account when specialists assumed clearance speed for more increase of traffic safety. It also turned out that the slope of the road does not affect in the clearance speed.

The conducted study gave grounds to determine the values of clearance speed as follows: straight ahead cars – 10 ms^{-1}; turning cars – 7 ms^{-1}; turning cars with curve radius smaller and 15 m – 5 ms^{-1}.

Author recommended these values be used in intergreen time determination with taking into account specific condition at a junction.

5. Acknowledgements

The author would like to thank the Research and Development Sector at the Technical University of Sofia for the financial support.

References

[1] Wolfermann A 2009 *Influence of Intergreen Times on the Capacity of Signalised Intersections* Dissertation ISSN 1613-8317 (Darmstadt: Technische Universität Darmstadt)
[2] Retzko HG, Boltze M 1987 *ITE Journal* p 23
[3] Hebenstreit C, Fellendorf M, etc. 2014 ***Final*** STREP ICT Call 8FP7-ICT-2011-8 *COLOMBO Deliverable 2.2 – Policy definition and dynamic policy selection algorithms* p47
[4] Keshuang T, Takeshi O, Kuwahara M, Tanaka S 2013 *J. Adv. Transp.* **47**:704–719
[5] Guberinic S, Senborn G, Lazic B 2011 *Optimal traffic control urabn intersections* (New York: CRC Press/Taylor & Francis Group) p 116
[6] Transport Management Centre 2010 *Traffic Signal Operation (New South Wale, Australia)* (Strawberry Hills: Road and traffic authotity of NSW) RTA-TC-106 p 23
[7] HristovV, Damyanov I, Savova-Mratsenkova M, Palagachev G and Mladenov G, 2017 *Proc. Conf. TechCo – Lovech 2017* ISSN 2535-079X (Lovech: Gabrovo University press) p11
[8] HristovV, Damyanov I, Savova-Mratsenkova M, Palagachev G and Mladenov G, 2017 *Proc. Conf. TechCo – Lovech 2017* ISSN 2535-079X (Lovech: Gabrovo University press) p17
[9] Damyanov I, Mladenov G, Savova-Mratsenkova M, Palagachev G and HristovV 2017 *Proc. Conf. BulTrans – Sozopol 2017* ISSN 1313-955X (Technical University Academic Publishing House) p118
[10] Damyanov I, Mladenov G, HristovV, Savova-Mratsenkova M and Palagachev G 2017 *Proc. Conf. BulTrans – Sozopol 2017* ISSN 1313-955X (Technical University Academic Publishing House) p122

Determination of inaccuracy by milling taking into account the phenomenon of technological heredity

H C Metev[1] and K K Krumov[2]

[1] Technical University of Gabrovo, department of Mechanical Engineering and Technology, BG

E-mail: hmetev@abv.bg

[2] Technical University of Gabrovo, department of Mechanical Engineering and Technology, BG

E-mail: kalin_krasimirov_krumov@abv.bg

Abstract: For the determination of the total inaccuracy in milling using analytical and calculation method taking into account the phenomenon of technological heredity in development are obtained for fixed (deterministic) component and composite, taking into account the mutual influence of various inaccuracies depending on technological conditions for machining. It was created based on the level dependencies, allowing increased efficiency of the method for determining the total inaccuracy of milling. A comparative analysis of the accuracy with and without taking into account the phenomenon of technological heredity.

1. Introduction

The main task of machine-building production is to improve the quality, reliability and durability of production, i.e. its user properties. The solution of this problem is carried out by rational selection of the appropriate sequence of technological operations for the workmanship of the pieces (articles), where special attention should be paid to ensuring the geometric accuracy and physico-mechanical properties of the piece surface layer.

To establish objective patterns, a comprehensive study of the accuracy obtained is necessary, taking into account the phenomenon of *"technological heredity"*, both within the specific technological operations and throughout the technological process.

To determine the expected inaccuracy in milling, taking into account the phenomenon of technological heredity at the transition level, a method has been proposed, allowing for the most complete description of the mechanism of formation of component inaccuracies and their mutual influence [1,2], and a methodology for determining the milling transformation coefficients [3], allowing the determination of the expected accuracy when considering the reciprocal influence of the component inaccuracies.

In determining the total inaccuracy, both the fixed (deterministic) constituents and the constituents are required, taking into account the mutual influence of the various inaccuracies in the operation.[1,2]. This necessitates the need to obtain dependencies to determine the transformation coefficients and the component inaccuracies of the technological conditions, which is the purpose of this work.

2. Exhibition

The disadvantage of the dependencies for determining the total inaccuracy by the analytical and computational method is that most elementary inaccuracies are determined by table or statistical data processing. To eliminate this deficiency, there is a mathematical apparatus that allows determination of elementary inaccuracies, taking into account the physico-mechanical characteristics of the machined and tool material, the type of machine tools, the parameters of the cutting tool, the stability of the workpiece, etc. In this case, the total inaccuracy obtained in the machining process by cutting is determined by the expression (1) [1,2], taking into account only the inaccuracies associated with the cutting process:

$$\varepsilon_{\Sigma} = \varepsilon_j + \varepsilon_{td_u} c + \varepsilon_{td_w} + \varepsilon_{w_c} \qquad (1)$$

where ε_j is an inaccuracy of insufficient stability of the technological system „machine - fixture - tool - piece" (MFTD"), ε_{td_c} - inaccuracy of temperature deformations of the teeth of the milling cutter in axial direction; ε_{td_w} - inaccuracy in temperature deformation of the workpiece; ε_{w_c} - inaccuracy associated with the wear of the teeth of the milling cutter.

Using the approach to determine total processing inaccuracy, the component inaccuracies in (1) are obtained

$$\begin{aligned}
\varepsilon_j &= (\varepsilon_j)_D - a_{1,\varepsilon_{td_c}} (\varepsilon_{td_c})_D - a_{1,\varepsilon_{td_w}} (\varepsilon_{td_w})_D + a_{1,\varepsilon_{w_c}} (\varepsilon_{w_c})_D, \\
\varepsilon_{td_c} &= -a_{2,\varepsilon_j} (\varepsilon_j)_D + (\varepsilon_{td_c})_D + a_{2,\varepsilon_{m\delta_\partial}} (\varepsilon_{m\delta_\partial})_D - a_{2,\varepsilon_{w_c}} (\varepsilon_{w_c})_D, \\
\varepsilon_{td_w} &= -a_{3,\varepsilon_j} (\varepsilon_j)_D + a_{3,\varepsilon_{td_c}} (\varepsilon_{td_c})_D + (\varepsilon_{td_w})_D - a_{3,\varepsilon_{w_c}} (\varepsilon_{w_c})_D, \\
\varepsilon_{w_c} &= a_{4,\varepsilon_j} (\varepsilon_j)_D - a_{4,\varepsilon_{td_c}} (\varepsilon_{td_c})_D - a_{4,\varepsilon_{td_w}} (\varepsilon_{td_w})_D + (\varepsilon_{w_c})_D,
\end{aligned} \qquad (2)$$

where $(\varepsilon_j)_D$, $(\varepsilon_{td_c})_D$, $(\varepsilon_{td_w})_D$ and $(\varepsilon_{w_c})_D$ are the fixed (deterministic) values of the inaccuracies ε_j, ε_{td_c}, ε_{td_w} and ε_{w_c}, occurring at the executing transition (passage); a_{i,ε_j} - coefficient of transformation of the transition (passage) of inaccuracy ε_j in the corresponding component of the total inaccuracy $(\varepsilon_{td_c};\varepsilon_{td_w};\varepsilon_{w_c})$, $(i=2,3,4)$; $a_{i,\varepsilon_{td_c}}$ - coefficient of transformation of the transition (passage) of inaccuracy ε_{td_c} in the compounding of the total inaccuracy $(\varepsilon_j;\varepsilon_{td_w};\varepsilon_{w_c})$, $(i=1,3,4)$; $a_{i,\varepsilon_{td_w}}$ - coefficient of transformation of the transition (passage) of inaccuracy ε_{td_w} in the compounding of the total inaccuracy $(\varepsilon_j;\varepsilon_{td_c};\varepsilon_{w_c})$, $(i=1,2,4)$; $a_{i,\varepsilon_{w_c}}$ - coefficient of transformation of the transition (passage) of inaccuracy ε_{w_c} in the compounding of the total inaccuracy $(\varepsilon_j;\varepsilon_{td_c};\varepsilon_{td_w})$, $(i=1,2,3)$.

The values of the quantities $(\varepsilon_j)_D$, $(\varepsilon_{td_c})_D$, $(\varepsilon_{td_w})_D$, $(\varepsilon_{w_c})_D$ are identical to the values of the quantities of the corresponding inaccuracies obtained by using the classical analytical-calculation method.

- *Dependencies for determining the coefficients of transformation*

Using the developed methodology [3] a database was created for the transformation coefficients of the components of the total inaccuracy in the face milling of steel workpieces 12XH2, $\sigma_B = 785$MPa, graphical dependences of the same on the cutting speed v_c and feed per tooth f_z are constructed and approximate dependencies are obtained by type [3]

$$a_{\Delta x,\Delta y} = C\, f_z^{\,a}\, v_c^{\,b} \qquad (3)$$

where $a_{\Delta x, \Delta y}$ is the coefficient of transformation of inaccuracy Δy into inaccuracy Δx; C, a, b - coefficient and exponents obtained when processing the results of the calculations [3] (see table 1).
The created base of power dependences for determining the transformation coefficients allows to increase the efficiency of the analytical-calculation method for determining the total inaccuracy in milling.

Table 1. Coefficients and exponents in dependences on the coefficients of transformation on the cutting condition at milling of workpieces of material 12XH2 and material of the active part of the tool P10

Elementary inaccuracies Δx	$a_{\Delta x,\Delta y} = C f_z^a v_c^b$											
	Elementary inaccuracies Δy											
	ε_j			$\varepsilon_{td,w}$			$\varepsilon_{td,c}$			$\varepsilon_{w,c}$		
	C	a	b	C	a	b	C	a	b	C	a	b
ε_j	-	-	-	1,096	0,081	0,046	0,25	0,003	-1,27	0,25	0,41	1,528
$\varepsilon_{td,w}$	0,078	0,024	-0,3506	-	-	-	0,024	0,007	-0,33	0,01	0,029	1,14
$\varepsilon_{td,c}$	1,16	0,042	-0,4974	2,14	0,179	0,55	-	-	-	0,46	0,006	-1,17
$\varepsilon_{w,c}$	0,94	0,006	-0,3069	2,26	0,17	0,29	0,41	0,0012	-1,11	-	-	-

- *Dependencies for the determination of elementary inaccuracies in milling with regard to the phenomenon of technological heredity*

With the help of the developed methodology used for obtaining the dependences for the transformation coefficients [3], dependences for the elementary inaccuracies from the technological processing conditions were obtained

$$(\Delta x)_{\Delta y} = C f_z^{\,a} v_c^{\,b} \tag{4}$$

where the coefficient C and the exponents a, b obtained by milling workpieces of material 12XH2 and material of the active part of the cutting tool K30 are given in table 2.

Table 2. Coefficients and exponents in dependences on the component inaccuracies of the technological conditions of processing at milling of workpieces of material 12XH2 and material of the active part of the tool P10

Elementary inaccuracies $(\Delta x)_{\Delta y}$	$(\Delta x)_{\Delta y} = C f_z^{\,a} v_c^{\,b}$											
	Elementary inaccuracies Δx											
	ε_j			$\varepsilon_{td,w}$			$\varepsilon_{td,c}$			$\varepsilon_{w,c}$		
	C	a	b	C	a	b	C	a	b	C	a	b
$(\Delta x)_D$	79,81	0,65	-0,355	2,65	-0,178	0,0285	251,12	1,165	0,478	55,88	0,635	0,545
$(\Delta x)_{\varepsilon_j}$	-	-	-	3,65	-2.93	0,15	230,82	1,5535	0,378	78,12	0,544	0,355
$(\Delta x)_{\varepsilon_{td,w}}$	90,15	0,677	0,421	-	-	-	254,35	1,17	0,388	8353	0,259	0,333
$(\Delta x)_{\varepsilon_{td,c}}$	95,52	0,557	0,355	3,45	-0,159	0,043	-	-	-	82,40	0,489	0,535
$(\Delta x)_{\varepsilon_{w,c}}$	79,25	0,552	0,291	3,95	-0,282	0,0555	221,33	1,332	0,288	-	-	-

With the help of (4), the determinants and values taking into account the mutual influence of the elementary inaccuracies in the milling of a plane surface of a workpiece of steel material 12XH2 are determined upon the following output data:

- material of the active part of the cutting tool – P10;

- cutting geometry – $\alpha = 12°$, $\gamma = 5°$, $\chi = 45°$, $\chi' = 15$, $r = 0{,}5mm$;
- machine tool - universal milling machine ФУ321;
- dimensions of the machined surface: b = *200mm*, l = *300mm*;
- depth of cut - $a_p = 2{,}5mm$;
- the range of variation of the cutting conditions: $f_z = 0{,}1 \ldots 0{,}3$ *mm*; $v_c = 100 \ldots . 300$ *m/min*.

Calculations have been made and graphical dependencies have been constructed (see figure 1), the analysis of which allows us to draw conclusions:

- Dependencies (4) allow the determination of elementary inaccuracies, taking into account their mutual influence, without the need of labor-intensive dependencies (2).
- By increasing the feed f:

- inaccuracies caused by insufficient stability of the technological system ε_j, determined by taking into account the influence of other inaccuracies increase, which is due to the increase in the constituent forces of the cutting force;

- inaccuracies caused by temperature deformations of the workpiece ε_{td_w} determined by taking into account the impact of other inaccuracies decrease, which is explained by the increase in the speed of movement of the heat source relative to the workpiece, as a result of which most of the heat is removed from the chips;

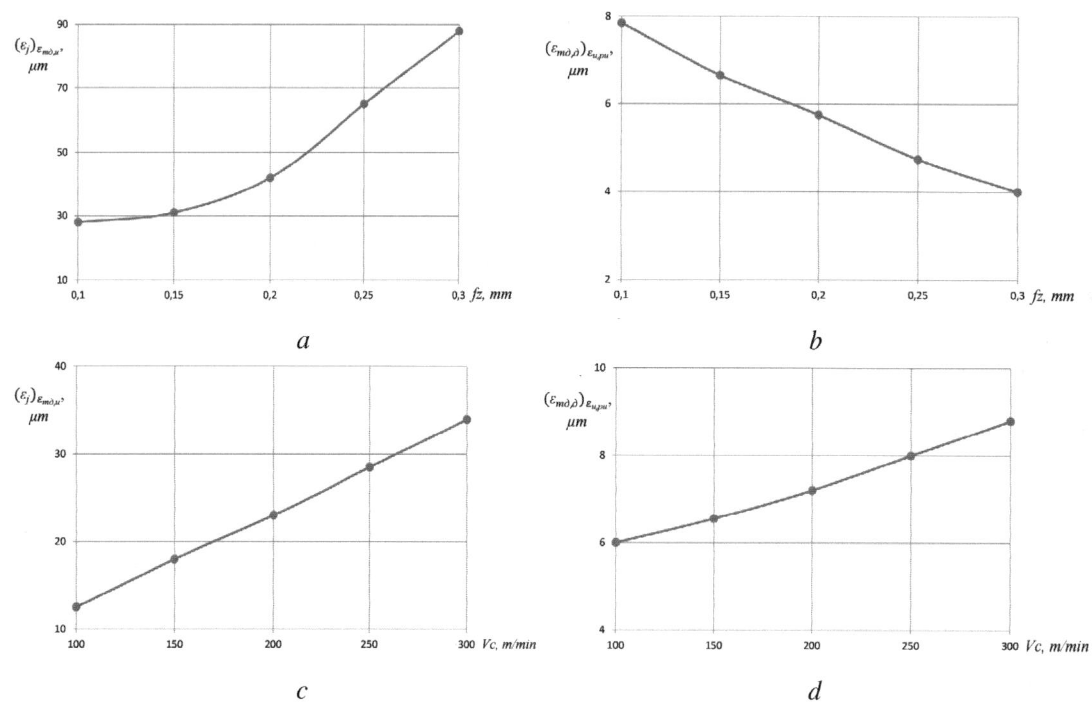

Figure 1. Graphic dependencies of elementary inaccuracies $(\varepsilon_j)\varepsilon_{td,c}$ and $(\varepsilon_{td,w})\varepsilon_{w,c}$ for face milling from the feed f_z (a,b) and cutting speed v_c (c,d), of processing at milling of workpieces of material 12XH2, $\sigma_B = 785$MPa and material of the active part of the tool P10

- inaccuracies of temperature deformations of the teeth of the milling cutter in axial direction ε_{td_c}, determined by taking into account the influence of the remaining inaccuracies increases, which is explained by the increase in the temperature in the cutting zone.

- With increasing cutting speed v_c:

- inaccuracies in temperature deformation of the workpiece ε_{td_w} and the teeth of the milling cutter in the axial direction ε_{td_c}, determined by taking into account the influence of the remaining inaccuracies increases, which is explained by the increase in the temperature in the cutting zone;

- inaccuracies of insufficient stability of the technological system, determined by taking into account the impact of other inaccuracies decrease, as a result, the cutting forces are reduced, and their reduction is much smaller than the increase in ε_{td_c}.

- *Comparisons of milling inaccuracies with and without reporting the phenomenon of technological heredity*

With the same output data, a comparative analysis of the accuracy with and without accounting for the technological heredity phenomenon was made, for the purpose calculations and experimental studies [4] in the sequence were carried out:

- Using the dependencies (3), the transformational coefficients of elementary inaccuracies are calculated for the selected range of change in technological conditions $a_{\Delta x,\Delta y}$.
- Using the dependencies (4), the determined values of elementary inaccuracies are calculated for the selected range of change of technological conditions $(\varepsilon_j)_D$, $(\varepsilon_{td_w})_D$, $(\varepsilon_{td_c})_D$ and $(\varepsilon_{w_c})_D$.
- With the determined deterministic values of elementary inaccuracies the total inaccuracy is determined, without taking into account the mutual influence of elemental inaccuracies [1]

$$\varepsilon_{\Sigma} = (\varepsilon_j)_D - (\varepsilon_{td_c})_D - (\varepsilon_{td_w})_D + (\varepsilon_{w_c})_D \qquad (5)$$

- By using deterministic values of elementary inaccuracies $(\varepsilon_j)_D$, $(\varepsilon_{td_c})_D$, $(\varepsilon_{td_w})_D$, $(\varepsilon_{w_c})_D$ and coefficients of transformation $a_{\Delta x,\Delta y}$, obtained with the help of the deduced dependencies (3) and (4), the values of the total milling inaccuracy were determined, taking into account the mutual influence of the elemental inaccuracies, using the dependences (2).
- The total inaccuracies are compared with and without taking into account the mutual influence of the elemental inaccuracies with the results of the experiments.

In order to confirm the obtained results, a study was carried out on the processed materials: steel 12XH2, σ_B = 785MPa and steel 40XH, σ_B = 980MPa, in the first case the material of the active part of the cutting tool is P10 and in the second - K20.

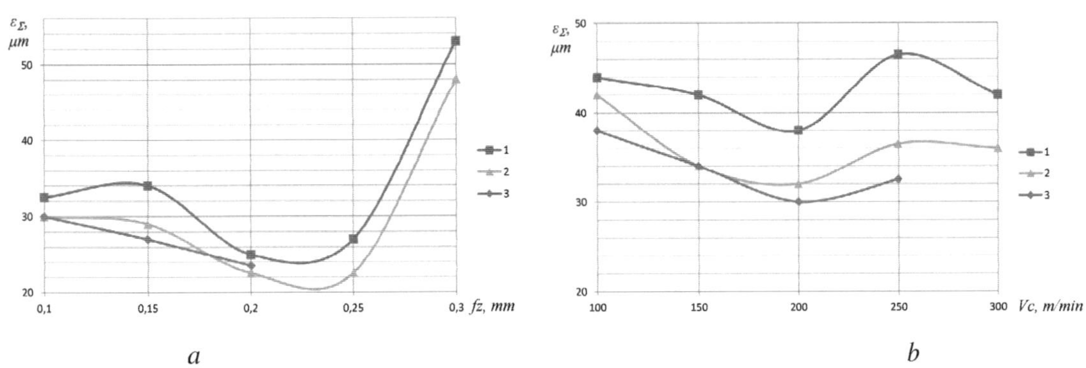

Figure 2. Graphic dependencies of total inaccuracies for face milling ε_{Σ} from the feed f_z (a) and cutting speed v_c (b), of processing at milling of workpieces of material 12XH2, σ_B = 785Mpa and material of the active part of the tool P10

IOP Conf. Series: Materials Science and Engineering **878** (2020) 012049 doi:10.1088/1757-899X/878/1/012049

With the results obtained, graphic dependences for the total inaccuracy are built, without taking into account the influence of the elementary inaccuracies (1), taking into account the influence of the elementary inaccuracies (2) and with the values of the total inaccuracy obtained from the experimental data. (3) (see figure 2 and 3).

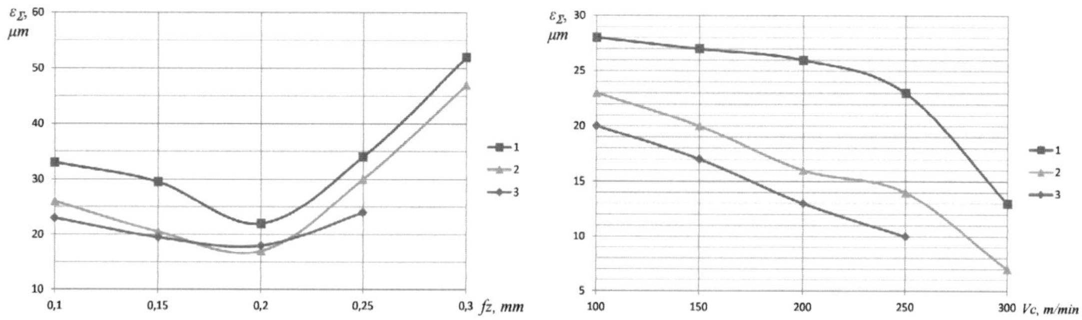

a *b*

Figure 3. Graphic dependencies of total inaccuracies for face milling ε_Σ from the feed f_z (a) and cutting speed v_c (b), of processing at milling of workpieces of material 40XH, $\sigma_B = 980$MPa and material of the active part of the tool K20

The analysis of the obtained results (see figure 2 and 3) allows to make *conclusions*:

- The nature of the dependencies of the total inaccuracy ε_Σ on the cutting conditions is the same for both pairs of machined and tool material. The quantitative difference in values is due to the different physico-mechanical characteristics of the materials under consideration.
- By increasing the feed f_z the absolute values of the total inaccuracy ε_Σ in face milling, determined without taking into account the mutual influence of the elemental inaccuracies, are greater than those determined by taking into account their mutual influence. The results of the studies show that, taking into account the influence of the other elementary inaccuracies, ε_j and ε_{w_c} they also slightly exceed, and ε_{td_c} and ε_{td_w} are, significantly smaller than their deterministic values. All this in accordance with the dependencies given in [3] explains the results obtained.

With the increase of feed f_z the difference between the total inaccuracies ε_Σ obtained with and without consideration of the mutual influence of the elemental inaccuracies remains approximately constant. This can be explained by the fact that the dependences on the elementary inaccuracies of the feed f_z obtained with and without taking into account their mutual influence are practically identical [4], and their difference is within a range of 5÷10 µm.

- The total inaccuracy without taking into account the mutual influence of elementary inaccuracies is greater than the same, determined by taking into account their mutual influence for the whole range of variation of f_z and v_c. This makes it possible to optimize cutting conditions so as to increase productivity.
- At a cutting speed of v_c from 150 to 250 m/min, the values of the total inaccuracy ε_Σ obtained by taking into account the mutual influence of the elemental inaccuracies are significantly smaller than those without taking into account their influence, such that at $v_c = 200$ m/min extreme (minimum) is observed, more pronounced in the first pair of treated and tool material. The reason for this is that in the range $v_c = 150÷200$ m/min the inaccuracies related to the thermal deformations of the workpiece and the teeth of the milling cutter sharply increase, with $v_c = 200$ m/min being dominant In the range $v_c = 200÷250$ m/min, the influence of these inaccuracies weakens and the inaccuracies due to

insufficient stability of the technological system and dimensional wear of the teeth of the milling cutter become dominant.

- The values of the total inaccuracies obtained from the experimental data and those obtained from the dependencies taking into account the mutual influence of the elemental inaccuracies are close (the difference does not exceed 10%). This confirms the reliability of the obtained dependencies and the need to take into account the impact of technological heredity when determining the total inaccuracy in milling.

References

[1] Metev H, Kuzmanov T, Stanev S and Krumov K 2015 Analysis and evaluation of the total inaccuracies and mutual influence of its compilation with reporting the phenomenon of technical heredity *„Mechanical Engineering and Science“* **vol.23** pp. 3-7 Varna ISSN 1312-8612.

[2] Metev H, Kuzmanov T and Krumov K 2015 Determination of accuracy considering the phenomenon technical heredity *Journal of the Technical University of Gabrovo* **vol.49** pp. 21-28 ISSN 1310-6686.

[3] Metev H, Krumov K and Zaharieva A 2018 Determination of transform coefficients by milling considering the phenomenon of technical heredity *Youth forums "Science, technology, innovation, business"* Plovdiv pp. 58-63 ISSN 2367-8569.

[4] Metev H, Krumov K and Zaharieva A 2018 Incompatibility consideration for milling with and without reporting the technological heredity *Youth forums "Science, technology, innovation, business"* Plovdiv pp. 30-33 ISSN 2367-8569.

IOP Conf. Series: Materials Science and Engineering **878** (2020) 012050 doi:10.1088/1757-899X/878/1/012050

Crack propagation evaluation using virtual prototyping techniques

K H Kamberov[1], Ts T Ivanov[1], B N Zlatev[1]

[1]Technical University – Sofia, FIT, TMMM, Laboratory "CAD/CAM/CAE in Industry", 8, "Kl. Ohridski" Blvd., 1797 Sofia, Bulgaria

corresponding author's e-mail address: gdt@tu-sofia.bg

Abstract. A methodology for crack propagation using virtual prototyping (VP) is demonstrated. It is based on a classical approach to assess crack propagation, improved with finite element analysis tools. It enables to examine in detail all aspects of crack formation and to evaluate the probability of its occurrence. Performed demonstration is based on Al2O3 substrate laser scribing process and its assessment in direction of possible crack occurrence. Presented study gives basic directions on application of virtual prototyping in crack initiation and propagation prediction. This approach gives the opportunity to monitor various parameters as stresses and to predict influence of various design changes over examined physical process output parameters.

1. State of the Art

Aluminum oxide substrates account for about 10% of the overall electronics printed circuit board (PCB) market. In many practical cases such as in automotive electronics, cellular radio transmitters, and aerospace systems, alumina is the substrate of choice, where its ruggedness in the face of mechanical and thermal shocks provides unique practical advantages. Because of the high absorption of alumina at 10.6 m, carbon dioxide lasers have been in common use since the early 1970s to drill vias, make slots, and produce scribe snap-lines in Al O substrates. [1]

Al_2O_3 substrate is widely used in automotive sensors industry, as a main support component. The current manufacturing technology uses the laser scribing process to divide substrate plate into several separate components. Laser scribing has a great advantage on the Al_2O_3 lines patterning process and thus becomes an important developing industry. Laser scribing ceramic substrates has been increasingly popular since its introduction to the industry nearly 50 years ago. The process involves pulsing a CO_2 laser through an optical configuration, through a focusing lens, focusing the beam through an air assisted nozzle assembly, and finally onto the substrate's work surface. Most Laser scribing systems utilize a CNC or PC based motion control system, moving the substrate below the fixed beam assembly at speeds as fast as 10 in./sec, or faster. Each pulse of the Laser provides an efficient and effective evacuation of molten and ablated ceramic substrate particulates. Well executed scribes allow for subsequent controlled Breaks, or "Snapping", of the substrate to its final size. [2]

At same time, there are new opportunities, which can accelerate and improve the product development process substantially by using virtual prototyping. Virtual Prototyping (VP) is a software-based engineering approach, which involves three-dimensional modelling, visualization and simulation of the behaviour of parts, assemblies or systems in close proximity to the actual operating conditions and subsequent iterative optimization and prediction of design and construction in a virtual form before drawing up any physical prototype of the product. Virtual prototyping is already widely used to predict the deformations and stresses and their evolution during various manufacturing processes. Nevertheless, still, experiments are necessary to verify the calculated results in order to be sure that the results obtained from virtual prototyping are reliable and accurate enough from both quantitative and qualitative perspective. [3] [4] [5] [6]

2. Motivation

Current study is focused on the laser scribing of substrate and its possible influence over its force-deflection behaviour under work loads. Slots over the side surface of the substrate are observed due to this process. Sample for the traces is shown on figure 1 below.

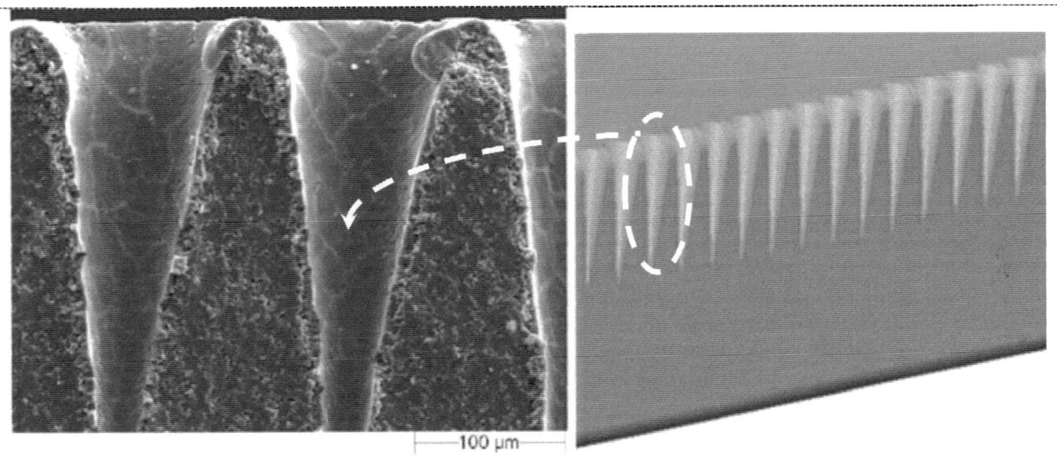

Figure 1. Laser scribing traces on the side substrate surface.

The environment examined in the presented study is at conditions of mounting of wire cushion to the housing and the substrate – with included dimensional tolerances – and under thermal loads. Main target is to demonstrate possibilities to evaluate two major possible issues:

- Initiation of a crack
- Crack growth

This is performed using fracture mechanics methodology and is based on behaviour of the structure under above described loads. Fracture propagation is a contemporary engineering analysis tool that is under intense development recent years. [7] [8] Thus, the analyses will deal with local geometry features – laser scribing slots – and their influence over stress/strain condition. [9] Detailed model of several slots over the substrate will be examined, with included surface pressure – as it is preliminary defined.

The study will be performed in two steps:

- Step 1: Analysis of the substrate with included slots, but without crack – to evaluate possibility for crack initiation;
- Step 2: Analysis/es of the substrate with included slots and crack – to evaluate crack growth.

3. Fracture analysis methodology

Initially, the initiation of a crack is evaluated through the principal stresses levels at certain surfaces as to obtain the possibility for crack generation and its direction.

Next step includes modelling of a possible crack and second analysis of the structure. The most important region in a fracture model is the region around the edge of the crack. In linear elastic problems, it has been shown that the displacements near the crack tip (or crack front) vary as $\sqrt{r}$, where r is the distance from the crack tip. The stresses and strains are singular at the crack tip, varying as $1 / \sqrt{r}$. To pick up the singularity in the strain, the crack faces should be coincident, and the elements around the crack tip (or crack front) should be quadratic, with the midside nodes placed at the quarter points. Such elements are called singular elements.

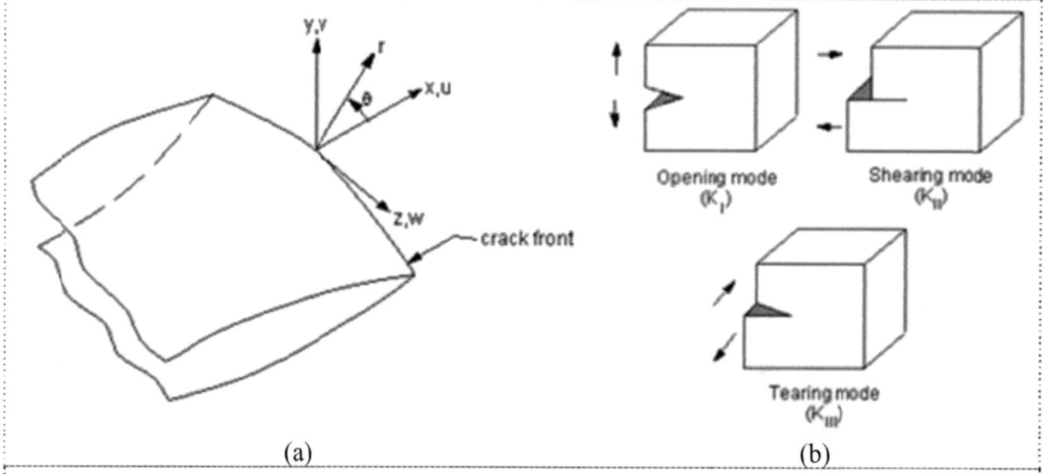

Figure 2. Crack measures and basic modes of fracture: (a) coordinate system orientation; (b) the Three Basic Modes of Fracture.

Typical fracture parameters of interest are stress intensity factors, the J-integral, and the energy release rate. [10] Current study will use stress intensity factors as reference values to determine crack growth. The stress intensity factors at a crack for a linear elastic fracture mechanics analysis may be computed using a fit of the nodal displacements in the vicinity of the crack. Calculation of these factors is performed according to the given expressions for no symmetry conditions (full model of crack):

$$K_I = \sqrt{2*\pi} * \frac{G}{1+\kappa} * \frac{|\Delta v|}{\sqrt{r}};$$

$$K_{II} = \sqrt{2*\pi} * \frac{G}{1+\kappa} * \frac{|\Delta u|}{\sqrt{r}};$$

$$K_{III} = \sqrt{2*\pi} * \frac{G}{1+\kappa} * \frac{|\Delta w|}{\sqrt{r}},$$ where:

u, v, w – displacements in a local Cartesian coordinate system as shown on figure 2;
Δu, Δv, Δw – motions of one crack face with respect to the other;
G – shear modulus;
$\kappa = 3 - 4*\nu$ - for plane strain or axisymmetric;
$\kappa = \frac{3*\nu}{1+\nu}$ - for plane stress;
ν – Poisson's ratio.

4. Virtual prototype

All overall dimensions and geometry of the substrate are according to used ones in previous models, but this model includes also laser scribing slots, as it is shown on figure 1, based as dimension on the measured on a microscope probe ones. Examined possible crack is placed at the bottom of one of these slots.

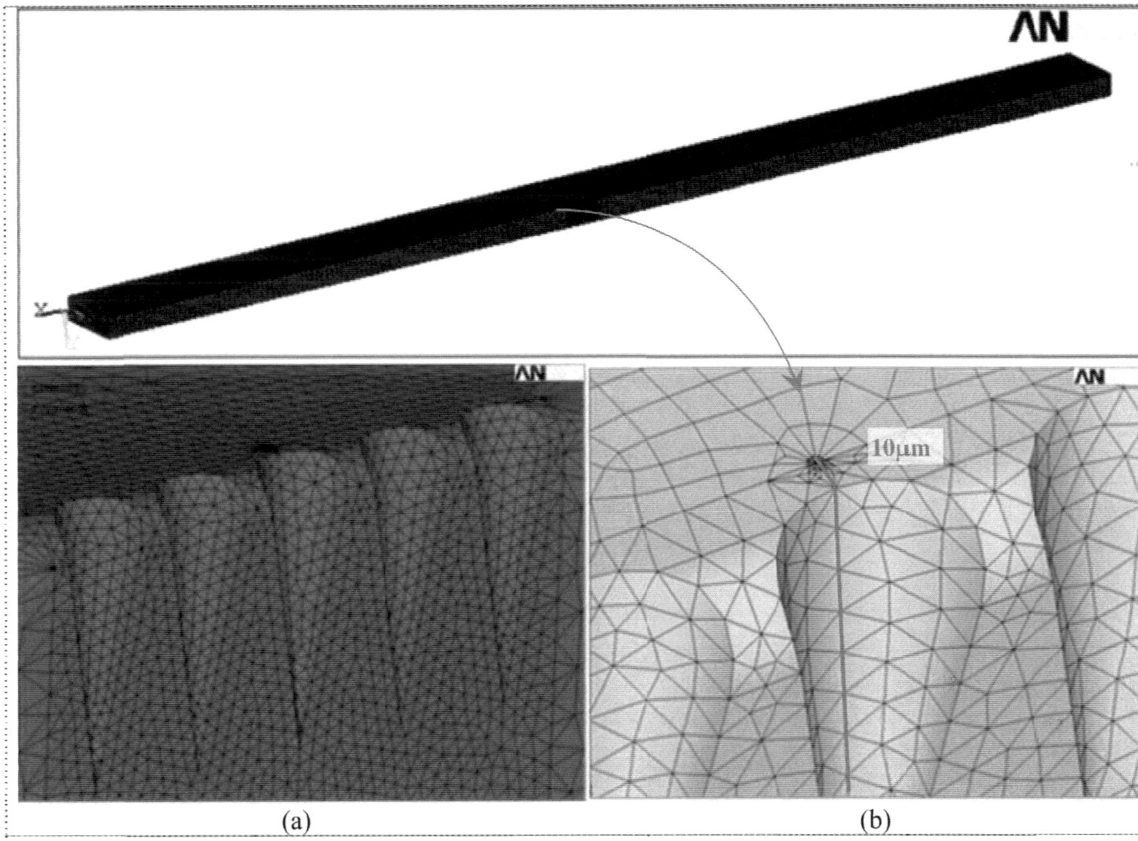

Figure 3. Meshed models: (a) model with no crack included; (b) model with crack.

The meshed structures – for the model with no crack and the model with crack included – are shown on figure 3. Each model contains approximately 75 000 nodes and 370 000 elements. The model with crack has defined space of 10 micrometers between two opposite sides of the crack as it is shown on figure 3 (b) – marked by red lines.

The material properties, used in the model, are common for Al_2O_3. The applied boundary conditions are shown on figure 4 – according to the analysis technology described above. A half model is examined – due to the structural and applied boundary conditions symmetry. Model symmetry is represented by constrained displacements in direction, normal to symmetry plane as it is shown on figure 4. The model is constrained for translation along the other two axes at the sliced surface, placed far enough as not to interfere with simulation results. Applied pressure zone is determined based on calculations of local contact of the substrate to another component in the sensor design and is defined prerequisite of current engineering analysis.

IOP Conf. Series: Materials Science and Engineering **878** (2020) 012050 doi:10.1088/1757-899X/878/1/012050

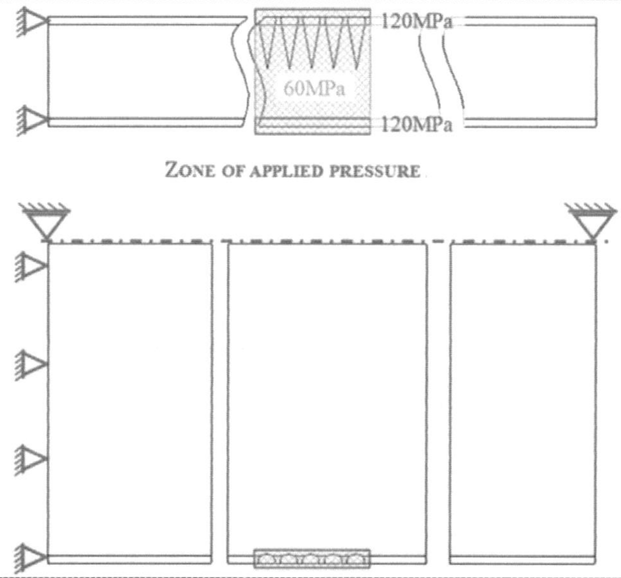

Figure 4. Applied boundary conditions.

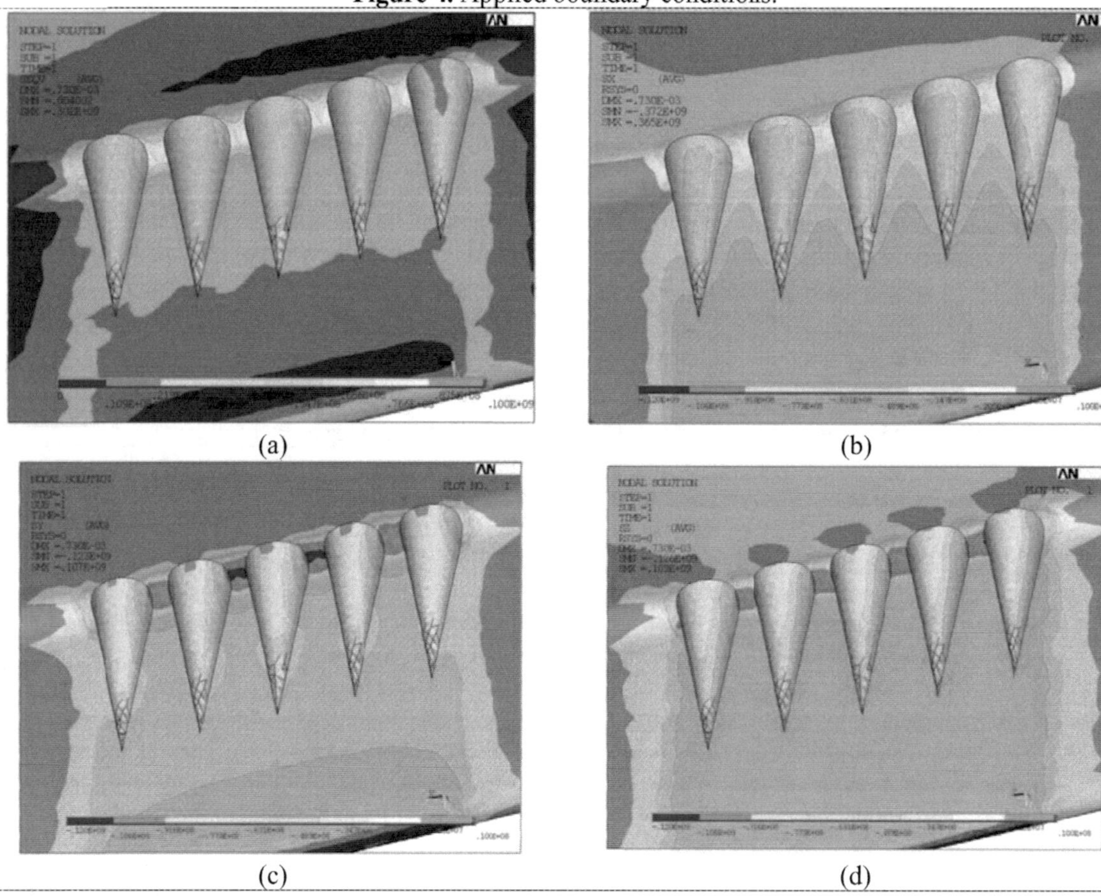

Figure 5. Model without crack: Stress distributions in the zone of laser scribing cavities, MPa: (a) equivalent; (b) normal, along X axis (longitudinal); (c) normal, along Y axis (transverse, vertical); (d) normal, along Z axis (transverse, horizontal).

5. Simulations results

5.1. Crack initiation – model with no crack

Initially performed structural analysis involves geometry with laser scribing slots, but no cracks. Its results are shown as stress distributions – equivalent and normal along different axes – on figure 5.

Equivalent stresses shows that the concentrations are on the edges, between neighbor laser scribing cavities. This is due to the character of the applied loads – normal pressure on the side surface and rounds of the edges. It leads to compressive character of the local behaviour at all. Another point is that calculated values of stresses do not show crack initiation. Maximal values are bellow 100MPa.

Nevertheless, possible crack in transverse direction is analyzed in next step.

5.2. Model with crack

The results are presented in similar way – to preview in detail the behaviour in the zone of laser scribing cavities and crack. Initially, the equivalent stresses are shown on figure 6 (a) below. It is seen that major stresses occurs at the bottom end of the crack. In fact, stresses at the top are redistributed among the crack and cavities. Further simulation results are presented again as normal stress distribution fields on figures 6 (b), (c) and (d).

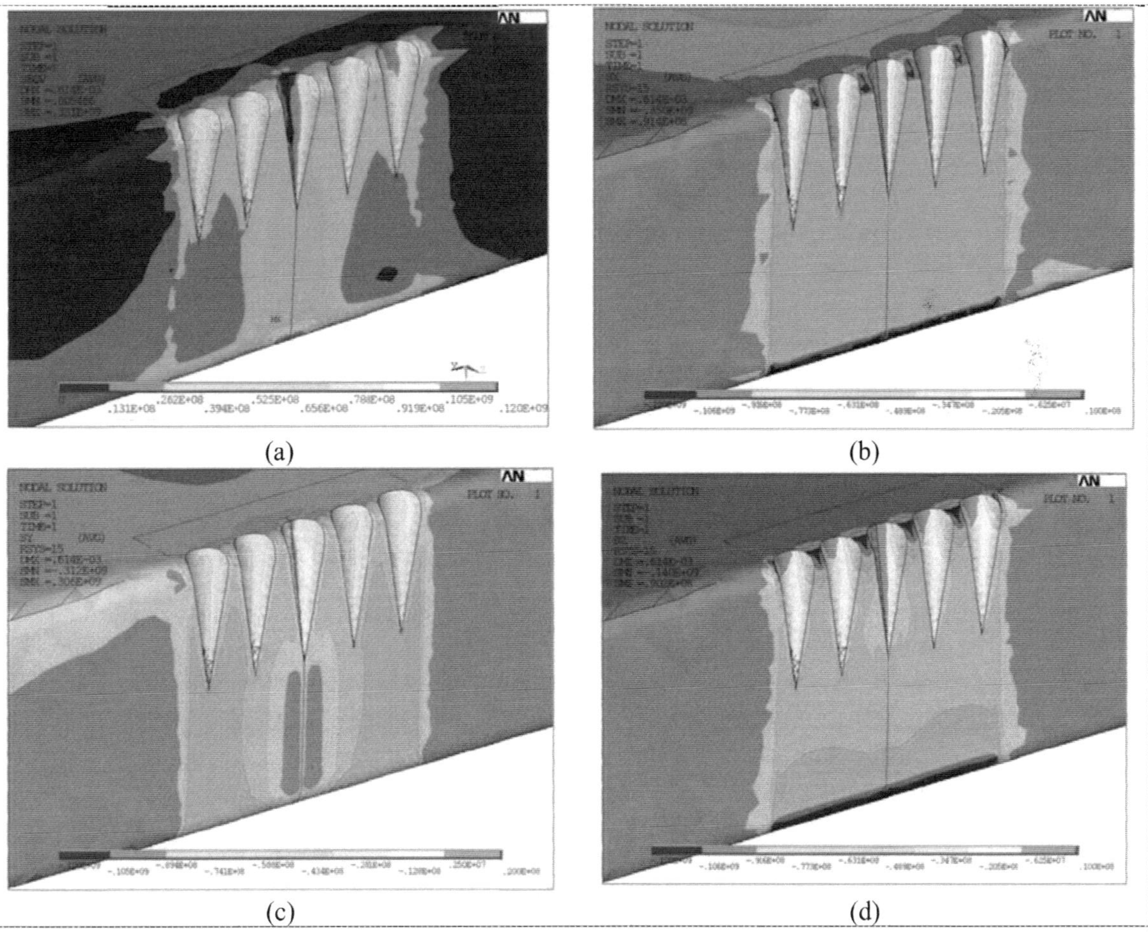

Figure 6. Model with crack: Stress distributions in the zone of laser scribing cavities, MPa: (a) equivalent; (b) normal, along X axis (longitudinal); (c) normal, along Y axis (transverse, vertical); (d) normal, along Z axis (transverse, horizontal).

5.3. Stress intensity factors

Both up side and down side ends are evaluated through calculations of the stress intensity factors. The calculation, as it was described in chapter 3, requires 5 nodes: one tip node, two "top" nodes and two "bottom" nodes – shown on figure 7 below, together with the crack reference coordinate system orientation. Top and bottom nodes coincide, but are from opposite sides of the crack.

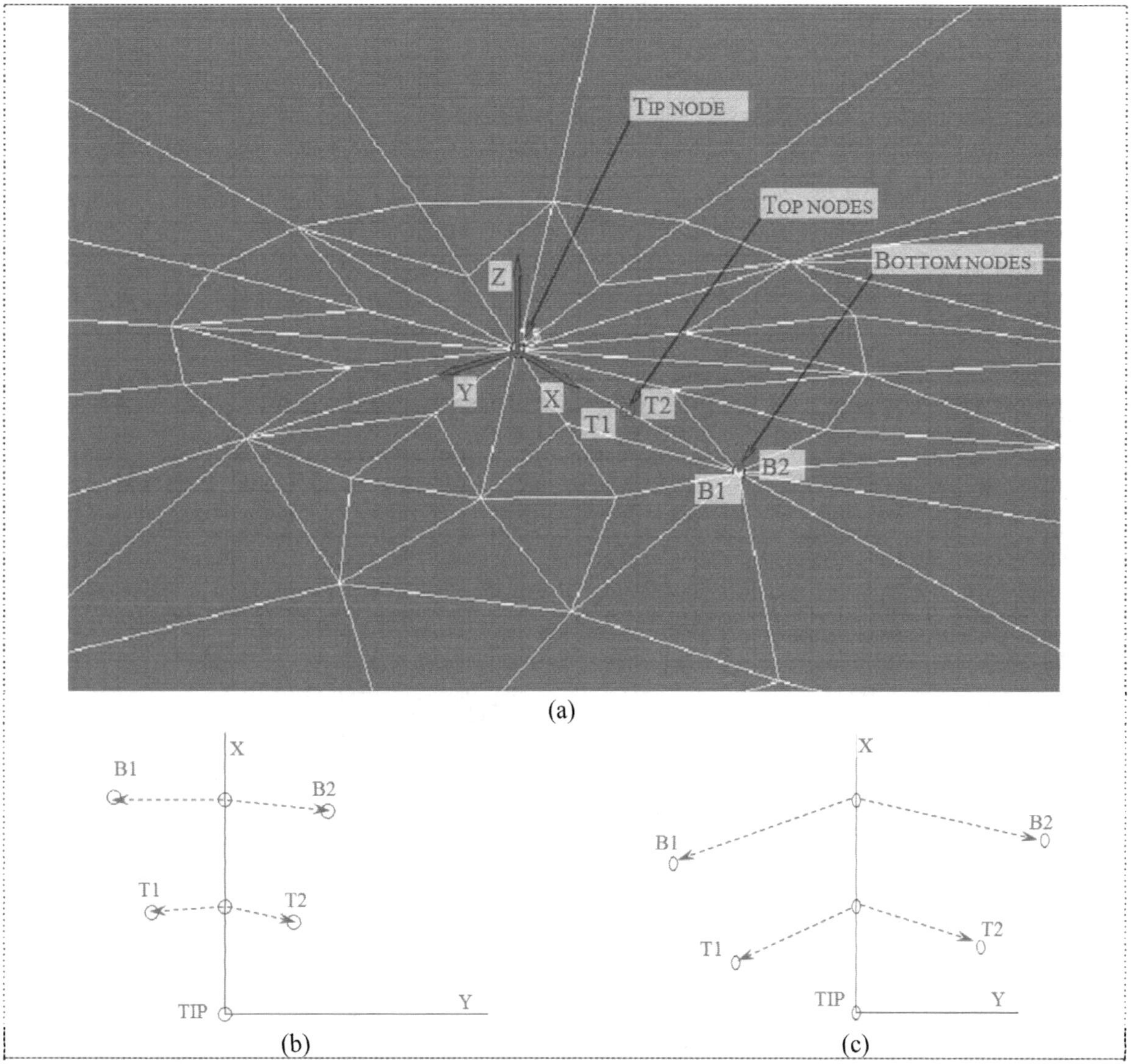

Figure 7. Crack propagation: (a) used nodes for stress intensity factor calculations; (b) relative crack edges displacements at upside edge; (c) relative crack edges displacements at down side edge.

The results are summarized on table 1 bellow.

Above shown results are presented graphically on figures 7 (b) and (c), for better understanding. The figure shows relative displacements of the top and bottom nodes to the tip. This could orient about the general behaviour of the examined crack.

The up side edge is definitively in opening mode – as it is seen on figure 7 (b). Stress intensity factor at Y axis direction (KI) has also highest value for the edge.

The down side edge is also in opening mode, even more as its stress intensity factor KI is higher, due to the highest stresses in this zone.

Tabl.1 Stress intensity factors calculation results

Side	Point	Radius to tip, r, μm	Displacements, relative to the tip deformations ($U_{(POINT)} - U_{(TIP)}$), Δ, μm * 1000			Stress intensity factors, K, Pa* $\sqrt{m}$ *10^{-4}		
			X	Y	Z	KI	KII	KIII
Up side	T1	2.5	-0.132	-1.792	0.795	23.2	1.53	9.6
	T2	2.5	-0.356	1.671	0.592			
	B1	5	0.060	-2.705	1.503			
	B2	5	-0.268	2.487	1.178			
Down side	T1	2.5	-1.327	-4.734	-1.172	63.5	2.52	1.29
	T2	2.5	-0.950	4.887	-1.464			
	B1	5	-1.493	-7.176	-2.303			
	B2	5	-0.929	7.418	-2.797			

6. Conclusions

Several conclusions could be formed, based on the performed analyses:

- Initial simulation of the local behaviour of the structure under thermal and contact loads shows equivalent stresses under 100MPa, that is not expected to be critical for the material;
- The values of the normal stresses show mainly compression in all directions and no indication for crack initialisation;
- Nevertheless, cracks are probable due to the local thermal treatment during laser scribing process and thus, they are examined further in the current study;
- Further processing of the model is focused on crack examination – at the middle of a laser scribing slot. Determined equivalent stresses have higher maximal value – about 120MPa – locally at the tip of the crack, but on the opposite side of the cavities;
- Normal stresses once again shows dominant compressive stresses;
- A detailed study of the fracture mechanics through stress intensity factors evaluation shows that the crack works in opening mode;
- The values for the stress intensity factors as well as the relative displacements are not significant and does not show increasing of the crack under explored loads (due to the thermal and contacts between wire cushion and substrate) and environment;
- More detailed study could be performed, based on da/DN vs DK material curve (fatigue analysis) to determine crack growth (da) per cycle (DN) at certain growth of the stress intensity factor (DK);
- Generally, analyses results under thermal and contact loads are not too critical for the crack initiation and growth. Fracture processes are more probable in case of vibration loading, when certain micro cracks are available, formed during the laser scribing process.

Presented study demonstrates virtual prototyping application for evaluation of crack propagation. Advantages of VP are mainly in direction of reviewing and monitoring of such "internal" parameters as stresses (normal, principal, equivalent) as well as in the ease of changing various geometry parameters and exploration of their influence over structural behaviour.

Acknowledgments
This work was supported by the European Regional Development Fund within the Operational Programme "Science and Education for Smart Growth 2014 - 2020" under the Project CoE "National centre of mechatronics and clean technologies" BG05M2OP001-1.001-0008.

7. References

[1] Moorhouse C J Villarreal F Wendland J J Baker H J Hall D R Hand D P 2005 CO2 Laser Processing of Alumina (Al2O3) Printed Circuit Board Substrates *IEEE Transactions on Electronics Packaging Manufacturing* vol 28 no 3 pp 249-258 DOI: 10.1109/TEPM.2005.852232

[2] Chuang H Ch Lee Y C 2014 Laser Scribing on Silver Flexible Substrate with Beam Shaping Technology *Journal of Laser Micro / Nanoengineering* vol 9 no 3 pp 198-203

[3] van Driel W D Hauck T Grech A M Schmadlak Zhang X Orain S 2006 Virtual Prototyping based Design Optimization of the Substrate, Leadframe, and Flip Chip Package Families with Low-k Technology *7th International Conference on Thermal, Mechanical and Multiphysics Simulation and Experiments in Micro-Electronics and Micro-Systems EuroSime 2006* 1 - 6. 10.1109/ESIME.2006.1644013

[4] Todorov T Todorov G Romanov B 2019 Design and simulation of mould tools with multi-material structure for plastic injection moulding based on Additive Technology *International Conference on Creative Business for Smart and Sustainable Growth CreBUS 2019* 8840061

[5] Todorov G Sofronov Y Petkov A 2018 Innovative Joystick Virtual Prototype Ergonomy Validation Methodology by Physical 3d Printed Functional Model *International Conference on High Technology for Sustainable Development, HiTech 2018 – Proceedings* 8566241

[6] Todorov G Nikolov N Sofronov Y Gabrovski N Laleva M Gavrilov T 2019 Computer aided design of customized implants based on CT-scan data and virtual *prototypes Lecture Notes of the Institute for Computer Sciences, Social-Informatics and Telecommunications Engineering LNICST* vol 283 pp 339-346

[7] Wang Y Wang Z Zheng Y 2019 Analysis of Fatigue Crack Propagation of an Orthotropic Bridge Deck Based on the Extended Finite Element Method *Advances in Civil Engineering* article ID 6319821

[8] Bergara A. Dorado J. I. Martín-Meizoso A. Martínez-Esnaola J. M. 2019 Fatigue crack propagation at aeronautic engine vane guides using the extended finite element method (XFEM) *Mechanics of Advanced Materials and Structures* DOI 10.1080/15376494.2019.1602236

[9] Hellen T Becker A 2013 *Finite Element Analysis for Engineers – A Primer* (NAFEMS)

[10] Ricardo L C H Crack 2018 Propagation by Finite Element Method *Frattura ed Integrità Strutturale* vol 43 pp 57-78; DOI 10.3221/IGF-ESIS.43.04

IOP Conf. Series: Materials Science and Engineering **878** (2020) 012051 doi:10.1088/1757-899X/878/1/012051

Black box/white box hybrid method for virtual prototyping validation of multiphysics simulations and testing

G D Todorov[1], K H Kamberov[1]

[1]Technical University – Sofia, FIT, TMMM, Laboratory "CAD/CAM/CAE in Industry", 8, "Kl. Ohridski" Blvd., 1797 Sofia, Bulgaria

corresponding author's e-mail address: gdt@tu-sofia.bg

Abstract. A hybrid method has been developed to predict the level of confidence in the results of multiphysical behavioural simulations of the object by virtue of virtual prototyping (VP) together with physical PP testing based on the stage of the virtual prototyping process and feedback from physical parameters tests. The method is based on a three approaches to assessing the reliability of the results of virtual prototypes - initial - for verification of boundary conditions, intermediate - to improve the accuracy of the results obtained and final - to validate the obtained results. A new method has been developed to improve the efficiency and effectiveness of Virtual Prototyping called "WHITE BOX / BLACK BOX" for the study of complex objects by applying VP as a "white box" and VP as a "black box" in iterative connectivity to achieve fine calibration of VP. The novelty of the method lies in the iterative connectivity of the VP and the PP, and in particular the calibration of the VP by feedback from PP data via a micro-cycle. The "WHITE BOX / BLACK BOX" method and the three approaches to verifying and validating VP have been demonstrated in three successfully completed projects with international industrial partners used as demonstration test cases.

1. State of the Art

Efforts in recent years have been focused on developing methods for virtual prototyping of products presented as complex objects with relatively small production data, allowing for the accounting of additional characteristics of the objects (material properties, object weight, kinematic and dynamic factors). Virtual prototyping (VP) introduces new approaches to engineering activities such as a predictive and prevent approach to minimize errors and reduce the number of physical prototypes (PP) and, accordingly, the cost of eliminating them.

In this connection, one of the great advantages of virtual prototypes can be pointed out - they reach the users of the product even before it is actually created. This also provides a kind of "feedback" through which the needs of the users reach the product developers. This is very important as it has a positive effect on reducing the time and investment involved in developing a new product. Often, more factors than just product characteristics must be considered when a prototyping approach is to be chosen. Liu argues that virtual prototypes are best suited to situations where physical prototyping is impractical [1], impossible or ineffective, while [2] emphasizes that physical and virtual prototypes are not competitive but complementary technologies. Campbell [3] also states that physical and virtual prototyping are two important technologies that must be used together to form a combined toolkit for prototyping complex products.

A critical element in Virtual Product Development (PDP) technology is the level of reliability of the results of engineering analysis and VP simulations. It is defined by two basic concepts:

- Stability, or, how well does the computational model describe the real physics of the state and processes and what is the impact of the model uncertainty (structural or parametric) on the simulation results?
  - Uncertainty of baseline physical condition and boundary conditions (How well are the initial / boundary conditions known?)
  - Engineering (structural) reliability (Does the model accurately reflect the physical process?): for instance - mathematical models of turbulence or model of combustion;
- Parametric stability (How accurately are the model parameters defined?): Condition equations and replacement models; Degrees of reaction and relaxation time;
- Geometric Stability (Does the geometric model accurately reflect the real object)? [4]
- Calculation error, or, how accurate are the results calculated according to the approximations made by a numerical method? [5] [6]

2. Motivation

This study focuses mainly on sustainability. It also looks at how physical modeling and testing can be integrated to increase the reliability of virtual prototyping results through feedback from test results.

Virtual prototyping as a technology involves the creation and use of virtual models to test and analyze a product before creating real physical models or optimizing parameters in parallel with physical prototypes. The application of VP in the study of multiphysical processes can be represented as a set of thermal, structural, thermos-fluid and others. models to simulate coupled processes, see Figure 1.

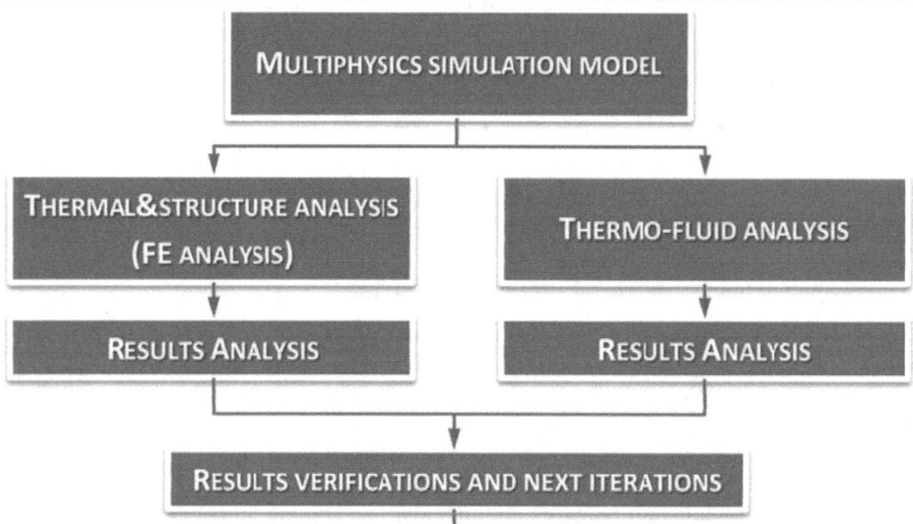

Figure 1. Simulation approach using Virtual Prototyping in multiphysics – connected processes as thermo structural or thermo fluid.

The practical applications of CFD analysis have been increasing significantly in recent years because it is already considered an industry-leading technology that can significantly contribute to the design, development and optimization of complex products incorporating fluid and / or thermos-fluidic systems. [7] [8]

This is due to the possibility of significantly reducing the cost of development and minimizing physical tests through the implementation of HR, which can also significantly reduce the time to reach the market and development costs.

3. New hybrid approach for virtual prototyping validation of multiphysics simulations

A new hybrid method is proposed here for assessment of the level of confidence in the results of analyzes of the behavior of an object under study through virtual prototyping (VP) using CFD, combined with physical testing on PP.

3.1. Alternative preliminary approach for VP validation by physical prototypes

Possible approaches for the implementation of physical prototyping for verification and validation of virtual prototyping results are considered. The following figure (figure 2) shows the different stages of the simulations and the location of the physical tests in the process.

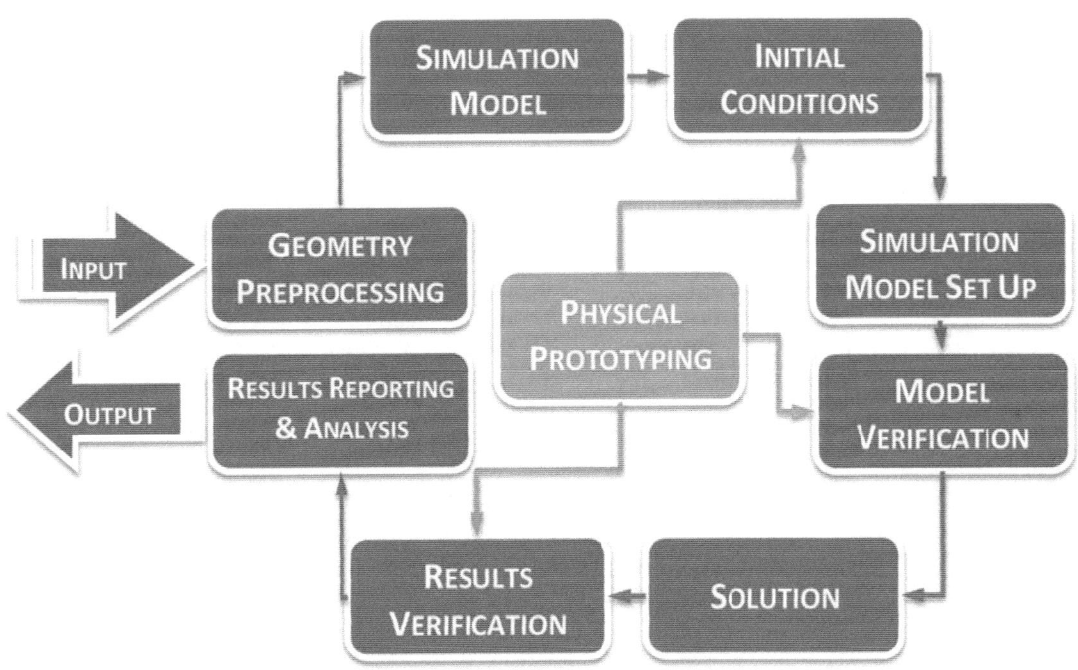

Figure 2. Stages of building a model of conducting an engineering analysis and a place of physical prototyping.

As can be seen from the figure, description and sequence of process steps, there are three possible approaches to applying physical tests to assist virtual prototyping - according to the stage of their use:

- Preliminary Verification Approach VP - the purpose in this case is to assist in defining / verifying boundary conditions - both in type and in parameters. This is a convenient approach to check the applied known boundary conditions, especially when using a hypothesis that does not fully cover the modelled physical case.
- VP Intermediate Validation Approach - mainly used to validate the simulation model constructed using intermediate measurements using the PP, very often on a partial physical prototype (of a process element or type) or an PP prepared from different material or different from the final production process;

- Approach for validation of VP through the results of PP. This type of assessment is integral to the behaviour of the prototyped object, as it is usually a "checkpoint" that proves not only the validity of the model but also the accuracy of the results.

Each of these approaches is demonstrated in the study through three examples of industrial projects undertaken by the author and team and demonstrated below.

3.2. Demonstration of an approach for pre-verification of VP

The specific example relates to the study of the design of a temperature sensor for the automotive industry in order to improve its operating parameters (response time and measurement accuracy) by means of a multiphysical engineering analysis (thermo-fluid). The example demonstrates the pre-physical testing approach. The reproduced physical process refers to an experimental setup in which the sensor's AF is placed in a flow of hot gas flowing at a controlled rate and temperature in a test tube - analogous to an ICE exhaust pipe (internal combustion engines). Figure 3 shows the experimental setup for physical testing.

The sequence of the experiment involves inserting the prototype of the sensor in the middle of the heated air flow into the tube in less than one second and starting the temperature measurement.

The purpose of the test is to determine the response time (dynamics) of the sensor by two main characteristics - a time to reach 63% (t63) and 90% (t90) at the final set temperature - that of the air flow (loss accounting). The VP via the computational model aims to replicate this experiment by simulation under the same conditions.

The main problem in modeling this experiment is related to the initial boundary conditions. The computational model includes only the tip of the sensor that is submerged in the air stream, while the rest, outside the tube, is not represented. In this way, the heat flow through the part outside the pipe cannot be correctly taken into account, nor can the "immersion" of the sensor itself into the pipe be reproduced. Although the introduction of the sensor into the heating zone occurs in a relatively short time, there is no real possibility to reflect this, nor to account for the volume of ambient fluid brought together into the sensor at this early stage of heating, which is also characterized by high intensity of the process.

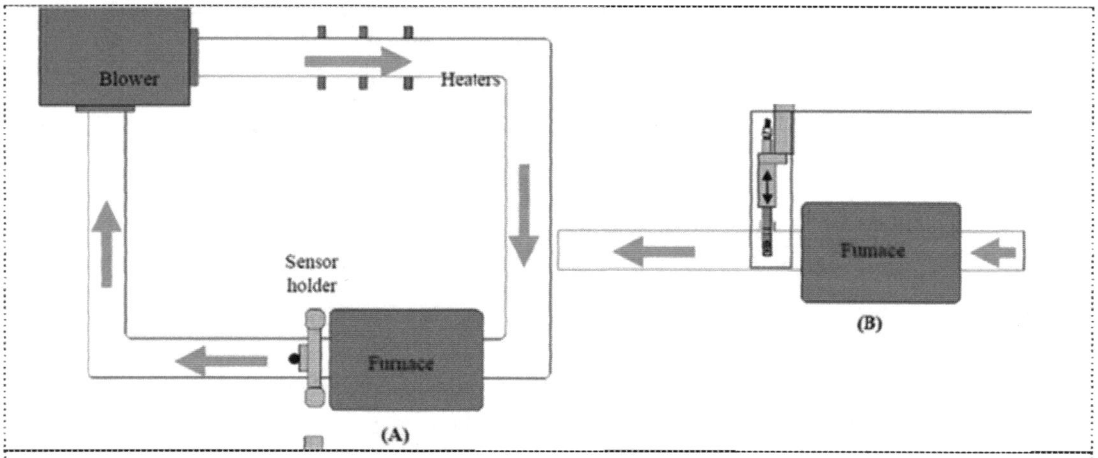

Figure 3. Experimental setup: (A) - Top view of a fan, heaters and sensor lowering system; (B) - Side view showing sensor transportation system.

In reality, there is no convenient way to reproduce this process and this is a source of uncertainty in its modeling.

The boundary conditions applied to the computational model are also shown on figure 4. They include convective heat exchange along the outer surface of the tube of the sensor, represented as a

IOP Conf. Series: Materials Science and Engineering **878** (2020) 012051 doi:10.1088/1757-899X/878/1/012051

convection coefficient to ambient temperature. In fact, there is insufficient data on this factor, which should reflect the heat dissipation outside the pipe.

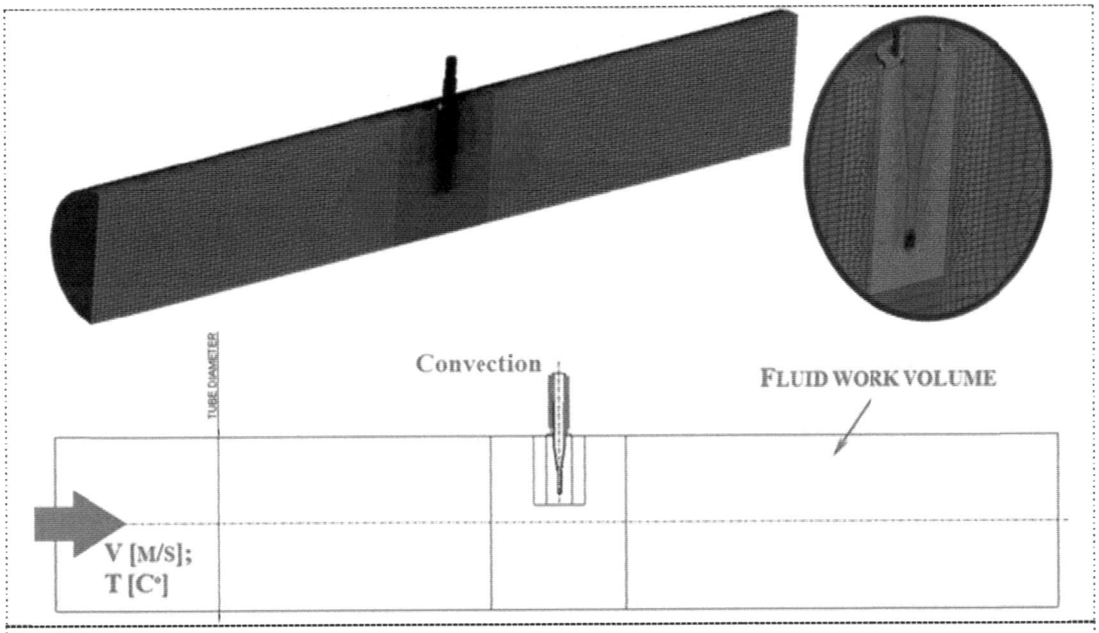

Figure 4. Computational model and set boundary conditions for temperature sensor (air flow and convective heat exchange on the outer surfaces of the sensor).

It is appropriate to use this preliminary approach to help determine the parameters of convective heat transfer. A physical test is performed using the experimental setup, and then a series of analyzes varying the convection parameters is made to iteratively reach their correct values. This is a good basis for further engineering analyzes on various design variants that use the already determined convective heat transfer parameters.

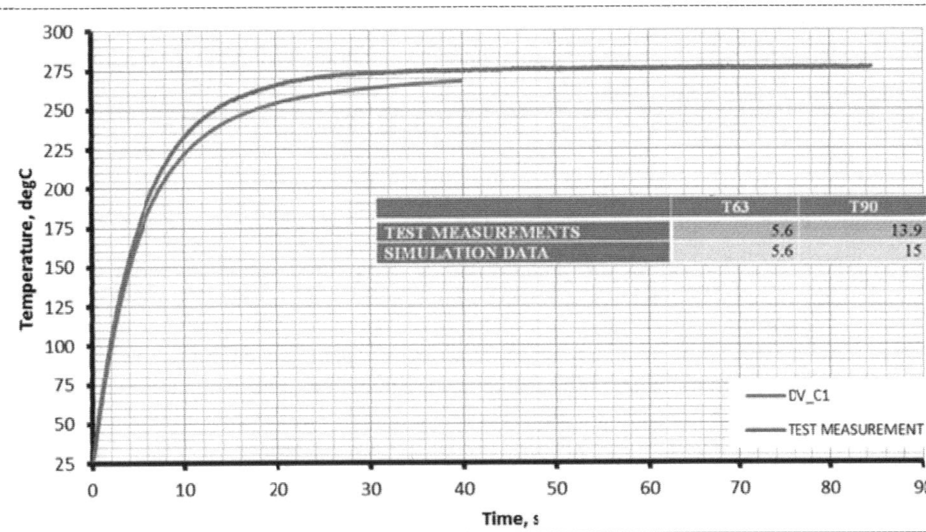

	T63	T90
TEST MEASUREMENTS	5.6	13.9
SIMULATION DATA	5.6	15

Figure 5. Comparison of data from a simulation model with test results with a physical prototype under the same conditions.

After validation of some VP parameters, which cannot be theoretically determined (e.g. convective heat transfer on the outer surfaces for a very short time) by comparison with the AF, a high degree of adequacy of the VP is achieved in a few steps. A comparison between the physical prototype test data and the virtual test data is shown in figure 5, together with the results for the calculated sensor response times. It shows good correspondence, which shows the adequacy of the preliminary approach applied.

Thus, VP can be used with a higher level of validity for testing various design modifications by conducting so-called planned virtual experiments with VP, which saves a lot of time and means of producing physical prototypes and testing them under the same conditions.

3.3. Demonstration of an VP Intermediate Validation Approach

The example given for the implementation of the proposed "Intermediate Validation Approach for VP" in practice is exemplified by a specific example of an industrial design for improving the functional parameters of a convection module, which is an important component of microelectronics production equipment of the tunnel kiln for heat treatment "Components - Solder" Surface Mounted Device (SMD) PCBs. This process equipment is part of a mass production line utilizing SMD technology that soldered the pins of the components to the circuit boards using pre-applied paste.

The process of SMD technology with lead-free formulations imposes high requirements on the thermal profile of the heating (mainly the duration, the rate of heating and the temperatures reached in the respective stages of the process), which is provided by this technological module.

In addition to the requirements for:

- precise heating rate (dynamics);
- allowable blowing speeds of heated objects (to avoid displacement);
- a very high uniformity of heating over the entire active area of the electronic circuit board is also required, taking into account the differences in their mass and size.

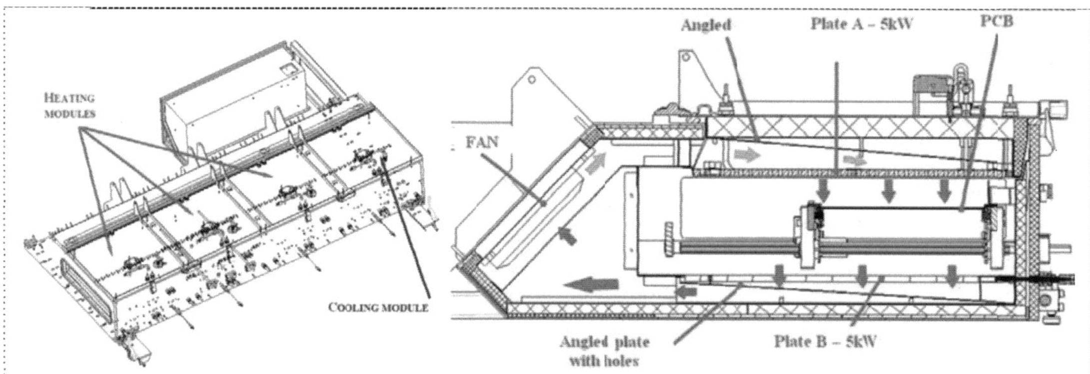

Figure 6. Examined design of a convective module, part of the technological equipment for SMD Line.

These high requirements have necessitated the use of virtual prototyping technologies that allow the evaluation of these parameters in a very large detail by applying a "dual" VP <> PP counter.

The end goal in this approach is again a reliable analysis of the thermo fluid process and the more accurate calibration of the model to adequately simulate the behavior of the structure in order to improve the required parameters and to achieve the best possible operating mode of the module in terms of the observed parameters. for convective heat transfer.

A schematic of the structure under study is shown in principle on figure 6.

The model under consideration is well defined as geometry, which allows one to construct a geometrically accurate computational model and to set boundary conditions adequate to the modeled physical process.

The challenge in this task comes from several significant nonlinearities related to the temperature dependent properties of the materials, which lead to a strong nonlinear behavior of the system under study and require a definite need for validation of the results obtained by physical tests. The calculation model used is presented on figure 7.

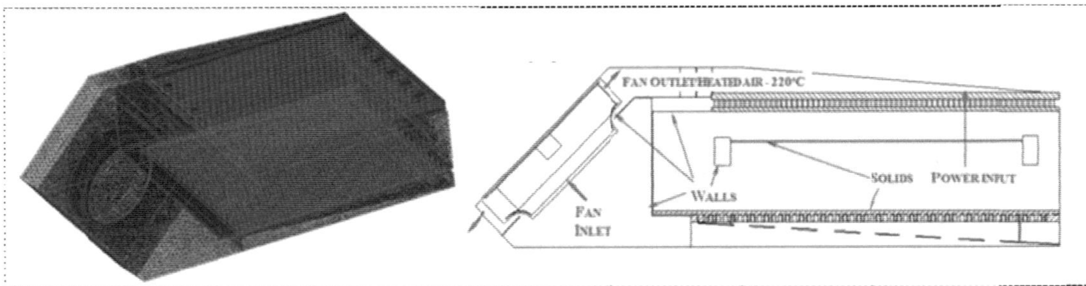

Figure 7. Calculation model (based on more than 1 700 000 finite elements) and boundary conditions for the convective module computational model (heated air outlet, plate heat and turbine flow).

Existing nonlinearity in the modeled object requires an intermediate verification of the results. This check is performed by comparing simulation data with physical test data on the functional prototype of the convective module. Air flow velocity measurements on the physical prototype are performed using calibrated high temperature anemometers.

The physical prototype designed for this purpose and the locations of the probes for measuring the flow velocity are shown on figure 8.

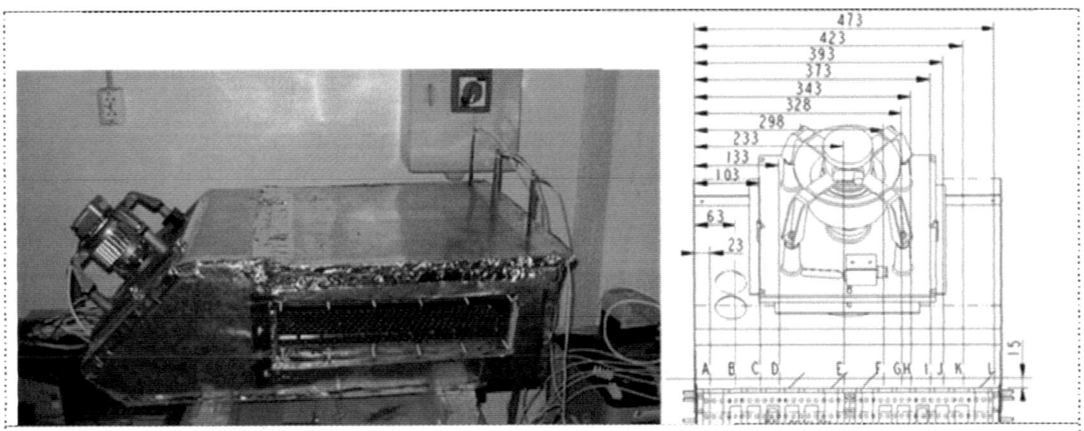

Figure 8. Physical prototype of the system and anemometers location for measuring air flow velocity at specific points of monitoring.

The results of the simulation thermal analysis using the virtual prototype created and the comparison with the measurements on the physical prototype are shown on figure 9.

The correspondence between the data of the physical prototype test and that of the simulation analysis of the virtual prototype is clearly visible. In fact, this correspondence is achieved through several iterations of model parameter values such as turbulence coefficients, as well as some boundary conditions adjustments (for example, a fan operating point that depends inversely on the system impedance).

These repetitions are performed at the "Model Verification" stage, as well as small modifications at the "Simulations" stage, and is a typical example of the approach described above to increase the reliability of the results of intermediate simulations.

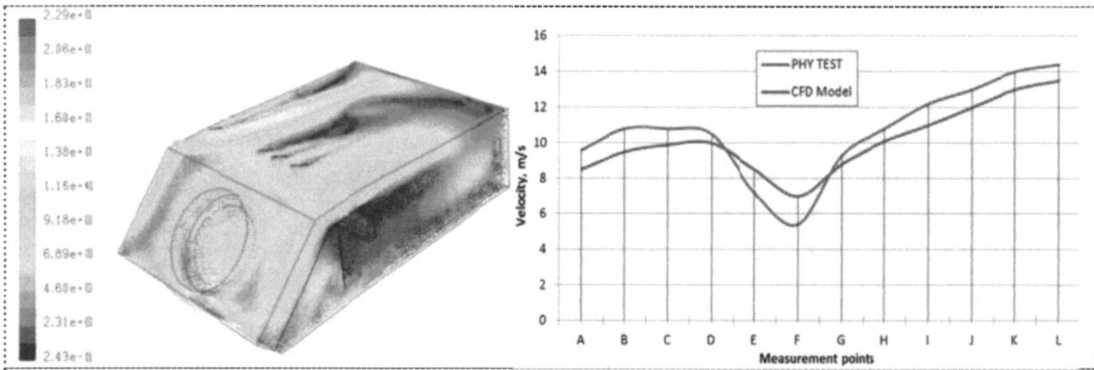

Figure 9. Virtual prototype results and comparison with measured values on a physical prototype.

The resulting parameters for the computational model setup are used to further optimize the design. The final result is tested on the physical prototype by directly measuring the basic operating parameter of the product - the heating temperature and its distribution on the surface of the circuit board.

4. A new white box / black box method, developed for practical applications and tests

This forms a well-structured approach for benchmarking, goal setting, model validation, parameter input, hybrid modeling, etc., which will contribute significantly to the overall goal of creating methods for faster, cheaper, and better development of new products.

On this basis, an innovative treatment of the ideology and philosophy of using PPs and VPs is formed as a dual method of implementing VPs and PPs in non-exclusive reciprocal application and with significantly improved quality and time indicators of the development process. The proposed new method for assessing the level of confidence in the results of virtual object prototype (VP) behavior tests and physical prototype testing is based on three approaches to applying physical tests to assist virtual prototyping - by stage of use. them:

- Approach for pre-validation of the HR - the objective in this case is to help define / verify the boundary conditions - both in type and in parameters. This is a convenient approach to test the known boundary conditions applied, especially when using a hypothesis that does not fully cover the modelled physical process. [9] [10]
- VP Intermediate Validation Approach - Mainly used to validate the simulation model constructed for the adequacy of the results obtained through intermediate measurements using the FP, very often on a partial physical prototype (of elements or process type) or PP prepared from different material. or with a different production process.
- Approach for verification of VP through the results obtained from the PP. This type of evaluation is integral to the behaviour of the prototyped object, as it is usually a "checkpoint", but is limited to be effective in the eventual negative results of the comparison of experimental data from the PP, which does not quickly lead to useful guidance on how to refine the simulation model.

From the point of view of the information that is entered and received accordingly from the prototype tests, they can be distinguished into two main categories: 'white box' - using the VP and 'black box' testing - using the PP.

The definition of "white box" is known as a tool mainly in the software industry and refers to systems for which structure and interconnections are known and can analyze the results based on the traceability of input / output processes [11]. The white box implies that we have full visibility of the internal workings of the system, in particular the logic and structure of the processes. Using white box testing techniques, the engineer can design test cases that are well investigated at the structural level, but very accurate quantitative results cannot be obtained due to the simulation type of virtual calculations.

White box testing (also called "comprehensible box" or "transparent box") assumes that the analyst designer knows and understands how the VP works - they can "see inside" in the simulation models, figure 10. Application of "white box" it is usually for designers as experience in modeling and simulation is required to create and use VPs.

Black box testing (also called "behavioral testing") is applied to physical tests. It requires the test specialist to understand what they need to do during the tests, but it is not necessary to understand exactly how the system under study works - the tests are not able to "see inside" the structure and logic of the processes in the AF. Here, the author proposes a new method for investigating complex objects by combining the application of HR as a "white box" and PP as a "black box" in an iterative connection.

Once the initial type of VP has been built, test simulations are conducted at points in the parameter space that can be adequately reproduced and tested with a physical prototype considered as "Black Box".

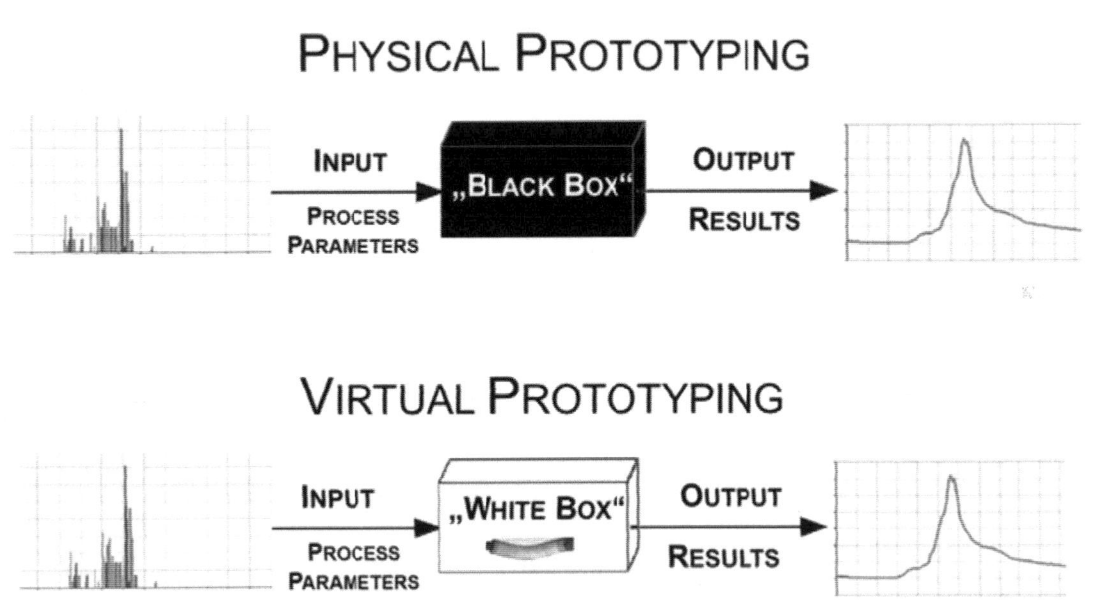

Figure 10. The method of combining the "WHITE BOX" using the VP and testing of the "BLACK BOX" using the PP.

The test results obtained from the AF are generally more accurate than the simulation accuracy. The quantitative results of the physical tests obtained are appropriately interpreted in the coefficients of the VP models, which are not well defined theoretically (e.g. friction coefficients, energy dissipation coefficients, contact effects, etc.). This process is repeated iteratively until a good level of agreement is obtained between the simulation results of the VP and the measurements from the PP tests. The test cases discussed above. prove the effectiveness and efficiency of this method.

Figure 11 shows the combination of the implementation of VP as a "white box" and the PP as a "black box".

The novelty of the method is the iterative connectivity of the VP and the AF, and in particular the calibration of the VP by feedback to the PP data through an internal micro-iterative cycle. This task is essentially a "reverse" - given exact output results and identical input effects, determine the parameters of the VP's computational models to achieve them. Multiple problem solving is applied to solve this inverse problem - the values of the corresponding coefficient in PP are set and the outputs of the simulations are monitored, and micro-iterations are applied within the VP to achieve good consistency of the obtained output results from the simulations, compared to those measured by the PP tests, with the same input parameters. Thus postulated, the method provides a great opportunity to achieve rapid and accurate research by iteratively matching the VP in the process of its construction and tuning, by rational and effective use of black box test data in the development phase of new products. In the presence of verified and validated VP, a large number of virtual simulations can be performed with it to investigate variants or to optimize different aspects of the product, while maintaining the structure and process identity within the validity range.

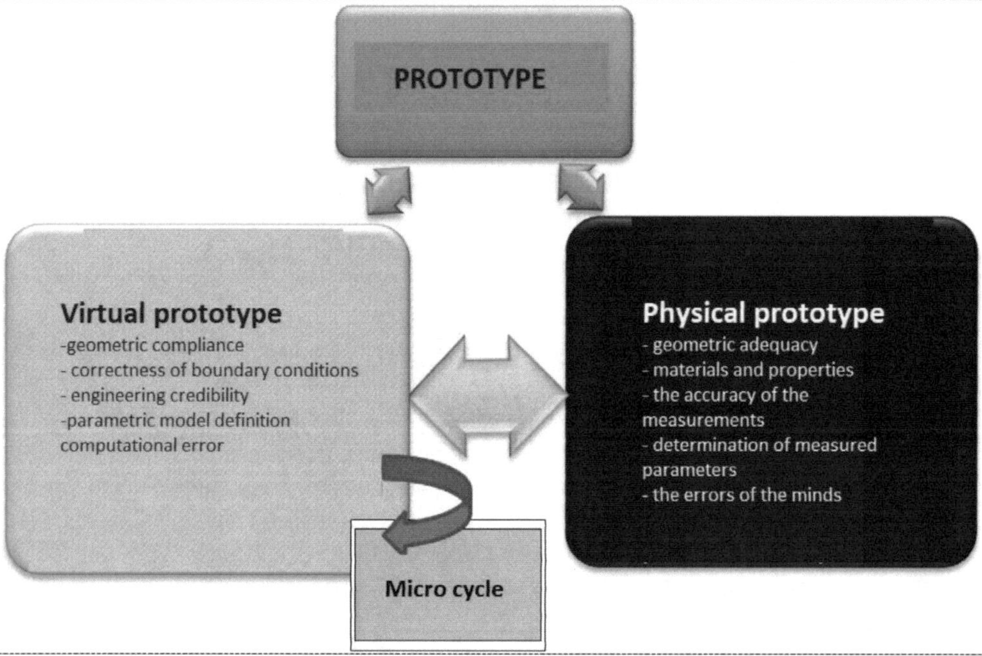

Figure 11. Combining the application of the VP as a "white box" and the PP as a "black box".

5. Conclusions

A hybrid method has been developed to predict the level of confidence in the results of multiphysical behavioral simulations of the object by virtue of virtual prototyping (VP) together with physical PP testing based on the stage of the virtual prototyping process and feedback from physical parameters tests.

Three approaches to assessing the reliability of the results of virtual prototypes - initial - for verification of boundary conditions, intermediate - to improve the accuracy of the results obtained and final - to validate the obtained results.

A new method has been proposed to improve the efficiency and effectiveness of Virtual Prototyping called "WHITE BOX / BLACK BOX" for the study of complex objects by applying VP as a "white box" and VP as a "black box" in iterative connectivity to achieve fine calibration of VP. The novelty of the method lies in the iterative connectivity of the HF and the PP, and in particular the calibration of the VP by feedback from PP data via a micro-cycle.

The "WHITE BOX / BLACK BOX" method and the three approaches to verifying and validating VP have been demonstrated in three successfully completed projects with international industrial partners used as demonstration test cases.

Acknowledgments
This work was supported by the European Regional Development Fund within the Operational Programme "Science and Education for Smart Growth 2014 - 2020" under the Project CoE "National centre of mechatronics and clean technologies" BG05M2OP001-1.001-0008.

6. References

[1] Liu B & Campbell R I 2008 Real time integration of user preferences into virtual *Undisciplined! Proceedings of the Design Research Society Conference 2008* (Sheffield UK)
[2] Grimm T A 2005 Virtual Versus Physical: Will Computer-Generated Virtual Prototypes Obsolete Rapid Prototyping *Time-Compression Technologies* vol 13 no 2
[3] Campbell R I 2004 A comparative study of virtual prototyping and physical prototyping *International journal of manufacturing technology and management* vol 4 no 6 pp 503-522
[4] Jivkov V Zahariev E Nikolov N 2018 Rotating cantilever beam with variable geometry – Frequencies and modal vectors Comptes Rendus de L'Academie Bulgare des Sciences vol 71 iss 8 pp 1108-1115
[5] Barth T 2011 *A Brief Overview of Uncertainty Quantification and Error Estimation in Numerical Simulation* (NASA Ames Research Center)
[6] Brinkgreve R 2013 *Validating Numerical Modelling in Geotechnical Engineering* (NAFEMS)
[7] Desai S 1999 Growth of CFD as an engineering tool for design and analysis of aerospace vehicles *Current Science* vol 77 no 10 pp 1283-1294
[8] Hellen T Becker A 2013 *Finite Element Analysis for Engineers – A Primer* (NAFEMS)
[9] Nedelchev K Kralov I 2019 Acoustic method for identification of railway wheel disc structural vibrations using COMSOL *Journal of the Balkan Tribological Association* vol 25 iss 3 pp 546-557
[10] Sofronov Y P Stoyanova Y P Kopralev N E Todorov G D 2019 Kinematic study of the articulated trucks operating layout of turn for articulated vehicles IOP Conference Series: Materials Science and Engineering vol 618 iss 1 article no 012044
[11] Beizer 1995 *Black-box testing: techniques for functional testing of software and systems* (New York: John Wiley & Sons, Inc.)

IOP Conf. Series: Materials Science and Engineering **878** (2020) 012052 doi:10.1088/1757-899X/878/1/012052

Martensitic transformations in Fe-Mn alloy with low stacking fault energy

S Parshorov[1], S Valkov[2,*], P Petrov[2]

[1] IMSETHC of Bulgarian Academy of Sciences, 67 Shipchenski prohod Blvd, Sofia 1574, Bulgaria

[2] Emil Djakov Institute of Electronics, Bulgarian Academy of Sciences, 72 Tzarigradsko Chaussee blvd., 1784 Sofia, Bulgaria

Abstract. Object of investigation is highly alloyed binary Fe-Mn alloys which in austenitic state are characterized with different stacking fault energy (SFE). Experimental results using the internal friction and X-Ray methods are obtained for the martensitic non isothermal transformation upon cooling and heating. It was determined that mainly the structural characteristics of the alloys define the type of the possible transformation mechanisms and the products of the martensitic reaction. The role of the SFE for the mechanisms of formation and growth of the martensite phase and the relation between the micro-structural parameters of the solid solution and the mechanism of non isothermal martensitic transformations is shown. On the basis of the existing models of dynamic relaxation associated with phase transformations, here are pivotal conclusions about the characteristics of the non-diffusion mechanism of transformation into a low SFE.

1. Introduction

Manganese is an austentabilizable element, it does not form carbides and nitrides in the steel and thus affects the properties of the iron solid solution. According to the state diagram, the phase composition of the dual alloy Fe-Mn depends on the manganese concentration and the alloy temperature as shown in Fig. 1. Conversions may be diffusion, straight and reverse reactions - γ-austenite ↔α-ferrite and may have a nondiffusional martensite character - γ-austenite ↔ε- or α-martensite [1-10].

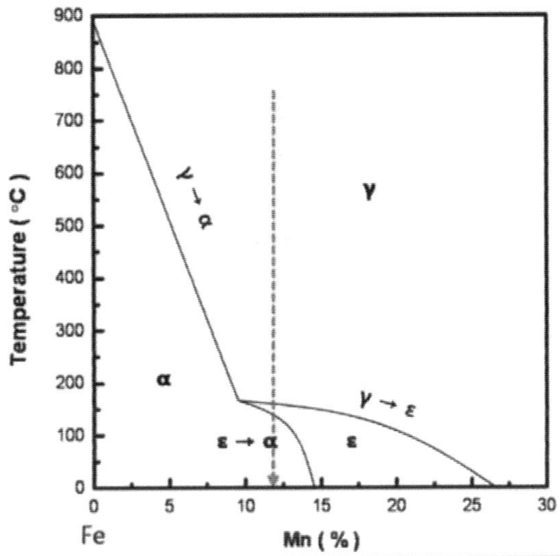

Figure 1. Fe-Mn alloy state diagram [2]

The reason for this "unusual" character of these type transformations lies in the specific dislocation structure of the initial austenite phase, characterized with high density of the low energy "surface" type of defects known as "stacking fault energy" (SFE) [3-10]. The defect lies right between the Shockley's partial dislocations- part of the main dislocations and is characteristic for the close- packed crystal structures (hexagonal an face-centered cubic) [11-14]. As measure for the SFE the elastic energy for a given area is adopted in the following form- erg/cm^2. The complete energy is a sum of the elastic energy of the partial dislocations and the SFE locked between them. As the SFE is decreasing the concentration and the size of the defects for a given volume are increasing. As long as SFE is connected to the atomic forces in the crystal lattice it is mainly determined by the type and the concentration of the alloying elements in the solid solution, the morphology of the high temperature phase, such as structural defects, size of the initial austenitic grain and temperature [11]. The general pattern of SFE dependence on manganese and carbon concentration, austenitic grain size and temperature are shown in Fig. 2 (a) - (d). Observed carbon and manganese concentration limits are in the range determined by the maximum solubility of each element in the defects in the arrangement, called the Suzuki-atmospheres. Another feature of the SFE is its strong temperature dependence. With decreasing of the temperature the SFE decreases, with the dependence being leapy - Fig.2 d). Plastic deformation increases the concentration and the width of stacking faults, thus causing phase transformation. This effect is typical for low SFE alloys and is known in the literature as TRIP- /Transformation Induced Plasticity/. These effect which is not subject of the current research but on its basis alloys with low SFE are found [5, 11,13].

The aim of the present work is to investigate the features of the martensitic phase transitions in the Fe-Mn double alloys related to their structural features by means of structure-sensitive methods.

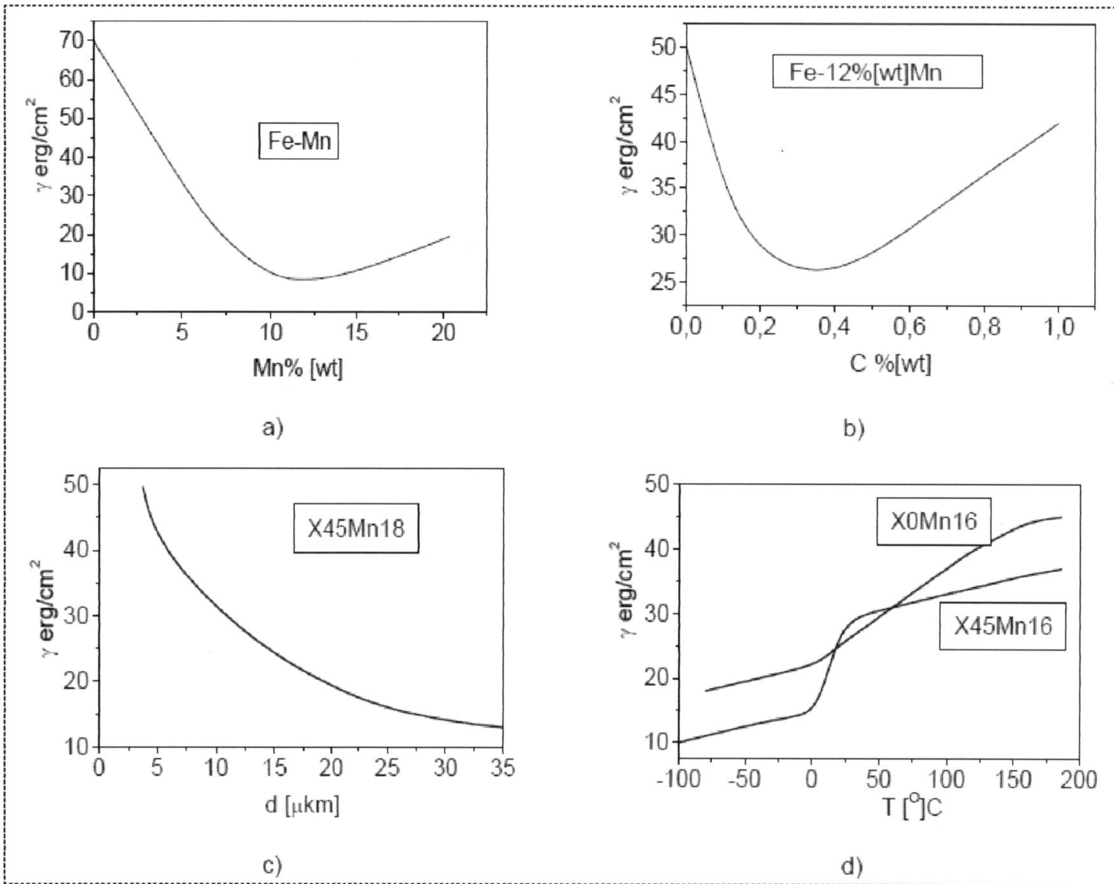

Figure 2. Dependence of SFE on: a) manganese concentration in austenite; (b) carbon concentration in austenite; (c) the size of the austenitic grain; d) temperature; [11]

2. Materials and methods

Object of investigation are alloys Fe-4% [wt] Mn, Fe-8% [wt] Mn and Fe-12% [wt] Mn, produced in the Institute of Metal Science. The specimens for investigation are manufactured by the following flow diagram: Casting of plates by vacuum furnace; Hot plastic deformation; The degree of the finally deformation was in range of 40%. Some of the plates were reduced to size of 5,5 mm and 1.2 mm. In order to reduce the internal stresses and remove the inherited structure obtained during the plastic deformation another set of samples were annealed for 10 hours at temperature of 7000 °C. In order of not changing the microstructure and the phase composition the cutting was made under water with electro-sparking machine. -The specimens are heat treated for finer structure - high temperature treatment - quenching from 1150 °C after 10 min of holding in water. The present investigations are led using tree structural-sensitive methods- X-ray diffraction-(specimens 15x15x5 mm), dilatometric-(specimens 5x50 mm) and the internal friction (specimens 50x1x1.2 mm) methods. In order of not changing the microstructure and the phase composition the cutting was made under water with electro-sparking machine. The X-ray structure analysis is performed by standard apparatus "Philips"using Co-Kα (λ=1,7880.10^{-8} cm) emission. The speed and the step of the goniometer are as follows-0,02o/min and 20 sec/step. On the base of standard calculations including the average value of the wave length for Co-radiation the unit cell parameter `a` was determined from the considered profile line $(211)_{\alpha}$.

The common integral profile line can be separated to its structural components (2.α_1 and 2.α_2-doublet) using the Gauss function for two maxima, by the programme „Origin". During the approximation, the values for the areas of the two maxima are considered to be constant in ratio of 2:1. Standard method for phase analysis based on the diffraction line $(110)_\alpha$ of the martensite is used. The microstructure parameters of the martensite solid solutions – the domain size - D and the micro strains of the crystal lattice - ε, are studied by the so called "single line analysis" shown in details in [15-17]. This method allows calculating the microstructure parameters of austenite and martensite on the basis of one diffraction line, for each of the phases. For reference sample of "armco" iron is used after full annealing. The dilatometric tests were carried out with standard equipment "NETZSCH- Geratebau GmbH". Linear expansion values of standard samples in micrometers are recorded as a function of temperature and time. The tests were conducted in the temperature range 20-1200°C with heating and cooling. Dilatation data is processed with the Origin program. The apparatus for the internal friction measurements is built on the principle of the reverse balanced pendulum also known as Ke`s pendulum. The operational frequency is in the range 1-1,3 Hz. The relaxation spectra of the specimens are recorded for the temperature range of 50-800K during heating and cooling in vacuum at rates of 20 oC/min. The degree of the relative deformation during the measurements is 5.10^{-5} -5.10^{-6}.

3. Results and discussion

According to the published data, during cooling from the austenite region diffusion reactions of $\gamma\rightarrow\alpha$ ferrite phase (Fe-4% [wt] Mn, Fe-8[wt] %Mn) or diffusion-free reactions of $\gamma\rightarrow\alpha$ - or $\gamma\rightarrow\varepsilon\rightarrow\alpha$ - martensite (Fe-12% [wt] Mn) type occur in Fe-Mn alloys in dependence of the manganese concentrations. Fig. 3 presents the X-ray structure spectra of the studied alloys for manganese alloys, respectively in the range of doubled reflection angles from 40 to 140o for Fe-8% [wt] Mn. and Fe-12% [wt] Mn after quenching. The result for Fe-4% [wt] Mn is analogous.

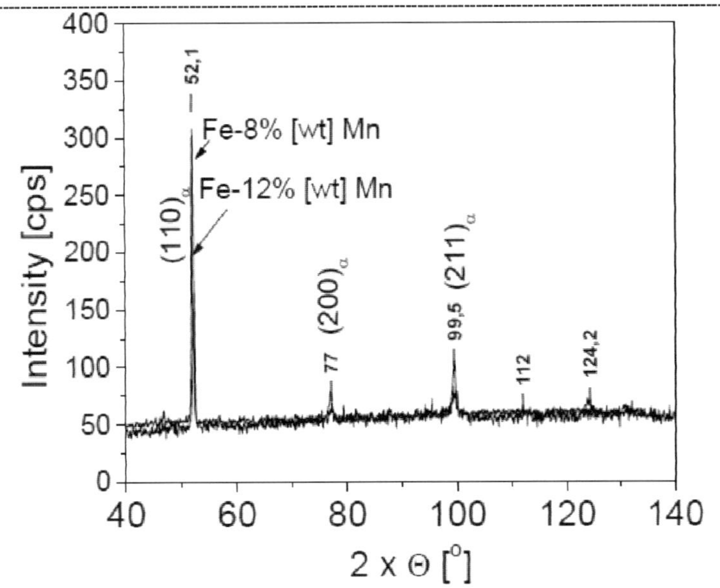

Figure 3. X-ray diffraction pattern of manganese alloys

The results show that the resulting spectrum corresponds to that of α-iron, irrespective of the concentration of manganese. Other authors find experimental presence of minimal quantities of ε-martensite that are not observed in our results [5]. This is probably due to the presence of ε-phase quantities in alloys, which are difficult to detect by the X-Ray method. The results obtained show (Fig. 3), that no alloys on Fe-Mn base form residual austenite. It is well known that the presence of residual austenite is caused by high internal stresses to which the austenite is subjected by the growth of the martensite lamellae and aciculae. The absence of residual austenite shows that the formation of ferrite or ε-martensite is connected with the occurrence of active relaxation processes resulting in decrease of the internal stress value. Verification of that is also the experimentally obtained by us very low value of the martensite lattice micro strains. Fig. 4 presents the microstructure parameters of the solid solutions of the martensite – size of the domains - D and the micro strains of the crystal lattice - ε. The results obtained show: (i) the concentration limit of the transition of the austenitic phase to ferrite or martensite is lower than that published in the literature. Martensitic phases are still formed at the manganese at 8% by weight; (ii) in the manganese alloys, the formation of epsilon-martensite in manganese concentrations of 8% [wt] reduces the microdeformations of the lattice.

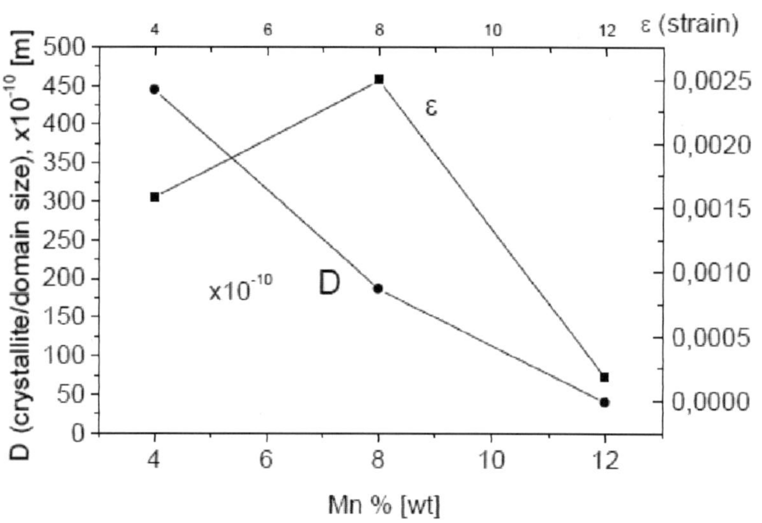

Figure 4. Dependence of domain size and micro strains on the manganese contents

The presence of two types of martensite transformations in the considered alloys is defined in the most general aspect by the thermo dynamical, crystallographic and other specific features of alloys. Determining parameter for that is mainly the energy of the austenite stacking faults from which the respective martensite phase has transformed. The published investigations [11] for the phase compositions and transformations in Fe-Mn alloys mainly show that Fe-12%Mn alloy is characterized by a very low energy of stacking faults, of the order of 18 erg/cm^2, which suggests a formation of ε-martensite.

In order to determine the influence of manganese concentration, hot plastic deformation and annealing on the structural state of austenite, the parameter of its crystal lattice was investigated according to the methodology shown above. After casting some of the samples were hot rolled to plates with thickness of 0.8 mm under the following conditions: preliminarily heating of the ingots to temperature of

1050°C; repeated hot rolling in the temperature interval of 1050- 980 °C; cooling down on air. The degree of the finally deformation was in range of 41- 43%. Some of the plates were reduced to size of 15x15x0.8 mm suitable for the further XRD measurements. In order to reduce the internal stresses and remove the inherited structure obtained during the plastic deformation another set of samples were annealed for 10 hours at temperature of 700° C. Fig. 5 shows the corresponding values for the unit cell parameter `a` of the lattice, calculated for the investigated alloys after hot rolling and annealing.

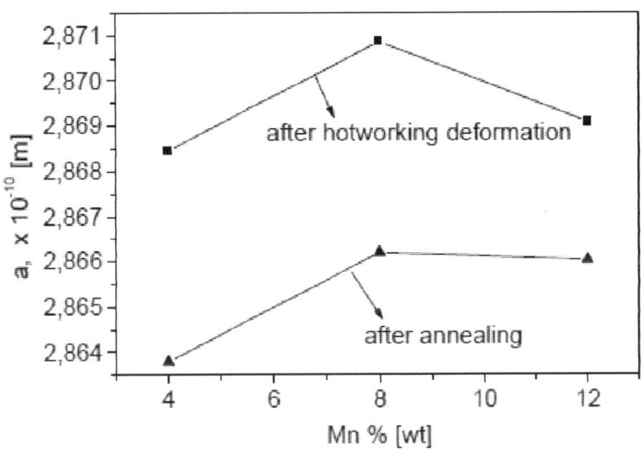

Figure 5. Dependence of domain size and micro strains on the manganese contents

Fig.6 a) shows the dilatation curves of Fe-12% [wt] Mn steel obtained under heating and cooling. Fig.6 b) shows enlarged cooling curve marked in a). According to the state diagram of alloy Fe-12% [wt] Mn, some published researches and our dillatometric investigations, the phase composition at room temperature is martensite [1]. On heating above 700 K the martensite transforms into austenite i.e. the reaction $\alpha \rightarrow \gamma$ takes place. On cooling below 400 K the austenite turns into ε-martensite, passing through the so-called. ε-martensite with hexagonal lattice. During cyclic heating and cooling below 700 K, the straight and the reverse martensite reactions of the type $\gamma \rightarrow \varepsilon$ and $\gamma \rightarrow \alpha$ take place- Fig.6 a) and b). The results showed that the $\gamma \rightarrow \varepsilon$ phase transformation is associated with minor dilatation changes in the alloy volume. That's why the internal friction method was used in our research, connected with the dynamic relaxation of the solid solution Fig. 7 presents the relaxation spectra of the Fe-Mn alloys at heating and cooling. It is characteristic that at room temperature they are martensitic [1] and on cooling, in compliance with the constitution diagrams, the martensite in them transforms into austenite. In case of cyclic heating and cooling in the alloy Fe-12%Mn runs the direct and reverse reaction of the martensite transformation of the type $\alpha \rightarrow \gamma$ and $\varepsilon \rightarrow \gamma$. These phase reactions are reflected as maxima on the temperature dependence of the internal friction. Fig. 8 shows the same are presented separately from the internal friction background after numerical integration and normalization. In practice, the numerical integration and normalization of the experimentally obtained maximums, shown in Fig. 8, represents the kinetics of the martensite reaction in non-isothermal conditions. The results show that the direct reactions of $\alpha \rightarrow \gamma$ and $\varepsilon \rightarrow \gamma$ transformations run close or in the high temperature background of the internal friction where the mobility of the dislocation structure is the highest, and the reverse $\alpha \rightarrow \gamma$ and $\varepsilon \rightarrow \gamma$ in the temperature interval before the high temperature background where the "pure", conservative dislocation creep/glide in an atomic plane is basic.

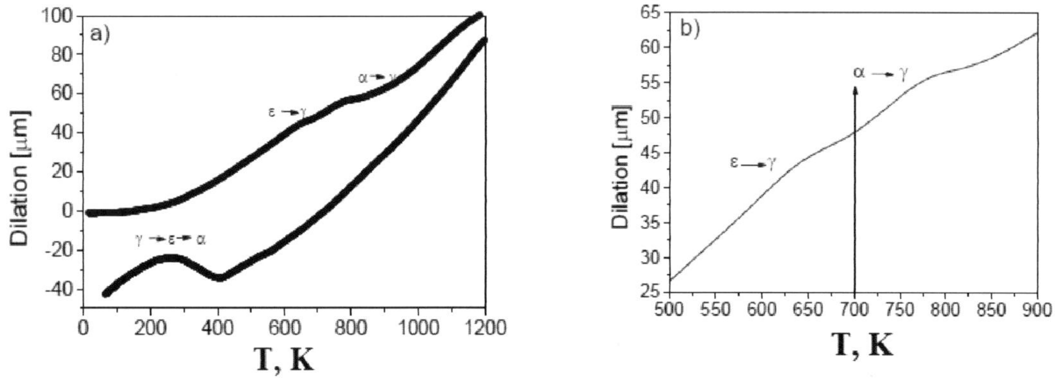

Figure 6. Dilatation curves at heating and cooling; b) The same enlarged heating curve marked in a) between 550 and 900 K

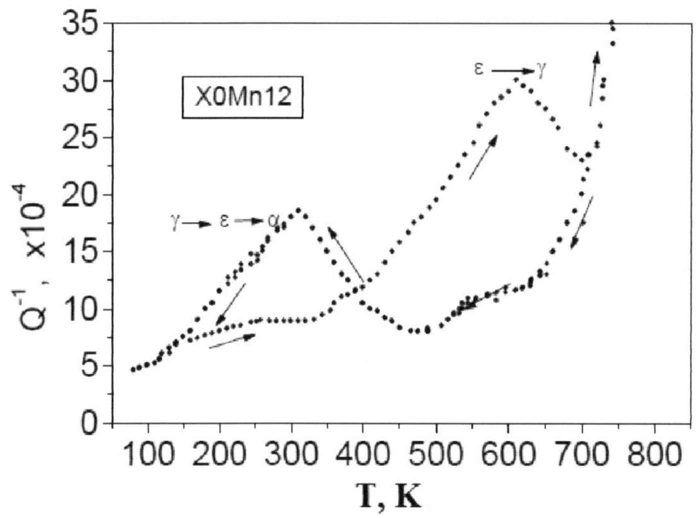

Figure 7. Relaxation spectra of alloy Fe-12% [wt] Mn (EN X0Mn12) at heating and cooling below 700 K

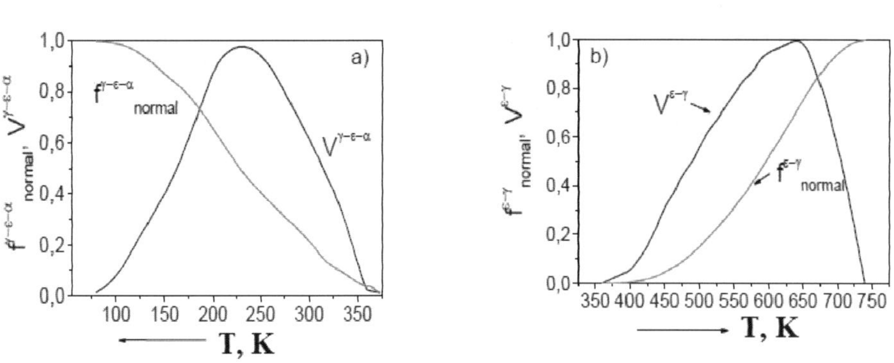

Figure 8. Numerical integration and normalization of experimentally obtained maxima after subtraction against the backdrop of internal friction:a) cooling; b) heating

According to published data, the alloy Fe-12% [wt] Mn characterizes by SFE which presumes formation of ε-martensite. In that sense, the dislocation structure of the manganese alloys is easily mobile and consists of widely sheared partial dislocations surrounding large areas of stacking faults. In the publications there are several models that describe the connection between the internal friction value and the phase transformation parameters that refer to the fact that: (i) the height of the internal friction maximum is proportional to the phase transformation rate and the heating rate; (ii) the height of the internal friction maximum is inversely proportional to the frequency of the measurement. In the base of these models [19] is the fluctuation mechanism of nucleation related to the process of nuclei thermal activation. It is assumed that the activation energy of nucleation is decreased by the presence of internal stresses which perform work in the nucleation process itself. It is proportional to the lattice non-elastic deformation, in this case the life time of the nucleus being much greater than the time of the process relaxation.

The dependence deduced by the authors allows by means of the area of the internal friction maximum to calculate the activation volume of the phase transformation. The values obtained by them are in good agreement with the number of atoms that normally enter in one critical nucleus during martensite transformations in some alloys, e.g. Co-Ni. We have made an attempt to calculate the nucleus critical size of the two alloys studied in the same way it is made by the above said authors. Unfortunately, our calculations in compliance with this theory do not give real values for the nuclei critical size; they are smaller than the volume of an elementary cell. In spite of all, V. Belko's theory shows that the presence of stresses in the lattice affects the size of the nuclei capable to grow.In respect of the boundary mobility in the model there are four possibilities according to the model of Dejonghe – Fig.9 [19].

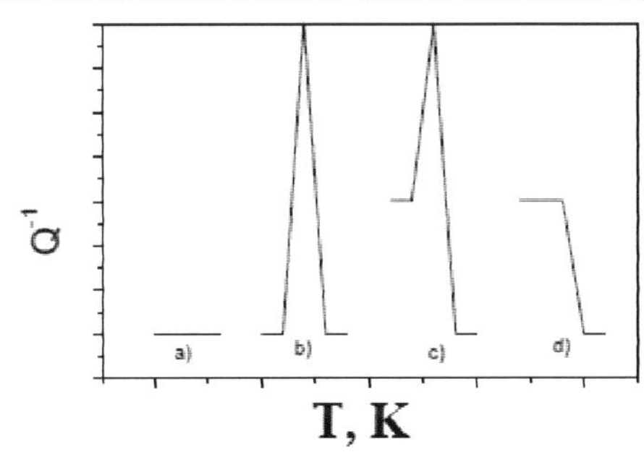

Figure 9. Scheme of addictions of the type of relaxation maximum of the temperature associated with movement of the boundaries of phase transformations

In respect of the boundary mobility in the model there are four possibilities: (i) absence of relaxation maximum when the applied stress is comparatively low and below the critical stress; (ii) presence of symmetrical relaxation maximum that shows that under the action of the applied stress the boundary between the new phase and martensite shifts; (iii) presence of symmetrical relaxation maximum that shows that under the action of the applied stress besides the boundary new phase/martensite the phase boundary martensite/martensite also shifts; (iv) occurrence of a step in the temperature dependence of internal friction - then the mobility of the boundary martensite/martensite is greater than the boundary new phase/martensite. It is obvious that for the studied by us $\alpha \rightarrow \gamma$ and $\varepsilon \rightarrow \gamma$ transformations the second case is relevant, i.e. the available relaxation maximum is caused by the motion of the phase boundary - new phase/martensite in both alloys studied by us. It is seen that there is a significant difference in the activation energies of the direct and reverse reactions of martensite type. The low values of the kinetic parameters of the austenite-martensite transformations in cooling show relieved mechanism of nucleation and growth of the martensite aciculae related with the low value of the SFE.

It is assumed that as martensite nuclei serve wide areas of the stacking faults surrounded by partial dislocations [11-13] which in crystallographic respect represent α-phase ready nuclei. The crossing of two packs of ε-phase results in α-phase nuclei that are ready nuclei for the heterogeneous mechanism of γ→α transformation. Up to now in the publications this mechanism is assumed to be the most probable and experimentally proved one. In our judgment basic moment in the nucleation mechanism is the temperature dependence of the austenite SFE. It is well known that it depends strongly on the temperature and changes at leaps by it. The temperature range of SFE change at leaps is below room temperature of the order of 280-230 K. The sharp SFE increase with the temperature increase results in high values of the activation energies of martensite phases transformation in austenite and in this connection a shift of the processes towards high temperatures, i.e. there is a strong temperature hysteresis of the direct and reverse martensite reactions that is observed also in our experimental results.

4. Conclusion

The results show that the complete X-ray structure spectrum of manganese alloys corespondes to that of α-iron irrespective of the concentration of manganese. No experimental presence of ε-martensite was observed. The X-ray diffraction measurements of Fe-Mn alloys showed different value for the unit cell parameter `a` of the crystal lattice depending on the concentration of the alloying element and the presence of plastic deformation. Probably these facts are due to a difference in the stacking fault energies of the two alloys. The concentration limit of the transition of the austenitic phase to ferrite or martensite is lower than that published in the literature. Martensitic phases are still formed at the manganese at 8% by weight. In the manganese alloys, the formation of epsilon-martensite in manganese concentrations of 8% reduces the microdeformations of the lattice. The temperature range of conversion of austenite to martensite is below room temperature in the range of 280-230 K, meaning that there is a process of leaping the EDP. The symmetry of the obtained maximums and the absence of a step of the internal friction temperature dependence show that the relaxation processes in these alloys are defined by the motion of the phase boundary new phase/martensite Presence of symmetrical relaxation maximum that shows that under the action of the applied stress besides the boundary new phase/martensite the phase boundary martensite/martensite also shifts. The results show that the direct reactions of α→γ and ε→γ transformations run close or in the high temperature background of the internal friction where the mobility of the dislocation structure is the highest, and the reverse α→γ and ε→γ - in the temperature interval before the high temperature background where the "pure", conservative dislocation creep/glide in an atomic plane is basic.

Acknowledgements: The authors are grateful to the financial support of Bulgarian National Science Fund at the Ministry of Education and Science, Contract No DN17/17 /12.12.2017/.

References:

[1] Hansen M 1958 Constitution of Binary Alloys (McGraw-Hill, New York).
[2] Nishiyma Z 1978 Martensitic transformation, (Academic Press, New York).
[3] Petrov Y 1978 Faults and non-diffusion transformation in steels, (Naukova dumka, Kiev, 1978).
[4] Witusiewicz V, Sommer F and Mittemeijer E 2004 Journal of Phase Equilibria and Diffusion **25(4)** 346–354.
[5] Kim J, Jeon J, Jung J, Um K and Chang Y 2014 Met. Mater. Int. **20(1)** 41-47.
[6] Hamada A, Sahu P, Ghosh S, Chowdhury L, Karjalainen J, Levoska and Oittinen T 2008 Journal, Metallurgical and Materials Transactions **A 39** 462-465.
[7] Sahu P, Hamada A, Ghosh R and Karjalainen L 2007 Metallurgical and Materials Transactions **A 38(9)** 1991-2000.
[8] Paidar V, Ostapovets A, Hardouin Duparc O and Khalfallah O 2012 Acta Physica Polonica A **122(3)** 490-492.
[9] Gupta S 2002 Solid state phase transformations, (Allied Publishers Pvt. Ltd. (APPL), New Delhi)

[10] Huang B, Wang X, Rong Y, Wang L and Jin L 2006 Materials Science and Engineering: A **438–440** 306-311.
[11] Petrov U 1971 UFJournal **16(9)** 1409.
[12] Cohen M 1972 Metall. Trans. **3** 1095.
[13] Olson G and Cohen M 1972 J. Less Common Met. **28** 107
[14] Olson G, Azrin M 1977 Transformation Behavior of Trip Steels, ARMY MATERIALS AND MECHANICS RESEARCH CENTER WATERTOWN MASS, (Pagination or Media Count : 21).
[15] Krumm S 1994 Acta Universitatis Carolinae - Geologica **38(2)** 253-261
[16] Balzar D 1999 International Union of Crystalography, 1-44.
[17] Ida T, Hibino H and Toraya H 2001 Journal of Applied Crystallography **34** 144–151.
[18] Krumm S, WINDOWS Software, http://www.geol.unierlangen.de/software/winsoft.html, krumm@geol.uni-erlangen.de.
[19] Novick A and Berry B 1972 Anelastic Relaxation in Crystalline Solids, Academic Press, New York 463.

Measuring the diameter of electron beam by rotating slit disc

M Tongov[1], T Hikov[2,*], I Angelov[2], P Petrov[2]

[1]Technical University, Sofia, Bulgaria

[2]Institute of electronics, Bulgarian Academy of Sciences, Sofia, Bulgaria

Email: thikov@ie.bas.bg

Abstract. A device for analysing the shape of the electron beam in electron beam welding apparatus has been developed. Experiments has been done with the device. The beam size at different focusing currents and constant beam current has been measured. The performance of the device is shown.

1. Introduction

Electron beam welding is widely used in industries that require high quality manufacturing. This asks for very a precise and stable parameter set. The process is often carried out in vacuum to avoid deflection of the beam and further oxidation of the treated materials.

There is a need for detailed information of the influence of parameters like cathode heating and wear, electron optical alignment, etc [1, 2]. These all influence the beam shape and dimensions and thus the focus conditions. This is especially sensitive when the equipment parameters are fixed within tight tolerances [3, 4].

One of the first techniques for characterization the electron beam consists of a single slit disc and Faraday cup below it. The slit is scanned by the beam and the resulting signal is gathered by the cup. The authors of [5] modified this method by adding a second slit to the disc, perpendicular to the first. The idea behind the two-slit disc is to being able to measure the electron beam in two dimensions simultaneously.

The method proposed by authors of [6] consists of a rotating disc with little pinhole in it. This way the electron beam is measured in "free fall" position and is not deflected. The pinholes are distributed angularly and situated around a mean radius. The signal gathered from the pinholes is relatively low due to their minute radius.

We propose equipment that is combination of the two methods. The idea is to avoid the scanning and deflection of the beam because this could yield additional aberrations which can't be eliminated after that [3]. That's why the beam is measured in "free fall" position while the disc is rotating.

The aim of the present study is to develop new equipment for measuring the diameter of the electron beam.

2. Materials and methods

Those kinds of measurements should be performed by either rotating disc or deflecting beam because of the high current density. The rotating disc scanner analyzes the beam in free fall position.

IOP Conf. Series: Materials Science and Engineering **878** (2020) 012053 doi:10.1088/1757-899X/878/1/012053

This equipment is based on a rotating disc with multiple slits in it (figure 1). The disc is turning at high speed (up to 3500 rpm). The heat load is relatively low due to the high volume of the disc which assures the evacuation of the incoming energy.

Figure 1. Electron beam chopper device.

There are 12 equally distant from one another radial slits with width 1mm. Below the beam is situated a small tungsten target in order to gather the signal. The signal after that is taken through a resistance and monitored at oscilloscope. A sample signal is shown in figure 2. The peaks are negative because the electron beam has negative potential when conducting current.

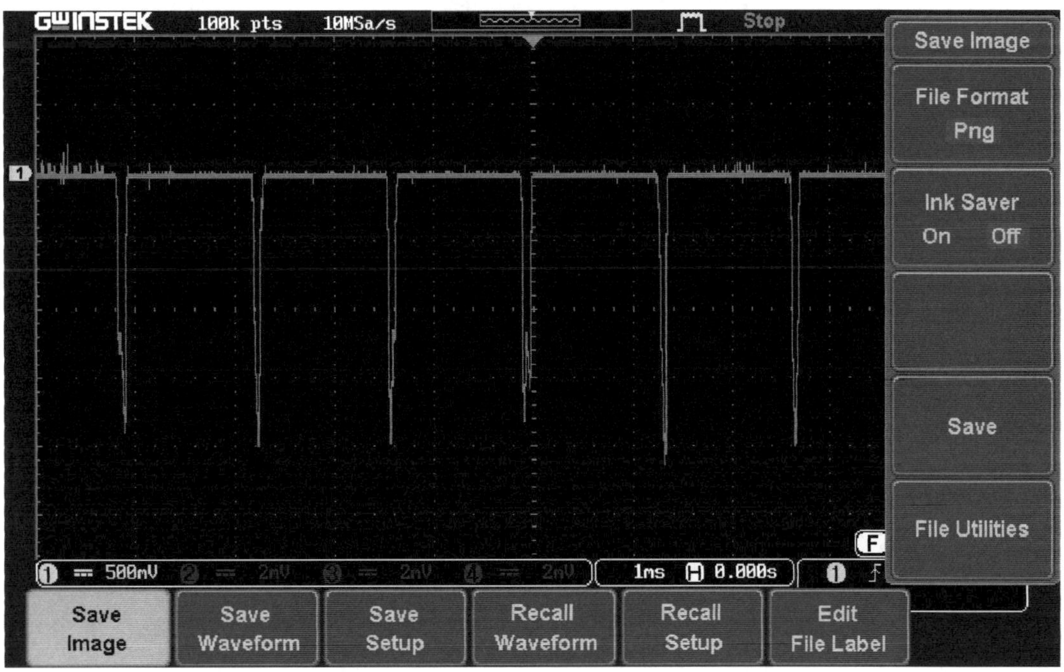

Figure 2. Sample signal from an oscilloscope. $I_{beam} = 5[mA]$; $I_f = 523[mA]$.

The electron beam diameter can be calculated using the law of uniform motion

$$d = t_p v - \delta \tag{1}$$

where d is the electron beam diameter, v is the velocity of the disc at the point where the beam is falling, t_p is the duration of a single impulse and δ is the width of a slit (figure 6). The velocity is calculated from the time needed for the disc to turn from one slit to the next via the equation

$$v = \pi * r/(6 * T) \tag{2}$$

where r is the measured distance between the center of the disc and the falling of the beam (r=37[mm]), T is the time between the start of two consecutive peaks, and π/6 is the angle between two consecutive slits in radians.

3. Results and discussion

The experiments are carried out with the same velocity with beam current I=5 mA and different focusing current shown in table 1. Images of the gathered signals are shown in figures 2 to 8.

Table 1. Focusing current for 7 different measured diameters

№	1	2	3	4	5	6	7
I_f [mA]	523	530	536	542	550	555	560

First, we have to calculate the time need for the disc to rotate between two consecutive slits. The period is T=1.35[ms] shown in figure 3. The turning velocity of the disc with respect to the electron beam is calculated to be v=14.35[m/s] by substituting in equation (2). The method for analyzing the experiment data is shown in figure 10. A typical peak, picked from the gathered signals, is graphed. This graph corresponds to the electron beam that has passed through a single slit. The starting point is chosen near the potential and is this potential difference is plotted against time (the absolute value of the voltage is used because it corresponds to the measured current). Then the duration of this impulse is measured t=0.13[ms] in the case if figure 5. The diameter of the electron beam in direction of the turning disc is calculated from equation (3).

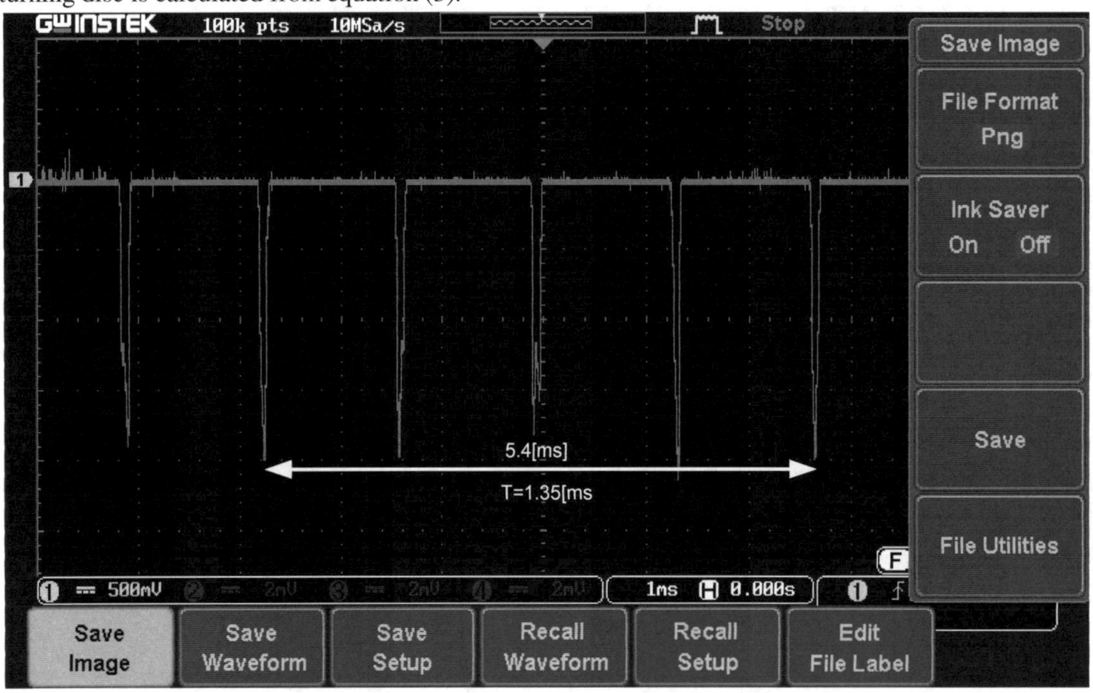

Figure 3. Calculating the time for the beam to pass two consecutive slits in the disc.

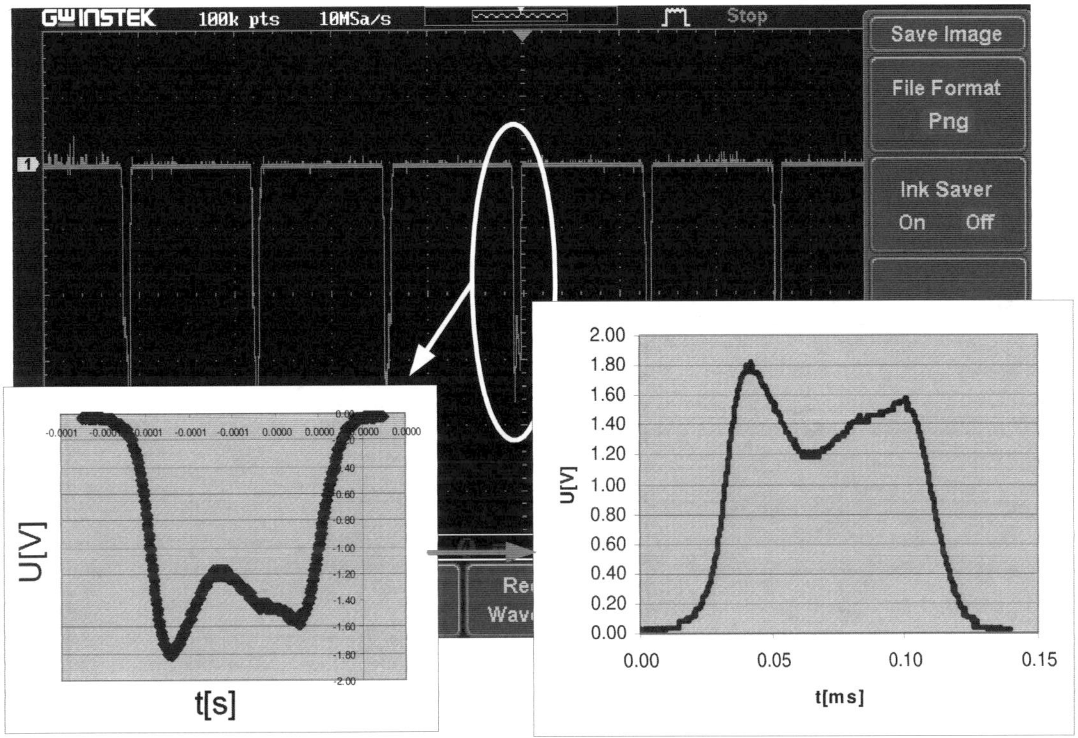

Figure 4. Different steps in processing the gathered signals.

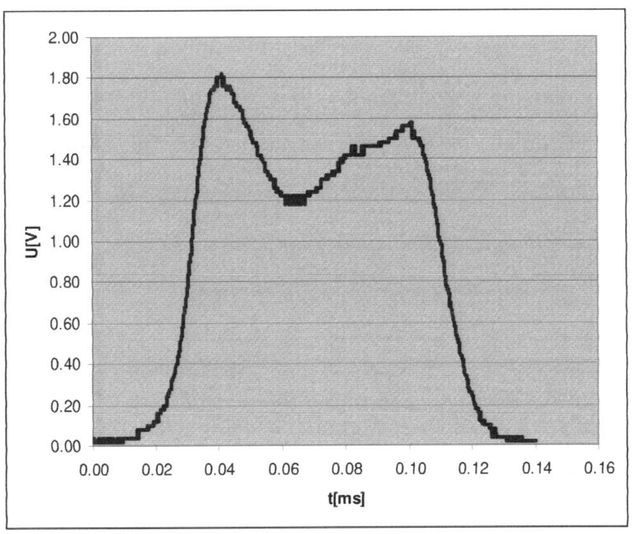

Figure 5. Duration of a single impulse.

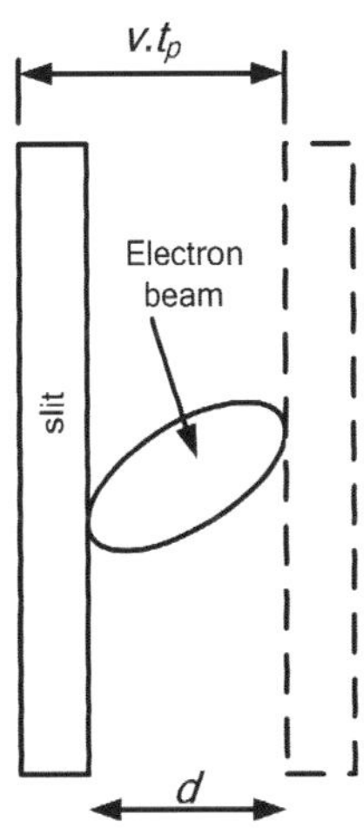

Figure 6. Image of the electron beam passing through a slit.

The calculated electron beam diameters with respect to the focusing current are shown in table 2 and are plotted in figure 7.

Table 2. Diameter of the electron beam with respect to the focusing current.

№	1	2	3	4	5	6	7
I_f [mA]	523	530	536	542	550	555	560
d[mm]	0.865	1.296	2.157	3.018	3.592	3.664	4.74

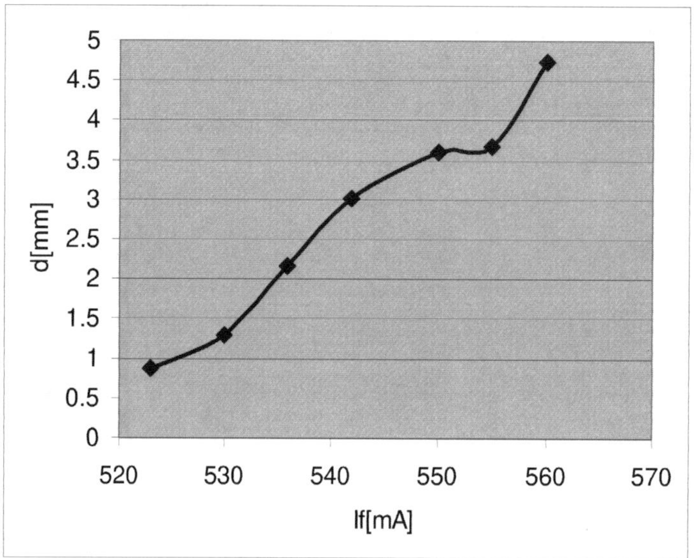

Figure 7. Graph of the electron beam current with respect to the focusing current.

4. Conclusion

A device (electron beam analyser) is developed to measure the diameter of the electron beam. The experiments that were carried out show that it is working as intended. The diameter of the electron beam with beam current of 5 [mA] in the plane of the detector is calculated with respect to the focusing current.

Aknowledgements
This work was supported by the Bulgarian National Scientific Fund under Grant DN 07/26.

References

[1] Longerich S and Metzler D 2015 Certification of an EB-welding machine for production of critical aerospace parts *3rd International Electron beam welding conference*, Chicago, November 10-11

[2] Dutilleul T, Priest J and Baufeld B 2016 Electron beam characterization *Electrotechnica & electronica: The union of Electronics* **175**

[3] Elmer JW, Palmer TA and Teruya AT 2012 An overview of the enhanced modified faraday cup electron beam power density distribution diagnostic *2nd International electron beam welding conference* Aachen pp 108-113

[4] Burgadt P, Pierce S and Hochanadel P 2015 Comparison of beam defocus and beam oscillation for Ta-10W EB welding *3rd international electron beam welding conference* Chicago November 10-11

[5] Kaur A, Ribton C and Balachandaran W 2015 Electron beam characterization methods and devices for welding *J of Mat Proc Tech* **221** pp 225-232

[6] Oving P 2012 Development of a new electron beam probing device and analyses method to ensure the weld quality in time, *2nd International electron beam welding conference* Aachen

Bead formation research in TIG welding of AISI 304 steel

Manahil Tongov[1,2], Rayna Dimitrova[1] and Konstantin Konstantinov[1]

[1]Technical University of Sofia, Bulgaria, Faculty of Industrial Technology, Department "Materials Science and Technology"
[2]Institute of Metal Science, equipment, and technologies with Center for Hydro- and Aerodynamics "Acad. A. Balevski" at the Bulgarian Academy of Sciences – (IMSETHC-BAS)

E-mail: tongov@tu-sofia.bg

Abstract. On the basis of the experimental results in depositing strips of molten metal onto AISI 304 steel using the D-optimal plan of the experiment and processing of the got results, regression equations were obtained describing the dependence of the bead width and the penetration depth depending on the welding speed and the value of the welding current. In parallel, a regression equation is obtained connecting the arc voltage and the welding current.

1. Introduction

High-alloy austenitic steel AISI 304 (1.4301) is widely used in various industries. Building regression models to determine the influence of process parameters on bead formation is a common practice for analysis and optimization [1÷10]. TIG welding is one of the main methods for welding of constructions from that steel. The application of technologies and tools for solving technological problems by simulation modeling of welding process is one of the most promising areas in this field. In order to be complete realized it is necessary to modeling the heat source with subsequent calibration, verification and validation of the built model.
In the present work, on the basis of experimental results, regression equations, which are applicable for the definition of the heat source in simulation modeling of the welding process, are obtained.

2. Materials and research method

The survey was performed using a consistent quasi-D-optimal plan. In this study, the magnitude of the welding current (in the range of 80 to 180 [A]) and the welding speed (12 to 30 [cm / min]) were varied. The arc length was constant and equal to 2 [mm]. The shielding gas flow rate was 6 [l / min] at a current value up to 120 [A] and 10 [l / min] at a higher value of the welding current. Thus, the number of factors in the experiment plan was two. During the experiment, strips were surfaced onto specimens 6x100x100 [mm] from AISI 304 steel at preliminary set values of welding current and welding speed. On each specimen several strips of molten metal were deposited after natural cooling the specimen to room temperature, measuring the values of the welding current, the arc voltage and the average welding speed. Samples for metallographic analysis were prepared from the specimens with deposited strips of molten metal. The preparation includes the following stages: sectioning of a sample, wet grinding in several stages, polishing and etching with a mixture of concentrated nitric and hydrochloric acid, optimally in a molar ratio of 1:3. Metallographic analysis with an optical metallographic microscope Neophot 21 "Carl Zeiss" were carried out. It was determined the geometric parameters for all samples – the penetration depth and the width of the welding bead.

The regression equations were sought as complete second-degree polynomials, and in the case of an inadequate model as third-degree complete polynomials. The plan of the experiment in relative coordinates is given in Table 1. In addition, seven experiments were performed in the center of the plan to verify the adequacy of the model. The value of the current was used as the first factor and the welding speed as the second factor. The processing of the obtained results was as follows:

The experiment plan matrix was specified, with each row representing the values of the factors for a particular surfaced stripe:

(1) $$\mathbf{X} = \begin{pmatrix} x_{1,1} & x_{2,1} \\ x_{1,2} & x_{2,2} \\ \dots & \dots \\ x_{1,21} & x_{2,21} \end{pmatrix}$$

Table 1. Experiment plan in relative coordinates.

№	X1	X2	№	X1	X2	№	X1	X2
1	1	1	8	1	-0.447	15	-0.447	-0.447
2	-1	-1	9	-0.447	0.447	16	0.447	0.447
3	1	-1	10	-0.447	-1	17	-1	1
4	-1	1	11	1	0.447	18	-1	-1
5	-1	-0.447	12	-0.447	1	19	1	-1
6	0.447	1	13	-1	0.447	20	1	1
7	0.447	-1	14	0.447	-0.447	21	0.447	1

The sought regression equation is of the type

(2) $$\hat{y} = b_0 + b_1x_1 + b_2x_2 + b_3x_1x_2 + b_4x_1^2 + b_5x_2^2 + b_6x_1^2x_2 + b_7x_1x_2^2 + b_8x_1^2 + b_9x_2^3$$

which can be written down as $y = \sum_{j=0}^{9} b_j f_j(\mathbf{x})$ and therefore the elements of the extended matrix of plan **F** are $f_j(\mathbf{x}_i)$, where j is the function number, i – the experiment number.

Fisher Information Matrix is $\mathbf{G} = \mathbf{F}^T\mathbf{F}$, and the sought coefficients were determined as

(3) $$\mathbf{b} = \mathbf{G}^{-1}\mathbf{F}^T\mathbf{y}$$

To verify the adequacy of the model according to the Fisher criterion, the ratio $F = s_{ost}^2 / s^2$ was used. Here s_{ost} and s are respectively the variation of adequacy and variance of experience. The variation of adequacy was determined by the obtained results y_i and the calculated value by equation (2) $\hat{y}_i$:

(4) $$s_{ost}^2 = \frac{\sum_{i=1}^{N}(y_i - \hat{y}_i)^2}{N-k} = \frac{\sum_{i=1}^{21}(y_i - \hat{y}_i)^2}{11}$$

where N is the number of experiments, a k - the number of coefficients in equation (2).

The variance of experience s was determined on the basis of further conducted experiments at the center of the plan such as:

$$s^2 = \frac{1}{n-1}\sum_{u=1}^{n}(y_u - \bar{y}_u)^2 \quad (5)$$

where $\bar{y}$ is the average of the measured values y_i.

Thus, the value of F was compared with the value of the Fisher distribution F_T at the significance level α and the number of degrees of freedom $v_{ost} = N - k = 11$ and $v = n - 1 = 6$. In order to consider the model adequate, it is necessary to observe the ratio $F < F_T$.

The Student's criterion was used to check the statistical significance of the coefficients of the regression equation. For its implementation, there is a matrix $\mathbf{C} = \mathbf{G}^{-1}$ (it has dimensions k x k) with diagonal elements c_{jj}. To make a coefficient significant, it is necessary to observe the condition

$$\frac{|b_j|}{\sqrt{s^2 c_{jj}}} > t(\alpha, v_{ost}) \quad (6)$$

where $t(\alpha, v_{ost})$ is the value of the Student distribution at the significance level α and degrees of freedom v_{ost}.

3. Results and discussion

The measurement of the geometric parameters – the bead width and the penetration depth is illustrated in Figure 1. The results of the experiments are shown in Table 2 in absolute values in the sequence in which the welds are performed. The expanded plan matrix and the inverted Fisher information matrix, respectively, are shown in Table 3 and Table 4. The results for the stripes in the center of the plan are shown in Table 5, and the values of the coefficients of the regression equations are shown in Table 6. The value of Fisher distribution at significance level 0.05, number of residual degrees of freedom 11 and degree of freedom 6 is 4.027442. The ratio is 2.0926827 for the penetration depth and 2.2872729 for the bead width. This indicates that the obtained models are adequate. Statistically significant coefficients are shown in Table 6 in bold.

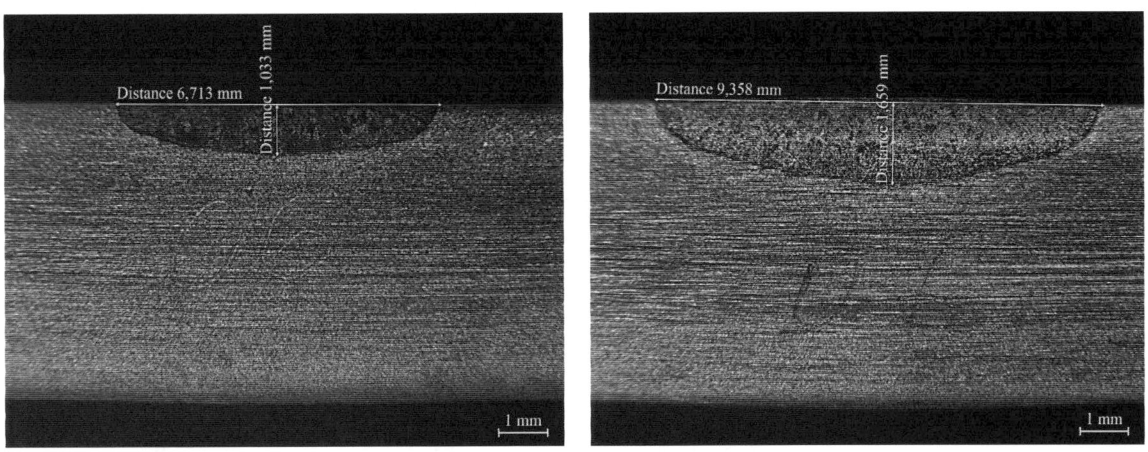

Figure 1. Measurement of the bead width and the penetration depth

Table 2. Experimental results.

№	X1 - welding current, [A]	X2 – welding speed, [cm/min]	bw - bead width, [mm]	hw - penetration depth, [mm]
1	**180**	**30**	7.609	1.24
2	**180**	**12**	11.518	2.112
3	**80**	**30**	2.801	0.438
4	**80**	**12**	4.025	1.006
5	**80**	**17**	3.552	0.767
6	**152**	**30**	6.713	1.033
7	**152**	**12**	9.358	1.659
8	**180**	**17**	9.242	1.517
9	**108**	**25**	5.305	1.017
10	**108**	**12**	6.322	1.377
11	**180**	**25**	8.324	1.221
12	**108**	**30**	4.719	0.804
13	**80**	**25**	2.942	0.538
14	**152**	**17**	8.326	1.221
15	**108**	**17**	5.295	1.046
16	**152**	**25**	7.267	1.074
17	**180**	**30**	7.542	1.23
18	**80**	**30**	1.92	0.449
19	**80**	**30**	2.307	0.325
20	**130**	**12**	7.543	1.283
21	**152**	**30**	6.51	1.111

Table 3. Expanded experiment plan matrix for penetration depth.

№	b0	b1	b2	b3	b4	b5	b6	b7	b8	b9
	1	X1	X2	X1*X2	X1^2	X2^2	X1^2*X2	X1*X2^2	X1^3	X2^3
1	1	180	30	5400	32400	900	972000	162000	5832000	27000
2	1	180	12	2160	32400	144	388800	25920	5832000	1728
3	1	80	30	2400	6400	900	192000	72000	512000	27000
4	1	80	12	960	6400	144	76800	11520	512000	1728
5	1	80	17	1360	6400	289	108800	23120	512000	4913
6	1	152	30	4560	23104	900	693120	136800	3511808	27000
7	1	152	12	1824	23104	144	277248	21888	3511808	1728
8	1	180	17	3060	32400	289	550800	52020	5832000	4913
9	1	108	25	2700	11664	625	291600	67500	1259712	15625
10	1	108	12	1296	11664	144	139968	15552	1259712	1728
11	1	180	25	4500	32400	625	810000	112500	5832000	15625
12	1	108	30	3240	11664	900	349920	97200	1259712	27000
13	1	80	25	2000	6400	625	160000	50000	512000	15625
14	1	152	17	2584	23104	289	392768	43928	3511808	4913
15	1	108	17	1836	11664	289	198288	31212	1259712	4913
16	1	152	25	3800	23104	625	577600	95000	3511808	15625
17	1	180	30	5400	32400	900	972000	162000	5832000	27000
18	1	80	30	2400	6400	900	192000	72000	512000	27000
19	1	80	30	2400	6400	900	192000	72000	512000	27000
20	1	130	12	1560	16900	144	202800	18720	2197000	1728
21	1	152	30	4560	23104	900	693120	136800	3511808	27000

Table 4. Matrix C (inverted matrix G) for penetration depth.

Ci	Ci0	Ci1	Ci2	Ci3	Ci4	Ci5	Ci6	Ci7	Ci8	Ci9
C0	597.577	-8.663	-37.715	0.1484	0.05583	1.3731	-0.0003	-0.0017	-0.0001	-0.0178
C1	-8.663	0.18749	0.1665	-0.0015	-0.0014	-0.0033	4E-06	1E-05	3E-06	3E-05
C2	-37.715	0.1665	4.81024	-0.0127	-0.0003	-0.196	2E-05	0.0002	-3E-07	0.0027
C3	0.14841	-0.0015	-0.0127	0.0001	1.9E-06	0.0002	-2E-07	-2E-06	8E-09	-2E-07
C4	0.05583	-0.0014	-0.0003	2E-06	1.1E-05	9E-06	-1E-08	2E-08	-3E-08	-2E-07
C5	1.37309	-0.0033	-0.196	0.0002	8.5E-06	0.009	-1E-07	-4E-06	-1E-08	-0.0001
C6	-0.0003	3.9E-06	1.6E-05	-2E-07	-1E-08	-1E-07	8E-10	3E-10	-2E-11	5E-10
C7	-0.0017	1.2E-05	0.0002	-2E-06	2.2E-08	-4E-06	3E-10	3E-08	-7E-11	7E-10
C8	-0.0001	3.2E-06	-3E-07	8E-09	-3E-08	-1E-08	-2E-11	-7E-11	7E-11	4E-10
C9	-0.0178	2.9E-05	0.00266	-2E-07	-2E-07	-0.0001	5E-10	7E-10	4E-10	2E-06

Table 5. Results of measuring the penetration depth and the bead width for the experiments carried out at the center of the plan I=130[A] и v_w = 21[cm/min].

u	1	2	3	4	5	6	7
hw, [mm]	1.08	1.142	1.127	1.033	1.095	1.002	1.095
bw, [mm]	6.697	6.869	6.572	6.494	6.369	6.353	6.588

Table 6. Regression equations coefficients values.

	for hw	for bw		for hw	for bw
b_0	**-0.7979**	0.40782	b_5	0.0089	**0.044**
b_1	0.08334	0.16434	b_6	**-7E-06**	**-2E-05**
b_2	-0.2401	-0.9219	b_7	**4E-05**	**0.0001**
b_3	0.0001	-0.0006	b_8	**2E-06**	2E-06
b_4	**-0.0006**	-0.0005	b_9	-0.0002	**-0.0009**

A graphical representation of the experimental and calculated results is shown on Figure 2 and Figure 3.

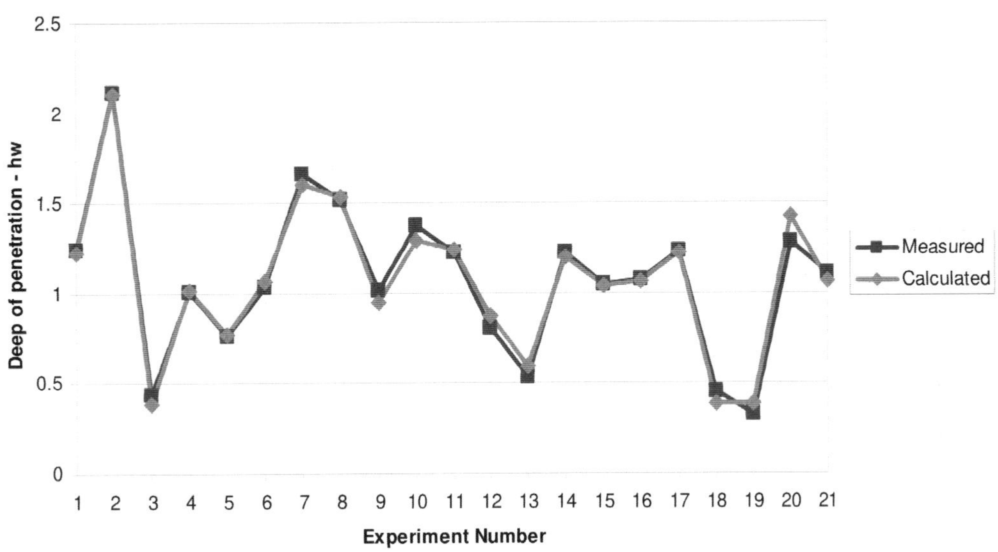

Figure 2. Comparison of experimental and calculated results for penetration depth.

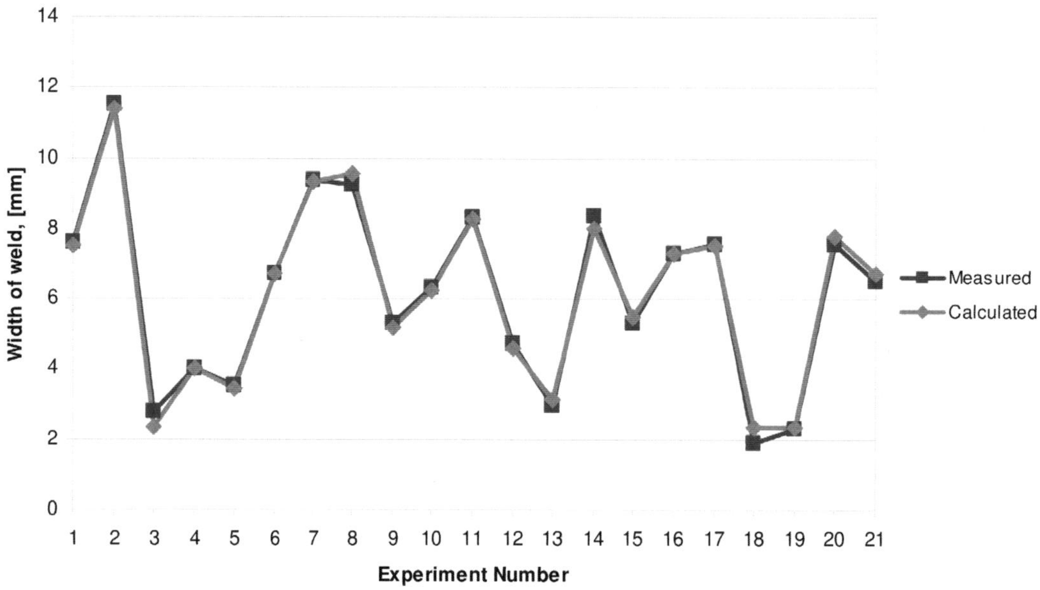

Figure 3. Comparison of experimental and calculated results for bead width.

To determine the dependence of the arc voltage on the current magnitude, a D-optimal plan for the second-degree full polynomial was used (Table 7). The processing of these data and taking into account only statistically significant coefficients the following arc voltage equation was obtained:

$$U_a = 7{,}0996 + 0{,}043345.I_w$$

Table 7. Arc voltage data.

№	Welding current, A	Welding speed, cm/min	Arc voltage, V	№	Welding current, A	Welding speed, cm/min	Arc voltage, V
1	180	30	13.7	**10**	180	30	12.8
2	180	12	13.7	**11**	180	12	13.3
3	80	30	10.1	**12**	80	30	9.9
4	80	12	10.1	**13**	80	12	10.1
5	130	30	11.8	**14**	130	21	11.4
6	180	21	13.2	**15**	130	12	11.6
7	130	12	12	**16**	180	30	13
8	80	21	10.1	**17**	80	30	10
9	130	21	12	**18**	180	12	13.2

The influence of the welding current magnitude on the penetration depth and the bead width for different values of the welding speed is shown on Figure 4 and Figure 5. As expected, with increasing welding current, the penetration depth is increasing. This increase is nonlinear depending on both the current magnitude and the welding speed (Figure 6). The bead width increases almost linearly with increasing current at a constant welding speed.

IOP Conf. Series: Materials Science and Engineering **878** (2020) 012054 doi:10.1088/1757-899X/878/1/012054

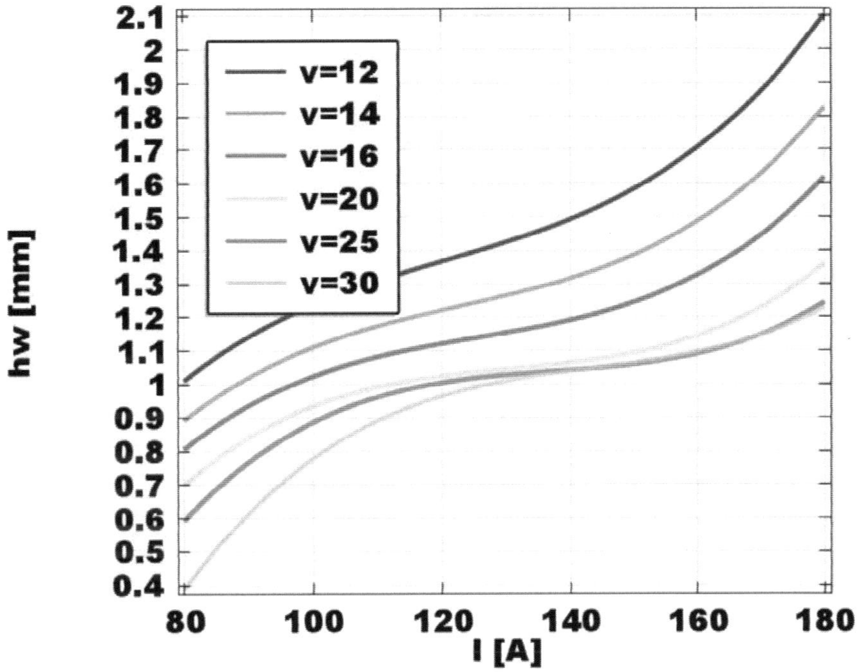

Figure 4. Influence of the current magnitude on the penetration depth for different values of the welding speed.

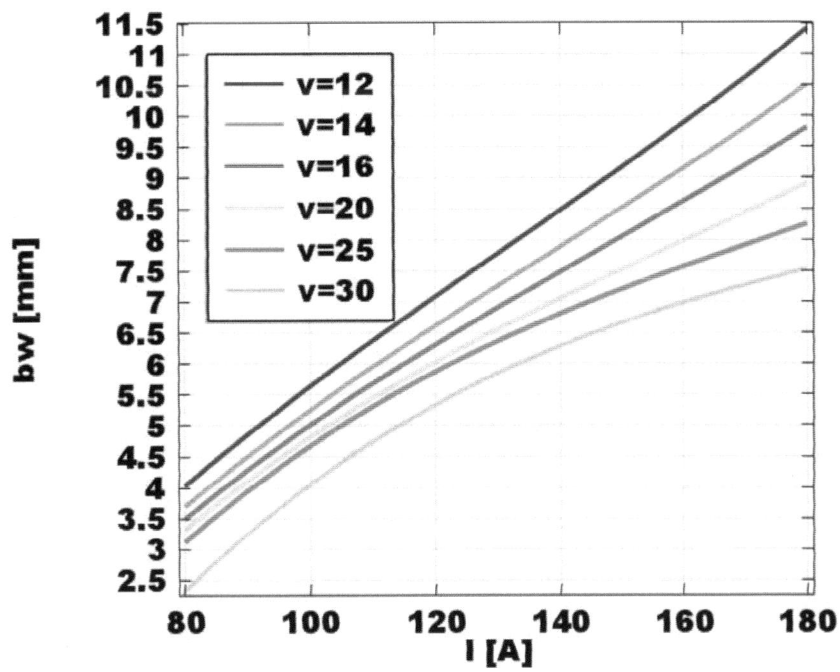

Figure 5. Influence of the current magnitude on the bead width for different values of the welding speed.

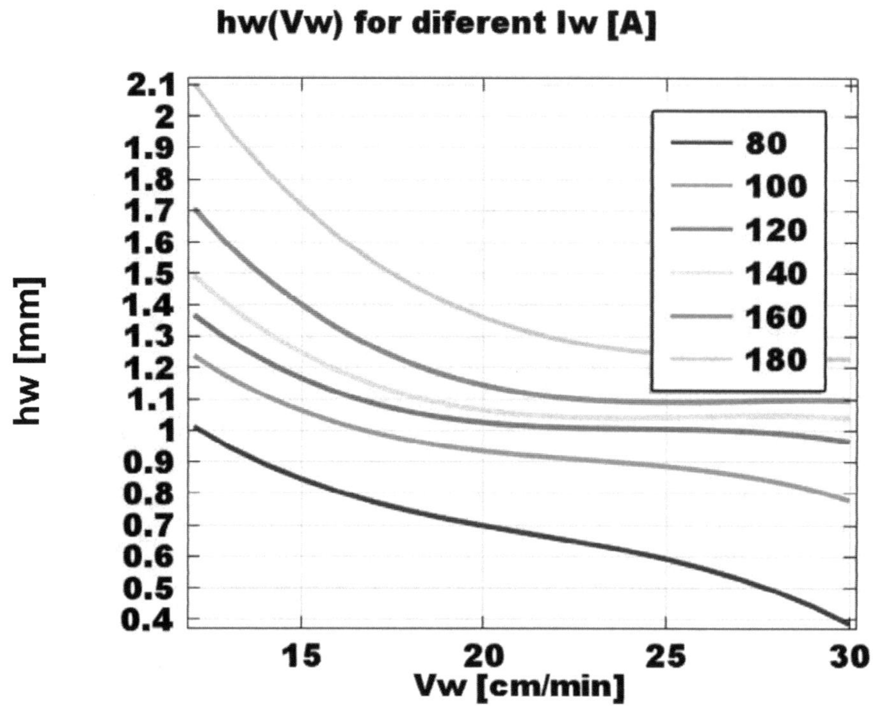

Figure 6. Influence of welding speed on penetration depth for different values of the welding current.

4. Conclusions

On the basis of experimental results for the conditions of TIG welding in argon and arc length 2 [mm], regression equations for the penetration depth, the bead width and the arc voltage were obtained. The penetration depth and the bead width depend on both the current magnitude/value and the welding speed, and these dependencies are described by third-degree polynomials. The arc voltage depends only on the current and the dependence is linear.

Aknowledgements

This work was supported by the Bulgarian National Scientific Fund under Grant DN 07/26.

5. References

[1] Nagesh, D.S., Datta, G.L., Genetic algorithm for optimization of welding variables for height to width ratio and application of ANN for prediction of bead geometry for TIG welding process, Applied Soft Computing Journal 10(3), pp. 897-907, https://doi.org/10.1016/j.asoc.2009.10.007

[2] Madadi, F., Ashrafizadeh, F., Shamanian, M., Optimization of pulsed TIG cladding process of stellite alloy on carbon steel using RSM, Journal of Alloys and Compounds 510(1), pp. 71-77, https://doi.org/10.1016/j.jallcom.2011.08.073

[3] Bodkhe, S.C., Dolas, D.R., Optimization of Activated Tungsten Inert Gas Welding of 304L Austenitic Stainless Steel, Procedia Manufacturing 20, pp. 277-282, https://doi.org/10.1016/j.promfg.2018.02.041

[4] Babu, A.V.S., Giridharan, P.K., Narayanan, P.R., Murty, S.V.S.N., Prediction of Bead Geometry for Flux Bounded TIG Welding of AA 2219-T87 Aluminum Alloy, ournal of

Advanced Manufacturing Systems 15(2), pp. 69-84, https://doi.org/10.1142/S0219686716500074

[5] Pichumani, S., Srinivasan, R., Weld profile optimization on weld speed & arc length parameters of TIG welding on Al-SiC composite, Journal of Advanced Research in Dynamical and Control Systems 9(Special issue 11), pp. 487-491, ISSN: 1943023X

[6] Jerold Jose, P.J., Anand, M.D., Optimization of weld bead profile parameters in TIG welding process for inconel 718 alloy using RSM and regression analysis, Journal of Chemical and Pharmaceutical Sciences 9(4), pp. 1953-1962, ISSN: 09742115

[7] Gopinath, V., Manojkumar, T., Sirajudeen, I., Yogeshwaran, S., Chandran, V., Optimization of process parameters in TIG welding of AISI 202 stainless steel using response surface methodology, International Journal of Applied Engineering Research 10(13), pp. 11053-11057, ISSN: 09734562

[8] Pavan, A.R., Arivazhagan, B., Vasudevan, M., Process Parameter Optimization of A-TIG Welding on P22 Steel, Lecture Notes in Mechanical Engineering pp. 99-113, DOI: 10.1007/978-981-13-8767-8_8, ISSN: 21954356, ISBN: 978-981138766-1

[9] Jung, S.H., Kim, J.-W., A study on the optimal welding condition for root-pass in horizontal butt-joint TIG welding, Transactions of the Korean Society of Mechanical Engineers, A 41(4), pp. 321-327, https://doi.org/10.3795/KSME-A.2017.41.4.321

[10] Deepak, P., Jualeash, M.J., Jishnu, J., Padmanaban, R., Thirumalini, S., Optimization of process parameters of pulsed TIG welded maraging steel C300, IOP Conference Series: Materials Science and Engineering 149(1),012007, https://doi.org/10.1088/1757-899X/149/1/012007

Comparative analysis of sorption characteristics of Bulgarian grape seeds and flours and flakes produced by them

A G Durakova[1], A L Bogoeva[2], A P Krasteva[3] and M L Temelkova[4]

[1]Department of Process Engineering, Technical Faculty, University of Food Technologies – Plovdiv, 26, Maritsa Blvd, 4002 Plovdiv, Bulgaria

[2]Department of Mechanical and Instrument Engineering, Faculty of Mechanical Engineering, Technical University-Sofia, Plovdiv Branch,25 Tsanko Diustabanov Str., Plovdiv 4000,Bulgaria

[3]Department of Technology of Grain, Fodder, Bread and Confectionery Products, University of Food Technologies – Plovdiv, 26, Maritsa Blvd, 4002 Plovdiv, Bulgaria

[4]Department of Language Training Centre, University of Food Technologies – Plovdiv, 26, Maritsa Blvd, 4002 Plovdiv, Bulgaria

The corresponding author's e-mail address: aldurakova@abv.bg

Abstract. The present paper contains a comparative characteristics of results obtained from statics of the sorption characteristics of grape seeds and biologically active products obtained from them – full-fatted flour, defatted flour and full-fatted flakes. The statics of the sorption processes is conducted for three temperatures - 10°C, 25°C and 40°C and eight water activities in the range from 0.112 to 0.868. For all of the examined products the temperature rise leads to a decrease of their sorption capacity. For the description of the equilibrium curves of the examined products the three parametric modified models of Oswin, Halsey, Henderson, Chung-Pfost are used. By the method of the smallest squares, using a programme for nonlinear regression "*Statistica*", the coefficients of the used models are calculated, as on the basis of universally used criteria of suitability, a relevant model is recommended for each of the four Bulgarian products. The calculation of the values of the monolayer moisture content (MMC) in water activities below 0.5 is done through linearization of the Brunauer-Emmett-Teller (BET) model. The obtained results for MMC vary within the range from 2.34% to 4.96% dry substance (d.b.) for the processes of adsorption and desorption.

Keywords: sorption characteristics, functional flours, grape seeds.

1. Introduction

Recently more and more studies are focused on the prospecting of natural plant species with high content of nutrients, which to be treated and proposed as products of high quality, satisfying the body needs [1].

The sorption characteristics (equilibrium and monolayer moisture content) of the nutrients afford an opportunity for proper methods of their treatment and preservation to be optimized. The knowledge about the equilibrium moisture enables the proper deciphering of the regimes and methods for the product's treatment, preservation and packaging [2]. The dependence between the equilibrium moisture (M) and the water activity (a_w) is given from a skillfully built under definite temperatures sorption isotherm. The isotherms afford the examination of the character and the quantity of the different types connected water. For the first time the term "water activity - a_w" is suggested by Scott in 1952 and it is widely used in the monitoring of the Critical Control Points (CCP) in the Hazard Analysis and Critical Control Point (HACCP) [3]. The equilibrium moisture calculated in different conditions of the nutrients can be prognosticated through a mathematic model. The use of empirical models for specifically examined products restricts their applicability. This imposes the search of models, allowing the description of the sorption isotherms of a larger number of nutrients grouped under similar indices [4]. Scientific teams suggest over 200 empiric and theoretical models for description of the equilibrium isotherms [2; 5-7]. A large part of the publications are concentrated over the use of two parametric methods, which eliminate the influence of the temperature in their equations. However, the investigations prove that the temperature affects the equilibrium moisture, which imposes the two parametric methods to be modified into three parametric methods containing the temperature effect [8]. The sorption isotherms of different nutrients modeled by a great number of scientific teams, proves that there is no universal model, i.e. to describe the isotherms of different products or groups of products. The first to reach this conclusion in 1989 are Chen and Morey, which necessitates that in the processing of the results obtained from an experiment, there must be used several models and according to certain criteria: average relative error, standard error and residues distribution, the most fitted to be chosen [9]. These criteria are adopted till nowadays [10]. In Standard 245D of American Society of Agricultural Engineers (ASAE) for description of the equilibrium isotherms of products of plant origin, three parametric modified models of Oswin, Henderson, Halsey и Chung-Pfost are suggested [11-12].

The model of Brunauer-Emmett-Teller (BET), based on the theory for the poly molecular adsorption is also one of the widely used [6; 13]. It is proved that the BET model is generally used in the calculation of the monolayer moisture content (MMC) for a large number of products in water activity not higher than 0.5. The moisture value corresponding to the monolayer moisture content (MMC) is a sorption characteristics, which affects the product's stability [14-17]. In a large number of investigations it is proved that reducing the product to moisture corresponding to MMC in conditions for storage, preservation of its quality indices is achieved [18].

In the detailed literature survey there is not data base of a summarized comparison of the sorption characteristics of grape seeds of Bulgarian origin and flours and flakes produced by them in cultivation, which gave us a reason for the present elaboration.

2. Materials and methods

2.1. Materials

The targets of the present studies are grape seeds of Bulgarian origin extracted after alcohol fermentation of wine of the sorts Mavrud, Cabernet Sauvignon, Syrah, Merlot, Dimyat and Sauvignon Blanc, and the obtained products - full-fatted flakes, full-fatted flour and defatted flour. The analysed samples are provided from an experimental base – village of Parvenets, Municipality of Rodopi, Plovdiv. They are selected on the base of limited and/or lack of data after a detailed literature survey.

2.2. Methods

Nutrients' equilibrium isotherms indicate the dependence between the product equilibrium moisture and the water activity of the environment in determinate temperature.

A static gravimetric method, recommended by Project COST 90 [19] and up-dated by Bell & Labuza, (2000) [14] is used for the purpose of the present paper. The study provides information about the sorption characteristics – adsorption and desorption in temperatures of 10°C, 25°C и 40°C and

IOP Conf. Series: Materials Science and Engineering **878** (2020) 012055 doi:10.1088/1757-899X/878/1/012055

eight different relative air moistures. The samples are preliminary dehydrated in desiccators over diphosphorus pentaoxide (P_2O_5) for 20 days with the purpose to investigate the process adsorption.

The samples are hydrated over distilled water (H_2O) for the same period – 20 days for the process of desorption. Aluminium weight containers, in which samples with a mass of 1±0.01g are measured, are used for the analysis. After that the containers are placed in hygrostats, in which relevant saturated solutions of salts are prepared in advance, upholding over their surface definite constant water activities (relative moistures of the environment) within the range from 0.112 to 0.868. In the hygrostats at a_w>0.6 thymol crystals are placed to prevent the product from microbiological damage for the whole period of the experiment.

2.2.1. Sorption isotherms modelling

The modified three parametric models of Oswin, Chung-Pfost, Halsey and Henderson are used for the description of the sorption isotherms.

Modified Chung-Pfost $$a_w = exp\left[\frac{-A}{t+B} exp(-CM)\right] \quad (1)$$

Modified Halsey $$a_w = exp\left[\frac{-exp\,(A+Bt)}{M^C}\right] \quad (2)$$

Modified Oswin $$M = (A + Bt)\left(\frac{a_w}{1-a_w}\right)^C \quad (3)$$

Modified Henderson $$1 - a_w = exp[-A(t + B)M^C] \quad (4)$$

where: *t* is temperature, °C; A, B and C - coefficients.

A programme for nonlinear regression by the method of the smallest squares "*Statistica*" (procedure "Nonlinear estimation") [20] is used to determine the coefficients of the models (A, B and C). Average relative error, standard error and residues distribution are the criteria applied for the estimation and comparison of the models:

Average relative error $$P = \frac{100}{N}\sum\left|\frac{M_i-\hat{M}_i}{M_i}\right| \quad (5)$$

Standard error $$SEM = \sqrt{\frac{\sum(M_i-\hat{M}_i)^2}{df}} \quad (6)$$

Residue $$e_i = M_i - \hat{M}_i \quad (7)$$

where: M_iи $\hat{M}_i$ are respectively the experimental and the predicted by the model equilibrium moisture; N – number of experiments; *df* – degrees of freedom (number of experiments minus number of coefficients).

The accidental and non-accidental distribution of residues is graphically represented. In model with the lowest values for P and SEM and accidental residues distribution, the relevant model can be recommended for description of the sorption isotherms of the examined product.

The model of Brunauer-Emmett-Teller (BET) (10) is transformed into linear type with the purpose of calculation of the monolayer moisture content values - MMC [21]:

Where: coefficients $$\frac{a_w}{(1-a_w)M} = P + Qa_w \quad (8)$$

$$P = \frac{1}{M_eC} \quad (9) \qquad \text{and} \qquad Q = \frac{C-1}{M_eC} \quad (10)$$

To receive the values of the coefficients P and Q of the linear equation, the experimental data at $a_w \leq 0.5$ and temperatures – 10°C, 25°C и 40°C are processed with the programme Excel by the method of the smallest squares. On the basis of the received results MMC of the examined products is calculated.

3. Results and discussion

The analysis of the sorption characteristics of the examined grape seeds and the biologically active components obtained from them indicates certain common objectives.

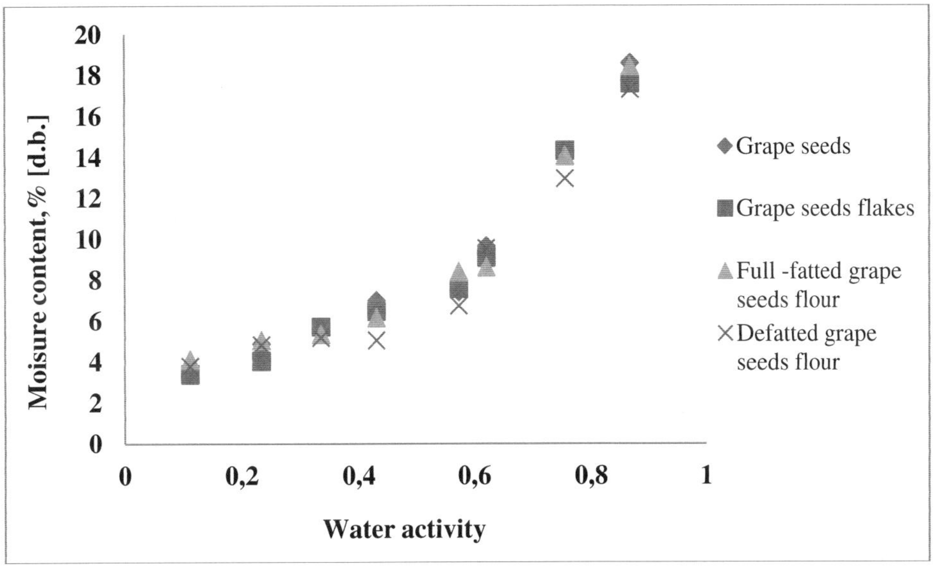

Figure 1. Adsorption isotherms of examined products at temperature of 10°C

From the analysis of the received results for temperatures of 10°C, 25°C and 40°C and water activities within the range from 0.112 to 0.868, it can be indicated that the temperature rise for all examined products leads to the reduction of their sorption capacity. The explanation of this fact is probably the activation of the water molecules' mobility, which are respectively held with difficulty on the hard surface. All isotherms have clearly indicated S-figured character, i.e. they are from 2nd type according to the classification of Brunauer et al., (1940), like more of the nutrient powdered products [22]. The comparative graphic dependencies of the adsorption and desorption isotherms of all examined products received for 10°C are presented in Figure 1 and 2.

In analysis of the results it is reported that the examined grape seeds and the bioactive products obtained from them at a_w approximately 0.6 have almost equalized adsorption and desorption capacity. An interesting fact is that at $a_w > 0.8$ for the adsorption process the sorption capacity which is most strongly defined is measured for whole grape seeds, followed by full-fatted flour, full-fatted flakes and defatted grape seeds flour. For the desorption process we measured the most strongly defined sorption capacity for full-fatted grape seeds flakes, followed by defatted flour, full-fatted flour and the initial material – the whole grape seeds remain with the lowest capacity. For water activity $a_w = 0.6$ at fig. 2 the four examined products are with equal sorption capacity. In the comparison sorption isotherms for all examined products (grape seeds, full-fatted flakes, full-fatted flour and defatted flour) lower or more clearly expressed hysteresis is defined. As a summary of the different theories explaining the hysteresis effect, we came to the conclusion that it is a result from many factors interacting between each other. The received results for the other two temperatures – 25°C and 40°C for adsorption an desorption processes are similar too.

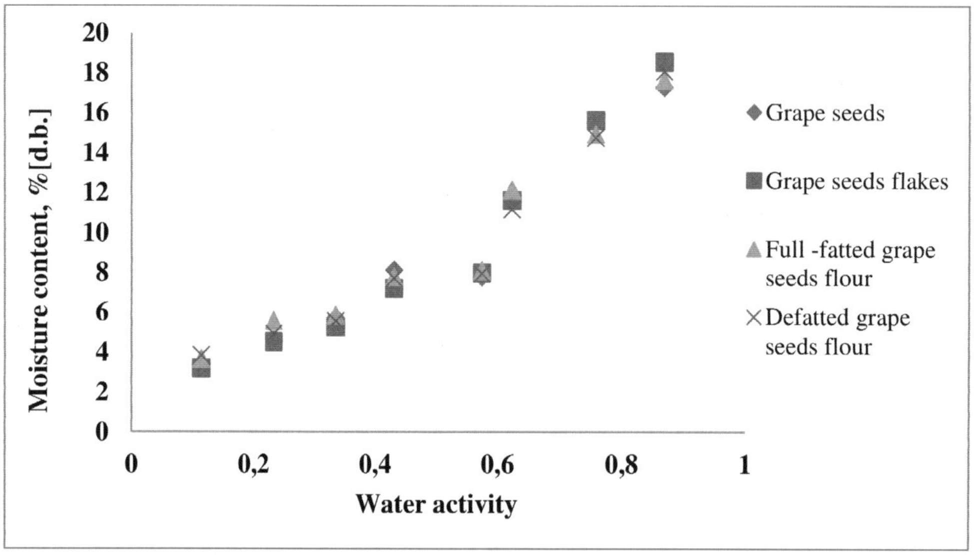

Figure 2. Desorption isotherms of examined products at 10ºC

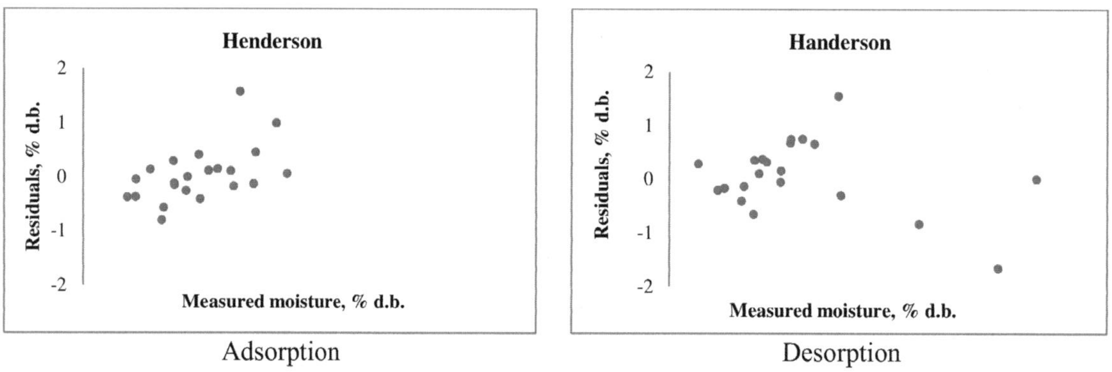

Figure 3. Grape seeds: Residues distribution
* Residues larger than ± 2 are not presented

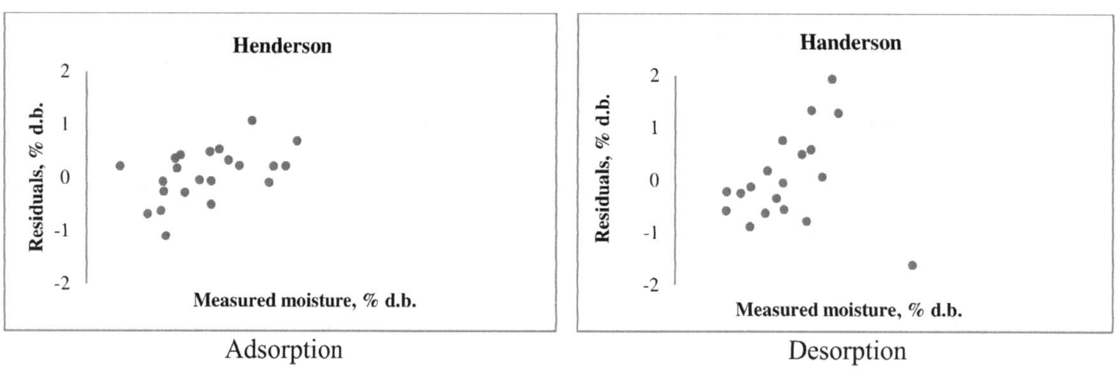

Figure 4. Grape seeds flakes: Residues distribution
* Residues larger than ± 2 are not presented

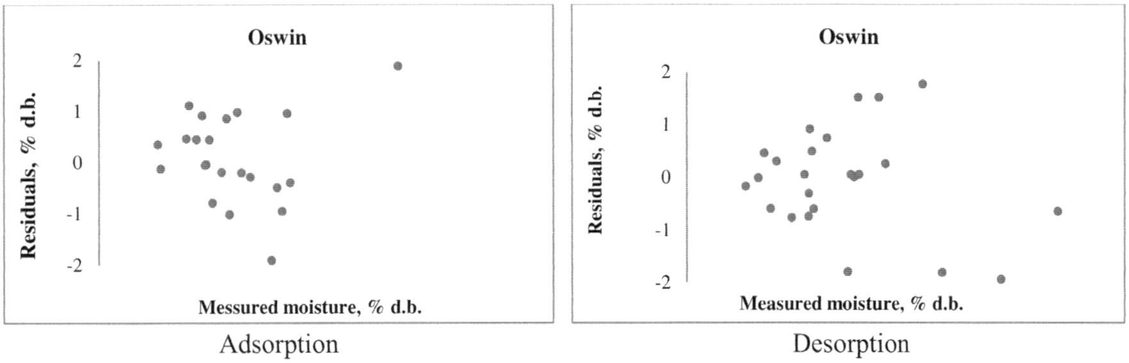

Figure 5. Full-fatted grape seeds flour : Residues distribution
* Residues larger than ± 2 are not presented

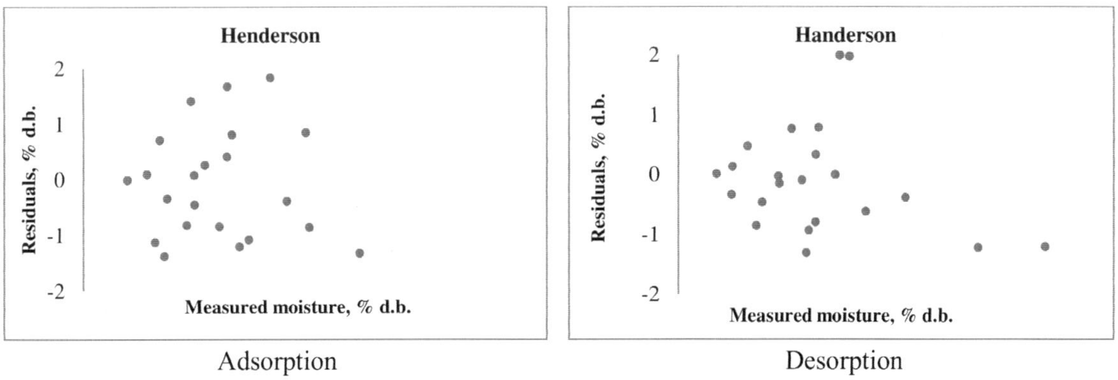

Figure 6. Defatted grape seeds flour: Residues distribution
* Residues larger than ± 2 are not presented.

On the basis of statistic processing three parametric modified models for description of the four examined products sorption isotherms are recommended, namely: whole grape seeds, full-fatted grape seeds flour, full-fatted grape seeds flakes and defatted grape seeds flour.

Table 1. Coefficients (*A, B, C*) of recommended models for sorption isotherms description

Product	Models	Process	*A*	*B*	*C*
Grape seeds	Henderson	Adsorption	0.000959	9.608507	1.737924
		Desorption	0.001089	8.806570	1.610936
Grape seeds flakes	Henderson	Adsorption	0.000830	14.97965	1.725401
		Desorption	0.000672	17.99822	1.712982
Full-fatted grape seeds flour	Oswin	Adsorption	547.982	0.33870	61.6750
		Desorption	9.68658	-0.10298	0.35869
Defatted grape seeds flour	Henderson	Adsorption	0.000291	95.37602	1.564150
		Desorption	0.000696	18.61505	1.63518

The models are recommended on the basis of commonly accepted criteria for the lowest values for average relative error, standard error (table 1) and accidental residue distribution (figures 3,4,5 and 6).

The presented models confirm the Chen & Morey, (1989) statement that there is no universal model to be recommended for description of all nutrients sorption isotherms, including the obtained as derivatives of initial materials [5-6; 9].

The numerical values of the monolayer moisture content (MMC) presented in table 2 are calculated through linearization of the BET model and the received calculations. It is clear from the data in the table that the highest value of MMC is reported at temperature 10°C, after which its values for temperatures 25°C and 40°C lower with approximately 1.3% for adsorption and approximately 1.5% for desorption process. The lowest values for the experiment's conditions are reported at temperature 25°C. The full-fatted flakes from pressed grape seeds have the lowest values of MMC at temperature 25°C for the both processes. The highest values 4.33% d.b. is calculated for desorption process at 10°C for this analyzed product, too. For the three temperatures the values of MMC for desorption process are higher than its values for adsorption process. The reported data for grape seeds full-fatted flour show clearly defined hysteresis; for the three temperatures are higher for desorption than adsorption with from 0.2% d.b. до 1.2% d.b. We report the lowest values for adsorption and desorption at temperature of 25°C for grape seeds Bulgarian full-fatted flour. Hysteresis is available for defatted flour but only for two temperatures – 10°C and 25°C. For temperature of 40°C the received results are approximately equal and similar to value with the results for desorption for 10°C.

Table 2. BET monolayer moisture content, % d.b. for adsorption and desorption processes of examined products for temperatures 10°C, 25°C and 40°C.

Product	Process	10°C	25°C	40°C
Grape seeds	Adsorption	4.03	2.46	2.59
	Desorption	4.96	2.34	3.03
Grape seeds flakes	Adsorption	3.95	2.62	2.67
	Desorption	4.33	2.78	3.23
Full-fatted grape seeds flour	Adsorption	3.34	2.48	3.11
	Desorption	4.59	2.70	3.55
Defatted grape seeds flour	Adsorption	3.41	2.54	4.53
	Desorption	4.58	2.75	4.25

As a result of the analysis we can summarize that the received results for all examined products indicate that the temperature does not affect the values of the monolayer moisture content. For all products an interesting fact is that the lowest values are calculated for temperature 25°C. The MMC values for examined whole grape seeds and powdered products obtained from them vary within the range from 2.34% d.b. to 4.96% d.b. for adsorption and desorption.

The sorption characteristics examination gives an important information for treatment, transportation and preservation which guarantees the nutrients' quality. The hygro-thermal equilibrium between relative air humidity and the examined products of Bulgarian origin – grape seeds, full-fatted grape seeds flakes, full-fatted flour and defatted flour for temperatures of 10°C, 25°C and 40°C and aw = 0.10÷ 0.90, is studied for the first time.

The numerical values of the proposed models for description of the equilibrium isotherms of the examined products, make them usable for calculation of equilibrium moisture in the engineering practice.

The moisture value corresponding to the monolayer moisture content (MMC) is a sorption characteristic which influence positively the stability of the product [16-17]. The MMC calculated values are a good precondition for an extensive research about regimes and terms of preservation of the analyzed products.

4. Conclusions

Equilibrium isotherms are received for the processes adsorption and desorption for temperatures 10°C, 25°C and 40°C and relative moistures within the range from 0.112 to 0.868 for grape seeds, full-fatted grape seeds flakes, full-fatted grape seeds flour and defatted grape seeds flour. It is established that the isotherms of all products are with *S*-graphic character, i.e., they are from 2nd type according to the classification of Brunauer et al., (1940). It is established that the temperature affects the sorption capacity of the grape seeds, full-fatted grape seeds flakes, full-fatted grape seeds flour and defatted grape seeds flour. With its increase at constant water activity the equilibrium moistures lower. On the basis of statistic processing and in accordance with the accepted criteria three parametric modified models for description of the four examined products sorption isotherms are recommended:

- For grape seeds, full-fatted grape seeds flakes and defatted grape seeds flour – modified model of Henderson;
- Full-fatted grape seeds flour – modified model of Oswin;

The values of MMC for the processes adsorption and desorption are calculated in the range within:

- Grape seeds: 2.34% ÷ 4.96%
- Full-fatted grape seeds flakes: 2.62% ÷ 4.33%
- Full-fatted grape seeds flour: 2.48% ÷ 4.59%
- Defatted grape seeds flour: 2.54% ÷ 4.58%

5. References

[1] Ghouila Z, Laurent S, Boutry S, Vander Elst L, Nateche F, Muller R, Baaliouamer A. 2017 *J Fundament Appl Sci* **9** (1) pp 392-420.
[2] Al-Muhtaseb A H, McMinn W A M, Magee T R A 2002 *Food Bioprod Process* **80** (2) pp 118-128.
[3] Sandulachi E 2012 *Technical University of Moldova* pp 40-48
[4] Kulchan R, Boonsupthip W, Suppakul P 2010 *J. Food Eng* **100** (3) pp 461-467
[5] Hinz C 2001 *Geoderma* **99** (3-4) pp 225-243
[6] Lemus R M 2011 *Vitae* **18** (3) pp 325-334
[7] Peleg M 1988 *J. Food Sci.* **53** (4) pp 1216-1217
[8] Muangrat R, and Nuankham C 2018 *Food ScI Nutr* **6** (3), pp 585-593
[9] Chen C-C and Morey R V 1989 *Transactions of the ASAE* **32** (3) pp 983-0990
[10] Bennamoun L, Harun N Y, Afzal M T 2018 *AJSE* **43** (3) pp 1195-1203
[11] Igathinathane C, Womac A, Sokhansanj S, Pordesimo L 2005 *Transactions of the ASAE* **48** (4) pp 1449-1460.
[12] Zhang L, Sun D-W, Zhang Z 2017 *Crit Rev Food Sci Nutr* **57** (5) pp 1052-1058
[13] Arthur E, Tuller M, Moldrup P, Greve M H, Knadel M, de Jonge L W 2018 *Eur. J. Soil Sci.* **69** (2) pp 245-255
[14] Bell L and Labuza T 2000 *St. Paul: American Association of Cereal Chemists*, Inc., pp 33-36.
[15] Labuza T P 1985 *In Properties of water in foods*, Springer, Dordrecht pp 421-445
[16] Timmermann E O 2003 *Colloid Surf A Physicochem Eng Asp* **220** (1-3) pp 235-260.
[17] Troller, J., 2012, *Water activity and food* Elsevier ISBN 032315901X.
[18] Decagon A L 2011 *Decagon Devices*, Pullman, WA Google Scholar.
[19] Wolf W, Spiess W E L, Jung G, 1985, COST-Project 90 and 90 bis, *Properties of Water in Foods in Relation to Quality and Stability Martinus Nijhoff*, Dordrech, pp 661-679.
[20] Menkov N, Dinkov K, Durakova A, Toshkov N 2009 *Bulg J Agric Sci* **15** (4) pp 281-285
[21] Brunauer S, Emmett P H, Teller E 1938 *JACS* **60** (2) pp 309-319
[22] Brunauer S, Deming L S, Deming W E, Teller E 1940 *JACS* **62** (7) pp 1723-1732

Sorption characteristics of Bulgarian Einkorn flakes

A L Bogoeva[1], A G Durakova[2] and N G Toshkov[2]

[1] Department of Mechanical and Instrument Engineering, Faculty of Mechanical Engineering, Technical University-Sofia, Plovdiv Branch, 25 Tsanko Diustabanov Str., Plovdiv 4000, Bulgaria

[2] Department of Process Engineering, Technical Faculty, University of Food Technologies – Plovdiv, 26, Maritsa Blvd, 4002 Plovdiv, Bulgaria

The corresponding author's e-mail address: adelina.bogoeva@abv.bg

Abstract. The current scientific research is focused on the Einkorn flakes of Bulgarian origin. The aim is to determine their moisture adsorption and desorption isotherms. The flakes are obtained of milling Einkorn grains with a reconstructed mill roll. The applied pressure was between 200 and 280 tons. The experimental database of equilibrium moisture content (EMC) are determined at three different temperatures – 10°C, 25°C and 40°C through the standard static gravimetric method in the compulsory presence of water activity within the range from 0.1 to 0.9. The obtained adsorption and desorption isotherms have an *S*-shape profile i.e. they are from the IInd class typical of the majority of food products. The results confirm a regularity that the sorption capacity decreases with increasing of the temperature in the presence of constant water activity which is valid for the both processes. The mathematical three-parametric modified models of Chung–Pfost, Oswin, Halsey and Henderson are evaluated for the description of the obtained sorption data. The monolayer moisture content (MMC) of Einkorn flakes is calculated through the linearization of Brunauer-Emmett-Teller equation. The values varied from 3.26% to 4.92%.

Keywords: sorption isotherms, adsorption, desorption, monolayer moisture content, Einkorn, *Triticum monococcum*, flakes

1. Introduction

Consumption of cereals as a staple food containing essential bioactive compounds contributes to the good and the normal human body functioning. Modern food production offers neglected regional high-nutrition energy products on consumers to meet their needs [1-3].

According to literature reviews, the *Triticum monococcum* as well-known as Einkorn is a variety of wild wheat species and is considered as one of the most ancient Bulgarian wheat [4-5]. It has a low gluten content compared to ordinary modern wheat that makes it suitable for people with celiac disease [6-7]. Einkorn grain is a good source of micro and macronutrients including vitamins A, B and E, minerals such as zinc, magnesium, manganese, phosphorus, lipids, fructans, and high content of protein, carotenoids and tocopherols as well as some antioxidant compounds [8-10]. Numerous scientific investigations are focused on the creation of new products with therapeutic and prophylactic properties based on the Einkorn flour [11]. Furthermore, Einkorn's products are recommended against the risk of coronary heart disease, cholesterol, chronic and intestinal diseases as well as their prevention [9, 12-13].

Nakov et al. 2018 are created biscuits based on Einkorn flour. The study is confirmed a high content of polyphenols, carotenoids, antioxidants and beta-glucans of the new product. Based on the results obtained the authors propose to consider the new Einkorn biscuits as functional food [14]. The positive consumer continuity leads to an increase in searching for biscuits and bread based on the whole Einkorn flour on the market. Izambaeva et al. 2016 are studied the Einkorn flour and bread. The presence of antioxidant activity in the investigated products is proven by three analytical methods ABTS, FRAP and CUPRAC, the lutein content of the Einkorn flour is 207.4 $\mu g.100g^{-1}$ and in the Einkorn bread is 226.8 $\mu g.100g^{-1}$[15].

The varied opportunities for Einkorn flour applications give new perspectives to the food industry. For example - the production of flakes - an integral part of modern balanced ready-to-eat breakfast cereals consumed daily by adolescents and adults in countries such as the United States, Germany, France and Bulgaria [16]. Einkorn flour and flakes are a hygroscopic material. The essential sorption characteristic is the equilibrium moisture content (EMC). EMC provides important information on the product which in turn is an important indicator determining storage regimes [17-19].

2. Materials and methods

2.1. Raw Material
Einkorn flakes were delivered by experimental institute located in Parvenets, Bulgaria. They were milled with a reconstructed mill roll [using 2 smooth shafts with a differential difference of 0.05/each roller rotates at different speeds, the difference between them for 1 rpm (rotation per minute) is 1.5 mm to 5 mm]. The applied pressure was between 200 and 280 tons.

2.2. Method

2.2.1. Physicochemical parameters
Moisture, (%) of the product studied - standard method via drying of 5g of flour at 105°C to constant weight, according to AOAC, 1990 [20]; Protein – Direct Kjeldahl method analysis (determination of nitrogen content/ nitrogen determination method) at 25±2°C, according to BDS EN ISO 14431:1978 [21]; Fat content, (%) – Soxhlet method via solvent extraction with petroleum ether at 105±2°C. BDS EN ISO 6997:1984 [22]; Carbohydrates - at 25±2°C, according to BDS EN ISO 7169:1989 [23]; Ash, (%) – The method used for the determination of ash content in flour is according to BDS EN ISO 7646:1982 at 580±2°C [24]; Energy value, kcal/kJ - Regulation EU 1169:2011, Appendices XIV; XV [25]. All physico-chemicals analyzes were carried out in the Food Testing Laboratory at The Institut of Food Preservation and Quality – Plovdiv, Bulgaria.

2.2.2. Sorption characteristics database

2.2.2.1 Preparation of the sample
The adsorption and the desorption isotherms are studied at 10°C, 25°C and 40°C using a gravimetric - static method [26-28]. One part of the samples is dehydrated in a vacuum desiccator over P_2O_5 before beginning the adsorption analysis. Simultaneously, the other part of the samples is hydrated in a desiccator over distilled for the desorption analysis. Eight saturated salt solutions (LiCl, CH_3COOK, $MgCl_2$, K_2CO_3, $MgNO_3$, NaBr, NaCl, KCl) providing constant water environment activities are prepared in a glass jar for each temperature and for each process used [29]. The full explanation of the method used is presented in an article of Bogoeva, 2020 [30].

2.2.2.2. Mathematical modelling of the data

According to a literature review four different modified three-parametric models were selected for fitting of the sorption isotherms data, namely modified Chung-Pfost, modified Halsey, modified Oswin, modified Henderson, recommended by American Society of Agricultural Engineers (ASAE) [31]:

Modified Chung-Pfost $$a_w = exp\left[\frac{-A}{t+B}exp(-CM)\right] \quad (1)$$

Modified Halsey $$a_w = exp\left[\frac{-exp\,(A+Bt)}{M^C}\right] \quad (2)$$

Modified Oswin $$M = (A + Bt)\left(\frac{a_w}{1-a_w}\right)^C \quad (3)$$

Modified Henderson $$1 - a_w = exp[-A(t + B)M^C] \quad (4)$$

where:
M is the average moisture content, %;
a_w is the water activity, decimal;
A, B and C are coefficients;
t is the temperature, °C.

The different modified models were fitted through a non-linear least-squares regression program "*Statistica*". According to Lomauro et al. (1985) one of the basic criteria for determination of the models' equations is the average relative error (P, %). If the value P is ≤ 10% the model is considered fit [32]. Moreover, Chen and Morey's research included two more criteria to compare the suitability of the models, namely the standard error of moisture (SEM) and the randomness of residuals [33]. Nowadays, the P (%), SEM and the randomness of residuals are taken into consideration in order to determine the suitability of the models [29-30]:

The average relative error $$P = \frac{100}{N}\sum\left|\frac{M_i-\widehat{M}_i}{M_i}\right| \quad (5)$$

The standard error of moisture $$SEM = \sqrt{\frac{\sum(M_i-\widehat{M}_i)^2}{df}} \quad (6)$$

The distribution of residuals $$e_i = M_i - \widehat{M}_i \quad (7)$$

where:

M_i and $\widehat{M}_i$ are experimentally observed and predicted by the model value of the equilibrium moisture content;

N is the number of data points;

df is the number of degree of freedom (number of data points minus number of parameters in the model).

According to a literature review the monolayer moisture content (MMC) is calculated through the linearization of Brunauer-Emmett-Teller (BET) equation (Eq. 8) including water activities over 0.45 for the three selected temperatures at 10°C, 25°C and 40°C [34-35]:

$$M = \frac{M_eCa_w}{(1-a_w)(1-a_w+Ca_w)} \quad (8)$$

where:

M is the equilibrium moisture content, %;

Me is the MMC, %;

a_w is the water activity, decimal;

C is the coefficient.

To our knowledge, this is the first scientific research considering the determination of sorption characteristics of the Einkorn flakes of Bulgarian origin and its physicochemical parameters and sorption characteristics.

IOP Conf. Series: Materials Science and Engineering **878** (2020) 012056 doi:10.1088/1757-899X/878/1/012056

3. Results and discussion

The food label demonstrates the physicochemical percentage distribution of the basic nutritional substances in the product. Initial analysis of the Einkorn flakes begins with the determination of proteins, carbohydrates, moisture, fat, ash content and energy values. The results are as follows 13.1%; 72.8%; 8.7%; 3.4%; 2.0% and 95.41kcal /399.20kJ, respectively. According to the literature review, our obtained results are approximative to the reported data [3-4, 14].

The initial moisture content of the sample before starting the adsorption process is 5.73%, for the desorption process, it is 17.22%, respectively. The results obtained from the Bulgarian Einkorn flakes for both processes are presented in table 1 and table 2.

Table 1. Equilibrium moisture content EMC (%) of the Bulgarian Einkorn flakes for adsorption process at different water activities (a_w) and at different temperatures *t* (°C).

	10°C			25°C			40°C		
Sel	a_w	**EMC**[a]	**SD**[b]	a_w	**EMC**[a]	**SD**[b]	a_w	**EMC**[a]	**SD**[b]
LiCl	***0.113***	3.99	0.18	***0.113***	3.37	0.09	***0.112***	3.00	0.10
CH_3COOK	***0.234***	4.27	0.07	***0.225***	4.96	0.16	***0.201***	4.38	0.07
$MgCl_2$	***0.335***	6.10	0.08	***0.328***	5.39	0.11	***0.316***	5.67	0.16
K_2CO_3	***0.431***	7.31	0.04	***0.432***	6.60	0.04	***0.432***	6.59	0.16
$MgNO_3$	***0.574***	8.35	0.17	***0.529***	7.61	0.08	***0.484***	7.31	0.09
NaBr	***0.622***	8.97	0.03	***0.576***	8.34	0.13	***0.532***	7.68	0.12
NaCl	***0.757***	10.98	0.11	***0.753***	10.57	0.17	***0.747***	10.31	0.19
KCl	***0.868***	14.21	0.16	***0.843***	13.07	0.23	***0.823***	13.07	0.11

[a] Mean values based on three replications
[b] Average standard deviation values based on three replications

Table 2. Equilibrium moisture content EMC (%) of the Bulgarian Einkorn flakes for desorption process at different water activities (a_w) and at different temperatures t (°C).

	10°C			25°C			40°C		
Sel	a_w	**EMC**[a]	**SD**[b]	a_w	**EMC**[a]	**SD**[b]	a_w	**EMC**[a]	**SD**[b]
LiCl	***0.113***	4.91	0.17	***0.113***	4.14	0.06	***0.112***	3.48	0.08
CH_3COOK	***0.234***	5.96	0.15	***0.225***	5.79	0.06	***0.201***	5.18	0.03
$MgCl_2$	***0.335***	6.98	0.08	***0.328***	6.60	0.05	***0.316***	5.99	0.15
K_2CO_3	***0.431***	8.30	0.12	***0.432***	7.80	0.06	***0.432***	7.81	0.12
$MgNO_3$	***0.574***	9.51	0.09	***0.529***	8.21	0.05	***0.484***	8.20	0.04
NaBr	***0.622***	10.26	0.14	***0.576***	9.11	0.06	***0.532***	7.89	0.09
NaCl	***0.757***	10.89	0.21	***0.753***	10.83	0.10	***0.747***	10.46	0.18
KCl	***0.868***	13.54	0.09	***0.843***	11.71	0.14	***0.823***	12.47	0.37

[a] Mean values based on three replications
[b] Average standard deviation values based on three replications

The obtained values of equilibrium moisture content for both processes are in the range of 3.00% to 14.21%. If we were selected the storage regime under condition relative air humidity from 45% to 55% and temperature from 18°C to 25°C, we could recommend a moisture content from 6.59% to 9.51%. Furthermore, our results confirmed the thesis of many similar conclusions that with the increase of temperature, the EMC decrease in the exclusive presence of constant water activity [29-30, 36-37].

It was performed the graphical comparison between the adsorption and desorption isotherms at 10°C shewed in Figure 1.

These sorption isotherms have an *S*-shape profile as showed in Figure 1, i.e. they are from the IInd class typical of the majority of food products according to Brunauer's classification [38].

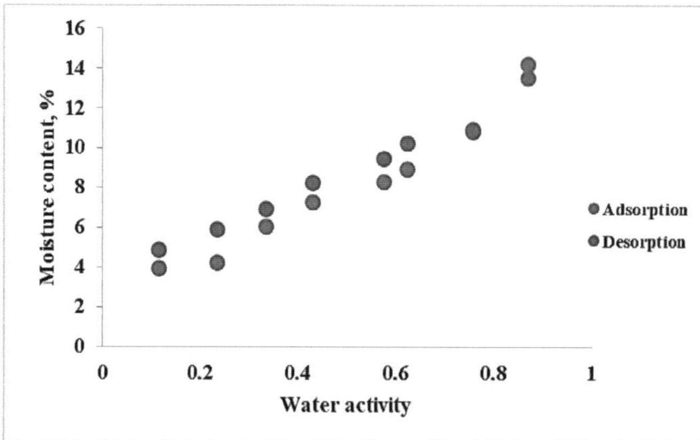

Figure 1. Comparison for both sorption process at 10°C

Coefficients of the three-parameter modified models (A, B, C), the corresponding mean relative error (P, %), standard error of moisture (SEM), residuals distribution and correlation coefficient (R^2) are presented in Table 3 and Table 4 for the two sorption process.

Table 3. Bulgarian Einkorn flakes: model coefficients (*A, B, C*), mean relative error (P, %), standard error of moisture (SEM), distribution of the residuals and correlation coefficient (R^2) for adsorption process.

Models	*A*	*B*	*C*	P	SEM	Residuals	R^2
Chung-Pfost	2067.646	372.129	0.268294	0.85	0.37	Random	0.9906
Oswin	***7.451963***	***-0.004156***	***0.351038***	***0.68***	***0.32***	***Random***	***0.9906***
Halsey	3.345982	-0.00161	1.861107	7.56	0.97	Non-random	0.9723
Henderson	0.000258	70.06473	1.662762	9.92	0.87	Non-random	0.9541

Table 4. Bulgarian Einkorn flakes: model coefficients (*A, B, C*), mean relative error (P, %), standard error of moisture (SEM), distribution of the residuals and correlation coefficient (R^2) for desorption process.

Models	*A*	*B*	*C*	P	SEM	Residuals	R^2
Chung-Pfost	1186.18	118.2324	0.301697	4.21	0.43	Random	0.9837
Oswin	***8.536256***	***-0.017735***	***0.266573***	***5.10***	***0.49***	***Random***	***0.9694***
Halsey	4.698046	-0.006131	2.360615	8.04	1.05	Non-random	0.9567
Henderson	0.000103	5.247957	2.421689	19.07	2.54	Non-random	0.6130

Based on the measuring and on the criteria suitability of the models, we could be recommended for satisfactory description of the Einkorn flakes sorption isotherms – the modified Oswin model for both processes. According to Chen and Morey, (1985) it does not exist a uniform model for the description of the sorption isotherms for all food products [33].

Monolayer moisture content (%) at 10°C, 25°C and 40°C for adsorption process are 4.50%, 4.36% and 3.26%, respectively for desorption processes – 4.92%, 3.96% and 3.68%.These results confirm the best condition for the stability of the Einkorn flakes. The values are calculated owing to the linearization of the Brunauer-Emmett-Teller equation and are showed in Figure 2 and Figure 3.

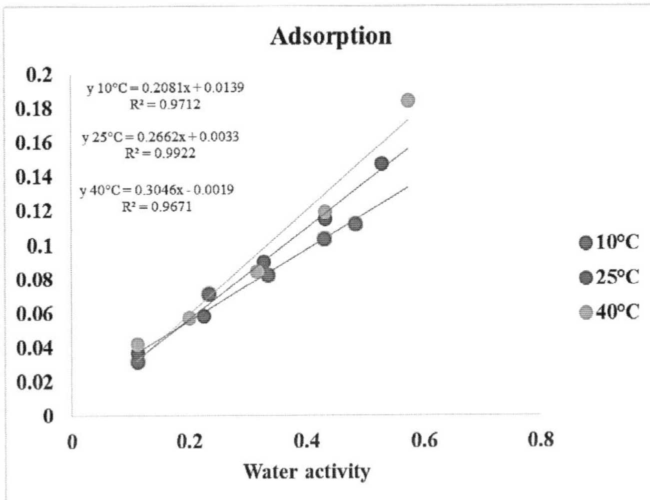

Figure 2. Bulgarian Einkorn flakes: linearization of BET equation for adsorption process.

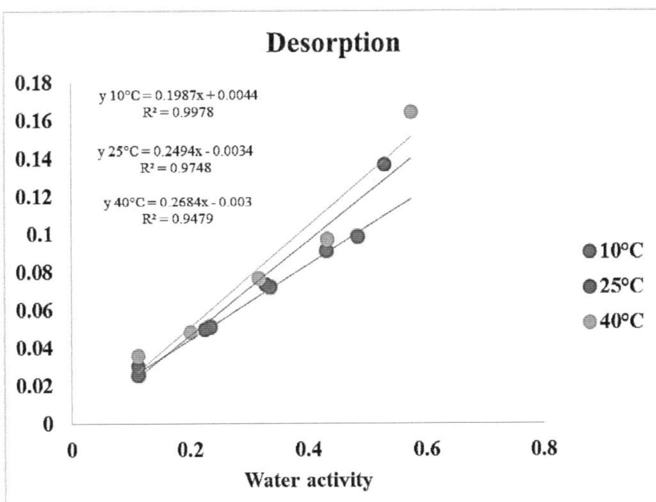

Figure 3. Bulgarian Einkorn flakes: linearization of BET equation for desorption process.

Based on the obtained results we could continue our experiment with confirmation and determination of the storage regime of Bulgarian Einkorn flakes. We will select the conditions imitated the condition on the store or market – relative air humidity from 44% to 55% and temperature between 18°C and 24°C where the Einkorn flakes will must have the equilibrium moisture content from 6.60% to 9.51%, according to table 1 and table 2. We will compare the duration of storage under MMC and of storage under the condition of EMC as in a study of Durakova et al, 2019 [39].

4. Conclusions

It was determined the equilibrium moisture content (EMC) of the Bulgarian Einkorn flakes (*Triticum monococcum*) for adsorption and desorption process at 10°C, 25°C and 40°C under the condition of different water activity in the range of 0.1% to 0.9%. We were graphical demonstrated that all obtained isotherms have the *S*-shape profile i.e. they are from the IInd class, according to Brunauer's classification.

According to the criteria for the suitability of the modified model we were recommended the modified Oswin model for the description of the obtained sorption data. The monolayer moisture content (MMC) was calculated through the linearization of the Brunauer-Emmett-Teller equation. MMC for the adsorption process is in the range from 3.26% to 4.50% and respectively for the desorption process – from 3.68% to 4.92%.

5. Acknowledgments
This study was part of the National Scientific Programme "Young Researchers and Postdocs" 2020 at the Technological faculty, University of Food Technologies – Plovdiv, Bulgaria.

6. References
[1] Barone F, Laghi L, Gianotti A, Ventrella D, Saa T, Bordoni A, Forni M, Brigidi P, Bacci M L, Turroni S, 2019 *Nutrients* **11** (1) p 16.
[2] Hidalgo A and Brandolini A 2014 *J Sci Food Agric.* **94** (4) pp 601-612.
[3] Goranova Zh , Baeva M, Stankov S. 2014 *Scien Works of UFT* **LXI** pp 891-895
[4] Dimov I and Stamatovska V, 2018 *Ukr. J. Food Sci.* **6** (1) pp 13-19
[5] Zencirci N, Ulukan H, Ordu B, Aslan D, Mutlu H T, Örgeç M, 2019 *IJSM* **6** (2), pp 113-128.
[6] Brandolini A and Hidalgo A 2011 *Flour and breads and their fortification in health and disease prevention* chapiter **8** pp 79-88 ISBN 978-0-12-380886-8 Elselvier Academic Press.
[7] Geisslitz S, Longin C F H, Scherf K A, Koehler P 2019 *Foods* **8** (9) pp 409.
[8] Hidalgo A, Ferraretto A, De Noni I, Bottani M, Cattaneo S, Galli S, Brandolini A *2018 Food chem* **240** pp 799-807
[9] Abdel-Aal E S and Hucl P 2002 *J Food Compost Anal* **15** (6) pp 737-747.
[10] Akar T, Cengiz M F, Tekin M 2019 *Qual Assur Saf Crop* **11** (2) pp 191-200.
[11] Hidalgo A, Lucisano M, Mariotti M, Brandolini A 2019 *J Food Process Pres* **43** (9) e14079.
[12] Abdel-Aal E-S M, Young J C, Wood P J, Rabalski I, Hucl P, Falk D, Fregeau- Reid J, 2002 *Cereal Chem.* **79** pp 455-457
[13] Arzani A and Ashraf M 2017 *Compr. Rev. Food Sci. Food Saf.* **16** (3) pp 477-488.
[14] Nakov G, Brandolini A, Ivanova N, Dimov I, Stamatovska V 2018 *J. Cereal Sci.* **83** pp 116-122
[15] Izambaeva A, Bozadjiev B, Gogova T, Durakova A, Dessev T, Koleva A, Krasteva A, 2016 *Bulg J Agric Sci* **22** (2), pp 331-338
[16] Fast R B 1990 *Breakfast cereals and how they are made* pp 15-42.
[17] Guo W, Wang S, Tiwari G, Johnson J A, Tang J 2010 *LWT-Food Sci Technol.* **43** (2) pp 193-201
[18] Labuza T P 1980 *Food Technol* **34** (4) pp 36-41
[19] Muangrat R, and Nuankham C 2018 *Food ScI Nutr* **6** (3), pp 585-593
[20] AOAC 1990 Official methods of analysis. Assoc Anal Chem.
[21] BDS EN ISO 14431:1978 *Bulgarian Institute for Standardization*
[22] BDS EN ISO 6997:1984 *Bulgarian Institute for Standardization*
[23] BDS EN ISO 7169:1989 *Bulgarian Institute for Standardization*
[24] BDS EN ISO 7646:1982 *Bulgarian Institute for Standardization*
[25] European Commission 2011 Regulation (EU) No 1169/2011 *Official Journal of the European Union* **54** pp 18-61
[26] Wolf W, Spiess W E L, Jung G, 1985, COST-Project 90 and 90 bis, *Properties of Water in Foods in Relation to Quality and Stability Martinus Nijhoff*, Dordrech, pp 661-679.
[27] Bell L and Labuza T 2000 *St. Paul: American Association of Cereal Chemists*, Inc., pp 33-36.
[28] Feng C, Janssen H, Wu C, Feng Y, Meng Q 2013 *Build Environ* **69** pp 64-71
[29] Durakova A 2004 *Bulg J Agric Sci* **10** pp 95-98.
[30] Bogoeva A 2020 Sorption characteristics of floury mixtures enriched with grape seeds flour of Bulgarian and French raw materials *JCEA* (*Preprint* ID 2711)
[31] Igathinathane C, Womac A, Sokhansanj S, Pordesimo L 2005 *Transactions of the ASAE* **48** (4)

pp 1449-1460.
[32] Lomauro C J, Bakshi A S, Labuza T P 1985 *LWT-Food Sci Technol* **18** (2) pp 118-124
[33] Chen C-C and Morey R V 1989 *Transactions of the ASAE* **32** (3) pp 983-0990
[34] Brunauer S, Emmett P H, Teller E 1938 *JACS* **60** (2) pp 309-319
[35] Timmermann E O 2003 *Colloid Surf A Physicochem Eng Asp* **220** (1-3) pp 235-260.
[36] Al-Muhtaseb A H, McMinn W A M, Magee T R A 2002 *Food Bioprod Process* **80** (2) , pp 118-128.
[37] Lemus R M 2011 *Vitae* **18** (3) pp 325-334
[38] Brunauer S, Deming L S, Deming W E, Teller E 1940 *JACS 62* (7) pp 1723-1732
[39] Durakova A, Bogoeva A, Yanakieva V, Gogova Tzv, Dimov I, Krasteva R, Dosheva K 2019 *Bulg J Agric Sci* **25** (6) pp 1287-1292

IOP Conf. Series: Materials Science and Engineering **878** (2020) 012057 doi:10.1088/1757-899X/878/1/012057

Tensile testing of Inconel 600 wire at high temperatures

N Kuzmanov[1,3], B Borisov[1] and I Muhtarov[2]

[1]Technical University – Sofia, "Strength of Materials" Department, Bulgaria
[2]VIB Engineering, Bulgaria

[3]E-mail: nkuzmanov@tu-sofia.bg

Abstract. This paper presents how to determine some mechanical properties of wires subjected to tensile testing. Additional equipment was added to an existing tensile testing machine for wires at high temperatures. A tensile testing procedure for wires at high temperature is described. Tensile tests of Inconel 600 wire were conducted at 700, 800 and 900 °C. The ultimate strength, the yield strength and the percentage elongation after fracture are determined. Analysis of the gathered results is done.

1. Introduction

Composite materials are becoming more widely used in different industries. Their share rapidly increases in automotive, boatbuilding, aircraft and military industries, building of skis and other sports equipment, as well as in production of construction materials. The composite materials are also known as reinforced materials because they often consist of reinforcing fibers (wires) inside polymer, wooden, ceramic, metal or other matrix. The composites are materials of which lightweight parts can be made, having high strength at the same time.

One of the major applications of Inconel wire is as reinforcing component in composite materials.

Stress-stain analysis of parts made from composite materials is most often done using numerical analysis. This kind of analysis demands the mechanical properties of the materials of which the matrix and the reinforcing fibers are made at the working conditions of the relevant part.

The mechanical properties of Inconel 600 at room temperature vary in wide ranges depending of the production technology – table 1 [1]. This table shows mechanical properties of Inconel 600 wires with diameter varying from 1,6 mm to 6,4 mm.

The study of material behaviour at temperatures higher than 600 °C is accompanied by overcoming significant difficulties set by the reliability of used devices and maintaining specified test conditions for a long period of time. Therefore, little data is available in literature for the materials behaviour at high temperatures.

The elasticity modulus, the shearing modulus and the Poisson's ratio of Inconel 600 at temperatures up to 1000 °C are given in table 2.

Figure 1 shows the mechanical properties of cold-drawn rod of Inconel 600, determined at high temperatures tensile tests.

The literature does not provide data for the yield strength, the tensile strength and the strain of Inconel 600 wires at high temperatures.

The following problem arises – to determine the yield strength, the tensile strength and the strain of Inconel 600 wire at 700 °C, 800 °C and 900 °C. A cold-drawn wire with diameter of 0,25 mm is used. The chemical composition of the wire material is given in table 3.

Table 1. Mechanical properties of Inconel 600 at room temperature.

Form and Condition	Tensile Strength		Yield Strength (0.2% Offset)		Elongation, %	Hardness, Rockwell
	ksi	MPa	ksi	MPa		
Rod and Bar						
Cold-Drawn						
Annealed	80-100	550-690	25-50	170-345	55-35	65-85B
As-Drawn	105-150	725-1035	80-125	550-860	30-10	90B-30C
Hot-Finished						
Annealed	80-100	550-690	30-50	205-345	55-35	65-85B
Hot-Finished	85-120	585-830	35-90	240-620	50-30	75-95B
Plate						
Hot-Rolled						
Annealed	80-105	550-725	30-50	205-345	55-35	65-85B
As-Rolled	85-110	580-760	35-65	240-450	50-30	80-95B
Sheet						
Cold-Rolled						
Annealed	80-100	550-690	30-45	205-310	55-35	88B max.
Hard	120-150	830-1035	90-125	620-860	15-2	24C min.
Strip						
Cold-Rolled						
Annealed	80-100	550-690	30-45	205-310	55-35	84B max.
Spring Temper	145-170	1000-1170	120-160	830-1100	10-2	30C min.
Tube and Pipe						
Hot-Finished						
Hot-Finished	75-100	520-690	25-50	170-345	55-35	-
Annealed	75-100	520-690	25-50	170-345	55-35	-
Cold-Drawn						
Annealed	80-100	550-690	25-50	170-345	55-35	88B max.
Wire[b]						
Cold-Drawn						
Annealed	80-120	550-830	35-75	240-520	45-20	-
No. 1 Temper	105-135	725-930	70-105	480-725	35-15	-
Spring Temper	170-220	1170-1520	150-210	1035-1450	5-2	-

Table 2. Elasticity modulus, shearing modulus and Poisson's ratio of Inconel 600.

Temperature	Young Modulus	Shear Modulus	Poisson's Ratio
°C	GPa	GPa	
22	214	80.8	0.324
100	210	79.6	0.319
200	205	78.0	0.314
300	199	76.2	0.306
400	193	74.2	0.301
500	187	71.9	0.300
600	180	69.2	0.301
700	172	65.9	0.305
800	164	62.1	0.320
900	154	57.9	0.330
1000	143	53.4	0.339

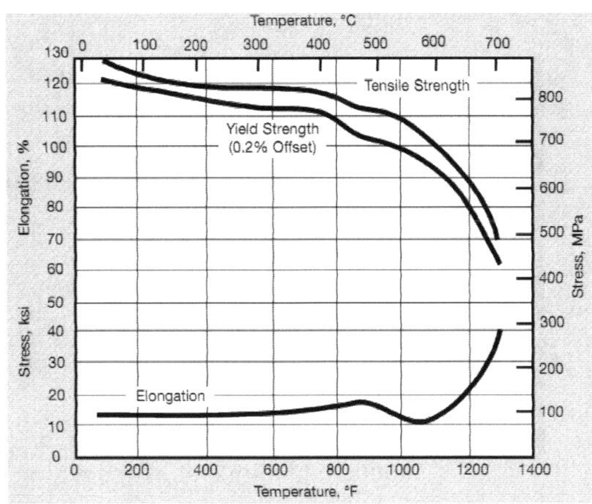

Figure 1. High-temperature tensile properties of Inconel 600 cold-drawn rod.

Table 3. Chemical composition of the tested wire (in %).

Ni	*Cr*	*Fe*	*Si*	*Co*	*Mn*	*Cu*	*C*	*S*
≥ 72,0	15,5	8	0,5	< 1	< 1	< 0,5	< 0,15	< 0,015

2. Metallic materials. Tensile testing. Method for tensile testing at high temperature

2.1. Test pieces with diameter less than 4 mm
The initial gauge length (L_o) must be (200 ± 2) mm or (100 ± 1) mm [2]. The distance between the grips of the machine must be no less than (L_o + 20) mm. If test pieces with small diameter are tested, the distance between the grips of the machine may be equal to L_o [3].

2.2. Testing conditions
The test piece must be heated to a predetermined temperature and it must be held at this temperature for no less than 10 minutes.

The strain rate of test piece working area should be 0,001 min^{-1} to 0,005 min^{-1} from the beginning of the test until the yield strength is reached. After the yield strength is reached strain rate of 0,20 min^{-1} is allowed [4].

2.3. *Determination of proof strength, plastic extension for given percentage elongation*
Proof strength, plastic extension (R_p) for given percentage elongation is determined from the force-extension curve of the material by drawing a line parallel to the linear portion of the curve at distance from it equivalent to the given percent of the percentage elongation, for example 0,2 %. The ordinate of the point at which this line crosses the stress-percentage elongation curve corresponds to the proof strength, plastic extension. An index showing the percentage of elongation is added to the designation of the yield strength for example $R_{p0,2}$.

2.4. Determination of percentage elongation after fracture
The two broken parts of the test piece must be carefully fitted back together so that their axes lie in a straight line. The final gauge length is determined (L_u).

The elongation after fracture (L_u - L_o) to the nearest 0,25 mm or better, should be determined using a measuring device with sufficient resolution.

The percentage elongation after fracture (A) is calculated using:

$$A = \frac{L_u - L_o}{L_o} 100, \ \% .$$

Index to the designation A must be added witch shows the used original gage length in mm, for example: $A_{100\,mm}$ is percentage elongation for original gauge length (L_o), equal to 100 mm.

3. Adding of additional equipment to existing tensile testing machine of wires at high temperatures

3.1. Available machine
In “Strength of materials” department at Technical University Sofia, a tensile testing machine at high temperatures is available – figure 2.

3.1.1. Tensile testing machine – it has the following features:

- The loading module of the machine is made using stepper motor, reduction gear and lead screw. To ensure the smooth and precise application of the force created by the lead screw, a reduction gear with large transfer ratio ($i = 50$) and stepper motor with large number of steps (200) per revolution are used;
- The grips of the machine (called upper and lower stem) are designed to test cylindrical test pieces with 6 mm diameter of the working area;
- The load cell of the machine is designed to measure loads up to 10 kN;

- An elastic body is present which assures smooth loading and minimizes the shock after the test piece fractures;
- The tensile testing machine complies with the requirements of EN ISO 7500-1 standard.

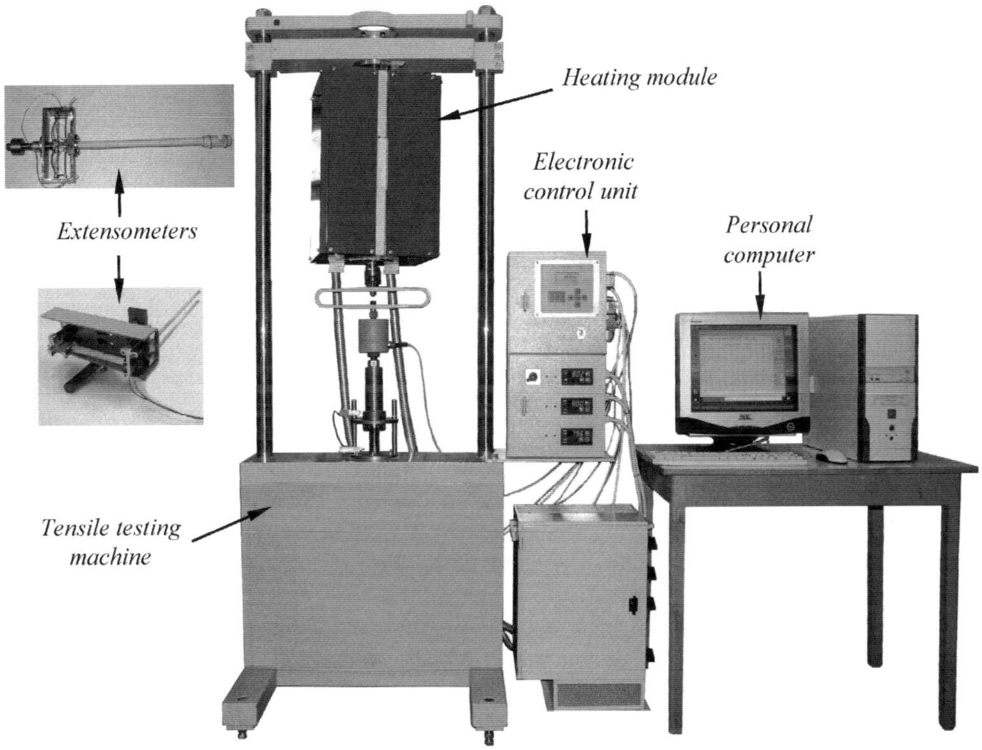

Figure 2. Available tensile testing machine at high temperatures.

3.1.2. Heating module – it has the following features:
- It assures heating of the test piece up to 1100 °C;
- To assure even distribution of temperature field the heaters are located in three sections each of which is controlled by a separate thermocouple and separate thermo-regulator [5]. Each section is comprised of two heaters (one situated in the left part of the module and the other - in the right part);
- The heating module complies with the requirements of BDS EN ISO 6892-2 standard [6].

3.1.3. Extensometers – they are designed to be attached to cylinder test pieces with diameter of the working are of 6 mm.

3.1.4. Electronic control unit – it controls the testing parameters: force and temperature [7].

3.1.5. Personal computer – the electronic control unit sends in real time all of the measured signals to the computer. The computer records processes and visualizes the received information using specialized software. The parameters processed by the software are: assigned force and real measured force, longitudinal strain, transverse strain, temperatures in all three sections of the heating module and the number of steps made by the stepper motor.

3.2. New additional equipment

To realize tensile testing of wires at high temperatures with the available testing machines, the following is required:

- To design and build new load cell, which allows accurate measurement of loads up to 0,1 kN;
- To design and build gripping device to test wires at tensile loading at high temperatures;
- To design and build new stems (upper and lower), which allow to attach the new gripping devices to the testing machine;
- The available extensometers cannot be attached to the wire under the test. It is planned the longitudinal strain to be measured using the stepper motor. To do this, the elastic body of the tensile testing machine should be removed and the lower stem of the machine should be extended with additional stem, which is connected directly to the load cell.

Figure 3 shows 3D model of the designed strain gauge load cell. It is comprised of body *1*, cap *2* and electrical plug *3* to send the measured signal to the testing machine. The load cell body is also used as elastic element and strain gauges *4* are mounted on it. The strain gauges are connected in “full bridge” circuit. This assures compensation for temperature influence on the measurement and manufacturing inaccuracies of the elastic body. The load cell body is made of 1.7045 steel.

Figure 4 shows 3D model of the gripping device for tensile testing of wires at high temperatures. The device comprises of two holding pieces – upper holding piece *1* and lower holding piece *2*. Each holding piece consists of two halves, pin *3* and screws *4*. The pins have diameter of 5 mm and radial holes are drilled through which the wire passes. A horizontal and vertical channels are machined in the holding pieces. The horizontal channels are needed to mount the pins. The vertical channels allow the wire to pass without touching the holding pieces. The two halves of the holding pieces are attached to each other using four screws M6x20. The griping device is made of temperature-resistant alloyed steel with high ultimate strength.

The shape and the size of the upper holding piece of the grip are designed taking into account the already available upper stem of the tensile testing machine (designed for testing 6 mm test pieces). It is necessary to build only one new lower stem for the tensile testing machine – figure 5. The upper holding piece is attached to the upper stem of the testing machine using attachment head. The lower holding piece is attached to the lower stem of the machine using thread connection. The machine’s lower stem is made of temperature-resistant alloyed steel with high ultimate strength.

Figure 5 shows 3D model of the extension stem with which the lower stem of the machine is extended. The connection between the load cell and the lower stem is made using thread connection. For this purpose, an adapter is needed, also shown on figure 5. The lower stem and adapter are build using common carbon steel S355JR.

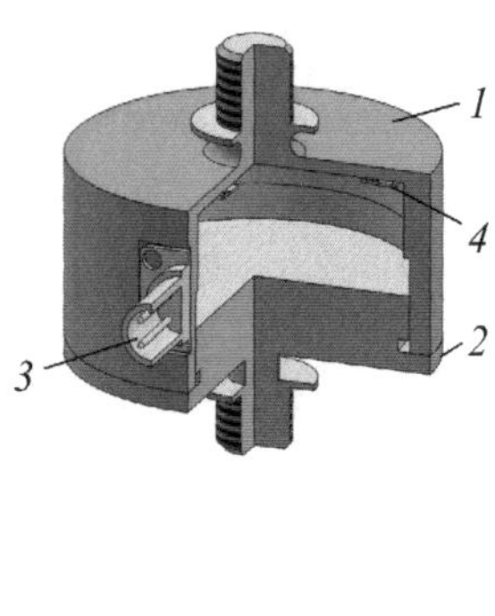

Figure 3. 3D model of the load cell.

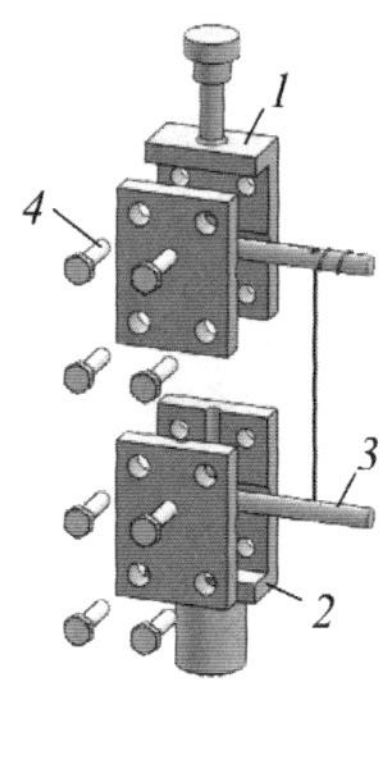

Figure 4. 3D model of the gripping device for tensile testing of wires.

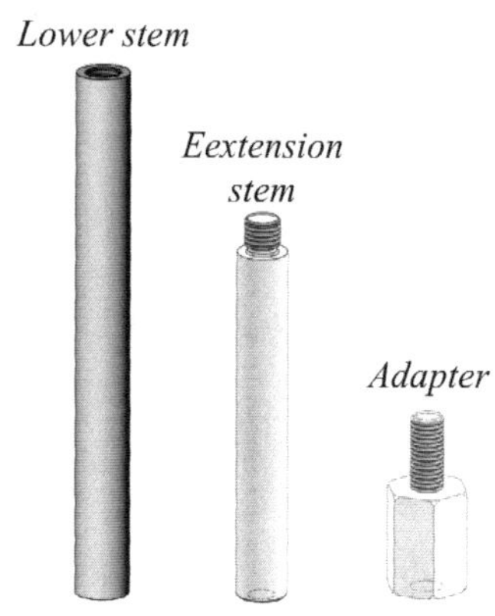

Figure 5. 3D models of the lower stem, the extension stem and the adapter.

Figure 6 shows a photograph of the machined and mounted additional equipment needed for tensile testing of wires at high temperatures with the available tensile testing machine.

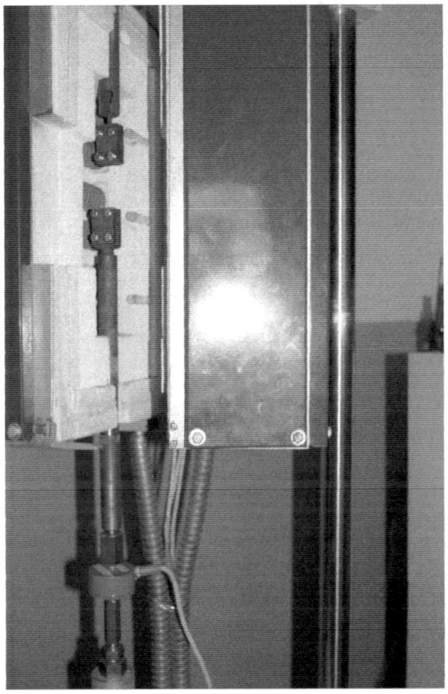

Figure 6. Photograph of the machined and mounted additional equipment.

3.3. Sequence for tensile testing of wires at high temperature

3.3.1. The wire is mounted in the pins – figure 7. Steps:

- A wire with 200 mm length is cut;
- The wire is passed through the radial hole of the first pin in such a way that it extends beyond the pin by 45 mm;
- The hole of the pin is filled with Pyro-Paint, while making sure that the wire doesn't touch any edges of the hole. It is necessary to wait for 2 hours for Pyro-Paint to dry out. This procedure allows to avoid stress concentration caused by the hole edge during tensile testing.
- The previous two steps are made for the other side of the wire;
- The wire with the two pins attached, is heated to 93 °C for 2 hours (this procedure is recommended by Pyro-Paint manufacturer);
- The two parts of the wire with 45 mm length each are wound to each of the pins;
- On each of the wound wires a few drops of stain gauge adhesive X60 are dropped (this assures easy mount of the pins in the holding pieces);
- The wire is straightened and the length of the wire between the pins is measured (the original gauge length L_o is needed to calculate the percentage elongation after fracture A) – L_o = 100 mm.

3.3.2. The pins with the wire are mounted in the holding piece and the attaching screws are tightened – figure 8.

3.3.3. The grips are mounted in special plastic brackets to prevent deformation of the wire during its mount in the heating module – figure 9.

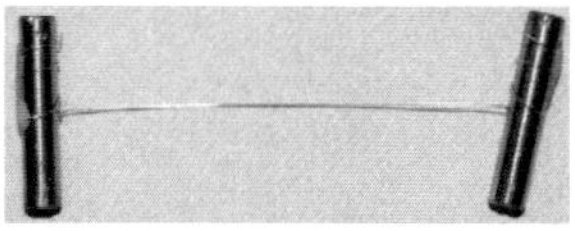

Figure 7. Mount of the wires in the pins.

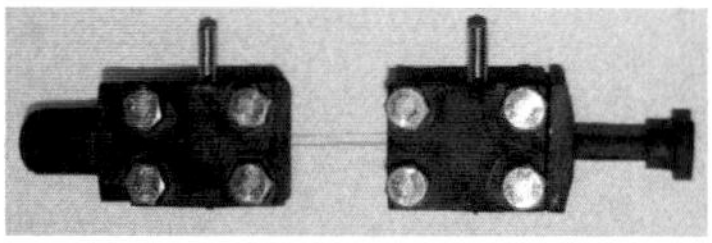

Figure 8. Mount of the wire and pins to the grips.

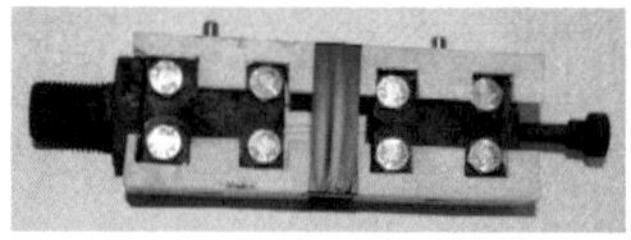

Figure 9. Mount of the grips to the plastic brackets.

3.3.4. The grips with the wire are mounted to the stems of the tensile testing machine and the brackets are removed – figure 6.

3.3.5. The desired testing temperature is set and the heating module is switched on, it is necessary to wait for the heating module to reach the set temperature after which additional 10 min have to pass.

3.3.6. The measuring equipment is zeroed.

3.3.7. The wire under test is subjected to pure tension until its fracture – during the loading, continuously the values of the force and the number of the steps made by the stepper motor are recorded (the number of steps is used to measure the strain of the test length on the wire).

3.3.8. The stress – percentage elongation curve of the material is drawn and the values of $R_{p0,2}$, R_m and A_{100mm} are determined.

4. Tensile testing

Twelve wires were tested (3 at room temperature, 3 wires at 700 °C, 3 at 800 °C and 3 at 900 °C).

Figure 10 shows the gathered stress-percentage elongation curves. Table 4 presents the values of the determined mechanical properties are given.

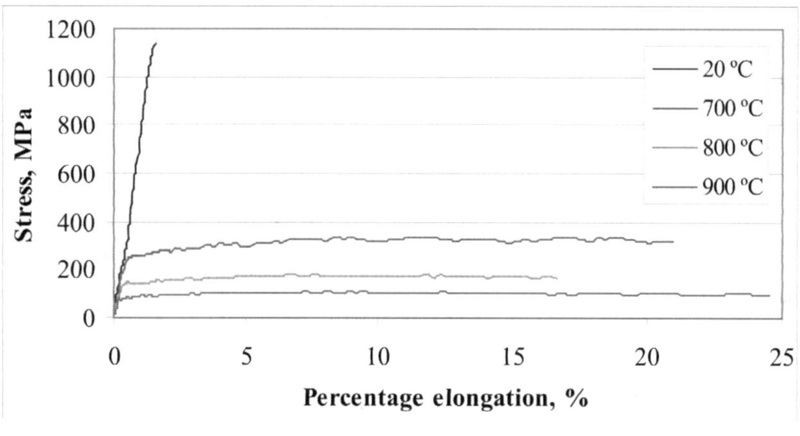

Figure 10. Stress-percentage curves.

At room temperature the Inconel 600 wire has the behaviour of brittle material, while at high temperatures it has the behaviour of ductile material.

As expected, the values of the stress properties decrease with the rise of the temperature. It is interesting to note that at 800 °C the longitudinal percentage elongation is less than the longitudinal percentage elongation at 700 °C, while at 900 °C we have the largest value of longitudinal percentage elongation. It is highly probable that this is due to structural change in the material. Such decrease followed by rise of the longitudinal percentage elongation with respect to the rise of the temperature is gathered in [1] – figure 1 – for Inconel 600 cold-drawn rod.

Table 4. Mechanical properties.

T, °C	Wire №	R_m, MPa	$R_{p0,2}$, MPa	A_{100mm}, %
20 °C	1	1136	-	2,2
	2	1118	-	1,8
	3	1124	-	2,4
	Average	**1126**	-	**2,1**
700 °C	1	338	260	20,4
	2	318	256	23,1
	3	324	249	21,9
	Average	**327**	**255**	**21,8**
800 °C	1	180	142	16,1
	2	190	150	14,2
	3	184	138	15,5
	Average	**185**	**143**	**15,3**
900 °C	1	102	80	27,2
	2	108	78	24,3
	3	104	74	25,1
	Average	**105**	**77**	**25,5**

5. Conclusion

Some mechanical properties of Inconel 600 are determined at 700, 800, 900 °C. The collected results shall be used for:

- Stress-strain analysis of composite materials (materials reinforced with Inconel 600 wires) exposed to high temperatures for short period of time.
- Determination of the impact of heat treatment of the Inconel 600 wire on its stress properties.
- Further study of the behaviour of Inconel 600 wire exposed to high temperatures for prolonged period of time.

The authors would like to thank the Research and Development Sector at the Technical University of Sofia for the financial support.

References

[1] Special Metals Corporation 2008 Inconel alloy 600 *Copyright © Special Metals Corporation* SMC-027

[2] EN ISO 6892-2:2018 Metallic materials - Tensile testing - Part 2: Method of test at elevated temperature.

[3] EN 10002-5:2000 Metallic materials - Tensile testing - Part 5: Method of testing at elevated temperature

[4] Tsonev V 2010 Determination of mechanical characteristics of a new alloy steel at high temperatures and static load *Proceedings of BulTrans-2010 (Sozopol, Bulgaria)* 209-211

[5] Nikolov N, Tsonev V, Lazov L and Georgiev V 2009 Design of a heating device for long material testing at high temperature *International Scientific Conference of the Faculty of Machine and Power Engineering* **1** 187-192

[6] Tsonev V and Nikolov N 2009 Experimental study on heating device for testing materials, intended for a long work at increased temperatures *Proceedings of RaDMI 2009 (Vrnjačka Banja, Serbia)* **1** 401-406

[7] Nikolov N, Tsonev V and Lazov L 2010 Controlling of stand for materials testing at high temperatures *Proceedings of AMO '2010 (Sv. Konstantin i Elena, Bulgaria)* 315-319

Application machining robots in processing operations

S Guergov, N G Nikolova and G Stefanov

Technical University - Sofia, FIT, FA, 8 Kl. Ohridski Blvd., 1000 Sofia, Bulgaria
sguergov@tu-sofia.bg, ninan@tu-sofia.bg, gommes@abv.bg

Abstract - An approach for the realization of various technological operations (drilling, milling, boring, grinding, etc.) on the basis of a specially designed execution technological system and an industrial robot has been developed. In the end effector of it, instead of a complex and large-sized tool turret, the workpiece is established. The method of producing geometrical surfaces of the workpiece is through the following formative motions: the feed motions are performed by the robot and the primary cutting motions by the tools, which are clamping in the turrets of the technological modules.

Keywords — machining robots, machine tools, industrial robots, processing operations.

1. Introduction

Created to be auxiliary, the idea of using robots to perform specific technological operations such as welding, assembly, coating, etc. is enforced [1,2,3].

This possibility is further enhanced by the rapid development of computer equipment, which creates prerequisites for the synchronization of the motions of the end effector, the kinematic and the dynamic capabilities of the robots when performing complex motions and even to perform some single technological operations, such as milling, drilling, polishing, grinding, deburring, sheet cutting, etc. [4.5].

Using a single robot as a technological one should not mean that only the robot holds the tool. Many applications work better when the robot holds the workpiece, bringing it to a stationary tool. This increases flexibility by allowing the robot to perform many different operations [6].

2. Formulation the problem

There are two approaches of using a single robot to perform multi-operational machining operations:

1) using a complicated end effector (analogous to a tool turret) to locate the cutting tools 'figure 1';

2) using a universal gripper for clamping the workpiece and the working motions to be performed by separate executive technology modules/machines 'figure 2'.

The advantage of the second approach is the reduction or elimination of the number of bases of the workpiece. This reduces errors, as well as the cycle time needed to process the workpiece.

During multi-operation machining, through using various tools (such as lathe tools, milling cutters, drills, grinding wheels, etc.), it requires a multifunctional end effector analogous to the DST-tool turret [7] 'figure 1'. The figure shows that the use of such end effector will greatly increase the requirements for the robot in terms of load capacity, stability, dynamic characteristics, etc.

For the above reasons, the second approach is preferred, with the following formatting movements being performed to obtain the geometric surfaces of the workpiece: the feed motions are performed by the robot and the primary cutting motions by the tools, which are clamping in the turrets of the technological modules.

Figure 1. Tool turret

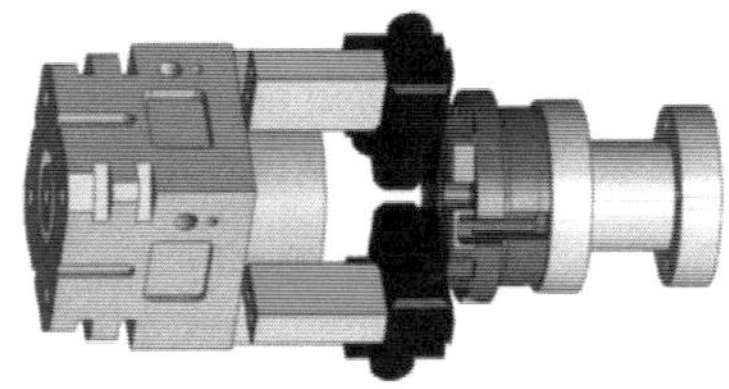

Figure 2. End effector

The decision to use industrial robot and specialized technological modules through optimal control in on-line mode with the ability to maintain the set quality of the workpiece is distinguished by significantly lower value of inputs, higher flexibility with possibilities for dynamic technological reconfiguration when performing various technological operations.

This article aims to develop the hardware and software environment of a robotic system for performing various technological operations (drilling, milling, boring, grinding, etc.) on the basis of a specially designed execution technological system and robot (in this case M430i-A/4FH) 'figure 3'.

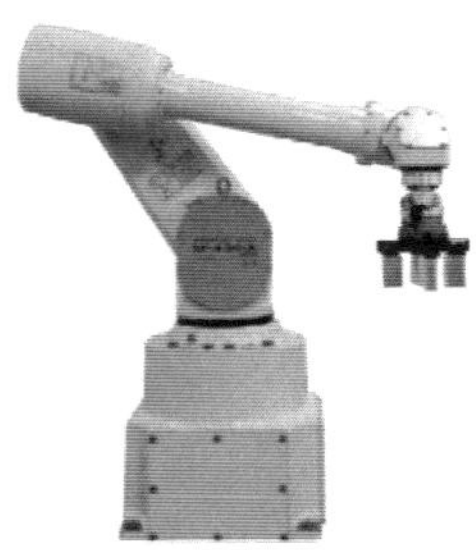

Figure 3. Robot Fanuc M430i-A/4FH

3. Operating environment

The executive technological system 'figure 4' includes: three working positions providing rotary motion, in which tools such as drills, milling machines, grinding discs, countersinks, grooves, etc. can be established; one position (turning knife) for turning and grinding operations and one rotating high-speed position with rotary motion (up to 30000 min^{-1}) for finishing operations, etc.

The specialized software package of FANUC Roboguide, which has very good hardware and software compatibility, flexibility and versatility, is used for software and information security of the entire system.

The package has a rich library consisting of a wide variety of robot models, static objects, obstacles, conveyors, metal cutters and more. It allows instructions to be generated for managing peripherals devices, defining states for logical structures execution, and communication with sensors, etc. Furthermore, its software compatibility creates preconditions for integrating external CAD models (e.g. through Solid Works) and optimizing constructive and technological solutions (e.g. through MATLAB).

The custom software application enables possibility for detailed modelling of the work environment, creating and simulating work movements, computer animation which supports validation and verification of the generated process, calibration of the virtual environment to the actual technology, communication with the robot control system, etc.

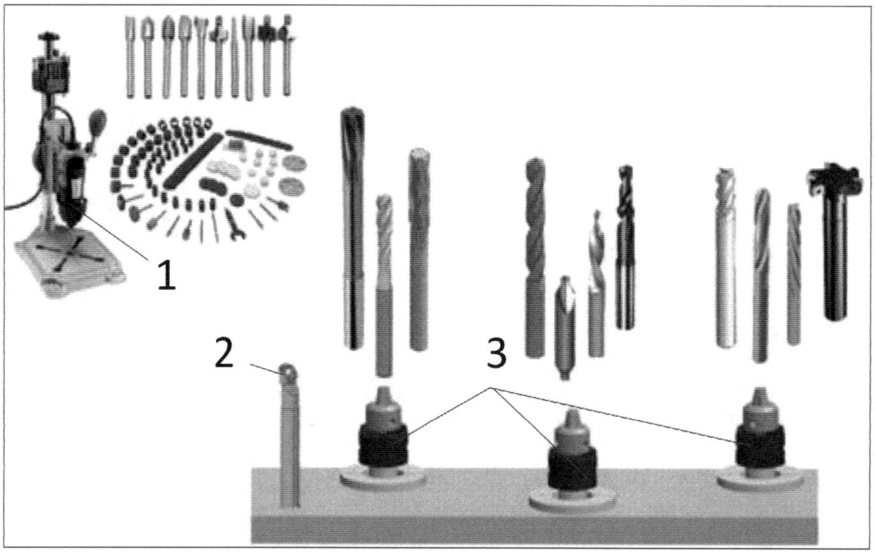

Figure 4. Executive technology system

Figure 5 shows the general architecture of the entire system, which includes the user software, interface and the elements of the multifunctional robotic technological system.

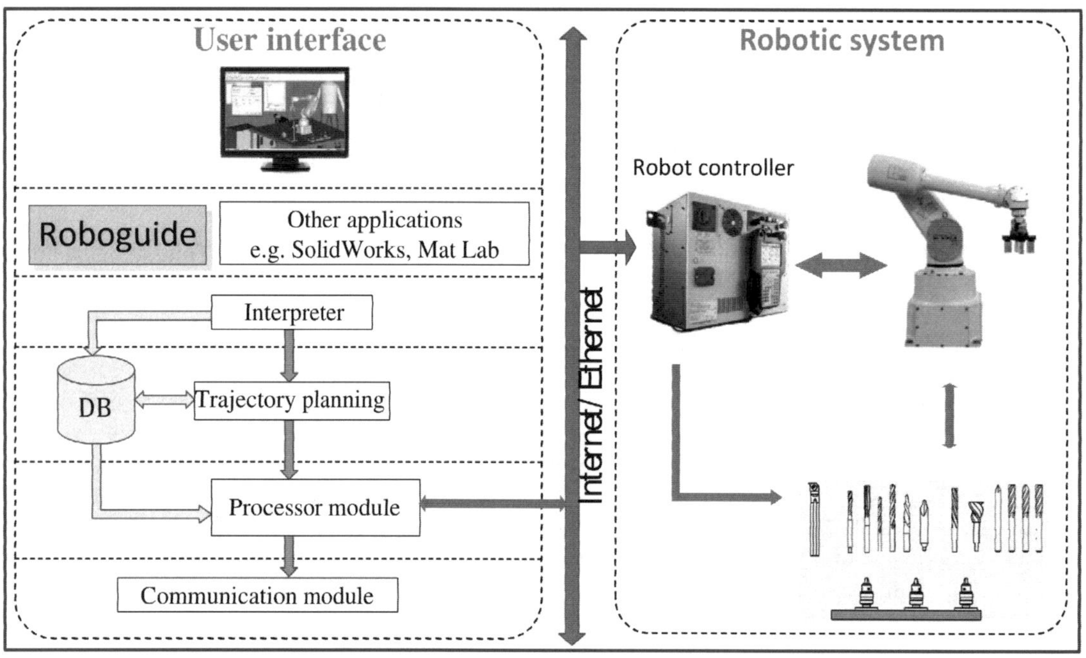

Figure 5. System architecture

The sequence for realizing the workflow is as follows:

- creating a virtual model 'figure 6b' of the working scene 'figure 6a' in Roboguide environment;
- performing a procedure to calibrate the virtual model;

- determining the geometric elements and processing modes of the workpiece from the technological documentation;
- analysis of trajectory movements, creation of logical control structures, while taking into account the forces arising from cutting processes;
- defining process variables;
- training of job positions 'figure 7';
- simulation and testing 'figure 6c';
- adjustments making;
- workflow in a real environment 'figure 6d'.

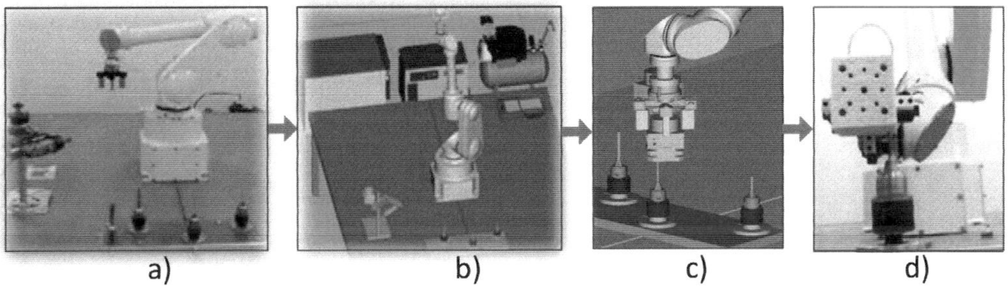

Figure 6. Workflow logic

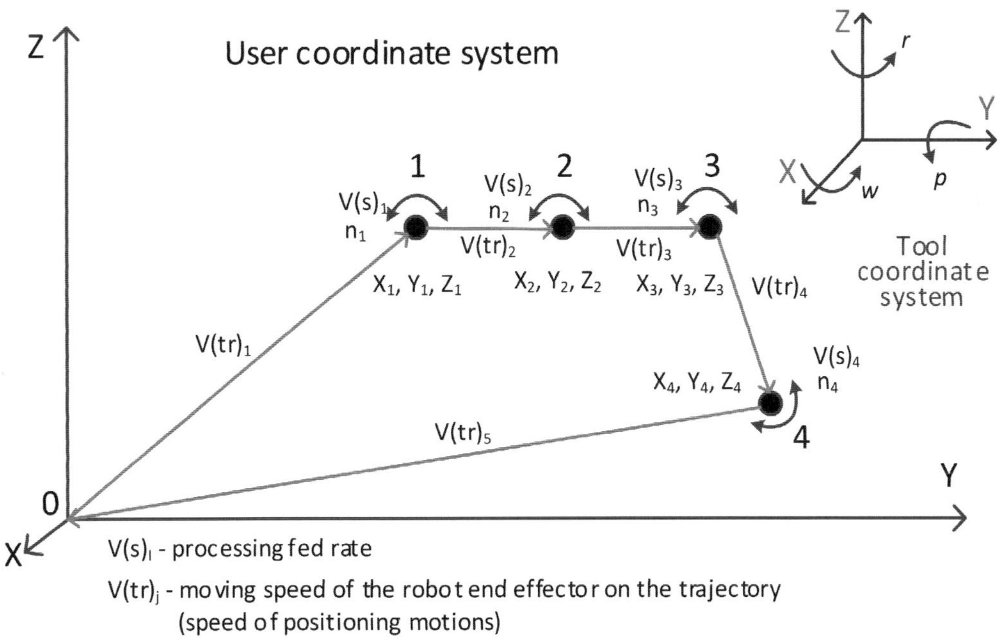

Figure 7. Trajectory and motions of the robot end effector

For machining the workpiece 'figure 2', which has holes, bevels, chamfers, grooves, surfaces etc., different technological operations (drilling, milling, etc.) are required, as well as control programs for workpiece production are developed, various auxiliary subroutines for activation and adjusting movements of the gripper and a main program that determines the sequence of execution of the technological operations.

In 'figure 8' (as an example) is given the control program as well as a fragment of the machined surface by radial channels milling.

```
    1: !FANUC Robotics America;
    2: !ROBOGUIDE Generated This
TPP;
    3: !Run SimPRO.cf to setup
frame and;
    4: UTOOL_NUM[GP1]=1;
    5: UFRAME_NUM[GP1]=4;
    6:              PR[24,1:STEP1
STP]=R[8:CHAMF1 X];
    7:              PR[25,1:STEP1
EP]=R[8:CHAMF1 X];
    8:      PR[26,1:STEP1    STP
ABOVE]=R[8:CHAMF1 X];
    9:      PR[27,1:STEP1     EP
ABOVE]=R[8:CHAMF1 X];
    10:J P[1] R[11:Vfast]% FINE;
    11:L  P[2]  R[11:Vfast]mm/sec
FINE;
  .......................................

  P[3]{
    GP1:
      UF : 4, UT : 1,
      CONFIG : 'F U T, 0, 0, 0',
      X =  -76.250  mm,      Y =  -
      30  mm,     Z =  200.000  mm,
      W =    0.000 deg,P =    90.000
deg,  R =    0.000 deg
  };
  P[4]{
    GP1:
      UF : 4, UT : 1,
      CONFIG : 'F U T, 0, 0, 0',
      X =  -76.250  mm,      Y =  -
178.830  mm,      Z =  117.890  mm,
      W =   90.000deg,P =  -45.000
deg,  R =   90.000 deg
  };
  /END
```

Figure 8. Control program for milling of radial channels

Conclusions

For machining the workpiece 'figure 2', which has holes, bevels, chamfers, grooves, surfaces etc., different technological operations (drilling, milling, etc.) are required, as well as control programs for workpiece production are developed, various auxiliary subroutines for activation and adjusting movements of the gripper and a main program that determines the sequence of execution of the technological operations.

In 'figure 8' (as an example) is given the control program as well as a fragment of the machined surface by radial channels milling.

This solution to use a technology robot and specialized modules, optimized on-line control with the ability to maintain the defined quality of the workpiece, has a significantly lower investments value, more flexibility with opportunities for dynamic technological reconfiguration when performing various technological operations.

Acknowledgment

The authors would like to thank the Bulgarian National Science Fund for the financial support by project КП-06-Н37/17.

References

[1] Waurzyniak P 2009 Machining with Robots SME Manufacturing Engineering magazine https://advancedmanufacturing.org/machining-with-robots/
[2] J&AP Scott Automation Ltd Robotic Milling/Machining Systems 0918581 https://www.scottautomation.com/products/milling-machining/
[3] Dikov R 2002 Simulation model of the machining of plane profile surfaces by traveling over *Proc. Int. Conf. on ICMEN* (Kassandra-Chalkidiki Greece) pp 281-289
[4] Iglesias I, M A Sebastián, J E Ares 2015 Overview of the State of Robotic Machining: Current Situation and Future Potential MESIC 2015 *Procedia Engineering* 132 pp 911– 917
[5] Dimitrov K, Y Velchev 2018 Infrared Array Sensor Wireless Network, HiTech 2018- *Proceedings 8566308 CFP18Q62-ART* Code 143562
[6] https://www.acieta.com/automation-application/material-removal/
[7] https://www.directindustry.com/prod/pragati/product-33450-1908350.html

IOP Conf. Series: Materials Science and Engineering **878** (2020) 012059 doi:10.1088/1757-899X/878/1/012059

Modeling and study of a novel electrothermal oscillator based on shape memory alloys

R P Mitrev [1*], E I Atamas [2], O I Goncharov [3], T S Todorov [4], and I I Iatcheva [5]

[1] Department of Logistics Engineering, Technical University of Sofia, Sofia 1000, Bulgaria
[2] Department of Nonlinear Dynamical Systems and Control Processes, Lomonosov Moscow State University, Moscow, Russia
[3] Department of Nonlinear Dynamical Systems and Control Processes, Lomonosov Moscow State University, Moscow, Russia
[4] Department of Theory of Mechanisms and Machines, Technical University of Sofia, Sofia 1000, Bulgaria
[5] Department of Theoretical Electrical Engineering, Technical University of Sofia, Sofia 1000, Bulgaria

[*]rosenm@tu-sofia.bg

Abstract. The paper deals with mathematical modelling and experimental investigation of a novel electrothermal oscillator based on Shape Memory Alloys (SMA). The operation of the oscillator is based on a heated by DC electric current SMA wire, mounted with a pre-tension between the free ends of two symmetrically arranged cantilever beams. Crystallographic changes in the SMA wire, consisting in the transition between austenitic and martensitic phases due to the periodic heating and cooling, lead to its intermittent shortening and stretching and thus mechanical and electrical vibrations are generated. The mathematical model of the system is derived using Lagrange-Maxwell formalism, Joule–Lenz law and Newton's law of cooling. Simulation results show two types of periodic motion – the first one with low frequency is due to the temperature variation in the wire and the second one is with much more higher frequency and is due to the mass and elastic properties of the system. Experimental validation of the mathematical model shows its suitability for description of the considered system dynamic behaviour.

1. Introduction

The modern smart materials allow a precise control of their physical properties by applying external influences. This unique feature makes it possible, by utilizing simple physical principles, to build a variety of technical devices with relatively complex behavior. Particularly widespread in the past two decades have become the Shape Memory Alloys (SMA). They are used for a considerable variety of practical purposes, some of which are vibration damping in mechanical devices [1], driving various devices by actuators and artificial muscles with reduced weight [2,3,4], natural frequency tuning of civil structures [5,6] and intensive application in the modern medicine and dentistry [7,8]. The nonlinear behavior of the SMA is determined primarily by the presence of a hysteresis and by the reversible crystallographic changes in the material due to the heating or cooling. SMA are particularly suitable for designing mechanical devices with cyclic behavior - so called oscillators. The authors [9,10,11] adequately demonstrate the possibility of using SMA to create oscillators that produce periodic, quasi-periodic or chaotic vibrations. Combining systems from different physical domains into one system is especially promising, mostly from the point of view of building technical devices with simple structure and reach dynamic behaviour. For example, combining an electrical, thermal and mechanical system in [12] makes it possible to develop a DC-driven electrothermal oscillator with a simple design and stable vibrations. The same approach is used in [13] to develop a mechanical oscillator with frequency range from 38 Hz to 1200 Hz with electrothermal actuation and without using any electrical components.

The principal aim of the present paper is to develop a conceptual design and mathematical model of a novel nonlinear electrothermal oscillator based on Shape Memory Alloys.

2. Conceptual design of the studied electrothermal oscillator

Figure 1 shows the developed design of the oscillator. It is composed of two symmetrically arranged elastic cantilever beams (pos.1a), fixed on an unmovable base 2, and a SMA wire (pos.3a), mounted with pre-tension between the free ends of the beams. The electric current, generated by the external power supply 5, is submitted to the SMA wire through the electrode 4, which contacts the wire in its initial position (pos.3a). The generated due to the flow of the electric current heat in the wire leads to crystallographic changes in the SMA material – a transition from martensitic to austenitic phase, resulting in shortening of the wire. This causes a symmetric deformation of the beams free ends (pos.1b), accompanied by an upward motion of the wire (pos.3b). Opening of the contact between the electrode and the wire terminates the flow of electric current through the wire whereby it begins to cool by convection. This provokes a reverse crystallographic change in the wire - the transition from austenitic to martensitic phase, resulting in a stretching of the wire, reverse deformation of the beams (pos.1a) accompanied by a downward motion of the wire (pos.3a) resulting in a restoration of the contact between the electrode and the wire whereby the heating process is repeated. The resulting motion of the free ends of the cantilever beams has a periodic behavior with a certain frequency. Deformation of the beams is used to produce electrical energy by two piezoelectric transducers 6, glued on the beams deformed surfaces near to their fixed ends. It should be specified that the electrode 4 is mounted in height so as to create a certain pre-tension of the wire in its initial position 3a.

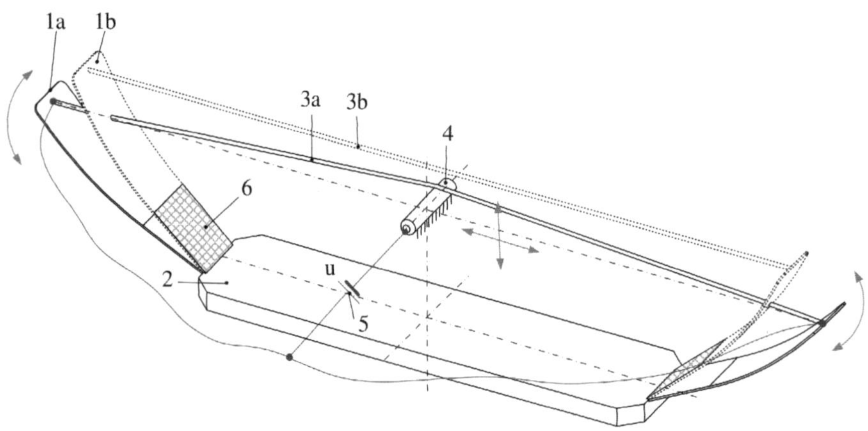

Figure 1. Conceptual design of the studied electrothermal oscillator.

3. Mathematical model of the oscillator

The electrothermal oscillator consist of three subsystems, belonging to different physical domains – mechanical, electrical and thermal. Every subsystem has its own dynamics, which affect the dynamics of the overall system.

3.1 Mathematical model of the mechanical subsystem

Figure 2 depicts the mechanism layout for an arbitrary wire length *s*, obtained for arbitrary wire temperature *T*. The following notations are used: δ_c is the beam deflection; δ_{y1} and δ_{x1} are the projections of the beam deflection on the *y* and *x*-axes correspondingly; F_S is the force in the SMA wire and F_S^t and F_S^n are its tangential and normal projections correspondingly.

For the determination of the forces and deflections, it was assumed that before the SMA wire mounting, the undeformed cantilevers are in the initial positions OA_0 and CB_0. The deformed cantilever beams positions OA_1 and CB_1 correspond to the mounted in the cold martensitic condition wire. The following assumptions are adopted: 1) Deflections of the cantilevers beams are small; 2) The free end of the beam during the deformation is moving along the perpendicular to its undeformed position; 3) The Hook law is valid for the wire in low temperature. If the initial length of the wire in the undeformed condition is s_0 then the mounting requirement has the form:

$$s_0 < A_0 B_0 \tag{1}$$

Mounting of the wire in the cold condition leads to its extension and its length is denoted by $s_1 = A_1 B_1$.

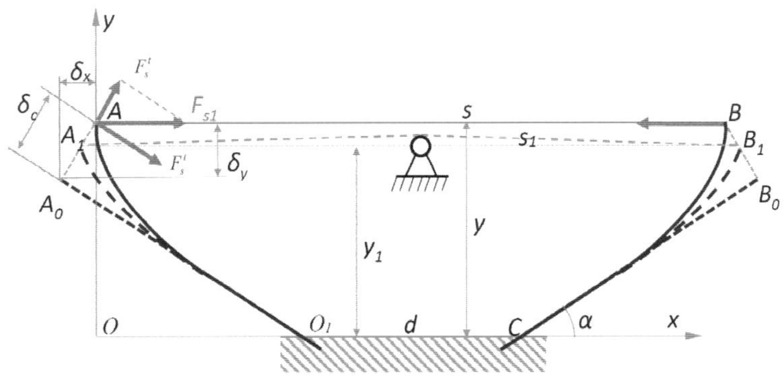

Figure 2. Layout of the mechanical structure of the oscillator.

Considering some geometric and force conditions, for the positions *OA* and *CB* corresponding to the coordinate *y* of the wire, the tensile force in the wire is determined as

$$F_s = E_m A_s \varepsilon_s = E_m A_s \frac{3EI(2l\cos\alpha + d)}{2E_s A_s l^3 \sin^2\alpha + 3sEI} \tag{2}$$

In equation (2), the following notations are used: E_m is Young modulus of the SMA wire in martensitic phase; A_s is the area of the wire cross-section; ε_s is the wire strain; l is the length of the cantilever beam; α is the angle of inclination of the beams; *d* is the distance between the beams fixed ends; $E_S = E_S(T)$ is the Young modulus of the SMA wire for an arbitrary temperature *T*, computed according to the SMA model, presented in [14]; E is the Young modulus of the beam; I is the area moment of inertia of the beam cross-section.

The following expressions for the *y*-coordinate of the wire and for its length *s* are obtained:

$$y = l\sin\alpha + \frac{3EI\left(l^2\sin 2\alpha + dl\sin\alpha\right)\cos\alpha}{2E_s A_s l^3 \sin^2\alpha + 3EIs} \tag{3}$$

$$s = \varepsilon_s s_0 = \frac{3EI\left(2l\cos\alpha + d\right)s}{2l^3 E_s A_s \sin^2\alpha + 3EIs} \tag{4}$$

For the arbitrary temperature *T*, the vertical translation Δy of the wire according to the initial mounting position is given by the following equation:

$$\Delta y = y - y_1 = 3EI\left(l^2\sin 2\alpha + dl\sin\alpha\right)\cos\alpha\left(\frac{1}{2l^3 E_s A_s \sin^2\alpha + 3EIs} - \frac{1}{2l^3 E_m A_s \sin^2\alpha + 3EIs_0}\right) \tag{5}$$

3.2 Mathematical model of the piezoelectric subsystem

The electromechanical piezoelectric subsystem is considered as a system with two degrees of freedom (DOF). The first DOF is mechanical, representing the deflection of the beam along the x-axis – figure 3(a). For the electrical generalized coordinate is chosen the electric charge q of the piezoelectric transducer circuit – figure 3(b) and figure 3(c). The mass of the cantilever beam and the half of the SMA wire mass are represented as a concentrated mass m, positioned at the free end of the beam at point A. The electric circuit consists of a piezoelectric transducer with capacitance C and load resistance R_L .

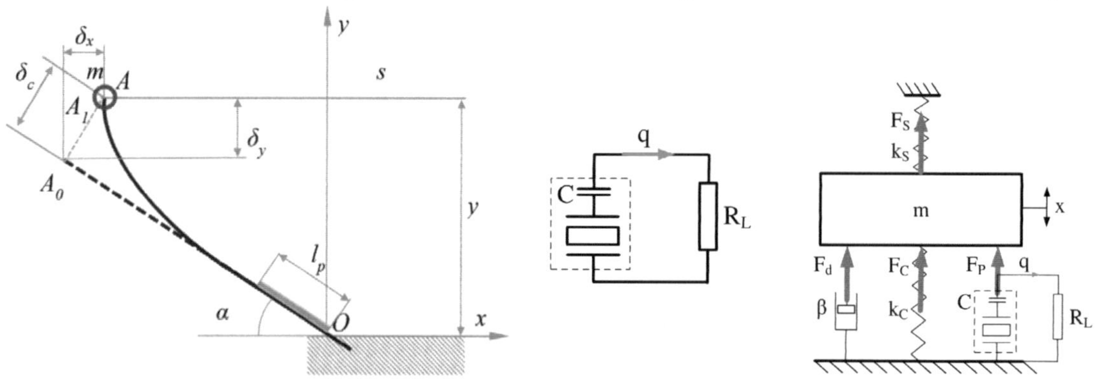

(a) (b) (c)

Figure 3. Dynamical model of the cantilever piezoelectric beam: (a) mechanical model; (b) electric circuit; (c) electromechanical model

Close to the beam fixed end three layers are applied – for the bottom electrode, for the piezoelectric film and for the top electrode and both electrodes are being connected by the load resistance. Also, it is assumed that the polarization of the piezoelectric layer is along the longitudinal axis of the beam. Since along the same axis is the strain of the piezoelectric layer, the mode of operation of the piezoelectric transducer is d_{33}.

Interaction of the mechanical and electrical variables is represented by the Lagrangian function:

$$W_L = W_K - W_P - W_{PT} \tag{6}$$

where W_k is the kinetic energy of the mass, and W_P is its potential energy. The third member W_{PT} is the full electromechanical energy of the piezoelectric transducer.

The cantilever beam and the SMA wire are considered as springs connected in parallel with stiffness depending on the deformations and the wire temperature. Stiffness of the cantilever beam along the x – axis is

$$k_c = \frac{3EI}{l^3 \sin\alpha} \tag{7}$$

Using the Lagrange equations of the second kind and the Lagrangian (6) one obtains the following system of differential equations:

$$\begin{cases} m\ddot{x} + k_c\left(x+\delta_{x_1}\right) - \dfrac{C'}{2C^2\left(1-k_p^2\right)}q^2 + \dfrac{d_{33}K_a p_x C'}{C^2\left(1-k_p^2\right)}q\left(x+\delta_{x_1}\right) - \dfrac{d_{33}K_a p_x}{C\left(1-k_p^2\right)}q + \dfrac{K_a p_x}{\left(1-k_p^2\right)}\left(x+\delta_{x_1}\right) = -\beta\dot{x} + F_S \\ \dfrac{1}{C\left(1-k_p^2\right)}q + \dfrac{d_{33}K_a p_x}{C\left(1-k_p^2\right)}\left(x+\delta_{x_1}\right) = -R_L\dot{q} \end{cases} \tag{8}$$

where:

$$C' = \frac{dC}{dx} = \frac{\varepsilon w_p p_x}{g_p} \tag{9}$$

$$p_x = \frac{3h_p l_p}{l^3 \sin\alpha}\left(l - \frac{l_p}{2}\right) \tag{10}$$

K_a is the stiffness of the piezoelectric transducer with short-circuited electrodes; k_p is the electromechanical coupling factor; d_{33} is the piezoelectric constant; C is the capacitance of the transducer; β is the damping factor; δ_{x_1} is the deformation of the beam in the initial mounting position; ε is the dielectric constant of the piezoelectric material; g_p is the distance between the electrodes of the piezoelectric layer and w_p is the width of the piezoelectric layer; l_p is the length of the piezoelectric layer; h_p is the distance between the beam neutral axis and the middle layer of the piezoelectric transducer. The force in the SMA wire F_s is presented in the form

$$F_s = \begin{cases} \left[E_a - (E_a - E_m)R_m(T)\right]\varepsilon_s A_s, & 0 \le \varepsilon_s \le \varepsilon_m^y \\ \left[E_a - (E_a - E_T)R_m(T)\right]\varepsilon_s A_s - (E_T - E_m)\varepsilon_m^y R_m(T) A_s, & \varepsilon_m^y \le \varepsilon_s \le \varepsilon_m^d \\ \left[E_a - (E_a - E_d)R_m(T)\right]\varepsilon_s A_s - (E_T - E_m)R_m(T)\varepsilon_m^y A_s - (E_d - E_T)R_m(T)\varepsilon_m^d A_s, & \varepsilon_m^d \le \varepsilon_s \end{cases} \tag{11}$$

where E_a, E_T, and E_d are the Young's modules at austenite phase, partly twined martensite, and detwinned martensite respectively, ε_s is the strain of the SMA wire, ε_m^y is the yield strain of twinned martensite, ε_m^d is the minimum strain of detwinned martensite. The hysteretic behavior of the SMA wire is presented by the martensitic fraction $R_m(T)$, calculated according to the model, presented in [14].

It is assumed that the position x of the mass is equal to 0 when the SMA wire has length s_1 (see figure 2), which corresponds to the preliminary deformation of the beam δ_{x1}:

$$\delta_{x_1} = \frac{\left(l^2 \sin 2\alpha \sin\alpha + dl \sin^2\alpha\right) E_m A_s l^2}{2E_m A_s l^3 \sin^2\alpha + 3s_0 EI} \tag{12}$$

3.3 Mathematical model of the thermal subsystem

In the considered system for the heat generation is utilized the flow of an electric current through a resistance, i.e. the Joule heating is present. Under the assumption of convective cooling and taking into account the Joule–Lenz law and Newton's law of cooling, one obtains the differential equation, describing the change of the temperature in the SMA wire:

$$\dot{T} = \frac{1}{\rho_s V_s c_p}\begin{cases} Ri^2 + A_s h_c (T_\infty - T), & if \quad x \le x_T \\ A_s h_c (T_\infty - T), & if \quad x > x_T \end{cases} \tag{13}$$

where x_T is the horizontal position of the beam free end, corresponding to the wire height y_T, for which the contact between the electrode and the wire is lost. In (13) the following notations are used: R is the electric resistance of the wire; i is the electric current in the wire; T_∞ is the temperature of the environment; ρ_s is the density of the wire material; V_s is the volume of wire; c_p is the specific heat of the wire; A_s is the surface area of the wire; h_c is the convection heat transfer coefficient.

The following additional simplifying assumptions are made: 1) The inclination of the SMA wire relative to the x-axis when the electrode and the wire are in a contact is neglected; 2) The change of the resistance in the contact point between the wire and the electrode due to the change of the contact pressure is neglected; 3) The change of the resistance of the SMA wire due to the increase of the temperature is neglected.

4. Numerical solution of the equations and discussion of the results

The final system of differential equations is composed by equations (8) and (13), represented as a system of first-order equations. The backward Euler method with a suitable step is used for the solution of the system for a time period of 2 s. Initial conditions for the variables are: $x(0) = 0$, $\dot{x}(0) = 0$, $T(0) = T_\infty$. At time t=0 the charge q has a value, computed from the second equation of (8) for $\dot{q} = 0$:

$$q(0) = -d_{33} K_a p_x \left[x(0) + \delta_{x_1} \right] \tag{14}$$

The obtained graphs for the output characteristics are shown in figure 4.

Figure 4(a) shows the graph of the horizontal translation x. As one can see, after a short transient period of 0.05 s the system switches to a steady-state mode with two types of periodic motion. The first periodic motion has a period $\tau_x = 0.591$ s and frequency $f_{Tx} = 1.692$ Hz. A comparison with the figure 4(b) graph shows that this frequency is equal to the frequency of the temperature variation. As one can see, all output characteristics of the system are modulated by this variation. The translations with this frequency are nonlinear and are caused by the temperature variation due to the consecutive heating and cooling of the SMA wire. They depend on the temperature convection time constant $\tau_T = \rho_s V_s c_p / (A_s h_c)$ and the gradient of the temperature $k_T = Ri^2 / (\rho_s V_s c_p)$. For that reason, f_{Tx} is denoted as a temperature-dependent frequency or macro frequency. The analysis of the graphs shows that except the macro frequency exists another much more higher frequency (micro frequency), which depends mainly on the mass and elastic parameters of the mechanical subsystem. The exponential decrease of the amplitude of the micro vibrations is caused by the assumed vicious damping. This vibration is modulated by the macro frequency.

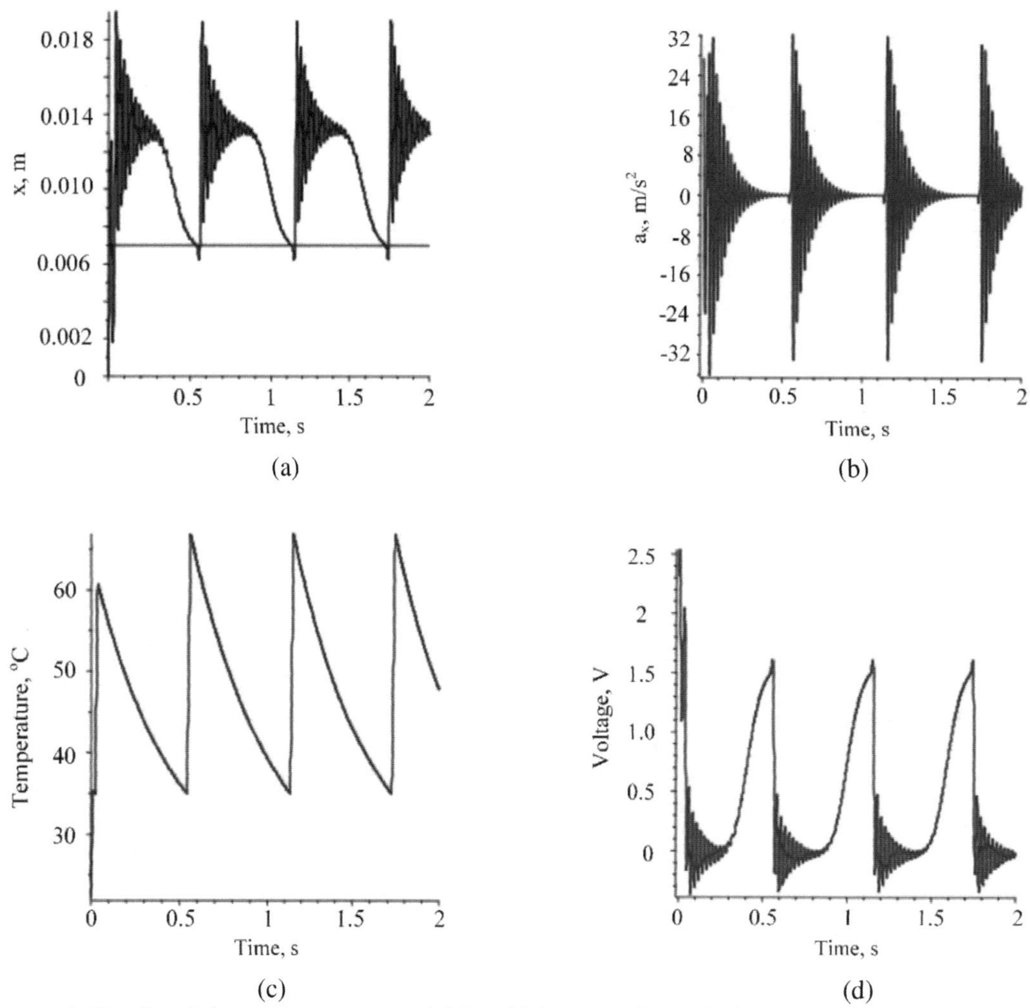

Figure 4. Graphs of the system output variables: (a) horizontal translation x of the beam free end; (b) horizontal acceleration; (c) temperature T variation; (d) voltage, generated by the piezoelectric transducer

5. Experimental setup and validation of the mathematical model

Validation of the mathematical model is conducted by a comparison between the simulation and the experimental data. Figure 5(a) depicts the experimental setup, whose main elements are the fixed base 1, the cantilever beams 2 with piezoelectric transducers 3, the SMA wire 5, the electrode 6 and the power source 9. It is equipped with sensors for measuring the acceleration 7 (ADXL 335) and the SMA wire temperature 4. Analog outputs from the sensors and from the piezoelectric transducers are converted to digital by an analog-to-digital converter (National Instruments 8 inputs, 14 bit) and sent to a computer measurement system 8, where a computer program LabView is used for the initial processing and visualization of data [15].

The experimental data for the acceleration a_x of the beam end in x-direction, for the generated voltage from the piezoelectric transducers and for the temperature of the SMA wire are shown in figure 5(b), figure 5(c) and figure 5(d). Although some differences between the experimental and simulated data shown in figure 4(b), figure 4(c) and figure 4(d) are present, the degree of coincidence is satisfactory and it can be considered that the developed mathematical model with good accuracy represents the real system. It should be noted that the difference between the experimental (figure 5(d)) and theoretical (figure 4(c)) temperatures is larger, but this is due to the high inertia of the thermocouple used.

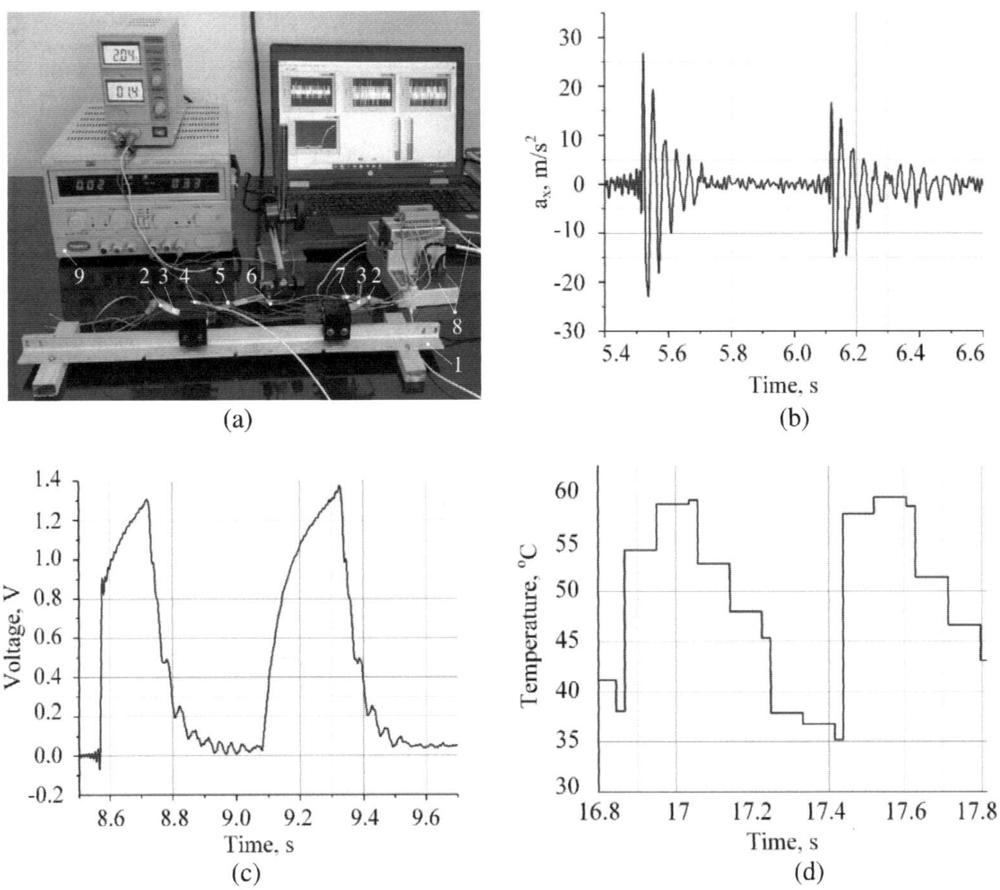

Figure 5. Experimental and simulation results: (a) layout of the experimental setup; (b) acceleration a_x of the beam end in x-direction; c) generated voltage from the piezoelectric transducers; d) temperature of the SMA wire

6. Conclusions

This paper studied issues related to the development and investigation of a novel electrothermal oscillator. The design of such a system, containing subsystems from different physical domains, namely mechanical, electromechanical and thermal, is difficult and the modelling method must take into account their interaction and its influence on the overall system behaviour. The developed and validated mathematical model considerably facilitates the oscillator design process. The conducted investigation clearly demonstrates the possibility to generate a periodic motion of the system using Shape Memory Alloy wire by supplying the system input with DC electric current. Such type of system is extremely sensitive to changes in system parameters, that's why by tuning the system parameters a considerable variety of alternative operating dynamic modes could be achieved. The primary direction for improvement of the system performance is its miniaturization which will cause an increase in its frequency, also its stability.

Acknowledgements

This research has been supported by the Bulgarian National Science Fund, under grant No. KP-06 – Russia/21.

References

[1] Lagoudas D and Hartl D 2007 Aerospace applications of shape memory alloys *Proc Inst Mech Eng G J Aerosp Eng* **221** 535-552

[2] Webster J 2006 High integrity adaptive SMA components for gas turbine applications *Smart Structures and Materials 2006: Industrial and Commercial Applications of Smart Structures Technologies* pp. 61710F

[3] Duerig T, Melton K and Proft J 1990 Wide Hysteresis Shape Memory Alloys *Engineering Aspects of Shape Memory Alloys* (Elsevier) pp. 130–136

[4] Nespoli A, Besseghini S, Pittaccio S, Villa E and Viscuso S 2010 The high potential of shape memory alloys in developing miniature mechanical devices: A review on shape memory alloy mini-actuators *Sensors and Actuators A: Physical* **158** 149-160.

[5] Song G, Ma N, Li H 2006 Applications of shape memory alloys in civil structures *Engineering Structures* **28** 1266-1274

[6] Wang L and Melnik R 2012 Nonlinear dynamics of shape memory alloy oscillators in tuning structural vibration frequencies *Mechatronics* **22** 1085-1096

[7] Karimi S and Konh B 2019 3D Steerable Active Surgical Needle *Design of Medical Devices Conference* 1-6

[8] Mereau T and Ford T 2006 Nitinol Compression Staples for Bone Fixation in Foot Surgery *Journal of the American Pediatric Medical Association* **96** 102–106.

[9] Savi M and Pacheco P 2002 Chaos and hyperchaos inshape memory systems *International Journal of Bifurcation and Chaos* **12** 645–657

[10] Bernardini D and Rega G 2005 Thermomechanical modelling, nonlinear dynamics and chaos in shape memory oscillators *Mathematical and Computer Modelling of Dynamical Systems* **11** 291–314

[11] Rusinek R and Rekas J 2018 Modelling of shape memory alloy oscillator and its application to middle ear structural reconstruction *AIP Conference Proceedings* **1922** 100017-1 - 100017-6

[12] Chen S, Yeh C and Yang Y 2017 DC-driven acrylate-copolymer-based electrothermal oscillator and its application in thermopneumatic diffuser pump *2017 IEEE 30th International Conference on Micro Electro Mechanical Systems (MEMS)* 825-828

[13] Udeshi K and Gianchandani Y 2004 A DC-powered, tunable, fully mechanical oscillator using in-plane electrothermal actuation *17th IEEE International Conference on Micro Electro Mechanical Systems. Maastricht MEMS 2004 Technical Digest* 502-505

[14] Madill D and Wang D 1998 Modeling and L2-stability of a shape memory alloy position control system *IEEE Transactions on Control Systems Technology* **6** 473-481

[15] Mitrev R 2007 Mathematical modelling of translational motion of rail-guided cart with suspended payload *Journal of Zhejiang University: Science A* **8** 1395-1400

Effect of P modifiers, SiC nanomodifiers and combinations them on the structure of a hypereutectic aluminum-silicon AlSi18 alloy

Iv Panov[1], B Dochev[2], V Manolov[3], A Velikov[4], V Dyakova[5], P Kuzmanov[6]

[1] Head Assistant, PhD - TU Sofia, Plovdiv Branch
[2] Assistant - TU Sofia, Plovdiv Branch
[3] Associate Professor, PhD - IMSETHC-BAS
[4] Head Assistant - IMSETHC-BAS
[5] Head Assistant, PhD - IMSETHC-BAS
[6] Head Assistant, PhD - IMSETHC-BAS

[1] specialista57@abv.bg
[2] boyan.dochev@gmail.com
[3] v.manolov@ims.bas.bg
[4] anmabg@abv.bg
[5] v_diakova@ims.bas.bg
[6] pawel_71@abv.bg

Introduction. Studies have been carried out with the following types of an aluminum-silicon AlSi18 alloy: unmodified, modified by a standard P modifier, modified by a SiC nanomodifier, and modified by a combination of both types if modifiers. The paper discusses the influence of the different types of modifiers and the combination of them on the shape and size of the primary Si crystals and their distribution in the structure of the alloy, as well as on the shape and size of the Si crystals in the eutectic composition.

Keywords: aluminum alloy, modification, nanomodifier, primary Si crystals, Si crystals in the eutectic composition

Exposition. The primary silicon crystals in hypereutectic aluminum-silicon alloys are large in size - from 80 to100 μm - and have an irregular shape. This is the reason why they have a considerable notching action on the structure of the hypereutectic aluminum-silicon alloys and worsen their mechanical and operational properties. In order to improve the properties of the alloys, the primary silicon crystals need to be refined, their uniform distribution in the structure of the alloy ensured and a compact shape given, thereby reducing their notching action. It is believed that this can be achieved by strengthening the eutectic matrix and refining it by modifying the primary silicon crystals [12]. Strengthening of the eutectic matrix can be achieved by alloying and/or grinding the silicon and alpha crystals in the eutectic composition. This is the aim of the recent experiments with novel types of modifiers - nanomodifiers. Nanomodifiers are ultra-fine nanosized powders with a particle size of 4-100 nm and a high melting point. The problem with their use is their difficult absorption by the melt and their homogeneous distribution in it. In order to obtain wettability and to facilitate the homogeneous distribution of the nanoscale particles as they are introduced into the molten metal, they need to be pre-coated with different metals [1,2,3,4,5]. Due to their small size and large specific surface area, the nanoparticles are characterized by extremely high sediment resistance and high surface stress; they are in constant motion and do not precipitate or float under the action of the gravitational forces [6]. By acquiring wettability (after cladding) and because of having high sediment resistance, these particles can serve as crystallization centers in heterogeneous crystallization when the melt solidifies.

Experiments with hypoeutectic and eutectic aluminum-silicon alloys and some grades of cast iron and steel only predominate in the literature [7, 8, 9, 10, 11].

Aim: The present work aims at investigating the influence of a standard modifier - P, a nanomodifier - SiC and their combinations, on the shape and size of the crystals of primary Si and their distribution in the melt, as well as on the shape and size of the Si crystals in the composition of the aluminum-silicon AlSi18 alloy eutectics.

1. Tested alloy and used modifiers

The object of the presented study was a two-component hypereutectic aluminum-silicon AlSi18 alloy with a chemical composition shown in Table 1.

Table 1. Chemical composition of the alloy AlSi18, wt.%

Si	**Fe**	**Cu**	**Mn**	**Mg**	**Cr**	**Ni**	**Zn**	**Pb**	**Al**
17,55	0,120	0,025	0,047	0,001	0,001	0,005	0,102	0,01	rest

A nanomodifier SiC 0.1 wt.%, passivated with oleic acid for protection against weathering was used. Cu, applied by a currentless method, served as a metal protector of the nanomodifier. Table 2 shows the SiC nanomodifier parameters.

Table 2. Parameters of the SiC nanomodifier

Composition of the nanomodifier	Specific Surface, m^2/g	Average Particle Size, nm	Particle Shape	Crystallographic phases
SiC + Cu (SiC:Cu-1:0,1)	35 ± 5	50 ± 5	cubic	β SiC

The used standard modifier P 0,4 wt.% was introduced into the melt by the ligature CuP10.

2. Methodology of the study

The study of the aluminium-silicon alloy consisted in conducting experiments with unmodified, modified by a P modifier, modified by a SiC nanomodifier, and simultaneously modified by both a P modifier and a SiC nanomodifier alloy.

Metallographic sections were prepared to study the microstructure of samples from the hypereutectic aluminum-silicon AlSi18 alloy (unmodified, modified by P, modified by SiC nanomodifier, and modified by P plus a nanomodifier at the same time). The size of the primary silicon crystals, their roundness, as well as the size and roundness of the silicon crystals in the eutectics were determined.

The influence of the different modifiers (the standard one, the nanomodifier, and the combination of them) on the structure of the investigated aluminum-silicon alloy AlSi18 was established.

3. Results

The results from the conducted microstructural studies of the hypereutectic aluminum-silicon alloy AlSi18 - unmodified, modified by phosphorus, modified by a SiC nanomodifier and modified by the combination of both the standard modifier and the nanomodifier - are shown in Table 3.

Table 3. Results from the microstructural studies of AlSi18 alloy

Alloy	Modifier	Arbitrary average diameter of the primary Si crystals D [μm]	Refinement %	Size of the Si crystals in the eutectic composition [μm]	Refinement %
AlSi18	-	92,4		250-260	
AlSi18	SiC	Not measured *		4-5	98
AlSi18	P	55,7	40,8	115-135	54
AlSi18	SiC + P	45,18	51,1	57-59	85

The microstructure of the unmodified AlSi18 alloy samples is composed of eutectics and individual primary silicon crystals (Figure 1). The shape of the primary silicon crystals is different: straight-walled polygons, well-formed plates, which in the plane of the microscopic observation resemble needles or irregularly shaped plate-type crystals (Figure 2). The average diameter of the primary silicon crystals in the unmodified AlSi18 alloy is in the range 87.2-97.6 μm.

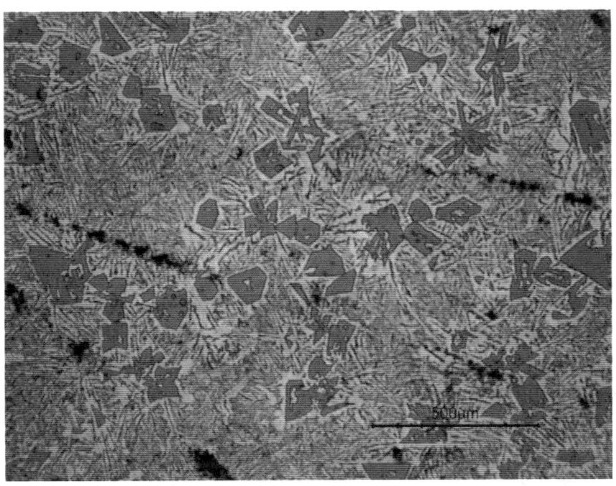

Figure 1. Unmodified AlSi18 (Si crystals)

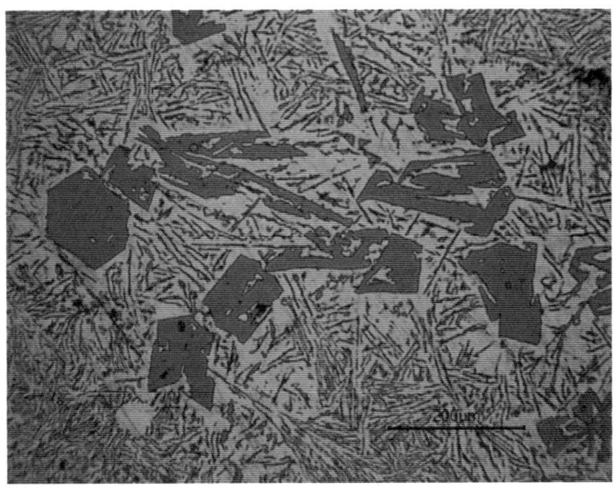

Figure 2. Unmodified AlSi18 (Si crystals)

Several types of zones are observed in the eutectics of an unmodified AlSi18 alloy. The first type has well-shaped, oblong-looking needle-like plates, measuring up to 250-260 μm in length (Figure 3). In the second type zone, small silicon crystals of several microns in size are formed, which group together or remain adjacent to each other, and under small microscopic magnification resemble a broken needle. In the third type zone, fine silicon crystals resembling a "fish bone" are observed (Figure 4).

IOP Conf. Series: Materials Science and Engineering **878** (2020) 012060 doi:10.1088/1757-899X/878/1/012060

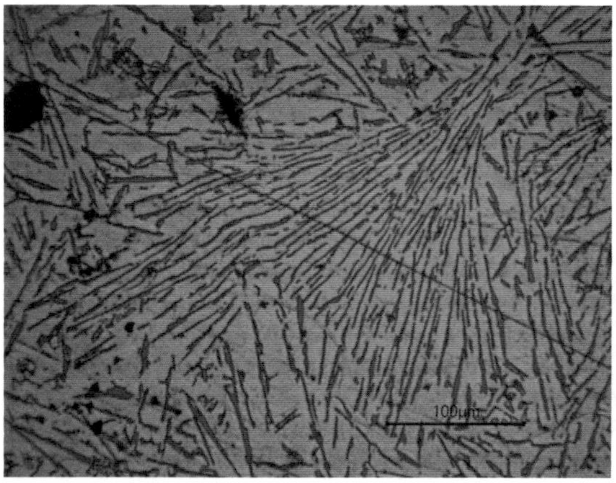

Figure 3. Unmodified AlSi18 (Si in eutectic)

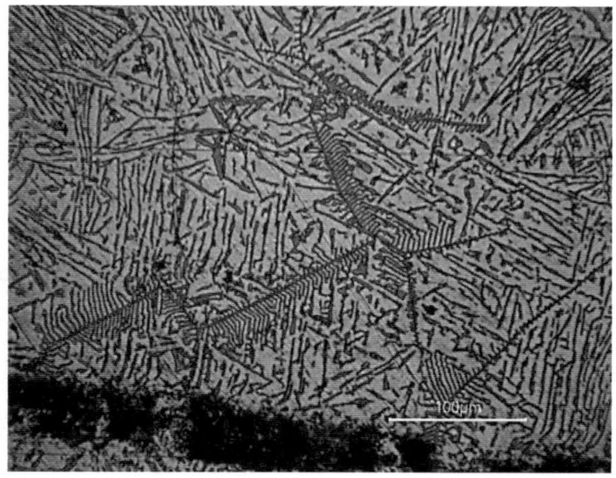

Figure 4. Unmodified AlSi18 (Si in eutectic)

In the AlSi18 alloy modified by the SiC nanomodifier (Fig. 5, Fig. 6), almost all primary silicon crystals have an irregular-star shape. Therefore, the arbitrary average diameter of the primary silicon crystals is not measured or calculated.

* The longest dimension of the primary Si crystals is measured about 250 μm.

.

Figure 5. AlSi18 modified by SiC (Si crystals)

Figure 6. AlSi18 modified by SiC (Si crystals)

In the AlSi18 alloy, modified by the SiC nanomodifier, the eutectics is highly refined, whole needles rarely occur - they are a series of stacked small silicon crystals not longer than 4-5 μm in length (Fig. 7, Fig. 8).

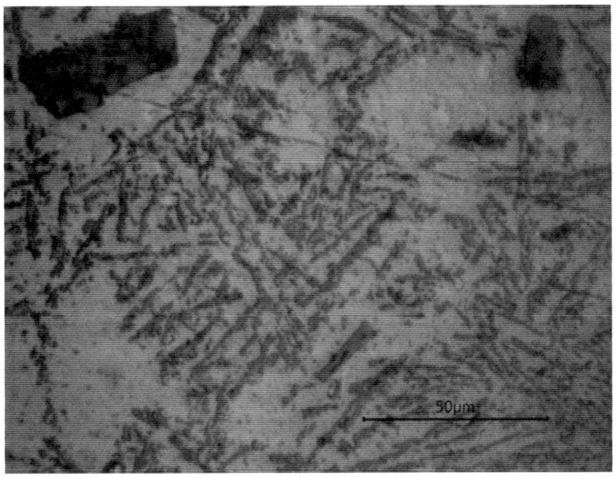

Figure 7. AlSi18 mod.by SiC (Si in eutectic)

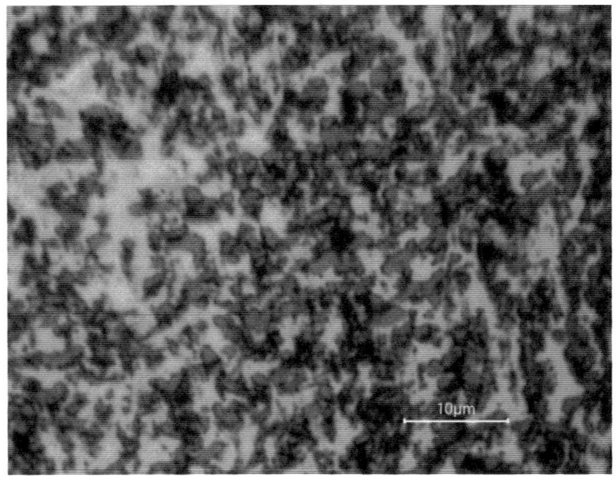

Figure 8. AlSi18 mod.by SiC (Si in eutectic)

The microstructure of the samples made of modified by phosphorus hypereutectic aluminum-silicon alloy AlSi18 consists of plate-like eutectics and primary silicon crystals, but the crystals differ in shape and size from those in the unmodified alloy (Fig.9, Fig.10). The arbitrary average diameter of the primary silicon crystals ranges from 52.1-59.2µm. Crystals in the form of straight-walled polygons predominate and no primary silicon crystals in the form of plates are observed.

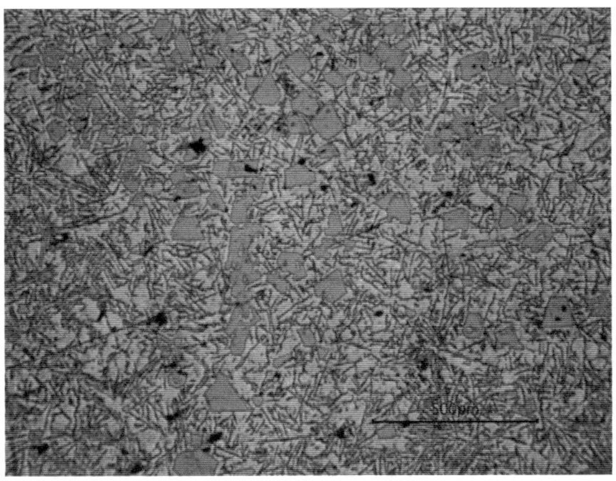

Figure 9. AlSi18 mod.by P (Si crystals)

Figure 10. AlSi18 mod. by P (Si crystals)

The eutectics in the modified by phosphorus hypreutectic aluminum-silicon AlSi18 alloy is refined. The silicon needles in it reach sizes of 115-135µm (Fig.11, Fig.12). Small round silicon crystals, which look like clumps or resemble broken needles, are observed here.

IOP Conf. Series: Materials Science and Engineering **878** (2020) 012060 doi:10.1088/1757-899X/878/1/012060

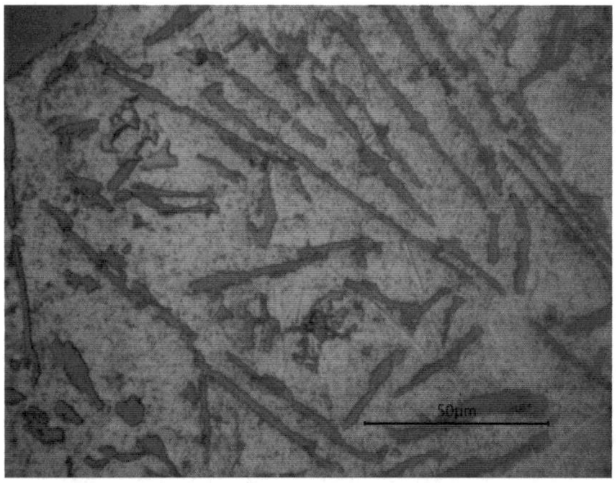

Figure 11. Mod. by P AlSi18(Si in eutectic)

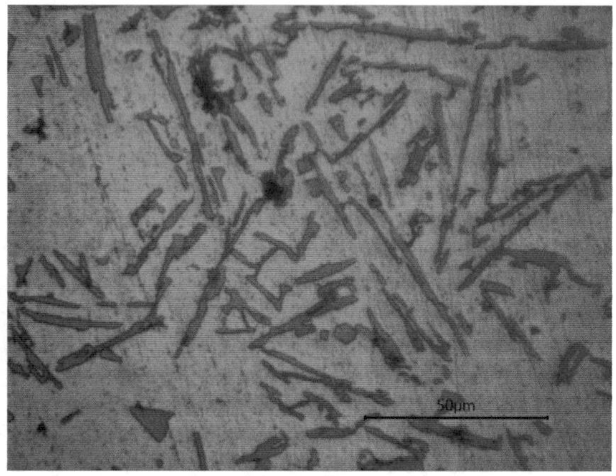

Figure 12. Mod. by P AlSi18 alloy(Si in eutectic)

The microstructure of the modified by both a standard modifier (P) and a nanomodifier (SiC) hypereutectic aluminum-silicon AlSi18 alloy is composed of eutectics and individual primary silicon crystals (Fig.13). The amount of primary silicon crystals having an irregular shape in the alloy structure is insignificant, and the primary silicon crystals in the form of polygons predominate (Figure 14). The arbitrary average diameter of the primary silicon crystals in the AlSi18 alloy, modified by both the nanomodifier (SiC) and the standard phosphorus modifier (P) in combination is 45.18 µm.

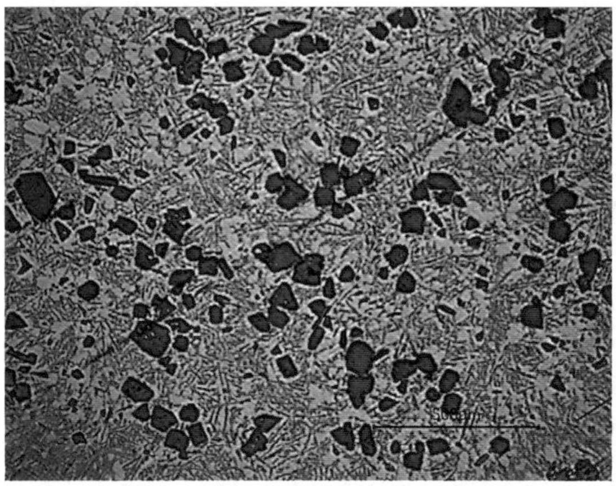

Figure 13. Mod. by P+SiC AlSi18 (Si crystals)

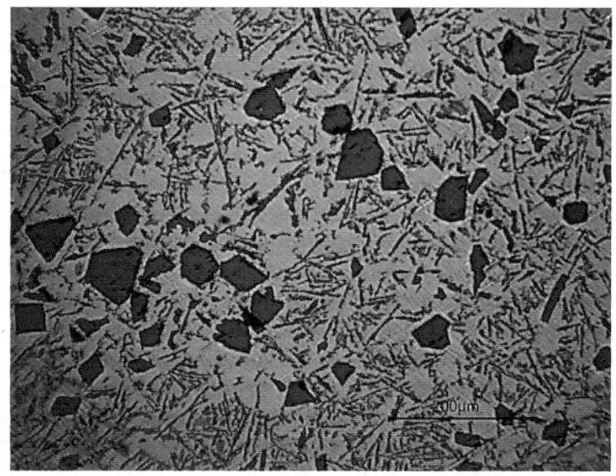

Figure 14. Mod. by P+SiC AlSi18 (Si crystals)

The eutectics of the hypereutectic aluminum-silicon alloy AlSi18 modified by the combination of a standard modifier (P) plus a nanomodifier (SiC), is made of well-formed wafers of eutectic silicon (having a needle-like shape in the plane of the section) around which small silicon crystals adhere (Fig. 15,16). The average maximum length of the "needles" in this structure is in the range 57-59µm.

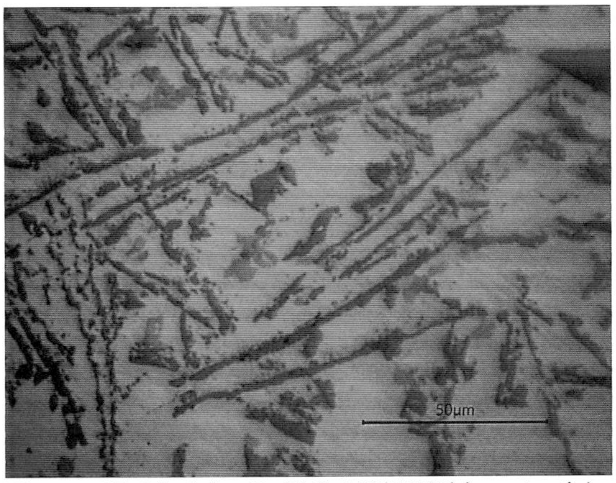

Figure 15. Mod. by P+SiC AlSi18(Si in eutectic)

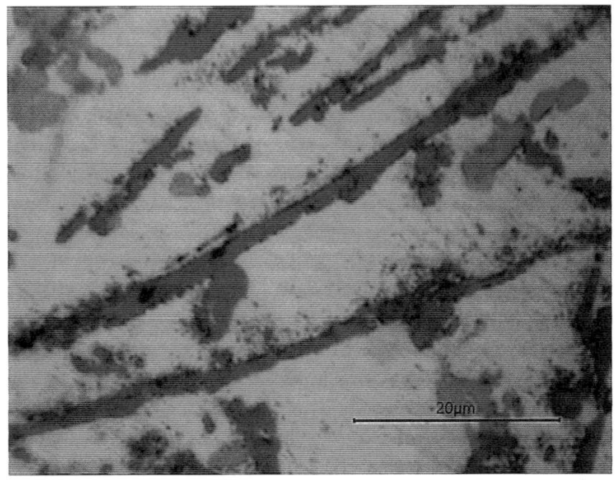

Figure 16. Mod. by P+SiC AlSi18(Si in eutectic)

4. Conclusions

• The SiC nanomodifier (in the concentrations we used) has a modifying effect on the size of the silicon crystals in the AlSi18 eutectic composition. However, it has no modifying effect on the deposited Si crystals of primary silicon.

• Since SiC has a positive effect on the structure of the hypoeutectic and eutectic aluminum-silicon alloys, but does not have a modifying effect on the primary silicon crystals in the hypereutectic aluminum-silicon alloy AlSi18, it cannot be recommended as a single and stand-alone modifier for this type of an alloy.

• The results from the microstructural analysis of the AlSi18 alloy, simultaneously modified by a standard phosphorus (P) modifier and a nanomodifier SiC show refinement of both the primary silicon crystals and the silicon crystals in the eutectic composition. Thence, it can be assumed that, with the use of a combination of a first-order modifier (SiC) and a second-order modifier (P), the formation of the AlSi18 alloy proceeds simultaneously according to both the adsorption and the overcooling theories.

Acknowledgements: The authors would like to thank the Research and Development Sector at the Technical University of Sofia for the financial support.

References:

[1] Kalinina N.E., Kalinin V.T., *Nanomaterials and nanotechnologies: production, structure, application.* Monography. Dnepropetrovsk: Makovetskij, Y.V, Publishing House, 188

[2] Cherepanov A.N., managing editor Fomin V.M., 1995, Plasma-chemical synthesis of ultradisperse powders and their application in metal and alloy modification, *Low-temperature plasma,* volume 12, Novosibirsk, Science, Syberian Publishing Company RAN, 344

[3] Manolov V., Cherepanov A., Lazarova R., and Konstantinova S., March, 22-25, 2011, Properties of the aluminium alloy AlSi7Mg, modified by refractory nanopowders, Interaction of highly concentrated energy flows with materials in promising technologies and medicine, IVth all-Russan conference, Novosibirsk, 186-190, ISBN 9785-93089-034-1

[4] Kalinina A.P., Cherepanov A.N., and Borisov O.V., 1999, The effect of linear tension on the rate of heterogeneous nucleation on dispersed particles, *Metals,* №5, 40-43

[5] Popov S., Manolov V., and Cherepanov A., *Mathematical modelling of metal alloys crystallization,* http://www.nanomodifiers-ims.com/botl.htm

[6] Krushenko, G.G, M.N. Filkov, Крушенко Г. Г., Фильков М. Н. Application of titanium nitride

nanopowder to obtain complex loaded aluminum-silicon alloys with the required mechanical properties, Nanotechniques, 2008, №3, 77-70 (in Russian);

[7] Lazarova R, Bojanova N, Dimitrova R, Panov I and Manolov V, *Influence of Nanoparticles Introducing in the Melt of Aluminum Alloys on Castings Microstructure and Properties*, https://link.springer.com/article/10.1007/s40962-016-0033-7

[8] Lazarova R, Dimitrova R, Bojanova N, Stanev S, Velikov A and Manolov V, *Modifying of luminium alloys using nanosized powders*, Project TK 01/076 funded by the National Fund "Scientific Researches'', published on http://www.nanomodifiers-ims.com/

[9] Manolov V., Cherepanov A., Lazarova R., and Konstantinova S., Influence of nanopowder inoculars on the structure and the properties of the AlSi7Mg alloy, *Foundry*, M., № **11**, 11-14, ISSN: 0024-449X, IF 0,088

[10] Stanev S., Lazarova R., Konstantinova S., and Manolov V., September, 2010, Study of the effect of nanopowder modification of AlSi7Mg alloy by thermal and metalograpic analyses, *Proceedings of the XXVIth international scientific conference "65 years Mechanical and technological Faculty at TU – Sofia* (Sozopol, Bulgaria), 277-282.

[11] Kuzmanov P, Dimitrova R, Lazarova R, Cherepanov A and Manolov V, March 2014, Investigations of the structure and mechanical properties of castings of alloy AlSi7Mg, cast iron GG15 and GG25 and steel GX120Mn12, modified by nanosized powders, *Journal of Nanoengineering and Nanosystems,* vol. **228** no. 1, pp. 11-18

[12] Stroganov, G.B., V.A Rotenberg, G.B Gershman, Aluminum alloys with silicon. M., Metallurgy, 1977,(in Russian).

IOP Conf. Series: Materials Science and Engineering **878** (2020) 012061 doi:10.1088/1757-899X/878/1/012061

Influence of a P modifier, a nanosized SiC modifier, and a combination of them on the mechanical properties of a hypereutectic aluminium-silicon alloy AlSi18

Iv Panov[1], B Dochev[2], V Manolov[3], A Velikov[4], V Dyakova[5], P Kuzmanov[6]

[1] Head Assistant, PhD - TU Sofia, Plovdiv Branch
[2] Assistant - TU Sofia, Plovdiv Branch
[3] Associate Professor, PhD - IMSETHC-BAS
[4] Head Assistant - IMSETHC-BAS
[5] Head Assistant, PhD - IMSETHC-BAS
[6] Head Assistant, PhD - IMSETHC-BAS

[1] specialista57@abv.bg
[2] boyan.dochev@gmail.com
[3] v.manolov@ims.bas.bg
[4] anmabg@abv.bg
[5] v_diakova@ims.bas.bg
[6] pawel_71@abv.bg

Introduction. The hypereutectic aluminium-silicon alloy AlSi18 has been modified by: a standard P modifier, introduced into the melt by means of the ligature CuP 0,4wt.%; a nanomodifier SiC 0,1wt.%; and a combination of these two. The influence of the different types of modifiers and their combination on the mechanical propertis of the alloy has been discussed. The mechanical properties of an unmodified AlSi18 alloy have been used as basic data to compare the obtained results with.

Key words: aluminium alloy, modification, nanomodifier, mechanical properties.

Exposition. In recent years, experiments have been carried out to modify aluminum alloys by new types of modifiers, namely, by nanomodifiers. Nanomodifiers are ultra-fine nanopowders with a particle size of 4-100 nm, and a high melting point. The main problem preventing the use of the nanoscale powders is that the nanoscale insoluble particles are difficult to get wet (or they do not get wet at all) by liquid metals and alloys. Therefore, their homogeneous distribution when introduced into the molten metal is a very difficult task [1]. The decision to pre-coat the nanoparticles has brought the resolution of this problem. One way to obtain active and melt-wetted nanomodifiers is by the method of their mechanical-chemical treatment, consisting in using cladding metals in centrifugal planetary mills [2]. When the modifiers are introduced into the melt, the coating layer dissolves and facilitates the process of wetting - thus the activated surface of the nanoparticles serves as a crystallization center upon which crystallites grow [5]. Another way to prepare and activate the surface of the nanoparticles is to chemically coat them in order to obtain a suitable metal cladding by various metals [6, 7].

Most of the existing studies have been performed with only hypo-eutectic and eutectic aluminum-silicon alloys, as well as with some grades of cast iron and steel [8, 9, 10, 11, 12].

Aim: The present work aims at studying the effect of a standard modifier - P, a nanomodifier - SiC and a combination of these, on the mechanical properties of a hypereutectic aluminum-silicon alloy AlSi18.

IOP Conf. Series: Materials Science and Engineering **878** (2020) 012061 doi:10.1088/1757-899X/878/1/012061

1. Tested alloy and used modifiers

The subject of this study was a two-component AlSi18 aluminum-silicon alloy. Its chemical composition is shown in Table 1.

Table 1. Chemical composition of the AlSi18 alloy, wt.%

Si	Fe	Cu	Mn	Mg	Cr	Ni	Zn	Pb	Al
17,55	0,120	0,025	0,047	0,001	0,001	0,005	0,102	0,01	rest

A nanomodifier SiC 0.1 wt.%, passivated with oleic acid for protection against weathering was used. Cu served as a metal protector of the nanomodifier, and the nanomodifier was applied by a currentless method to the surface of the nanomodifier. Table 2 shows the parameters of the coated SiC nanomodifier.

Table 2. Parameters of the SiC nanomodifier

Composition of the nanomodifier	Specific Surface, m^2/g	Average Particle Size, nm	Particle Shape	Crystallographic phases
SiC + Cu (SiC:Cu-1:0,1)	35 ± 5	50 ± 5	cubic	β SiC

The used standard modifier P 0,4 wt.% was introduced into the melt by the ligature CuP10.

2. Methodology of the study

The study of the hypereutectic aluninum-silicon alloy AlSi18 consisted in conducting experiments with unmodified, modified by P, modified by a nanomodifier SiC, and sinultaneously modified by both P and SiC modifiers alloy.

The melting of the alloy was carried out in an electrical resistance laboratory furnace using pre-cleaned and dry stock materials. The melting process takes place under a layer of refining flux in the amount of 0.5 wt.% from the charge material amount. After melting, the melt is vigorously stirred for removal of non-metallic inclusions out of it, then the slag is removed and degassing of the alloy is stsrted. The degassing is carried out at 760 ° C by purging with argon for 3min, cleaning the melt surface and casting samples.

When working with a P-modified alloy, the modification is carried out at 760 ° C before degassing. It is typical for treating the hypereutectic aluminum-silicon AlSi18 alloy by a SiC nanomodifier, that the process of modification is performed after refining and degassing, at 760 ° C again. The calculated amount of the SiC nanomodifier is mounted on the blade of a titanium stirrer, which is part of the stirring device. The nanomodifier is introduced near the bottom of the crucible. Mechanical stirring is then started for 3 min a lower number of revolutions per minute at 120 - 130 min-1, aiming at a uniform dispersion of the nanomodifier particles throughout the melt volume.

In order to produce from the tested alloy types (unmodified, modified with phosphorus and nanomodified) test specimens for carrying out mechanical tests and workpieces for metallographic microsections, samples were cast – see the "wedge castings" in Fig. 1. Metal die casting equipment (steel mould) with vertical dividing surface was used. Prior to casting, the metal mould was heated, coated with mould coating, and heated until it reached the temperature for pouring, e.g., 200°C. The temperature of the form is measured by a contact thermocouple, and all the experiments were performed at the same temperature of the metal form used.

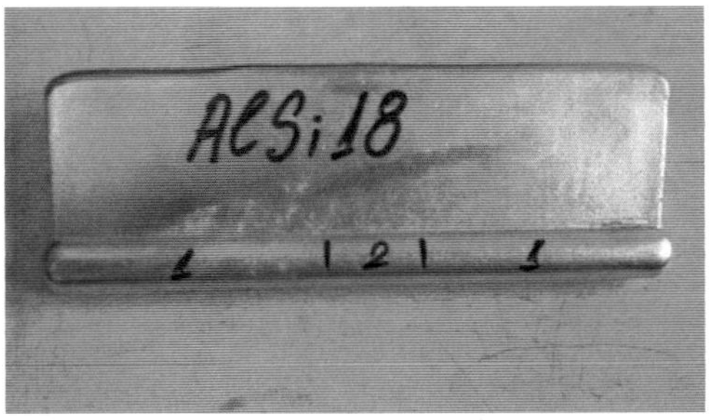

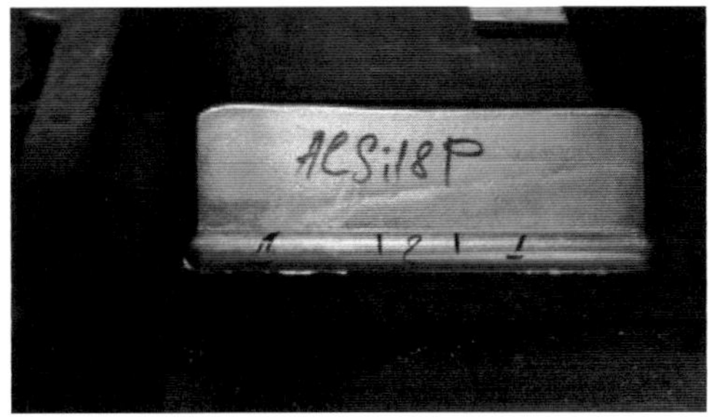

Figure 1. Wedge type castings with markings for test specimens

To determine the mechanical parameters of the AlSi18 aluminum-silicon alloy under study, short test samples were made from the marked by the number 1 places in the wedge casting. The samples were tested on a "Zwick/Roell Z 250" tensile test machine and both their tensile strength Rm and relative elongation A5 were established. The values of the mechanical characteristics were averaged over the test results from 3 ÷ 4 samples. The macro-hardness of the tested samples was also measured.

3. Results and discussion

The results from the mechanical tests of the different aluminum-silicon AlSi18 alloys - unmodified, modified by phosphorus, modified by a SiC nanomodifier, and modified by a combination of a standard P modifier and a nanomodifier SiC - are shown in Table 3.

Table 3. Results from the mechanical tests of the AlSi18 alloy

Alloy	Modifier	Rm/Mpa	A_5 /%/	HB2,5/62,5/30
AlSi18	-	108	1,4	65
AlSi18	SiC	82	0,8	65
AlSi18	P	128	1,6	64
AlSi18	SiC + P	133	1,2	69

Figures 2, 3 and 4 present the recorded tensile strength, relative elongation and macro-hardness of the test samples in the form of diagrams.

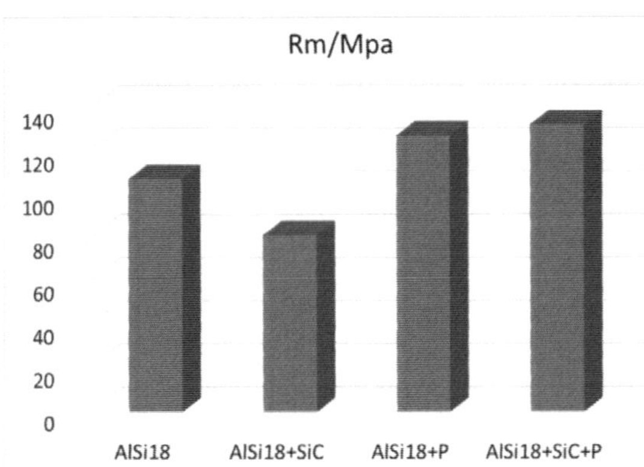

Figure 2. Change of R_m

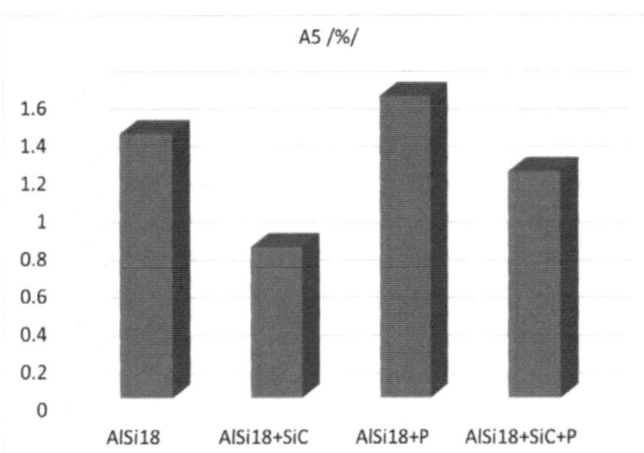

Figure 3. Change of A_5

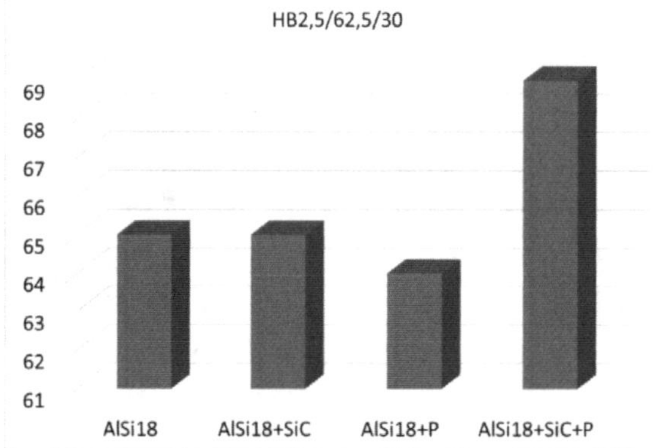

Figure 4. HB change

The results from the mechanical tests of the unmodified aluminum-silicon alloy AlSi18 are the following: tensile strength (Rm) 108 MPa, relative elongation (A5) 1.4% and hardness 65 HB (HB2.5 62.5/30). During the conducted microstructural analysis, the dimensions of the primary silicon crystals (92.4 μm) and the silicon crystals in the eutectic composition (250-260 μm) were measured.

The results from the mechanical tests of the modified by a standard modifier (P) AlSi18 alloy show an increase in its tensile strength (Rm) by 18.5% and also an increase in its relative elongation (A5) by 14.3%. The hardness decreased by 1.6%. The mechanical properties of the alloy, as well as its structure have changed in result of the modifying treatment. The microstructural analysis reveals that the primary silicon crystals in modified by a standard modifier (P) test samples have been refined by 39.7%, and the silicon crystals in the eutectics have been refined by 48.1% to 54% compared to the unmodified alloy. It has been proved that the modifying effect of phosphorus (P) on the primary silicon crystals comes not only from the AlP compound, but also from the formation of solutions and their adsorption at the boundaries of the grains [11]. Considering the fact that the modifiers most often have double effects [12], we assume that (P) could also be adsorbed along the grain boundaries of the alpha phase (a solid solution of silicon in aluminum), thus preventing the alpha phase grains from growing, i.e. it influences the structure the way it does a first-order modifier as well. By preventing the growth of the alpha phase grains, the distances between the branches of the dendrites remain small, and this is where the separated silicon crystals distribute when the eutectics is formed. In our view, this is a prerequisite for the effect of phosphorus (P) on the size of the silicon crystals in the composition of the eutectics of the alloy. The modification of the primary silicon crystals and the silicon crystals in the eutectic composition leads to an increase in the hardness and plasticity of the alloy under investigation.

The microstructural analysis of the AlSi18 alloy modified by a nanomodifier (SiC) reveals that the silicon crystals in the eutectic composition refine by 98.1% to 98.4%, compared to the unmodified alloy. This modifier in the concentrations we use does not affect the shape or the size of the primary silicon crystals. The notching action of the primary silicon crystals on the structure of the alloy results in reduction of its tensile strength (Rm) by 24.1% and decrease in the relative elongation (A5) by 43%. The hardness of the tested sample stays the same, as in the unmodified alloy, despite the significant refinement of the silicon crystals in the eutectic composition. The nanomodifier (SiC) has been shown to have a modifying effect on the hypoeutectic aluminum-silicon alloys [6, 7], though, in the used concentrations, it does not affect the shape or the size of the primary silicon crystals in the hypereutectic aluminum-silicon alloy. This fact confirms the view that in the pointed concentrations, the SiC nanomodifier acts only as a first-order modifier. It can be assumed that it is adsorbed along the grain boundaries of the alpha phase, thus inhibiting the growth of the alpha crystals, and many silicon crystals are formed on the branches of the dendrites as a result of the concentration fluctuations, which also do not have the potential to increase significantly in dimensions. This is the most likely reason for the considerable refinement of the silicon crystals in the eutectic composition, which is confirmed by the results from the microstructural analysis of the alloy.

The mechanical tests results from for a hypereutectic aluminum-silicon alloy modified simultaneously by both a standard modifier P and a nanomodifier SiC, are as follows: increase in the tensile strength (Rm) by 23.2%, relative elongation (A5) by 14.3%, and increase in hardness by 6.2%. The increase in the mechanical properties of the alloy (tensile strength and macro-hardness) is due to a change in its structure. In this modification, the primary alloy crystals refine by 51.1% and the silicon crystals in the eutectic composition - by 77.2%-77.3%, compared to the unmodified alloy. The measured arbitrary average diameter of the primary silicon crystals is 45.18 μm, and the sizes of the silicon crystals in the eutectic composition are in the order of 57-59 μm. From the results of all conducted experiments with a hypereutectic aluminum-silicon AlSi18 alloy it becomes clear that when the alloy is modified by the combination of a standard modifier (P) and a nanomodifier (SiC), the refinement of the crystals of primary silicon is the most significant and the highest values of tensile strength and hardness of the alloy are measured.

The microstructural analysis of the AlSi18 alloy, simultaneously modified by a standard phosphorus (P) modifier and a nanomodifier SiC shows refinement of both the primary silicon crystals and the silicon crystals in the eutectic composition. We can assume that, with the use of one first-order modifier (SiC) and one second-order modifier (P), the structure formation of the AlSi18 alloy proceeds in accordance with the adsorption theory and the overcooling theory at the same time. The successful two-stage modification treatment (P + SiC) proves the assumption that the modification of the eutectic silicon, as well as of the silicon crystals in the eutectic composition will have a positive effect on the properties of the hypereutectic aluminum-silicon AlSi18 alloy.

4. Conclusions

• The use of the SiC nanomodifier as a stand-alone and sole modifier for the AlSi18 aluminum-silicon alloy degrades its mechanical properties and is not recommended in this concentration.
• Due to the inability of the SiC nanomodifier to etch and change the shape of the primary Si crystals (ie, to avoid their notching action on the structure of the alloy), its application does not lead to increasing the hardness characteristics of the AlSi18 alloy, despite the considerable refinement of the eutectics of the alloy.
• The complex modification of the AlSi18 alloy melt by both a standard P modifier and a SiC nanomodifier results in the highest values of the hardness parameters (Rm, HB), compared to the other compositions tested and could be recommended for practical application.

Acknowledgements: The authors would like to thank the Research and Development Sector at the Technical University of Sofia for the financial support.

References:

[1] Kalinina N.E., Kalinin V.T., *Nanomaterials and nanotechnologies: production, structure, application.* Monography. Dnepropetrovsk: Makovetskij, Y.V, Publishing House, 188

[2] Cherepanov A.N., managing editor Fomin V.M., 1995, Plasma-chemical synthesis of ultradisperse powders and their application in metal and alloy modification, *Low-temperature plasma,* volume 12, Novosibirsk, Science, Syberian Publishing Company RAN, 344

[3] Manolov V., Cherepanov A., Lazarova R., and Konstantinova S., March, 22-25, 2011, Properties of the aluminium alloy AlSi7Mg, modified by refractory nanopowders, Interaction of highly concentrated energy flows with materials in promising technologies and medicine, IVth all-Russan conference, Novosibirsk, 186-190, ISBN 9785-93089-034-1

[4] Kalinina A.P., Cherepanov A.N., and Borisov O.V., 1999, The effect of linear tension on the

rate of heterogeneous nucleation on dispersed particles, *Metals,* №5, 40-43

[5] Popov S., Manolov V., and Cherepanov A., *Mathematical modelling of metal alloys crystallization,* http://www.nanomodifiers-ims.com/botl.htm

[6] Lazarova R, Bojanova N, Dimitrova R, Panov I and Manolov V, *Influence of Nanoparticles Introducing in the Melt of Aluminum Alloys on Castings Microstructure and Properties,* https://link.springer.com/article/10.1007/s40962-016-0033-7

[7] Lazarova R, Dimitrova R, Bojanova N, Stanev S, Velikov A and Manolov V, *Modifying of luminium alloys using nanosized powders*, Project TK 01/076 funded by the National Fund "Scientific Researches", published on http://www.nanomodifiers-ims.com/

[8] Manolov V., Cherepanov A., Lazarova R., and Konstantinova S., Influence of nanopowder inoculars on the structure and the properties of the AlSi7Mg alloy, *Foundry*, M., № **11**, 11-14, ISSN: 0024-449X, IF 0,088

[9] Stanev S., Lazarova R., Konstantinova S., and Manolov V., September, 2010, Study of the effect of nanopowder modification of AlSi7Mg alloy by thermal and metalograpic analyses, *Proceedings of the XXVIth international scientific conference "65 years Mechanical and technological Faculty at TU – Sofia* (Sozopol, Bulgaria), 277-282.

[10] Kuzmanov P, Dimitrova R, Lazarova R, Cherepanov A and Manolov V, March 2014, Investigations of the structure and mechanical properties of castings of alloy AlSi7Mg, cast iron GG15 and GG25 and steel GX120Mn12, modified by nanosized powders, *Journal of Nanoengineering and Nanosystems,* vol. **228** no. 1, pp. 11-18

[11] Stroganov, G.B., B.A.Rotenberg, G.B.Gershman, Alloys of aluminum with silicon, M., Metallurgy, 1977 (In Russian);

IOP Conf. Series: Materials Science and Engineering **878** (2020) 012062 doi:10.1088/1757-899X/878/1/012062

The effect of the type of protective suit on the thermophysiological comfort of surgeons in an operating room

Radostina A Angelova, Rositsa Velichkova

Bulgaria, 1000 Sofia, * Kl. Ohridski Bld, Technical University of Sofia

joy_angels@abv.bg

Abstract. Surgical gowns and medical clothing have to protect the surgeons from the patient and vice versa. However, all players in the operating room (OR) have to be in a state of thermophysiological comfort. The comfort of the surgeons is of special importance, as they are exposed to stressful situations, which could influence their performance negatively. The study presents an analysis of the thermophysiological comfort of the surgeons in an OR, using protective clothing ensembles with different clothing insulation and protection: a surgical scrub suit (SSS) and a clean air suite (CAS). The analysis is based on the calculation of PMV-PPD indexes, following Fanger's model. Different combinations of thermal environment parameters and clothing insulation are tested and analysed.

1. Introduction

The thermophysiological comfort of people is based on the complex interaction between the human body and the environment. Several parameters influence this interaction, such as air temperature and humidity, air velocity, human activity, clothing insulation and many others. Some of them are time-dependent (activity, stress); others can be considered as constant (clothing insulation).

The original function of the surgical cloths is to protect the patient from a possible impact of the surgical team. Later on, with the development of the knowledge on blood-borne pathogens, additional functions were added to the surgical clothing: to protect the surgeons from the patient himself [1].

The ISO 7730 [2] standard, does not make a difference between types of indoor spaces in terms of thermal comfort. Therefore, surgical scrub suits (SSS) or clean air suits (CAS), used in the OR, have to assure the thermophysiological comfort of the surgeons as the business suit in an office room. SSS are applied not only in the ORs, but they are used as a kind of uniform outside the OR. CAS are specially designed for surgeons and belong to Class I medical devices (Medical Devices Directive 93/42/EEC, amended by 2007/47/EC [3])

Clothing is the most significant difference between the medical players in the OR. The environmental conditions are also different: the surgeons that use the most insulated protective clothing and devices (including SSS and CAS), stay under the surgical lights, where the temperature is the highest, due to the radiation.

The problem of the thermophysiological comfort of the surgeons in an OR has been investigated from different points of view: the ventilation system [4, 5], the flow around the human bodies [6], the clothing insulation [7-9]. Some of the studies were based on the calculation of the Predicted Mean Vote (PMV) or Predicted Percentage of Dissatisfied (PPD) indexes [10,11], which are used in the present study s well

The developed by Fanger in 1972 [12] Predicted Mean Vote (PMV), is the most applied model for assessing the thermal comfort in the indoor environment. It is based on the calculation of the joint influence of the heat balance of the human body, sweat rate and mean skin temperature. The related index, namely the Predicted Percentage of Dissatisfied (PPD) [12] allows predicting the quota of the occupants that would feel uncomfortable in the particular thermal environment.

Our study applies the PMV-PPD indexes as a point for an analysis of the thermophysiological comfort of surgeons, protected by two types of surgical ensembles: a surgical scrub suit (SSS) and a clean air suit (CAS). Different environmental conditions were tested. A computational program for the prediction of the heat strain, following ISO 7933:2004 [13], was applied to calculate the PMV-PPD indexes [14].

2. Methodology

Three groups of factors were taken into account when calculating the PMV-PPD indexes for the surgeons in an OR: clothing, environmental factors and body activity.

2.1 Clothing

Two types of clothing ensembles were considered for the determination of the thermophysiological comfort of the surgeons in the OR:

- a reusable surgical scrub suit (SSS) – Fig. 1a;
- a reusable clean air suit (CAS) – Fig. 1b.

The composition of SSS is 67% polyester (PES) and 33% cotton fibres, with thermal insulation of 1.39 clo. The composition of CAS is 99% PES and 1 % carbon fibres, with thermal insulation of 1.31 clo [8].

a/ b/

Figure 1. Surgical clothing: a/ reusable surgical scrub suit (SSS); b/ reusable clean air suit (CAS)

2.2 Environmental factors

The ASHRAE standard 170-2013 [15] suggests the air temperature of an OR (Class B and C) to be from 20 °C to 24 °C. The air temperature under the lights could be with up to 3 °C higher [16], but the measurements in [17] did not find such a difference. The temperature during the calculations was changed from 18 °C to 24 °C with a step of 1 °C. The relative humidity (RH) of the OR was set to be 40%, 50% and 60% [15, 17]. The air velocity was set to 0.02 m/min [17].

2.3 Factors related to human body activity

Moderate activity (145 W/m^2) of the surgeons was considered for the calculations. An average male body was taken with height 175 cm, weight 85 kg and body surface area 2.0 m^2 [18]. The time of surgery was set to 1 h 40 min, which corresponds to a kidney operation [17].

2.4 Cases

Forty-two cases were simulated all together.

3. Results and Discussions

The thermal sense scale of the PMV index varies from -3 to +3, as shown in Table 1. The PPD index changes from 0% to 100%. ISO 7730 [2] international standard suggests the limits -0.5 < PMV < 0.5 and the PPD index below 10% for a moderate thermal environment. These are, in fact, the limits of the thermophysiological comfort.

Though a temperature interval of 20 °C to 24 °C is suggested in an OR (RH 60%) [4], our first predictions showed that for the clothing insulation provided by the two surgery ensembles, the discomfort would appear very quickly in this thermal interval. Therefore our predictions were made for a larger temperature interval: from 18 °C to 24 °C and a larger interval of relative humidity: from 40% to 60%.

Table 1. PMV Index sense scale

Value	Thermal sense
-3	cold
-2	cool
-1	slightly cool
0	neutral
+1	slightly warm
+2	warm
+3	hot

Figures 2-4 summarize the PMV index for surgeons, dressesd in the two clothing ensembles, for RH 40%, 50% and 60%, respectively. The PMV index increases with the temperature, relative humidity and clothing insulation. The results show that 21 °C is the critical temperature in the investigated air temperature interval. The thermophysiological comfort of the surgeons would depend strongly on the relative humidity of the air and the type of protective clothing. A surgeon would feel comfortable during the operation, dressed in CAS (1.31 clo), if the RH is 40% or 50%. If the relative humidity increases to 60%, the surgeon would feel slightly warm, leaving the zone of the thermophysiological comfort. A surgeon, dressed in SSS (1.39 clo) would not be in a state of a thermophysiological comfort at a temperature, higher than 20 °C for all investigated cases of RH.

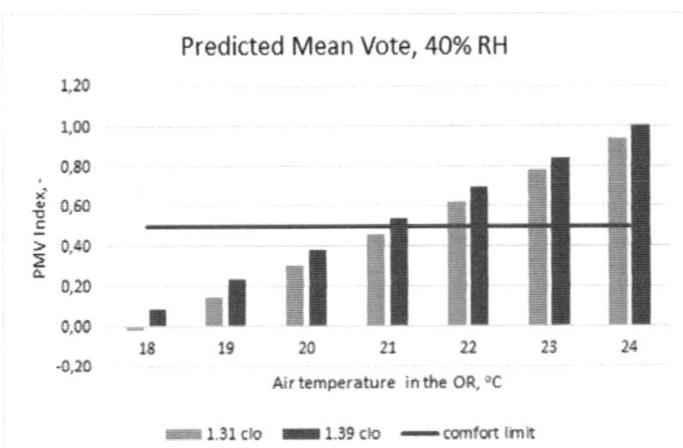

Figure 2. PMV index for surgeons, dressed in CAS (1.31 clo) and SSS (1.39 clo): air RH 40%

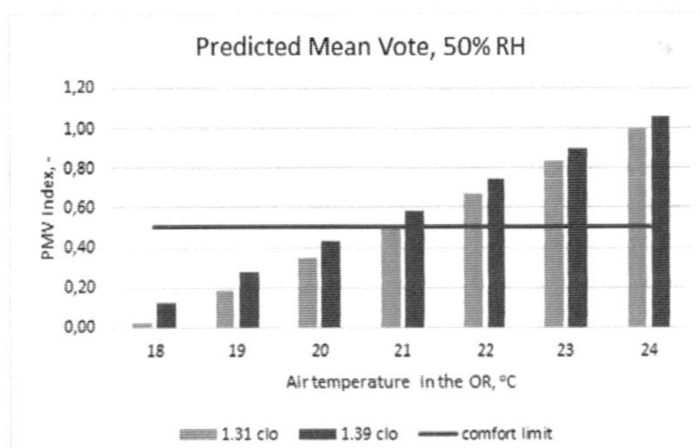

Figure 3. PMV index for surgeons, dressed in CAS (1.31 clo) and SSS (1.39 clo): air RH 50%

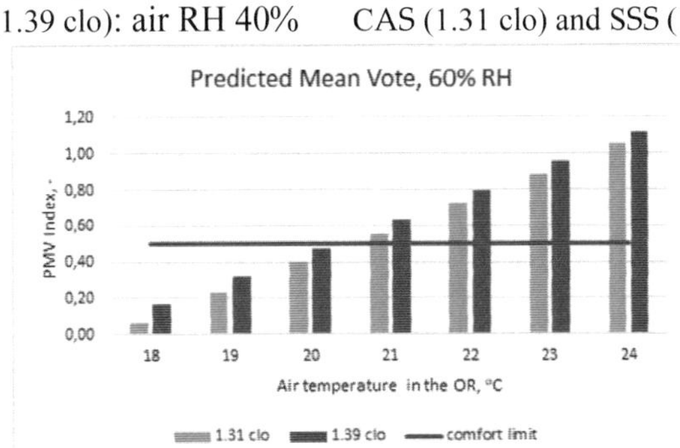

Figure 4. PMV index for surgeons, dressed in CAS (1.31 clo) and SSS (1.39 clo): air RH 60%

To avoid the influence of the type of protective clothing, the temperature in the OR should be up to 20 °C. This coincides with the discussed in [2] comfort temperature interval (20-22 °C). However, the discomfort could appear even if the temperature of the OR is up to to 20 °C, as the local air temperature under the OR's lights, where the surgeons stay, could be higher.

Figures 5-7 present the results for the PPD index for the three investigated values of RH: 40%, 50% and 60%, respectively. It can be seen that for 24 °C, which is still in the suggested temperature interval for an OR, approx. 25% of the surgeons (for RH 40%, Fig. 5) and approx. 30% of the surgeons would complain (for RH 60%, Fig. 7) from the thermal environment.

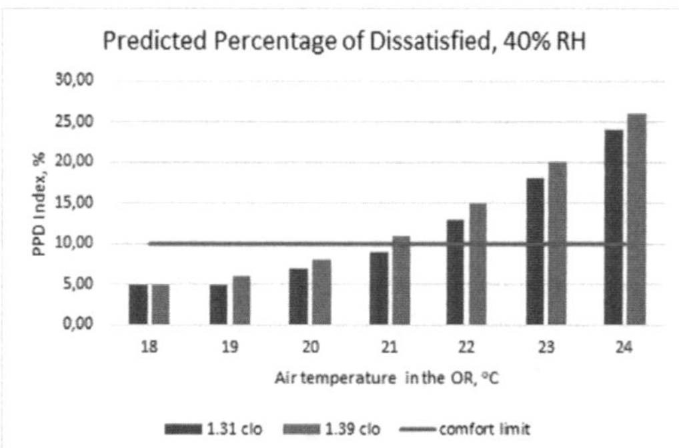

Figure 5. PPD index for surgeons, dressed in CAS (1.31 clo) and SSS (1.39 clo): air RH 40%

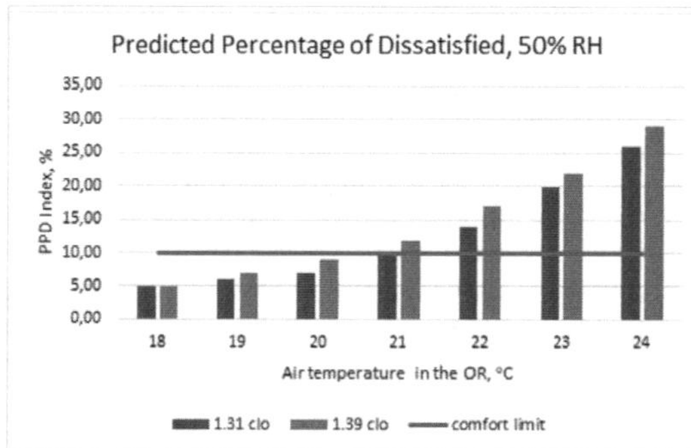

Figure 6. PPD index for surgeons, dressed in CAS (1.31 clo) and SSS (1.39 clo): air RH 50%

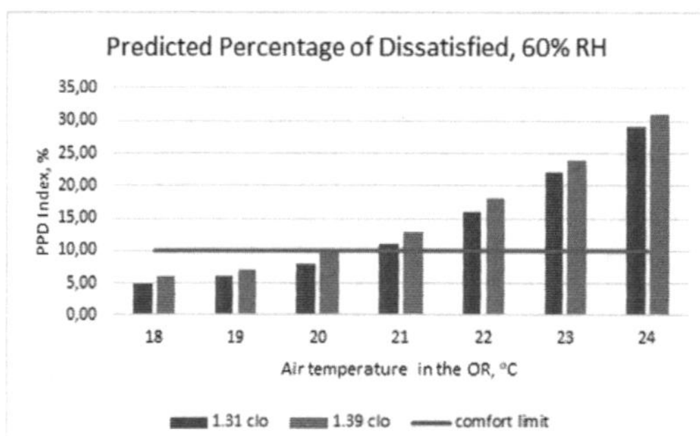

Figure 7. PPD index for surgeons, dressed in CAS (1.31 clo) and SSS (1.39 clo): air RH 60%

The analysis of Figures 2-7 shows that temperature up to 20 °C is favourable for the thermophysiological comfort of the surgeons in the OR when using the investigated types of protective clothing. However, the decrement of the temperature is unfavourable for the patient, who is under anaesthesia, with lower insulation over a naked body. An option is to use local methods for warming the patient. Another option is to play to a certain level with the air humidity in the OR.

In fact, the human body does not have a sensory organ for air humidity [19]. The RH decrement is related to upper airways discomfort and irritation in eyes [20], but these symptoms usually appear after prolonged exposure to RH in the room below 30%. As for the work performance, it is barely affected, especially for short exposures (below 5 hours): dry air was found to have decreased the rate of performance by 3%–7% [21].

The predicted values for the PMV and PPD show that the thermophysiological comfort of a surgeon would be assured in an OR in the following thermal environment (the determination of the interval is done for the combination between lowest to highest temperature and from lowest to highest RH):

- If dressed in CAS (1.31 clo): from 18 °C, 40% RH (PMV = -0.02) to 21 °C, 50% RH (PMV = +0.5).
- If dressed in SSS (1.39 clo): from 18 °C, 40% RH (PMV = +0.08) to 20 °C, 50% RH (PMV = +0.43).

4. Conclusions

This study was performed to compare the effect of two protective clothing ensembles in an OR on the thermophysiological comfort of the surgeons. The difference between the clothing was the level of clothing insulation and the type of protection: a surgical scrub suit (SSS) and a clean air suite (CAS). The calculation of the PMV and PPD indexes was used to assess the thermophysiological comfort of the surgeons in the temperature interval from 18 °C to 24 °C and an interval of the air relative humidity from 40% to 60%.

The results obtained and their analysis showed that the surgeons would stay in the condition of thermophysiological comfort when the temperature of the OR is up to 20 °C. The relative humidity for 18 °C and 19 °C can change from 40% to 60% without affecting the comfort of the surgical staff. For a temperature of 20 °C, the humidity should be lower than 60%, if SSS is used. Clean air suits could also be used when the temperature of the OR is 21 °C, but the RH should be lower than 60%.

A recommendation for future work is to compare the thermophysiological comfort of different staff in an OR: surgeons, anaesthetists, nurses. The effect of the duration of the operations should also be investigated.

References

[1] Angelova RA 2015 *Textiles and human thermophysiological comfort in the indoor environment* (NY, CRC Press)
[2] International Organization for Standardization 2005 *ISO 7730*
[3] Metcalfe D 2001 *Medical Textiles* (Manchester, Woodhead Publishing)
[4] Balaras CA, Dascalaki E and Gaglia A 2007 *Energy and Buildings* **39**(4) 454-470
[5] Tacutu L, Nastase I, Bode F, Dogeanu A and Croitoru C 2019 *E3S Web of Conferences* (Vol. 111). EDP Sciences
[6] Cao G, Storås MC, Aganovic A, Stenstad LI, Skogås JG 2018 *Am. J of Inf. Control* **46**(10) 1115-1122
[7] Bogdan A, Sudoł-Szopińska I and Szopiński T 2011 *FTEE* **19** 65-69
[8] Abreu I, Ribeiro P and Abreu MJ 2017 *Dyna* **84**(200) 234-239
[9] Kuivjogi, H. (2019). Field investigation of thermal comfort level of patients and surgical staff in operating rooms at St Olavs hospital (Master's thesis, NTNU)
[10] Van Gaever R, Jacobs VA, Diltoer M, Peeters L and Vanlanduit S 2014 *Building and Environment* **81**(11) 37-41
[11] Pourshaghaghy A and Omidvari M 2012 *Applied ergonomics* **43**(6) 1089-1095
[12] Fanger PO 1972 *Thermal comfort analysis and applications in environmental engineering* (NY McGraw-Hill)
[13] International Organization for Standardization 2004 *ISO* 7933
[14] Malchaire J 2016 *Computation of the PMV-PPD, WBGT and PHS indexes*
[15] ASHRAE Standard 170-2013, ASHRAE, 2013
[16] Chow T, Kwan A, Lin Z and Bai W 2006 *J. of Hosp. Inf.* **64** 371-378
[17] Zwolińska M and Bogdan A 2013 *Int J Occup Saf Ergon* **19**(3) 443-453
[18] DuBois D and DuBois EF 1915 *Arch Int Med* **15**(5_2) 868-881
[19] Wolkoff P 2018 *Int J Hyg Environ Heal* **221**(3) 376-390
[20] Bluyssen PM, Roda C, Mandin C, Fossati S, Carrer P, De Kluizenaar Y, Mihucz VG, de Oliveira Fernandes E and Bartzis J 2016 *Indoor Air* **26**(2) 298-317
[21] Wyon DP, Fang L, Lagercrantz L and Fanger PO 2006 *Hvac&R Research* **12**(2) 201-213

IOP Conf. Series: Materials Science and Engineering **878** (2020) 012063 doi:10.1088/1757-899X/878/1/012063

Classification of parts used in mechatronic products and produced by permanent-mold casting methods

V Zaharinov, I Malakov, S Nikolov, R Dimitrova and G Stambolov

Technical University Sofia, blvd. Kliment Ohridski 8, Sofia, Bulgaria

vzaharinov@tu-sofia.bg

Abstract. The paper presents the results from a research on parts used in mechatronic products and produced by the permanent-mold casting methods of gravity die casting and die casting. An overview and analysis of different definitions for mechatronic systems and examples of mechatronic products is made. As a result, a new definition for a mechatronic product is proposed. Various classifications of mechatronic products are analysed and deemed too general for the purposes of the current research. Permanent-mold casting methods are analysed and their advantages and disadvantages are pointed out. Arguments as to why these methods are suitable for producing mechatronic product parts are given. Based on the proposed definition for mechatronic product, the analysed classifications, and the overview of the permanent-mold casting methods a new classification is proposed. The classification is used for classifying permanent-mold cast parts used in mechatronic products. The parts are classified in six groups with regard to their application in the mechatronic product. In addition, the classification gives information regarding permanent-mold casting method used, material, and mechatronic product that includes the part.

1. Introduction

According to the first definition of mechatronics, proposed by an engineer at Yaskawa Electric Corporation, called Tetsuro Mori [1], and the definitions that were brought up later [2, 3, 4, 5, 6, 7], a great deal of the engineering products designed and manufactured in the last forty years, that integrate mechanical and electrical systems, may be classified as mechatronic systems. Some examples of such systems are industrial robots, CNC processing machines, car control systems like ABS, ESP, automated measuring systems, etc. [4, 6, 7, 8, 9].

Before classifying anything related to mechatronic products it must be clarified what is a "mechatronic product". There are a number of publications [6, 7, 10, 11,] that give definitions for a "mechatronic system", and use the term interchangeably with "mechatronic product". There are authors [4] that make distinction between a "mechatronic system" and a "mechatronic product", but do not give a clear definition of the latter. Therefore, due to this discrepancy, there is a need for clarification and definition of the term.

The choice of a new product for manufacturing, the choice of a test object for research in a particular manufacturing process, or starting a business of offering manufacturing services, is related to an extensive research on the process itself, the parts that are suitable for production with said process, and the products in which these parts are used as components. One tool for directing such searches is existing classifications of the types of parts, types of products, and the manufacturing processes used. Surprisingly enough, although mechatronic products today are widely used, and mechatronic design is a strong tendency in engineering design, there are relatively small number of classifications proposing some kind of taxonomy of mechatronic systems [4, 12, 13, 14]. In addition, to the authors' knowledge there are no existing classifications aiding in the search of parts used in mechatronic products and produced by gravity die casting and/or die casting. The development of such a classification will aid future searches in the area of permanent-mold cast parts and mechatronic product components.

The aim of the current paper is to present the results from a research on permanent-mold cast parts used as mechatronic product components. To that end, a definition for the term "mechatronic product" is proposed, existing classifications of mechatronic products are analyzed, an overview of gravity die casting and die casting processes is made, and an open classification scheme summarizing the research, and enabling future expansion is developed. The presented work is part (subtask) of a larger research effort for the development of optimization methods and algorithms for automated permanent-mold casting processes through optimization of the robotic servicing of a permanent-mold casting machine.

2. Definition of a mechatronic product

Despite the variety of definitions for mechatronics, their analysis shows, that the authors agree in one thing: mechatronics unite at least several scientific and/or engineering fields. Because the nature of mechatornics is interdisciplinary, finding a widely accepted answer to the question "Which product is mechatronic?" is difficult.

At first, mechatronics referred to systems including only mechanical and electrical components - no computational system was included. Examples of such systems are: automated sliding door, vending machines, and automated garage doors. Nowadays, practically all mechanical devices include electronic components and some kind of computer monitoring or control. Therefore, the term "mechatronic" encompasses a huge number of devices and systems [15].

The term "mechatronic product" very often [6, 7, 10, 11] is used together with "mechatronic system" interchangeably, without clarification as to why. The authors [5, 7, 15, 16] instead of giving definitions for a "mechatronic product", prefer to give examples of mechatronic systems. In [7] a block diagram is proposed of a typical mechatronic system, including the following main components: user interface; controller; drive circuits; actuators; mechanical system; sensors; signal conditioning. In [15] also a schematic of the main components of a mechatronic systems is proposed which agrees with the previously mentioned components.

The mechatronic systems are characterized by the following features [4]: integration of mechanical and electronic components; software that defines functions; new design instruments for concurrent design; synergistic effects.

Some authors [4] suggest a certain degree of nuance between a mechatronic system and a mechatronic product: "While not a mechatronic product on the order of a camcorder, it is (home/office heating/cooling system) a mechatronic system because of its combination of mechanical, electrical, and computer components."

In an attempt to clarify the term "mechatronic product", and for the purposes of the paper, a definition of a "mechatronic product" is proposed, as follows: mechatronic is a product, that realizes full integration of mechanical systems, drive systems, and control systems into one aggregate technical system through the synergistic interaction between two main flows - energetic and informational.

3. Classifications of mechatronic products

For the systematic study of mechatronic products that include parts produced by permanent-mold casting methods, the study of existing classifications of mechatronic products is beneficial.

The rarity of mechatronic products' classifications in the specialized literature could be explained, to some extent, with the great variety of applications in which mechatronic products are used.

In [12] maybe one of the first classifications of mechatronic products is presented, developed in the late seventies of the twentieth century. The classification divides the mechatronic products in four classes:

- Class I: mainly mechanical products with included electronics for increased functionality. Examples are NC machines and velocity control drives in processing machines;
- Class II: traditionally mechanical systems with considerably modernized structure which includes electronics. The user interface is not changed. Examples are modern sewing machines and automated production systems.
- Class III: systems that keep the functionality of traditional mechanical systems but their components are replaced with electronics. The example given is a digital watch;
- Class IV: products designed with mechanical and electronic systems through synergistic integration. Examples are photocopiers, intelligent washing machines and dryers, automated cooking appliances and automated ovens.

According to [4] mechatronic products can be divided into mechatronic systems, mechatronic machines, mechatronic vehicles, precision mechatronics, and micro mechatronics.

In [13] examples are proposed of various mechatronic systems, which are divided into separate groups as follows: mechatronic machine components, mechatronic motion generators, mechatronic power producing machines, mechatronic power consuming machines, mechatronic automobiles, mechatronic trains.

In [14] the mechatronic systems are divided into conventional mechatronic systems, micromechatronic systems, and nanomechatronic systems.

These classifications, although some [12, 13] give examples of classified products, are very general and hardly can be used when searching for particular mechatronic products, their main components or manufacturing processes used. For the aims of the research laid out in the current paper, these classifications do not give enough information, and do not aid significantly the information search process.

4. Gravity die casting and die casting as processes for production of parts for mechatronic products

Some authors [4] point out, as a reason for the advent of mechatronic products, the desire for increasing the ratio of performance over costs, and the creation of new products, situated at the intersection point of the traditional engineering disciplines, informatics, and the natural sciences. A lot of today's mechatronic products are integrated systems from which high performance parameters are expected. Obtaining high performance parameters and enhanced functions puts a lot of pressure on the design of such products, and the use of new and complex methods for design optimization becomes mandatory. Advanced optimization techniques like shape and topology optimization generate complex geometric structures, that can be produced with economic benefit only through casting processes and/or 3D printing. As application of metal 3D printed parts outside the prototyping field gets more and more common [17], still the majority of metal parts with high importance to system performance and with complex shape are produced through casting processes. Moreover, 3D printer technology cannot provide the production rate that a casting process can. Some of the advantages of the metal casting processes, such as [18]: creation of complex external and internal geometry, and part production without the necessity of additional processing for obtaining the end product geometry and dimensions, also make them suitable for the production of mechatronic system components.

Gravity die casting, and die casting, are metal casting processes that use permanent-molds (metallic molds or a combination of metallic molds and sand cores) [18, 19]. Furthermore die casting is performed in two types of technological processes: hot-chamber die casting, and cold chamber die casting.

The main advantages of permanent-mold casting are [18]: good part surface quality; high precision; due to the faster solidification when using metal molds, structure with finer granularity is produced, leading to better strength characteristics compared to expendable-mold casting processes. Additionally, both gravity die casting and die casting have the following advantages [20, 21]: suitable for mass production; high precision of the dimensions (higher for die casting); good quality and details of the part surface (excellent for die casting); high production rates.

The main disadvantages of permanent-mold casting are [18]: the process is limited to metals with lower melting temperatures; simpler geometry compared to expendable-mold castings, due to the need for opening the mold; price of the mold. Additionally, both gravity die casting and die casting have the following disadvantages [20, 21]: not suitable for large castings; high surface hardness and internal stresses of the casting due to fast cooling; not suitable for small production quantities.

Both processes, gravity die casting and die casting, are mainly used for casting of non-ferrous metals like aluminium, zinc, magnesium, copper, lead, and low melting point alloys of said metals [18, 19, 22]. Some example parts produced with these processes are: carburettors housings, pump housings, connecting rods, internal combustion engine pistons, hydraulic break cylinders, parts for fridges and washing machines, gears and gear housings, parts for aircraft and rocket projectiles.

5. Classification of parts used in mechatronic products and produced by gravity die casting and die casting

A classification of parts produced by gravity die casting and die casting, and used as mechatronic product components is proposed. The classification divides the parts by their application in the mechatronic product in six groups, as follows: mechanical system, optical system, sensors, actuators, control system, and user interface (figure 1). These groups account for the main components of the mechatronic product with the addition of the optical system, which accounts for opto-mechatronic products [23].

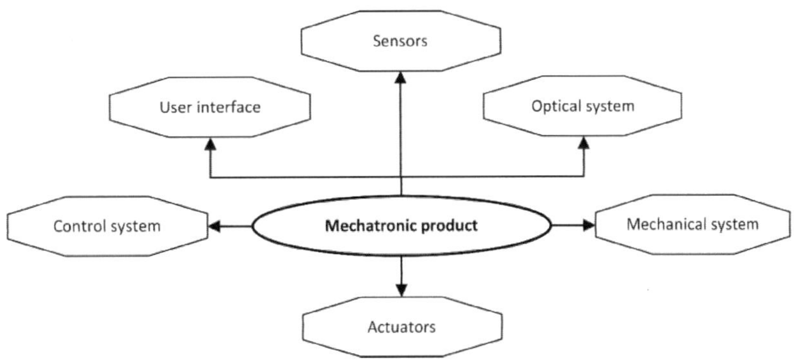

Figure 1. Classification groups

The classified parts in each group are shown in figure 2, figure 3, figure 4, figure 5, figure 6, and figure 7. Additionally, for each part classified, information regarding the manufacturing process and material used is given. These are noted on the left and right sides of the block for the part accordingly, where DC means die cast; GC - gravity die casting; Al - aluminium; Zn - zinc; Mg - magnesium; Br - bronze. More than one manufacturing method, and more than one material can be utilized (attached to the block of a part) for the production of the same part. Under each part's box are given examples of mechatronic products in which it is used.

Twenty-nine parts are classified in the shown classification. From the parts included in the latter it can be seen that the gravity die casting and die casting manufacturing processes are used mainly for parts that function as housings for components. In addition, from the classification can be seen that a large number of the parts are made from aluminium and its alloys, although zinc and zinc alloy parts are also commonly used.

As noted before, mechatronic products encompass a huge number of devices and systems, and the classification of all existing parts by the classification criteria given, is a labour intensive and long process. Therefore, the classified 29 parts shown here do not make by any means an exhaustive list. There are also the space limitation constraints for the paper to be considered.

The given classification is just a beginning and gives a base that one can build upon. That is why the classification scheme is open, and easy to expand.

The proposed classification can be used as a tool in aiding the search for: mechatronic products, parts that build mechatronic products, permanent-mold processes used in manufacturing parts for mechatronic products, and commonly used materials for these parts.

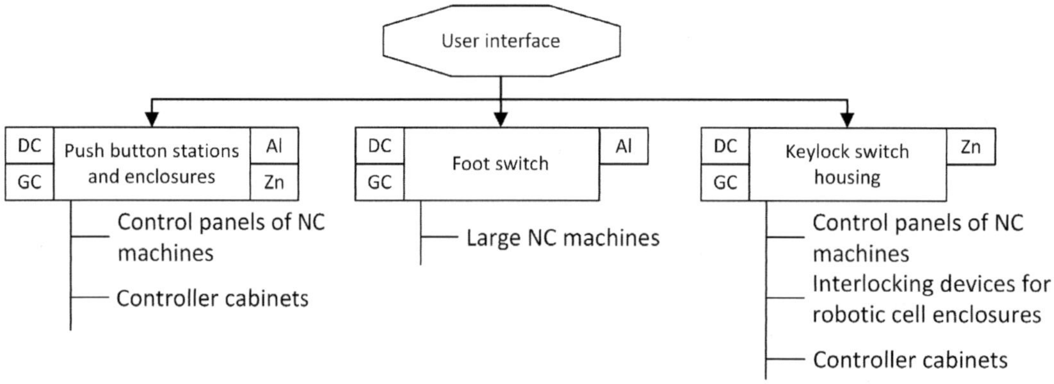

Figure 2. User interface group

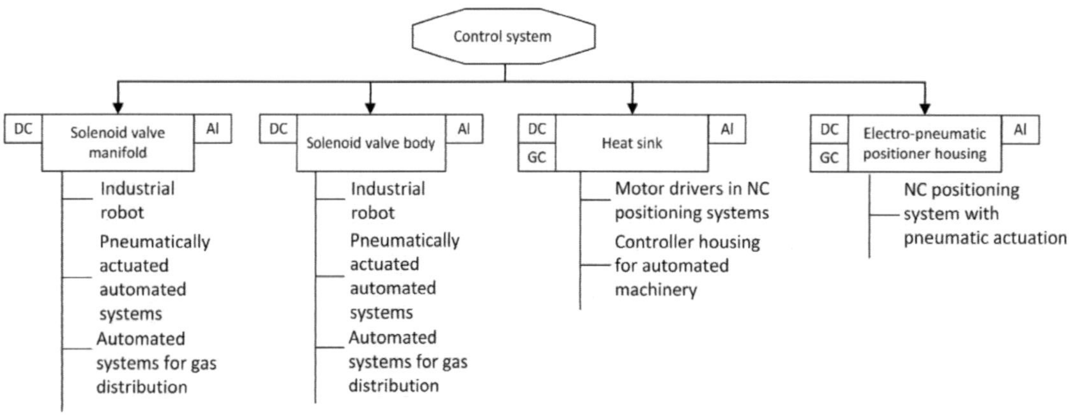

Figure 3. Control system group

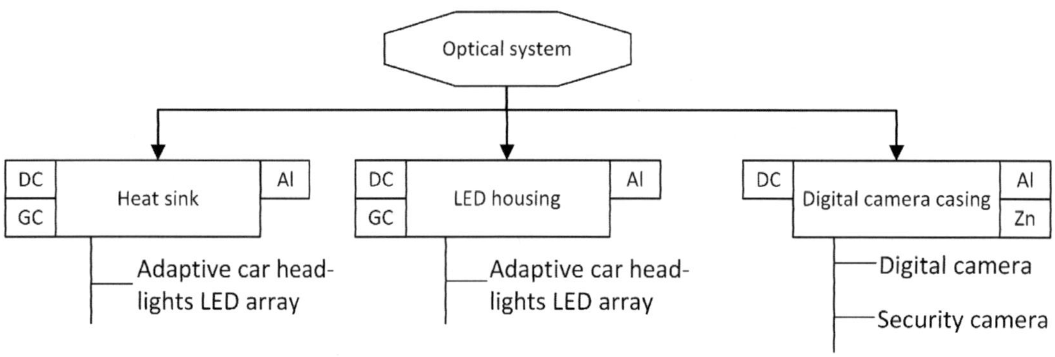

Figure 4. Optical system group

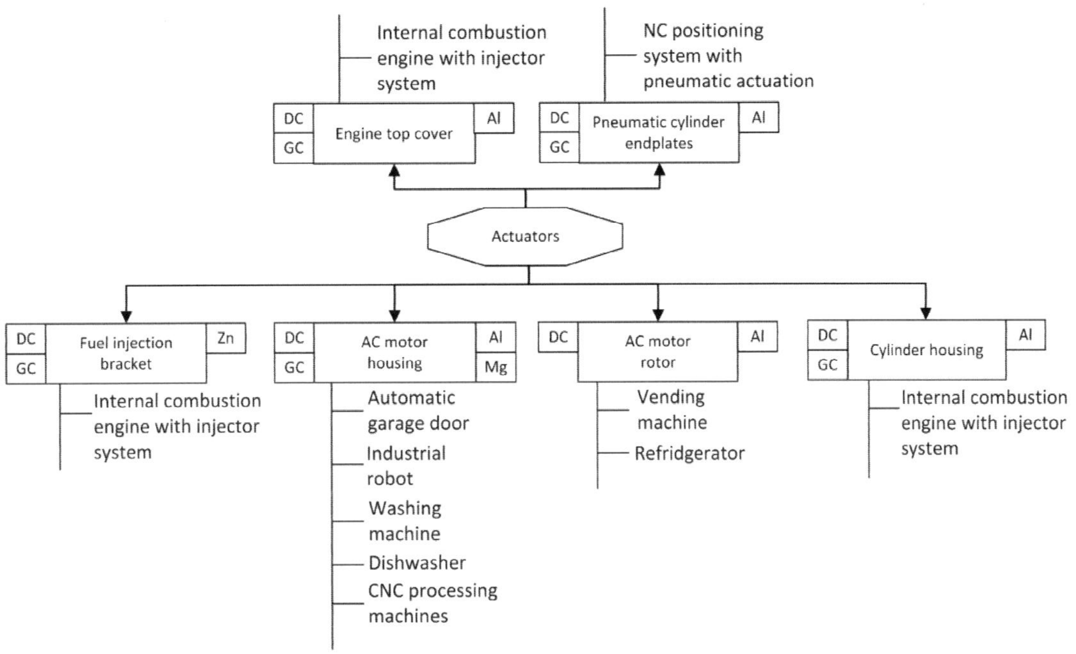

Figure 5. Actuators group

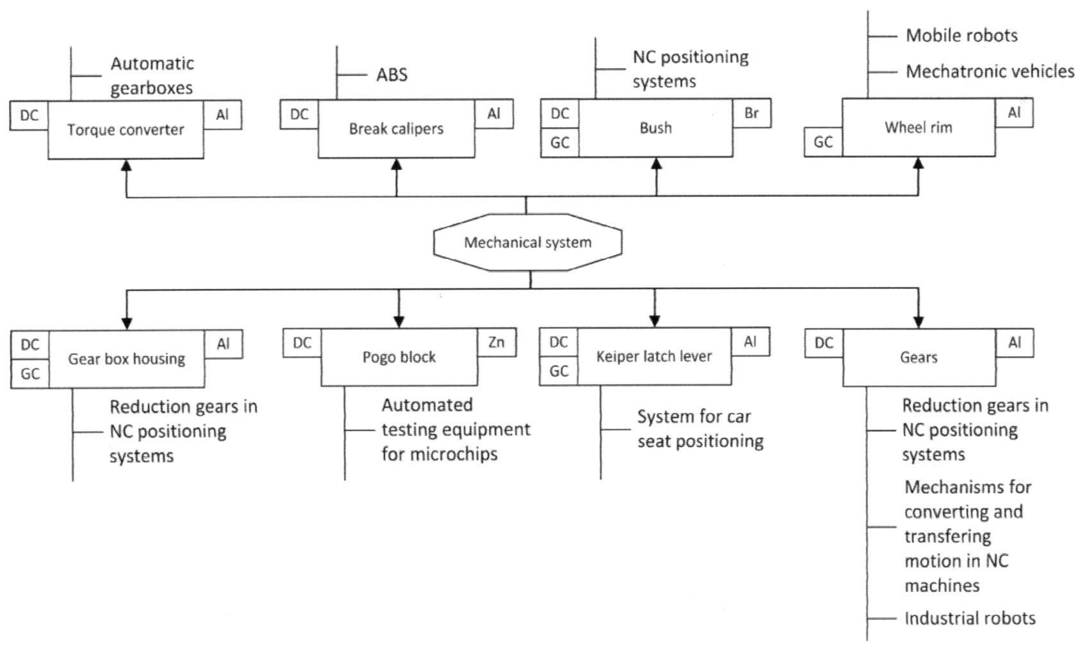

Figure 6. Mechanical system group

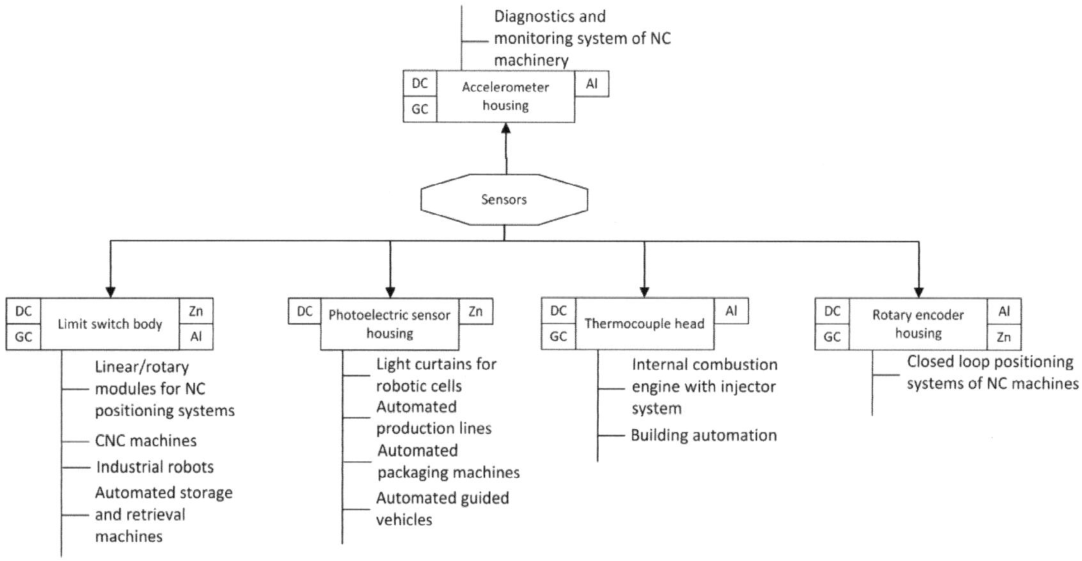

Figure 7. Sensors group

6. Conclusion

From the research laid out in the paper the following important results are obtained:

- A definition of mechatronic product is proposed;
- An open classification scheme for parts used in mechatronic products and produced by gravity die casting and die casting is developed;
- Twenty-nine parts conforming to the classification scheme criteria are classified, showing how the proposed classification is built, and how it can be used for classification of parts;
- The most of the classified parts are used as housings;
- Out of the classified parts, the most commonly used materials are aluminium/aluminium alloys, and zinc/zinc alloys;
- Examples are given of mechatronic products in which the classified parts are used.

The proposed definition for mechatronic product is based on definitions from literature for mechatronic systems and it helps clarify the relation between a mechatronic product and a mechatronic system.

The developed classification can be used as a research aid tool, when information is sought regarding parts, processes, and materials used for manufacturing of mechatronic product parts.

The classification uses an open scheme which allows for expansion, and thus further development of the classification is related to adding more classified parts to it and/or adding more classification groups. Additionally, as the presented research is a part of a larger scope research, the classification will be used in the development of design and optimization methods for two automated robotic systems servicing a die casting and a low-pressure casting machine.

Acknowledgments

This work was supported by the European Regional Development Fund within the Operational Programme "Science and Education for Smart Growth 2014 - 2020" under the Project CoE "National center of mechatronics and clean technologies" BG05M2OP001-1.001-0008

References

[1] Mori T 1969 "Mechatronics, Yasakawa Internal Trademark Application Memo 21.131.01"
[2] Harashima F, Tomizuka M and Fukuda T 1996 *IEEE/ASME Transactions on Mechatronics* **1** 1-

4 ISSN 1941-014X
[3] Auslander D M 1996 *IEEE/ASME Transactions on Mechatronics* **1** 5-9 ISSN 1941-014X
[4] Bishop R 2002 *The Mechatronics Handbook - 2 Volume Set* Mechatronics Handbook 2e (CRC Press) ISBN 9781420042450
[5] Onwubolu G 2005 *Mechatronics Principles and Applications* (Butterworth-Heinemann) ISBN 9780750663793
[6] Shetty D and Kolk R A 2010 *Mechatronic System Design* (Cengage Learning) ISBN 143906198X
[7] Jouaneh M 2013 *Fundamentals of Mechatronics* (Cengage Learning) ISBN 9781111569013
[8] Karastoyanov D, Stoimenov N and Gyoshev S 2019 *IFAC-PapersOnLine* **52** 539-542 ISSN 2405-8963 19th IFAC Conference on Technology, Culture and International Stability TECIS 2019
[9] Dichev D, Koev H, Diakov D, Panchev N, Miteva R and Nikolova H 2017 *Proceedings Elmar – International Symposium Electronics in Marine* **2017-September** 219-224
[10] Bolton W 2015 *Mechatronics: Electrical Control Systems in Mechanical and Electrical Engineering* (Pearson) ISBN 9781292076683
[11] Isermann R 2005 *Mechatronic Systems: Fundamentals* (Springer-Verlag London) ISBN 9781852339302
[12] Kyura N and Oho H 1996 *IEEE/ASME Transactions on Mechatronics* **1** 10-15 ISSN 1941-014X
[13] Isermann R 2008 *Control Engineering Practice* **16** 14-29 ISSN 0967-0661 URL http://www.sciencedirect.com/science/article/pii/S0967066107000603
[14] Zambare A, Lande P, Walke P, Soudagar S and Gawali R 2016 *IJRASET* **4** 49-54 ISSN 2321-9653
[15] Alciatore G D and Histand M B 2012 *Introduction to Mechatronics and Measurement Systems* (McGraw-Hill) ISBN 9780073380230
[16] Billingsley J 2006 *Essentials of Mechatronics* (John Wiley and Sons) ISBN 9780471723417
[17] Savov I, Todorov G, Sofronov Y and Kamberov K 2019 *IUTAM* Bookseries **33** 149-163
[18] Groover M P 2010 *Fundamentals of Modern Manufacturing: Materials, Processes, and Systems* (John Wiley and Sons) ISBN 9780470467008
[19] Frolov K V et al. 2005 *Machines and Equipment used in the Forging, Stamping, and Casting Manufacturing (Russian)* (*Mechanical Engineering. Encyclopedia in 40 volumes* vol IV-4) (Mashinostroenie) ISBN 521702416X
[20] Singh R 2006 *Introduction to Basic Manufacturing Processes and Workshop Technology* (New Age International) ISBN 9788122423167
[21] Gupta H, Gupta R and Mittal A 2009 *Manufacturing Processes* (New Age International) ISBN 9788122428445
[22] Tonchev N 2005 Technology of metal materials (bulgarian) (*Preprint* www.castingarea.com)
[23] Cho H et al. 2003 *Opto-mechatronic systems handbook: techniques and applications* (CRC Press) ISBN 0849311624

IOP Conf. Series: Materials Science and Engineering **878** (2020) 012064 doi:10.1088/1757-899X/878/1/012064

Impact of atmospheric corrosion on the mechanical properties of B235 steel rods

Veselin Tsonev[1], Nikolay Nikolov and Kamen Penkov

Technical University – Sofia, Strength of Materials Department, 8 Kliment Ohridski Blvd., 1000 Sofia, Bulgaria

[1] E-mail: tzonev@tu-sofia.bg

Abstract. It is well known that atmospheric corrosion greatly affects the stress-strain properties of reinforcing steel. To be able to assess and forecast the remaining service life of constructions which have been subjected to atmospheric corrosion, it is necessary to have good knowledge of this effect. Research is often complicated to do due to the lack of accurate information on the initial dimensions of the steel parts and the properties of the material of which those parts are made. In this paper an attempt is made to overcome this obstacle. Reinforcing steel rods are tested. Those rods have been subjected to natural atmospheric conditions in a temperate climate zone for 25 years, in vertical orientation. To determine the initial stress-percentage extension curve of those rods test pieces have been made by machining on a lathe – to remove the corrosion layer and to set a standard diameter. The chemical composition and the density of the material are determined by appropriate experiments. The yield strength, the ultimate strength, the percentage elongation after fracture are determined by standard tensile testing. To determine the final stress-percentage extension curve, tensile tests are carried out with the corroded rods without any treatment. A comparison of both curves and analysis of the atmospheric corrosion effect on the stress-strain behavior is made.

1. Introduction

It is often necessary to assess the remaining service life of corroded constructions after their prolonged service in an aggressive environment. This assessment is often difficult to make due to the lack of information about the initial properties of the materials and about the exact shape and dimensions of the parts.

The inconvenience caused by the lack of information about the initial properties of the materials can be easily overcome by removing the corrosion layer from the corroded parts followed by machining of standard test pieces which are subjected to a tensile test in standard conditions.

When the mechanical properties of real corroded constructions must be assessed the only possible approach is to cut off the test pieces from the investigated construction. Usually the reinforcing rods are extracted from concrete constructions which have stayed for long period of time (20-40 years) in an atmospheric or an artificial aggressive environment [1, 2] or parts are cut off from steel constructions which have worked in similar conditions [3].

When the goal is to determine the amount of corrosion and the rate of corrosion in order to predict how long the construction may be used in an aggressive environment, it is necessary to know the initial shape and dimensions of the parts. The inability to measure them with great accuracy is compensated by some approximation using data from the accompanying technical documentation. If such information is not available the problem of determining the rate and the amount of corrosion is impossible to solve.

The object of this research are corroded smooth rods of reinforcing steel with nominal diameter of 6,5 mm. The rods have stayed outdoors for 25 years (atmospheric corrosion) oriented vertically in a temperate climate zone. The chemical composition (table 1) is studied in a specialized laboratory. The steel is determined as grade B235 in accordance with the acting Bulgarian standard BDS 4758:2008 [4]. This steel with some approximation can be accepted as an analog for the widely used structural steel S235JR [5]. The European standard dedicated to reinforcing steels is EN 10080:2005 [6]. The standard does not discuss specific grades of steel or compositions, but the upper limits for the quantities of some chemical elements are given. The composition of the major alloying elements is taken into account by calculating the equivalent amount of carbon using the following formula, in which the values from table 1 are entered:

$$C_{eq} = C + \frac{Mn}{6} + \frac{Cr + Mo + V}{5} + \frac{Ni + Cu}{15} = 0.2 + \frac{1.18}{6} + \frac{0.12 + 0 + 0}{5} + \frac{0.28 + 0.62}{15} = 0.48. \quad (1)$$

The value of C_{eq} is lower than the standard upper limit of 0.50 [6] which means that the amount of alloying elements in the tested steel complies with the requirements of the standard.

Table1. Chemical composition of the tested rods.

Element	Amount, %	No grade, EN 10080:2005, %	B235 BDS 4758:2008, %	S235JR EN 10025-2:2004, %
C	0.2000	≤ 0.24	≤ 0.20	≤ 0.19
C_{eq}	0.48	≤ 0.50		
Mn	1.1800			≤ 1.50
Cu	0.6200	≤ 0.85		≤ 0.60
Ni	0.2800			
Si	0.2300			
Cr	0.1200			
As	0.0585			
S	0.0450	≤ 0.055	≤ 0.040	≤ 0.045
P	0.0240	≤ 0.055	≤ 0.040	≤ 0.045
N	0.0075	≤ 0.014	≤ 0.014	≤ 0.014

The goal of this paper is to make an assessment of the impact of atmospheric corrosion on the mechanical properties of reinforcing steel B235, analog of structural steel S235. To investigate this effect two sets of tensile tests have been done. The first set is with cleaned test pieces cut off from the corroded rods and without additional treatment. The second set is with standard test pieces machined on lathe to a diameter of 4 mm.

2. Test pieces for tensile testing

On figure 1 the different stages of preparation of the test pieces are shown. The corroded rod in its original appearance is shown in figure 1 (a). Its surface is covered with a dense layer of corroded products with high mechanical strength, which is hard to remove. Figure 1 (b) shows a rod after several cycles of chemical cleaning according to EN ISO 8407:2014 standard [7]. The uneven nature of corrosion can be observed with multitude of pittings and patches, and a large amount of residual corrosion products. On figure 1 (c) a photograph of a fully cleaned rod is shown. A combination of mechanical cleaning (scrubbing with brush) and chemical cleaning is used. The surface texture is preserved as much as possible. The piece is ready for measurement, weighing and mechanical testing. On figure 1 (d) a standard test piece for tensile testing [8] is shown. The test piece is made from the corroded rod by machining it to a diameter of 4 mm. The purpose of this test piece is to get the stress-percentage extension curve of the uncorroded material.

(a) (b) (c) (d)

Figure 1. Preparation of the corroded test pieces for tensile testing:
(a) original appearance; (b) partial chemical cleaning; (c) fully cleaned using mechanical and chemical process; (d) standard test piece machined from the corroded rod.

3. Properties of the test pieces

3.1. Density of the steel

As the initial mass (before the beginning of the corrosion) of the rods is not known it is necessary to know the exact density of the material to be able to calculate the mass of the rods knowing their dimensions.

The density of the steel is determined experimentally. A cylindrical piece with accurate dimensions – 4 mm diameter and 10 mm length – is machined using a lathe. The final dimensions of the same piece are measured with a micrometer and its volume is calculated. It is also weighed with 0,001 g accuracy. Finally the steel density is calculated to be 7745 kg/m^3.

3.2. Initial dimensions and mass

The initial dimensions and mass of the studied corroded rods are unknown. It is known that these are reinforcing rods manufactured in accordance with BDS 4758:1984 standard, which is a predecessor to [4]. This means that the initial diameter of the rods is d_1 = 6.5 mm. If the diameter and the density are known, it is very easy to calculate the initial parameters of the test piece with a 100 mm length – volume V_1 = 3318.31 mm^2; surface area S_1 = 2108.4 mm^2; mass m_1 = 25.7 g.

3.3. Final dimensions and mass

The diameter of a set of three pieces of cleaned corroded test pieces with a length of 100 mm as shown in figure 1 (c) is measured with a micrometer in four cross-sections. The average diameter d_{av} and the final mass m_2 of the test pieces are determined. The results are given in table 2.

Table 2. Determination of average diameter d_{av} and average final mass m_2 of the corroded test pieces.

№	Measured diameter *d*, mm				d_{av},	m_2,
	section 1	section 2	section 3	section 4	mm	g
1	6.25	6.21	6.21	6.17	6.210	22.888
2	6.24	6.18	6.20	6.18	6.200	22.483
3	6.25	6.18	6.18	6.21	6.205	22.479
				Average:	*6.205*	*22.617*

3.4. Mass loss

Mass loss for the 25-year period of corrosion in natural atmospheric conditions is calculated in three different forms:

- absolute mass loss of a 100 mm long rod:

$$\Delta m = m_1 - m_2 = 25.7 - 22.617 = 3.083\,\text{g}; \tag{2}$$

- specific mass loss:

$$\frac{\Delta m}{S_1} = \frac{3.083}{2108.4.10^{-6}} = 1462.25\frac{\text{g}}{\text{m}^2}; \tag{3}$$

- percentage mass loss:

$$\frac{\Delta m}{m_1}100 = \frac{3.083}{25.7}100 = 12\,\%. \tag{4}$$

3.5. Corrosion rate

The corrosion rate k is one of the most important parameters of the corrosion behavior, because it allows the remaining service life prediction in case of operation in corrosive environment. In our case:

$$k = \frac{\Delta m}{S_1 t} = \frac{3.083}{2108.4 \times 10^{-6} \times 9125} = 0.16\frac{\text{g}}{\text{m}^2\text{days}}, \tag{5}$$

where $t = 25 \times 365 = 9125$ days is the number of days in 25 years.

4. Tensile testing

The tensile tests are made using enhanced universal testing machine ZD10 with a lead screw loading mechanism which makes it possible to use precise digital control of loading [9] in accordance with EN ISO 6892-1:2016 standard [8].

A set of three proportional test pieces of each type, shown in figure 1 (c) and (d), are tested. The machined test piece has the behavior of non-corroded material. To measure extension, extensometers with gauge length equal to the original gauge length of the test pieces are used, as given in table 3. The mechanical properties upper yield strength R_{eH}, ultimate strength R_m and percentage elongation after fracture A are taken from experimentally obtained stress-percentage extension curves, like ones shown on figure 2. The stress rate was 6 MPa/s.

Table 3. Mechanical properties in accordance with [8].

Type of test piece	Nominal diameter, mm	Gauge length, mm	№	R_{eH}, MPa	R_m, MPa	A, %
Non-corroded (machined from corroded)	4	20	1	314	391	27.8
			2	312	390	29.9
			3	314	395	28.5
			Average	*313*	*392*	*28.7*
Corroded (25 years atmospheric corrosion), percentage loss of mass 12%	6.5	32.5	1	272	346	20.2
			2	266	340	21.4
			3	268	345	20.8
			Average	*269*	*344*	*20.8*

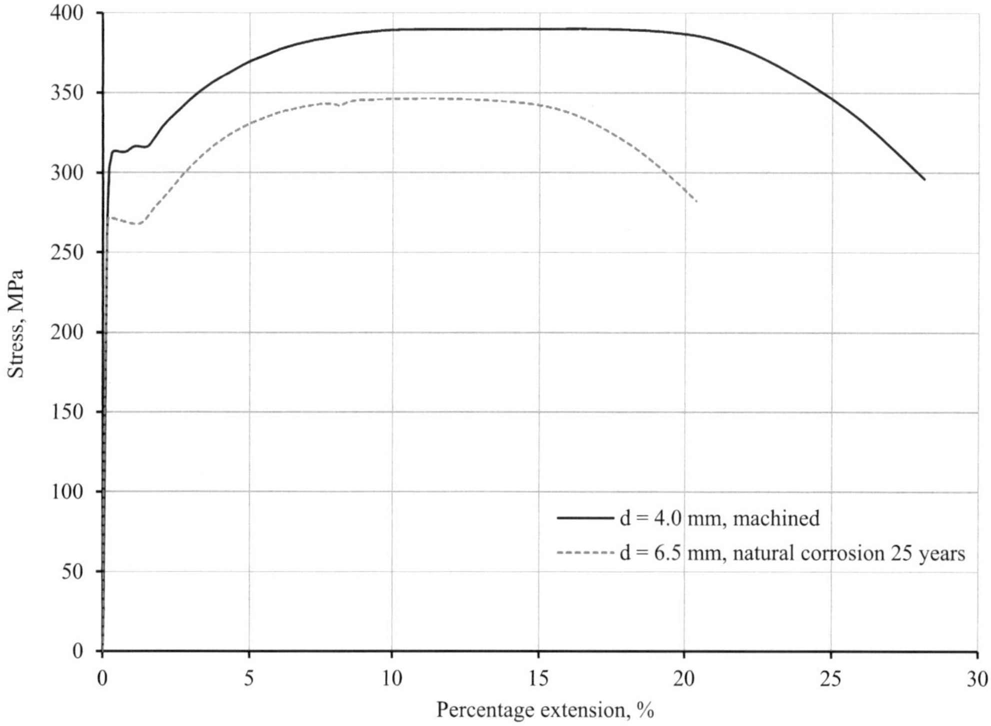

Figure 2. Stress-percentage extension curves – test piece №1 of each set.

In accordance with [4], for B235 steel the yield strength $R_{eH} \geq 235$ MPa, the ultimate strength $R_m \geq 370$ MPa and the elongation after fracture $A_5 \geq 25$ %. In accordance with [5] R_{eH} of S235JR steel must be greater than 235 MPa, the ultimate strength R_m must be between 360-510 MPa, and the elongation after fracture A_5 – 24-26%.

5. Analysis of the obtained results

Corrosion leads to loss of material and loss of area of the carrying cross sections. As a result the strength properties of the corroded piece decrease. In our case we can observe the expected decrease in the yield strength R_{eH} by 14.06% and the ultimate strength R_m – by 12.24% in the corroded test pieces. At 12% mass loss, the measured yield strength is above the minimal value set in standards [4, 5], but the ultimate strength is less than the lower limit set in the same standards.

The corrosion worsens the quality of the surfaces in contact with the aggressive environment. A large number of surface defects with complex and irregular shape and varying depth occur. In tensile testing these defects are stress concentrators, from where cracks are easily developed. This worsens not only the strength, but also the deformational properties of the corroded piece. In our case a significant decrease of plasticity of the examined test pieces is observed – at 12% mass loss the elongation after fracture A is reduced by 27.53% to a value which is significantly below the value prescribed in [4, 5].

The results presented in this paper are going to be compared with results from an accelerated corrosion test.

Acknowledgements
The authors would like to thank the Research and Development Sector at the Technical University – Sofia for the financial support.

References

[1] Apostolopoulos C A and Papadakis V G 2008 Consequences of steel corrosion on the ductility properties of reinforcement bar *Construction and Building Materials* **22(12)** 2316–2324
[2] Apostolopoulos C A, Papadopoulos M P and Pantelakis S G 2006 Tensile behavior of corroded reinforcing steel bars BSt 500s *Construction and Building Materials* **20(9)** 782–789
[3] Kaita T, Appuhamy J M R S, Itogawa K, Ohga M and Fujii K 2011 Experimental study on remaining strength estimation of corroded wide steel plates under tensile force *Procedia Engineering* **14** 2707–2713
[4] BDS 4758:2008 Steel for reinforcement of concrete. Weldable reinforcing steels B235 and B420
[5] BDS EN 10025-2:2019 Hot rolled products of structural steels – Part 2: Technical delivery conditions for non-alloy structural steels
[6] BDS EN 10080:2005 Steel for reinforcement of concrete – Weldable reinforcing steel – General
[7] EN ISO 8407:2014 Corrosion of metals and alloys – Removal of corrosion products from corrosion test specimens
[8] BDS EN ISO 6892-1:2016 Metallic materials – Tensile testing – Part 1: Method of test at room temperature (ISO 6892-1:2016)
[9] Tsonev V, Kuzmanov N, Borisov B and Penkov K 2019 System for materials testing at static loading *IOP Conference Series: Materials Science and Engineering* **618** 012048

IOP Conf. Series: Materials Science and Engineering **878** (2020) 012065 doi:10.1088/1757-899X/878/1/012065

The influence of magnesium alloy AE42 test specimens deformation magnitude on their microstructure

Y Marcheva [1,2], M Loukaycheva [1], V Tsonev [1] and B Dimova [1]

[1]Technical University – Sofia, Bulgaria

[2]E-mail: ysm@tu-sofia.bg

Abstract. It is known that the properties of metals and their alloys depend on their microstructure. An important aspect of microstructure study is the ability to predict the properties and the behavior of the material in the process of its exploitation taking into account determined structural changes and their relationship with the material properties. The goal of this work is to study the effect of the magnitude of the plastic deformation on the microstructure of magnesium alloy AE42 test specimens. This paper describes the actions to get test specimens with various magnitude of residual deformation. The microstructure of those test specimens is studied. The gathered results are discussed.

1. Introduction

The magnesium alloys find increasingly wider application in automotive industry, aircraft building industry, production of electronic devices, etc. [1].

The magnesium is the lightest structural metal, much lighter than steel, which is the most used material with industrial application. Unfortunately the magnesium alloys, available to the industry, are expensive. To produce automotive structural components, forming operations at high temperatures are required due to ductility problems of the magnesium alloys. This is the reason for the high cost of magnesium automotive parts. After a lot of resources were invested in research in the last twenty years, it has been discovered that the addition of rare-earth elements in small quantities can be a potential solution to achieve better ductility of magnesium alloys [2]. The most commonly used magnesium alloys which contain rare-earth elements are AE42 and AE44 [3].

The use of magnesium alloys is often limited by their high electrochemical activity and low corrosion resistance. The corrosion behavior of the metallic materials is highly dependent on their microstructure. The change of this microstructure under the influence of various factors may lead to changes in the corrosion behavior of a given alloy due to a change in the dimensions of metal grains, the borders of the grains, change in the distribution of intermetallic particles and others. The effect of the change of the microstructure of magnesium alloys on their corrosion behavior hasn't been systemically studied [1].

The microstructure of metal alloys can be affected by the influence of mechanical deformations, especially during the processes of forming metal parts using plastic deformation and in the cases when the material is subjected to stress higher than the elastic limit stress during its service life at ambient temperature. In the case of ductile materials, the yield stress R_{eH} ($R_{p0,2}$) can be accepted for allowed stress. That is why it is interesting to study the influence of the magnitude of plastic deformation on the microstructure of AE42 magnesium alloy test specimens as a part of wider study of corrosion resistance of magnesium alloys.

2. Experimental

AE42 is the representative of Mg-Al-RE system, which is in the group of high-pressure die casting (HPDC) alloys and are particularly attractive for mass production [4]. The chemical composition of the studied alloy is given in table 1.

Table 1. Chemical composition of the studied AE42 magnesium alloy, (wt %).

Al	*Ce*	*Mn*	*Zn*	*Cu*	*Ni*	*Fe*	*other*
3,6 – 4,4	max. 2,5	min. 0,1	max. 0,02	max 0,04	max 0,001	max 0,004	0,01

2.1 Preparing of test specimens with various magnitude of residual deformation

In order to determine the mechanical properties of the studied magnesium alloy, tensile tests have been carried out at ambient temperature using modernized testing machine ZD10 with lead screw loading mechanism-and numeric control [5], which assures precise loading in accordance with EN ISO 6892-1:2016.

Four test specimens were subjected to pure tension until their fracture, while continuously the values of the applied force and longitudinal deformation were recorded. The fracture of the test specimens occurs outside their gauge length – figure 1a. In table 2 the gathered results for the following parameters are given:

- Tensile strength R_m – the stress corresponding to the maximum force, which the test specimen can bear during testing;
- Proof strength, plastic extension $R_{p0,1}$ – the stress, under which the relative percentage extension is equal to 0,1%;
- Proof strength, plastic extension $R_{p0,2}$ – the stress, under which the relative percentage extension is equal to 0,2%.

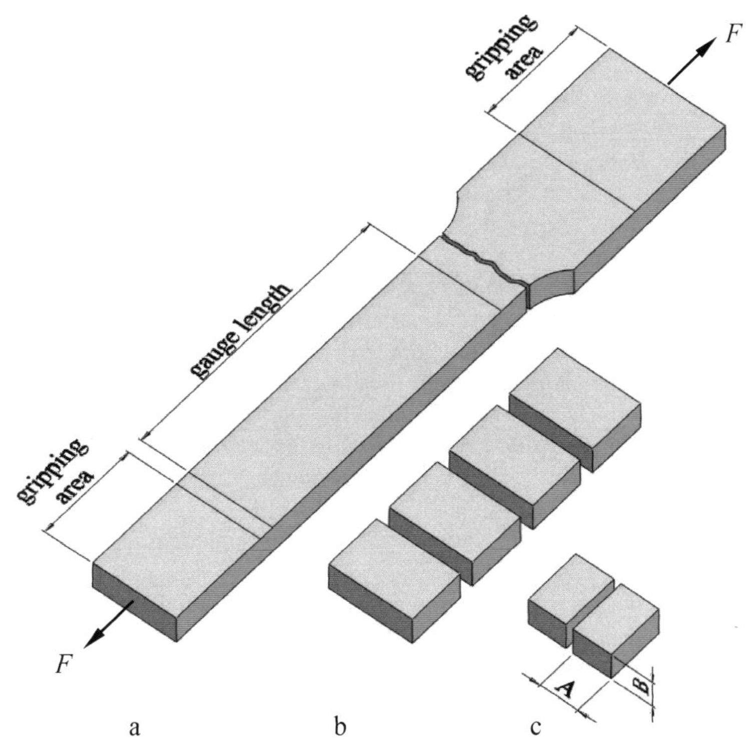

Figure 1. Test specimens.

Table 2. Mechanical properties of the studied AE42 magnesium alloy.

Test specimen №	R_m, MPa	$R_{p0,1}$, MPa	$R_{p0,2}$, MPa
1	203	98	121
2	209	94	118
3	205	97	119
4	210	99	123
Average	**207**	**97**	**120**

Four test specimens are loaded until $R_{p0,1}$ is reached, after that the load is removed. Four test specimens are loaded until $R_{p0,2}$ is reached, then the load is removed. Four test specimens are not subjected to any load.

Four sets of 4 test specimens each are produced. The first set contains 4 non-loaded test specimens, the second set contains 4 test specimens loaded up to $R_{p0,1}$, the third set contains 4 test specimens loaded up to $R_{p0,2}$ and the fourth set contains 4 test specimens loaded up to R_m.

From the gauge length of each test specimens 4 smaller test specimens are cut (figure 1b). Thus from each set 16 test specimens are collected (15 test specimens for corrosion tests and one for microstructure study).

Each test specimen, intended for microstructure study is split in 2 smaller test specimens – figure 1c.

2.2 Microstructure study

Test specimens with various magnitude of deformation are packed in acrylic resin. The studied cross-sections have been cleaned up with a series of sand papers followed by mechanical polishing. The microstructure is developed for 5s at room temperature using an etching solution with the following composition: 20ml CH_3COOH, 1ml HNO_3, 60 ml ethylene glycol and 20 ml H_2O.

The microstructure was studied by optical observations using inverted metallographic microscope Optika XDS-3MET with built-in camera for digital photography.

3. Results and discussion

Figs. 2-4 show the typical etched microstructure of unloaded and loaded to various magnitude of deformation test specimens. It can be concluded, that a serious difference exist in the microstructure in the core of the specimen and in the vicinity of the surface of the test specimen, even in the case of unloaded specimens (figure 2). A narrow band that follows a contour parallel approximately to the surface divides the specimen into two different regions.

3.1. Regions in the vicinity of the edges of the test specimens

In the vicinity of edge A (close to the surface) of the cross-sections (figure 1) of the test specimens, the size of the grains decreases. Areas with different microstructure are seen (figure 2). Bands with dendritic structure can be found as well as segregations of intermetallic phase.

The occurrence of areas with different microstructure is a feature of high-pressure die castings of Mg alloys and is called *skin effect* [6, 7]. Due to the very high rate of heat extraction (rapid cooling), the surface layer (skin) of Mg die-castings has a very fine microstructure with very small (<0.5 μm) α-Mg grains and a higher volume fraction of the intergranular eutectic β-$Mg_{17}Al_{12}$ phase. This results in a hard surface layer, which can have a significant effect on the mechanical properties of the casting [6].

Changes due to the various degree of loading are apparent mainly in the areas near the edges B of the cross-sections of the test specimen (figure 3). In the test specimen loaded up to $R_{p0,1}$ elongation of the α-phase grains can be seen. In the test specimen loaded up to $R_{p0,2}$ this elongation is greater, while in the test specimen loaded up to R_m micro-fractures are seen. The fracture is trans-crystal and probably starts from the line between α and β phases.

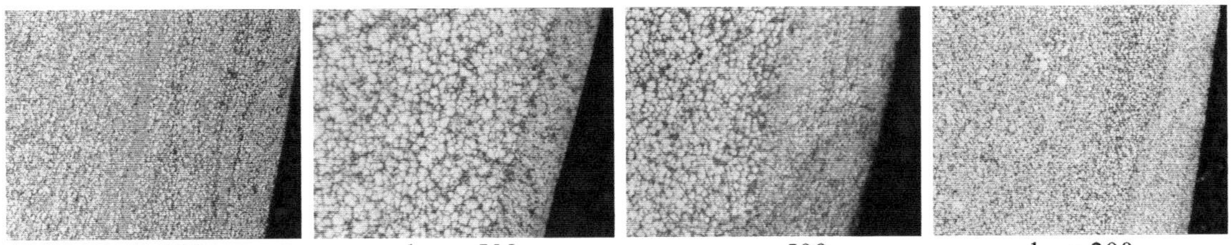

a – x 200 b – x 500 c – x 500 d – x 200

Figure 2. Microstructure along edge A of the cross-section of the test specimen: (a) non-loaded; (b) loaded up to $R_{p0,1}$; (c) loaded up to $R_{p0,2}$; (d) loaded up to R_m.

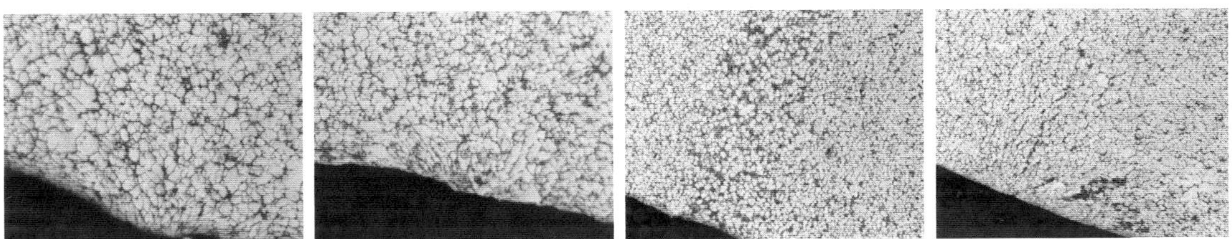

a – x 500 b – x 500 c – x 200 d – x 200

Figure 3. Microstructure along edge B of the cross-section of the test specimen: (a) non-loaded; (b) loaded up to $R_{p0,1}$; (c) loaded up to $R_{p0,2}$; (d) loaded up to R_m.

3.2. Areas of the core of the studied specimens

The microstructure of non-loaded specimen consists of two phases: α-Mg matrix and eutectoid (intermetallic) phase [3, 8 - 11]. In the core of the cross section the structure is homogenous in respect to the grain size. The grains of the α-phase are equiaxial and 5 to 10 μm in diameter. The eutectic distribution along the α-phase grain boundary is uniform, but separate aggregations (about 5 μm in diameter, figure 4a) are present.

According to the literature, in the presence of rare-earth (RE) elements in the alloy, even though the $Al_{11}RE_3$ phase, mainly located at the grain boundaries of the (Mg) matrix, potentially, there are many different phases that may form, such as the $Mg_{12}RE$, $Mg_{17}RE_2$, Mg_3RE or Mg_2RE phases. However, it is not clear which one, or in which form, these phases will nucleate under given solidification conditions [12]. The intermetallic phase could content not only $Mg_{17}Al_{12}$, but also Mg-Ce [3, 10], or $(Al, Mg)_2Ce$ and $Al_{11}Ce_3$ [3], or Al_2Ce and $Al_{11}Ce_3$ [11]. The eutectic has mostly lamellar structure. The lamellas consist of α and β-phases, but there are also areas with prevailing β-phase in which α-phase grains are enclosed.

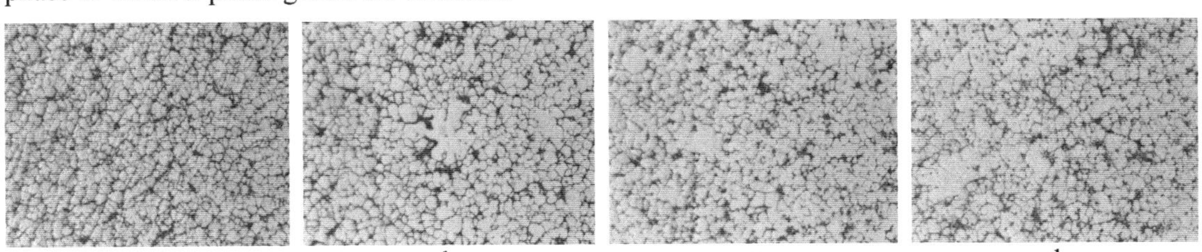

a b c d

Figure 4. Microstructure (x 500) in the core of the cross-section of test specimen: (a) non-loaded; (b) loaded up to $R_{p0,1}$; (c) loaded up to $R_{p0,2}$; (d) loaded up to R_m.

In the core of the test specimen loaded up to $R_{p0,1}$, the microstructure consists of α and β-phases. Most of the α-phase grains are equiaxial, but bigger grains with irregular shape are also present (figure 4b). The intermetallic phase is formed along the α-phase grain boundary forming separate bigger clusters compared to the non-loaded test specimen.

In the test specimen loaded up to $R_{p0,2}$ the structure is not significantly changed due to the loading and consists of α and β-phases. Here, formation of clusters of the α-phase with irregular shape also can be seen. They are larger when compared with the test specimen loaded up to $R_{p0,1}$ (figure 4c).

Generally, the microstructure of the loaded to R_m specimen is similar, the formation of larger grains (clusters, about 20 μm in diameter) is apparent. The intermetallic phase grains also recombine in a bigger clusters.

4. Conclusion

With the increase of the magnitude of deformation changes in the microstructure of the alloy occur.

Next to the surface, the skin effect is clearly seen. In the core of the studied cross-sections of the test specimen segregation of the metallic grains of α-phase occurs and formation of larger clusters with the increase of the deformation magnitude. Full explanation of the observed effects could be obtained with more detailed study and using of other methods of examinations.

The authors would like to thank the Research and Development Sector at the Technical University of Sofia for the financial support.

References

[1] Song G and Xu Z 2012 *Corrosion Science* **54** 97-105
[2] Imandoust A, Barrett CD, Al-Samman T, Inal KA and El Kadiri H 2017 *Journal of Materials Science* **52** 1-29
[3] Birbilis N, Easton MA, Sudholz AD, Zhu SM and Gibson MA 2009 *Corrosion Science* **51** 683-689
[4] Ghali E, Dietzel W and Kainer KU 2004 *Journal of Materials Engineering and Performance* **13** 7 -23
[5] Tsonev V, Kuzmanov N, Borisov B and Penkov K 2019 *IOP Conference Series: Materials Science and Engineering* **618** 012048
[6] Zhang J, Leng Z, Zhang M, Meng J and Wu R 2011 *Journal of Alloys and Compounds* **509** 1069-1078
[7] Padfield TV 2004 *Metallography and Microstructures ASM Handbook* Vol 9, ed. G. F. Vander Voort (ASM International) p. 801–815
[8] Zhang J, Zhang D, Tian Z, Wang J, Liu K, Lu H, Tang D and Meng J 2008 *Mater Sci Eng A* **489** 113-119
[9] Nakatsugawa I., Kamado S., Kojima Y., Ninomiya R. and Kubota K., 1998 *Corrosion Review* **16** 139 - 157
[10] Zheng L, Nie H, Liang W, Wang H and Wang Y 2016 *Journal of Magnesium and Alloys* **4** 115-122
[11] Shi Z, Cao F, Song G, Liu M and Atrens A 2013 *Corrosion Science* **76** 98-118
[12] Chia T.L., Easton M.A., Zhu S.M., Gibson M.A., Birbilis N. and Nie J.F. 2009 *Intermetallics* **17** 481–490

IOP Conf. Series: Materials Science and Engineering **878** (2020) 012066 doi:10.1088/1757-899X/878/1/012066

Classification analysis of tensile strength of alloyed steels

G L Dobrev[1] and I P Iliev[1]

[1] Technical University Sofia – Branch Plovdiv, 25 Tzanko Diusstabanov Street, 4000 Plovdiv, Bulgaria
E-mail: dobrev.1975@abv.bg

Abstract. Based on experiment data about 63 steels, a statistical models and a methodology are developed to assess the influence of chemical composition on the tensile strength of steels. Using the methods of factor and cluster analysis, the independent quantities are grouped according to correlation and similarity criteria into factors and clusters. The degree of influence of the chemical composition on the tensile strength of some alloyed steels is determined.

1. Introduction

The construction of machine components poses special physical, chemical and technical requirements. Ductility, tensile strength, and resistance to deformation depend to a large degree on the right combination of chemical elements, the presence of phases, which cause strengthening, the emergence of compounds, that are a factor for the shift in the fatigue value.

Modern statistical and mathematical methods are actively used to study the status and to predict the behavior of various metals and alloys. This makes it possible to find new dependencies when studying the behavior of the examined quantity, which cannot be discovered using any other experimental or theoretical method.

Statistical prediction is performed in [1] for the average tensile strength of 316L stainless steel using the Weibull distribution method. Paper [2] presents the results of mechanical testing at room temperatures and elevated temperatures of 17-7PH stainless steel with modified chemical composition. Regression analysis is performed, revealing the statistically significant variables for predicting tensile strength and yield strength of 17-7PH steel with modified chemical composition. The regression analysis describes the dependency between the content of chromium, nickel, aluminum and their interactions, and the variables R_m and $R_{p0.2}$ (the tensile strength and yield strength of the steel 17-7PH). Second order equations are obtained. [3] describes the results of tensile strength testing of AISI type 304, deformed at room temperature and elevated temperatures. It has been found that the parameters of the Weibull distribution show a trend towards reduction with temperature elevation and the coefficient of variance tends to increase as the temperature is elevated. In [4], regression models are used as a tool to reduce the time and costs related to the development and selection of new metal alloys. A multiple regression model is developed, which can predict accurately the tensile strength of high-strength low alloy steel production based on its chemical composition and processing parameters. A linear regression model is developed to include the chemical elements C, Mn, Cr, Ni, Mo, Cu, N, V, plate thickness, solution processing and aging temperatures.

Classification analysis is used to qualify the variables according into specific target groups. When these target groups are known in advance, the methods of logistic regression and discriminant analysis are used, and if the groups are unknown – factor and cluster analysis. Classification analysis is also used to study the state and to predict the behavior of complex systems, which are dependent on a large number of independent variables [5]. The basis for classification analysis is the availability of a lot of experiment data.

As it is known, the tensile strength of steels depends on multiple independent variables. The objective of this paper is to study the influence of chemical composition on this property of steels. The use of classification analysis to study the tensile strength of steels enables the solution of the following problems: 1) classification of independent variable quantities (chemical composition of steels) into macro categories and their grouping into several significant groups; 2) determining the degree of influence each variable has over the dependent variable – tensile strength; 3) determining the position of each independent variable in a general hierarchy; 4) determining the distance of each independent variable to the dependent variable – tensile strength. The resulting classifications and classification groups of independent variables can serve as the basis for subsequent regression models. These models will enable the forecasting of the behavior of the studied dependent variable in order to plan the experiment and to create new steels with enhanced tensile characteristics. Classification analysis allows the expression of new properties and dependencies, which cannot be found using known theoretical and experimental methods.

This paper uses the results for 63 steels with regard to the influence of chemical composition on tensile strength [6, 7].

The statistical studies are performed using the SPSS software package [8].

2. Results of multivariate factor analysis

The main objective of factor analysis is: 1) Reduction in the number of independent variables (reduction of data dimensionality); 2) Classification of variables, determining the relationships between independent variables, as well as between independent variables and the dependent variable. The most commonly used method for extracting the significant factors and factor structure is the method of principal component analysis (PCA).

The classification through factor analysis is based on the presence of multicollinearity between the variables. The objective is to predetermine the known number of macrovariables (factors), which group within themselves the input variables according to their mutual correlation. The resulting factors are mutually independent, which allows them to be used subsequently to build regression models and to predict the behavior of systems.

The classification analysis in this paper includes 10 independent parameters: 9 chemical elements: Mn, Cr, Ni, Si, S, P, Al, Mo, C (% content) and the diameter of the specimen for testing tensile strength: d (mm). The tensile strength is considered as the dependent variable, *MPa*.

The factor analysis of the 10 considered independent parameters needs to be analyzed for admissibility. Following the factor analysis process in SPSS, the results are given in Table 1.

The results show that the Kaiser-Meyer-Olkin measure of sampling adequacy (KMO) is equal to 0.657 (this indicator needs to be between 0.5 and 1) and Bartlett's Test of sphericity is significant (Sig.= 0.000). These results demonstrate the applicability of factor analysis in this particular case.

Table 1. Results for testing of admissibility of factor analysis.

KMO and Bartlett's Test		Value
Kaiser-Meyer-Olkin Measure of Sampling Adequacy		0.657
Bartlett's Test of Sphericity	Approx. Chi-Square	275.249
	Df	45
	Sig.	0.000

The next step is to determine the number of components or factors, into which the 10 independent parameters are grouped. Using the PCA method, all components (factors) are obtained with aggregation of experiment data as a percentage, Table 2.

From this table, 10 independent parameters are selected to be grouped in 5 factors. In this case, 85.3 % of experiment data are described. We consider this process to be acceptable for subsequent studies.

Table 2. Total variance explained using PCA.

Component	Total	% of Variance	Cumulative %
1	3.506	35.060	35.060
2	1.887	18.867	53.927
3	1.520	15.204	69.131
4	1.011	10.109	79.240
5	0.606	6.060	85.300
6	0.505	5.050	90.350
7	0.442	4.419	94.769
8	0.225	2.246	97.015
9	0.179	1.790	98.805
10	0.119	1.195	100.000

Next, the 10 independent variables are grouped into 5 factors, in keeping with the principle that the variables in each factor are strongly correlated and the variables in the different factors are not correlated or correlate weakly. The results are obtained using the PCA and Oblimin rotation method with Kaiser Normalization as shown Table 3.

Table 3. Grouping of predictors into factors[a].

Variable	Factors				
	F1	F2	F3	F4	F5
Mn	0.902				
Cr	0.735				
Ni	0.718				
Si	0.665				
S		-0.945			
P		-0.871			
Al			0.916		
Mo			0.884		
d (mm)				0.998	
C%					0.930

[a] Loadings smaller than 0.5 are omitted.

Table 3 shows that the first factor groups the quantities Mn, Cr, Ni and Si. This means that they are strongly correlated and describe 35% of the sample variance in the experiment data, Table 2. The second factor groups together with a negative sign the quantities P and S, which influence 18.9% of experiment data, Table 2. Because they are in one group (factor) they are also correlated. The third factor includes quantities Ai and Mo, which influence 15.2 % of experiment data. They are also correlated. The last two factors consist of a single quantity, d and C, which do not correlate with any of the other 9 quantities and influence experiment data by 10.1% and 6% respectively. The clear differentiation of factors confirms the adequacy of this factor analysis.

3. Cluster analysis

Cluster analysis (CA) is the name given to a group of various calculation procedures used to classify objects. Usually in CA, the number of groups is known in advance and is determined by the researcher during the study. Formed groups (clusters) need to be mutually homogeneous within and mutually heterogeneous among each other according to specified characteristics. The quantitative assessment of the term "similarity" is related to the term "metric".

In engineering and technical sciences, and physics, clusters are usually defined using hierarchy agglomeration methods. The application of these methods produces a dendrogram (tree diagram).

Table 4 shows the distribution of 10 independent quantities into 4 to 8 clusters using the between-groups linkage method.

Table 4. Grouping independent variables into clusters.

Variable	8 Clusters	7 Clusters	6 Clusters	5 Clusters	4 Clusters
P	1	1	1	1	1
Mo	2	2	2	2	2
Al	3	2	2	2	2
d (mm)	4	3	3	3	3
C%	5	4	4	4	3
Mn	6	5	5	5	4
Si	7	6	6	5	4
Cr	7	6	6	5	4
Ni	8	7	6	5	4
S	1	1	1	1	1

The table shows that with 4 clusters, the grouping of chemical variables is the same as that in factors, Table 3. Clusters are clearly differentiated. The first cluster includes the variables S and P, the second cluster – Mo and Al, the third – d, the fourth – C, the fifth –Mn, Si, Cr and Ni. Even though the grouping is done according to different criteria – correlation in the factor analysis and proximity in the cluster analysis, the results are identical, which is an indication of their reliability.

The following Fig. 1 shows a dendrogram built using the between-groups linkage method. The horizontal axis visualizes the distance between the variables in relative units, expressed as the squared Euclidean distance.

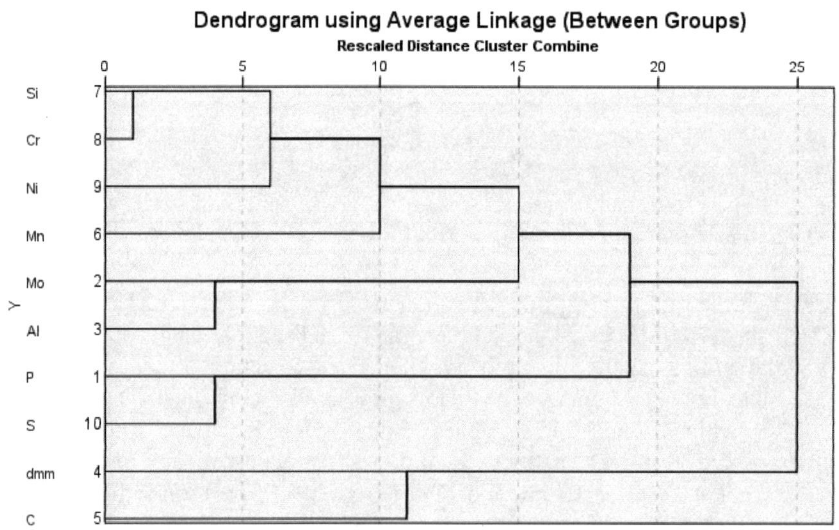

Figure 1. Dendrogram of the independent variables.

The next step is to indicate the place of the dependent variable – tensile strength – among the 10 independent variables. This is given in Table 5.

The grouping of the 10 variables into 5 clusters is maintained, with the dependent variable σ (Mpa) grouped in the second cluster together with quantities S and P. In both cases, tensile strength σ (Mpa) is significantly affected by the variables S and P.

The next figure 2 shows a dendrogram based on Table 5. It clearly shows that tensile strength σ (Mpa) is grouped in a cluster together with variables S and P.

Table 5. Distribution of independent variables and dependent into clusters.

Variable	8 Clusters	7 Clusters	6 Clusters	5 Clusters	4 Clusters
C (%)	1	1	1	1	1
σ (Mpa)	2	2	2	2	2
Mn	3	3	3	3	2
Si	4	4	3	3	2
Cr	4	4	3	3	2
Ni	5	4	3	3	2
S	6	5	4	2	2
P	6	5	4	2	2
Mo	7	6	5	4	3
Al	7	6	5	4	3
d (mm)	8	7	6	5	4

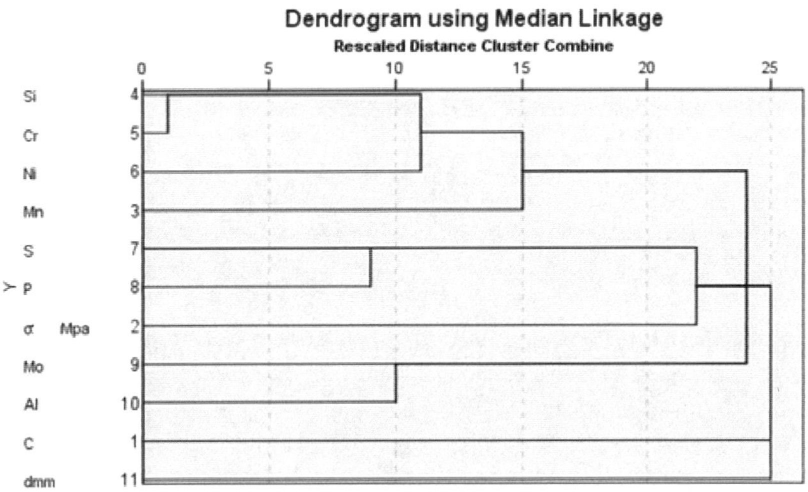

Figure 2. Dendrogram of the independent variables and dependent.

4. Analysis of results from factor and cluster analysis

The grouping of 9 chemical elements into 4 factors allows for the reduction of dimensionality of the problem when performing experiment studies in order to develop new steels with enhanced strength properties. The nine independent parameters, chemical elements Si, Cr, Ni, Mn, P,S, Mo, Al and C are replaced by 4 new latent independent variables (factors) F_1, F_2, F_3, F_4, F_5. The reason for this is that in each factor the independent variables are grouped together due to high mutual correlation. This means that the variables in each factor should change **at the same time**. In this manner, they lose their identity and start to represent a single independent variable, that of the factor.

The performed cluster analysis is confirmatory. It validates the results of the factor analysis.

In accordance with the results in Table 3, the coefficients for impurities S and P have a negative sign, which means that the values of S and P need to be reduced. As it is well-known, sulfur in steel produces FeS or MnS compounds and causes hot brittleness; therefore, its content should not exceed 0.003% to 0.005%. Where sulfur forms FeS, there is hot brittleness and cracks which form in the temperature interval (1000–1250) 0C. Phosphorus causes cold brittleness, which makes shaping more difficult and for this reason its content should not exceed 0.03%-0.04%. In accordance with the results in Table 3, alloy elements Mn, Cr,Ni and Si are found in one factor. This means that they are strongly correlated and should be modified together. They increase deformation resistance and hardness more strongly with higher carbon content. Si content from 0.17% to 0.35% and of Mn between 0.5% and 0.8% do not influence ductility, however over these values ductility deteriorates. This further reduces the dimensionality of the problem of predicting new experiments.

The results obtained by factor and cluster analysis are significant when performing regression analysis. One of the crucial requirements for regression models is for predictor quantities to be mutually independent, i.e. they should not be strongly correlated. In this case, the factor analysis indicates strong correlation between the initial quantities and the use of linear regression analysis is statistically reliable. The development of regression models [4] in our opinion should always be preceded by factor analysis in order to establish the collinear dependence of the predictors. The development of regression models of second or higher order excludes the presence of multicollinearity between predictors and better reflects the nonlinear properties of processes in nature. For this reason, the second order regression model developed in [2] provides very good fit between experiment data and predicted results.

5. Conclusion

Factor and cluster analysis are performed on 9 independent quantities and the dependent variable – tensile strength. Independent quantities are grouped into 5 factors according to their degree of correlation. The results of the cluster analysis are confirmatory. It is found that the linear regression analysis with independent quantities cannot be statistically reliable due to the lack of strong correlation between some of these. It is recommended to perform regression analysis of second or higher order to improve fit with experiment results.

Acknowledgement

This work was supported by the Research and Development Sector at the Technical University of Sofia, Bulgaria.

References

[1] An Z L, Chen T, Cheng D L, Chen T H and Wangl Z Y 2017 Statistical analysis and prediction on tensile strength of 316L-SS joints at high temperature based on Weibull distribution 1st International Workshop on Materials Science and Mechanical Engineering IOP Publishing. IOP Conf. Series: Materials Science and Engineering **281** 012062 doi:10.1088/1757-899X/281/1/012062

[2] Fakić B, Ćubela D, Burić A and Horoz E 2019 Regression analysis of tensile strength testing results steel 17-7 PH with modified chemical composition 14th International conference on Accomplishment in Mechanical and Industrial Engineering-DEMI 2019, Banja Luka, 24-25 May 2019, 691–697

[3] SungHo P, NoSeok P and JaeHoon K 2010 A statistical study on tensile characteristics of stainless steel at elevated temperatures 15th International Conference on the Strength of Materials (ICSMA-15) IOP Publishing *Journal of Physics: Conference Series* **240** 012083 1–4 doi:10.1088/1742-6596/240/1/012083,

[4] Golodnikov A, Macheret Y, TrindadeA A, Uryasev S and Zrazhevsky G 2005 Statistical modelling of composition and processing parameters for alloy development IOP Publishing *Modelling Simul. Mater. Sci. Eng.* **13** 633–644 doi:10.1088/0965-0393/13/4/013

[5] *Atlas Specialty Metals Technical Handbook of Bar Products* 2005 www.atlasmetals.com.au

[6] Bringas J E (ed) 2004 *Handbook of Comparative World Steel Standards* ASTM Intern. DS 67B.

[7] Izenman A J 2008 *Modern Multivariate Statistical Techniques. Regression, Classification, and Manifold Learning* (New York: Springer)

[8] SPSS IBM Statistics https://www.ibm.com/analytics/spss-statistics-software

Corrosion behaviour of aluminium alloys EN AW-6026 and EN AW-6082 in a sulphuric acid medium

Kalina Kamarska
Department of Mathematics, Physics, Chemistry, Technical University–Sofia, Branch Plovdiv, Bulgaria
E-mail: kamarska@tu-plovdiv.bg

Abstract. The corrosion behaviour of aluminium alloys EN AW-6026 and EN AW-6082 in 0,1 M and 1 M sulphuric acid solutions at different temperatures has been investigated by using gravimetric techniques and a digital microscope. The study revealed that the corrosion rate increases with higher temperature and concentration of H_2SO_4. The surface attack increased with the increasing of the concentration of sulphuric acid and the temperature. Alloys EN AW-6026 and EN AW-6082 have a higher corrosion resistance at 25 °C.

1. Introduction

Corrosion is the destructive attack on a material by reaction with its environment and the serious consequences of the corrosion process have become a problem of worldwide significance [1]. Aluminium and its alloys are widely used in engineering applications and are very reactive metals, predisposed to corrosion. They frequently come in contact with acids, such as sulphuric acid, during pickling, de-scaling or electrochemical etching of aluminium.

Dilute sulphuric acid solutions, up to 10% in concentration, cause-some attack on aluminium alloys, but the action is not sufficiently rapid at room temperature in order to prevent them from being used in special applications [2].

The studied aluminium alloys EN AW-6026 and EN AW-6082 refer to the series of 6xxx alloys and magnesium and silicon are the principal alloying elements.

Aluminium–magnesium–silicon wrought alloys usually show some susceptibility to intergranular corrosion. With a balanced magnesium–silicon composition that results in the formation of Mg_2Si constituent, the intergranular attack is minor. When the 6xxx alloy contains an excessive amount of silicon, the intergranular corrosion increases because of the strong cathodic nature of the insoluble [2]. What makes the Mg_2Si phase unique is that it is an ionic compound and is not only anodic to aluminum but also reactive in acidic solutions [3].

The 6*xxx* alloys offer moderately high strength and a very good resistance to corrosion which make these alloys extremely suitable for various structural, building, marine, machinery, and process equipment applications [3].

Most of the reported studies of corrosion have been conducted on EN AW-6082 in chloride solution [4-6] and under simulated acid rain conditions [7] but there has been little investigation into the reaction of alloy 6026 in different media.

The present work is focused on the corrosion behaviour of aluminium alloys EN AW-6026 and EN AW-6082 in 0,1M and 1M H_2SO_4 solutions at different temperatures. The study was conducted by using gravimetric techniques and a digital microscope.

2. Experiments

2.1 Gravimetric technique

One of the main methods that was used to express the corrosion rate is the measurement of weight loss per unit area and unit time.

The weight loss experiments were carried out with cylindrical samples of aluminium alloys EN AW-6026 and EN AW-6082 with an exposed area of 8.40 cm^2. Before immersion in the solution, the experimental samples were initially abraded with fine emery paper, washed in detergent for five minutes and then rinsed with distilled water and dried. The dried samples were weighed with an analytical balance Acculab ATILON to an accuracy of ± 0,0001g and then immersed in test solutions of 0,1 M and 1M H_2SO_4 for 4 hours. The experiments were carried out at temperatures of 25 °C and 50 °C (± 0,5 °C). After 4 hours of total time of exposure, the specimens were taken out, washed with distilled water, dried and weighed accurately.

The corrosion rate (CR), in terms of weight loss measurements, were calculated using the following equation:

$$CR = \frac{(m_1 - m_2)}{S.t} \quad (1),$$

where m_1 is the weight of the sample, g; m_2 is the weight of the sample after the corrosion test, g; S is the area of the sample, m^2; and t is the test time, h.

2.2. *Surface analysis*

The structures of the corroded and the uncorroded surfaces of EN AW-6026 and EN AW-6082 were examined with a digital microscope (Digi Micro Scope 2,0M Pixels).

3. Results and discussion

3.1. *Gravimetric technique*

The comparative plots of the corrosion rates versus exposure time in 0,1 M and 1 M H_2SO_4 solutions for EN AW-6026 and EN AW-6082 at 25 °C are shown in figure 1 and figure 2. Observation of the plots shows an initial increase in the corrosion rate with respect to time. During the first two hours of immersion in the solutions, the graphs reach the maximum level followed by a decrease in the corrosion rate of these alloys.

The behaviour of this process can be attributed to the formation of a protective layer on the alloy surface and the increase in the corrosion rate with respect to time. The increase in the corrosion rate ceases with the formation of a protective oxide layer [8]. The results obtained in this work also showed that in the conditions employed in the present study EN AW-6082 at 25 °C has a higher corrosion rate at both concentrations.

The corrosion rates depend strongly on other factors, such as surface state, electrolyte flow rate, geometry and temperature [9].

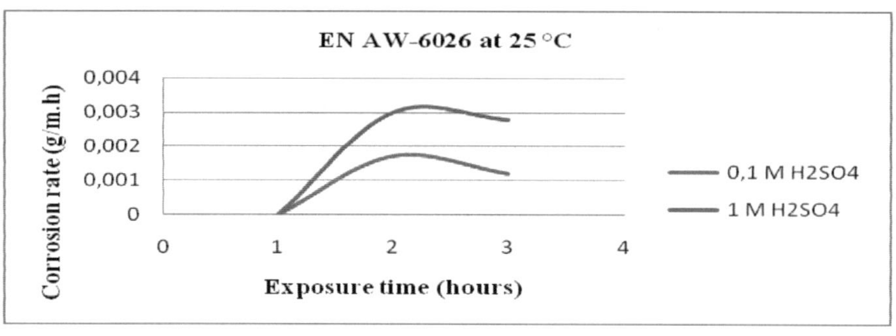

Figure 1. Corrosion rate of EN AW-6026 in 0,1M and 1M H_2SO_4 at 25 °C after 4 hours.

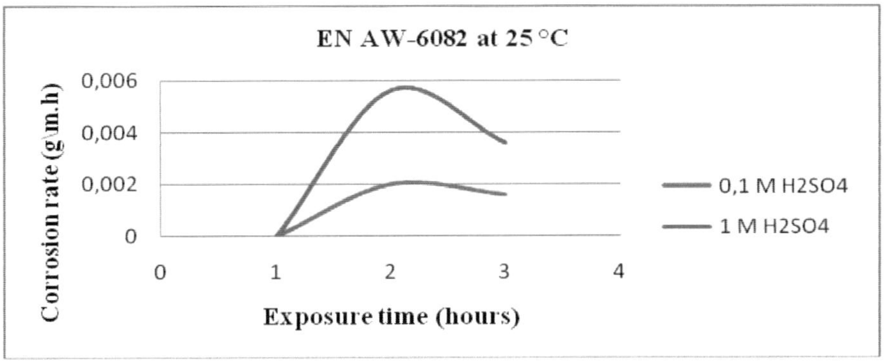

Figure 2. Corrosion rate of EN AW-6082 in 0,1M and 1M H_2SO_4 at 25 °C after 4 hours.

The corrosion rates of these alloys immersed in 0,1M and 1 M H_2SO_4 solutions at 50 °C are presented in figure 3 and figure 4. The results show that the corrosion rate increased with the increase of temperature. The aluminium alloys tested demonstrated different behaviour at 50 °C (± 0,5 °C) after the fourth hour of immersion. With the increase of the temperature in the 0,1 M H_2SO_4 solution, EN AW-6026 had a higher weight loss than EN AW-6082. At the same temperature in the 1 M H_2SO_4 solution, EN AW-6026 demonstrated a higher corrosion resistance and EN AW-6082 had a significantly higher corrosion rate in these conditions.

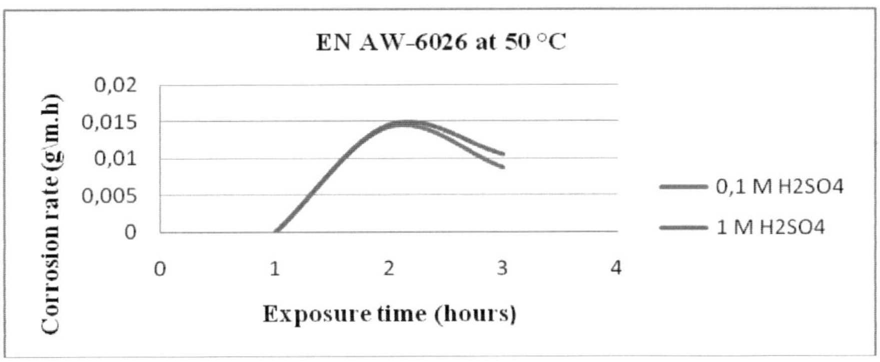

Figure 3. Corrosion rate of EN AW-6026 in 0,1M and 1M H_2SO_4 at 50 °C after 4 hours.

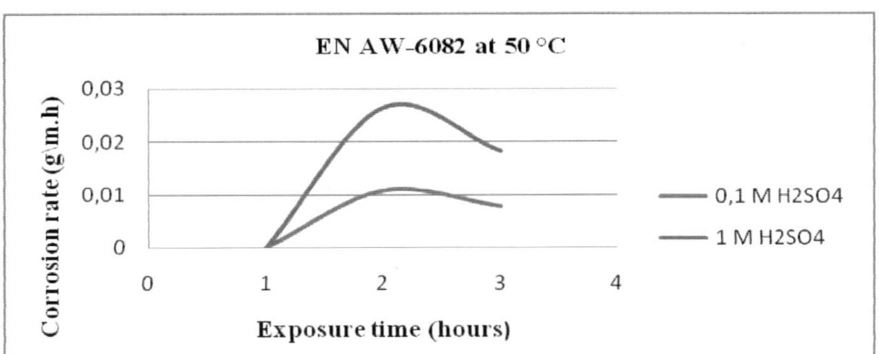

Figure 4. Corrosion rate of EN AW-6082 in 0,1M and 1M H_2SO_4 at 50 °C after 4 hours.

3.2. *Surface analysis*

The structures of the uncorroded surfaces of both alloys are shown in figure 5a and figure 6a, respectively.

(a) x 200 (b) x 200

Figure 5. Stuctures of the surface of EN AW-6026: a - before immersion; b - after 4 hours of immersion in 1M H_2SO_4 at 50 °C.

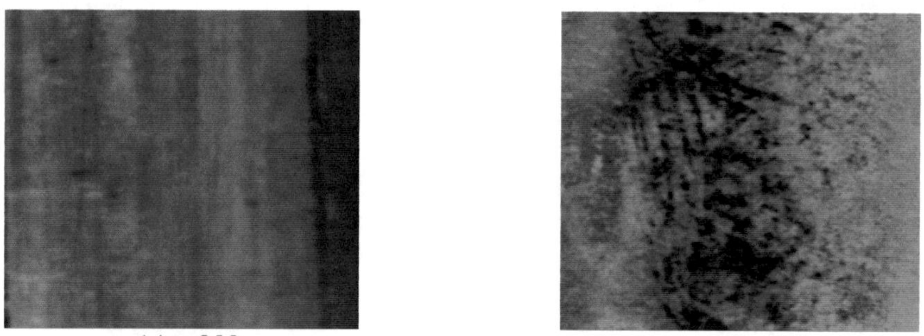

(a) x 200 (b) x 200

Figure 6. Structures of the surface of EN AW-6082: a - before immersion; b - after 4 hours of immersion in 1M H_2SO_4 at 50 °C.

The corrosion rate of aluminium alloys depends on the specific ions present in the solution. Furthermore, acidic medium could activate the alloy surface. When exposed to an environment with aggressive sulphate ions, the oxide layer on the alloy surface breaks down at definite places causing corrosion on the aluminium surface. The structure of the surface of both samples was changed in the presence of sulfuric acid solutions and the surface attack increased with the increasing of the temperature and the concentration of H_2SO_4. Large pits were observed on the surface of EN AW-6082 (figure 6b) with corrosion products after 4 hours of immersion in 1M H_2SO_4 at 50 °C.

4. Conclusion

From the investigation of the corrosion behaviour of aluminium–magnesium–silicon alloys can be concluded that EN AW-6082 at 25 °C is more susceptible to corrosion attack in 0,1 M and 1 M H_2SO_4 than EN AW-6026. Comparison of the corrosion rate of these alloys from the weight loss analysis showed that EN AW-6026 and EN AW-6082 both have a higher corrosion resistance at 25 °C.

The structure of the surface of both samples was changed in the presence of sulfuric acid solutions. The surface attack increased with the increasing of the temperature and the concentration of H_2SO_4.

IOP Conf. Series: Materials Science and Engineering **878** (2020) 012067 doi:10.1088/1757-899X/878/1/012067

Acknowledgments
The author would like to thank the Research and Development Sector at the Technical University of Sofia for the financial support.

References

[1] Roberge P R *Handbook of Corrosion Engineering* (New York: McGraw-Hill Education)
[2] Wisdom J, Kelechi I D and Ugochukwu J 2018 *IRJET* **5** 121
[3] Kaufman J G 2005 *Corrosion of Aluminum and Aluminum Alloys* (Ohio: ASM International)
[4] Panagopoulos C, Georgiou E and Gavras A 2009 *Tribology Int.* **42** 886
[5] Cicolin D, Trueba M and Trasattim S 2013 *Electrochim. Acta.* **124** 27
[6] Xhanari K and Finšgar M 2017 *Int. J. Electrochem. Sci.* **12** 5845
[7] Gerengi H, Slepski P, Ozgan E and Kurtay M 2015 *Mater. and Corrosion.* **66** 233
[8] Tolulope R and Adeleke A 2016 *J Fail. Anal. and Preven.* **16** 874
[9] Bardal E 2003 *Corrosion and protection* (London: Spring)

New thermodynamic investigation of some solid phases of Sn-Ti phase diagram

V Gandova[1] and I Petrova[2]

[1] University of Food Techniligies, Analytical and Physical Chemistry Department 4002 Plovdiv, Bulgaria
[2] SofiaTechnical University, branch Plovdiv, Mathematics, Physics and Chemistry, 4002 Plovdiv, Bulgaria

e-mail: ivalinapetrova.com@abv.bg

Abstract. Thermodynamic description of the part of Sn-Ti phase diagram is presented using the CALPHAD approach. The thermodynamic parameters of the some phases of system modifying according new experimental phase equilibrium and thermodynamic data. Some phases described as stoichiometric, but another with sublattice models. Ti_3Sn, TI_5Sn_3, TI_6Sn_5 are modelled with three sublattice modes. Better agreement is obtained by the present optimization. Calculated results are compared with the original experimental data to demonstrate the successfulness of this assessment. The enthalpies of formation of solid phases are calculated and compared by two methods.

1. Introduction

Electronic industry production relies to a great extent on the soldering process. Nevertheless, it is well known that the classical solders (lead-tin based alloys) represent a serious health and environmental risk. For this reason, the use of lead-containing solders was recently restricted [1]. Sn based alloys were appropriate to change the lead-containing ones. Binary Sn-Ti phase diagram was investigated experimentally several times [2, 3]. Pietrokowsky et al. [4] investigated the binary system by metallographic, X-ray and thermal analysis. They found four intermetallic phases: Ti_3Sn, Ti_6Sn_5,Ti_2Sn and Ti_5Sn_3. Ti_3Sn and Ti_6Sn_5 are congruently melted. Ti_2Sn and Ti_5Sn_3 are formed by peritectic reaction and found a eutectoid reaction, (β-Ti) ->(α-Ti) + Ti_3Sn at 864 ^{0}C. Using X-ray diffraction, they found that alloys containing more than 45 at.% Sn are mixtures of Ti_6Sn_5 and (β-Sn). Eremenko et al. [5] determined the Sn-rich liquidus and (β-Sn) eutectic temperature by differential thermal analysis. They also identified an exothermic reaction at 790 ^{0}C with a polymorphic transformation in Ti_6Sn_5. According McQuillan [6] (β-Ti) transforms congruently to (α-Ti) at 6.7 at.% Sn and 842 ^{0}C. McQuillan et al. [7] and Glazova et al. [8] confirmed congruent transformation of (α-Ti) to (β-Ti), but they not found evidence for a two-phase (α-Ti) + Ti_3Sn field. The start temperature of the martensitic (β-Ti) ⇔ (α-Ti) transformation was measured by Sato et al. [9]. In the range 2 to 9 at.% Sn, it is 740 ^{0}C to 760 ^{0}C, apparently independent of composition.

Enthalpies of formation of the binary compound phases Ti_3Sn, Ti_2Sn, Ti_5Sn_3 and Ti_6Sn_5 in the system Sn-Ti by differential scanning calorimetry were measured by Pondarevskaya et al. [10]. Meschel et al. [11] determined enthalpy of formation of phase Ti_6Sn_5 by high-temperature direct synthesis calorimetry. First optimization of the system the Sn-Ti is given by Hayes [12]. Liu et al. [13] assessed the system using a two sublattice model $(Sn,Ti)(Sn,Ti)_3$ for phase Ti_3Sn. They treated phase

Ti_3Sn as phase with homogeneity range and all the other compounds as stoichiometric phases. Some thermodynamic investigations of system by calculations were done using TC AB software [14].

The main objective in these studies is to investigated and described some solid phases in the Sn-Ti phase diagram.

In Table 1 are presented crystal structures of binary system and ternary system Sn-Ti.

Table 1. Crystal structures of pure components, binary and ternary phases in system Sn-Ti

Phase name	Pearson symbol	Prototype	Space group	Struktur-berich designation	Wyckoff positions	Referens
(βSn)	*tI4*	Sn	*I41/amd*	A5	b	[2]
(αTi)	*hP2*	Mg	$P6_3/mmc$	A3	c	[2]
(βTi)	*cI2*	W	*Im3m*	A2	a	[2]
Ti_3Sn	*cP4*	Ni_3Sn	*Pm3m*	$D0_{19}$	ac	[2]
Ti_2Sn	*hP6*	$InNi_2$	$P6_3/mmc$	$B8_2$	acd	[2]
Ti_2Sn_3	*oS40*	...	*Cmca*	...	e^2f^3	[3]
Ti_5Sn_3	*hP16*	Mn_5Si_3	$P6_3/mmc$	$D8_8$	dg^2	[2]
Ti_6Sn_5	*hP22*	Ti_6Sn_5	$P6_3/mmc$	...	$acgh^2$	[2]

2. Method and modeling

The Gibbs energy (${}^0G^{\phi}_i$) of pure component i in phase ϕ was expressed relative to the enthalpy of the component in its stable phase at 298.15 K. $L^{\phi}_{i,j}$ was a binary parameter describing the interaction between components i and j in phase ϕ. For these parameters, ${}^0G^{\phi}_i$ was a function of temperature and $L^{\phi}_{i,j}$ can be functions of temperature and composition.
Gibbs energy of formation of the binary compounds was expressed as:

$$ {}^0G^{\phi}_{A:B} = a\,{}^0G^{\alpha}_{A} + b\,{}^0G^{\beta}_{B} + A + B^{*}T \quad (1)$$

The solid binary phases belonging to the Sn-Ti binary system were described with the respective sublattice models. Three phases were remodelled Ti_3Sn, TI_5Sn_3, TI_6Sn_5 with three sublattice modes.

3. Results and discussion

The thermodynamic description of the binary systems and Sn-Ti system is presented in Table 2.

Table 2 . The optimized parameters and models describing the phase relations of the Sn-Ti system.

Phase	Model	Thermodynamic parameter
Liquid	(Sn,Ti)	${}^0L^{L}_{Sn:Ti}=-91598.9-0.9416T$
		${}^1L^{L}_{Sn:Ti}=45682.64-12.1045T$
Bct	(Sn,Ti)	${}^0L^{Bct}_{Sn:Ti}=50000$
Ti_2Sn	(Sn,Va)(Ti,Va)	${}^0G^{Ti2Sn}_{Sn:Ti}-2\,{}^0G^{Hcp}_{Ti}-{}^0G^{Bct}_{Sn}=-122644.77+6.0034T$
		${}^0G^{Ti2Sn}_{Va:Ti}-2\,{}^0G^{Hcp}_{Ti}=10000$
		${}^0G^{Ti2Sn}_{Sn:Va}-{}^0G^{Bct}_{Sn}=5000$
		${}^0G^{Ti2Sn}_{Va:Va}-{}^0G^{Bct}_{Sn}=300000$
		${}^0L^{Ti2Sn}_{Sn,Va:Ti}=-33085.17$
		${}^0L^{Ti2Sn}_{Sn:Ti,Va}=-49803.91+24.471T$
Ti_2Sn_3	(Sn)(Ti)	${}^0G^{Ti2Sn3}_{Sn:Ti}-2\,{}^0G^{Hcp}_{Ti}-3\,{}^0G^{Bct}_{Sn}=-173931.75+7.82593T$
Ti_3Sn	(Ti)(Ti,Sn)(Ti,Sn)	${}^0G^{Ti3Sn}_{Ti:Ti:Ti}-2\,{}^0G^{Hcp}_{Ti}-{}^0G^{Bct}_{Sn}=-122644.77+6.0034T$

		${}^{0}G^{Ti3Sn}_{Ti:Ti:Ti}-{}^{0}G^{Hcp}_{Ti}=100$ ${}^{0}G^{Ti3Sn}_{Ti:Sn:Ti}-0.75{}^{0}G^{Hcp}_{Ti}-0.25{}^{0}G^{Bct}_{Sn}=-35090.44+0.1698T$ ${}^{0}G^{Ti3Sn}_{Ti:Sn:Sn}-0.5{}^{0}G^{Hcp}_{Ti}-0.5{}^{0}G^{Bct}_{Sn}=-7477.04-2.8T$ ${}^{0}G^{Ti3Sn}_{Ti:Ti:Sn}-0.75{}^{0}G^{Hcp}_{Ti}-0.25{}^{0}G^{Bct}_{Sn}=24927.12-5.2572T$ ${}^{0}L^{Ti3Sn}_{Ti:Ti:Ti,Sn}={}^{0}L^{Ti3Sn}_{Ti:Sn:Ti,Sn}=-18458.96$ ${}^{0}L^{Ti3Sn}_{Ti:Ti,Sn:Ti}={}^{0}L^{Ti3Sn}_{Ti:Ti,Sn:Sn}=4000$
Ti_5Sn_3	(Ti)(Ti,Sn,Va) (Ti,Sn,Va)	${}^{0}G^{Ti5Sn3}_{Ti:Ti:Ti}-{}^{0}G^{Hcp}_{Ti}=1344.86+4.11T$ ${}^{0}G^{Ti5Sn3}_{Ti:Sn:Ti}-{}^{0}G^{Bct}_{sn}-7{}^{0}G^{Hcp}_{Ti}=147086+15.53066T$ ${}^{0}G^{Ti5Sn3}_{Ti:Ti:Sn}-3{}^{0}G^{Bct}_{sn}-5{}^{0}G^{Hcp}_{Ti}=-329876.45+5.3066T$ ${}^{0}G^{Ti5Sn3}_{Ti:Sn:Sn}-4{}^{0}G^{Bct}_{sn}-4{}^{0}G^{Hcp}_{Ti}=-309763.261+9.404T$ ${}^{0}G^{Ti5Sn3}_{Ti:Sn:Va}-{}^{0}G^{Bct}_{sn}-4{}^{0}G^{Hcp}_{Ti}=129320.576+18.352T$ ${}^{0}G^{Ti5Sn3}_{Ti:Va:Sn}-3{}^{0}G^{Bct}_{sn}-4{}^{0}G^{Hcp}_{Ti}=129320.576+18.352T$ ${}^{0}G^{Ti5Sn3}_{Ti:Ti:Va}-5{}^{0}G^{Hcp}_{Ti}=136430.937+3.908T$ ${}^{0}G^{Ti5Sn3}_{Ti:Va:Ti}-7{}^{0}G^{Hcp}_{Ti}=136430.937+3.908T$ ${}^{0}G^{Ti5Sn3}_{Ti:Va:Va}-4{}^{0}G^{Hcp}_{Ti}=60008.541+12.332T$ ${}^{0}L^{Ti5Sn3}_{Ti:Ti,Sn:Ti}={}^{0}L^{Ti5Sn3}_{Ti:Ti,Sn:Sn}={}^{0}L^{Ti5Sn3}_{Ti:Ti,Sn:Va}=-28746.5+10.7T$ ${}^{0}L^{Ti5Sn3}_{Ti:Ti,Va:Ti}={}^{0}L^{Ti5Sn3}_{Ti:Ti,Va:Sn}={}^{0}L^{Ti5Sn3}_{Ti:Ti,Va:Va}=-142832.9+2.8T$ ${}^{0}L^{Ti5Sn3}_{Ti:Sn,Va:Ti}={}^{0}L^{Ti5Sn3}_{Ti:Sn,Va:Sn}={}^{0}L^{Ti5Sn3}_{Ti:Sn,Va:Va}=-92831.1+27T$ ${}^{0}L^{Ti5Sn3}_{Ti:Ti:Ti,Sn}={}^{0}L^{Ti5Sn3}_{Ti:Ti:Ti,Va}={}^{0}L^{Ti5Sn3}_{Ti:Ti:Sn,Va}=-37108.8+15.4T$ ${}^{0}L^{Ti5Sn3}_{Ti:Sn:Ti,Sn}={}^{0}L^{Ti5Sn3}_{Ti:Sn:Ti,Va}={}^{0}L^{Ti5Sn3}_{Ti:Sn:Sn,Va}=0$ ${}^{0}L^{Ti5Sn3}_{Ti:Va:Ti,Sn}={}^{0}L^{Ti5Sn3}_{Ti:Va:Ti,Va}={}^{0}L^{Ti5Sn3}_{Ti:Va:Sn,Va}=40108.6+8.84T$
Ti_6Sn_5	(Sn)(Ti)(Va)	${}^{0}G^{Ti6Sn5}_{Sn:Ti:Va}-5{}^{0}G^{Bct}_{Sn}-6{}^{0}G^{Hcp}_{Ti}=-468938.25+5.372$

The Sn-Ti system contains five intermetallic compounds. Ti_5Sn_3, Ti_6Sn_5 and Ti_2Sn_3 are treated as stoichiometric compounds. The sublattice models of $(Sn,Va)(Ti,Va)_2$ and $(Sn,Va)(Ti)_3$ are used to describe the homogeneity ranges of Ti_2Sn and Ti_3Sn, respectively.

Fig. 1 presented Sn-Ti phase diagram calculated with parameters obtained in this work. The phases Ti_3Sn, TI_5Sn_3, TI_6Sn_5 are modeled with three sublatice modes (Table 2).

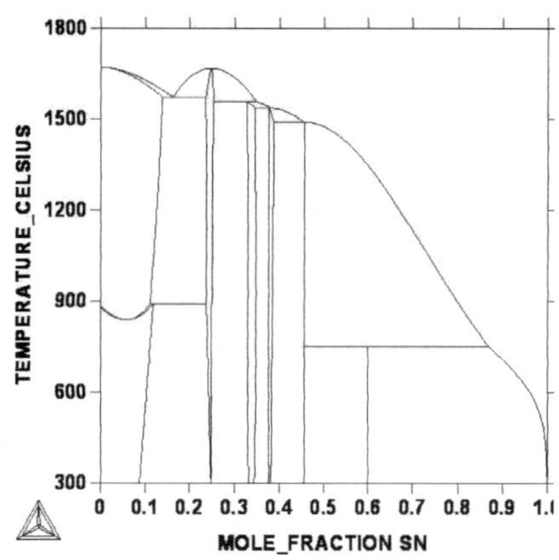

Figure 1. Calculated Sn-Ti phase diagram with parameters obtained in this work.

In Fig. 2 seen diagram compared with another results calculated and experimental found in literature. In the figure good correlation was observed in equilibrium point in diagram.
Comparative revue in Fig. 3 was done between enthalpies of formation of all solid phases in the binary system with AB-initio calculations and Thermocalc calculations. Good agreement was observed in obtained results.

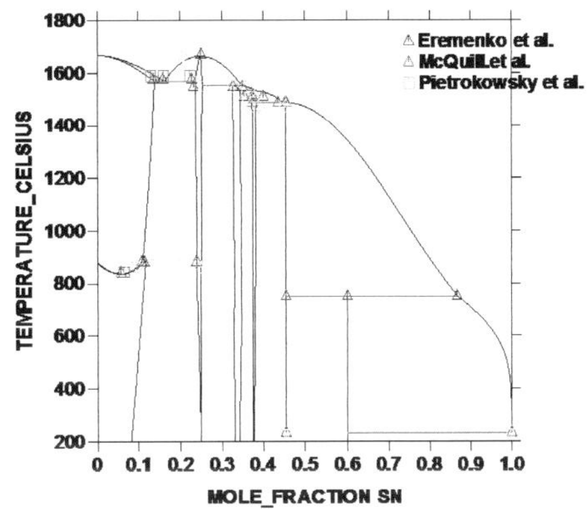

Figure 2. Calculated Sn-Ti phase diagram with parameters obtained in this work compared to experimental and calculated dates.

Table 3 shows the calculated invariant points for the binary Sn-Ti system in this work. The calculated invariants were compared by another taken from literature. All obtained temperatures appeared in close results.

Table 3. Calculated invariant points from this work of the binary Ni-Sn, Sn-Ti system. ϕi – equilibrium phase

$\phi 1$ - $\phi 2$ - $\phi 3$	Temp (^{o}C)	$C_B^{\phi 1}$ (at%)	$C_B^{\phi 2}$ (at%)	$C_B^{\phi 3}$ (at%)	Ref.
		Sn-Ti			
L=(βTi)+Ti_3Sn	1571	16.2	13.5	24.2	This work
	1574	15.9	13.6	22.8	[15]
	1606	23.0	19.5	25.0	[4]
L= Ti_3Sn	1668	25.0	25.0	-	This work
	1670	25.3	25.3	-	[15]
	1670	25.0	25.0	-	[4]
L+Ti_3Sn=Ti_2Sn	1552	34.7	22.8	32.6	This work
	1548	35.0	23.2	32.7	[15]
	1550	33.0	25.0	-	[4]
L+Ti_2Sn=Ti_5Sn_3	1517	42.3	34.7	37.3	This work
	1511	40.1	35.7	37.5	[15]
	1490	37.5	-	-	[4]

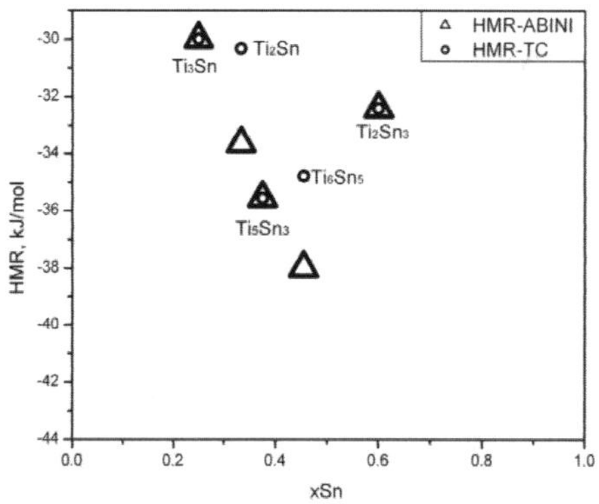

Figure 3. Comparative review of enthalpies of formation in solid phases of the binary Sn-Ti system calculated with AB - initio method and with Calphad method.

4. Conclusions

In this work three binary phases Ti_3Sn, TI_5Sn_3, TI_6Sn_5 are remodelled. It was found transition reaction between α and β-Sn. New topological and thermochemical data are available for the Sn-Ti systems. The good agreement in binary system is obtained between calculated and experimental phase equilibria data.

Acknowledgements

The authors would like to thank the Research and Development Sector at the Technical University of Sofia for the financial support.

References

[1] Official Journal of the European Union L37 2003 *http://europa.eu.int/eurlex/lex/JOHtml.do?uri=OJ*:L:2003:037:SOM:EN:HTML.
[2] Massalski T, Okamoto H, Subramanian P and Kacprzak L 1996 Binary alloy phase diagrams, in: *ASM International*
[3] Nash A and Nash P 1985 *Bull Alloy Phase Diag* **6** 350
[4] Pietrokowsky P and Frink E 1958 *Trans. ASM* **49** 339
[5] Eremenko V and Velikanova V 1962 *Russ. J. Inorg. Chem.* **77** 902
[6] McQuillan A 1954-1955 *Journal of the Institute of Metals* **83** 181
[7] McQuillan M and Member M 1955-1956 *Journal of the Institute of Metals* **84** 307
[8] Glazova V and Kurnakov N 1960 *Dokl. Akad. Nauk SSSR* **134** 1087
[9] Sato T, Hukai S and Y. Huang Y 1960 *J. Austr. Inst. Met.* **5** 149
[10] Pondarevskaya O, Petrenko O, Sudavtzova V, Lisnyak V and Stus N 2002 *Journal of Thermal Analysis and Calorimetry* **67** 649
[11] Meschel S and Klepp O 1998 *Thermochim Acta* **314** 205
[12] Hayes F, Ansara I, Dinsdale A and Rands M 1998 Thermochemical Database for Light Metal Alloys, *COST 507, vol. 2, Luxembourg* pp 284-287
[13] Liu C, Klotz U, Uggowitzer P and Loffler J 2005 *Monatsh. Chemie* **136** 1921
[14] Tedenac J C, Fartushna I, Bulanova M, Gandova V, Colinet C, Jund P 2012 Phase stabilities and phase diafram in ternaries systems based on Titanium: Experimental investigations, AB initio calculations and Calphad modelling, CALPHAD XLI, Berkeley, USA, Thm6.
[15] Yin F, Tedenac J C, Gascoin F 2007 *Calphad* **31** 37

IOP Conf. Series: Materials Science and Engineering **878** (2020) 012069 doi:10.1088/1757-899X/878/1/012069

Digitalization of HR activities in industrial enterprises

T B Mihova[1] and I M Ivanova[2]

[1] Faculty of Mechanical Engineering, Department of Industrial Management, Technical University of Sofia, Plovdiv Branch, 25 Tsanko Diustabanov Str., Plovdiv, 4000, Bulgaria

[2] Faculty of Mechanical Engineering, Department of Industrial Management, Technical University of Sofia, Plovdiv Branch, 25 Tsanko Diustabanov Str., Plovdiv, 4000, Bulgaria

E-mail: expert2009@abv.bg, iva_1982@abv.bg

Abstract. The report substantiates the need for digital transformation of human resources management in industrial enterprises in order to be adequate to the dynamic development of technologies in the modern digital world. Based on the conducted research, the current state of the human resources management system in industrial enterprises and the degree of introduction of digitalization of HR activities were analyzed. The main problems and difficulties related to their digitalization are formulated, as well as guidelines for improving the management of human resources in the studied enterprises.

1. Introduction

Effective human resource management (HRM) is a decisive factor for the overall effective management and development of industrial enterprises. Therefore, a number of scientific publications are devoted to this issues. Some of them explore the human resources (HR) in the knowledge triangle and the development of the Bulgarian economy, others focus on the challenges of human resources in the digital age, and third explore the possibilities for increasing the qualification [1, 2, 3]. In general, the studies conducted at the present time in this area are very limited in the specialized literature, lacking extensive and in-depth analyses, which is a prerequisite for studying the role of digitalisation in the activities of Management of human resources.

The analyses of the company reality in Bulgaria, at this stage, show that the wider use of digital technologies in the management of human resources is rather a privilege and practice of foreign and high-tech companies, but increasingly Enterprises are in the process of developing a strategy for digital transformation of the human resources management system or have introduced some form of Eplearning. In this regard, it is a challenge for the authors to explore the actual state of the management system of the most important resources in industrial enterprises, which activities have been digitized and what are the main difficulties and problems of digital Transformation of HR activities.

2. Exposition

IOP Conf. Series: Materials Science and Engineering **878** (2020) 012069 doi:10.1088/1757-899X/878/1/012069

The dynamic development of technology in the modern digital world leads to changes in the business models, which determines a radical change in the management of human resources. We are witnessing a continuous change of the function of the specialists in Human resources management, depending on the entry and implementation of the new technologies. In modern conditions it is necessary for them to know perfectly the business of the enterprise, with a view to establishing the relationship between their own work and the stakeholders in such a way as to add value to the business by retaining talents, leadership and organisation of work. In the article "5 Human Resources Models Every HR Practitioner Should Know", the author argues the role of human resources in business, indicating 5 practical models explaining how the HR adds value to the business and vice versa, how the business affects the Human resources [4, 13]. Our position on the models discussed in the article is that the popular HR value chain is the most clearly expressed the HR-business link shown in Figure 1. According to him, everything that is done in the HR field can be divided into two categories: HRM activities and HRM outcomes.

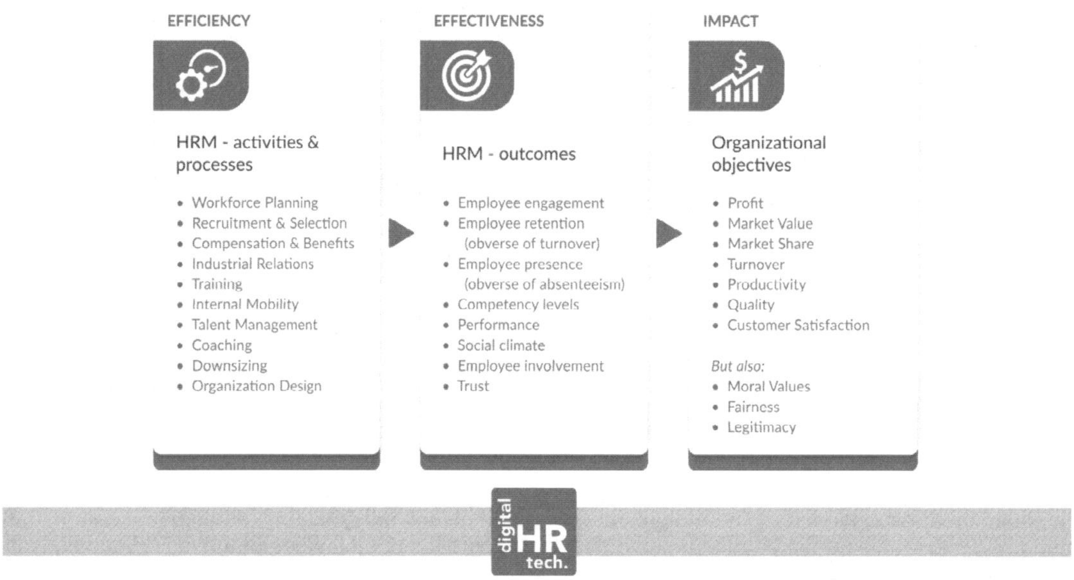

Figure 1. HR value chain

The more companies focus on measuring the effectiveness of HR activities, the more it will reduce the cost of carrying them out [12]. The focus should be on the results of HR activities, which will lead to the fulfillment of company goals. Of interest is the feedback – successful companies have always more effective HRM, as invest more in HR programs, including HR Software and learning & development opportunities for their people [6, 13]. The digitalization of HR activities was investigated by the world renowned specialist in HR theories and practice Prof. Dave Ulrich. It justifies its assertion that HR practices respond to changing external conditions (e.g. social, technological or political trends) and external stakeholders. Shifting the logic of the HR activity "out" requires HR professionals to truly understand the business of the organization so that they can make the connection between their own work and the added value for the stakeholders. The main conclusion that he does is that digitalization in HR work will not replace human judgement and ability in terms of talent, leadership and organization that are needed for success on the market [5,12].

The literature review on the issues in question gives us the opportunity to discover a number of researchers on the role of HRM in the digital environment, the amendment of HRM activities, in order to be resilient companies in the digital age, the requirements for individual and Organisational competencies in the new conditions as well as the mechanisms of cooperation between "human beings" and "smart Things" [6,7,8].

The overview of the specialized Bulgarian sources shows that there are no in-depth analyses and studies of the problems related to the digital transformation of human resources management, but rather separate articles from HR consultants and HR experts [9, 10,11].

They explain the steps of the digital transformation of HR activities, the attitudes of managers for its implementation, as well as the competitive advantages that companies will have after digitization of HR activities. In a synthesized form, these are the main highlights of the review of the specialized literature on the digitalization of HR activities.

2.1 Methodology of the study

The survey was carried out in thirty eight industrial enterprises in Bulgaria in the period September 2019 – February 2020 year. Of these, fourteen are large enterprises, and twenty-two are medium-sized enterprises. The nineteen of the enterprises are foreign management (Germany, France, Austria, Sweden, Denmark, Italy, Belgium, etc.) and nineteen are Bulgarian enterprises. The companies surveyed employ 9873 people. The survey participants were 184 HR managers and specialists. Of these, 92 work in foreign management companies. Conditionally denote these participants in the study with Group A. The other 92 respondents are HR managers and specialists in Bulgarian industrial enterprises. These are the group B participants.

The survey covers five phases: the development of surveys; Conducting inquiry; HR activities; Determining the degree of digitalization of HR activities; conducting structured interviews. The questionnaire sets out the issues related to the evaluation of the digitalization of 15 HR activities, and a scale is made to assess the degree of their digitisation from 1 to 5, as follows: 1 – fully digitised; 2 – largely digitalised; 3 – There are elements of digitization; 4-the initial degree digitized; 5-unitalised. The participants assess the extent to which each HR activity is digitized by indicating the corresponding degree in the range from 1 to 5. As a result, the average value of the indicator is calculated and the activity is performed. The survey is conducted for the participants of Group A and Group B. Fifteen human resource management activities can be grouped into 4 groups: routine activities (12, 13,14 and 15); Training (6.7 and 8); Organisation of Continuity and development (9 and 10); Recruitment and selection of human resources (1, 2,3, 4 and 5); The main idea is to establish and compare the degree of digitalization of HR activities in the two groups of enterprises. The interviews were conducted with 20 HR managers, of which 10 from Group A and 10 from group B. The only question is related to the problems and difficulties encountered in the digital transformation of human resources management.

2.2 Results of the study

As a result of the survey, we received the following results for Group A and group B participants (table 1). HR managers and specialists from industrial enterprises with foreign management assess the degree of digitization of human resources management activities as follows:

- routine activities related to the preparation of reports, missions, holidays, organization of meetings with the shellfish are in the scale between "fully digitised" and "largely digitized";
- activities relating to the human resources training process are also in the scale between ' fully digitised ' and ' largely digitised ';
- the activities of the Organization of continuity and development fall within the scale between ' largely digitised ' and ' there are elements of digitisation ';
- recruitment and selection of human resources are also on the scale between "largely digitised" and "there are elements of digitisation".

The analysis of the results obtained from group B participants ' questionnaires shows that:

- routine activities related to the formation of missions, leave, organizing meetings with the management are in the scale between "in the initial degree digitized" and "non-italised";

- the activities relating to the human resources training process are also in the scale between ' largely digitised ' and ' there are elements of digitisation ';
- the activities of the Organization of continuity and development fall within the scale between "C" there are elements of digitization "and" in the initial degree digitised ";
- recruitment and selection of human resources are also in the scale between "C" there are elements of digitization "and" in the initial degree digitised ";
- the activities of the Organization of continuity and development fall within the scale between "C" there are elements of digitization "and" in the initial degree digitised ";
- recruitment and selection of human resources are also in the scale between "C" there are elements of digitization "and" in the initial degree digitised ".

Table 1. Results of the survey

№	HR activities	Group A Average	Ranking	Group B Average	Ranking
1.	Post Job vacancies across multiple sources with one click	2,26	10	3,45	8
2.	Schedule for interviews	2,42	11	3,28	6
3.	Access to the applicant's profile for all participants in the recruitment campaign	2,12	9	3,87	9
4.	Automatic notification of managers and colleagues about the emergence of new staff recruitment	2,92	15	4,48	12
5.	Organizing meetings of the employee with the manager and his automatic planning	2,52	13	4,62	13
6.	Open access of managers to the training desires of each employee	1,26	5	2,34	1
7.	The ability to quickly create and view a training schedule for each employee	1,28	6	2,45	2
8.	Access to feedback on results in order to further plan employee training	1,34	7	2,84	3
9.	Registration of the platform for career desires and achievements of employees	2,43	12	3,26	5
10.	Automatic update of the Personnel reserve catalogue	2,68	14	3,84	7
11.	Document management System	1,82	8	4,27	10
12.	Organizing meetings with the management	1,14	2	4,42	11
13.	Request and confirmation of leave	1,18	3	4,68	14
14.	Shaping travel missions	1,20	4	4,86	15
15.	Shaping of reports	1,12	1	3,25	4

The ranking of activities, depending on the average assessment of their digitalization, shows that in Group A, the first places are occupied by routine activities related to the preparation of reports, missions, leaves, meetings, followed by training activities, Development and recruitment and selection of human resources.

In group B, the first places are training activities, followed by routine activities, recruitment and selection and organization of continuity and development.

The ranking of activities, depending on the average assessment of their digitalization, shows that in Group A, the first places are occupied by routine activities related to the preparation of reports, missions, leaves, meetings, followed by training activities, Development and recruitment and selection of human resources.

In group B, the first places are training activities, followed by routine activities, recruitment and selection and organization of continuity and development.

The analysis of the results of the study shows the relatively large difference in the assessments of the degree of digitalization of HR activities in the two groups, which we will summarize in the conclusions of the survey.

2.3 Conclusions from the study

- In industrial enterprises with foreign management, most of the activities related to the management of human resources are largely digitized and fully digitised. The activities of organization of continuity and development are largely digitized and there are elements of digitization.
- In the Bulgarian industrial enterprises the routine activities are in the initial degree of digitization and non-unitalised; The recruitment and selection activities, the organization of continuity and development are in the initial stage digitised and have elements of digitisation; The training activities are largely digitised.
- The comparative analysis of the results between the two groups of participants shows that the digital transformation of human resources management can be mainly spoken in industrial enterprises with foreign management. Still, in the Bulgarian industrial enterprises, in large part digital are only activities related to the training of human resources. Other activity groups have initial digitization or digitization elements.
- In a summarized and synthesized form, the conclusions of the analysis of the results of the interviews conducted on the difficulties and problems in the digital transformation of human resources management are as follows:
- In some of the Bulgarian industrial enterprises there is still mindset of underestimation of HR activities and in a relatively small part of them – 3% has developed a strategy for the Digitabizaciâ of human resources management.
- In the enterprises of both groups it is difficult to find creative, innovative and emotionally intelligent workers. Of interest is the response of HR managers that they are easier to discover with technological skills than with "soft skills" – criticality, innovativeness, teamwork skills, creativity and adaptability to the digital environment.

The conclusions of the survey give us the opportunity to summarize that in Bulgaria the digitalization of HR activities is still in its initial phase. It is therefore necessary to start from the development of a digital transformation strategy for HRM and a programme for its implementation. The practical result of the survey is that we have attempted to direct the concrete steps for digitization to the four main groups of activities, namely routine activities; Recruitment and selection; Training Organisation of continuity and development.

For this purpose it is necessary to select the teams with the necessary qualities – technological and soft skills. It is precisely at this initial stage of digital transformation of human resources management that the leading role of HR managers and specialists in industrial enterprises is of paramount importance.

3. Conclusion

In the report, we have tried to draw attention to the results of the survey, which show that serious efforts are needed in industrial enterprises on the digitisation of HR activities, so that they can gain competitive advantages in three Directions:

- digital Workplace – it will provide freedom for workers to carry out their work tasks, not only from the office but also from home or their preferred place. In this way, labor reality can become an exciting experience, as a result of which employees will be more efficient and creative;
- digital Learning – it must be carried out constantly, in relation to the changing conditions in various forms, so that it is highly effective and provides concrete results for the company;
- digital Talent – it is assumed that the modern digital environment will contribute to attracting and especially to the retention of talents in industrial enterprises, as well as to providing a positive experience for employees.

Acknowledgements
The author/s would like to thank the Research and Development Sector at the Technical University of Sofia for the financial support."

References

[1] Konarev A and Konstantinova Sn 2014 *Choveshkite resursi v triagalnika na znanieto i razvitieto na ikonomikata na Balgariya* Nauchna konferentsiya: Osnovni tendentsii v razvitieto na choveshkite resursi (Plovdiv: UI P Hilendarski) Sbornik s dokladi pp 20-29 ISSN 978-619-90178-4

[2] Shatarova D and Zlatanova-Pazheva E 2018 *Human resources in the digital age – trends and challenges* Scientific researches of the Union of Scientists in Bulgaria – Plovdiv series BNatural Sciences and the Humanities Vol XVIII ISSN-1311-9192 (Print) ISSN-2534-9376 (On-line)

[3] Anguelov K Ferdov A and Mihova T 2018 *Research of Interactions between Universities and High-Technological Enterprises on Employees Qualification* 9th National Conference with International Participation ELECTRONICA 2018 Proceedings (2018)

[4] Vulpen, Erik Van, 2013 *5 Human Resources Models Every HR Practitioner Should Know*, https://www.digitalhrtech.com/

[5] Ulrih D *Digitalizatsiyata na HR rabotata nyama da zamesti choveshkata pretsenka* https://www.capital.bg

[6] El-Khoury D 2017 *Digital transformation and the world-class HR difference*, Strategic HR Review, Vol. 16 No. 2, pp. 86-88. https://doi.org/10.1108/SHR-01-2017-0001

[7] *HRM 4.0 For Human-Centered Organizations*, 2020 Advanced Series in management, Volume 23, Howard House, Wagon Lane, Bingley BD16 IWA, UK, ISBN:978-1-78973-536-9 (Print), ISBN:978-1-78973-535-2 (Online)

[8] Trost A *Human Resources Strategies*, 2020 e-book, ISBN 978-3-030-30592-5

[9] Petrov Sv *Digitalnata transformatsiya na upravlenieto na choveshkite resursi* http://ww2.business-club.bg/

[10] Savov V 2019 *HR transformatsiyata zapochva ot nas* https://www.manager.bg/

[11] *Digitalizatsiyata na biznesa ne znachi roboti, a e shansza HR da poeme vodeshtata rolya* https://jobtiger.tv/hr-industriya/

[12] Nikolova-Alexieva V 2019 Background of Bulgarian bioeconomy and opportunities for its development", Monograph, ISBN 978-619-876-73-6, "Intelekspert-94,", pp. 43-45

[13] Mihova T and Nikolova-Aleksieva V 2014 *Novi tendentsi i na trudoviya Pazar prez* 2014 *godina* Mezhdunarodna nauchna konferentsiya-UNITEH-Gabrovo Tom IV, 21-22 Noemvri (Gabrovo: Universitetsko izdatelstvo Vasil Aprilov)

IOP Conf. Series: Materials Science and Engineering **878** (2020) 012070 doi:10.1088/1757-899X/878/1/012070

Business communities - a factor of industry and bioeconomy development

T B Mihova[1] and V N Alexieva[2]

[1] Faculty of Mechanical Engineering, Department of Industrial Management, Technical University of Sofia, Plovdiv Branch, 25 Tsanko Diustabanov Str., Plovdiv, 4000, Bulgaria
[2] Department of Industrial Business and Management, University of FoodTechnologies-Plovdiv, Bulgaria

E-mail: expert2009@abv.bg, valentina_nikolova@abv.bg

Abstract. The population of Europe and the world is constantly growing, and so is its need for more food. This leads to rapid depletion of natural resources, and environmental impacts and related climate change challenges are increasing.It is therefore essential to move to a new way of economic growth that is compatible with the protection of the environment and the sustainable use of scarce natural resources, while guaranteeing a much higher standard of living reducing poverty. The bioeconomy strategy and its action plan become the basis for a more innovative, efficient and competitive society that combines food security, sustainable use of renewable industrial resources and environmental protection.In recent years, the role of business associations has increased significantly, in connection with the rapidly changing business environment of ever-increasing corporate competition.The authors aim to reveal the goals of these business communities and the means to achieve them.The report outlines the specific results of the team's own study of the competitive advantages of industrial enterprises created by business associations.

1. Introduction

The sustainable development of the industry and the bioeconomy depends on the business environment in which businesses operate. According to the annual ranking Doing Business for 2020, prepared by the World Bank, Bulgaria gave two seats to the 61 position among 190 economies around the world [1]. The country stays back to other countries around the world when comparing the procedures needed to start a business, to obtain building permits, to credit and access electricity, to recognise insolvency and to pay Taxes. The main reason indicated by the specialists is that the country has not changed anything in regulations and requirements compared to last year. A number of authors analyse the factors of the internal and external business environment, allowing for industrial growth, with the emphasis in recent years mainly on the digital transformation of enterprises [2].

In this report, the team of authors aim to consider the role of business communities as a factor in the development of industrial enterprises. This issue is relatively little explored, creating a challenge for us.

IOP Conf. Series: Materials Science and Engineering **878** (2020) 012070 doi:10.1088/1757-899X/878/1/012070

2. Exposition

Long-term forecasts show that, without radical political changes, the current trends in world economic growth and development will have a major impact on the natural resources and the ecosystem. The population of Europe and the world is constantly growing, and so does its need for food, more and more natural resources are exhausted, the environmental impact and the related climate change challenges are increasing. It is therefore essential to move to a new way of economic growth that is compatible with environmental protection and sustainable use of limited natural resources, while ensuring a much higher standard of living reducing poverty. The development and application of innovative biotechnology methods and processes in the agriculture, health, chemistry and energy sectors has recently been seen as one of the solutions to accelerate sustainable growth and development. By focusing more on scientific research and innovation, new products from biomass and new services needed for the development of the bioeconomy would be created, helping to reduce climate change, waste and create new jobs .

The bioeconomy encompasses the production of renewable biological resources and the conversion of these resources and waste streams into value added products, such as food, feed, bio-based products and bioenergy. The bioeconomy includes the sectors of agriculture, forestry, fisheries, food and pulp and paper production, as well as parts of chemical, biotechnological and energy industries. Bio-based products are products that are wholly or partly derived from materials of biological origin, excluding materials embedded in geological formations and/or fossilized [15]. The bioeconomy relies on life sciences, agronomy, ecology, food science and social sciences, biotechnology, nanotechnology, information and communication technologies (ICT), and engineering[1].

Comparative analysis of the definition of bioeconomy in the EU, OECD and the EU member state documents allowed determining that bioeconomy is a part of economics associated with the following: *sustainable production of bio-resources and their processing into value added products*; *land and water* (plants, animals and micro-organisms) and bio-based products; *types of economic activities* (in other words – economic sectors) *related to the manufacture, processing or recycling and use of such as plants, animals, micro-organisms and their products*; *the use of advanced technologies, such as biotechnology, nanotechnology, information and communication technologies, in addition to traditional technologies* [15].

Thus the definition of bioeconomy according to the Communication from the European Commission "Innovation for Sustainable Growth: A Bioeconomy for Europe" reflects the essence and the content of bioeconomy, covering all the specified material elements of bioeconomy. According to this Communication, "*.....the bioeconomy encompasses the production of renewable biological resources and the conversion of these resources and waste streams into value added products, such as food, feed, bio-based products and bioenergy*"[15]. This description of bioeconomy is clear, easily understandable and suitable for defining the bioeconomy in Bulgarian legislation. It should be added that this definition of bioeconomy has recently been often quoted in various EU bioeconomy studies.

In the specialized literature there are different names such as business communities, business associations, business associations, business federations, business structures and others, among which there are some similarities and differences. What unites them mostly is that, by their essential characteristics, in themselves, they are among the most important forms of social capital [3]. As the most famous authors, published theories about social capital, we can point out sociologists Pierre Bourdieu, Robert Putnam and James Coleman. According to Bourdieu, the social capital represents the sum of resources (real or virtual) that individuals or groups accumulate because of the existence of a permanent network of more or less institutionalised acquaintance and acceptance relationships [4]. Complementary emphasis in the understanding of social capital puts Robert Putnam "Social Capital is a system of interactions with Nejerarhičen, networked nature. It is deployed as a result of obŝovzprieti norms and mutual trust. Putnam focuses on the traits of social organization such as trust, norms and networks. They can improve the efficiency of society by facilitating coordination of actions [5]. According to Coleman's views, social capital is defined by its functions. And it consists of a multitude of different units with two common characteristics: they have traits of social structure and support certain actions of the individuals who belong to this structure [6].

The survey of sources by Bulgarian authors, examining the essence of the social capital makes it possible to highlight the definition of Ivan Bondokov "the social capital is all the resources

that a person acquires through personal and professional contacts and relations. These are for example: information, ideas, business opportunities, influence, trust, emotional support, friendship and cooperation. The social capital resource is built and located in networks of relationships and relations with others "[7]. Other Bulgarian researchers analyse in depth various definitions of social capital, which reveal different aspects of its nature [8, 9, 10]. For our development for a working definition we can accept the understanding of Robert Putnam, which reveals the social capital as a system of interactions with Nejerarhičen, networking nature. It is deployable as a result of obŝovzprieti norms and mutual trust [11,16]. The analysis of the different definitions allows us to summarize that some of the main features of social capital are:

- a system of interactions with non-heretic, networked nature;
- the result of continuous work on creating, developing and strengthening social relations based on trust;
- associated with group membership.
- having already pointed out that business communities are a form of social capital, we will look at the objectives of three of them operating on the territory of the South Central planning region.
- in summary, they are limited to:
- to protect the rights and legitimate interests of its members;
- to promote dialogue between business representatives, individual investors, experts and NGOs on the one hand, and municipal and regional administration on the other;
- to undertake and implement initiatives to attract national and international investments;
- to unite and promote the efforts of its members to regulate professional ethics in the field of entrepreneurship and investment and to prevent unfair competition;
- to bring together entrepreneurs to protect their economic, social and legal interests by interacting with state and local authorities to create conditions conducive to the development of private enterprise;
- to support the Bulgarian business and economy [12, 13, 14].
- the means to achieve these objectives are:
- organizing and conducting events related to the fulfilment of the objectives;
- participation in the development of investment projects of local, regional and national importance;
- organizing and conducting initiatives (workshops and discussions) in support of the territorial-spatial, architectural and economic development;
- attracting professionals and experts to solve business issues [16];
- promoting cities and areas as favourable economic zones for investment.

The review of the objectives and the means for their realization of business communities that they possess the identified distinctive characteristics of social capital. In a synthesized form, these are the results of the literature review on the essence of business communities.

3. Methodology and results of the study

Methodology of the study

The survey was conducted in seventy three industrial enterprises in Bulgaria, which are members of business communities in the period September 2019 – February 2020 year. Of these, seventeen are large enterprises, thirty seven are medium-sized and nineteen are small enterprises. Industrial enterprises are from different industries: "Automobiles and Autoparts", "Information technology", "Lighting", "Transport services", "Machinery and Machine building", "Polygraphy and packaging", "Metals and metalworking", “Bio-fuel","Bio-chemical" and „Bio-pharmaceutics“. Twoof them and one of the companies are foreign management (Germany, France, Austria, Sweden, Denmark, Italy,

Belgium, etc.) and fifty two are Bulgarian enterprises. The companies surveyed employ 19325 people. The participants in the survey are 231 managers at higher and middle level.

The study covers five phases: conducting structured interviews to determine the benefits for the development of industrial enterprises; developing surveys; Conducting inquiry; benefits; Processing of the results.

The first stage relates to the conduct of structured interviews which contain a question relating to the identification of the most significant benefits for the development of industrial enterprises from their participation in business communities. They were scheduled for an interview with twenty business managers who are members of three businesses communities, or a total of sixty managers to answer the only question. Of the sixty managers planned for the structured interview, 9 declined to participate, which follows that the respondents were fifty-one. The above-mentioned benefits are over twenty, but the comparatively most support gets the first eight, and they define them as "an essential factor for the operation and development of industrial enterprises".

Based on the results of the interview, questionnaires were developed – the second stage of the study.

The questionnaire sets out questions related to the assessment of the benefits for the development of the companies from belonging to the business communities, making a scale for the assessment of the degree of 1 to 4 as follows: 1 – High level; 2 – Medium degree; 3 – Low level; 4 – Zero degree. The participants assess how much each benefit contributes to the development of the companies

Results of the study

As a result of the survey, we received results illustrated in the table. (Table 1). Interviewed managers indicated as the most significant benefits:

- -Business information;
- Business opportunities;
- Prestige
- Influence
- Cooperation
- Business ideas
- Consultations with experts;
- Emotional support.

Table 1. Results of the study

№	**Benefits for the development of the enterprises**	**1 high degree %**	**2 middle degree** %	**3 low degree %**	**4 zero degree** %
1.	Information for businesses	85	8	5	2
2.	Consultations with experts	62	23	12	3
3.	Business ideas	58	24	13	5
4.	Cooperation	37	33	21	9
5.	Business opportunities	76	14	7	3
6.	Prestige	71	15	12	2

7.	Influence	68	24	6	2
8.	Emotional support	28	19	38	15

The results of the questionnaires relate to the assessment of the benefits of the business communities on a four-tier scale and are the following:

-85% of respondents are of the opinion that the information about the business is useful in "high grade", 8%-"Medium grade", 5%-"low grade" and 2%-"zero degree";

-76% assess the significance for the development of the companies of "high grade" business opportunities, 14%-with "medium grade", 7%-with "low grade" and 3%-with "zero degree";

-71% of respondents assessed the ' high-grade ' prestige, 15% with ' Medium grade ', 12% with ' low grade ' and 2% ' nil '.

These are the benefits mentioned in the first three places for the development of industrial enterprises by the survey participants. The rest are shown in table 1. On the basis of these results, their ranks were performed (table 2).

Table 2. Ranking position of the benefits

№	**Benefits for the development of the enterprises**	**Ranking**
1.	Information for businesses	1
2.	Consultations with experts	5
3.	Business ideas	6
4.	Cooperation	7
5.	Business opportunities	2
6.	Prestige	3
7.	Influence	4
8.	Emotional support	8

Conclusions from the study

1. Business communities are significant for the development of high-level industrial enterprises with information about business, business opportunities and prestige and, to a lesser extent, consulting experts, providing influence and business ideas.
2. Business communities are not very successful in helping to achieve cooperation and emotional support for their members.

In a summarized and synthesized form, the analysis of the results show that business communities have their place and are a factor for the successful operation and development of industrial enterprises and the bioeconomy at all.

Recommendations

The conclusions of the survey give us the opportunity to summarize that in Bulgaria the business communities as a form of social capital are useful for the development of the industry, but they need to improve their activities with a view to creating better conditions for Cooperation and emotional support of its members. In this strand, there is good European practice, which could be addressed by business community manuals, raising awareness among members and building techniques to enhance the usefulness of business ideas, cooperation and emotional Support.

4.Conclusion

In the report, we tried to uncover some of the essential characteristics of business communities and argue that they are one of the forms of social capital. Through the research we have established the benefits of these communities for the effective development of industrial enterprises.

This report is funded from the National Science Program "Healthy Foods for a Strong Bioeconomy and Quality of Life" of the Ministry of Education and Science, approved by decision of the Council of Ministers №577 / 17.08.2018 , contract No 68 / NSP under Work Package 4.2 "Regional ecosystems for bio-economy"

References

[1] Ignatova L 2019 *Usloviyata za biznes v Balgariya se vloshavat za chetvarta poredna godina*, https://www.capital.bg/politika_i_ikonomika

[2] Gigova T and Valeva K and Nikolova-Alexieva V 2019 *Digital Transformation – Opportunity for Industrial Growth,* 2019 International Conference on Creative Business for Smart and Sustainable Growth (CREBUS), Sandanski, Bulgaria, INSPEC Accession Number: 18994787, DOI: 10.1109/CREBUS.2019.8840102

[3] Bian Y 2002 *Social Capital of the Firm and its Impact on Performance: A Social Network Analysis.* Kluwer Academic Publishers. Boston, pp. 275-297

[4] Burdiyo P 2005 *Prakticheskiyat uset* Izd. „Kritika i Humanizam", S.

[5] Coleman J S *Social Capital in the Creation of Human Capital* The American Journal of Sociology, 94:95

[6] Putnam R and Leonardi R and Nanetti R Y 1993 *Making democracy work: civic traditions in modern Italy*. Princeton, N.J.: Princeton University Press

[7] Bondokov I 2018 *Kakvo e sotsialen kapital i kolko e vazhen za biznesa i demokratsiyata*, https://www.dnevnik.bg/biznes/

[8] Bakardzhieva M 2013 *Mezhdulichnostnata komunikatsionna mrezha kato komponent na sotsialniya kapital,* Nauchni trudove na Rusenskiya Universitet, seriya 5.2

[9] Georgieva T 2012 *Sotsialniyat kapital v kooperativnite organizatsii: prilozhenie na kontseptsiyata i izmervane*, Upravlenie i ustoychivo razvitie 1/2012 (32)

[10] Pachev T 2012 *Sotsialniyat kapital i vliyanieto mu varhu ikonomicheskoto razvitie*, sp.„Ikonomicheski alternativi", br.3

[11] Putnam R D *1995 Bowling Alone: Amerika's deslining social capital* Journal of Democraty, 6 (1), p.p. 10-12

[12] Sdruzhenie s nestopanska tsel "Biznesat za Plovdiv ",https://www.businessforplovdiv.com/

[13] Targovsko-promishlena kamara – Plovdiv, https://pcci.bg/new/

[14] Sdruzhenie na chastnite predpriemachi v Pazardzhik, http://businesspz.org/

[15] European Commission 2017 *The Bioeconomy Strategy*. Research & Innovation: Bioeconomy

IOP Conf. Series: Materials Science and Engineering **878** (2020) 012071 doi:10.1088/1757-899X/878/1/012071

Diffusion of innovation in the Bulgarian dairy industry

Valentina Nikolova-Alexieva[1], Irina Krasteva[2]

[1] University of Food Technologies, Plovdiv, Bulgaria

e-mail: valentina_nikolova@abv.bg

Abstract. This report presents a study of the factors that influence the diffusion of innovation in Bulgarian dairy businesses. Maintaining high innovation activity and positive innovation potential is the result of the effective functioning of the innovation system. The different types, functions and factors, prerequisites and mechanisms for implementing the process of innovation diffusion are explored. It is proved that the efficiency of investment is determined by external factors, primarily by the level of economic concentration of the market of innovative products. As a results the external factors have a relatively greater impact on the economic efficiency of investing in innovation and and hence a significant direct impact on the diffusion of innovation in the dairy industry. The «highest» level of correlation is determined with the factor of economic concentration of the industry and the market of innovation project.

1. Introduction

Keeping innovative activity creates the conditions for businesses to profit and maintain competitive advantages. Diffusion is of great economic importance, since only the widespread diffusion of an innovation can have a noticeable economic effect [1]. The existence of mechanisms for dissemination of innovation results and their subsequent refinement alters the targeted scope of innovation and, from a factor relevant to enterprises, transforms them into an engine of social development and the creation of competitive economies.

The **purpose of this article** is to present the essence and characterize the mechanisms for diffusion of innovation in enterprises and their influence on their innovation activity, efficiency and competitiveness.

The object of research is the nature and factors influencing on the diffusion of innovations in the national dairy industry.

The methodology of research - the task of correlation analysis was set to identify the factors that affect the performance indicator - the level of diffusion in innovative projects. The authors identified 8 internal and 4 external factors, hypothetically determining the profitability of investments in innovative projects. The choice of hypothetical factors is determined by the following criteria: the frequency of allocation in the scientific literature; measurability; relative independence from the scale of projects; the possibility of comparison in projects; availability of management accounting documents (business plans, investment projects and reports on marketing research).

The sample of the study is based on a survey of the retrospective of 241 innovative projects of enterprises from food industry, implemented in the period 2010-2019. To clarify individual quantitative assessments and justify the nature of economic efficiency, in-depth interviews were conducted with the leaders of 27 innovative projects.

The practical side **of the result** can be formulated as a criterion for selecting markets or types of activities for investing in innovations by national enterprises from dairy industry [2]. The results of the research are in solidarity with the scientific conclusions about the dominance of external factors in the formation of economic efficiency of investment in innovation: The formulation of the criterion: as optimal are determine the markets, where the level of economic concentration at the time of the start of sales of the innovative product is 15-45%, which characterizes the stage of "growth".

2. Literature review

In most studies, innovation is viewed in a broad and narrow sense [3]. In a broad sense, it is presented as a complex phenomenon of mutually conditioned and connected creative actions of people in the formulation of an idea, creation, implementation and use of something (product, service, process, etc.), by which they improve and transform the opportunities and conditions for business and your environment. In a narrow sense, it is considered as an act of actually introducing and commercializing new goods, services, technology, technology and organization. In both cases, innovation can be viewed in dynamics and in statics.

The theory of the diffusion and perception of market innovations was originally launched by Everett Rogers [10]. It defines diffusion as "a process in which information about innovation is disseminated through members of a social system" [5] and identifies four specificities of dissemination:

- This is a specific communication aimed at disseminating innovative ideas;
- The novelty of ideas creates uncertainty in the diffusion process - in terms of projections, structure, information, etc.
- Dissemination of information leads to a reduction of uncertainty in situations requiring the choice of alternative behaviors.
- Diffusion of innovation is a type of social change associated with changes in existing structures and functions.

Two interpretations of the term "innovation diffusion" have become commonly accepted in the economic literature. According to the first understanding, it is defined as a process with strictly defined parameters and boundaries in terms of its ability to be controlled. Researchers refer to this type of diffusion as an active spread that is planned, formal, often centralized and hierarchical [8, 9]. The second group of economists view diffusion as spontaneous, unplanned, informal, decentralized, and mainly through horizontal connections, the dissemination of new ideas. This is the so-called. "Pure diffusion". Practice shows that the implementation of diffusion processes is a combination of planned actions in certain situations and spontaneous ad hock decisions - in others.

At its core, diffusion of innovation is the dissemination of successfully implemented innovations in the processes or products of other enterprises in the industry and / or in other areas of social practice [4]. Therefore, diffusion is "a macroprocess and is the dissemination of innovation from its source - the enterprise, to consumers (the target market) through communication channels (mass media, distributors, traders, etc.) over a period" [7]. It consists of four elements [6]: innovation; communication channels; time / period; social system.

The main function of diffusion is to drive and reach innovation to the maximum number of users. Several additional functions can be specified as derivatives: *economic function, social function, incentive function and psychological function.*

Table 1. Factors and prerequisites for diffusion of innovation

External: by industry affiliation of the enterprise	Internal: the average value of the indicator in the sample during the implementation of the innovative project
Market share by main business activity	The growth rate of the revenue of an industrial enterprise
Average market growth rate	Profitability of operating activities
Coefficient of economic concentration by the sum of 3 (CR3)	Proportion of staff engaged in R&D
Science intensity of the industry (average for OECD groups)	Margin coefficient
Information system for the supply and demand for innovation;	Science-intensive: costs of R&D in revenue
Reliable legal protection of technological transfer facilities;	The share of R&D costs in the budget of the innovation project
Efficient regulatory framework and court arbitration and contractual practice;	The ratio of innovation costs to annual revenue
Reliable measures of the relative advantages of the first successful innovation;	The share of innovation costs in total investment

Closely related to the term "innovation diffusion" is the term "reinvention". According to Rogers, it reflects the extent to which innovation is altered or modified by the consumer and in the process of its adaptation or use. For Ivland, reinvention shows the extent to which the use of a new idea is moving away from the basic version of innovation originally marketed [8].

The dissemination of an innovation cannot be accomplished without the parallel flow of another important process - the adoption of innovation. According to Rogers and Schumaker, perception can be defined as "a decision to use it fully or to choose a new idea as the best course of action" [9]. The adoption process has two main components: the decision to adopt / reject the innovation and the period of use [10]. Diffusion and the perception of innovation differ (see tabl 1). While diffusion refers to whether the proposed new or improved product (service) will be accepted by the market, the perception is related to how quickly the product (service) will be accepted (implemented) by the market. The diffusion mechanisms are [3]:

- **Indirect diffusion** - an increase in the scale of production or a new process in a company that is the first implementer in the world at all (Figure 1).
- **Direct diffusion** - Adaptation of already implemented innovation by related companies.
- **Mixed diffusion** - combines direct and indirect diffusion. Examples of mixed diffusion are the transfer of innovations from the military to the civilian sphere, from the chemical to the food industry.

Each innovation system is characterized by a certain degree of development. It can be tracked, analyzed and reported through the level of innovation and the accumulated innovation potential (IIA) of the economic system that determine the innovation (IA) and economic (EA) activities. In this regard, it is important to give a brief description of some basic concepts related to new trends, requirements and processes [6].

Research Potential (RP) is related to the capacity of the scientific and research system to generate a scientific product in a different form. Innovation potential (IP) is the accumulated economically significant innovations in the economic system and its opportunities for their development and implementation in the economic practice. According to Kerchev [3], innovation shows the possibility and the achieved level of the innovation potential of the innovation system, ie it is an indicator for the effective use of research and innovation potential. Innovation activity (IA) takes into account the dynamics and capabilities of the innovation system to feed production with ready-to-use innovations. Innovation's economic activity (EA) characterizes the capabilities and flexibility of the production system with the available resources to perceive and absorb innovations and to offer them as products in the form of goods, services or technologies.

The analysis of the relationship between IA and EA and their tracing over time gives an idea of the trends and the degree of development of the innovation potential [13]. When IA has values higher than Ca, an innovation reserve (IR) is formed. Otherwise, there is an innovation gap (IG). Charts 1 through 4 show the relationship between EA and IA in time (t). Many conditions and changes in the relationships between them are possible, but several are the basic ones.

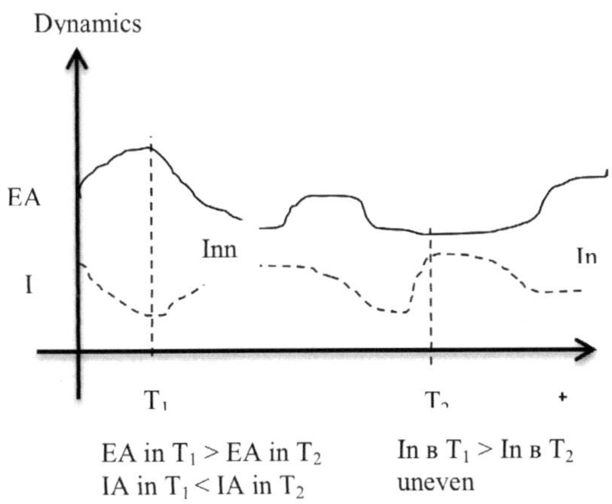

Chart 1. Imbalance in favour of the business activity

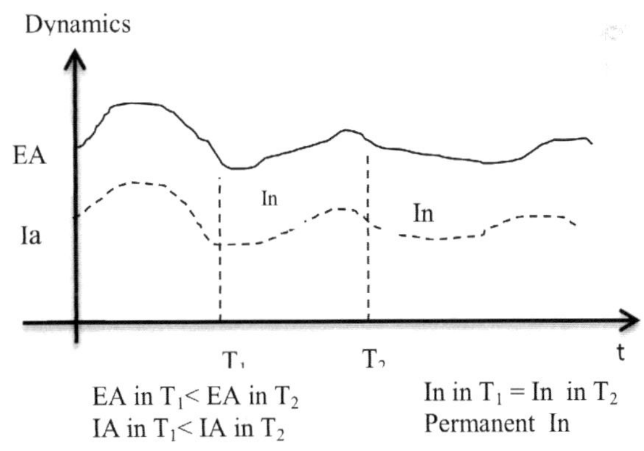

Chart 2. Synchronous negative imbalance

1. Imbalance in favor of economic activity (Chart 1). The innovation potential has been negative throughout. The difference between EA and IA is always in favor of production, with larger (T1) or smaller (T2) differences between them.

2. Synchronous negative imbalance (Figure 2). In this case there is a correspondence between EA and IA for each period of time and the innovation potential is constant (permanent) but negative.

3. Balance and imbalance (Chart 3). Viewed in time, the development of production and the availability of innovation are very uneven, ie. the project receives positive and negative values4. Positive imbalance (Chart 4). If the innovation policy manages to keep in sync with IA and EA with a

positive IP, this is the best option for the company. On the one hand, production and innovation develop in relation to each other, and on the other, there is always an II that can be used in production.

From these basic relationships between IA and EA, three conditions for a dynamically progressing relationship between the innovation and production activity of the business, of the economic life and of the whole economic system are deducted:

1) EA in T2> EA in T1 ; 2) IA in T2> IA in T1; 3) IA>EA for any meaning of T> 0.

As a result of active R&D, there is a positive IP and an accumulation of IR. They create the conditions to satisfy the needs of the economic system for innovative products.

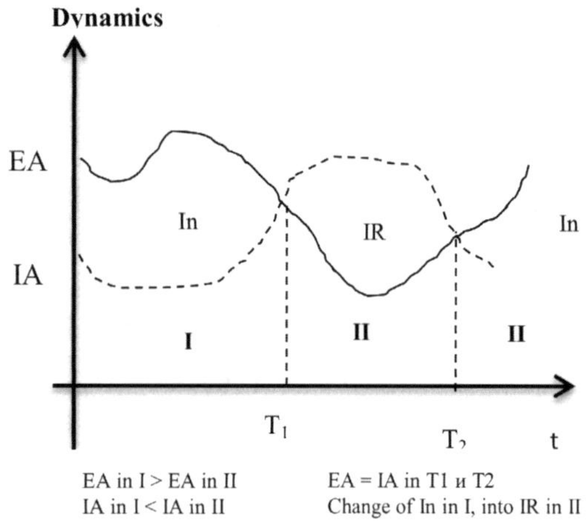

Chart 3. Balance and imbalance

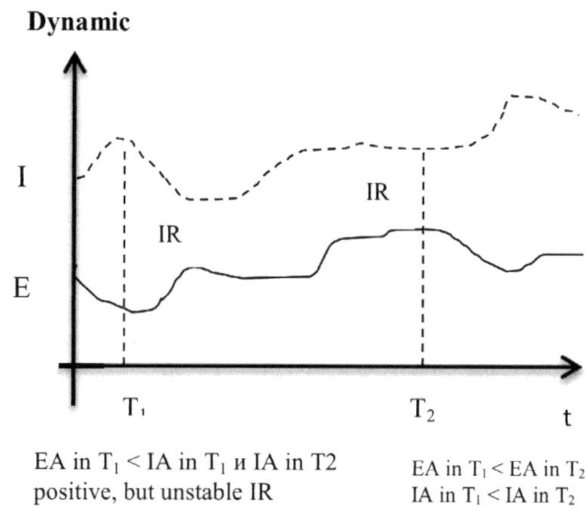

Chart 4. Positive balance

3. Object and Methododlogy of research

The subject of the research: the nature and factors influencing on the diffusion of innovations in the national food industry.

The object of research: Bulgarian enterprises from the dairy industry.

Method of research: correlation analysis. The task of correlation analysis was set to identify the factors that affect the performance indicator - the level of diffusion in innovative projects. The authors identified 4 main external factors (see table 1), hypothetically determining the profitability of investments in innovative projects in dairy industry. The choice of hypothetical factors is determined by the following criteria: the frequency of allocation in the scientific literature; measurability; relative independence from the scale of projects; the possibility of comparison in projects; availability of management accounting documents (business plans, investment projects and reports on marketing research).

The sample of the study is based on a survey of the retrospective of 241 innovative projects of enterprises from food industry, implemented in the period 2010-2019. The population statistics was obtained from Bulgarian Chamber of Commerce and Confederation of the Employers and Industrialists in Bulgaria – CEIBG. A stratified random sample was a useful blend of randomization and categorization, which enabled both a quantitative and qualitative process of study to be undertaken. Statistics on internal investment indicators of projects are supplemented with information on external indices of the industry obtained from business plans of innovation projects, marketing research and official statistics (in particular, NSI and Eurostat). To clarify individual quantitative assessments and justify the nature of economic efficiency, in-depth interviews were conducted with the leaders of 27 innovative projects. The study used a structured questionnaire in data collection.

The questionnaire was carefully designed and administered to the respondents. The questionnaire was designed on a four point Likert-Scale which ranged from strongly agree (4 points), agree (3 points), disagree (2 points) and strongly disagree (1 point). The items were structured to capture information on the dependent variable (diffusion of innovations) and the independent variables (Average market growth rate; Coefficient of economic concentration by the sum of 3 (CR3); Science intensity of the industry (average for OECD groups); Market share by main business activity).

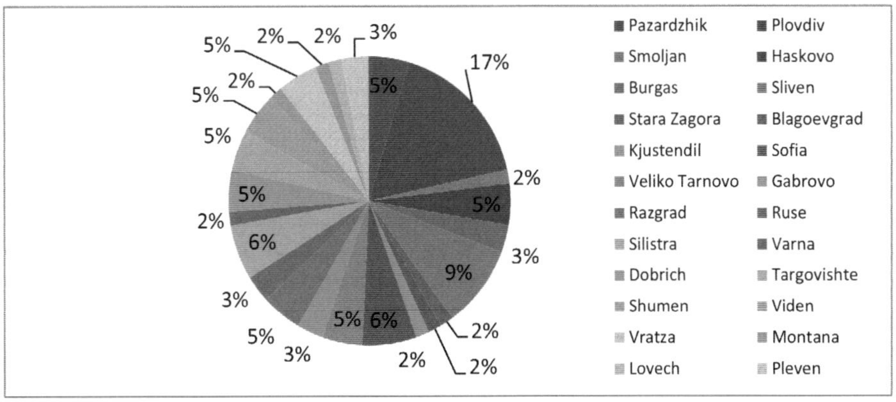

Figure 1. Territorial distribution of enterprises

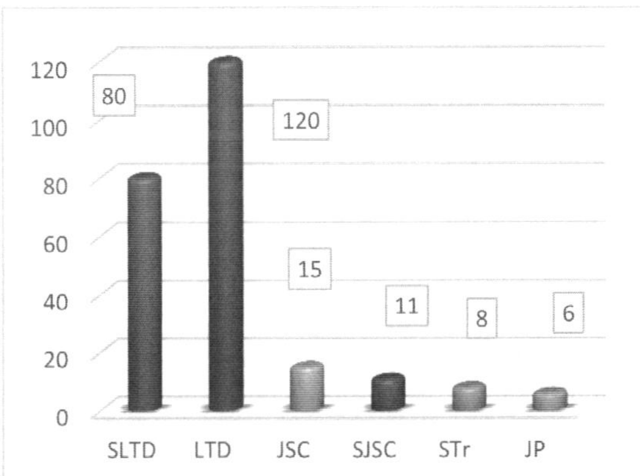

Figure 2. Legal status of companies

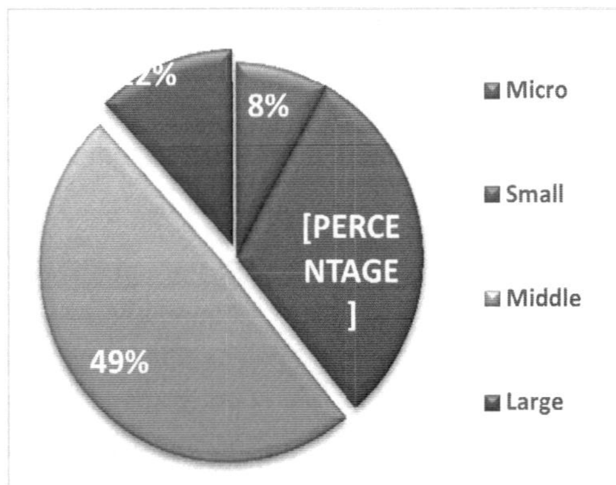

Figure 3. Distribution of enterprises by size

4. Analysis and results

The territorial distribution of the surveyed dairy enterprises in Bulgaria is shown in Figure 1. ***This distribution indicates that the sample is sufficiently representative for this survey, as enterprises from all regions of Bulgaria are included.*** The analysis of the legal status of the surveyed enterprises shows that 80 of the enterprises are Sole LTD, 120 of them are LTD, 15 are joint stock companies, and at least the enterprises with legal status are general partnership - 6 piece (see fig.2). 49% of the surveyed enterprises are small, 31% are medium-sized, large enterprises are 12% and micro-enterprises are 8% (see fig.3).

Pearson's measures the strength and direction of the linear relationship between variables. From the results, a significant relationship exists between the variables (table 2). *Science intensity of the industry* was shown to contribute 76.3% of the diffusion of innovations as indicated by the correlation coefficient value of 0.763 which is significant at $\alpha = 0.01$. *Coefficient of economic concentration by the sum of 3 (CR3)* rate was positively correlated to innovation's diffusion as indicated by correlation coefficient value of 0.571 indicating that the Coefficient of economic concentration by the sum of 3 (CR3) was a significant factor and contributed up to 57.1% of the change in diffusion of innovations. *Average market growth rate* was also shown to contribute 44.7% of the change in diffusion of dairy innovations as indicated by the correlation coefficient value of 0.447 which is significant at $\alpha = 0.01$.

Table 2. Correlations statistic for relationship between variables

Variable	*Diffusion of innovation*	*Average market growth rate*	*Coefficient of economic concentration by the sum of 3 (CR3)*	*Science intensity of the industry (average for OECD groups)*	*Market share by main business activity*
Diffusion of innovation	1				
Average market growth rate	*0,447*	1			

Coefficient of economic concentration by the sum of 3 (CR3)	*0,571*	*0,432*	1		
Science intensity of the industry (average for OECD groups)	*0,763*	*0,321*	*0, 421*	1	
Market share by main business activity	*0,482*	*0,341*	*0,398*	*0,432*	1

* *Correlation is significant at the 0.01 level (2-tailed).*

The correlation for *coefficient of economic concentration by the sum of 3 (CR3)* showed that 43.2% of the change in diffusion of dairy innovations was significantly accounted for by Coefficient of economic concentration by the sum of 3 (CR3) as shown by correlation coefficient value of 0.432 (significant at $\alpha = 0.01$). This paves way for multiple regression analysis.

Table 3. Multicollinearity analysis test for independent variables

Diffusion of innovation	***Tolerance***	***VIF***
Average market growth rate	*0,821*	*1,576*
Coefficient of economic concentration by the sum of 3 (CR3)	*0,865*	*1,099*
Science intensity of the industry (average for OECD groups)	*0,787*	*1,321*
Market share by main business activity	*0,421*	*1,730*

Results in table 3 showed that the VIF value for all the estimated parameters was found to be less than 4 and the tolerance values are more than 0.2 which indicate the absence of multi-collinearity among the independent variables of the study. This implies that the variation contributed by each of the independent factors was significant independently and all the factors should be included in the prediction model.

Table 4. Multiple regression model

R	***R^2***	***Adjusted R^2***	***Std. error of the estimate***	***Durbin Watson***
0.931*	0.828	0.771	0.843	1.626

**Predictors: (Constant)*

The study assessed the contribution of the independent variables on dependent variable. The findings of the study in table 4 illustrates multiple regression model had a coefficient of determination (R2) of about 0.828. This means that 82.8% variation of diffusion of dairy innovations performance is explained by joint contribution of innovativeness, creativity, business alertness and risk-taking. The findings are supported by ANOVA (F test) results that the model was fit or none of the parameters was equal to zero hence significance adjusted R square ($F = 26.721$, $\rho<0.05$).

Table 5. F test results (ANOVA)

	Sum of squares	***Df***	***Mean square***	***F***	***Sig***
Regression	*8,435*	*4*	*9,421*	*26,721*	*0,000***
Residual	*9,002*	*187*	*7,87*		
Total	*17,437*	*191*			

* *Dependent variable: diffusion of innovations*

In addition, Durbin Watson test had value less than two indicating minimal autocorrelation with no effect on the study output (Watson value = 1.626). The rule of thumb was applied in the interpretation of the variance inflation factor which states that a principle with broad application that is not intended to be strictly accurate or reliable for every situation. The results of survey of the sample of innovative projects are presented as a table 6 of correlation coefficients (r) of hypothetical factors with diffusion of innovations as a resultant indicator. The analysis allows to formulate the following conclusions:

- External factors have a relatively greater impact on the economic efficiency of investing in innovation and hence a significant direct impact on the diffusion of innovation in the dairy industry.
- The «highest» level of correlation is determined with the factor of economic concentration of the industry and the market of innovation project (coefficient by the sum of 3, CR3).

Table 6. Multiple regression model

Variable	***Coefficient***	***Std. error***	***Beta***	***t-statistics***	***Sig.***	***Tolerance***	***VIF***
Constant	*3,419*	*0,499*		*4,211*	*0,198*		
Average market growth rate	*0,436*	*0,001*	*0,303*	*3,291*	*0,006*	*0,821*	*1,576*
Coefficient of economic concentration by the sum of 3 (CR3)	*0,554*	*0,087*	*0,345*	*3,239*	*0,003*	*0,865*	*1,099*
Science intensity of the industry (average	*0,386*	*0,122*	*0,331*	*2,367*	*0,032*	*0,787*	*1,321*

for OECD groups)							
Market share by main business activity	*0,281*	*0,158*	*0,243*	*2,876*	*0,029*	*0,421*	*1,730*

* *Dependent variable: diffusion of innovations*

5. Conclusion

Hypothesis one (H01) estimated that *average market growth rate* has no significant effect on diffusion of innovations. However, research findings showed that average market growth rate had coefficients of estimate which was significant based on β1 = 0.303 (p-value = 0.001 which is less than α 0.05) implying that we reject the null hypothesis stating that there is no significant effect of average market growth rate on diffusion of innovations in the companies from dairy industry in Bulgaria. This indicates that for each unit increase in the positive effect of *average market growth rate*, there is 0.303 units increase in diffusion of innovations. Furthermore, the effect of *average market growth rate* was stated by the t-test value =3.291 which implies that the standard error associated with the parameter is less than effect of the parameter.

Hypothesis two (H02) stated that *coefficient of economic concentration by the sum of 3 (CR3)* has significant effect on diffusion of innovations. Findings showed that *coefficient of economic concentration by the sum of 3* had coefficients of estimate which was significant based on β2 = 0.345 (p-value = 0.003 which is less than α 0.05) hence we reject the null hypothesis and conclude that *coefficient of economic concentration by the sum of 3 (CR3)* has significant effect on the diffusion of innovations of dairy industry in Bulgaria.

Hypothesis three (H03) stated that *Science intensity of the industry (average for OECD groups)* has no significant effect on organizational performance. Research findings indicated we reject the null hypothesis and conclude that this factor has significant effect on the diffusion of innovations of dairy industry in Bulgaria.

Hypothesis four (H04) stated that *market share by main business activity* has no significant effect on organizational performance. However, findings of the study revealed that the factor had coefficients of estimate which was significant based on β2 = 0.243 (p-value = 0.024 which is less than α 0.05) hence we reject the null hypothesis and conclude that market share by main business activity has significant effect on organizational performance in the diffusion of innovations of dairy industry in Bulgaria.

References

[1] Nikolova-Alexieva V and Alexieva I 2019 *Impact of innovation – supportive culture for an effective bio-economy in Bulgaria* DOI:10.15547/ictte.2019.07.073; ICTTE 2019 Bulgaria pp 514-523
[2] Konarev A Konstantinova S and G V - Georgieva 2019 *Technological innovations and total factor productivity in dairy industry* IOP Conference Series: Materials Science and Engineering (Published: IOP Publishing Ltd) Volume 618 conference 1 pp 1-6
[3] Kerchev Kr., 2015 *Innovation and innovation system* (University Publishing Economy), pp 49
[4] Konstantinova S 2017 *New models for product planning* Trakia Journal of Sciences (Stara Zagora: Trakia University) Volume 15 Supplement 1 pp 195-198
[5] Gadrdner C A Acharya T and Yach D 2007 *Technological and Social Innovation: a Unifying New Paradigm in Global Health* (Publisher: Health Affairs) Vol 26 (4) p 1052-1061
[6] OECD 2005 *Oslo Manual* European Commission Eurostat (Brussels, Belgium) p 47
[7] Roger E M 2003 *Diffusion of innovation* 5 ed (New York: Free press) p 5
[8] Varamezov L Panteleeva I 2009 *Innovative management* (Svishtov: AP Chenov) p 132
[9] Eveland J et al 1977 The innovation process in public organizations p 17
[10] Rogers E and Shoemaker F 1977 *Communication of innovation: a crosscultural approach* (New York: Free Press) p 5
[11] Black W 1983 *Discontinuance and Diffusion: Examination of the Post Adoption Decision Process* Advanies in Consumer Research Vol 10, p 360
[12] Greenhalgh T., Robert G., Bate P. et al 2005 *Diffusion of Innovation in Health Service Organisations : A Systematic Literature Review* (Oxford, Blackwell) p 145
[13] Nikolova-Alexieva V 2019 *Opportunities for raising the Entrepreneurial culture in Bulgaria* ISBN 978-619-7220-33-9 (Publisher: Intelexpert-97)

IOP Conf. Series: Materials Science and Engineering **878** (2020) 012072 doi:10.1088/1757-899X/878/1/012072

Corporate growth benchmarking in the food industry

Snezhinka Konstantinova[1], Gergana V. – Georgieva[1]

[1] University of Food Technologies, Plovdiv, Bulgaria

E-mail: sks_ko@abv.bg

Abstract. This study presents results of benchmarking of corporate growth of public companies in food industry. Security and profitability parameters from growth financing were selected as the main benchmarks. For the benchmarking have been selected results of the best public industrial companies in other sectors. For information were used audited annual reports on an individual basis. The comparative analysis shows that parameters of corporate growth management in food industry are significantly inferior to those of leading industrial companies.

1. Introduction

Benchmarking is process of comparing the performance of a company or group of companies with industry best practices or best practices from other industries [1, 2].

The name of "benchmarking" method is derived from English words "bench" and "mark" and are interpreted differently: "benchmark", "reference marker", etc. According to benchmarks, the following types of benchmarking differ [2]:

1. Internal benchmarking – a comparison is made within the organization.
2. External benchmarking – conducted outside the organization.
3. Competitive benchmarking.
4. Functional benchmarking.
5. Generic benchmarking.

The following types of benchmarking differ according to goals and objectives for improving productivity [3, 4].

1. Strategic;
2. Product-oriented;
3. Process oriented.

The strategic needs of companies to explore best practices have turned to this specific benchmark, which began in the 1950s in Japan and United States, and turn into a widely used toolbox around the world. For this purposes, strategic common (general) benchmarking was used to analyze, evaluate, and compare corporate growth of food industry companies with the leading industrial companies [5, 6, 7].

2. Methodology and methods

Corporate growth benchmarking is based on two criteria:

- security of financing (reciprocal of creditors' risk);
- profitability of financing.

The term “balance of growth” is used to refer to the degree of growth that a company can maintain with its own operating cash flows.

When the growth of the company is financed with equity, cash flows of the company are considered to be in equilibrium and growth is called “self-financing growth”. The growth equilibrium model contains three parameters [8].

On this basis, formula for self-financing growth is determined as follows:

$$C'_g = \frac{R}{T} \quad (1)$$

, where
C'_g is self-financing growth;
R – a parameter that represents the ratio of profit after taxes for a given year and net sales revenue for the same year;
T is a parameter that represents the ratio of equity and net sales revenue.

Self-financing growth is recommended in terms of taking a lower risk from foreign lenders as the company develops. In this context, the following levels of risk can be assessed in terms of financing growth:

1. Option one: self-financing growth – zero risk from foreign lenders;
2. Option two: 100% financing growth with foreign capital – high risk from foreign lenders.
3. Option three: Combined financing with equity and foreign capital.

To determine and compare the level of corporate growth, a sample of two groups of public industrial companies was sampled:

Group I – companies from food industry;
Group II – companies from other industrial sectors.

3. Results and dictation

Using results of food companies for 2017 and the previous year, Table 1 was compiled.

Table 1. Real and self-financing growth of food companies

№	Companies	Real growthC_g (%)	Self-financing growth C'_g (%)	Securitization$S_f = C'_g/C_g$
1.	Zarneni hrani Bulgaria AD	1.12	1.23	1.10
2.	Zaharni zavodi AD	1.05	2.55	2.42
3.	Gradus AD	11.80	4.88	0.37
4.	Lavena AD	29.07	12.62	0.43
5.	Balgarska Roza Sevtopolis AD	1.03	1.57	1.52
6.	Aroma Kosmetiks AD	0.35	0.18	0.51
7.	Bulgartabak-holding AD	8.98	3.89	0.49
8.	Dupnitsa-BT AD	9.88	3.35	0.34
9.	Nikotiana-BT AD	5.48	-1.83	-0.33
10.	Haskovo-Tabak AD	15.86	-2.18	-0.14
11.	Shumen-Tabak AD	3.54	-7.10	-2.01
12.	Yuri Gagarin AD	8.27	7.06	0.85
	Average	8.03	2.15	0.46

The data in Table 1 shows that the corporate growth (C_g) of whole group of food companies is 8.03%. It can be assumed that this is a moderate corporate growth. It is formed mainly by the high growth rate of the cosmetic company Lavena AD, Haskovo-Tabak AD and Gradus AD. The other two

IOP Conf. Series: Materials Science and Engineering **878** (2020) 012072 doi:10.1088/1757-899X/878/1/012072

companies in food industry have low corporate growth. Following three have moderate corporate growth: Yuri Gagarin AD; Bulgartabak-holding AD; Dupnitsa-BT AD.

Other companies have low corporate growth.

The own cash flows allow for low corporate growth, which is confirmed by average for whole group self-financing growth $C_g' = 2.15\%$. Only Lavena AD and Yuri Gagarin AD have moderate levels of this growth. In all other companies, the potential for financing growth through own cash flows is low. In three tobacco companies, (C_g') is negative, which means that the cost of using foreign capital outweighs its own profits and corporate growth generates losses and uncertainty.

The "risk of creditor" criteria is reciprocal of security which was formed by ratio between the two types of growth:

$$S_f = \frac{C_g{}'}{C_g} \quad (2)$$

, where
S_f – security from financing.

The analysis of data in Table 1 shows that average security level for the whole group of companies in the food industry is 0.46. It can be concluded that the managers of these companies realize corporate growth within the limits of moderate (about 50%) risk from creditors.

This average contains a wide variety of values for S_f. Three companies achieve corporate growth at very high risk from creditors: Shumen-Tabak AD (S_f = -2.01); Nicotiana-BT AD (S_f = -0.33); Haskovo-Tabak AD (S_f = -0.14). This is probably main reason that they lose their publicity status in the years after the economic crisis. The growth security of four other companies is low: Dupnitsa-BT AD, Gradus AD, Lavena AD and Bulgartabak-holding AD. There is a group of companies that have high potential risk-free corporate growth, but due to commodity, capacity or other resource and market constraints, have not been able to use this potential: Zarneni hrani Bulgaria AD (S_f = 1.10); Zaharni zavodi AD (S_f = 2.42); Balgarska Roza Sevtopolis AD (S_f = 1.52).

In order to apply benchmarking approach, results of food companies were compared with those of the best public companies in other sectors.

The selection of the companies is purposeful and the aim is to research the best performing industrial companies of BSE AD – Sofia in the following way of grading:

Level I: All industrial companies included in calculation of basic index of the stock exchange – SOFIX: Sopharma AD; M+C Hydraulic AD; Sirma Group AD.

Level II: Premium segment on main market: Monbat AD; Korado Bulgaria AD.

Level III: Standard Segment on Main Market – all other companies.

Using the companies' results for 2017 and the previous year, Table 2 was compiled:

Table 2. Real and self-financing growth of companies from other sectors of the industry

№	Companies	Real growth C_g (%)	Self-financing growth C_g' (%)	Securitization $S_f = C_g'/C_g$
1.	Sopharma AD	24.14	8.81	0.36
2.	Sirma Groups AD	50.49	4.35	0.07
3.	Korado Bulgaria AD	30.01	25.85	0.86
4.	Hydraulic Elements and Systems AD	19.93	13.40	0.67

5.	Alcomet AD	14.20	8.96	0.63
6.	Elhim Iskra AD	6.13	1.15	0.19
7.	M+C Hydraulics AD	11.11	18.70	1.68
8.	Monbat AD	2.72	4.31	1.58
9.	EMKA AD	18.14	28.22	1.56
10.	Fazerles AD	1.75	2.28	1.30
11.	KRZ Odesos AD	11.14	1.96	0.18
12.	Tchaikafharma AD	5.65	11.01	1.95
	Average	16.28	10.75	0.92

The data in Table 2 show that corporate growth (C_g) for the whole group of companies is 16.28%. This is a relatively high annual corporate growth, but it is formed by a very high level of growth of the IT company Sirma Group AD and Korado Bulgaria AD. The growth rate of three other companies is high (Sopharma AD, Hydraulic Elements and Systems AD and EMKA AD), while the rest of companies is moderate and low.

Moderate level of growth can be achieved, as the average for the whole group of annual self-financing growth (C'_g) is 10.75%. It should be noted that this level is formed by the high potential for self-financing growth in three companies: Korado Bulgaria AD (25.85%) M+C Hydraulic AD (18.70%) and EMKA AD (28.22%). In other companies, this potential is moderate and more often low.

The study shows that the average security level for the entire group of companies is 0.92, which means that managers realize corporate growth within low risk of lending. This average contains a wide variety of S_f values.

In addition to “security of finance” criterion, the profitability (efficiency) of growth financing must be analyzed and managed. Attracted capital has the ability to reduce or increase the return on equity. The ability of borrowed capital to change return on equity is known as “leverage effect” (leverage), which is defined as follows [9]:

$$L_E = \frac{RSE.PAT+[(1-RSE).PAT-I]}{RSE.PAT} \quad (3)$$

, where symbols have the following meanings:
L_E – leverage effect;
RSE – relative share of equity;
PAT – profit after tax;
I – interest.

Studies and calculations made on evel of leverage in food companies made it possible to compile Table 3:

Table 3. Profitability and security in financing of the companies in the food industry

№	Companies	Profitability *Le*	Security *Sf*	Multiplication *Mes*
1.	Zarneni hrani Bulgaria AD	0.75	1.10	0.83
2.	Zaharni zavodi AD	1.55	2.42	3.75
3.	Gradus AD	1.21	0.37	0.45
4.	Lavena AD	1.41	0.43	0.61
5.	Bulgarska Roza Sevtopolis AD	1.17	1.52	1.78
6.	Aroma Kosmetiks AD	1.09	0.51	0.56
7.	Bulgartabak-holding AD	1.10	0.49	0.54
8.	Dupnitsa-BT AD	1.47	0.34	0.50

9.	Nikotiana-BT AD	0.90	-0.33	-0.30
10.	Haskovo-Tabak AD	0.48	-0.14	-0.07
11.	Shumen-Tabak AD	0.50	-2.01	-1.00
12.	Yuri Gagarin AD	1.19	0.85	1.01
	Average	1.07	0.46	0.72

The analysis of data in Table 3 shows that most of companies have a favorable leverage effect. The level of this effect is relatively high for three companies: Zaharni zavodi AD, Lavena AD and Dupnitsa-BT AD. The profitability level of five more companies is good: Gradus AD, Yuri Gagarin AD, Aroma Kosmetiks AD, Bulgartabac-holding AD and Bulgarska Rosa Sevtopolis AD. The other four companies have an unfavorable leverage factor, with a relatively low level in two companies in tobacco industry. The leverage effect is good for whole group: $L_e = 1.07$.

The multiplication presented in Table 3 is determined by the following model:

$$M_{ES} = S_f . L_E \tag{4}$$

where M_{ES} is a multiplication of efficiency and security.

The study of the level of leverage effect in leading industrial companies made it possible to compile Table 4:

Table 4. Profitability and security in financing the corporate growth of companies in other industrial sectors

№	Companies	Profitability *Le*	Security *Sf*	Multiplication *Mes*
1.	Sopharma AD	1.19	0.36	0.43
2.	Sirma Groups AD	1.15	0.07	0.08
3.	Korado Bulgaria AD	1.26	0.86	1.08
4.	Hydraulic elements and systems AD	1.20	0.67	0.80
5.	Alcomet AD	1.46	0.63	0.91
6.	Elhim Iskra AD	1.05	0.19	0.20
7.	M+C Hydraulic AD	1.19	1.68	1.99
9.	Monbat AD	1.25	1.58	1.97
11.	EMKA AD	2.98	1.56	4.65
12.	Fazerles AD	0.98	1.30	1.27
13.	KRZ Odesos AD	0.98	0.18	0.17
15.	Tchaikafharma AD	1.14	1.95	2.21
	Average	1.23	0.92	1.31

The table data shows that the majority of companies have a favorable leverage effect. Only two of the companies have an unfavorable leverage effect (Fazerles $-L_E = 0.98$; KRZ Odessos $-L_E = 0.98$). The leverage effect of EMKA AD is high ($L_E = 2.98$), the effect level in Alcomet AD ($L_E = 1.46$) and Korado Bulgaria AD ($L_E = 1.26$) is very good. For the rest of companies, the level of leverage is good, which is confirmed by the average level for whole group ($L_E = 1.23$).

4. Conclusion

Benchmarking results for corporate growth of companies in food industry, compared with leading other sectors allow to draw following comparison:

1. Corporate growth (C_g):

a) companies in food industry – 8.03%;

b) leading companies – 16.28%;
c) ratio (a : b) – 0.49.
2. Self-financing growth (C_g'):
a) companies in the food industry – 2.15%;
b) leading companies – 10.75%;
c) a ratio of 0.20.
3. Security (S_f):
a) companies in food industry – 0.46%;
b) leading companies – 0.92%;
c) a ratio of 0.50.
4. Yield (L_E):
a) companies in food industry – 1.07%;
b) leading companies – 1.23%;
c) ratio – 0.86.
5. Multiplication (M_{es}):
a) companies in the food industry – 0.72%;
b) leading companies – 1.13%;
c) a ratio of 0.64.

In summary, the comparison shows that parameters of corporate growth management in food industry are significantly inferior to those of leading industrial companies.

In general, benchmarking has six main stages, five of which are presented in this report:
- Choosing a benchmarking process – corporate growth for companies in food industry;
- Determination of the main benchmarks – security and profitability of its financing;
- Choosing etalon for comparison – the best publicly traded industrial companies;

Collection of information – official audited annual reports on an individual basis;
- Analysis and comparison – presented in table and text in detail.

Due to the limited volume, sixth chapter in adaptation and implementation of best practices will be presented in other scientific publications.

References

[1] Robert J and Boxwell Jr 1994 Benchmarking for Competitive Advantage (New York & McGraw – Hill)
[2] Camp R C 1995 *Business Process Benchmarking: Funding and Implemementing Best Practices* (Milwaupkee, Wiscosin: ASQ Quality Press)
[3] Mertins K and Holger K 2009 *Benchmarking: Leufaden für den Vergleich mir den Besten* – (Herausgeber: Symposion Publishing) Ausgabe 2
[4] Chipriyanov M 2008 *Product benchmarking research*, (*E-Journal Dialogue*) 3 (Svishtov: Tsenov Academy).
[5] Loginova E 2016 *Benchmarking – tool for developing competitive advantages* (Moscow: Scientific book)
[6] Longbottom D 2000 *Benchmarking in the UK: an empirical study of practitioners and academics* (Benchmarking: An International Journal) 7 (2) pp 98-117
[7] Berezin A and Kovalenko A 2014 *Benchmarking in the system of competitive actions of entrepreneurial structures (E-Journal: Journal of Modern Competition)* 5 (47) pp 117-129
[8] Konarev A and Konstantinova S 2017 *Analysis of the Economics Activity of Industrials Companies* Chapter 12 Analysis of corporate growth 1st edition (Plovdiv: Publishing house KSI) pp 222-231
[9] Konarev A and Konstantinova S 2015 *Finance and financial management of the company* Chapter 12 Asset and capital management (Plovdiv: Public house Janet-45) p 146

IOP Conf. Series: Materials Science and Engineering **878** (2020) 012073 doi:10.1088/1757-899X/878/1/012073

Investment in intangible assets and corporate growth in the industrial companies

Snezhinka Konstantinova[1], Asen Konarev[1, 2]

[1] University of Food Technologies, Plovdiv, Bulgaria
[2] Technologies and Innovations Corporation AD, Plovdiv, Bulgaria

E-mail: sks_ko@abv.bg

Abstract. This report examines the impact of investment in intangible assets on corporate growth of public industrial companies. An intangible asset growth indicator model has been developed. A new conceptual approach for measuring corporate growth has been implemented, using an integral indicator determined by four single indicators using weighted geometric mean model. Based on built models, a study has been conducted to show that the companies have achieved moderate corporate growth, accompanied by faster growth in intangible assets. Investments in intangible assets are proven to be a stable factor for corporate growth. Bulgarian industrial companies do not sufficiently utilize the intangible assets of leading companies. The investment strategy of Bulgarian companies should be optimized primarily to achieve a spillover effect.

1. Introduction

The economy of tangible assets are constantly growing – an economy symbolically called "weightless world" [1]. Haskel and Westlake emphasize that economy does not rely "solely on investment in tangible assets" [2]. The fundamental changes that take place in modern business with different names among which "digitalization of the economy" affect all aspects of company's business, including corporate growth.

Under International accounting standards (IAS) 38 intangible asset is an identifiable non-monetary asset without physical substance.

Like any asset, an intangible asset is a resource, controlled by entity as a result of past events that are expected to receive future economic benefits. In industrial companies, the intangible assets are patents, licenses, copyrights, trademarks and more. Their growth is an indication of an increase in company's competitive potential [3]. In balance sheets of companies, these assets are a separate group called "intangible assets", to which four articles are distinguished:

1. Property rights.
2. Software products.
3. Development products.
4. Other intangible assets.

It is normal to expect that investments in intangible assets are a powerful factor in the development of industrial companies. To examine the impact of these investments on corporate growth, a sample of

15 publicly traded industrial companies was selected, as the selection is targeted to be sampled the best companies:

- three companies included in the calculation of the basic SOFIX index of the BSE AD – Sofia Sopharma AD, M+C Hydraulic AD and Sirma Group AD;
- two companies from Premium segment of the main market: Monbat AD and Korado Bulgaria AD;
- eight companies in the Standard segment on the main market;
- one company from the alternative market of BSE – Sofia: Saharni zavodi AD;
- a company that goes out of the status of publicity in 2016, but has traditions in investments in intangible assets: Biovet AD.

The period 2007 – 2018 was selected for analysis for the following reasons:

- 2007 is one of the good pre-crisis years;
- 2018 forms the boundaries of a relatively long 12-year period, sufficient for serious changes in the studied parameters of industrial companies.

2. Methodology and methods

The levels and dynamics of intangible assets are defined as follows:

- the balance sheet value is used as a level indicator, by which the asset is recognized in statement of financial position after deduction of accumulated depreciation and impairment losses;
- the growth indicator of intangible assets is used as an indicator of dynamics, which is defined as follows:

$$G_{ia} = \frac{IA_t}{IA_{t-n}}.100 - 100 \qquad (1)$$

, where
G_{ia} is growth of intangible assets;
IA_t – book value of intangible assets in t-th year;
IA_{t-1} – the carrying amount of intangible assets during the base year.

In some studies [4, 5], net sales revenue was used as an indicator of enterprise growth and corporate growth was determined by the formula:

$$G_g = \frac{NSR_t}{NSR_{t-n}}.100 - 100 \qquad (2)$$

, where
G_g – corporate growth;
NSR_t – net sales revenue for the t-th year;
NSR_{t-n} – net sales revenue for the base year.

There are significant differences in the economic literature regarding firm size indicators. It is believed that the company's growth analysis may be different because there are different ways to measure it. This diversification is sometimes due to the goals of each author, but more often due to lack of data [6]. Some researchers [7, 8] have identified the following basic indicators used in the empirical literature:

- financial or stock exchange value;
- the number of employees;
- sales and revenue;
- production capacity;

- production value;
- the added value.

To these indicators we can add the following:

- the value of total assets;
- the net value of the property;
- the permanent capital of the company;
- the value of non-current assets (called the "supporting structure" of industrial companies), etc.

Taking the research findings [9] as valid, that some of the presented indicators are interdependent and expendable, we believe that each indicator can paint a different picture of the company's growth.

For the purposes of this study, we suggest a new approach and new indicators for measuring company size and growth. The available economic literature all studies are based on a single indicator of the size.

- most often the number of employees (up to 2000);
- in periods of financial stability, the stock exchange or financial value;
- after 2000, mostly through the value of total assets;
- after 2010, most often through the turnover (total revenue or net sales revenue).

The problem is that the only indicator, however perfect it may be, has some drawbacks. In addition, researchers are often criticized for choosing the metric that is most appropriate to prove their own thesis. For this reason, we propose a new conceptual model for measuring of the size and growth, For this reason, we propose a new conceptual model for measuring of the size and growth, respectively, not through one metric a single indicator but through an integral indicator. After a thorough study of the qualities of the single indicators, the possibilities for collecting information, the specificity of the analyzed period and the diverse industry affiliation of the companies, we offer the following models for an integral indicator for the size of an industrial company:

a) Model of unweighted the arithmetic mean average

$$S_c^i = \frac{NSR + VA + TA + LE}{4} \tag{3}$$

where

S_c^i is an integral indicator of company size;
NSR – net sales revenue;
VA – added value;
TA – total assets;
LE – labor costs (labor as measured by the company's salary and social security costs).

b) Model of weighted average arithmetic mean

$$S_c^i = \frac{aNSR + bVA + cTA + dLE}{4} \tag{4}$$

where a, b, c, d are the coefficients of significance (relative weights) of the individual indicators involved in the integral indicator model.

c) Model of unweighted average, quadratic value:

$$S_c^i = \sqrt[4]{NSR + VA + TA + LE} \tag{5}$$

, where the symbols have the same meaning.

d) Model of weighted average, quadratic value:

$$S_c^i = \sqrt[4]{aNSR.bVA.cTA.dLE} \tag{6}$$

In the models of integral indicator involving four single indicators. Thus, the size of the company is more comprehensive and objective. Preference of authors of this report for geometric mean over the arithmetic mean or other mean is due to the property of geometric mean better describes the harmony in variation of single indices. Significance coefficients a, b, c, d has the role of ranking single indicators by expert or econometric methods. For the purposes of this study are identified through expert assuming the following meanings: a = 0.25, b = 0.35, c = 0.20, d = 0.20.

Depending on the objectives of the survey and the possibilities for collecting information, the number of single indicators involved in the integrated indicator model may be increased or decreased. Therefore, we propose the following principle model of an integral indicator of company size:

$$S_c^i = \sqrt[n]{\prod^{n} a_i X_i} \tag{7}$$

where
n is the number of unit indicators;
a_i – coefficient of significance (relative weight) of the i-th unit indicator (i = 1… n);
X_i – level of the i-th unit indicator (i = 1… n);
Π is a mathematical symbol for a multiplication.

Corporate growth is determined by the following model:

$$C_g^j = \frac{S_{ct}^{ij}}{S_{ct-n}^{ij}}.100 - 100 \tag{8}$$

where
C_g^j is the corporate growth of the jth industrial company;
S_{ct}^{ij} – integral indicator of the size of the j-th industrial company in the t-th year;
S_{ct-n}^{ij} – an integral indicator of the size of the same company in the base year.

3. Results and discussion

Based on a specific study of the actual data of the surveyed industrial companies and the content of the models (1) and (6) were compiled Table 1:

Table 1. Dynamics of intangible assets and corporate growth of industrial companies (2007 – 2018)

Companies	Intangible assets growth		Corporate growth
	thousand BGN	%	(%)
Alcomet AD	721	270.51	81.54
Sopharma AD	4 433	591.07	87.66
Monbat AD	223	437.25	143.14
Neochim AD	65	382.35	8.99
Yuri Gagarin AD	79	146.75	88.16
Hydraulic Elements and Systems AD	175	336.54	46.11
Zaharni zavodi AD	2	9.09	37.54

M+C Hydraulic AD	-146	-56.37	83.31
Lavena AD	468	754.84	588.82
EMKA AD	-7	-32.58	58.77
Elhim Iskra AD	-23	-32.40	25.39
Korado Bulgaria AD	-15	-63.22	122.56
Sirma Group AD	80 616	899.13	379.94
Tchaikafharma AD	-350	-10.86	90.86
Biovet AD	29 397	186.66	310.88
Average	7 709	267.62	143.58

The analysis of the results presented in Table 1 shows that all companies achieved corporate growth during the period 2007 – 2018. Corporate growth in three companies is high: Lavena AD, Sirma Group AD and Biovet AD. At the same time, all three companies have increased both in absolute value of intangible assets and high growth modeled on (1). Only Sopharma AD, which has a significant increase in intangible assets, has achieved moderate corporate growth. The other pharmaceutical company Tchaikapharma AD achieved the same growth with a significant decrease in intangible assets. In addition, with four other companies, corporate growth was accompanied by a decrease in these assets. This reduction is very small in absolute terms, but one of the companies (M+C Hydraulic) which are blue chip on the stock exchange, this decrease is significant.

The analysis of the average results for the entire aggregate of 15 companies showed the following:

- the companies achieved moderate corporate growth – 11.96% annual average increase in size;
- this growth is accompanied by a faster growth of intangible assets – 22.30% annually;
- each percent of corporate growth is accompanied by an increase of intangible assets of nearly BGN 54 thousand.

The proposed integral metric used for indicator to measure the size of the company is an average quadratic value. It's known, that the averages value are those summarizing characteristics, which reflect the typical in phenomenon under study. In this sense, it is unambiguous to conclude that the growth of intangible assets is a stable factor for corporate growth. But within the typical there are deviations that cause serious disturbance. For example, in five companies, corporate growth is accompanied by a decrease in intangible assets, which represents one third (33.13%) of total. If, in three of these companies, this reduction is insignificant in absolute value, for two of the companies, which is considered for successful (M+C Hydraulics AD and Tchaikapharma AD), the decrease is significant. Specific analysis shows that the decrease in intangible assets in the engineering company M+C Hydraulics AD is mainly due to the reduced value of the used software products (-106 thousand BGN) and other intangible assets (-72 thousand BGN), while property rights (industrial and intellectual) are growing significantly.

In the pharmaceutical company Tchaikapharma AD, property rights are also increasing, but the following three articles are substantially reduced:

- software products (-293 thousand BGN);
- R&D products (-128 thousand BGN);
- others (-142 thousand BGN).

Development by international standards is the application of research findings or other knowledge to a plan or design for the production of new or substantially improved materials, devices, products, processes or services prior to commencement of commercial production or use. In our opinion, development is a typical manifestation of the so-called "spillover effect" [2]. With this effect, one company takes advantage of the private investment of another company to create a certain innovation. R&D keywords are "implementation of discovery". Practice has shown that some intangible assets have an extremely high spillover effect, which means that other companies can relatively easily benefit from intangible investments that they have not made and, through development, to apply foreign discoveries, with or without improvement. The main feature of intangible assets, called "spillover effect", is of great importance for the development of Bulgarian industrial companies. These are companies, who do not have the potential for independent research to create innovation. Therefore, the innovation strategy of these companies in the development of intangible assets should be to achieve a spillover effect through the direct application or adaptation of unprotected innovations of developed companies. And this effect can be obtained most quickly through R&D. The analysis shows that only three companies report ownership of R&D products in their balance sheets: Sirma Group AD, Tchaikapharma AD and Biovet AD. Research by international standards is an original and planned study undertaken to gain new scientific or technical knowledge and understanding. The spillover effect can be obtained not only through development, but also through the use of research results from other companies. This process is successfully characterized by Thomas Jefferson through the following symbolism: “He, who receives an idea from me, receives instruction himself without lessening mine; as he who lights his taper at mine, receives light without darkening me” [10, 11].

Spillover effect can occur not only in creating new products and services ideas of foreign companies, but also in design, training, marketing, brand and many other intangible assets.

In the practice of the Bulgarian industrial companies there are no separate articles on the intangible assets of the research activity, and they are usually reported in the complex article “other intangible assets”. The analysis shows that this article has reported assets in the balance sheets of 12 companies in the sample surveyed. Significant intangible assets of scientific value are created by the company Biovet AD, which has its own research institute. The biggest products in absolute value are the products of the software company Sirma Group AD. The two pharmaceutical companies (Sopharma AD and Tchaikapharma AD) also have significant intangible assets from research. In other companies, these assets are completely or almost missing.

4. Conclusion

The following important conclusions can be drawn from this study:

1. Investments in intangible assets of industrial companies are part of the fundamental changes that occur in modern business, characterized by the name “economy of intangible assets”.
   In the balance sheets of Bulgarian companies these assets are reported in a separate group with four items, and there are serious problems with their proper accounting. In addition, an entire category of intangible assets called “economic competences” such as training, marketing research and branding, restructuring, etc. is not considered as an investment.
2. For an indicator of the level of intangible assets, it is appropriate to use the balance value, by which the asset is recognized in the statement of financial position after deduction of accumulated depreciation and impairment losses. For an indicator of the dynamics of these assets, a model has been drawn up that represents the ratio of the book value during the surveyed and base year.
3. A new conceptual approach to measuring the size of a company and, respectively, its growth through an integrated indicator has been applied to study corporate growth. Four specific and one principle model of integral indicator are formed, formed by the individual levels of four single indicators. The authors recommend the use of the model of weighted average geometric value because it better characterizes the harmony in the variation of individual indicators and enables the ranking (weighting) of the individual indicators depending on the situation, research objectives, development prospects or other considerations.
4. Based on the constructed models of indicators for the growth of intangible assets and corporate growth, a survey was conducted in 15 public industrial companies, which are considered to be the most successful and best listed on BSE AD – Sofia. The analyzed period is 2007 – 2018 – before and after the crisis. The analysis of the average results for the whole aggregate shows that the companies have achieved moderate corporate growth, which is accompanied by a faster growth of the intangible assets, with each percentage of corporate growth corresponding to an increase of the intangible assets by BGN 54 thousand.
5. The study of common characteristics convincingly shows that investing in intangible assets is a stable factor for corporate growth. But within the typical deviations occur – in 5 companies

corporate growth is accompanied by a decrease in intangible assets, which represents one third (33.33%) of the aggregate. The specific analysis shows that in three of the companies this decrease is insignificant, but in the other two companies the decrease is mainly due to the protected intangible assets – property rights and software.

6. The analysis clearly shows that industrial companies surveyed do not use enough unprotected intangible assets of leading companies called economic "spillover effect". This basic characteristic of intangible assets is a great importance for the development of Bulgarian industrial companies that do not have the human and financial potential for their own research and innovation. The investment strategy of these companies in intangible assets should be optimized primarily to achieve a spillover effect by directly applying or adapting unprotected innovations of developed companies to the purchase of protected assets.
7. Spillover effect can be obtained most quickly through R&D. The analysis shows that only three companies account for R&D in their balance sheets. Spillover effect can also occur through research on ideas from foreign companies. This activity is reflected in the comprehensive article "other intangible assets". The analysis shows that this article contains intangible assets in the balance sheets of 12 companies in the sample. Significant intangible R&D assets are owned by the most successful companies: Sirma Group AD, Biovet AD and two pharmaceutical companies: Sopharma AD and Tchaikapharma AD.

References

[1] Coyle D 1997 *The Weightless World* 1st edition The Mit Press, 1999 (Cambridge, Massachusetts) books.google.com

[2] Haskel J and Westlake S 2018 Capitalism without Capital: The rise of the intangible economy (Sofia: Publishing house East-West) p 12

[3] Konarev A and Konstantinova S 2017 *Analysis of the economics activity of industrials companies* Chapter 12 Analysis of corporate growth 1 st edition (Plovdiv: Publishing house KSI)

[4] Konstantinova S and Konarev A 2017 *Corporate growth total factor productivity in industrial companies* Trakia Journal of Sciences 15(1) pp 191-194

[5] Konarev A Konstantinova S and Georgieva-V G 2019 *Technological innovations and total factor productivity in dairy industry* Published under licence by IOP Publishing Ltd. IOP Conference Series: Materials Science and Engineering Volume 618 conference 1, pp 1-5

[6] Correa A Acosta M Conzales L A and Medina U 2003 *Size, age and activity sector on the growth of the small and medium firm size* Small Business Economics, 21, (Journal Article) pp 289-307

[7] Ardishvili A Cardozo S Harmon S and Vadekath S 1998 *Towards a theory of new venture growth* Paper presented at the 1998 Balson Entrepreneurship Research Conference (Ghent, Belgium)

[8] Dalmer F Davidsson P and Garrtner W 2003 *Arriving at the high-growth firm* Journal of Business Venturing 18 (2) pp 189-216

[9] Kirchhoff A B 1992 *Entrepreneurship's Contribution to Economics* Research Article, Volume 16 Issue 2 pp 93-112

[10] Aghion P. and Durlauf N St 2005 *Handbook of Economic Growth* 1st edition (Publisher: Elsevier) Volume 1B p 1069

[11] *The Writings of Thomas Jefferson* 1905 Edited by Andrew A. Lipscomb and Albert Ellerg Bergh (Thomas Jefferson Memorial Association) 13 pp 333-335

IOP Conf. Series: Materials Science and Engineering **878** (2020) 012074 doi:10.1088/1757-899X/878/1/012074

Why do Bulgarian wine producers innovate?

D Pastarmadzhieva[1] and M Angelova[1]

[1]Faculty of Economics and Social Sciences, University of Plovdiv Paisii Hilendarski, 4000 Plovdiv, 24 Tzar Asen str, Bulgaria

daniela.pastarmadzhieva@uni-plovdiv.bg

Abstract. The positive effects of the innovations on the performance of the companies is inevitable. However, these effects can be numerous and it is important the entrepreneurs to be aware what they can expect. Thus, the *aim* of the current study is to identify the results, expected by the Bulgarian wine producers from the application of innovations and if their choice is associated to some factors. The *object* of the study is Bulgarian SMEs wine producers. The *focus* is their opinion as regards the expected results from the application of innovations. The results are obtained through survey, conducted from December 2019 to January 2020. The results show that the most expected results by the Bulgarian wine producers are the increase of competitiveness and efficiency of their enterprises.

1. Introduction

It is inevitable that the innovations have positive effects on the performance of the companies. However, these effects can be numerous and it is important the entrepreneurs to be aware what they can expect. Thus, the *aim* of the current study is to identify the results, expected by the Bulgarian wine producers from the application of innovations and if their choice is associated to some factors. The *object* of the study is Bulgarian SMEs wine producers. The *focus* is their opinion as regards the expected results from the application of innovations. The results are obtained through an online survey, conducted in the period December 2019 - January 2020 through structured interviews using own questionnaire. The collected data is processed with IBM SPSS Statistics and frequencies and cross tables are used.

The article is structured as follows: following the introduction, in the second part is described a conceptual framework of innovations and the effects of their application in particular. This analysis paves the way for the third part, in which is highlighted the methodology of the study. The fourth part introduces the obtained results of the survey and discussion based upon the figures. The paper closes with conclusions and recommendations for future research, regarding the reasons for application of innovations in wine industry.

2. Literature review

The innovations can lead to the creation of new product, service, process, or form of organization. The relevance of innovations for the development has been examined since 19th and gained popularity in the beginning of 20th [1]. It was Joseph Schumpeter [2] who defined the notion of innovations as an amendment to introduce and use new types of consumer goods, new production and transport means, markets and forms of organization in production.

In the Oslo Manual 2018 (4th edition) OECD innovations are classified in two major groups – product innovation and business process innovation. Product innovations can be innovations of goods and of services. Business process innovations include innovations in production of goods or services, distribution and logistics, marketing and sales, information and communication systems, administration and management and product and business process development [3]. Innovations are essential for the development of wine industry as well [4]. However, in order to implement innovations the wine producers have to be motivated to do so by expecting some positive results.

First of all it should be noted that the results of innovations application may vary in accordance with the specific innovation – "from product development, methods of management, ways of doing works and etc." [5]. Furthermore, innovations are related to number of factors, which are vital for their impact. Among them are the legal framework and national policies, access to technologies, market demands, interactions between stakeholders and to some extent the skills of the entrepreneur [6]. As concerns the latter it is very important to mention that the innovative thinking is more an intuitive one, rather than based on previous knowledge [7]. Nevertheless, in the context of the current study we want to measure the overall expectations of the wine producers as regards the application of innovations in general and there are numerous studies, which examine the effects of the application of innovations.

Some of the scholars focus on the role of the innovations for *increasing the competitiveness of the enterprise*. According to Ciocanel and Pavelescu [8] "innovation, when defined as a process that allows companies to produce more with the same amount of resources or produce as much with a smaller amount of resources, leads to sustainable competitive advantage". Furthermore, the focus on ICT innovations is one of the main prerequisites to the increase of competitiveness of the enterprises Therefore, the competitiveness of companies and countries depends on their capabilities to make innovations and their orientation to technology and information [9]. And the competitiveness produced by the innovations is relevant not only on the level of the company, but on national level as well. Innovation is a key condition in the context of the need to enhance the competitiveness of the Bulgarian economy [10].

Alongside, the innovations have the potential to *increase efficiency* [11]. Some authors even refer to the innovations as "fundamental to making sustainability possible and improving efficiency" [12]. Scholars reach the conclusion that if entrepreneurs want to increase the efficiency of their work and of the investments in particular, they should search and implement innovative solutions [13].

A significant area to which the innovations can contribute is the *increasing of customer satisfaction*. The latter is of essential importance for every organization and in industry. Customers and the covering of their expectations are crucial for the survival and sustainability of the companies [14]. The innovations are crucial for the satisfaction of the consumers especially in the area of services [15] [16].

Further positive effects of the innovation implementation is the *increase of productivity* [17]. According to an analysis of the World Bank, the "productivity growth is about innovation" [18].

The studies on the effects of innovations also show that sometimes they may lead to the *increase in the employees' satisfaction*. However, the satisfaction of the employees is related to the types of innovations, which are implemented. The results show that "enterprises who introduced organizational innovation and no other innovation reported higher effects on employee satisfaction compared to enterprises that did marketing innovation and no organizational innovation" [19].

In addition to these wide spread effects of the innovations, the latter may also cause *reduction of the costs* [20] [21], *better management* [22], *higher profit* [8], *environmentally friendly economy* [12] and other.

3. Methodology of the study

The empirical results are obtained through a structured interview, conducted in the period December 2019 - January 2020. The survey consisted of four sections: general information about the respondent, information about the enterprise, innovations section and competitiveness section. The section on the innovations consists of a list of fifty (50) innovations. One of the questions is about the results, which the wine producers expected of the application of innovations. They were able to make multiple selection among the following options: Increasing the competitiveness of the enterprise; Increasing efficiency; Increasing customer satisfaction; Increasing productivity; Increasing employee satisfaction; Reduce costs, shorten product life-cycle at the expense of more efficient use of resources; Better interpretation of the activities and more transparency of information flows, and Clear allocation of responsibilities and duties between management and executive staff.

In the current paper are presented some of the results of the survey. In the latter all the SMEs wine producers were invited to take part in the online survey by e-mail and some of them were personally contacted. As a result, 156 online questionnaires were sent by e-mail and 57 of them participated in the survey which represents 37% response rate, thus the collected data provides relevant results. The statistical methods, used in the study are frequencies and cross tables. The latter is used in order to identify if there is an association between the expected results and various factors. The examined factors are: years of functioning of the enterprise, size of enterprise, additional activities, markets for production realization, sources of financing and sources of raw material.

4. Results and discussion

The SMEs wine producers have various expectations about what innovations will contribute to their work. As figure 1 indicates they believe that the innovations are mostly relevant for the competitiveness (20%) and efficiency (19.3%) of their enterprise. They are followed by the expectations for rise in the customers' satisfaction (15.6%) and in the productivity (15.6%). The least expected result is the clear allocation of responsibilities and duties between management and executive staff (3.7%).

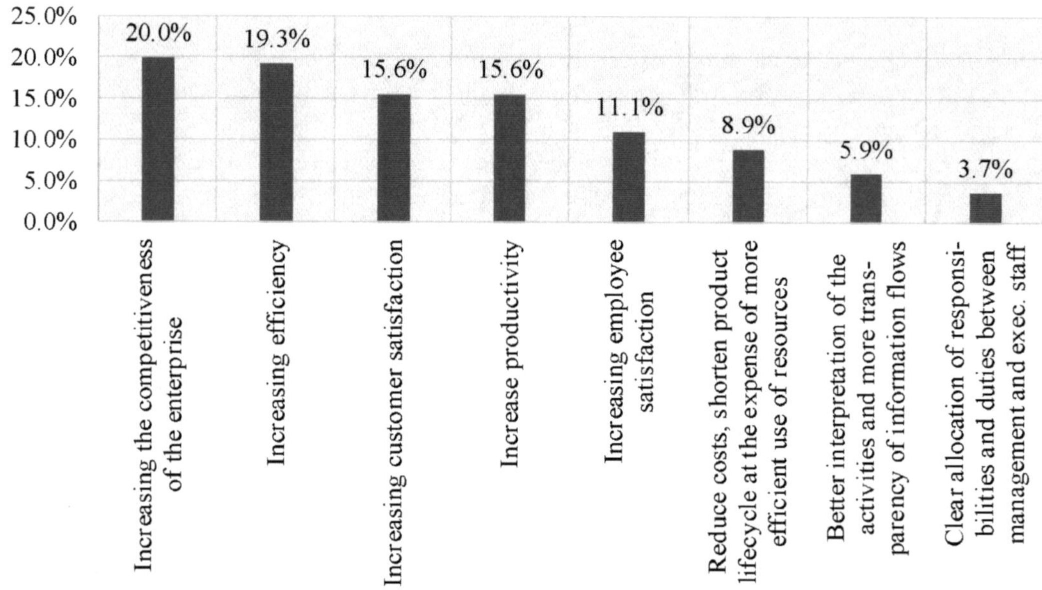

Figure 1. Expected results of the application of innovations among SMEs wine producers, 2019

Source: Own data

The results on figure 2 show that the years of enterprise functioning are relevant to some of the results. None of the most experienced SMEs expected the innovations to contribute to the satisfaction of their employees. Their share is bigger compared to other two groups as concerns the expectation the innovations to produce clear allocation of responsibilities and duties between management and executive staff (47%), alongside to reduce costs, shorten product life-cycle at the expense of more efficient use of resources (47%). Here we should mention that as concerns the latter innovation, only 13% of those who selected it are functioning 11-15 years. Overall, there is a relatively even distribution of attitudes on this indicator.

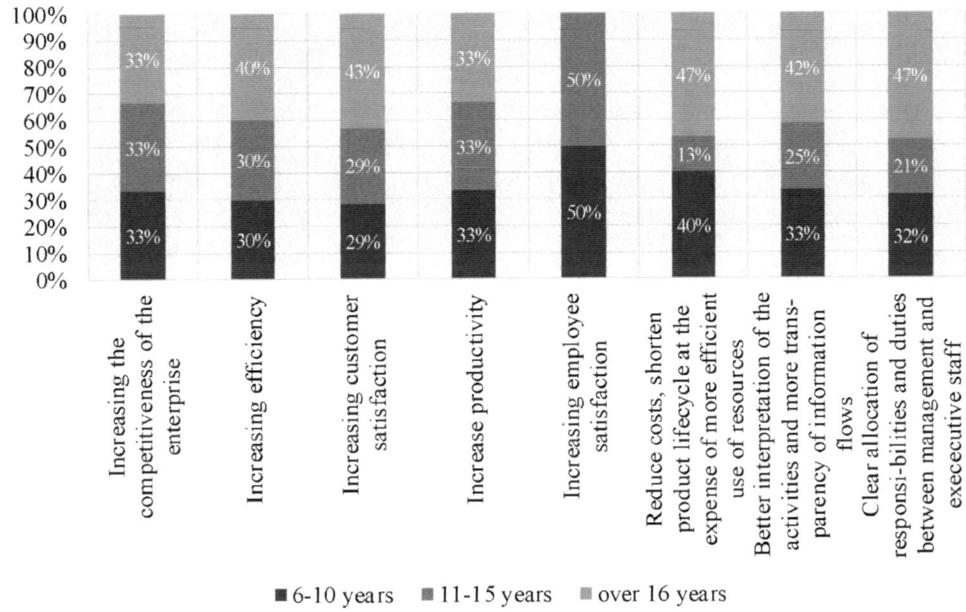

Figure 2. Years of functioning of the enterprise and expected results of innovations application, 2019

Source: Own data

As concerns the size of the enterprise the results show that the micro enterprises have biggest share as regards the expectations innovations to lead to clear allocation of responsibilities and duties between management and executive staff (52%), alongside to reduce costs, shorten product life-cycle at the expense of more efficient use of resources (54%). They have equal share with the medium-sized enterprises as regards the increase of employees' satisfaction (40%). The small enterprises are predominant about the expected results increasing customers' satisfaction (50%).

IOP Conf. Series: Materials Science and Engineering **878** (2020) 012074 doi:10.1088/1757-899X/878/1/012074

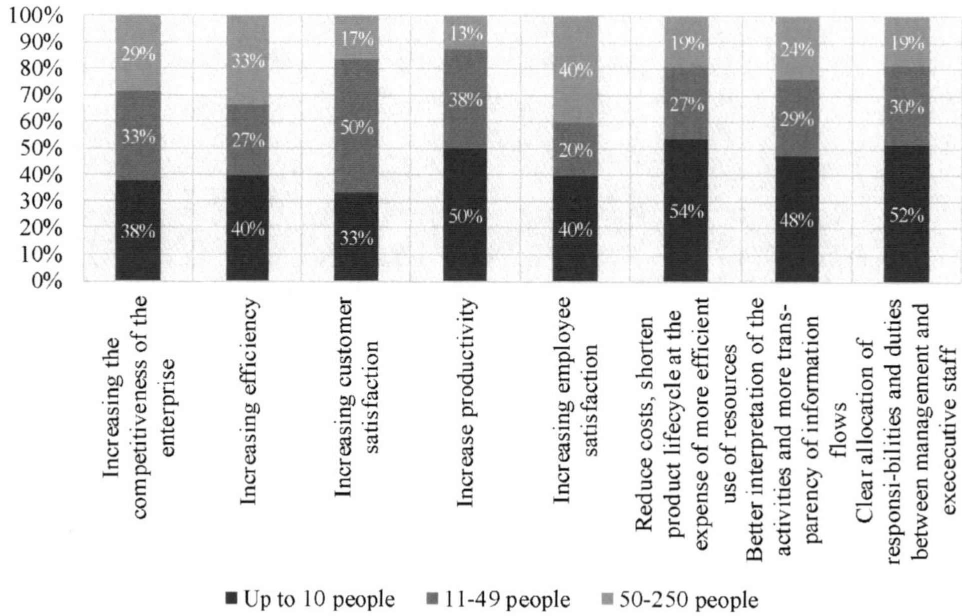

Figure 3. Size of the enterprise and expected results of innovations application, 2019
Source: Own data

It seems that additional activities, performed by the wine producers are not very relevant for their expectations about the results of innovations. Such statement is based on the fact that there is a relatively even distribution of attitudes on this indicator. The only result that is worth mentioning is the increase of productivity. Only 20% of the entrepreneurs, who selected this result, do wine tourism along with the wine production. (Fig. 4)

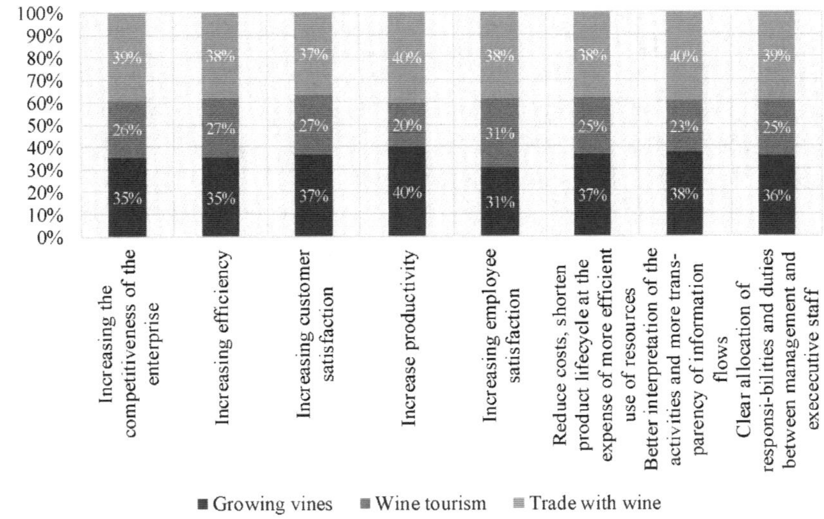

Figure 4. Additional activities and expected results of innovations application, 2019

Source: Own data

As concerns the markets for the realization of the production, we identified a dependency. The highest share for each expected result belongs to those, who sell to Bulgarian markets. They are followed by the wine producers, who sell on European markets and last are the one, who sell worldwide. Such dependency is explainable as regards the share of each in the overall number of wine producers. Thus, here we also identify even distribution of the opinions. (Fig. 5)

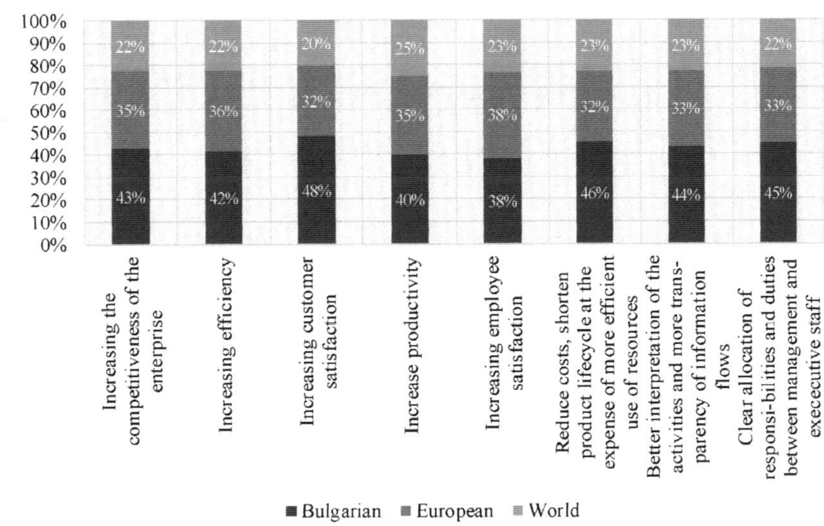

Figure 5. Markets for product realization and expected results of innovations application, 2019

Source: Own data

A similar dependency was observed as concerns the funding of the SMEs wine producers. The largest share belongs to the entrepreneurs, who rely on their own funds. They are followed by the ones, who use bank credits or European Funds. The wine producers, who rely on National programs have the smallest share in each of the expected results. These corresponds to their overall share among the respondents. (Fig. 6)

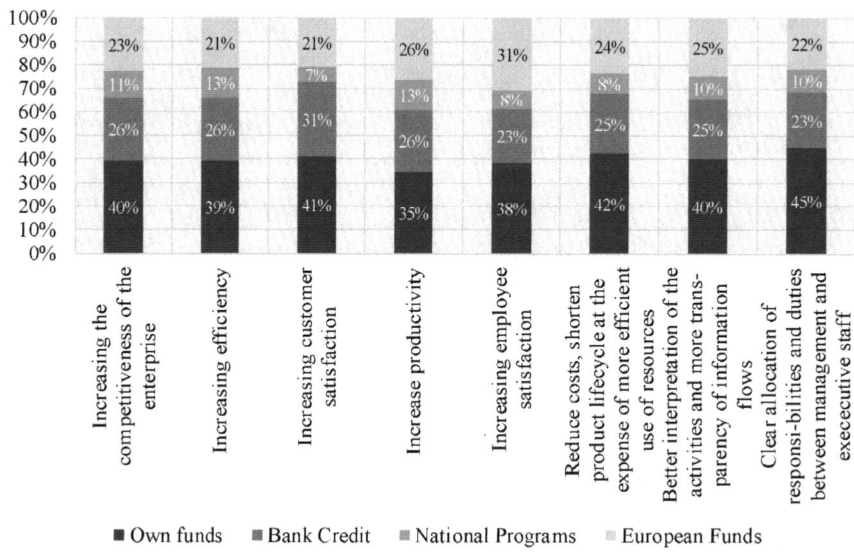

Figure 6. Sources of financing and expected results of innovations application, 2019

Source: Own data

The distribution of opinions as concerns the source of raw material also corresponds to the share of the latter in the sample. Only as regards the increase of productivity those with own vineyards are highly represented (58%). (Fig. 7)

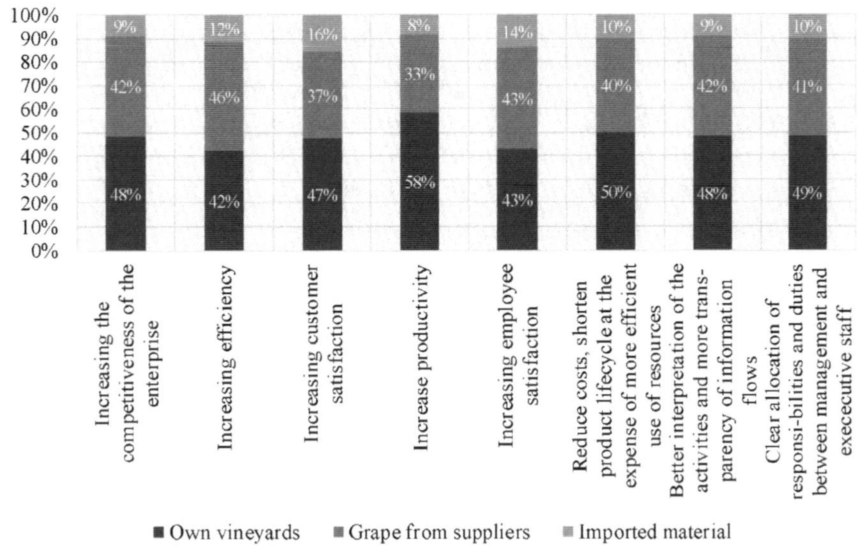

Figure 7. Sources of raw material and expected results of innovations application, 2019

Source: Own data

Conclusion

Application of innovations can lead to numerous contributions to the SMEs. The SMEs wine producers believe that the innovations are mostly relevant for the competitiveness and efficiency of their enterprise. The empirical results show that only two of the examined factors are extent relevant for the results expected by the wine producers. These are the years of functioning of the enterprise and its size. They are mostly relevant for the expected result clear allocation of responsibilities and duties between management and executive staff, alongside to reduce costs, shorten product life-cycle at the expense of more efficient use of resources. We also identified two exceptions. One of them is related to the additional activities, as only 20% of SMEs, who do wine tourism along with wine production expect increase of productivity. The other exception is about the source of raw material. As regards the increase of productivity those with own vineyards are predominantly the ones, who expect such result. However, these findings require further investigation and more efforts in order to identify the reasons for the results.

References

[1] Schumpeter J A 1949 *Economic theory and entrepreneurial history,* in Center for Research in Entrepreneurial History, Harvard University (ed.), *Change and the Entrepreneur. Postulates and Patterns for Entrepreneurial History,* (Cambridge, MA: Harvard University Press, pp. 63-84)

[2] Schumpeter, J A 1946 *The Future of Private Enterprise in the Face of Modern Socialistic Tendencies*, in Swedberg R. (ed.) Joseph A. Schumpeter. *The Economics and Sociology of Capitalism*, (Princeton: Princeton University Press, 1991 (1st edition 1946), pp. 401- 405)

[3] OECD/Eurostat 2018 *Oslo Manual 2018: Guidelines for Collecting, Reporting and Using Data on Innovation*, 4th Edition, The Measurement of Scientific, Technological and Innovation Activities, OECD Publishing, Paris/Eurostat, Luxembourg. https://doi.org/10.1787/9789264304604-en

[4] Karagiannis D and Metaxas T 2018 *Innovation in Food & Wine Tourism Related Businesses: Turning Ashes to Gold*, in *Management and Marketing of Wine Tourism Business: Theory, Practice, and Cases*, Marianna Sigala, Richard N. S. Robinson (Eds), (New York: Springer) pp. 345-363

[5] Tohidi H and Jabbari M M 2012 The important of Innovation and its Crucial Role in Growth, Survival and Success of Organizations, *Procedia Technology 1*, pp. 535-538

[6] Technopolis Group & Mioir 2012 *Evaluation of Innovation Activities. Guidance on methods and practices*. Study funded by the European Commission, Directorate for Regional Policy

[7] Horth D and Buchner D 2014 *Innovation Leadership How to use innovation to lead effectively, work collaboratively, and drive results*, Center for Creative Leadership.

[8] Ciocanel A B and Pavelescu F M 2015 Innovation and competitiveness in European context, *Procedia Economics and Finance 32*, pp. 728 – 737

[9] Doğan E 2016 The Effect of Innovation on Competitiveness, *Ekonometri ve İstatistik Sayı:24*, pp. 60-81

[10] Yosifov T 2019 Competitiveness Through Innovation, *Sotsialno-ikonomicheski analizi*, 2, pp. 71-81

[11] Kogabayev T and Maziliauskas A 2016 The Role of Innovation for Efficiency of Investments, *Science and Studies of Accounting and Finance: Problems and Perspectives, 10*(1), pp. 85-91

[12] Schilirò D 2019 Sustainability, innovation, and efficiency: A key relationship, In *Financing Sustainable Development: Key Challenges and Prospects*, Magdalena Ziolo and Bruno S. Sergi (eds.) (83-102), Cham, Switzerland: Palgrave Macmillan.

[13] Sandal J-U and Ejsmont A 2018 Innovativeness and Efficiency of Investments Undertaken by Networking Companies, *Zeszyty Naukowe Wyższej Szkoły Bankowej w Poznaniu, 81*(4), pp 13-23

[14] Daragahi G A 2017 The Impact of Innovation on Customer Satisfaction: A Study of the Cosmetics Producer in Tehran, *International Review, 1-2*, pp. 121-132

[15] Diaw B and Asare G 2018 Effect of Innovation on Customer Satisfaction and Customer Retention in the Telecommunication Industry in Ghana, *European Journal of Research and Reflection in Management Sciences, 6*(4), pp. 15-26

[16] Mahmoud M A, Hinson R E and Anim P A 2017 Service innovation and customer satisfaction: The role of customer value creation, *European Journal of Innovation Management*, https://doi.org/10.1108/EJIM-09-2017-0117

[17] Woltjer G, Galen M van and Logatcheva K 2019 Industrial Innovation, Labour Productivity, Sales and Employment, *International Journal of the Economics of Business*, DOI: 10.1080/13571516.2019.1695448

[18] World Bank 2017 *Innovation for Productivity and Inclusiveness*, Washington, USA: The International Bank for Reconstruction and Development/THE WORLD BANK

[19] Jiřinová K and Vrbová L 2017 Do management innovations yield in higher employees´ satisfaction?, *Proceedings of Innovation Management, Entrepreneurship and Sustainability (IMES 2017),* pp. 371-377

[20] Carlan V, Sys C and Vanelslander T 2019 Innovation in Road Freight Transport: Quantifying the Environmental Performance of Operational Cost-Reducing Practices, *Sustainability, 11*(8), https://doi.org/10.3390/su11082212

[21] Bertermann D, Müller J, Galgaro A, Cultrera M, Bernardi A and Di Sipio E 2016 Cheap-GSHPs, an European project aiming cost-reducing innovations for shallow geothermal installations. Geological data reinterpretation, *Geophysical Research Abstracts, 18*, EGU2016-13086

[22] Tsonev N and Kaleychev S 2018 Innovative Practices as a Key for a Better Management in Tourism Industry, *European Journal of Economics and Business Studies, 4*(1), pp 8-16

Acknowledgments

This article is a part of a project № KP-06- M 25/5/17.12.18, financed by National Science Fund, Ministry of Education and Science, Bulgaria.

IOP Conf. Series: Materials Science and Engineering **878** (2020) 012075 doi:10.1088/1757-899X/878/1/012075

Application of Management Innovations in the Bulgarian Wine Industry

D Pastarmadzhieva[1] and M Angelova[1]

[1]Faculty of Economic and Social Sciences, University of Plovdiv Paisii Hilendarski, 4000 Plovdiv, 24 Tzar Asen str, Bulgaria

daniela.pastarmadzhieva@uni-plovdiv.bg

Abstract. The innovations are of crucial importance for the contemporary business development. Management innovation is one of the groups of innovations that can be applied and according to the scholars it has the potential to enhance the competitiveness of the enterprise. As the wine industry is a key sector of the Bulgarian economy, it is essential for the wine producers to select a relevant type of innovations for their enterprise. Thus, the *aim* of the current study is to identify if the Bulgarian wine producers are ready to implement management innovations and if this decision is associated to some factors. The *object* of the study is Bulgarian SMEs wine producers. The *focus* is their activity, regarding the management innovations. The results are obtained through an online survey, conducted from December 2019 to January 2020. The results show that the priorities of Bulgarian wine producers are the management innovations, related to processes and resources, the employees and the knowledge.

1. Introduction

The role of innovations for the development is a topic, which has been examined since 19th century [1]. However the most recent development in the concepts of innovations alongside with the increasing role of ICTs transforms and enhance the approaches to the implementation of innovations. The innovations are relevant for every area – personal, economic and business and for the states as well. Nowadays countries adopt special innovation strategies for various areas in order to achieve competitive advantages.

Wine industry is a traditional sector of the Bulgarian economy with high potential for development. It is also a leading sector in Bulgaria. In the International Organization of Vine and Wine (OIV) Statistical Report on World Vitiviniculture in 2018, Bulgaria was ranked 21st in the world, regarding wine production, and 24th as for the area of vineyards [2]. However, regarding the production, sales, consumption, import and export indicate that there is a necessity of imposing a policy for promotion and realization of the Bulgarian wines on more foreign wine markets. Given that there is a need to limit production capacity and the need for technical and technological upgrading for the majority of producers of goods, given the availability of more funds. Thus, the sector requires more efforts regarding the marketing, product and production innovations [3].

Alongside, the management of a company is of essential importance for its innovativeness and competitiveness. The innovations in the area of management can enhance its effectiveness and support the functioning of the whole enterprise. Thus, the *aim* of the current study is to identify if the Bulgarian wine producers are ready to implement management innovations and if this decision is associated to some factors. The *object* of the study is Bulgarian SMEs wine producers. The *focus* is their activity, regarding the management innovations. The results are obtained through an online survey, conducted in the period December 2019 - January 2020 through structured interviews using own questionnaire. The collected data is processed with IBM SPSS Statistics and frequencies and cross tables are used.

The article is *structured* as follows: following the introduction, in the second part is described a conceptual framework of innovations, innovations in wine industry and the management innovations in particular. This analysis paves the way for the third part, in which is highlighted the methodology of the study. The fourth part introduces the obtained results of the survey and discussion based upon the figures. The paper closes with conclusions and recommendations for future research in the field of investigation the management innovations in wine industry.

2. Literature review

The very concept of innovation appeared for the first time in scientific research in the 19th century. The concept was gained in the early 20th century by the Austrian scholar Joseph Schumpeter [1]. According to Schumpeter [4], the term "innovation" means an amendment to introduce and use new types of consumer goods, new production and transport means, markets and forms of organization in production. The end result of the process is creating a new product, service, process, or form of organization.

Nonetheless, the innovation activity is related to various factors starting from personal to national level. The overall environment plays significant role on the entrepreneurial and innovation activity [5]. To some extent the countries like Bulgaria, who transformed their economy from central planning into a market one have their own challenges in policy-making [6], respectively in the implementation of innovations.

However, during the last decade, the Bulgarian economy has achieved macroeconomic stability and growth. Various measures were implemented in order to improve the environment for doing business especially for SMEs. Entrepreneurs exploit new opportunities and are associated with disturbing the market equilibrium. They often revolutionize industries overturning long-established technologies, business models and dominant companies [7]. In order to do so, they innovate and take risks.

In the Oslo Manual 2018 (4th edition) OECD innovations are classified in two major groups – product innovation and business process innovation. Product innovations can be innovations of goods and of

IOP Conf. Series: Materials Science and Engineering **878** (2020) 012075 doi:10.1088/1757-899X/878/1/012075

services. Business process innovations include innovations in production of goods or services, distribution and logistics, marketing and sales, information and communication systems, administration and management, and product and business process development [8]. Thus, the management innovations are in the second group. Innovations are essential for the development of wine industry as well [9]. The scholars, who examine the innovations in wine industry identify that they apply mostly product and marketing innovations [10] [11] [12].

Although management innovations are not in the top list of the wine producers, they are crucial for the sustainable development of a company [13]. Management innovations can be defined as "implementation of new management practices, processes and structures that represent a significant departure from current norms" [14]. The examination of management innovations allows scholars to identify the dynamics of the way a company is managed, its progress and if the used approach enhances its effectiveness over time [15].

Management innovations' impact on the organizational performance is predominantly positive. One of the positive results is related to improvement of the company's competitiveness. The latter is related to the fact that the effective management can boost the effect of technical and technological innovations. Management innovations are also related to the competitiveness as they has impact on "productivity, delivery time, quality and flexibility". Furthermore, the studied type of innovations transforms the daily routines of the company and enhance its overall performance. Nevertheless, the wine producers have to implement the management innovations in an adequate way, because in some cases they may have negative effect [16].

3. Methodology of the study

The methodological framework consists of a structured interview, conducted in the period December 2019 - January 2020. The survey includes four sections: general information about the respondent, information about the enterprise, innovations section and competitiveness section. Section Innovations is the most important and it consists of a list of fifty (50) innovations. For each of them the respondents are invited to select one of the following answers: *1-Not applicable to us*, *2-We know about this opportunity, but we would not apply it*, *3-We know about this opportunity and we would apply it*, *4-We apply it now*, and *5-We applied it in the past*.

The current paper presents some of the results of the survey. In the latter all the SMEs wine producers were invited to take part in the online survey by e-mail and some of them were personally contacted. As a result, 156 online questionnaires were sent by e-mail and 57 of them participated in the survey which represents 37% response rate, thus the collected data provides relevant results. The statistical methods, used in the study are frequencies and cross tables. The latter is used in order to identify if there is an association between the application of management innovations in the company and various factors. The examined factors are: years of functioning of the enterprise, size of enterprise, additional activities, markets for production realization, sources of financing and sources of raw material. The figures display the share of answer *3-We know about this opportunity and we would apply it* and answer *4-We apply it now.*

4. Results and discussion

As figure 1 shows a significant share of wine producers apply or would apply some kind of management innovations. The most applied innovation is the optimization of business processes as all of the SMEs wine producers apply (55.6%) or would apply (44.4%) such. It is followed by more efficient use of resources (95.8%), further education and advanced training of employees (92.3%) and exchange of knowledge regarding production (92.2%). Thus, the priorities for the wine producers are processes and resources, the employees and the knowledge. The least preferred management innovation is the use of private consultants as 40.9% apply it and only 22.7% would apply it.

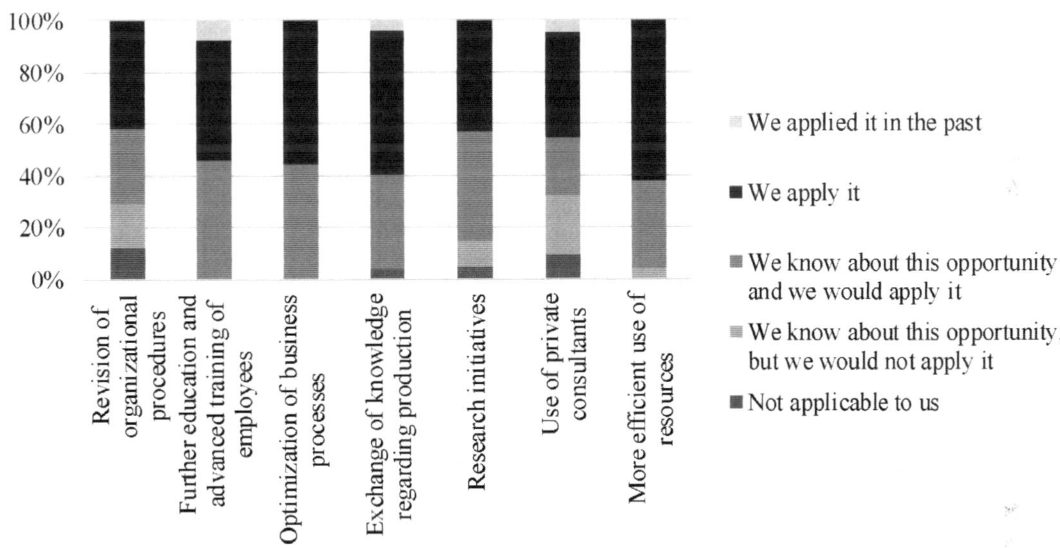

Figure 1. Application of management innovations among SMEs wine producers, 2019

Source: Own data

As concerns the revision of organizational procedures, they are applied mostly by the medium-sized enterprises (50-250 employees), who use finances mostly by National programs and use imported material for their production. The oldest SMEs are least likely to implement revision of organizational procedures (only 43% of them). Such result is explainable as they are experienced and know what works for their enterprise. (Fig. 2)

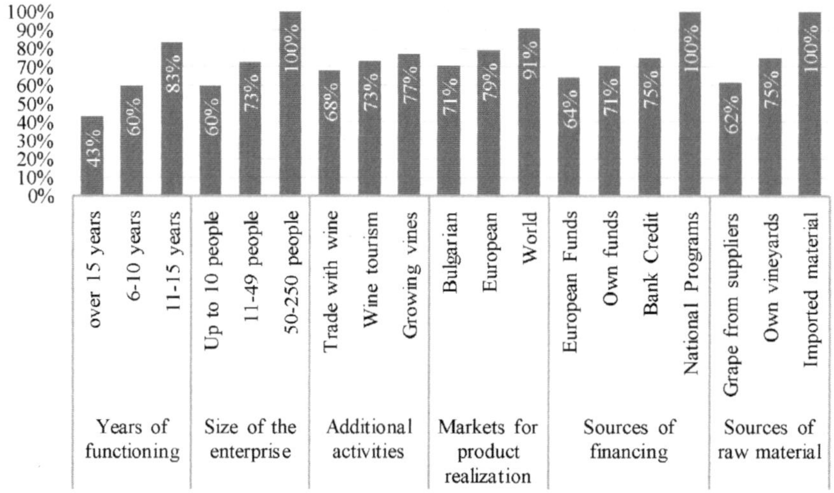

Figure 2. Association between revision of organizational procedures and a selection of factors, 2019

Source: Own data

The further education and advanced training of employees as management innovation is applied predominantly by medium-sized enterprises (50-250 employees), which have functioned for over 10 years. They are implemented by enterprises, which rely on national programs for financial resources and use imported material for their production. The newest enterprises are less ready to make further education and training for their employees. (Fig. 3)

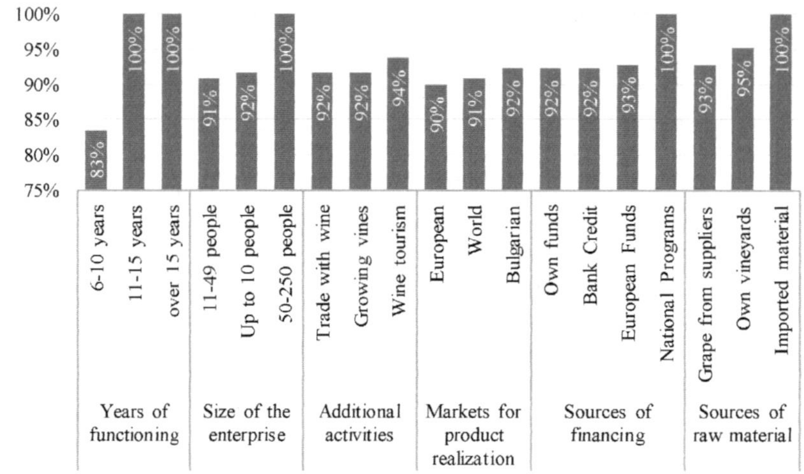

Figure 3. Association between further education and advanced training of employees and a selection of factors, 2019

Source: Own data

However, it should be noted that in almost every of the examined groups the share of those who implement or would implement such innovations is over 90%. It is 83% only for the enterprises, who function for 6 to 10 years. (Fig. 3)

We already mentioned that exchange of knowledge, regarding the production is one of the most implemented management innovations. It is preferred by micro (up to 10 people) and medium-sized (50-250 people) enterprises. These are enterprises, which have been functioning for 6-10 years or over 15 years. Again, all of the SMEs, who use financial resources from national programs implement this kind of innovations. The group of the enterprises, who produce with imported material is the most innovative as regards the exchange of knowledge, regarding the production. The SMEs, who are least ready to apply the latter are the ones, which function for 11-15 years (67%). (Fig. 4)

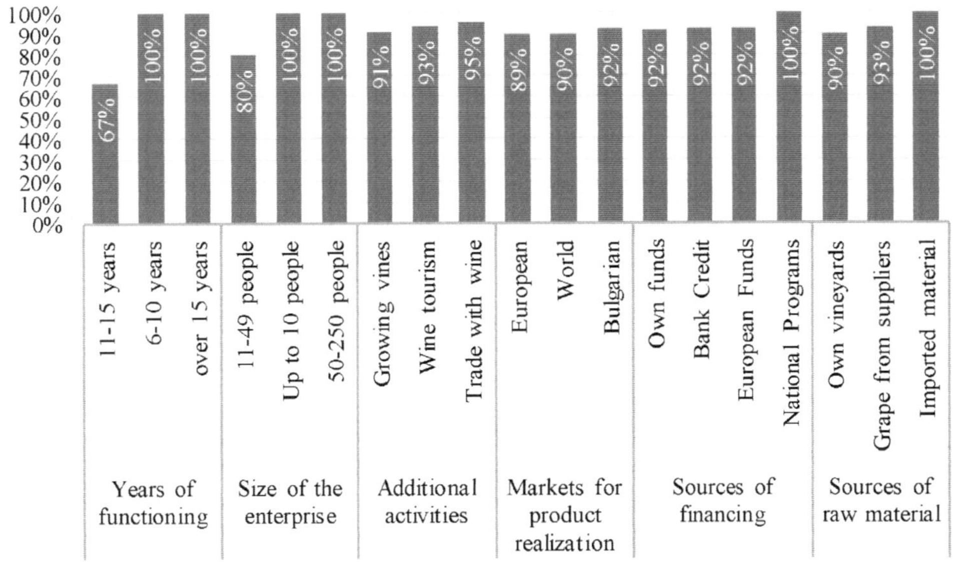

Figure 4. Association between exchange of knowledge regarding production and a selection of factors, 2019

Source: Own data

All of the medium-sized enterprises (50-250 people), who sell their production worldwide make research initiatives as part of their management innovations. The same is also valid for those, who use grape from suppliers and imported material. Those SMEs, who rely on National programs for financing are least likely to implement such innovations. (Fig. 5)

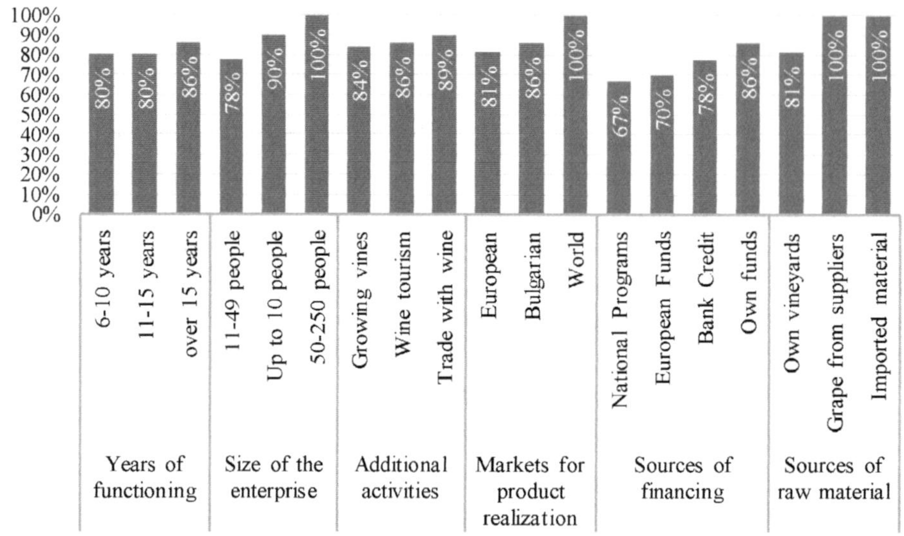

Figure 5. Association between research initiatives and a selection of factors, 2019

Source: Own data

The use of private consultants is the least applied innovation. However, 100% of the medium-sized enterprises implement such, as well the companies, who use financial resources from National Programs. (Fig. 6)

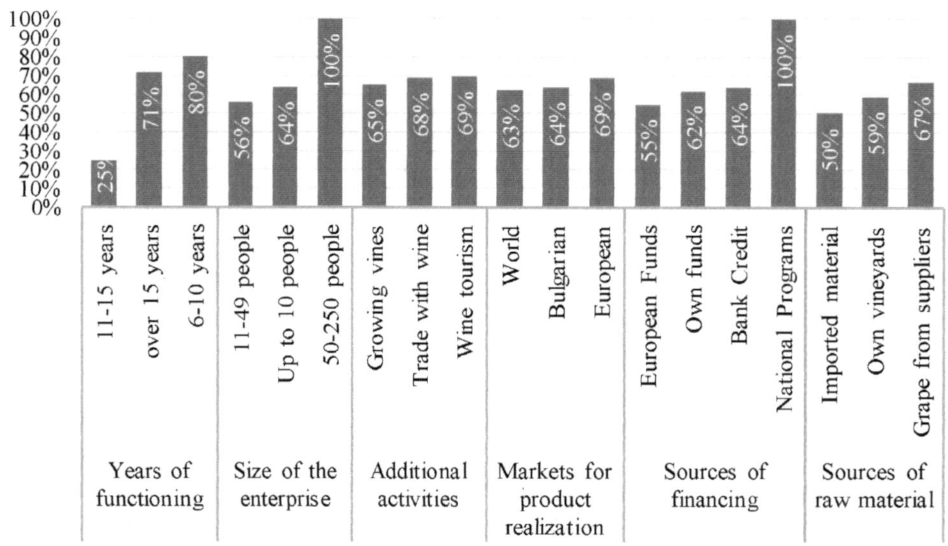

Figure 6. Association between use of private consultants and a selection of factors, 2019

Source: Own data

More efficient use of resources is one of the most preferred innovations. It is applied by all micro (up to 10 people) and medium-sized (50-250 people) enterprises, alongside with those, which have functioned for over 10 years. 100% of the wine producers, who sell their production worldwide also apply them. The same is valid for the ones, who use imported material and grape from suppliers. The SMEs, who rely on national programs are least likely to think about more efficient use of resources. (Fig. 7)

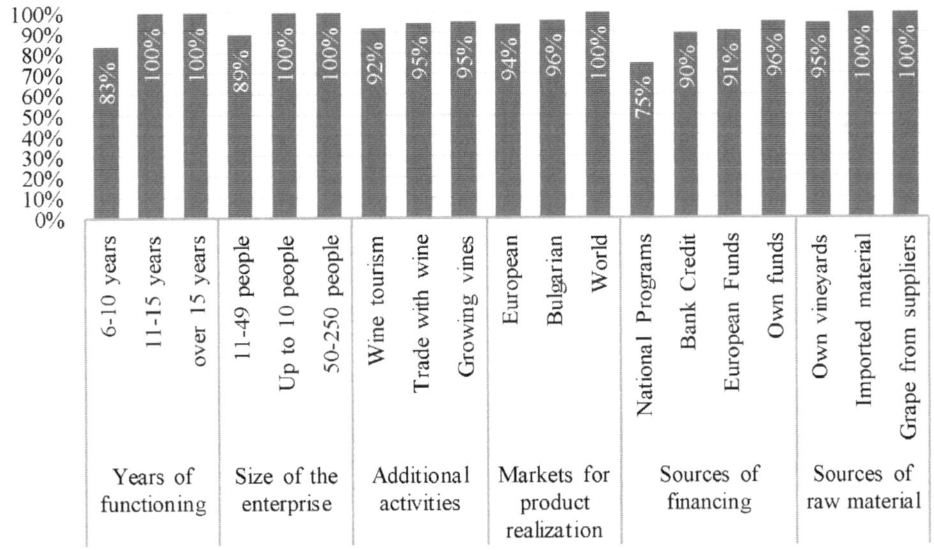

Figure 7. Association between more efficient use of resources and a selection of factors, 2019

Source: Own data

Conclusion

Innovations and innovation activity are a booster of the competitiveness of a company. Bulgarian wine industry as a significant sector in the economy needs to enhance its performance. The innovations can contribute to high extent for the latter and the management innovation also can add value to this process. The obtained results indicate that a significant part of the Bulgarian SMEs wine producers implement or are ready to implement management innovations. The latter is a positive indication and can lead to the improvement of the competitiveness of the sector. The most applied management innovations are the optimization of business processes, more efficient use of resources, further education and advanced training of employees and exchange of knowledge regarding production. The least preferred management innovation is the use of private consultants. The implementation of management innovation is mostly spread among the SMEs, who are medium-sized and work for over 15 years. Alongside, an interesting result is that almost all these type of innovations are applied by all the wine producers, who rely on national programs for funding, who sell their production on world markets and who use imported raw material for the production of wines. However, these findings require further investigation and more efforts in order to identify the reasons for the results.

References

[1] Schumpeter J A 1949 *Economic theory and entrepreneurial history,* in Center for Research in Entrepreneurial History, Harvard University (ed.), *Change and the Entrepreneur. Postulates and Patterns for Entrepreneurial History,* (Cambridge, MA: Harvard University Press, pp. 63-84)

[2] International Organization of Vine and Wine 2018 *World Vitiviniculture Situation*, OIV Statistical Report on World Vitiviniculture

[3] Angelova M, Pastarmadzhieva D, Dimitrova G and Georgiev P 2019 Innovative Practices in Wine Industry: Opportunities for Competitiveness Enhancement in Bulgaria, *IEEE*, International Conference on Creative Business for Smart and Sustainable Growth (CREBUS), Sandanski, Bulgaria, doi: 10.1109/CREBUS.2019.8840046

[4] Schumpeter, J A 1946 *The Future of Private Enterprise in the Face of Modern Socialistic Tendencies*, in Swedberg R. (ed.) Joseph A. Schumpeter. *The Economics and Sociology of Capitalism*, (Princeton: Princeton University Press, 1991 (1st edition 1946), pp. 401- 405)

[5] Pastarmadzhieva D and Angelova M 2019 Optimization of Bulgarian Political Culture as a Prerequisite for Enhancement of the Entrepreneurship, *IEEE*, International Conference on Creative Business for Smart and Sustainable Growth (CREBUS), Sandanski, Bulgaria, doi: 10.1109/CREBUS.2019.8840081

[6] Vilisov M V 2017 Political strategies of Belarus, Kazakhstan and Russia: Problems of formation and practical implementation, *Political science (RU), Special issue*, pp. 41-62

[7] Hisrich R, Peters P and Shepherd D A 2017 *Entrepreneurship (10th ed.)* (New York: McGraw-Hill Education)

[8] OECD/Eurostat 2018 *Oslo Manual 2018: Guidelines for Collecting, Reporting and Using Data on Innovation*, 4th Edition, The Measurement of Scientific, Technological and Innovation Activities, OECD Publishing, Paris/Eurostat, Luxembourg. https://doi.org/10.1787/9789264304604-en

[9] Karagiannis D and Metaxas T 2018 *Innovation in Food & Wine Tourism Related Businesses: Turning Ashes to Gold*, In *Management and Marketing of Wine Tourism Business: Theory, Practice, and Cases*, Marianna Sigala, Richard N. S. Robinson (Eds), (New York: Springer) pp. 345-363.

[10] Wood E and Kaplan D 2005 Innovation and performance improvement in the South African wine industry, *Int. J. Technology and Globalisation, 1*(3/4), pp. 381-399.

[11] King L and Forbes S L 2013 *Exploratory analysis of marketing innovations in the New Zealand wine industry*, Working Paper, Faculty of Commerce, Lincoln University, Lincoln, New Zealand,

[12] Luca P de and Penco P 2006 *The Role of Packaging in Marketing Communication: an Explorative Study of the Italian Wine Business,* 3rd International Wine Business Research Conference, Montpellier, 6-7-8 July, 2006

[13] Zhang Y, Khan U, Lee S and Salik M 2019 The Influence of Management Innovation and Technological Innovation on Organization Performance. A Mediating Role of Sustainability, *Sustainability, 11*(2), 495-516

[14] Birkinshaw J and Mol M 2006 How Management Innovation Happens. *MIT Sloan Management Review, 47*(4), pp. 81-88

[15] Mol M J and Birkinshaw J 2014 The Role of External Involvement in the Creation of Management Innovations, *Organization Studies, 35* (9): 1287-1312

[16] Lin H 2018 *Adoptive Management Innovation* (Singapore: Springer Nature Singapore)

Acknowledgments

This article is a part of a project № KP-06- M 25/5/17.12.18, financed by National Science Fund, Ministry of Education and Science, Bulgaria.

IOP Conf. Series: Materials Science and Engineering **878** (2020) 012076 doi:10.1088/1757-899X/878/1/012076

Digital competitiveness of Bulgarian enterprises – 2019

T B Gigova[1] and A R Arabadzhieva[2]

[1] Faculty of Mechanical Engineering, Department of Industrial Management, Technical University - Sofia, Plovdiv Branch, 25 Tsanko Diustabanov Str., Plovdiv, 4000, Bulgaria

[2] Faculty of Mechanical Engineering, Department of Industrial Management, Technical University - Sofia, Plovdiv Branch, 25 Tsanko Diustabanov Str., Plovdiv, 4000, Bulgaria

E-mail: t_gigova@abv.bg, anet2003@abv.bg

Abstract. This paper examines the position of Bulgaria with respect to the Digital Economy and Society Index by analysing the following indicators: connectivity, human capital, use of Internet services, integration of digital technology and digital public services. Digital competitiveness of European Union member states has been measured by the Digital Economy and Society Index (DESI). The main barriers to the use of digital technologies and processes in Bulgarian enterprises have been identified on the basis of research.

1. Introduction

Essential characteristics of the industries of the future are human spirit, ingenuity and enterprise. Facing new dynamic business models and increasing competition in the respective market sectors, many industrial enterprises encounter the need of thorough reconsideration of their current strategies. Entrepreneuers are confronted with the challenge of building and developing digital enterprises to make their businesses sustainable and efficient in the forthcoming decades. Digitisation holds a key position in this scenario. Digital technologies and the opportunities they provide will be a key driver for development in an ever-increasingly digital future.

The term „digitisation" derives from the English word „digital" (which means „based on digits") and is referred to as the process of converting information from analogue source material (text, sound, videosignal, telephone and pulses) into a numerical format by meaqns of electronic devices, i. e. converting analogue information into digital. This allows for digital environment processing, storage and transmission of information to users regardless of their location via computer networks, satellites, the Internet, social networks, etc.

Digitisation provides greater possibilities for integration, processing, structuring, categorization, usage and transmission of heterogeneous information flows and access to information on the global network to many users simultaneously without time limits.

Below is a summary of the current technological trends in digital transformation of business models [1]:

- Internet of Things (IoT);
- Blockchains;

IOP Conf. Series: Materials Science and Engineering **878** (2020) 012076 doi:10.1088/1757-899X/878/1/012076

- Big Data Calculations;
- Machine Learning.

The main goals of digitisation [2, 3] are:

- preservation of analogue information resources and their long-term storage in the form of digital copies;
- provision of access to these digital copies through digital devices and networks;
- collection of digital copies in digital libraries.

The pillars of digital transformation are:

- thinking outside the box, i.e a digital way of thinking;
- a digital transformation company roadmap, i.e. a digital strategy;
- appropriate skills;
- technologies;
- evolution.

The impact of new technologies on production and business processes in Bulgarian industry (Industry 4.0) [4] can be summarized as follows:

- Creation of new products and services with built-in intelligence, innovative business models and opportunities to personalise and adapt them to customer needs.
- Digitisation of the entire production cycle; research and development acceleration through digital prototyping and virtual production; flexible organisation of the manufacturing process.
- Miniaturisation as a trend in production of microchips, electronic devices, implants, etc.

2. Exposition

The study on the degree of digital competitiveness of Bulgarian enterprises has been carried out by analysing and summarising data and information gathered by ICAP Agency for Market Research, the European Commission, the German-Bulgarian Chamber of Commerce and Industry, EUROSTAT, etc. The research is also based on data derived from a number of surveys, strategies, conclusions and work guidelines from national and regional programmes.

The Digital Economy and Society Index (DESI) [5] tracks the evolution of EU member states in digital competitiveness.

Although for the year 2017 Bulgaria has made progress in expanding its broadband infrastructure and in developments with freely available data relevant to the previous yeaar, it ranks 27th in the Digital Economy and Society (DESI) Index Ranking for 2017. What is worse, in 2019 Bulgaria ranks last (28th) among EU member states overall in the European Commission Index Ranking (Figure 1) [5].

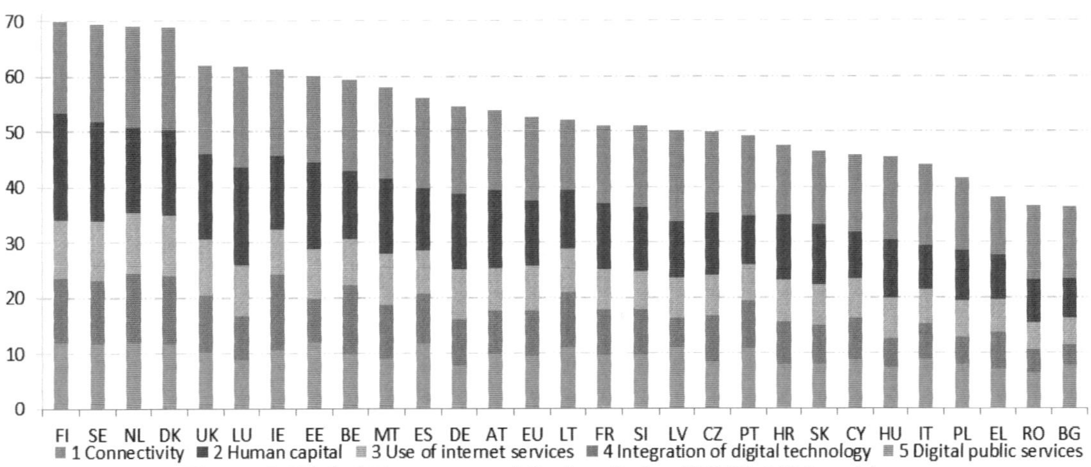

Figure 1. Digital Economy and Society Index (DESI) 2019 ranking

The DESI Index comprises five indicators: connectivity, human capital, use of Internet services, integration of digital technology and digital public services. Scoring one hundred points is the highest rating which means that a particular country takes utmost advantage of digital technologies for business and public development. The total score of Bulgaria is 36.2 compared to 35.5 for the previous year. The EU average is 52.5 compared to 49.8 for 2018. Bulgaria's ranking is lower both due to the limited results in some of the reported DESI dimensions and the better performance of the other member states by some of DESI dimensions (Figure 2). In terms of the human capital dimension (with a score of 28.5 relevant to an EU average of 59.3) and the integration of digital technology dimension (with a score of 18.1 relevant to an EU average of 41.1), Bulgaria ranked 28th in 2019. Regarding the use of Internet services dimension the country ranked 27th (with a score of 32.5 relevant to an EU average of 53.4) and with respect to the connectivity dimension – 25th (with a score of 51.6 relevant to an EU average of 59.3). Bulgaria's poor performance regarding digital skills, digitization of business and public services appears to be a hindrance to the further development of its digital economy and society. Thus, the country falls into the cluster of poor performing countries.

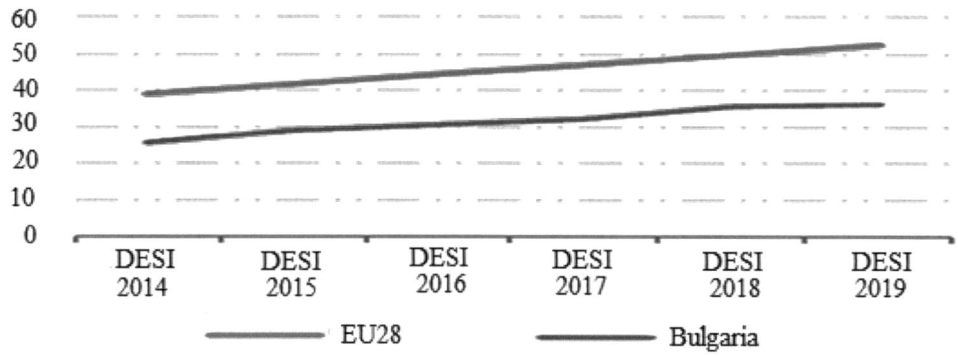

Figure 2. Evolution overtime of Digital Economy and Society Index, 2014 ÷ 2019

The connectivity profile of Bulgaria (Figure 3) shows that it lags behind among EU member states overall by the following indicators: fixed broadband take-up, 4G network coverage, fixed broadband coverage, broadband price index, utrafast broadband take-up and fast broadband (NGA) coverage and that the country nearly reaches the EU average in terms of the fast broadband take-up, mobile broadband take-up and ultrafast broadband coverage indicators.

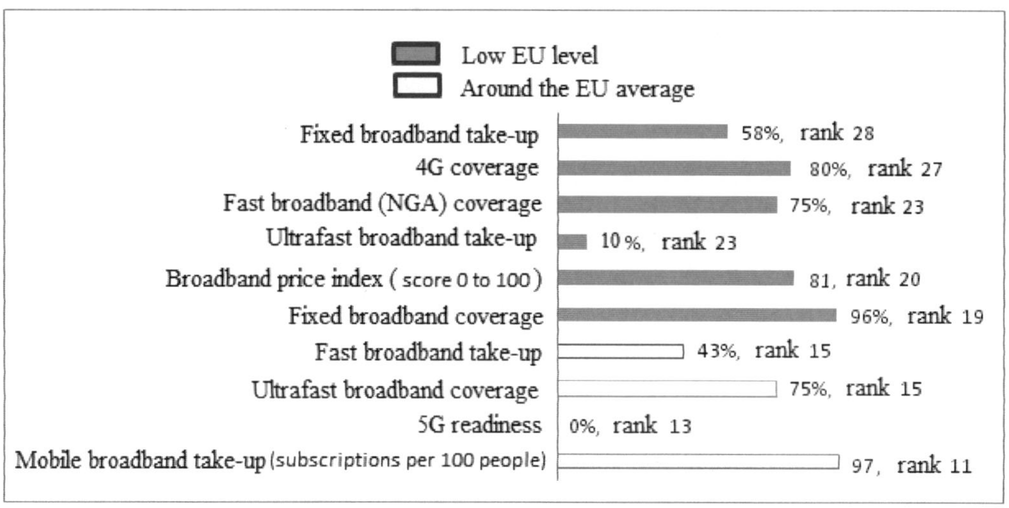

Figure 3. Connectivity profile of Bulgaria

The human capital dimension (Figure 4) shows that by the Female ICT Specialists indicator only Bulgaria ranks 15th, thus reaching the EU average, while by the other indicators the country falls behind among the other EU member states, with an overall level of digital skills being among the lowest in the EU. The share of people having at least basic digital skills is approximately 29% relevant to an EU average of 57%. Only 54% of the young people aged between 16 and 24 have at least basic skills, while the EU average is 81%. The share of people with above basic digital skills amounts to 11% of the total number, which is less than one-third of the EU average of 31%.

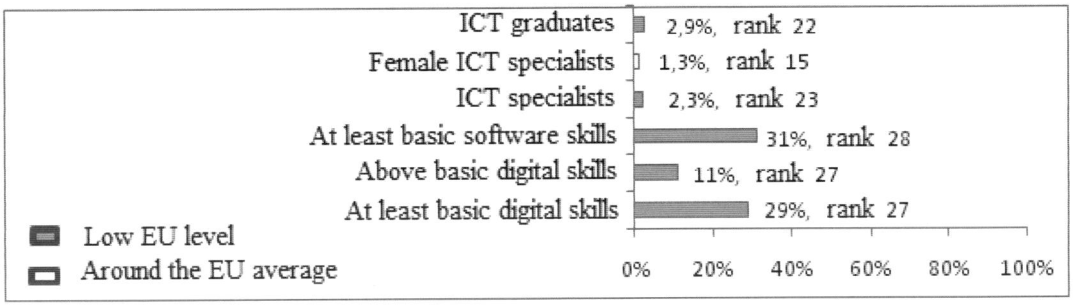

Figure 4. Human capital profile of Bulgaria

Regarding the use of Internet services dimension Bulgaria has improved its performance, reaching a score of 32.5 compared to 29.9 for the previous year, but still falls below the EU average by this indicator. (Figure 5). The country stands well above the EU average in terms of video chatting, however. Bulgarian Internet users make the largest number of video calls and thus rank 1st among all EU Internet users. In terms of activity on social networks Bulgaria ranks 7th which is above the EU average as well. The percentage of Internet users reading news online (79%) is also close to the EU average (65%). On the other hand, Bulgarian Internet users are less interested in e-banking, shopping and selling online, doing online courses, video on demand, online consultations and voting.

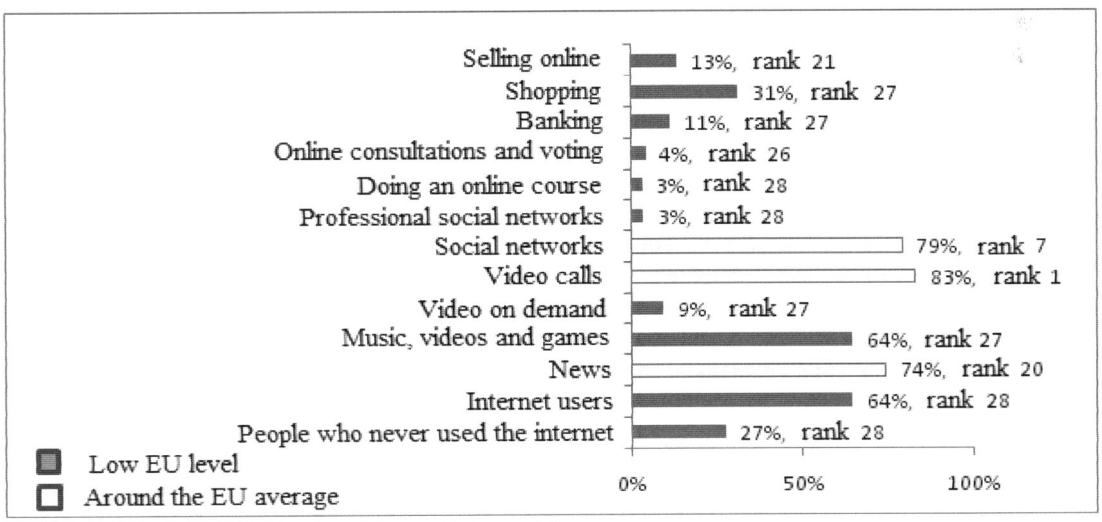

Figure 5. Use of Internet services

With respect to the integration of digital technology dimension, Bulgaria scores 18.1, thus being well below the EU average score which is 41.1 for 2019. The country nearly reaches the EU average (Figure 6) in terms of big data sets (7% relevant to an EU average of 12%) and electronic information sharing (23% relevant to EU average of 34%), but it falls below the EU average by the rest of the indicators. Bulgarian enterprises encounter significant difficulties in realising the potential of e-Commerce. Only 6% of all SMEs sell online; 3% of them realise cross-border sales and their e-Commerce turnover only amounts to 2%. Isolated 9% of Bulgarian enterprises use social media to promote their business relevant to an EU average of 21%.

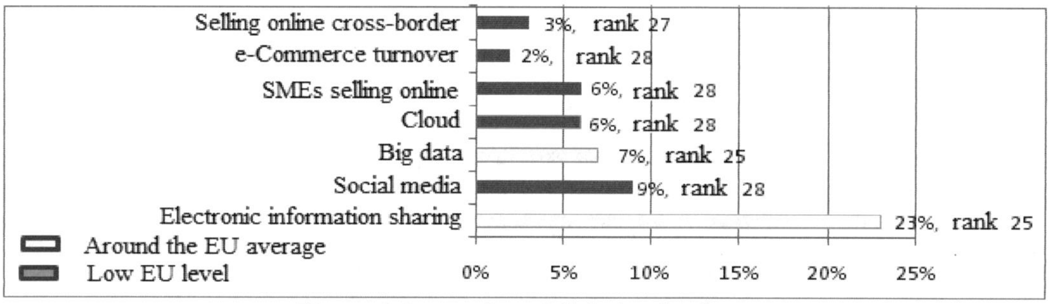

Figure 6. Integration of digital technology

According to data provided by Siemens and the German-Bulgarian Chamber of Commerce [6] the low level of digital maturity can be accounted for by certain key barriers encountered by Bulgarian enterprises which are as follows: inadequate personnel qualification (50%), amount of investment (46%), insufficient technology maturity (31.6%), other priorities (27.6%) and unclear economic benefits (23.7%) (Figure 7).

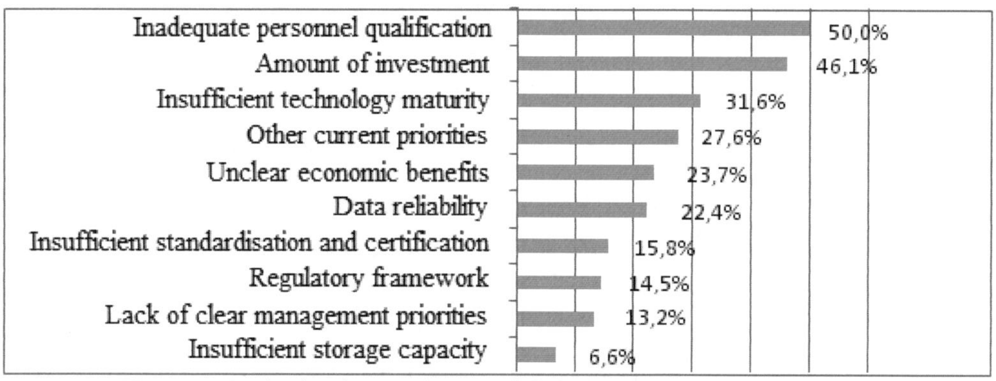

Figure 7. Major barriers to the use of digital technology and processes

The analysis results show that digital transformation streamlines business processes, decision-making and interaction with value chain partners and end customers.

Digitization is not an unfamiliar concept for Bulgarian companies. However, while most enterprises in Bulgaria are aware of the digitization potential for development and expansion of their businesses, nearly a third of them admit that they do not have a comprehensive digital strategy.

Bulgarian enterprises haven't yet reached an adequate level of digital maturity.

Bulgaria considerably lags behind relevant to the overall EU trend toward evolving digital society, and it is therefore necessary for the country to adopt specific measures and invest efforts to overcome the lag [7] and increase digital competitiveness of Bulgarian enterprises.

Digital transformation is a many-sided process that can bring benefits to enterprises in the following respects:

- Users – service improvement and creation of unique user experience.
- Operations – optimisation of business processes, improvement of work efficiency and decision-making process.

- Products – renovation of existing products and services, creation of new products and services and introduction of innovations.

Bulgaria lags behind the other EU member states in terms of fixed broadband take-up, ultrafast broadband take-up and 4G network coverage.

Bulgarian Internet users fall below the EU average in using Internet services, doing online courses, using professional social networks, e-banking, online professional consultations and voting, and have the lowest overall level of digital skills.

Bulgarian businesses encounter serious difficulties in realizing e-Commerce potential and do not exploit the social media potential to promote their business.

Inadequate personnel qualification, small amount of investment and insufficient technology maturity appear to be the key barriers faced by Bulgarian business to further integration of digital technology and processes.

3. Conclusion

Technological development is a natural reaction to the business pursuit of optimisation, i.e. growing more efficient, more productive, and more flexible. Digital transformation will have a positive impact on the efficiency and formation of competitive advantages for Bulgarian industrial enterprises. Digitisation is a prerequisite for gaining strategic advantages over competitors, opening up new markets and enabling businesses to become proactive.

Acknowledgements

The authors would like to thank the Research and Development Sector at the Technical University of Sofia for the financial support.

References

[1] Mihova T and Chukalov K 2019 *Digital business models in industrial enterprises* IOP Conference Series: Materials Science and Engineering, Volume 618, 2019 IOP Conf. Ser.: Mater. Sci. Eng. 618 012074 https://doi.org/10.1088/1757-899X/618/1/012074

[2] Zlatanova - Pazheva E and Shatarova D 2017 *Internet Marketing – Certain Types and New Trends*, In: Conference proceedings, Vol. I, of XV International Scientific Conference „Management and Engineering '17", June 25-28 2017, Sozopol, pp. 275 – 284. ISSN 1310-3946, ISSN 1314-6327.

[3] Nikolova-Alexieva V and Angelova A 2018 *Opportunities For Raising The Entrepreneurial Culture A Factor For Competitiveness Of The Bulgarian Economy*, In: International Journal of Entrepreneurship and Small Business, Vol. 1, Issue 1, 2018, pp. 196 – 245. [online]. doi.org/10.1504/ijesb.2019.10014160.

[4] Gigova T and Nikolova-Alexieva V and Valeva K 2019 *Digital Transformation of Bulgarian Enterprises in the Conditions of Industry 4.0*, 2019 International Conference on Creative Business for Smart and Sustainable Growth (CREBUS), Sandanski, Bulgaria, 18-21 March 2019, INSPEC Accession Number: 18994760, INSPEC Accession Number: 18994760, DOI: 10.1109/CREBUS.2019.8840065

[5] European Commission 2019 *Digital Economy and Society Index* (DESI) Country Report Bulgariaq, https://ec.europa.eu/digital-single-market/en/desi

[6] Siemens Bulgaria & Deutsch-Buigarische Industrie – und Handelskammer 2018 *Study on the level of digitalization in Bulgaria*, Sofia, Bulgariaq, https://bulgarien.ahk.de/fileadmin/AHK_Bulgarien/News/Digitalization_Survey_Bulgaria-BG.pdf

[7] Topleva S 2016 *Economic aspects of GM-foods: consumer perceptions of welfare*, 1901 – In: Scientific Journal L'Association „SEPIKE", France, 2016, № 11. pp. 244 – 247. ISSN 2196-95312.

IOP Conf. Series: Materials Science and Engineering **878** (2020) 012077 doi:10.1088/1757-899X/878/1/012077

Algorithm for business process reengineering in industrial enterprises

T B Gigova[1] and N S Geshanova[2]

[1] Faculty of Mechanical Engineering, Department of Industrial Management, Technical University of Sofia, Plovdiv Branch, 25 Tsanko Diustabanov Str., Plovdiv, 4000, Bulgaria
[2] Faculty of Mechanical Engineering, Department of Industrial Management, Technical University of Sofia, Plovdiv Branch, 25 Tsanko Diustabanov Str., Plovdiv, 4000, Bulgaria

E-mail: t_gigova@abv.bg, n_geshanova@gmail.com

Abstract. The present article delivers an analysis of the specific stages of reengineering implementation according to reengineering proponents and examines the strengths and limitations of the methodology developed by them. The analysis of the outlined stages of business process reengineering provides a detailed sequence of the stages of reengineering project implementation. The algorithm for business process reengineering is depicted in a block diagram which presents the summarized conceptual and structural components and the logical connections between the distinct steps of business process reengineering methodology which incudes seven stages: initiative for business process reengineering; diagnostics of existing business processes; business process redesign; implementation of redesigned business processes; summary and evaluation of business process reengineering efficiency; analysis and regulation.

1. Introduction

Considerable structural changes in the industrial enterprise in the course of reenginering – fundamental and revolutionary BP (business process) restructuring coupled with transition to the new principles of enterprise building call for a special project development and implementation as well as creation of reengineering team including both enterprise staff members and invited consultants. After accomplishing the project goals and objectives, the enterprise is supposed to transition to a new stage of development called business enhancement involving slight modernization in the course of the ongoing work.

Reengineering implementation is indispensable even after the possibilities for gradual improvement have been exhausted, but in this case the project is only intended for particular subdivisions of the industrial enterprise. Thus, the continuous modification activities become part of the enterprise everyday life as a reaction of the ever-changing external environment factors: market, technologies, customer preferences, competition, etc.

2. Exposition

Different reengineering methodologies are available. The strengths and limitations of Hammer and Champi's, Davenport's, Manganelli and Klein's and Kodak's methodologies are surveyed on the basis of an analysis of the particular stages of BPR execution [1].

According to Hammer and Champi, the goal of the reengineering process lies in dramatic increase of business process efficiency and the tool for achieving it is creation and implementation of original, 'irrational' ideas capable not only of improving the process, but of completely overhauling it. In this case, information technologies play an improtant, but not a key role. In his work, Hammer [2] argues that excessive fascination with IT is what can cause the reengineering project to be merely reduced to automation of all business process stages.

The lack of a specific project time frame can be deemed a limitation of the proposed methodology. On one hand, if the ideas are conceptualized and do not raise doubts, then project success is more than likely guaranteed and its implementation deadline is short enough. On the other hand, if ideas turn out to be not revolutionary enough and the proposed model – unviable, then it is necessary to invest additional efforts into conceptualisation and validation of new ideas.

According to Davenport the goal of the reengineering process lies in dramatic increase of business process efficiency and the tool for achieveing it is implementation of new IT although Davenport recognizes the great significance of organisational and staffing issues.

According to Davenport, culture may appear as a grave hindrance in case innovative processes do not fit the cultural environment. With regard to management, Davenport focuses on traditional functions such as planning, leading, management, control, decision-making and communication [3, 4]. Unlike Hammer, Davenport views the reengineering project as part of a package of measures intended to improve industrial organisation including other unrevolutionary approaches, such as TQM [4, 5] rather than as a single activity. In respect to this, activity efficiency analysis is made on a continual basis.

Davenport's methodology presupposes detailed description of existing business processes – for large enough organization description takes from six to nine months. On one hand, detailed description reveals where major issues lie, but on the other hand, such an approach only pressuposes correction of existing errors and solution of existing problems rather than process reconceptualisation. The focus on IT, on one hand, significantly streamlines tasks for reconceptualisation pressuposes increased awareness of the ways in which implementation of new IT will affect processes. On the other hand, problems often lie in process logic itself rather than in insufficient automation.

According to Manganelli and Klein the goal of the reengineering project lies in dramatic increase of competitiveness and the tool for its achievement is implementation of new IT.

Manganelli and Klein's approach differs considerably from Hammer and Champy's and Davenport's approaches. Reengineering proponents also consider increase of competitiveness, but what they focus on in fact is increase of BP efficiency only [6]. Competitiveness is above all based on unique competitive advantages of industrial enterprise products and services. Achievement of a particular competitive advantage is what underlies industrial enterprise strategy. The lack of strategy validation in BP redesign causes the reengineering project to only increase efficiency i.e. reduce costs, time, increase customer satisfaction rate rather than contribute to the creation of unique competitive advantages. Strategy validation can result in substantially different processes.

According to Kodak, the goal of the reengineering process lies in fundamental increase of efficiency and the instrument for its accomplishment is implementation of new IT.

The main strength of Kodak's approach is a clearly designed system for project administration. The focus of methodology founders is rather on IT implementation than on development of 'irrational' reengineering ideas.

The algorithm for BPR depicted in figure 1 below in the form of a summarized block diagram is based on Hammer and Champi's approach and considers the need for relation to enterprise strategy proposed by Manganelli and Klein. The algorithm has been developed on the basis of practical expertise of enterprises which have carried out reengineering.

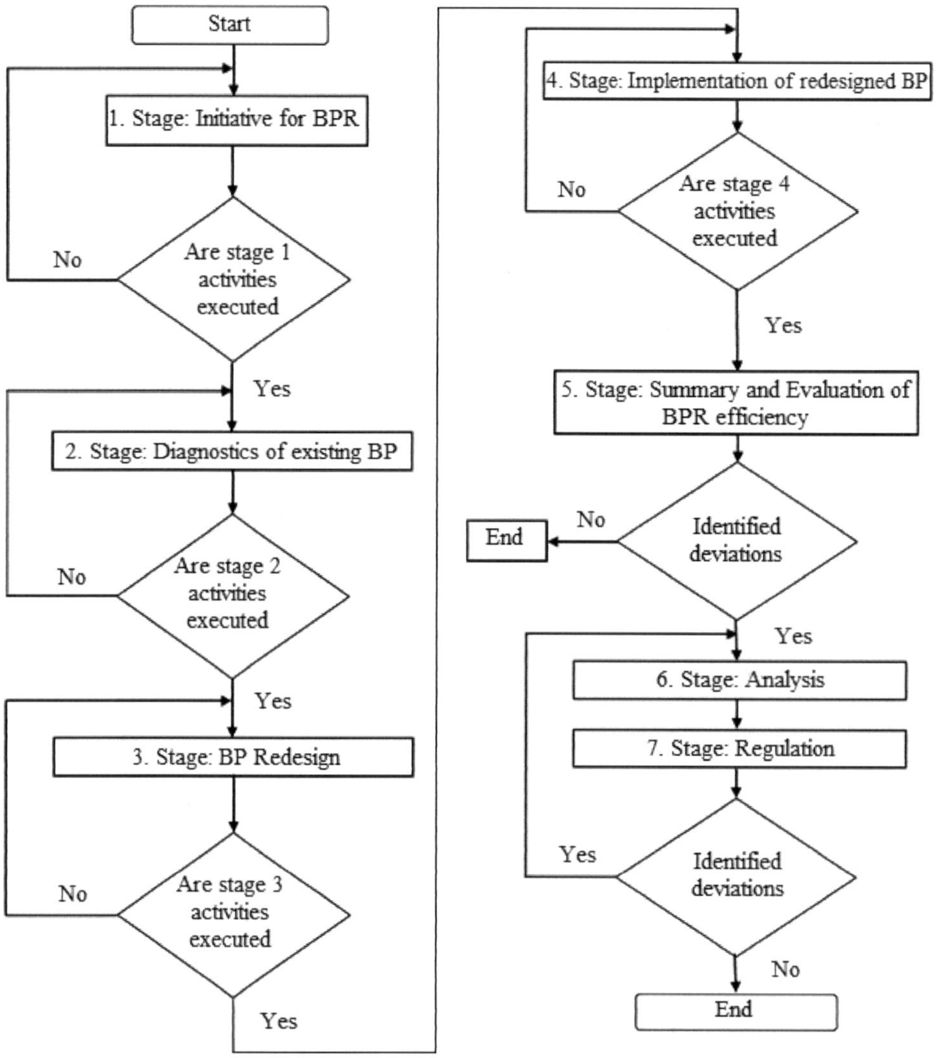

Figure 1. Block diagram of BPR methodology

The summarized block diagram of the proposed methodology for BPR represents the summarized conceptual and structural components and the logical connections between the distinct steps of the methodology including the following stages: initiative for BPR; diagnostics of existing business processes; business process redesign; implementation of redesigned business processes; summary and evaluation of BP efficiency; analysis; and regulation.

Stage: 1. Initiative for business process reengineering

In this stage, the BPR project is launched. Project launch timing is pivotal to the successful accomplishment of its goals and objectives. The sequence of the activities included in this stage is as follows:

- *1.1. Creation of BPR management system.*
- *1.1.1. Selection of participants in BPR.*

- *1.1.2. Allocation of roles and responsibilities.*
- *1.1.3. Team work organisation.*
- *1.2. Planning of BPR execution.*
- *1.2.1. Scheduling.*
- *1.2.2. Selection of implementation methods and tools.*

Team initiates action by selecting the methodology used in reengineering.

- *1.2.3. Selection of interaction tools.*

The fist stage appears essential to the success of BPR, but also extremely difficult due to an emerging inconsistency – on one hand, BPR is launched, project team is built, research work is initiated, but final outcome is not yet clear. Diagnostics can reveal that project launch timing is not right.

Stage: 2. Diagnostics of existing business processes

Different authors evaluate the stage of diagnostics of existing business processes in different ways. Davenport proposes detailed description of existing business processes [7], while Hammer suggests that that no efforts at all should be wasted on description [8]. What is deemed reasonable is an intermediate approach. A design concept can be developed without description of processes. As a result, however, it will appear too general – fit for preliminary discussion, but not for actual implementation. Wasting too much time on description of processes subject to radical overhaul is inexpedient. It is recommended that stage 2 – diagnostics of existing business processes should include the following activities:

- *2.1. Description of existing business processes.*

Once reengineering project launch is scheduled, it is necessary to create a process map at the highest possible level. Selection of processes which will be subject to reengineering presupposes an awareness of the processes which are currently ongoing in the enterprise. This is by no means an easy task for enterprises built on the basis of functional criterion. Processes are most often not identified, but scattered as distinct functions among different enterprise subdivisions. In case a single approach for process selection is not available, the task appears complicated even to experts dealing with the process approach.

It is necessary to select key (priority) processes which will undergo reengineering. Key factors for success (7-8 factors) such as duration, costs, after sales service, etc. are also determined. The number of processes depends on the quantitative composition of the project team and the prospective investments.

The description of existing business processes includes:

- *2.1.1. Creation of a summarized business model and identification of the most essential processes by means of expert evaluation such:*
  - processes which have greatest impact on customers;
  - processes which have lowest efficiency in comparison with analogous (similar) processes in leading companies;
  - processes the reengineering of which can yield positive future results.

Processes are to be arranged in order of importance.

- *2.1.2.* Distinct processes *require high-level description.*

Prototype building is an instance of high-level description of future processes. Consequences (forecasting of changes in process characteristics) are evaluated; both success and risk factors are determined.

- *2.1.3. Indicators (measurable properties) are measured*:
  - time indicators – average order processing time, productivity, etc.
  - cost-efficiency indicators – prices of goods, number of participants in the process, etc.
  - quality indicators – rejecton rate (customer refusals), level of service provided, etc.

A description of enterprise organisational structure is made.

- *2.2. Analysis of customer, supplier and partner requirements* which comprises:
- *2.2.1. Customer requiremenet analysis* made on the basis of surveys, interviews, market research. It involves:
  - classification of existing and potential customers;
  - a list of requirements which are met/not met by BP;
  - a description of the ideal (from customer point of view) BP;
  - customer evaluation of existing business processes through the use of indicators.
- *2.2.2. Partner and/or supplier requirement analysis* which encompasses:
  - evaluation of their roles in BP;
  - requirements for ideal partners and/or suppliers;
  - partner and/or supplier requirements for BP.
- *2.3. Evaluation of enterprise level.*

Evaluation of enterprise level is carried out by drawing a comparison between the particular enterprise and the best competitor organisations which:

- have high sales volumes;
- boast good reputation;
- produce high-quality goods.

Comparison is executed by means of indicators.

Enterprise level evaluation allows for identification of enterprise strengths and weaknesses in comparison with competitiors.

- *2.4. BPR goal setting.*

Specific project goals are set on the basis of results of customer, partner and supplier requirement analysis and enterprise level evaluation.

Goals are to include indicator values.

- *2.5. Documenting and verification.*

Stage: 3. BP redesign

In this stage it is expedient to again adopt a mixed approach involving BP redesign, defiance against old rules, and employment of 'revolutionary ideas' along with the widest possible application of IT capabilities.

Scenario planning of BP redesign involves goal achievement strategy development. Methods of expert evaluation can be employed in scenario efficiency evaluation. BP redesign plans are generally created by project teams. It is necessary that all process activities should be simulated by means of imitation models prior to actual redesign process.

The sequence of the activities included in the stage of BP redesign is as follows:

- *3.1. Creation of existing business process model 'as it is'.*

Initially, a model of the existing business process 'as it is' is created. Once the model is developed and trialed, its adequacy is validated. The model allows for identification of 'impediments' and evaluation of key parameters.

- *3.1.1. Creation of external model* which includes participant selection and description of interactions (diagram).
- *3.1.2. Event flow description* by means of IDEF0 diagrams, workflow diagrams, Gantt diagrams, simulation (imitation) models, etc.
- *3.1.3. Creation of external model* which includes object selection and diagram representation and interaction represenation (diagrams of sequence or collaboration).
- *3.1.4. Documenting and verification* – discussion of created model for the purpose of error identification, documenting and verification.
- *3.2. Analysis of existing BP model.*

The analysis of existing BP model involves:

- *3.2.1. Measure and evaluation of BP indicators.*

Process characteristics are measured through the use of indicators:

- value – techniques of functional-value analysis
- time – scheduling techniques (Gantt chart, network diagrams) and simulation modelling;
- quality – expert evaluations (quality points).

➢ *3.2.2. Identification of problems and limitations* – on the basis of measurement of BP quality attributes, evaluation of BP steps, as well as staff interviews and analysis of existing information systems.

The problem list is to be organized in a descending order of importance.

➢ *3.2.3. Correction of BPR goals* outlined in the stage of existing BP diagnostics.

The following step is creation of new BP model (what it is supposed to be like) which involves establishment of new organisational structure and new information system should the need arise.

➢ *3.3. Creation of new BP model (what it is supposed to be like).*

The creation of new BP model encompasses the following activities:

➢ *3.3.1. Development of innovative ideas.*

➢ *3.3.2. Development of alternatives (further options).*

Innovative ideas are considered in the process of specification of summarized scenarios (from goal specification). Reengineering principles are to be considered in scenario development process.

➢ *3.3.3. Creation of discrete models for each of the alternatives.*

Future process design is created by reengineering project team by means of tools for process modelling. The procedure entails the same activities as the ones involved in the process of creation of existing BP model.

➢ *3.3.4. Analysis of discrete models of alternatives and selection of the best option.*

Discrete models of alternatives are compared to one another as well as to existing BP.

As a final step, it is necessary to establish whether the measurable goals and objectives set in the diagnostic stage are accomplished.

The next step in stage 3 is *development of new organisational structure* which involves:

1) Determining project team composition.
2) Organising team work.
3) Defining enterprise employee rights and obligations.
4) Creating motivation and evaluation systems which reflect new beliefs and values adopted by the particular enterprise.

Stage: 4. Implementation of redesigned BP

In the stage of redesigned BP implementation it is reasonable to apply Hammer's three-phase implementation strategy which includes laboratory, pilot and circulation phase.

Laboratory stage involves simulation of a number of process models and verification of hypothesis viability.

Pilot testing involves launching a number of processes in 'real-life' environment, but within a limited range, i.e. trial run.

Circulation phase involves process launch on a large-scale basis within the particular enterprise.

The sequence of the implementation stage activities is as follows:

➢ *4.1. Verification of new BP* which includes:

➢ *4.1.1. Verification of new BP model.*

➢ *4.1.2. Verificataion of information maintenance system prototype.*

➢ *4.1.3. Trial run of integral business model on a small-scale basis within the particular enterprise.*

Trial run is better conducted by employing prototypes. Certain alterations and model specifications can be made on the basis of trial run results.

➢ *4.2. New BP implementation.*

As new business processes are implementated, it is necessary that existing processes continue functioning. Customers and partners are not supposed to feel uneasiness during integration of new processes in old ones.

IOP Conf. Series: Materials Science and Engineering **878** (2020) 012077 doi:10.1088/1757-899X/878/1/012077

A universal application method which would be feasible for each and every project is not available. Economic, political and cultural idiosyncrasies of the particular industrial enterprise are to be considered in the process of selection and rational combination of different events. Among the most determining factors are:

- total amount of reorganisation work;
- number of engaged employees;
- number and significance of affected processes;
- scale of change (revolution or evolution);
- statutory deadlines;
- prevailing management style.

Stage: 5. Summary and evaluation of business process reengineering efficiency

The main activities in this stage are:

- *5.1. Project summary.*
- *5.2. Evaluation of project efficiency.*

Following implementation BP is to be under continuous supervision and control for the purpose of obtaining statistical data on actual progress and outcome of executed activities and customer satisfaction. Comparison between planned performance and reported performance is drawn in order to identify possible deviations. Derived data allows for BP measurement in accordance with target criteria and indicators. In case of possible deviations, two more stages are to be considered – analysis and regulation.

Stage: 6. Analysis

This stage involves examining and analysing reasons for deviation occurrences and outlining respective solutions to eliminating them. The aforementioned components of the efficiency diagnostics on the basis of suggested factors can be used for this purpose.

Stage: 7. Regulation

In this stage, should need arise corective action plan is developed on the basis of monitoring results. In accordance with it BP model and respectively the ongoing (real) process undergo overhaul. Regulation continues until equilibrium is fully restored.

BPR presented above has been trialed in Bulgarian industrial enterprises. Business process redesign has led to positive outcomes. Reengineering of 'sales and customer service' process in one of the enterprises has resulted in reduction of process cycle time by about 2/3 on average. The main project goal has been accomplished – the number of accepted purchase orders has risen by 50% and sales revenue – by 20%; the number of points of contact has been reduced; the roles and participants in the process; the number of error occurrences in information conveyance and reflection has been reduced which has in turn contributed to achieveing information coherence and consistency; the opportunity for system update and improvement has been exploited.

3. Conclusion

What is specific of the reengineering approach is that it offers new models for organisation, management, planning and execution of processes, functions, activities and tasks. It is based on the integration of simple and elementary operations into a single integral process. The synergy between reengineering and modern information technologies can serve for analysis and optimisation of organisational structure, staff reduction, cycle time reduction, improvement of BP quality, new customer attraction and cost saving. Successful implementation of reengineering projects involves considerably greater efforts by comparison with implementation of technical aspects of new decisions.

Reengineering projects are unique, 'know-how' projects in the sense that they consider specific enterprise peculiarities which accounts for the development of a special project for each particular enterprise. It is practically impossible and therefore unnecessary to provide for all design details. What is essential is the fact that BPR methodology reveals goals and provides guidelines on their accomplishment. Enterprises which have launched reengineering can use methodology components, introduce certain corrections justified by production specifics which gives reengineering team opportunity for creative work and enables them to develop their own reengineering projects with minimal time and labour costs.

Acknowledgements
The authors would like to thank the Research and Development Sector at the Technical University of Sofia for the financial support.

References

[1] Gigova T 2013 *Reengineering projects – stages and aims*, XI International scientific conference 'Management and engineering '13', Technical University – Sofia, Faculty of Management, Conference proceedings, Sozopol, Bulgaria, Volume I, ISSN 1310-3946, ISSN 1314-6327, pp. 238-243.
[2] Hammer M *1990 Re-engineering Work: Don't Automate*, Obliterate, Harvard Business Review, pp. 104-112.
[3] Davenport T 1993 *Process Innovation: Reengeneering Work through Information Technology*, Boston, MA: Harvard Business School Press., ISBN: 0875843662 pp.1, 2, 4, 11, 5-8, 27-28, 71, 117, 141, p. 337.
[4] Davenport T N and Short J E 1993 *The New Industrial Engineering: Information Technology and Business Process Redesing*, Sloan Management Review, pp.46-49, summer 1990, Vol.31, pp.11-27.
[5] Davenport T and Jarvenpaa S and Beers M 1996 *Improving knowledge work processes*, In: Sloan Management Review.
[6] Manganelli R and Klein M 1996 *The Reengineering Handbook: A Step-By-Step Guide to Business Transformation*, New York: Amacom, p.318.
[7] Масааки И и Гемба Кайдзен 2014 *Путь к снижению затрат и повышения качества*, Альпина Паблишер, М. ISBN: 978-5-9614-1136-2, с. 345.
[8] Kamennov K 1998 *Management of Change*, Abagar Publishing House, Veliko Tarnovo.

IOP Conf. Series: Materials Science and Engineering **878** (2020) 012078 doi:10.1088/1757-899X/878/1/012078

Analysis of significant floods in Bulgarian Danube region till 2010 y

Rositsa Velichkova

Bulgaria, Sofia, 8 Kl. Ohridski Ave, Technical University of Sofia

rvelichkova@tu-sofia.bg

Abstract. Floods in Bulgaria are natural hazards that occur often and make huge damage. In current work the significant floods for Danube region are discussed. An analysis for the period till 2010 is made. The rivers which cause most damages in that region are shown.

1. Introduction

Floods are common natural disasters on the territory of the Republic of Bulgaria. They can cause huge damage as they affect urban territories, productive agricultural lands and woodland. According to the scientific classification for this type of hazards and according to the Water Act, the floods can be natural and technogenic - caused by other influences, such as damage to a hydrotechnical facility, which can lead to an accident as well as to prevent critical situations in hydrotechnical facility.

Floods over recent years have shown that the flood risk reduction activities that have been applied so far are not sufficient and a policy of comprehensive flood risk management should be pursued under the conditions of coexistence with them. Such an integrated approach - flood risk assessment and management is set out in the European Floods Directive, which was adopted in 2007.

Its requirements were introduced into the national legislation through the Water Act in 2010. The reduction of the flood risk is mainly due to the construction of hydrotechnical protection systems and facilities, good spatial planning of the territory and last but not least - increasing the readiness to prevent or reducing the negative effects of floods through preventive measures, training of the population, adequate preparation, planning of rescue activities, etc. Regular cleaning of the riverbeds and their protection in natural conduction conditions is necessary as well.

The aim of the present work is to analyse the significant floods of the rivers in the Danube region by 2010. (Fig. 1)

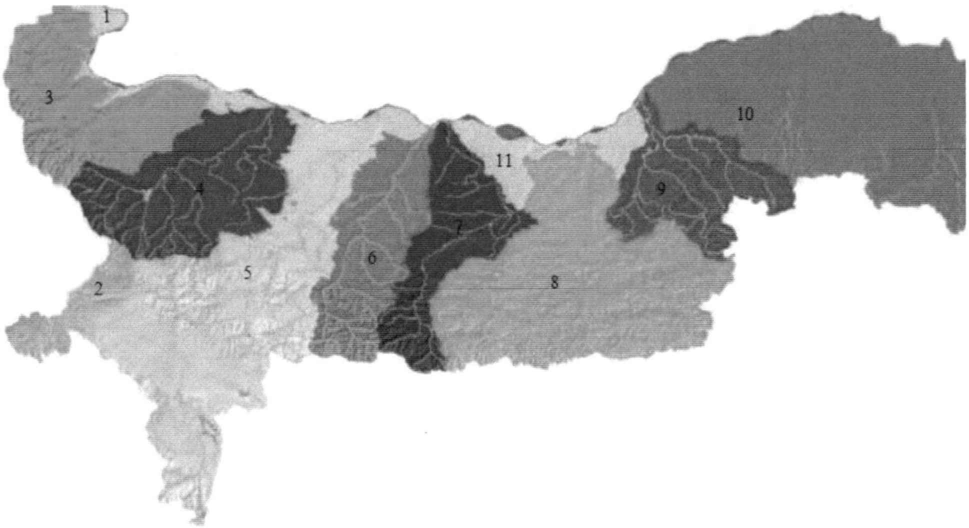

Figure 1. Main basins in the Danube region
1-basins of river Erna, 2-basins of river Nishava, 3- basins of rivers west form river Ogosta, 4- Basins of river Ogosta, 5- basins of river Iskar, 6- basins of river Vit, 7- basins of river Osam, 8- basins of river Yantra, 9- basins of river Rusneski Lom, 10- Danubian Dobrudjan rivers 11- basins of river Danube

2. Preliminary risk assessment

According to the recommendations of the Floods Directive, in the framework of the ex-ante flood risk assessment, past floods with significant impacts on the protected categories "Human health", "Environment", "Cultural heritage" and "Economic activities" which are likely to repeat in a similar form in the future (Article 4.2.6). For this purpose, significant floods of the past have to be described as far as significant negative consequences of future similar events can be expected (Article 4.2).

In order to make use of the knowledge of past floods and to identify areas with a significant potential flood risk, pursuant to Article 5 of the Floods Directive, the historical flooding limits need to be redeveloped in addition to the verbal description of past floods. In addition, historical flooding information is of great importance for the development of hazard maps and flood risk maps as they can be used to calibrate hydraulic computational models as well as to control the credibility of the computational results.

Depending on the region, the identification of flood-risk areas, flood risk assessment maps are prepared for different parts of the river network on the territory of the respective municipality or region.

Risk Assessment Cards are developed for the following scenarios:

- floods with low security or with a long period of repetition (extreme floods);
- floods with average collateral (<1%), respectively with an average replication period of 100 years);
- at the discretion and for floods with high security or with a small period of repetition

In accordance with the Directive 60/2007 / EC requirements, flood hazard maps must contain the following information:

- Flooding range / limits;
  Depth of flooding or discretionary water structures;
- at an estimated rate of flow or significant water quantity.

IOP Conf. Series: Materials Science and Engineering **878** (2020) 012078 doi:10.1088/1757-899X/878/1/012078

3. Project units in Danube region

The volume of activities in each project unit also depends on the existing flood risk. The determinate project units in our country correspond to the currents basins for the implementation of the Water Framework Directive. The project units determination includes:

- justification of initial territorial river basins for analysis;
- threshold values of the height of the water column;
- analysis of the affected areas and determination of the scale of the flood;
- identifying the importance of flood protection measures for the population and infrastructure.

Project units in the Danube area are listed in Table 1

Table 1. Project Units

Project Unit	Rivers
1-st	Timok
	Topolovets
	Voynishka
	Vidibol
	Archar
	Skomlya
	Lom
	Tsibritsa
2-nd	Ogosta and its influx
3-th	Iskar and its influx
4-th	Vit and its influx
5-th	Osam and its influx
6-th	Yantra and its influx
7-th	Rusenski Lom and its influx
8-th	Tsartsar
	Senkovets
	Kanagyol
9-th	Suha and its influx
10th	Vatova
	Shablenska

European Commission's guidelines set put in the document "Draft list of flood types and stand consequence" form 16.02.2011 distinguished five types of floods according of their source:

- river – caused by rain and snowmelt;
- slope-caused by rain ;
- lake-caused from rain water inflowing in lakes;
- sea – caused by sea storms;
- infrastructure – caused by insufficient conductivity of sewer and drainage systems due to rainfall and wave transferring seawater over protective embankments in the coastal areas of cities.

The collected information for floods in Bulgaria covers the period 2000-2010 and refers to the 2211 past floods (fig.2).

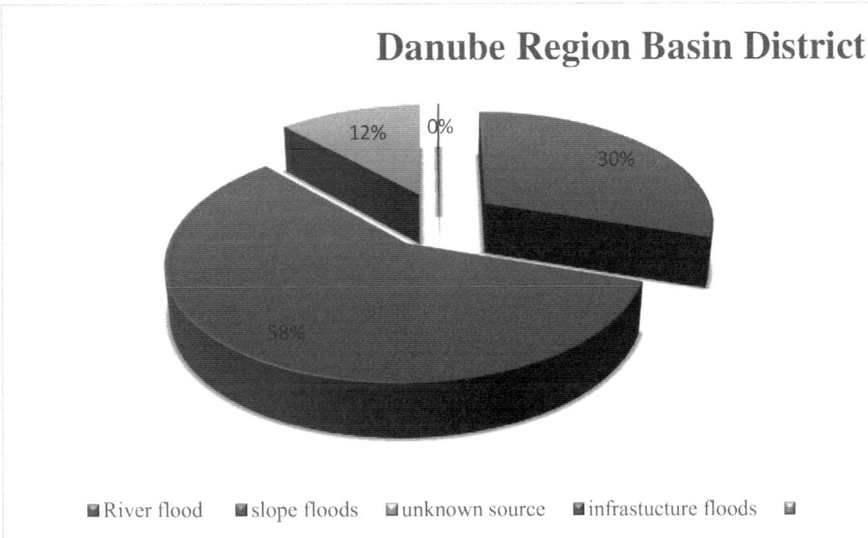

Figure 2. Information for Danube Region Basin District

The predominant mechanism for flood formation is the natural overflow of water over the river banks. Historical flood chronology describes one case of river floods as a result of the blocking of a river with icy blocks (Danube River, 1942), as a result of the rivers (Ogosta, Vit, Osam and Yantra, Rusenski Lom, 2006) from the high level of the Danube River and as a result of reduced conductivity (p. Yantra, 1991). The number of the river floods caused by the reduced possibility of the river beds is very large.

Catastrophic floods are monitored and documented with particular attention. Therefore, such floods have been thoroughly documented and described. These records are an important source of information on the existing flood threat to be used in the framework of the preliminary flood risk assessment, both for describing the flood threat and its consequences and for identifying areas with significant potential flood risks.

4. Analysis of significant floods at Danube region

In this part are made analysis for significant floods by districts in Danube region. It is given the significant floods at each river in this region.

At Danube region have 9 districts: Vidin, Montana, Vratsa, Pleven, Veliko Tarnovo, Rousse, Razgrad, Silistra, Dobirch

The rivers in this region are given in table 1.

On fig. 3-9 are given the significant floods in Danube region till 2010 year by districts and river is Danube region.

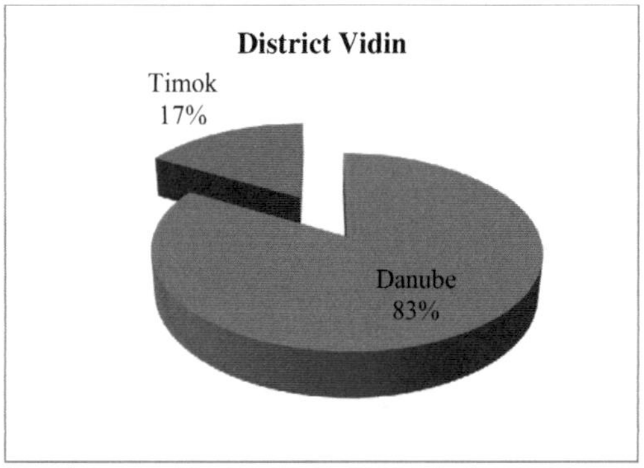

Figure 3. Significant floods in district Vidin

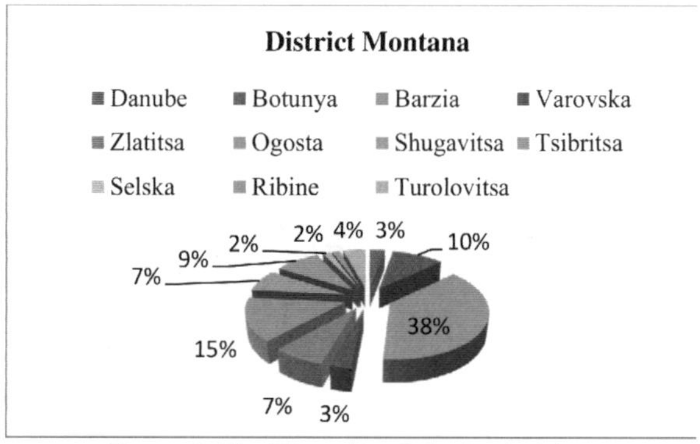

Figure 4. Significant floods in district Montana

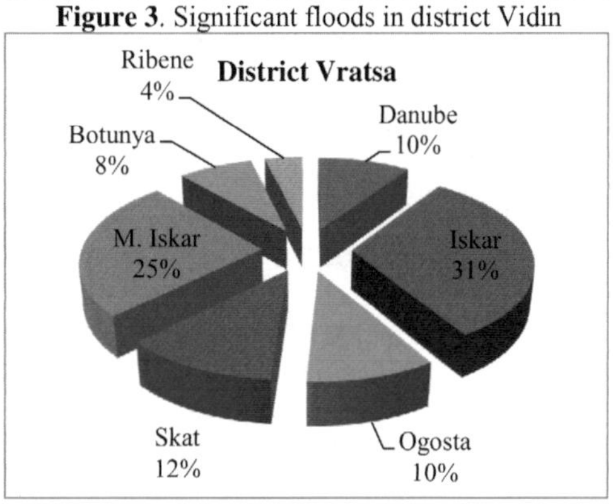

Figure 5. Significant floods in district Vratsa

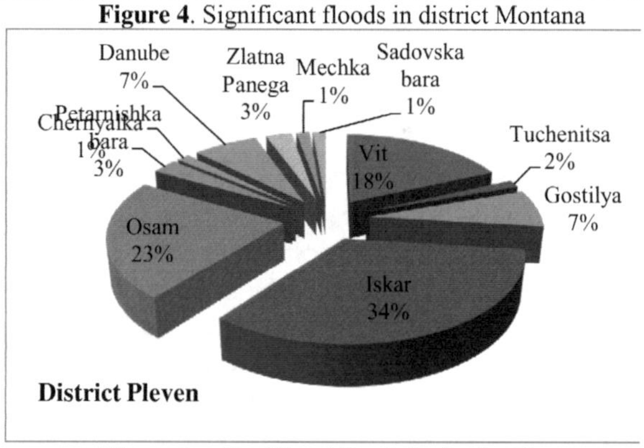

Figure 6. - Significant floods in district Pleven

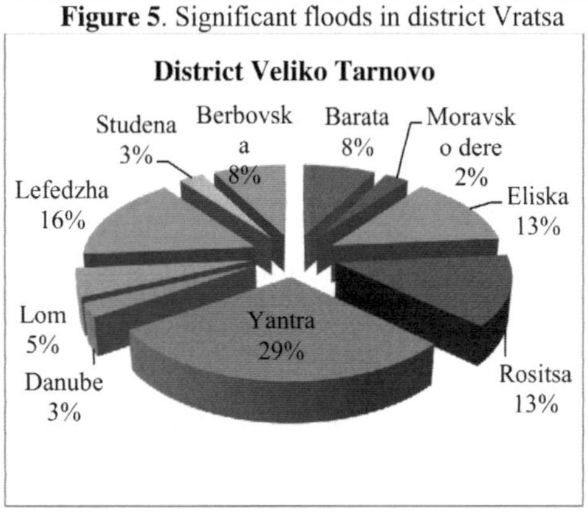

Figure 7. Significant floods in district Veliko Tarnovo

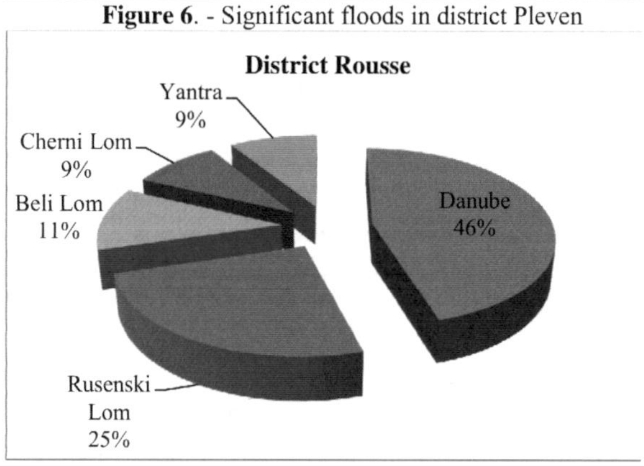

Figure 8. Significant floods in district Ruse

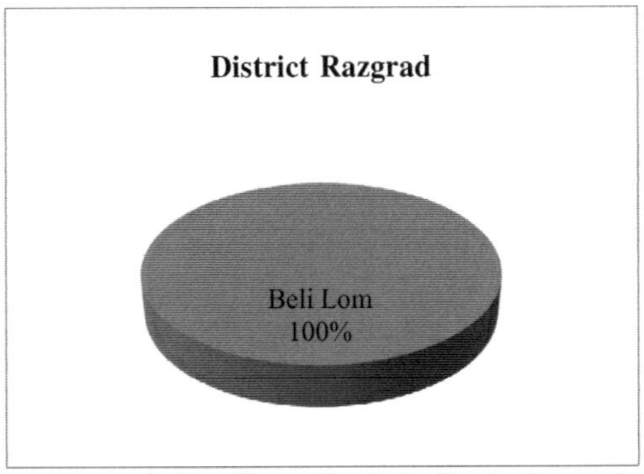

Figure 9. Significant floods in district Razgrad

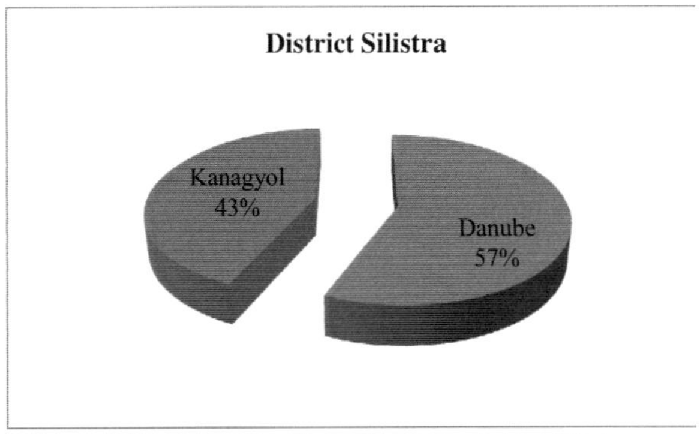

Figure 10. Significant floods in district Silistra

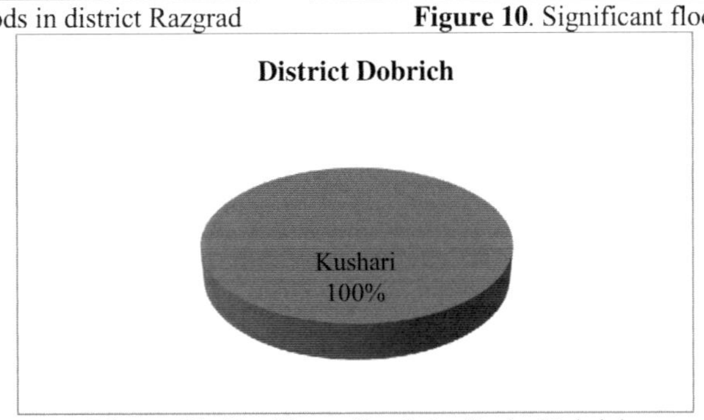

Figure 11. Significant floods in district Dobrich

In fig. 10 is shown the significant floods by rivers at Danube region

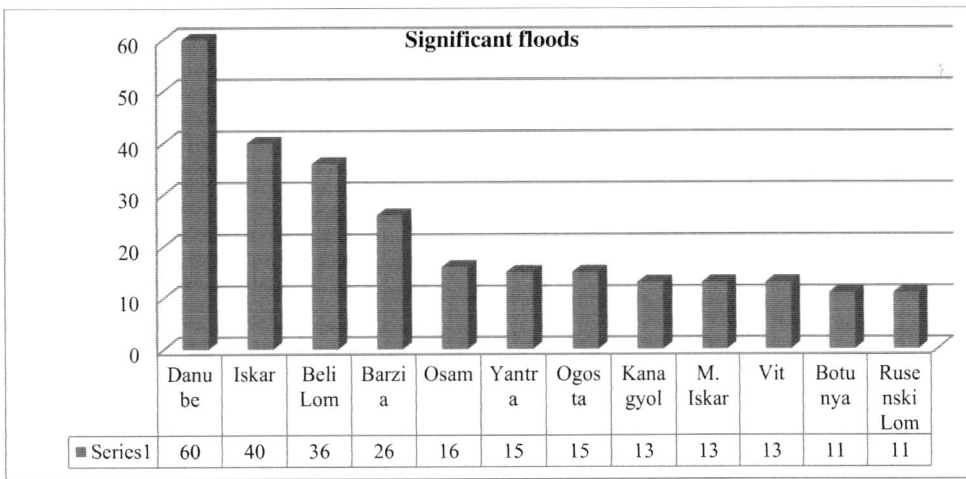

Figure 12. Significant floods by rivers at Danube region

IOP Conf. Series: Materials Science and Engineering **878** (2020) 012078 doi:10.1088/1757-899X/878/1/012078

5. Conclusion

In current paper the significant floods in the Danube plain are analysed. The floods which is happen in past year in the disrticts are given, as well as the significant floods on the rivers in this area. Based on this, areas with a significant potential flood risk can be identified and validated, focusing efforts on the mapping of the threat and risk of floods and the development of flood risk management plans.

Acknowledgment : This work has been carried out in the framework of the National Science Program "Environmental Protection and Reduction of Risks of Adverse Events and Natural Disasters", approved by the Resolution of the Council of Ministers № 577/17.08.2018 and supported by the Ministry of Education and Science (MES) of Bulgaria (Agreement № ДО-230/06-12-2018).

References

[1] DIRECTIVE 2007/60/EC OF THE EUROPEAN PARLIAMENT AND OF THE COUNCIL of 23 October 2007 on the assessment and management of flood risks (Text with EEA relevance) Official Journal of the European Union, 6.11.2007
[2] Garo Mardirossian, (Sofia 2009, prof. Marin Drinov Academic Publishing house), Natural disaster and ecological catastrophes, Study, prevention, protection, Second revised and supplemental edition
[3] Management plan for flood risk for Danube Region basin District
[4] Venelin Makakov, Rositsa Velichkova, Iskra Simova, Detelin Markov, FLOODS RISK ASSESSMENT IN BULGARIA, CBU International Conference on innovation of science and education, March 22-24.2017, Prague, Czech Republic pp 1253-1258
[5] http://www5.moew.government.bg/?page_id=23341

An approach to develop a collaboration platform for tool making industry

Pavel Vitliemov[1]

[1]University of Ruse, Department of Management and Business Development, Bulgaria

pvitliemov@uni-ruse.bg

Abstract. Critical point of the tool industry is data management considering that all applications are mostly based on the industrial data processing as well as the business process. Nowadays electronic collaboration platforms are unique environments that support synchronous and asynchronous communication through a variety of devices and channels within the toolmakers facilities This paper gives a brief overview of an approach to develop a collaboration platform for tool making industry focusing the goal to improve the collaboration along the entire life-cycle of both the tool and the customer's product on a cross-company and cross-country level of relationship.

1. Introduction

Electronic collaboration platforms are unique environments that support synchronous and asynchronous communication through a variety of devices and channels within the toolmakers facilities.

Also the collaboration platforms offer a set of software components and software services in order the managers to find the information they need and to be able to communicate and work together to achieve common business goals.

The objective of this material is to propose an approach to develop a concept of collaboration for European toolmakers. The main result of this work is estimated to be an internet-based platform for collaboration in global networks of toolmakers. These global networks should to comprise the collaboration along the entire life-cycle of both the tool and the customer's product on a cross-company and cross-country level of relationship.

The proposed collaboration platform is expected to be tool-centred, as the different tools link the single parties and allow collaboration between toolmaker, moulder and tool owner for each single unique tool.

The paper is structured as follows: In the sections in terms of related work are listed the basic requirements that need to be taken into consideration during the implementation and the crucial factors for successful collaboration in tool making industry. Section 2 shows the proposed approach for concept of collaboration platform including the architectural view and the functional view

2. Related work

In case the platform is web-based will means that an easy access will be guaranteed and will serve a collaboration and information to all relevant information on a detailed level as it stores and processes data that has so far been implicit or non-existent. The actors in that business model would have the possibility to select the information of interest out of the information provided.

The successful model of platform have to consist of six essential parts (Figure 1) in order to fulfil the specified requirements. [3]

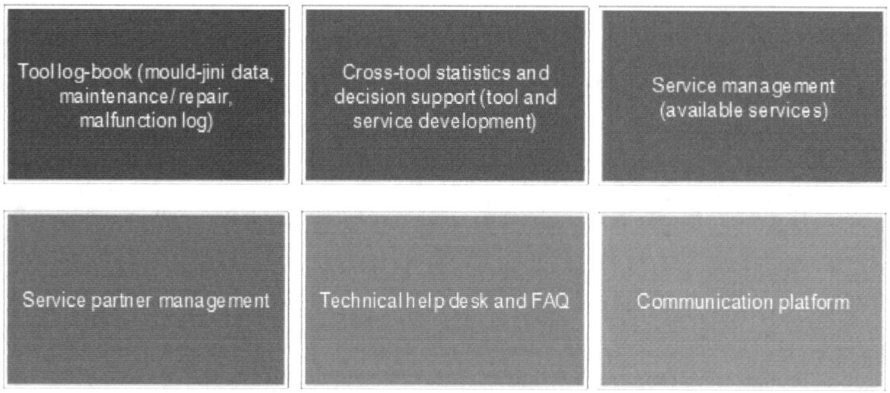

Figure 1. Features of the platform concept

The successful concept of collaboration for toolmakers have to allow collaboration between toolmakers, their customers and service providers and to deliver process data from in terms of enabling the toolmaker to offer an optimised preventive maintenance and to react promptly and adequately to specific problems. The concept of such collaboration platform is shown below. [6]

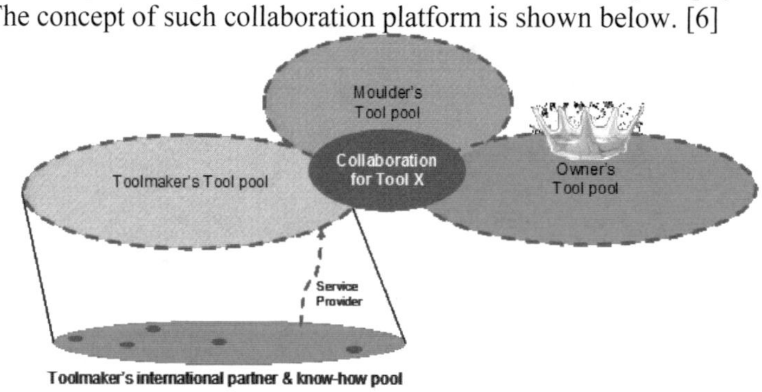

Figure 2. Concept of collaboration platform

Details about the organisation, rules, processes and requirements needed for the collaboration platform should to be additionally elaborated. Basic requirements and a framework can be extracted concerning collaborations in general and for the specific collaboration concept of European toolmakers. The basic requirements that need to be taken into account during the implementation are the following: [2]

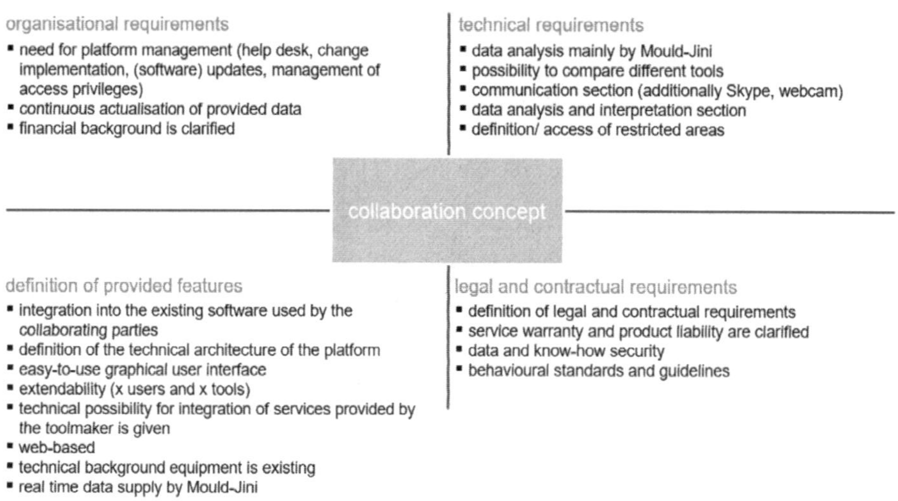

Figure 3. Requirements for the collaboration platform concept

Considering the analysis of the virtual factory and workshops, the main success factors have been depicted and were included into a criteria list for collaboration platforms, and are summarised in Figure 4. In general, criteria for choosing a collaboration partner are the same size, similar qualifications and the corporate culture. More specific elements of successful collaborations are the following: [3]

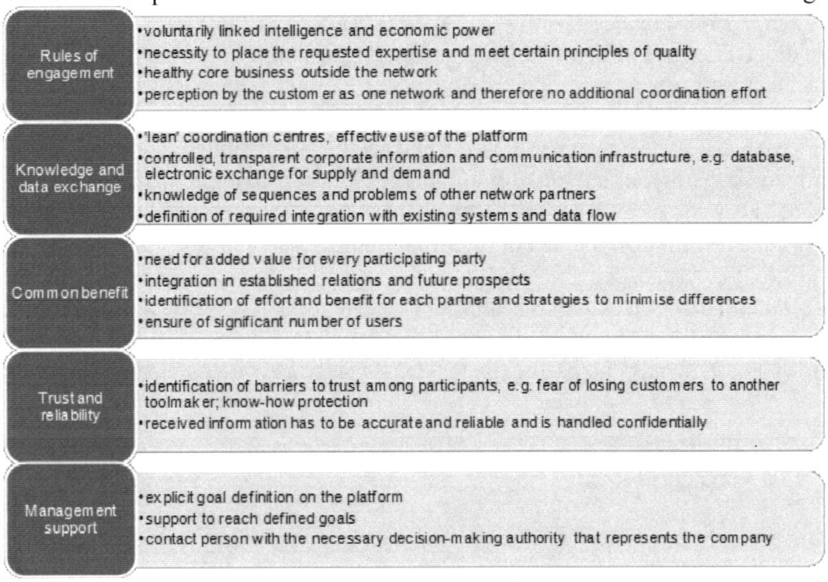

Figure 4. Criteria list for collaboration platforms

Besides these crucial factors for successful collaborations, further factors need to be considered to permit a good working collaboration. At first it has to be ensured that every collaboration partner has the necessary technical background to use the platform properly. [1]

In order to facilitate the selection of a partner for a specific task, a description of the partner's various competences has to be available. Furthermore, the mode of distributing the performance inside the network must be considered so that, on the long run, there is a use for every single member.

There has to be an interface for order processing that guarantees the document traceability and responsibility. Each company is responsible for the actuality and accuracy of its data. Another aspect is the monetary side of the collaboration. It has to be clarified how the expenses of the platform are divided and where the responsibilities for e.g. tendering and accounting are. [5]

The most important elements of such collaboration platform are messaging (email, calendaring and scheduling, and contacts), team collaboration (file synchronization, ideas and notes), and real-time collaboration and communication, and Social Computing tools. According to such definitions, it is possible to say that a collaboration platform can be classified on the basis of the services provided. [7]

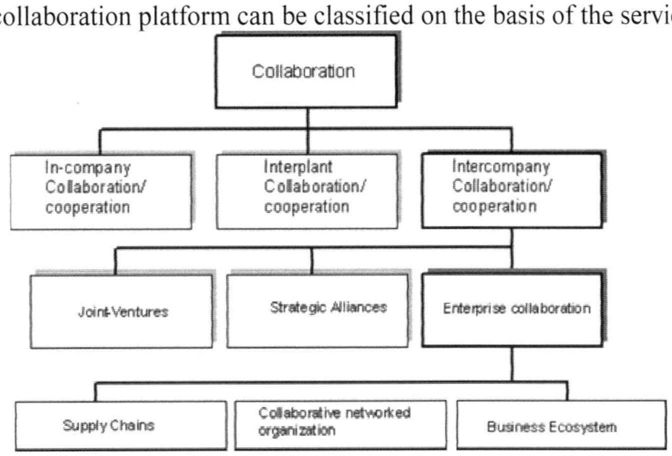

Figure 5. A collaboration platform organizational view

The collaboration platform can also be classified according to the self-explaining schema in Figure 5 from an organizational point of view, where the requirements can be identified with the 'collaborative networked organization' box.

3. A concept of collaboration platform for tool making industry

Management of machine manufacturing is the most complex case of production management. As a rule, machine manufacturing industry produces items with complex hierarchic structure. These items are usually multi-variant on all levels of production. In difference to the chemical industry where the product has its process route from the raw material to the end product, in machine manufacturing every assembly, part and semi-finished product have individual process routes. In other words, machine manufacturing industry produces complex multi-variant products based on complex multi-variant process routes.

A general view of proposed concept of collaboration platform is presented in Figure 6 with more focused on IT architectural aspects, where different layers are focused to specific technical and logical components. One can distinguish the user interface (UI) layer that will be provided through a web portal, some middleware layers for the management of the software services and the bottom layer devoted to handling data sources such as databases, file systems, real time process data generators, etc. [4]

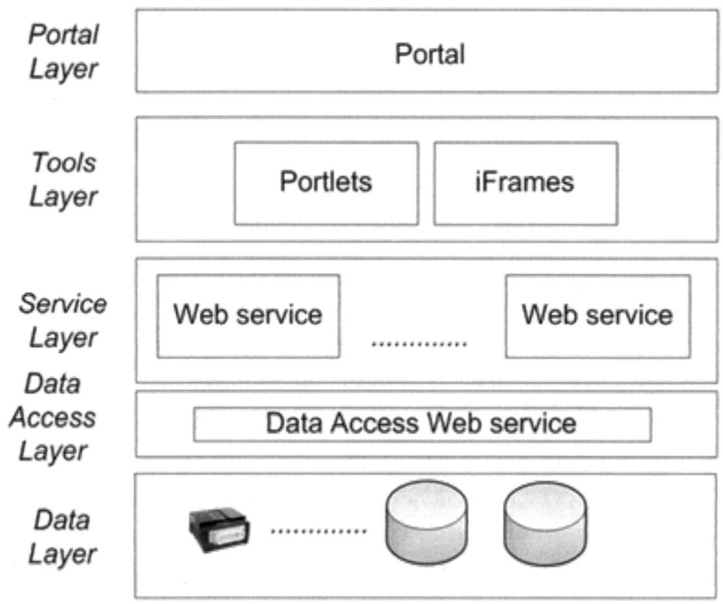

Figure 6. A collaboration platform architectural view

The proposed collaboration platform is showing the most relevant support tools that can be accessed from different types of users playing different roles within the toolmaker's business community. The different actors as it has been mentioned above in this collaboration are toolmakers, tool owners and service partners as well as moulders. They can access the collaboration environment from any internet access point and – provided the right authorization level – use the functions available with a simple web-browser.

An overview is presented in Figure 7 where functions and the information handled are given for the different systems and roles and the most general user working conditions are displayed.

IOP Conf. Series: Materials Science and Engineering **878** (2020) 012079 doi:10.1088/1757-899X/878/1/012079

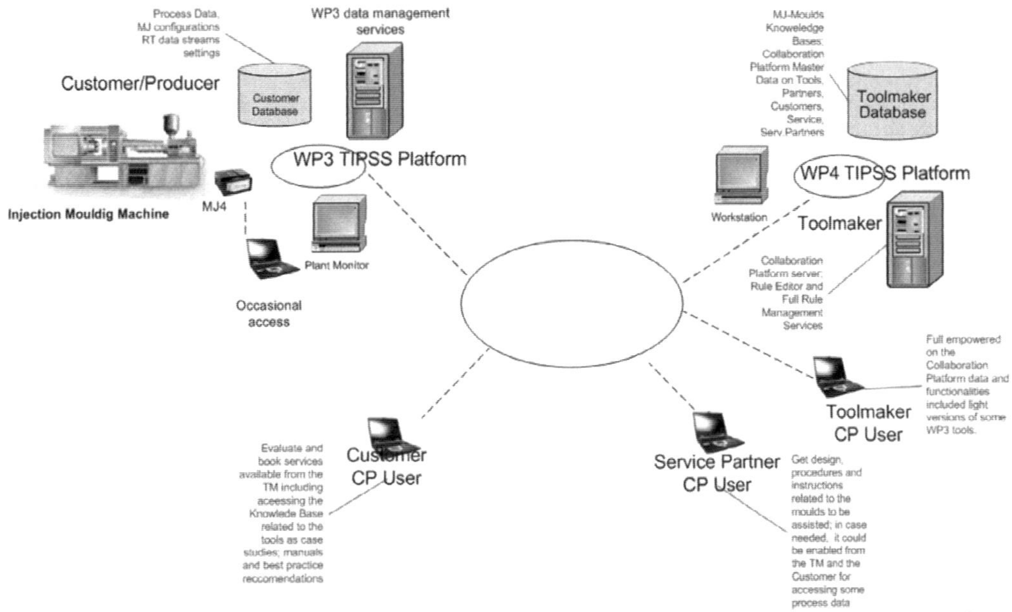

Figure 7. Collaboration platform functional view

On the customer/part producer side, the mould Jinis (data loggers) are the principal process data sources. The proposed platform needs to be installed in order to allow the access of the data logger devices and the provision of typical functions on a server basis which enables the connection of all such devices. Data related to the mould Jinis configurations is stored on the local customer databases as well. Database and mould jinis are connected and accessible by the data collection services through the customer intranet. These services are in turn accessible from an authorized user/application that requires the process data from any internet access point. Therewith the customers may use the platform for monitoring and analysing the situation of their installed injection moulds. They can also keep track of relevant events or check potential technical services available for their moulds and book the ones of interest. [8, 10]

The collaboration platform IT components as the database, for storing all master data used by the functionalities provided, and the server have to be installed on the toolmaker's side. The stored data is related to tools, customers, a service catalogue and service partners and can be accessed through the toolmaker's intranet, mainly for managing the web collaboration platform.

This distribution of roles is logical, since from the physical point of view the adopted IT technology allows to keep the software components also running in a distributed environment. With some of the resources running anywhere, the toolmaker's IT operators can still access and control them.

Finally, the service partners' role has been considered, too. For them the collaboration platform does not include specific functions. Since the exchange of technical documents has been considered as the most relevant requirement this service is available on the portal for all members of the community. [9, 11]

The IT architecture described in greater detail is based on a pool of IT resources installed within the toolmaker domain as most of the functions are in fact based on the use of data with the toolmaker as logical owner (toolmaker database). Since some of the functions available on the platform also require the use of process data (both historical and real time) from a specific mould as well as data logger data, a platform installed on the customer side is required too. Before using the functions based on process data they need to be properly configured. Here the identification and set-up of rules that are activated on a specific set of process variables for a given mould is required. This kind of set-up should be provided from the toolmaker platform administrator and requires good IT management capabilities as well as an adequate level of process engineering knowledge.

Conclusion

This paper presents an approach that can be applied to develop a collaboration platform for tool making industry. It is crucial furthermore the platform to be focused on surveillance in terms of ensuring solutions for online early detection of tool wear based on recent Internet of things technology and artificial intelligence. All that would be focused to present any damage from the tool increasing the tool reliability and efficiently.

In order to overcome the challenges of production in high-wage countries innovative services can be integrated within the collaboration platform in the area of after-sale to differentiate themselves from global competitors. A specific database can be designed for the collaboration platform that includes all the master data necessary for the characterization of the most relevant entities of the IT domain: toolmakers, partners and customers as well as moulds and technical services that can be provided.

References

[1] Berente, N., Yoo, Y., (2012); Institutional contradictions and loose coupling: post implementation of NASA's enterprise information system, Information System Research 23 (2): 376–396;

[2] L. K. Chung, J. Z. Shyu, K. Ding, (2017) "A Cross-Strait Comparison of Innovation Policy under Industry 4.0 and Sustainability Development Transition", *Sustainability*, vol. **9**;

[3] Günther Schuha, Martin Pitscha, Stefan Rudolfa, Wilhelm Karmanna,*, Martin Sommera (2014), Modular Sensor Platform for Service-oriented Cyber-Physical Systems in the European Tool Making Industry, Variety Management in Manufacturing. *Proceedings of the 47th CIRP Conference on Manufacturing Systems, Procedia CIRP* **17**, 374 – 379;

[4] M. Rath, J. Döring, W. Stark, G. Hinrichsen (200), Process monitoring of moulding compounds by ultrasonic measurements in a compression mould, *NDT E Int*, **33** (**2**), pp. 123-130;

[5] Milkusz M. (2014), Towards an Understanding of cyber-physical systems as industrial software-product-service systems; Procedia CIRP 16 pp385-389;

[6] Schuh G., Boss W., Kozlelski (2009); Life cycle cost-oriented service models for tool and ide companies; Proceedings of the 1st CIRP Industrial Product-Service Systems (IPS2) Conference, Cranfield University, pp.249-254;

[7] Schuh G., Salmen M., Kuhlmann, Wiese J. (2016), Life-cycle-oriented product service-systems in the toll and die making industry, Procedia CIRP 47 pp.555-560;

[8] XAware, XAware Community website, retrieved from http://www.xaware.org/introduction-to-xaware.html in February (2020);

[9] Distributed Transaction Processing: The XA Specification, retrieved from http://www.opengroup.org/onlinepubs/009680699/toc.pdf in January (2020);

[10] Distributed Transaction Processing, The Open Group model, retrieved from http://www.opengroup.org/bookstore/catalog/c193.htm in February (2020);

[11] Review SQL Server 2017, Replication Publishing Model Overview, retrieved from http://technet.microsoft.com/en-us/library/ms152567.aspx in February (2020).

IOP Conf. Series: Materials Science and Engineering **878** (2020) 012080 doi:10.1088/1757-899X/878/1/012080

Process mining in support of technological readiness level assessment

Veneta Markovska[1,*], **Stanimir Kabaivanov**[2]

[1] University of Food Technologies, Plovdiv, Bulgaria
[2] Plovdiv University "Paisii Hilendarski", Plovdiv, Bulgaria

* venetta@abv.bg

Abstract. We develop a framework for assessing technological readiness level using available data on business processes. By constructing a network of actors and linking process steps together it is possible to estimate the complexity of organizational structure, examine the bottlenecks and analyse whether advantages of available technology are fully utilized. Using publicly available data on business event logs we also test an automated process mining procedure and suggest a measurement to link our results to the TRL.

1. Introduction

Since its first steps in NASA in 1974 [1], technological readiness level (TRL) has gone a long way to become a wide spread tool for assessing adoption of new concepts and use of existing ones. Based on general, yet sound principles, it has proven to be useful tool for supporting management decisions [2], [3], as well as for assessment of innovations [4], [5]. Considering that some concepts may be more complex than others, the TRL research has been carried on to extend original classification, either by adding new levels [6] or by transforming it into a performance matrix [7]. While there is still an ongoing debate on advantages of specific extensions of the idea in different industries, it is particularly important for practitioners to have a reliable tools supporting their assessment. Due to the fact that real businesses need to take decisions under uncertainty and considering multiple factors that may not be explicitly taken into account in academic research, it is vital for these tools to be flexible.

Quite common approach for assessment of technological readiness is to conduct a survey with carefully selected questions and based on the answers to assign respondent descriptions to a set of predefined categories. Despite being a straightforward process, this method has several significant drawbacks:

- The number of respondents in a survey may be limited, based on the organization structure, as well as the complexity of the problems addressed. Trying to assess a technologically diverse organization with a limited data set is very challenging and error prone.
- Business processes tend to have different scalability properties which means that similar TRL assessments may end up with a completely unrelated results under modified environment conditions.
- Non-linear relationships between different processes and steps in the organization performance have to be taken into consideration.

We believe that careful analysis of business processes could help improve technological readiness assessment both in terms of required efforts and reliability of final output. Dependencies between business process understanding and TRL has been subject to multiple studies, like for example [8], [9]. Instead of building on a predefined and limited assumptions on the processes, we suggest to analyze information on procedures in the organization in order to build a map of current processes and inspect their complexity and depth. There are a lot of algorithms, aiming at automated business process discovery (ABPD), that offer a tradeoff between speed and accuracy [10], [11]. In this paper we use precedence and activity matrices in order to analyze business processes and then apply metrics that can assess their complexity. TRL and process analysis are important for non-business organizations [12], [13].

2. Model details

The model as shown on Figure 1, consists of five steps, that need to be repeated on demand or periodically in order to account for changes in the environment and internal organization. It should be pointed out that these five steps can be used with different TRL metrics and risk measures as well as with different complexity assessment procedures.

A sequence of events would typically contain observations that are part of different business processes. While there are no restrictions that a particular element of the input data should participate in a single process, it is useful to spend more effort analyzing those events that are part of multiple processes. That helps not only to identify potential bottlenecks in the operations but also to perform additional tests on what steps could be optimized or bundled together.

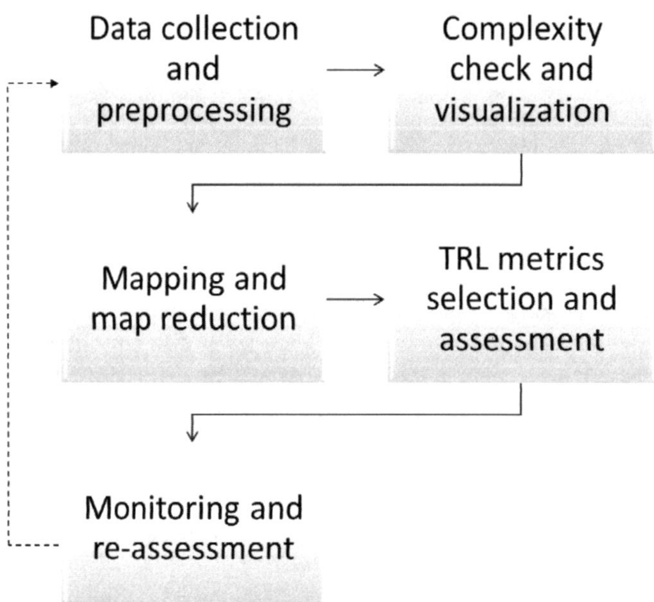

Figure 1. Business process discovery in support for TRL

2.1. Data collection and pre-processing

Data collection and preprocessing is the first step that needs to be considered. It is particularly important as cleaning up the inputs can reduce the dimensions of the data to be analyzed. Preprocessing is also essential for making sure that different characteristics of analyzed events are properly encoded and prepared for further analysis.

2.2. *Complexity assessment and visualization*

This step contributes to achieving two major goals. The first one is to provide an estimate on how complicated different processes are, as this allows to focus on those that are particularly tough to examine. It should be noted that complexity of a process is not the same as its business importance. Thus, the second goal of this step is to make sure that only relevant data is passed for further analysis. Considering possibilities for fully automated assessment, this part is particularly prone to errors and may require regular support and assistance from experts.

2.3. *Mapping and map reduction*

Mapping and map reduction of inputs plays important role for complex systems, where summary of similar steps can simplify further processing. This step contributes to the final results but building two intermediate output data sets – a map between events and logical processes and a set of combined (summarized) steps that can be used to simplify data handling.

2.4. *TRL metrics selection and assessment*

At this point typical TRL metrics like those suggested by Sauser et all [13] and Yaseri [14] can be extended with parameters indicating process complexity (length, cycles, inter-dependencies etc.). There are multiple possible metrics that can be used to estimate complexity of the business process, once it has been presented in a proper notation. While they are all relevant to the internals of the process, from management point of view the leading role should be attributed to one single metric – how much does this process contribute to reaching specified goals (for example to generating higher profit or serving more customers).

From an economic point of view, a particularly important characteristic of different processes is their throughput – time required to complete a specific process. When combined with step 2.5, changes in this parameter can be used to estimate the effects from particular management decisions as well as to account for the time needed to move to the next readiness level. To be able to measure economic impact we use the following complexity metric, applied to each step/node in the process:

$$C_i = \sqrt{n_i . x_i}, \tag{1}$$

where n_i is the number of inputs required to start this point and x_i is the number of outputs from this particular step, that are on their own turn used by next steps. There are other ways to measure the impact, including adapted metrics of cyclomatic complexity used in software analysis, although we prefer to keep calculations very simple and easy to understand. For the complete process we estimate the impact as weighted average of complexity metrics, using available free capacity (e.g. possible increase in intensity or workload on each node that would not require additional investment or changes in technology):

$$C = \frac{\sum w_i C_i}{\sum w_i}, \tag{2}$$

where w_i is the respective weight, calculated as the time throughput of the node relative to the total time required to complete the process, multiplied by the one, minus the percentage of available spare capacity:

$$w_i = \frac{t_i}{\sum t_i}(1 - fc_{\%i}) \tag{3}$$

Weight calculation puts emphasis on the fact that operations being able to easily cope with additional load are better prepared for changes and could be used to handle future changes and challenges. Once calculated weights could be further normalized so that then sum up to 1 in order to simplify the calculations. We prefer not to do this, as there is additional benefit for managers to see directly how weight values are influenced by individual factors.

IOP Conf. Series: Materials Science and Engineering **878** (2020) 012080 doi:10.1088/1757-899X/878/1/012080

2.5. Monitoring and re-assessment

Monitoring and re-assessment is essential to make sure that management decisions are continuously supported as business conditions change [15]. When all described steps are repeated that also gives possibility to use them for simulation and estimation of future development. Therefore, iterations are inherent part of the suggested model.

3. Application and results

We use publicly available data on EU Common Agricultural Policy, which is part of BPI Challenge 2018 [15]. Only 253K events have been analyzed in order to reduce the amount of processing time and only cases with refusal have been analyzed. This would correspond to an attempt to assess TRL for automatic check, validation and refusal of applications. We have analyzed data with R and visualization and case handling has been performed with bupaR package [16].

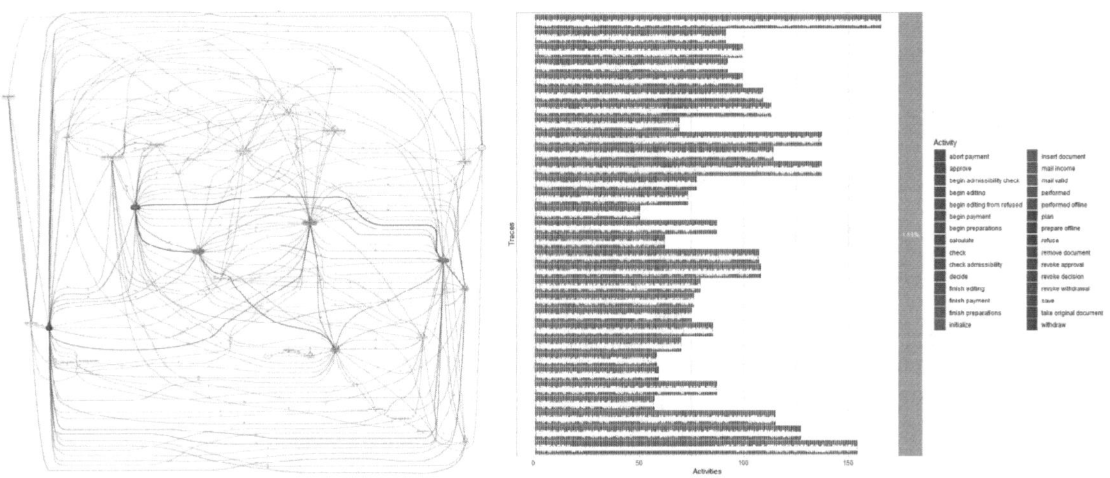

Figure 2. Process map and activity trace of the refusal cases

Activity trace and process map of the refusals shown on Figure 2 indicate that mapping and reduction can significantly reduce the amount of data being processed and can also help focus management attention to particularly important areas instead of just presenting specific activities. With mapping of the activities, instead of 30 different entries their number can be reduced to 8.

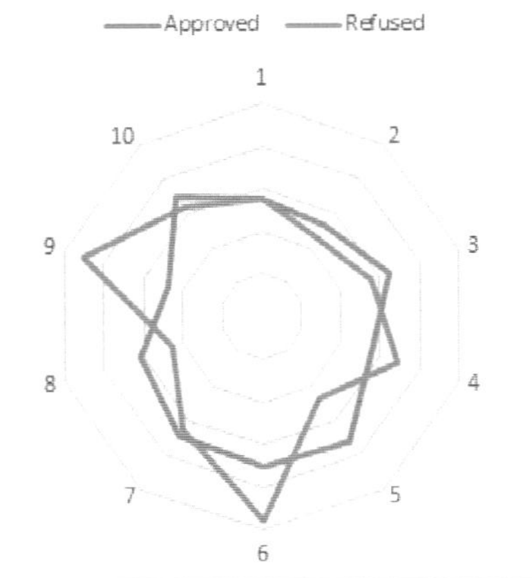

Figure 3. Mean throughput time

Differences in mean throughput time as shown on Figure 3 show how process features can be used in order to assess how technological changes will affect different steps. As data shows the refusals tend to have much more diverse timing requirements which is expected as there may be quite different reasons for rejecting a request as in this particular data set. Referring to TRL levels that would mean that much more detailed analysis would be needed when analyzing impact of management decisions on steps leading to refusal. Therefore, it would be more accurate to separate these steps and study their readiness separately.

When there is no huge difference between separate processes or cases, then economic impact metrics can support focusing on those steps and features that have highest contribution for achieving the goals set. Thus, if one step has 4 inputs required, single output and only 5% spare capacity its metric would be twice as large, as similar step having a single input and a single output.

4. Conclusion

Managers can take advantage of technological readiness level assessments in order to focus their attention on critical issues and opportunities in their organization. Despite being a useful tool that can combine diverse technical information into easy to understand and communicate way, TRL success depends on correct understanding of business processes and how different technologies fit in. We suggest an iterative procedure that can utilize process mining algorithms and give an overview of the actual business processes. In contrast to theoretically developed steps, event logs can shed light on the real time necessary to perform a specific action and about the actual relations between parts of the organization.

Using open data set we applied suggested steps in order to demonstrate how preprocessing and complexity checks can simplify further data processing and identify what are the most important steps in the underlying business activities. Process discovery and throughput analysis are important for correctly studying economic impact on different management decisions and importance of discovered processes and steps. Combination of process mining can therefore provide for better TRL assessment and decisions that are aligned with organization objectives.

*This report is funded from the National Science Program "Healthy Foods for a Strong Bioeconomy and Quality of Life" of the Ministry of Education and Science, approved by decision of the Council of Ministers No577 / 17.08.2018, contract No 68 / NSP under Work Package 4.*3 "New generation of value-added chains in bio-economy".

References

[1] J. C. Mankins, "Technology readiness levels," NASA, April 6 1995.

[2] HQ Office of the Chief Engineer/Steven Hirshorn and HQ Office of the Chief Technologist/Sharon Jefferies, "Final Report of the NASA Technology Readiness Assessment (TRA) Study Team," NASA Technology Readiness Assessment (TRA) Study Team, 2016.

[3] J. R. Lavoie and T. U. Daim, "echnology Readiness Levels Improving R&D Management: A Grounded Theory Analysis.," *Portland International Conference on Management of Engineering and Technology (PICMET),* pp. 1-9, 2017.

[4] C. A. Bates and C. Clausen, "Engineering Readiness: How the TRL Figure of Merit Coordinates Technology Development," *Engineering studies,* vol. Feb 19, pp. 1-30, 2020.

[5] S. Lokuge, D. Sedera, V. Grover and X. Dongming, "Organizational readiness for digital innovation: Development and empirical calibration of a construct," *Information & Management,* vol. 56, no. 3, pp. 445-461, 2019.

[6] J. Straub, "In search of technology readiness level (TRL) 10," *Aerospace Science and Technology,* vol. 46, pp. 312-320, 2015.

[7] J. Weber, "WEC Technology Readiness and Performance Matrix–finding the best research technology development trajectory.," in *Proceedings of the 4th International Conference on Ocean Energy*, Dublin, Ireland, 2012.

[8] A. Neda, A. Albadvi and Z. Ferdowsi, "Assessing readiness for business process reengineering," *Business Process Management Journal,* vol. 14, no. 4, pp. 497-511, 2008.

[9] M. Sanjay, A. Choudhury and K. Ganesh, "Framework for supporting'business process reengineering'-based business models.," *International Journal of Business Innovation and Research,* vol. 13, no. 4, pp. 451-474, 2017.

[10] A. Adriano, R. Conforti, M. Dumas, M. La Rosa and G. Bruno, "Automated discovery of structured process models: Discover structured vs. discover and structure.," *International Conference on Conceptual Modeling,* pp. 313-329, 2016.

[11] E. Marc, A. Koschmider and A. Oberweis, "Measuring similarity between semantic business process models.," *Proceedings of the fourth Asia-Pacific conference on Comceptual modelling-Volume,* vol. 67, pp. 71-80, 2007.

[12] V. Kazashka, M. Ruseva and P. Stoyanova, "Mapping and ranging of cultural and creative industries in the city of Plovdiv," in *CBU INTERNATIONAL CONFERENCE ON INNOVATIONS IN SCIENCE AND EDUCATION*, Prague, 2017.

[13] S. Kabaivanov and V. Markovska, "Making a Difference: Accounting for the Impact of Management Decisions in Environmental Management," *Scientific Annals of Economics and Business,* vol. 66, no. 2, pp. 131-139., 2019.

[14] B. Sauser, J. Ramirez-Marquez, D. Henry, D. Dimarzio and J. Rosen, "Methods for Estimating System Readiness Levels. The Schoolof Systems and Enterprises White Paper.," Stevens Institute of Technology, Hoboken NJ, 2007.

[15] S. Yasseri, "Subsea system readiness level assessmen," *International Journal of Underwater Technology,* vol. 31, pp. 77-92, 2013.

[16] T. Mihova and V. Nikolova-Alexieva, "Approaches and challenges in the application of high technology in the industry," *International scientific journal "Science. Business. Society",* vol. IV, no. 3, pp. 105-110, 2019.

[17] B. F. van Dongen, "BPI Challenge 2018," 4TU.ResearchData, 2018, 2018.

[18] G. Janssenswillen, B. Depaire, M. Swennen, M. Jans and K. Vanhoof, "bupaR: Enabling reproducible business process analysis," *Knowledge-Based Systems,* vol. 163, pp. 927-930, 20z19.

Opportunities for process reengineering in Bulgarian bread production (a case study)

Magdalena Damyanova-Bakardzhieva

Department of Industrial business and entrepreneurship,
University of Food Technologies, Plovdiv, Bulgaria

magidamyanova@gmail.com

Abstract. The aim of this paper is to outline the importance of business process reengineering and to prove that it could have great effects on the manufacturing industry. Sustainability becomes an important factor in production and the issue of energy efficiency is examined in the paper. The research is conducted in a bread producing factory. After reengineering, the company achieves better performance, cost minimization, increase in production and a significant energy efficiency improvement.

Keywords: Reengineering, Process, Energy Efficiency, CO2 emissions

1. Introduction

Companies need to be adaptable to change so that they are able to offer quality goods and services. The organizations that do not change according to environment will disappear from the markets [1-2]. In response to the competitive pressure, customer demands and ever-changing regulatory conditions, many companies are fundamentally rethinking the way they do business [3-5]. One of the most popular management tools for organizational change is the so called 'business process reengineering' (BPR).

BPR aims a successful market realization of newly-developed products, reduction of production costs and efficiency increase [6]. Hammer and Champy defined it as "the fundamental rethinking and radical redesign of business processes to achieve dramatic improvements in critical, contemporary measures of performance, such as cost, quality, service, and speed" [7]. According to T. Davenport BPR "encompasses the envisioning of new work strategies, the actual processes design activity, and the implementation of the change in all its complex technological, human, and organizational dimensions"[8]. Talwar emphasized on rethinking and reconstructing the organizational structure, workflow and value chain [9]. BPR is the redesign of business processes and the associated systems and organizational structure to achieve a dramatic improvement in business performance [10], [3]. Despite the variety of definitions and the differences between them, all authors agree on the importance of BPR.

BPR could be implemented everywhere, including manufacturing companies [11]. Manufacturing industry is one of the greatest energy consumers and carbon emitters globally [11]. That is why energy efficiency is vital for the company. According to the Global Energy Institute, industrial energy use is forecast to jump 26% by 2050. Energy-related carbon dioxide emissions decline except for the industrial sector [12]. The increasing price of energy and the current trend of sustainability have exerted new pressure on manufacturing enterprises that have to reduce energy consumption for both cost saving and environmental friendliness, as well as Life-Cycle Inventories initiatives [13-14]. Manufacturing companies have to find the balance between competition and environmentally friendly production. The environmental awareness leads the EU member states agreeing on the principle of "20/20/20 by 2020", i.e. a 20% reduction in greenhouse gases, a 20% share of renewable energies and a 20% increase in energy efficiency by the year 2020 as compared to 1990 indicators [15]. According to the International Energy Agency (IEA), it is vital that a reduction by two-thirds in the energy intensity of the global economy is achieved by 2050. This could be accomplished only by strict energy efficiency measures [16]. The prices of energy are continuously growing as a consequence of the increased demand for energy combined with a restricted supply in the market. This together with the dynamics in price setting generates uncertainties for organization schemes with respect to accurately calculated energy costs [17]. Energy savings are expected to be achievable from increasing both the energy efficiency of production [18] and the logistic processes, as well as innovative energy monitoring and management approaches [19], leading industries to a way of producing "more with less" [20].

This paper aims at outlining the opportunities of BPR. It shows that when used to redesign a process, BPR can give outstanding results. The research is conducted in a factory producing bread. What the management of the company aims at is not only cost minimization and increased production but most importantly – energy efficiency improvement. The main goal of the management of the organization is to improve the energy performance of the factory, i.e. to reach measurable results of energy efficiency, energy use and energy consumption.

2.Methodology

Appropriate instruments must be used so that the process of production is successfully reengineered. A lot of different methodologies related to BPR have been identified in the specialized literature. A consolidated one is presented here which according to the author reflects the most important aspects in a concise and clear way. The methodology consists of five phases [21].

The first step is the preparation for reengineering. Here it is vital for the managers to be sure that BPR is necessary and of great significance for the company.

The second step includes mapping and analysis of the existing process. This phase aims at identifying any obstacles that could occur through the whole progress.

The third activity is related to the design of the future process. The main objective here is the creation of alternatives to the current situation which must meet the company's goals.

What follows is the implementation of the reengineered process. This phase is the most difficult one as the efforts for reengineering meet opposition and resistance.

The last phase includes continuous improvement of the reengineered process. The focus here is on monitoring the progress of action as well as the outcome.

3.Implementation and Results

The research is conducted in a factory producing bread and confectionery products. It is medium-sized enterprise which was established in 1982. "Bread and Confectionery Company" Ltd. produces 38 types of bread and 40 types of bread products. At the moment the factory has two main bread lines, one special bread line and a pastry line. The lines have a capacity of 1200 pieces/hour (Line 1) and 2400 pieces/hour (Line 2). The weight of the produced bread is 700 g. The four bread lines, together with the silos, represent the main electricity consumers of the company. The management of the factory is constantly trying to adopt innovation and uses different strategies in order to improve the processes and increase profit. The company is certified according to the international standards ISO 22000:2005, ISO 9001:2008, ISO 14001:2004, BS OHSAS 18001:2007 and there is a plan for implementation of an energy management system and a certification according to ISO 50001. This proves that the management of the factory values not only quality but also safety work conditions.

Following the steps in the methodology presented above, we start with considering the need for reengineering. The company plans to increase production, as the current capacity does not correspond to the market demand. That is why in order to improve the energy efficiency of the existing process of production and to reduce production costs, the company plans to implement the following measure (later mentioned as Measure 1): Purchase of a bread line with a capacity of 2400 pcs/hour.

The second phase of the reengineering methodology includes analysis of the situation as it is at the moment. For the needs of its activity, the company buys two types of energy carriers - electricity and diesel fuel. The main energy carrier in "Bread and Confectionery Company" Ltd. is electricity. It is used by all facilities involved in the production and the provision of indoor climate.

The consumption of diesel fuel is complementary - only for refueling vehicles for the transport of goods and materials and for providing administrative needs. The two largest consumers are the tunnel furnaces of Line 1 and Line 2 with installed capacity of 630 kW and 770 kW. The furnaces are equipped with a modern control system and consume 450 kW and 400 kW in normal operation.

Electricity consumption for the last three years (2017-2019) is presented in Table 1. It could be summarized that the consumption was relatively stable. The average over the three years is 4 265 124.67 kWh/year, with deviation within ±2%. Diesel deviations are observed within ± 17% compared to the average of 839 441.23 kWh/year.

The third phase of the methodology is related to the design of the future processes. As it was mentioned above, the management of the company has decided to replace the old main bread line with a capacity of 1200 pieces/hour with a new one with a capacity of 2400 pieces/hour. New machinery need to be purchased and put into operation in view of the expansion of the capacity of the bread production plant. For that aim the enterprise makes a survey of the market for similar equipment from different manufacturers and suppliers and collects information about the technical data and price. The new line requires a change in the equipment and technological scheme of the silo. The technological scheme of the bread line will not change compared to the old one, presented in Fig. 1.

What follows next is of great importance for the company – calculation of the energy savings and comparison of the results before and after reengineering - Table 2. 2019 is taken for a representative year. First, work hours per day and daily consumption are calculated for each facility of the old bread line. The calculations for the new bread line are made according to the same algorithm but for a capacity of 48000 breads per day.

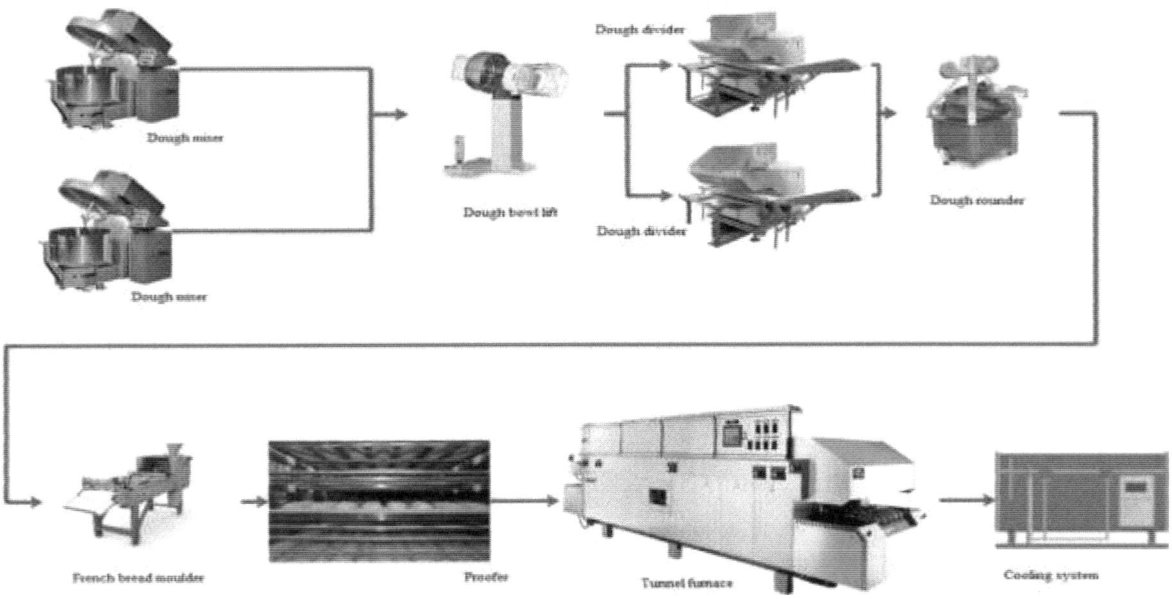

Figure 1. Technological scheme of the bread line.

Table 1. Annual energy consumption

ENENRGY		ANNUAL CONSUMPTION								
№	NAME	2017			2018			2019		
		t	kWh	lv.	t	kWh	lv.	t	kWh	lv.
1	Diesel fuel	59.89	698739	142253.81	84.87	990140	193535.63	71.10	829445	136451.44
2	Electrical energy		4168395	540429.39		4351721	501106.66		4275258	524575.59
	TOTAL:		4867134	682683.20		5341861	694642.29		5104703	661027.03

Table 2. Energy savings before and after Measure 1

Measure №	BEFORE MEASURE 1				AFTER MEASURE 1						
	ENERGY CONSUMPTION REQUIRED FOR MEASURE 1	PRODUCTION	SPECIFIC CONSUMPTION	CORRECTED ENERGY CONSUMPTION REQUIRED FOR MEASURE 1	ENERGY CONSUMPTION REQUIRED FOR MEASURE 1	PRODUCTION	SPECIFIC CONSUMPTION	ENERGY SAVINGS	ENERGY SAVINGS FOR A PRODUCT UNIT	ENERGY SAVING RATIO	SAVED CO_2EMISSIONS
	E_{in} kWh/y	N_{in} kg/y	E_{in}/N_{in} kWh/y	E_{in}/N_{in} x N_{exp} kWh/y	E_{exp} kWh/y	N_{exp} kg/y	E_{exp}/N_{exp} kWh/kg	kWh/y	kWh/kg	ESR %	t/year
1	3523672	6081600	0.58	7047344	3563030	12163200	0.29	3484313	0.29	49.4	2853.65

Table 2 presents the production converted to kilograms as its number is multiplied by the weight of one bread – 0.700 kg. Then, the energy saving ratio (ESR) for Measure 1 is calculated using the formula:

$$ESR=[(Ein \div Nin \times Nexp-Eexp)\div(Ein \div Nin \times Nexp)]x100 \quad (1)$$

ESR – Energy Saving Ratio;
Ein – Energy consumption before the implementation of
Measure 1 (3523672 kWh/year);
Eexp – Energy consumption after the implementation of Measure 1 (3563030 kWh/year);
Nin – Manufactured products before the implementation of Measure 1 (6081600 kg/year);
Nexp – Manufactured products after the implementation of Measure 1 (12163200 kg/year);
Ein/Nin – Specific consumption before the implementation
of Measure 1 (0.58 kWh/kg);
Eexp/Nexp - Specific consumption after the implementation of Measure 1 (0.29 kWh/kg).

The financial benefits from Measure 1 are presented in Table 3 where the energy savings could be examined. The payback period is calculated using the formula:

$$PB = I_0 \div B \ (2)$$

PB – payback period
I_0 – Investment
B – savings/year

Then follows a calculation of carbon emissions reductions. After implementation of Measure 1, the annual energy savings equal 3484.313 MWh/year. This means a reduction in CO2 emissions of 2853.65 t/year (at an emission factor for electricity of 0.819 t/MWh). The calculations are presented in Table 4.

Last but not least, is presented the information about the planned energy savings for the factory after the implementation of Measure 1:

Annual energy consumption: 5 104 702.81 kWh/year.

Corrected annual energy consumption for the year accepted as representative: 8 628 374.61 kWh/year.

Annual energy savings from the recommended package of measures: 3 484 313.35 kWh / year.

Planned energy savings after the implementation of Measure 1: 40.38%. They are calculated by dividing the amount of annual energy savings by the amount of corrected annual energy consumption, then the amount is multiplied by 100.

The last two phases – implementation in the factory and monitoring the progress of action will happen after the management examine the research in detail and approve the suggested steps. The results from the reengineered process are outstanding not only for the profit of the company but also for the environment.

Table 3. Financial benefits form Measure 1

Savings	Investments	Payback period	Life
lv./year	lv.	year	year
427526.42	1 203 352.00	2.81	10

Table 4. CO2 emissions before and after Measure 1

Indicator	Energy carrier	Ecological equivalent gCO_2/kWh	Energy consumption kWh/year	CO_2 emissions t/year
CO2 emissions before Measure 1	Electro energy	819	7047343,60	5771,77
CO2 emissions after Measure 1	Electro energy	819	30563030,25	2918,12
Reduced CO2 emissions after Measure 1	Electro energy	819	3484313,35	2853,65
CO2 after Measure 1			49.4%	

4.Conclusion

The reengineering model that was discussed in the paper shows the importance of BPR as the results are outstanding: better performance, cost minimization, increase in production and most significantly – considerable energy efficiency improvement. The latter must become a priority for industrial companies not only because of increase of energy costs and political demands but mainly because of the positive impact they could make on environment.

Acknowledgements

Throughout the entire research process, the author was fortunate to receive the support of Associate Professor Doctor Valentina Nikolova-Alexieva. Without her backing and encouragement, this work would not have been possible. The author would also like to acknowledge and thank "Bread and Confectionery Company" Ltd., for the extremely useful, up-to-date information and the opportunity of conducting the research in industrial conditions.

References

[1] Nisar Q. A., Ahmad S. and Ahmad U., "Exploring factors that contribute to success of business process reengineering and imact of business process reengineering on organizational performance: Aqualitative descriptive study on banking sector at Pakistan", Asian Journal of Multidisciplinary Studies, 2 (6), pp. 219-224, 2014.

[2] Habib M N and Shah 2013 A Business process reengineering: Literature review of approaches and applications, Proceedings of 3rd Asia-Pacific Business Research Conference, Kuala Lampur, Malaysia pp. 1-25

[3] Goksoy A, Ozsoy B and Vayvay O 2012 Business process reengineering: Strategic tool for managing organizational change an application in a multinational company International Journal of Business and Management 7 (2) pp.89-112

[4] Jamali G, Abbaszadeh M A, Ebrahimi M and Maleki T 2011 Business process reengineering implementation: Developing a causal model of critical success factors International Journal of e-Education, e-Business, e-Management and E-learning 1 (5) pp. 354-358

[5] Rahali E, Chaczko Z C, Agbinya J I and Chiu C C 2008 Business process reengineering in Saudi Arabia: A survey of understanding and attitudes International Journal of Artificial Intelligence and Interactive Multimedia, IEEE 1 (2) pp. 33-38

[6] Dakov I and Anguelov K 2001 Reengineering – a new method for industrial engineering and management vol. 4 pp. 26-32 ISBN 954-438-303-9, in Bulgarian

[7] Hammer M and Champy J 1993, Reengineering the corporation: A manifesto for business revolution, New York pp. 90-91

[8] Davenport T 1993 Process Innovation: Reengineering Work through Information Technology Boston, Harvard Business School Press

[9] Tawlar R 1993 Business re-engineering – a strategy-driven approach [doi: 10.1016/0024-6301(93)90204-S] pp. 22-40

[10] Saleem P. M 2012 Contemporary Office Handbook Kottayam, IN: DC Books

[11] Nikolova-Alexieva V Survey of the latest BPM concepts used by Bulgarian enterprises Scientific Researches of the Union of Scientist in Bulgaria, Plovdiv, Series B, Natural Sciences and the Humanities vol. XVI pp.216-221

[12] Global Energy Institute https://globalenergyinstitute.org/

[13] Kellens K, Dewulf W, Overcash M, Hauschild M Z and Duflou J R 2012 Methodology for systematic analysis and improvement of manufactirung unit process life-cycle inventory (UPLCI) – CO2PE! Initiative cooperative effort on process emissions in manufacturing, Part 1: Methodology description Int J Life Cycle Assess 17 pp.69-78

[14] Duflou J R, Kellens K and Dewulf W 2011 Unit process impact assessment for discrete part manufacturing: A state of the art CIRP Journal of Manufacturing, Science and Technology vol. 4 pp. 129-135

[15] Council of the European Union http://consilium.europa.eu/
[16] IEA: Energy Technology perspectives 2012, pathways to a clean energy system, executive summary http://iea.org/
[17] Jucker B, Leupp P and Sjokvist T 2008 Electrical energy – The challenge of the next decades Abb Review pp.8-13
[18] Bunse K, Vodicka M, Schönsleben P, Brülhart M and Ernst F O 2011 Integrating energy efficiency performance in production management – gap analysis between industrial needs and scientific literature Journal of Cleaner Production vol. 19, 6-7 pp. 667-679
[19] Weinert N, Chiotellis S and Seliger G 2011 Methodology for planning and operating energy-efficient production systems CIRP Annals – Manufacturing Technology vol. 60, 1 pp. 41-44
[20] Chryssolouris G, Papakostas N and Mavirikios D 2008 A perspective on manufacturing strategy: Produce more with less CIRP Journal of Manufacturing Science and Technology vol. 1, 1 pp. 45-52
[21] Muthu S, Whitman L, and Cheraghi S H 1999 Business process reengineering: a consolidated methodology, 4th Annual Conf. on Industrial Engineering Theory, Applications and Practice, Texas, USA

AUTHOR INDEX

Institute of Physics
Dirac House, Temple Back
Bristol BS1 6BE UK

ISSN: 1757-8981
ISBN 978-1-7138-4742-7

CURRAN ASSOCIATES INC.
proceedings.com